Bursting the Limits of Time

Martin J. S. Rudwick

Bursting THE LIMITS OF TIME

The Reconstruction of Geohistory in the Age of Revolution

Based on the Tarner Lectures delivered at Trinity College, Cambridge, in 1996

The University of Chicago Press · Chicago & London

Martin J. S. Rudwick is research associate in the Department of History and Philosophy of Science at the University of Cambridge and professor emeritus of history at the University of California, San Diego. He is the author of *The Meaning of Fossils, The Great Devonian Controversy, Scenes from Deep Time,* and *Georges Cuvier,* all published by the University of Chicago Press.

The University of Chicago Press, Chicago 60637
The University of Chicago Press, Ltd., London

Published 2005
Printed in the China
Published with the generous support of the Getty Grant Foundation.
14 13 12 11 10 09 08 07 06 05   1 2 3 4 5

ISBN: 0-226-73111-1 (cloth)

Library of Congress Cataloging-in-Publication Data

Rudwick, M. J. S.
Bursting the limits of time : the reconstruction of geohistory in the age of revolution / Martin J. S. Rudwick.
p. cm.
"Based on the Tarner Lectures delivered at Trinity College, Cambridge, in 1996".
Includes bibliographical references and index.
ISBN 0-226-73111-1 (cloth : alk. paper)
1. Science—Europe—History—18th century. 2. Geology—Europe—History—18th century. I. Title.
Q127.E8R83 2005
551'.094'09034—dc22

2004022007

∞ The paper used in this publication meets the minimum requirements of the American National Standard for Information Sciences—Permanence of Paper for Printed Library Materials, ANSI Z39.48-1992.

In memory of Susan Abrams
(1945–2003)

We admire the power by which the human mind has measured the movements of the globes, which nature seemed to have concealed forever from our view; genius and science have burst the limits of space, and observations interpreted by reason have unveiled the mechanism of the world. Would there not also be some glory for man to know how to burst the limits of time, and, by observations, to recover the history of this world, and the succession of events that preceded the birth of the human species?

Georges Cuvier, *Researches on Fossil Bones*, 1812

CONTENTS

ILLUSTRATIONS

ACKNOWLEDGMENTS

First and foremost, I thank the Master and Fellows of Trinity College, Cambridge, for appointing me Tarner Lecturer for 1994–97. No one who looks down the list of those who held this position in earlier years—starting with such giants as Alfred North Whitehead and Bertrand Russell—can accept the invitation without a profound sense of inadequacy. In my case this was deepened by the realization that no previous Tarner Lecturer had been primarily a historian, and in fact I almost decided to decline the honor. But what in prospect was daunting, even alarming, became immensely pleasurable in the event. I am particularly grateful to Sir Michael and Lady Atiyah for the warm welcome they gave me in the Master's Lodge, to Boyd Hilton for his moral support and efficient organization of all the practicalities of my lectures, and to him and many other old and new friends among the Fellows for making me feel at home in my original college.

Before and after giving the Tarner Lectures I had several valuable opportunities to try out my interpretation of this material in an even more condensed form, and I am very grateful for the response and criticism that I received on these occasions. Among them were the Faculty Research Lecture at the University of California San Diego, and a lecture at the Centre Alexandre Koyré (EHESS) in Paris, both in 1996; the 1998 Distinguished Lecture to the History of Science Society (meeting that year in Kansas City, Missouri); and lectures at the Universities of New South Wales and Melbourne, and the Australian National University, also in 1998. Moving back in time, I am very grateful to the President and Fellows of Clare Hall, Cambridge, for electing me to a Visiting Fellowship for 1994–95. At Clare Hall I found an exceptionally friendly and congenial environment in which to review and sort out much of the material for this book and to begin the task of actually writing it. My year back in Cambridge was made possible by the award of a fellowship from the John

Simon Guggenheim Foundation and of a President's Fellowship in the Humanities from the University of California: I am immensely grateful for both. On a longer timescale, indeed through most of my years in the United States, my research was generously supported by Scholars' Awards from the National Science Foundation (grant numbers SES-8705907/8896206, DIR-9021695, and SBR-9319955) and by grants from the Academic Senate of the University of California San Diego.

The argument of this book has been aided immeasurably by the advanced courses that I taught at, successively, Cambridge, Amsterdam, Princeton, and San Diego. I am extremely grateful to those who as students took part in often lively discussions during these seminars; their own suggestions, and their criticisms of my ideas, contributed much more than they may have realized to modifying and improving my interpretation. It would be invidious to single out any one of these seminars, but I owe a special debt to some of my Dutch students, first in Amsterdam and much later and more briefly in Utrecht. The internationalism that they took for granted in their academic work, and their almost casual multilingual competence, opened my eyes to my own unthinking insularity and made me resolve to try, in my own work, to transcend the overwhelmingly anglophone and anglocentric bias of much of the historiography of the earth sciences.

The research that lies behind this book has been spread over more than thirty years. It is impossible for me to acknowledge adequately all the friends and colleagues who have helped me on the way with their expert advice and information. Above all, however, they have encouraged me to believe that the project was important enough to deserve the time and effort that it has taken to bring it to some kind of completion. Certain scholarly periodicals in our field have recently adopted the shortsighted policy of excluding from the ranks of potential reviewers anyone who is thanked in print by an author, which in practice excludes most of those best qualified to give an informed judgment on a book. Nonetheless, at the risk of depriving this work of the reviews that both I and my readers would find most valuable, I must mention by name at least a few of the friends and colleagues who have supported me over the long haul (even if some of them were almost unaware of it), and others who have read and commented on specific parts of the text or who have given me invaluable help on specific topics. Among them are David Bloor (Edinburgh), Pat Boylan (London), Chip Burkhardt (Urbana), Albert and Marguerite Carozzi (Urbana and Geneva), Claudine Cohen (Paris), Pietro Corsi (Paris), Patricia Fara (Cambridge), David and Rosemary Gardiner (Cambridge), Mark Hineline (San Diego), Jim Kennedy (Oxford), Pat de Maré (London), Jack Morrell (Bradford), Ralph O'Connor (Cambridge), David Oldroyd (Sydney), Mark Phillips (Ottawa), Rhoda Rappaport (Poughkeepsie, N.Y.), Simon Schaffer (Cambridge), Jim Secord (Cambridge), Steve Shapin (Cambridge, Mass.), Phil Sloan (Notre Dame), Ken Taylor (Norman, Okla.), Hugh Torrens (Keele), and Ezio Vaccari (Varese). Other friends and colleagues who gave me much help and encouragement in earlier years have not, sadly, lived to see the results in print: among them were the late Gerd Buchdahl, Bill Coleman, François Ellenberger, Steve Gould, Brian Harland, Tom Kuhn, Roy Porter, Jacques Roger, and John Thackray.

I am grateful to all the curators, librarians, and archivists responsible for the collections that are recorded in my list of sources, for access to them, and in many cases for supplying the photographs that have been used for the illustrations in this book and for permission to reproduce them. The latter is acknowledged specifically in the captions (illustrations without such a note are of materials in my own collection, or are my own photographs, or maps or diagrams of my own design). However, two institutions deserve special mention, because I have made exceptionally heavy use of them over the years and have received so much courteous and patient help from their staff. They are the Rare Books Room at the University Library in Cambridge, for printed sources, and the Bibliothèque Central at the Muséum National d'Histoire Naturelle in Paris, mainly for manuscripts. Likewise, the photographers at two institutions deserve particular thanks, because I have made so many demands on their outstanding skills: they are those at the University Library in Cambridge and the British Library in London. I am also grateful to the Getty Foundation for their generous grant to the University of Chicago Press, which has made it possible for this book to be published with the density of illustration that I believe its subject demands.

Finally and above all, I am thankful for the life and work of Susan Abrams, to whose memory I dedicate this book. During the past twenty years, Susan's superb editorial work for Chicago has immeasurably improved the standard of scholarly publications on the history of the sciences. My own books have been among those to benefit hugely from her enthusiastic support and encouragement, but also from her penetrating yet always constructive criticism. I doubt that the present book would ever have reached completion without her constant interest and concern about its progress, which continued unabated right up to the time of her final illness. Above all, she was a wonderfully warmhearted friend, and is greatly missed.

A NOTE ON FOOTNOTES

The narrative of this book, and the interpretative thread that holds it together, can be followed in its entirety without looking at a single footnote. I hope that many readers will do just that, and enjoy the book in that way. But many others will want to glance at the footnotes, and some will need to scrutinize them in detail, for some very good reasons.

Footnotes are often scorned or ridiculed—sometimes by those who ought to know better—as pretentious, obfuscating, or just plain pedantic. However, I hope this book will be read not only by historians, who are familiar with the use of footnotes, but also by scientists and other readers, some of whom may not be. So it is worthwhile to state briefly why footnotes are indispensable in any work that embodies original historical research, and also to explain their more specific uses in this volume.

First and foremost, footnotes are used to give references to the sources that support and justify what is written in the main text: particularly the *primary sources* that date from the period under study, such as manuscripts and published articles and books, but also the books and articles by modern historians that constitute the *secondary sources* used. In this respect, footnotes function like the sections on methods and materials in a scientific paper: they enable the reader to follow the author's procedures and to evaluate the author's evidence. Full references, with exact page numbers for quotations, etc., are as important in historical research as accurate instrumental readings in a table or graph of experimental results: in both cases, the details—utterly tedious in themselves—are the only way of checking the adequacy of investigative standards and the reliability of conclusions.

The form of citation customary among scientists, e.g., just "Smith 1997b", is not helpful for historical purposes, because in most cases it requires the reader to turn to the list of references to find out which Smith, and which of the papers that he or she published in 1997, is referred to; and even then it gives no indication how the work has been used. Many historical works adopt an equally unhelpful system, giving full details the first time a source is cited, but thereafter requiring the reader to turn back page after page to find out what work an "op. cit." denotes. In this book a compromise system is used: sources are cited every time by author, year, and highly abbreviated title; this should usually be sufficient to enable readers familiar with the material to recognize the source; but for those who are not, or in case of doubt, the full title and other details are given in the Sources section at the end of the volume. That section also gives a key to the abbreviations used in the footnotes to denote the archives or libraries where manuscripts are held (e.g., "Paris-IF" for the library of the Institut de France) and the museums where historically decisive specimens are on display (see below). Printed sources are almost always cited in their original language (the form in which, with very few exceptions, I have studied them for this work); but modern English translations are noted in the bibliographies wherever I am aware of their existence.

In this volume, footnotes also have some more specific functions. I use them to refer to secondary sources that I think would be helpful as "further reading" for those who may want to pursue a particular topic in greater detail, whether or not I myself have derived my information or interpretative ideas from these particular sources, or indeed whether or not I agree fully with them. I also use footnotes to make comments or add information that would be out of place, or might be disruptive, in the main text. Such comments are of several kinds. Some are designed to explain historical terms that may not be familiar to readers who are not historians; or, conversely, to explain geological terms to those who are not geologists. Other comments are "translations" of scientific objects or phenomena into the terms of modern geology, which may help scientific readers to comprehend the kind of empirical evidence that the historical figures were confronting and trying to understand (this does *not* entail using what "we now know" as a yardstick for evaluating their reasoning, still less their intelligence). The footnotes also mention briefly how and where some of the same features can be seen at the present day, in the field or in museums: at the end of the volume I include—in addition to the customary lists of manuscripts and primary and secondary printed sources—a list entitled "Places and Specimens", which I hope will encourage some readers to utilize for themselves this important and woefully neglected source of historical evidence.

One further point is worth making here, although it does not refer only to footnotes. The main text is, I hope, wholly intelligible to readers without knowledge of the languages (other than English) used in many of the original sources. Almost all words, phrases, and longer quotations in other languages are translated into English (translations are my own, unless stated otherwise in the footnotes). The only untranslated words are those that are so similar to their equivalents in English that to translate them would insult the reader's intelligence. However, specific words that are particularly significant in the original texts are given (in brackets) after the translated ones, for the benefit of readers familiar with the language concerned. In the footnotes I make occasional comments that assume some knowledge of other languages; these too are of course directed at more specialized subsets of readers.

INTRODUCTION

TIME AND GEOHISTORY

Sigmund Freud claimed that three great revolutions had transformed what his generation—in blissful innocence of modern political correctness—often called "Man's Place in Nature". The first revolution, associated with Copernicus, was said to have toppled us from our privileged position at the center of the universe. The second, focused on Darwin, had demoted us to the status of naked apes. And it is not difficult to guess whom Freud had in mind as the genius behind the third and most recent revolution, which was supposed to eliminate human pretensions to rationality. Historians of the sciences are now uneasy about calling any such intellectual changes "revolutions", except perhaps to help sell their books; and many of those who study these developments closely would now be cautious about interpreting them in terms of successive assaults on human dignity. But anyway, as Stephen Jay Gould pointed out, Freud's list omitted one major historical change that certainly deserves a place in the same league. Compared to the other three, it has been grossly underexplored by historians, and neglected by those who popularize science and its history (with the honorable exception of Gould himself), perhaps because it cannot so easily be labeled with the name of any specific Dead White Male.[1]

Many years ago, in a pioneering survey of this fourth major change (the second in historical sequence), Stephen Toulmin and June Goodfield called it "the discovery of time". One could indeed start the story in the seventeenth century or earlier, with a taken-for-granted timescale of only a few thousand years for the whole history of the universe. Even as late as the nineteenth century, one minor English poet famously referred to the ruins of Petra as a "rose-red city half as old as time" and

1. Gould, *Time's arrow* (1987), 1–3.

probably meant it quite literally, being apparently ignorant of what by that time had long been a scientific consensus to the contrary. By contrast, the story leads eventually to the casual use of millions and even billions of years in the everyday work of modern scientists: the literally inconceivable expanses of the astronomers' "deep space" are matched by what John McPhee has aptly called the "deep time" of the geologists.[2]

However, I believe that a focus on the magnitude of the timescale has masked the much greater significance of the "deep history", as I shall call it, that fills up the vast tracts of deep time. Above all, it obscures an even more dramatic feature, namely the change from regarding human history as almost coextensive with cosmic history, to treating it as just the most recent phase in a far longer and highly eventful story, almost all of it *prehuman*. Back in the seventeenth century Sir Thomas Browne could remark, quite casually, "Time we may comprehend, 'tis but five days elder than ourselves"; and he would have meant it just as literally as the later line about Petra. But we, unlike Browne and his contemporaries, now take it for granted that even our most remote human ancestors were long preceded by an age when dinosaurs dominated the terrestrial fauna, periods when our present continents were split up or joined together in unfamiliar ways, an era when the atmosphere was as yet without oxygen, and moments when comets or asteroids crashed catastrophically into our planet.

This book is about how that highly eventful narrative of deep or prehuman history first began to be pieced together. At the time, it was not certain that any such reconstruction would ever be possible, except perhaps as a speculative kind of science fiction. The prospect of extending detailed and *reliable* history back into a *prehuman* world was so novel that it inspired one of the major scientific actors in this drama to write the eloquent prose that gives this book its title and serves as its epigraph. Georges Cuvier saw that he and his successors could hope to "burst the limits of time" by making prehuman history reliably *knowable* to human beings confined to the present, just as the astronomers had already "burst the limits of space" by making the movements of the whole solar system knowable with precision to human beings confined to one small planet.[3]

The scientific research that led to this dramatic change in our human perspective on time and history was focused on the earth, rather than the whole universe: it was the work of those who would now be called geologists, rather than astronomers or cosmologists. So in my subtitle I borrow the word *geohistory* from the modern earth scientists: geohistory is the immensely long and complex history of the earth, including the life on its surface (biohistory), as distinct from the extremely brief recent history that can be based on human records, or even the somewhat longer preliterate "prehistory" of our species.

HISTORICAL PARAMETERS

The reconstruction of geohistory goes on today in the work of earth scientists worldwide. But learning *how* to work out a reliable geohistory was something that their intellectual forebears achieved during just a few decades, or within the span of a single scientific career. The relevant period is more than amply enclosed at one end by the rebellion in 1776 by transatlantic colonials against His Britannic Majesty George III, and at the other end by the "year of revolutions" in Europe in 1848, with the most traumatic Revolution of all, in France in the years from 1789, in between. So for my subtitle I have borrowed another convenient phrase, "the age of revolution", this time from the political historians. In fact, the present volume is focused on just one part of the age of revolution, namely the period of the French Revolution and the subsequent Napoleonic wars, plus a few years before that turbulent time and a few years after it. But this is not to claim the causal primacy of the political over all other dimensions of society. The period covered in this book might also be defined, perhaps with no less relevance, as paralleling the early phase of the Industrial Revolution, or as stretching from Haydn to Schubert or from Fragonard to Goya.[4]

One further parameter is not explicit in either the title or the subtitle of this book. In tracing the reconstruction of geohistory during part of the age of revolution, I have tried to reflect the pervasive internationalism of the scientific world of that time. Even when the cosmopolitan culture of the late Enlightenment was replaced by the often nationalistic cultures of the early Romantic period, the outlook of those who worked in the natural sciences remained *in practice* highly international. Even when they felt most patriotic (or chauvinistic), they knew they had to pay very close attention to the work of their foreign counterparts. They knew themselves to be part of a scientific network—not always deserving to be called a "community"—that transcended national frontiers and linguistic boundaries. In this book I have tried to replicate that outlook by describing and analyzing work that was done in all the leading scientific nations of Europe, and indeed in what were then distant outposts of European culture such as Russia to the east and the United States to the west. The work I mention may seem heavily weighted towards sources in French; but this is because throughout this period French was the international language of the sciences, just as English is today, and because France itself was generally—if often enviously—treated as the center of the scientific world, just as the United States is today.[5]

Readers who are scientists will have no difficulty in feeling at home with this kind of internationalism. Geologists will know additionally that for their kind of

2. Toulmin and Goodfield, *Discovery of time* (1965); the best surveys to be published more recently are Richet, *L'Age du monde* (1999), and, in more popular style, Gorst, *Aeons* (2001). On "deep time", see McPhee, *Basin and range* (1981); for its more recent and routine usage by earth scientists, see for example Erwin and Wing, *Deep time* (2000).

3. Cuvier, *Ossemens fossiles* (1812), 1: "Discours", 3; Rudwick, *Georges Cuvier* (1997), 174–5, 185; see this volume, §9.3.

4. See, for example, Eric Hobsbawm's classic *Age of revolution* (1962), though this starts at 1789; or, closer to the present subject, Gascoigne, *Science in the service of empire* (1998), which deals with "the uses of science in the age of revolution".

5. I use the word "Europe" throughout this book in the traditional sense that includes my native island. It is only in recent years, as part of the baneful legacy of the hostility of the Thatcher regime towards all things beyond the Channel, that in British English—though not in American—"Europe" has come to mean what used to be called Continental or mainland Europe, or just "the Continent", so that for the British the Europeans have become "them" and no longer also "us".

science it is an intrinsic necessity: geological features do not respect national frontiers. By contrast, most current research on the history of the sciences is restricted to one particular national culture, overlooking or rejecting the internationalism that usually characterized the leading scientific figures of the time. I will not speculate on whether this has anything to do with linguistic limitations (or laziness). But any historian hoping to understand the scientific world in the age of revolution, while being unable to read sources in French, is being as foolish as a modern non-anglophone scientist who tries to contribute to his or her chosen field while being unable to read the scientific literature published in English. Instances of the latter kind of fool would be very hard to find, but not of the former.[6]

Readers who are professional historians will in fact recognize that this book is deeply unfashionable. Not only does it try to replicate the international outlook of the main historical figures; above all it is unfashionable in that it focuses on the details of the scientific work itself. Of course, scientific activity can be understood fully only in relation to its context in the culture and society of its time. But "context" has no meaning without "text": the political, economic, social, and cultural dimensions have little historical significance if their analysis neglects the precise claims to knowledge and epistemic goals that were the ostensible raison d'être of the scientific work. It is a regrettable fact that much current work on the history of the sciences treats those claims and goals as peripheral and their production and shaping as matters of secondary interest. Some historians seem to go to great lengths to avoid having to engage with the technical details of past scientific knowledge and of how it came into being. It is difficult not to conclude that in some cases this is due to intellectual laziness, dressed up in various items from the Emperor's ample wardrobe of postmodernist clothes.

It is high time for this fashion to be challenged. At least some historical writing should be focused on what the practitioners of the various sciences did in their scientific work, and "what the devil they thought they were up to". So this book is centered unfashionably on the ways in which specific concrete claims to reliable knowledge were formulated, argued over, and consolidated or rejected in the course of reconstructing geohistory during the age of revolution. Therefore, it is also centered on the activities of members of the scientific elite, among whom those claims were shaped most effectively, rather than on the beliefs and opinions of the literate public as a whole, let alone those of other social classes. It is good that "popular science", for example, is now getting the historical attention it deserves and that it is being treated as something actively constructed within popular culture as a whole, rather than just as material diffused with more or less distortion from its source in an elite. But there is now a danger that historians, anxious perhaps to demonstrate their democratic credentials, are neglecting or at least failing to understand the construction of scientific claims among the scientists themselves. Elites too deserve their histories. And this particular elite was not one based primarily on birth, wealth, or social class, but on hard intellectual work. Its achievements do not deserve to be trashed with the "anti-elitist" slogans of mindless modern populism.

However, although the scientific work that is the subject of this book was largely the product of an intellectual elite, much of it could be understood at the time by a wide range of educated people, and there is no reason why it should not now be equally accessible when treated historically. I have tried in fact to make this book intelligible not only to my fellow historians of the sciences, but also and equally to my friends and former colleagues in the earth sciences. I have tried to bear in mind, in effect, that whereas some readers will associate the term "Jurassic" only with a science-fiction movie, for others the word "Thermidor" may only evoke memories of haute cuisine at a seafood restaurant. By catering as well as I can to both groups, and by not assuming a lot of background knowledge in either direction, I hope this book will also be accessible to those who are not academic specialists of any kind. If any readers feel I am talking down to them by explaining in a simple way what they regard as too elementary to mention, I hope they will remember that others may be grateful not to have that background knowledge taken for granted.[7]

HISTORICIZING THE EARTH

What I shall be describing was a series of researches that first created the conceptual space, as it were, within which modern earth scientists can now do geohistory as a matter of routine and not even have to think about its possibility. But as often happens in "science studies" (which include the historical study of the sciences), what is generally taken for granted is just what most deserves scrutiny. In this case, the feasibility of acquiring reliable knowledge of geohistory was initially quite problematic, as already mentioned, and there was no inevitability about how, when, or even whether it would be achieved.

However, it was not problematic because of any simple conflict between geology and Genesis, or more generally between "Science" and "Religion". That older

6. Hardly less foolish is the tendency among anglophone historians to neglect or even to ignore the rich *secondary* literature published in other languages, most of which is never translated into English, or only after several years' delay.

7. I should explain at this point the relation between the present work and my earlier publications. My first historical book, *Meaning of fossils* (1972), set out a first sketch of some of the debates that I now see as central to the reconstruction of geohistory, but only insofar as they affected what came to be called paleontology (the science I had been practicing professionally when the book was conceived). Much more recently, *Georges Cuvier* (1997) offered an anthology of the work of one of the towering figures in the present story—often erroneously regarded as the villain of the piece—but in the present volume I try to relate his research much more fully to that of his contemporaries. In the quarter-century between these two books, a long series of articles, many reprinted in *New science of geology* (2004) and *Lyell and Darwin, geologists* (2005), dealt with the meaning of "the uniformity of nature" and the legitimacy of "catastrophist" explanations in geology; but these issues were explored mainly in relation to the work of Charles Lyell and his contemporaries at a slightly later period than that covered in the present volume. Likewise, *Scenes from deep time* (1992) illustrated—literally, in pictorial form—the imaginative reconstructions of "prehistoric monsters" that first made the new geohistory vividly credible to the general public, but this too was mainly a story of later decades. So was *Great Devonian controversy* (1985), which traced in technical detail the resolution of one crucial problem in geohistory; however, it did portray the kind of intensive argument that lay behind the reconstruction of every part of geohistory, so it too is indirectly relevant here. In short, the present volume draws on all this earlier work, but it offers a synthesis that uses a much wider range of sources and covers an earlier period than that of most of my other publications.

Manichaean style of describing the history of the sciences evaded the real issues. It reduced the problem to that of describing how one reified entity was liberated from the grip of another: once freed from the dead hand of Religion (it was said), Science could flourish, and its growth was simply the March of Truth. That kind of claim may be effective antireligious propaganda, but it makes lousy history. In fact, it has long been abandoned by historians, who rightly regard it as a product of specific social and political conditions in the late nineteenth century. But it continues to dominate the public perception of the sciences, and more damagingly that of popular science writers, television producers, and others in the media, long after those circumstances have been transformed almost beyond recognition. The fundamental defect of the conflict thesis is that it treats both "Science" and "Religion" as hypostatized and homogenized entities: as if there were a deep distinction in "essence" between them, and as if practices and beliefs, both scientific and religious, are and always have been uniform and unvarying. I hope this book will help to show, just by example, how religious and scientific practices and knowledge claims have interacted, in ways that have varied widely according to place, time, and, above all, social location. Such a modest and untidy conclusion will be deeply unsatisfying to crusading atheistic fundamentalists (and perhaps to religious ones too), but it does better justice to the historical realities.[8]

I shall argue that what was involved in the reconstruction of geohistory, far more importantly than any occasional and local conflict with religious beliefs, was a new and surprising conception of the natural world. Rather than being essentially stable and bound by unchanging "laws of nature"—ever since an initial act of creation, or else from uncreated eternity—one major part of nature, the earth itself, came to be seen as a product of *nature's own history*. Furthermore, geohistory turned out to be as contingent, as unrepeated, and as unpredictable (even in retrospect) as human history itself. This book, then, is about the *historicization* of the earth itself during the age of revolution. It can hardly be emphasized too strongly that this was a radically new feature on the conceptual landscape of the natural sciences: understanding and explaining the natural world began to be seen to entail its contingent past history as much as its directly observable present.[9]

This newly historical way of looking at the world of nature had two distinct but related sources. Both are perhaps unexpected and surprising, though for different reasons; but both are, I believe, substantiated by a wealth of historical evidence. The first of these sources was the contemporary practice of *human* history. Ideas, concepts, and methods for analyzing evidence and for reconstructing the past were deliberately and explicitly *transposed* from the human world into the world of nature, often with telling use of the metaphors of *nature's* documents and archives, coins and monuments, annals and chronologies. In one sense this transposition from culture into nature is unsurprising, for this was just the period in which human historiography was taking its modern shape, with a newly rigorous and critical evaluation of sources and a newly keen appreciation of the sheer otherness of earlier periods of history. If nonetheless the transposition seems surprising, it may be because the rigid modern distinction—particularly in the anglophone world—between the natural sciences and the humanities inhibits us from recognizing that in this instance "Science" has been radically transformed from outside its own sacred boundaries.

The second source for the historicizing of nature is perhaps still more surprising and may even be, to some, distasteful and unwelcome. But I believe there is strong evidence—some of which is set out in this volume—that the idea of nature having its own history was most congenial to those who already had a profoundly historical perspective, not only on their own human world but also on the cosmos as a whole and on the transcendent realm of divine initiative that they believed underlay it. In other words, those who were most attracted by the possibility of reconstructing an eventful past history of the earth were often also those who already understood their human place in the cosmos in terms of an unrepeated sequence of contingent events, suffused with divine meaning and intent, stretching from primal Creation through pivotal Incarnation towards ultimate Parousia. Within the intellectual framework of the Christian religion it made sense to try to understand the natural world, no less than the human, as part of this divine drama; and when the evidence for an immensely long geohistory became overwhelming, it made sense to try to construct a *history* for the vast tracts of prehuman time, and to *link* it on to the history recorded in more traditional and human ways. Historians no longer indulge in the sterile game of identifying goodies and baddies in the history of the sciences; but if, in this instance, we were to do so, "Religion" could certainly not be condemned, as it was so often in the past, for having invariably "retarded the Progress of Science".[10]

In any event the historicization of the earth, in what became the science of geology, was soon extended to other parts of the natural world, above all in Darwin's conception of the historical character of living organisms. In fact a further reason for the neglect of the theme of this book by historians is that the early phase of the reconstruction of geohistory has generally been treated as a mere prelude to the better-known story of the "Darwinian revolution". The two developments were indeed quite close in historical period, and they overlapped importantly in empirical material, particularly fossils. But I hope that this book will make plausible my claim that the reconstruction of a contingent geohistory was historically distinct from, as well as being an indispensable precondition for, the slightly later reconstruction of an equally contingent history of life in the perspective of Darwinian evolution.

8. For critiques of the simplistic "conflict thesis", see for example Brooke, *Science and religion* (1991), chap. 1; and more specifically, Rudwick, "Meaning of earth history" (1986), and Roberts (M.B.), "Geology and Genesis unearthed" (1998). Gillispie's classic *Genesis and geology* (1951) had a more subtle thesis than its title might suggest, and anyway it was focused mainly on the *popular* reception of geology in just one rather peculiar country at one specific period, namely Britain in the first half of the nineteenth century.

9. An excellent but brief survey, which offers an interpretation quite closely parallel to mine, but which reached me just too late to be used in the main text of the present volume, is Gohau, *Naissance de la géologie historique* (2003). The treatment of a related but even broader theme in Foucault, *Mots et choses* (1966), was based on decidedly slender empirical foundations; its "catastrophist" account of the historical change is one that few historians of the natural-history sciences would, I believe, now endorse.

10. I should put it on record that the conclusion summarized in this paragraph came to seem compelling only at a late stage in the writing of this book, as a result of my detailed research; it was not a guiding feature of my interpretation from the start. My own personal beliefs may have made me more open to the evidence in its favor than I might otherwise have been—I did not approach the sources with the usual knee-jerk hostility to all things religious—but I did not expect this conclusion, still less strive to demonstrate it.

The reconstruction of geohistory in the age of revolution was a major feature—even arguably the defining feature—of a new science of "geology", which embraced the subject matters of all the modern earth sciences. However, I should emphasize that this book is not a comprehensive history of geology, even within my chosen few decades. Rather, it is a history of a specific perspective within the sciences of the earth and of the practices and conclusions in which it was embodied. My choice of topics to analyse in detail has been governed throughout by that goal; my relative neglect of other topics does not imply that I think them unimportant, still less that I am unaware of them. This book focuses, then, on the history of one specific idea within what came to be called geology, but my approach could hardly be further removed from the traditional genre of "the history of ideas". I have tried to trace the development of a sense of the historicity of the earth, not as a disembodied idea, but as it became incarnate in the practical procedures and concrete conclusions of specific individuals and groups in particular historical circumstances, as they interacted in collaboration or controversy while exploring certain specific features and phenomena. Using this approach, however, a comprehensive survey of all the relevant scientific work would be impossibly onerous for the author and exhausting for his readers; so my policy has been to choose for closer description and analysis those cases that were either *innovative* or *exemplary* and to let them stand for a much larger mass of contemporary research.

TEXT AND ILLUSTRATIONS

My emphasis on the practical activities of the historical figures in this story also explains the relation between the text of this book and its illustrations. Most research by professional historians of the sciences is still rooted in *literary* traditions that have little or no comprehension of the central place of *visual* imagery in the life and work of most natural scientists, both past and present. Having been trained as a geologist, and having worked for many years in the earth sciences, I took the pervasive "visual language" of my science so much for granted that like M. Jourdain I was unaware of it, until in mid-career I tried to turn myself into a historian. I then found myself in a foreign country, where they spoke only the language of texts and where the use of illustrations was widely regarded as a mere prop for the feeble and unworthy of a true scholar. In more recent years the visual imagery of scientists has at last become an acceptable—even a fashionable—topic of special study among historians, sociologists, and philosophers of the sciences. But it is still not employed, as it should be, *routinely*—rather than with fanfares and song and dance—as an ordinary and indispensable part of any attempt to understand scientific work, past or present.[11]

In this book the illustrations are not merely decorative, nor are they included just to break up the pages of text; I did not select them after the rest of the book was completed, still less were they chosen for me by some picture researcher. They form an integral part of my narrative and my interpretation, just as they were considered essential by those whose work I describe and analyze. Their substantial number reflects the centrality of visual thinking and visual communication among the historical actors themselves. They are visual quotations that deserve to be studied with as close attention as any textual quotations, not merely glanced at or admired as examples of "art in science": hence the rather detailed explanatory captions that I have provided for many of them.

MAPS OF KNOWLEDGE

The making of a new science of "geology" in the decades around 1800 was just one aspect of what Thomas Kuhn, perceptive as ever, referred to as the *second* "Scientific Revolution". It is significant that the words "geology" and "biology" were both coined at this time, and even the much older words "physics" and "chemistry" underwent intense metamorphism, as it were, in their ranges of meaning. All these basic sciences of nature were shaped or reshaped into recognizably modern forms, reconstituted from mature preexisting branches of "natural history" and "natural philosophy". In effect, the implicit map of knowledge was radically redrawn; it was a period of major change in the mental topography on which we try to represent our conception of "the relations, or want of relations, between the different departments of knowledge".[12]

The phrase I have just quoted was that in which Edward Tarner expressed his intentions for the lectureship that he endowed at Trinity College, Cambridge, in 1916. In my Tarner Lectures in 1996, on which this book is based, I tried to fulfill his wishes by using the reconstruction of geohistory as a concrete example of a much broader argument. As the first historian to be appointed Tarner Lecturer—the position had previously been held by philosophers and scientists—I thought it important to use my own research as an illustration of the historical contingency of *all* our ways of mapping or classifying "the different departments of knowledge". The relations between the various natural sciences, and between them and the social sciences and humanities—all of them *Wissenschaften*, *sciences*, *wetenschappen*, *scienze* (etc.) in the broad sense still retained in European languages other than English—are not intrinsic to the natural and human worlds: all our maps of knowledge are themselves human constructions, embedded in the contingencies and specificities of history.

11. Rudwick, "Visual language" (1976), set out long ago my own plea for the use of pictorial sources in historical work on the sciences. This article was quite well received by scientists with historical interests, but it was almost totally ignored by professional historians of the sciences, although it was published in one of our leading periodicals. It was several years before it was "discovered" by the sociologists, and then still later by philosophers and historians, and—to my surprise and bemusement—assigned "classic" status retrospectively.

12. See Kuhn, "Function of measurement" and "Mathematical *vs* experimental traditions", reprinted in *Essential tension* (1979); also for example Hahn, *Anatomy of a scientific institution* (1971), 275, and Cannon, *Science in culture* (1978), chap. 4. Even larger claims have been made for the invention of "science" itself in this period: Cunningham, "Getting the game right" (1988); Cunningham and Williams, "De-centring the big picture" (1993). Kuhn's phrase has also been used, however, for events around the turn of the *twentieth* century: see for example Brush, *History of modern science* (1988).

For example, Jack Morrell and Arnold Thackray showed, in their classic study of the early years of the British Association for the Advancement of Science, how the modern anglophone meaning of "science" was shaped in specific circumstances in nineteenth-century Britain. The gentlemanly elite that gained power in the BAAS imposed its own ideas about just which sciences were fit for "advancement" in public support and merited the name of "science" *and which did not.* The modern restrictive definition of "science" was not just out there in nature, waiting to be recognized and adopted: it was enforced—not without opposition—at a particular time and place, for specific social and political purposes. It continues to exercise its cultural power, for good or (mainly) ill, more than a century and a half later and in quite different circumstances: for example, it now divides English-speaking intellectuals into the "two cultures" of "scientists" and "nonscientists", in a way that non-anglophones often find quite bizarre.[13]

In rather the same way, my analysis of the construction of "geology" is designed in part to show how this new kind of natural science, with historicity at its core, was no inevitable product of intrinsic scientific progress: it was the outcome of highly contingent events in Europe, and in outliers of Western culture beyond Europe, during the age of revolution. Above all, as already mentioned, I shall argue that it entailed and required the deliberate transposition of analogical and metaphorical resources from right outside the sacred boundaries of "Science", namely from the human sciences (*Wissenschaften*) of history and even theology. My story, like that of Morrell and Thackray, is therefore an illustration of the constructedness and contingency of the tacit mental maps on which we plot "the relations, or want of relations, between the different departments of knowledge". By exploring the origins of a new sense of *nature's own history* in the science of geology, and hence the origins of a quite new *kind* of natural science, I hope this book will serve as a reminder of the sheer diversity of all the sciences, both natural and human. Once we acknowledge that proper—and fascinating—diversity, all attempts to reduce the sciences to a single model or method, or to rank them in a linear pecking order (usually with mathematical physics as the domineering alpha male), become clearly illegitimate and indeed pernicious.

Living at the high tide of scientific triumphalism, Edward Tarner was disturbed by the cultural and epistemic hegemony of what he called "Physical science", or what the anglophone world now calls just "science". He wanted to claim a place for other ways of knowing, and above all for theology as the traditional "Queen of the sciences". He regarded the material world accessible to science as just one "cosmic province"; there might be others, he argued, equally important but incommensurable with the everyday material world, at least within the limitations of human existence. He claimed that once that possibility was accepted, even the eschatological concept of resurrection, for example, might become both intelligible and credible. Tarner recognized that Science with a capital *S* is intrinsically limited in its epistemic scope; or rather, as I would prefer to put it, the plural *sciences* are each limited by their diverse choices of empirical materials, practical methods, and epistemic goals. Those limitations make nonsense of hegemonic claims by the practitioners of any particular science, or of any group of sciences, that they provide us with the *only* valid form of human knowledge. Edward Tarner believed that there is more to our universe than was dreamt of by the scientists and philosophers of his day. Just possibly he was right.[14]

❧ ❧ ❧

This volume had its origin, as already mentioned, in the Tarner Lectures that I gave at Trinity College, Cambridge, in 1996. That invitation facilitated the writing of the book far more effectively than I had anticipated. I had to find a way to condense a sprawling mass of material and ideas, accumulated over many years, into a coherent argument that could be set out in just a few lectures. This imposed a much-needed discipline on what might otherwise have become a leisurely ramble through the early history of geology: rather than stopping to describe all my favorite trees in loving detail, it forced me to map out what I see as the shape of the whole wood. However, unlike some earlier Tarner Lecturers (but like some others), I have chosen not to publish my lectures in their original brief form. The main reason was that much of my source material, apart from the work of a few major figures, has never been adequately described or analyzed in the historical literature (which would not be the case if my topic had been, for example, anything to do with Copernicus, Darwin, or Freud): geology, and indeed the earth sciences as a whole, have long been a Cinderella among historians. So I have thought it right to put my own ideas on this one aspect of the history of geology into the public realm, with the degree of detail that I think the material requires and deserves. This massive enlargement of my original text has necessarily entailed a lengthy delay in publication; but I have retained the structure of my lectures, most of their illustrations and, I hope, the coherence of the overall argument.[15]

Readers who feel daunted by the sheer scale of this work will find each section summarized in its "Conclusion", and an even briefer summary of the argument of the whole book (with cross-references to the main text) in the "Coda" that brings it to an end. Alternatively, those who are visually literate—the unavoidable phrase reflects the primarily textual orientation of academic culture—could try just looking at the illustrations and their captions, which jointly summarize much of the textual argument.

13. Morrell and Thackray, *Gentlemen of science* (1981), chap. 5.

14. Tarner, *Athetic philosophy* (1916) and *Letter to the Vice-Chancellors* (1924). These and other pamphlets suggest that his views were idiosyncratic, even eccentric; nonetheless he had a valid point about the arrogant "scientism" of his time (and indeed of ours too).

15. In fact the narrative Part Two of this volume deals only with the earlier part of the historical period that I covered in the lectures. I hope in a sequel to describe and analyze the slightly later developments that built on the exemplary research with which the present volume is concerned, and which led to a reconstruction of geohistory that is recognizably congruent with the views of earth scientists in the twenty-first century.

Part One

UNDERSTANDING THE EARTH

Fig. 1.1. "Mr Saussure's climb to the summit of Mont Blanc in the month of August 1787": one of a pair of colored etchings, published at Basel in 1790 by Christian von Mecheln. It was the first time any "savant" had reached the highest point in Europe. The artist Henri L'Evêque did not depict the veils of black crêpe that Saussure and his guides and porters had used to protect themselves from snow blindness, which would have given them a somewhat sinister appearance.

CHAPTER 1

Naturalists, philosophers, and others

1.1 A SAVANT ON TOP OF THE WORLD

First ascents of Mont Blanc

At eleven o'clock on the morning of 3 August 1787, a party of twenty men climbed wearily through the snow onto the summit of Mont Blanc. All were wearing boots studded with iron nails to improve their grip on the snow, and all held long metal-tipped staffs to help them keep their balance. But one member of the party was distinguished from the rest by his elegant clothing and by the absence of a heavy load on his back. He was clearly a gentleman, one of the others was his personal servant, and the remainder were men from the village of Chamonix far below (Fig. 1.1).[1]

The gentleman was Horace-Bénédict de Saussure (1740–99), a forty-seven-year-old member of a prominent and wealthy family in the city of Geneva, some forty miles to the northwest. He was a "savant"—it would be anachronistic to call him a scientist—who was already famous throughout the scientific world for his *Alpine Travels* [*Voyages dans les Alpes*], of which a second handsome volume had been published in Geneva the previous year, with the promise of another still to come. In his *Travels* he was setting new standards for the scientific description of mountain regions, and providing others with a fine model to emulate. To set foot on what was believed to be the highest point in the Alps, and therefore in Europe, was not only a moment of personal triumph; it was also an event of great symbolic significance for the sciences of nature. It was an even more striking achievement in

1. Fig. 1.1 is reproduced from Mecheln, *Voyage de Saussure* (1790), pl. 1; the following account is based on Saussure, *Mont Blanc* (1787). What was depicted was in fact his expedition the following year to spend over two weeks on the Col du Géant, when he was accompanied by his son: the design clearly includes *two* gentlemen, both enjoying a greater degree of security than their peasant guides and porters. The discrepancy was pointed out by Freshfield, *Life of Saussure* (1920), 260–61; see also Carozzi, "Géologie" (1988), fig. 17.

Saussure's century than the first ascent of Everest was in the twentieth; as remarkable in its time as the first successful expeditions to the north and south poles.[2]

Saussure and his party were not quite the first to stand on the summit of Mont Blanc. He was guided there by Jacques Balmat, a villager from Chamonix, who had reached the peak the previous summer with the local physician, Michel-Gabriel Paccard. Balmat had already been dubbed "le Mont Blanc" for his achievement; and when Saussure wrote an account of his own ascent, he praised the local pair for their courage in climbing to the summit "without even being certain that men could live in the places they aspired to reach". There had in fact been a series of attempts on the mountain during the previous decade, by Saussure himself among others, and some of the villagers had become quite experienced at scaling the slopes of ice and snow. After his first success, Balmat had been commissioned by Saussure to try again in 1787, and he had duly made a second ascent in July, accompanied this time by two other villagers. Saussure had been on his way to Chamonix when they did so, but had not got there in time to go with them. Then the weather had closed in for almost a month, and he had had to wait in patience for another opportunity. When at last the clouds cleared, he had been ready to make it a truly scientific expedition.

Saussure and his party found the ascent arduous. After a relatively simple first day's climb over grassy slopes and easy rocks, they camped 779 *toises* or fathoms (about 1,500m) above Chamonix. But the second day took them up across a glacier, where they had to negotiate dangerous crevasses—one man fell in, but was saved by

A
B
C
Vue du Mont-Blanc et de la Route par la quelle on a atteint sa Cime,
A. Cime du Mt Blanc. B. Dome du Gouté. C. Aiguille du Gouté. D.E. Arve et Vallée de Chamouny. * * Places ou l'on a campé en montant.

Fig. 1.2. The north face of Mont Blanc, as seen from the far side of the valley of the Arve, in which Chamonix lies: an engraving based on a drawing by Marc-Théodore Bourrit, published in a later volume of Saussure's *Alpine Travels* (1796). His route to the peak (A) is shown, with asterisks to denote their camp sites on the way up. (By permission of the Syndics of Cambridge University Library)

being roped to others—and when they stopped for the night Saussure had to assure them that they would not die of cold if they camped on snow. At an altitude of 1,995 *toises* (about 3,900m) above sea level, even such fit men were exhausted by the effort of digging the snow to make level ground for the tents. Leaving them to such manual labor, Saussure the gentleman contemplated the scene around him: as he recalled later, "no living being was to be seen there, no trace of vegetation; it is a realm of cold and silence". After a fine clear night, during which they were alarmed by the sound of an avalanche, the party toiled slowly up steep slopes of snow, pausing frequently for breath; and two hours after they passed the last outcrop of solid rock they stood on the highest point in Europe (Fig. 1.2).[3]

Science on the summit

Once they were on the summit, a flag was unfurled; but this was no nationalistic exercise, and Saussure did not record what form the flag took. Its purpose was simply to be a sign of their successful arrival, visible from below. His first action on reaching the peak was to look down to Chamonix, where he knew his wife and her sisters would be watching the summit through a telescope. Then he looked out to the horizon, but it was too hazy to see the distant plains of Piedmont (now in northern Italy) to the southeast, or those of France to the northwest beyond Geneva and the hills of the Jura. Still less could he see the Mediterranean some 56 *lieues* or leagues (about 200km) away, although he later calculated that it was possible in principle to do so, even allowing for the curvature of the earth and for the Apennines rising above the coastline. However, as he recalled, any disappointment in that respect was amply compensated by the spectacular views of the Alps all around him:

> What I saw with the greatest clarity was the ensemble of all the high peaks, the arrangement of which I had for so long wanted to understand. I could not believe my eyes, and it seemed to me as if it were a dream . . . I was seeing their relationships, their connections and their structure; and a single glance relieved the doubts that years of work had not been able to resolve.[4]

In other words, the view enabled him to improve—at least in his mind's eye—the map of the western Alps that he had published the previous year and to gain a better understanding of the complex physical geography of the environs of Mont Blanc (Fig. 1.3).[5]

2. Saussure, *Voyages dans les Alpes* (1779–96); in the event, two further volumes were published, but only after a ten-year gap. The essays in Sigrist, *Saussure* (2001), give a comprehensive review of many aspects of his life and work. He was the great-grandfather of the famous linguistics scholar Ferdinand de Saussure (1857–1913).

3. Fig. 1.2 is reproduced from Saussure, *Voyages* 4 (1796) [Cambridge-UL: Mm.49.4], pl. 2. As a young man, a quarter-century earlier, Saussure had offered a prize for the first ascent, but there had been no takers at that time. The history of the early ascents is recounted in detail in Brown and De Beer, *Mont Blanc* (1957).

4. Saussure, *Mont Blanc* (1787), 15.

5. Fig. 1.3 is reproduced from Saussure, *Voyages* 2 (1786) [Cambridge-UL: Mm.49.2], part of frontispiece; see Sigrist, "Géographie de Saussure" (2001).

Fig. 1.3. The Mont Blanc massif, as depicted on Saussure's map of the western Alps, published in his *Alpine Travels* the year before his successful ascent. He acknowledged that the cartography was defective: the valleys had been surveyed quite accurately, but his mapmaker had to use conventional "molehills" to depict the mountains, and the snowfields and glaciers were as yet poorly understood. The area shown here is about 20km by 25km; Mont Blanc is near the center, with Chamonix ("le Prieuré") to the north, in the upper valley of the Arve. Crammont, his memorable viewpoint over the south face of the massif (see §4.5), is southwest of Courmayeur. (By permission of the Syndics of Cambridge University Library)

While Saussure was enjoying his waking dream, his porters were erecting a tent and a table nearby. Then he set to work with his battery of scientific instruments, for this was no mere sporting trip. The most important of his observations were with the two barometers he had brought: each nearly three feet of glass tubing filled with mercury, carried precariously to the summit on the backs of the porters (see Fig. 1.4). A third barometer was with his son down at Chamonix, and a fourth with his friend Jean Senebier, the city librarian at Geneva and a keen experimentalist; both men, he was confident, would be making equivalent measurements at the same time. Later, back in Geneva, Saussure would convert all the barometric readings into altitudes, though rival formulae for doing so left some margin of uncertainty; he would conclude that they confirmed measurements already made by triangulation, and that the true height of Mont Blanc above sea level was about 2450 *toises* (about 4,780m, close to the modern measurement of 4,807m). Anyway, even while on the summit he was able to confirm that Mont Blanc was certainly higher than any other mountain in the Alps, for its most likely rivals, the peaks of Monte Rosa and the Schreckhorn, were some thirty minutes of arc below the horizon.[6]

Other standard measurements followed, as swiftly as the rarified air allowed the exhausted Saussure to make them. In all it took him four and a half hours to make observations he had made in only three hours at sea level, during a trip to the Mediterranean coast earlier in the summer to establish a baseline for comparison with Mont Blanc. Two hair hygrometers—made in Geneva to his own well-known design (see Fig. 1.7)—gave him readings that he later computed as a humidity one-sixth that at Geneva, which seemed to account for the inordinate thirst they all experienced on the summit. Static electricity (no other kind had yet been recognized) was measured with an electrometer, yet another delicate precision instrument of his own design, and found to be surprisingly low. The air temperature in and out of the sun, and the boiling point of water, were measured with mercury thermometers, again with great precision. The intensity of the color of the sky was matched against a set of sixteen pieces of paper that he had prepared in advance, painted in graded shades of blue, again for comparison with observations being made simultaneously with identical papers at Geneva; on the summit, the sky's color fell between the two most intense shades. The wind was noted, and the declination of a magnetic needle found to be the same as at Chamonix. The effect of altitude on sound, and on the sense of hearing, was tested by firing a pistol. Even the party's meal was put to scientific use, as Saussure noted that the senses of smell and taste were unaffected by the climb; but they had little appetite, and their altitude sickness must have been aggravated by the wine and spirits that they drank as a matter of course with their meal. Finally, Saussure measured his own pulse, and that of his personal servant François Têtu and one of the four guides from the Balmat family, having done the same at Chamonix before they started out. The figures indicated that Saussure himself was the best acclimatized, which was hardly surprising: he was used to climbing high in the Alps, although never before as high as this.

Either on the way up or the way down, Saussure completed his observations by noting that the highest rock outcrops were all composed of granite, one showing what he took to be clear vertical stratification; and he noted that one conspicuous flat slab could serve as a benchmark to measure any future changes in the thickness of the snow cover. He also recorded that the highest flowering plant, at 1,780 *toises* (about 3,500m), was what the great Swedish botanist Carl Linnaeus (writing under his scholarly Latin name) had called *Silene acaulis*, or what the leading French botanist Jean-Baptiste de Lamarck called the "carnillet moussier" or moss campion; lichens ranged even higher. Two butterflies above the snow line were the highest sign of animal life.

Return to civilization

At half past three, the party left the summit and started back down the mountain. The sun had melted the snow, and the going was more difficult; the bright light also accentuated the alarming precipices below them. They spent a second night on snow, but much lower than on the way up; they were relieved to find that with the drop in altitude their appetites revived. On the fourth and last day of the expedition, they returned down the glacier, which the sun had made even more dangerous, with many new crevasses revealed (Fig. 1.4).[7]

6. From the direction of Mont Blanc the Schreckhorn (4,078m) lies behind the Finsteraarhorn (4,274m), the highest peak in the Bernese Oberland and probably the one he had in mind. Monte Rosa (4,538m), near the better known Monte Cervino or Matterhorn (4,477m), is much closer. See also Huta, "Jean Senebier" (1988).

7. Fig. 1.4 is reproduced from Mecheln, *Voyage de Saussure* (1790), part of pl. 2; the whole print is reproduced in Carozzi, "Géologie" (1988), fig. 16, and the rejected design in fig. 15. See also Ripoll, "Iconographie des *Voyages*" (2001), 323–26.

Fig. 1.4. Saussure's party descending from Mont Blanc: a detail from the companion print to Fig. 1.1. Saussure himself is depicted (rather inadequately) in the act of glissading down the slope; his guide Balmat, who is offering him a helping hand over a crevasse, carries a long slender load, probably one of the mercury barometers. The original version of this print showed Saussure being towed down the slope on his buttocks, but he rejected this as undignified and ordered the artist to change the design before it was published. In both prints the artist was also ordered to make Saussure himself more svelte and youthful than the portly middle-aged reality.

Eventually they regained solid ground, and were safely down in Chamonix in time for dinner. But that was not the end of the story. As soon as Saussure had completed his calculations, he published a booklet (on which this summary has been based) giving a brief account of his ascent, and his achievement was widely reported throughout the Western world. It spawned a small industry in Mont Blanc memorabilia. For example, at Saussure's instigation three-dimensional scale models of the Mont Blanc massif were designed by the director of mines for the Duchy of Savoy, in which Mont Blanc itself lay, and were advertised for sale less than a month later. A less expensive edition of Saussure's *Alpine Travels* was published the same year; this must have been already in press, but the author's enhanced fame would have helped its sales. Meanwhile, like NASA with the modern Moon rocks, Saussure had also distributed duplicates of his most important and precious specimens, such as the granite from near the summit, to some prominent institutions.[8]

Three years later Saussure issued a new edition of his account of the ascent, evidently to promote the sale of Mecheln's pair of prints (Figs. 1.1, 1.4). He endorsed them as giving an accurate impression of the climb and described how his Chamonix guides used a glissading technique "with astounding boldness and dexterity" to descend rapidly across slopes of snow. Above all, however, Saussure approved the prints as providing effective *proxies* for the scenes he had witnessed, so that others could share his arduous experience in the safety and comfort of their own homes:

> What was most difficult to depict, and what Mr Mecheln has portrayed as well as can be done in a colored print, is the appearance of these deserts studded with jagged rocks covered with snow and ice, and of the yawning chasms amidst the eternal ice. So, by means of the work of Mr Mecheln, those who cannot get there to admire these astonishing features will be able, without fatigue or danger, to furnish their minds with these great images.[9]

Meanwhile, one of the original conquerors of the mountain had less success in claiming the credit that was due to him. Paccard, the Chamonix physician, planned to publish his own account of the *first* ascent of Mont Blanc. Issuing a prospectus appealing for advance subscriptions, as was usual at the time, he noted that "crowds of travelers come every year to admire it and to walk on the glaciers that issue from it". Alpine tourism was on a tiny scale by modern standards, but it was already there: in the 1780 season about thirty visitors a day had managed to reach Chamonix. Paccard listed as his first subscribers seven foreigners who were staying in Chamonix at the time: three from Italy, two from Britain, and one each from Germany and Ireland. One young Englishman, Mark Beaufoy (later a prominent astronomer), had reached the summit with ten guides and porters less than a week after Saussure; another, William Woodley, made the fifth ascent the following summer. Paccard was evidently confident that an account of the very first ascent would arouse interest throughout Europe: his prospectus listed twenty-three book dealers in most of the important cities—from London to Rome, Paris to St Petersburg—who were ready to receive further subscriptions. But all this effort was in vain: Saussure, with his international reputation already established, had all the advantages, and the local physician's account never appeared.[10]

Conclusion

Saussure's ascent of Mont Blanc in 1787 will function in this book as the analogue of a modern geologist's "golden spike". This is a virtual marker, driven notionally into a sequence of strata at a convenient but not arbitrary point, to define consensually

8. The booklet, Saussure, *Mont Blanc* (1787), was being advertised by 1 September, less than a month after the ascent, probably to ensure that he gained full credit for it: see Brown and De Beer, *Mont Blanc* (1957), 8. Freshfield, *Life of Saussure* (1920), chap. 8, also quotes from his manuscript journal. One of the models, still in the possession of the Saussure family in Geneva, is illustrated in Delécraz, "Reliefs de montagne" (1998), 130–31. Another, bought by Martinus van Marum, is still on display in Haarlem-TM; see also Wiechmann and Touret, "Van Marum als verzamelaar" (1987), 129–30. A specimen of granite went, for example, to the École des Mines in Paris; see Decrouez and Lanterno, "Collection de Saussure" (1998).

9. Saussure, *Mont Blanc* (1790), avis de l'auteur; Mecheln, *Explication* (1791). Mecheln later issued a print depicting one of the models, giving a quasi-aerial perspective view of the Mont Blanc massif and making a sense of the Alpine topography even more widely accessible. Hineline, *Visual culture* (1993), first defined as "proxies" those scientific illustrations that try to replicate accurately and thereby *stand in* for firsthand experience, for example of sites in the field or of specimens in a museum, making them accessible to others at second hand; on "mobiles" in general, see Latour, "Drawing things together" (1990).

10. Paccard, *Premier voyage* [1786]; a copy of this rare prospectus is in Genève-BPU, Saussure MSS, 20/1,4. The text is printed in Brown and De Beer, *Mont Blanc* (1957), 408–13, a work that focuses on Paccard's ascent and chronicles the complex and unedifying story of the attempts by Bourrit, a Genevan author (and Saussure's artist: see Fig. 1.2), to deprive him of his glory; Bourrit made several attempts on Mont Blanc, but never reached the summit. Though not implicated in all this, Saussure was certainly determined to highlight his own achievement. On the early tourism, see Broc, *Montagnes* (1991), 247–49, and Wagner, "Gletschererlebnis" (1983).

the start of a specific named portion of geohistory (such qualitative subdivisions are generally more useful to geologists, even today, than quantitative radiometric dates). Saussure's successful climb will act here as a historian's golden spike, plunged into the seamless flow of human history. Rather than risking an almost infinite regress by having to trace and explain a long sequence of antecedents to the events and ideas to be described, this book begins (in the rest of Part One) with a synchronic *survey* of the sciences of the earth, as they were being practiced around the time of Saussure's climb. It reviews the kinds of work and the kinds of ideas that were, or could have been, familiar to him and to other leading savants with interests in the sciences of the earth, around 1787. The choice of baseline is not arbitrary, because Saussure's achievement conveniently exemplifies many aspects of those sciences during the late Enlightenment and in the last years of the Old Regime. In 1787 the political and cultural environment throughout Europe still seemed relatively stable, or anyway hardly less so than it had been at other times earlier in the century. The synchronic survey in Part One shows how a sense of the earth's historicity was also quite limited around this time; few savants used a truly geohistorical approach to understand the earth, and those few applied it only to a restricted range of physical features.[11]

Only two years later, in 1789, the outbreak of the Revolution in France began to disrupt the institutions of the sciences, and the lives of many individual savants, not only in France itself but also to varying degrees throughout Europe and even beyond it. In Part Two, a diachronic *narrative* traces how a geohistorical perspective developed, through the turbulent period of the Revolution and the Napoleonic wars and into the first years of the subsequent peace, until, less than four decades after Saussure's climb, it had transformed all the older sciences of the earth into the new science of geology.

1.2 THE REPUBLIC OF LETTERS AND ITS SUPPORTERS

Savants, professional and amateur

The historicization of the earth in the new science of geology was the work of human beings, not disembodied ideas. The first stage in setting the scene for the reconstruction of geohistory in the age of revolution must therefore be to sketch the "scientific community" of the time. However, that modern phrase misleadingly suggests a homogeneous interior composed of "scientists", and a sharp boundary separating them from the "nonscientists" outside. In the late eighteenth century (and, arguably, in the modern world) the social topography of the sciences was far more complex. It will be outlined briefly in this section, to indicate in general terms *who* was practicing the sciences, and particularly the sciences of the earth, and who was supporting their work.

Leading practitioners of all the sciences—in the broad sense of that word retained today in European languages other than English—considered themselves *savants.* They were recognized as being "savant" or learned, whether their expert knowledge lay in, say, chemistry or classics, physiology or theology. Those whose main interests were in the *natural* sciences were not regarded as fundamentally different from the rest (and are still not, outside the anglophone world). And contrary to an impression commonly held today, many savants who worked in the sciences two centuries ago were already what would now be called "professionals" in one way or another, while those who considered themselves "amateurs" did not treat that term as conceding any lesser status. Neither group can be called "scientists" without gross anachronism. The word was not coined until half a century later, and did not come into general use in English until the twentieth century; only very recently have equivalent words begun to be used in other languages. More important than the word itself, however, is that the narrowly conceived "professionalism" that it denotes is quite alien to the period with which this book is concerned.[12]

Those who earned their living from their practice of the natural sciences two centuries ago might, for example, be professors of "natural history", "natural philosophy", "physics" (three terms to be explained in §1.4), or medicine, in one of the many universities scattered across Europe in almost every city of any cultural importance. By modern standards, most universities were very small, and many professors taught a wide range of subjects. However, some of them enjoyed pan-European reputations, and students often migrated from one university to another in the course of their studies, to take advantage of the best lectures in each. But universities were primarily teaching institutions; they were places for the transmission of established knowledge. For most professors, research was a peripheral activity. If they did any, which most did not, they had to pay its expenses out of their salary, and they often received little credit from their colleagues for its published results. Academic life was certainly not yet dominated by the ethos of "publish or perish" (Fig. 1.5).[13]

By contrast, in most European countries (Britain was an exception) the universities were complemented by a national "academy of sciences". Each of these academies was regarded as a place for the making of new knowledge; their members received salaries for doing research, being in effect retained by the government to give expert advice when required. However, the number of savants employed in academies was very small; and even the most distinguished, the Académie Royale des Sciences in Paris, paid salaries that were hardly adequate without other sources of income. On the other hand, academies dispensed an important kind of support

11. The survey is not restricted to the precise year of Saussure's climb, but covers broadly the 1780s, with of course a retrospective review of earlier work that was still being treated as current during that decade; a few sources from the 1790s or even later are also used, for example where work done in the 1780s was not published until later.

12. See Jardine, "Uses and abuses of anachronism" (2000), for a judicious review of that issue. The word "savant" was used consistently in French, and often in English. The phrase "man of science", which became common in English in the early nineteenth century, is, like "scientist", too restrictive in meaning, though its gendered character is historically correct: see Barton, "Men of science" (2003).

13. Fig. 1.5 is compiled from Darby and Fullard, *Modern history atlas* (1970), 68–69, for universities; McClellan, *Science reorganized* (1985), appendices, for academies and societies; and other sources (in the interests of clarity, all varieties of academies and societies have been lumped together). Pyenson and Sheets-Pyenson, *Servants of nature* (1999), and McClellan, "Scientific institutions" (2003), survey the institutional context to be reviewed in this and the next section with an international perspective that is regrettably unusual among modern histories of the sciences. On universities, see also for example McClelland, *State, society, and university* (1980), and Heilbron, *Early modern physics* (1982), chap. 2.

Fig. 1.5. Scientific Europe at the end of the Old Regime: the distribution of universities, academies of sciences, scientific societies, and mining schools around the 1780s; they varied widely in size and quality. A.N., Austrian Netherlands (roughly, modern Belgium); S.C., Swiss Confederation (roughly, modern Switzerland); U.P., United Provinces (modern Netherlands). Political frontiers in central Europe and Italy were far too complex to be shown here, and only the larger states are named.

for scientific research, when they offered prizes—usually a substantial sum in cash, often with a gold or silver medal—for completed research on a specific topic. Since those who chose the prize topics were usually close to the corridors of state power, these competitions were often in effect a rudimentary form of governmental science policy, promoting research in some useful direction. In any case the chosen topics indicate the unsolved problems that the leading savants in a given country regarded as most important at a particular time. The overt purpose of the prizes was often that of national prestige or national economic welfare, but the competitions were generally advertised internationally. For example, when in 1785 the Russian academy in St Petersburg offered a substantial sum for an improved classification of rocks, it was awarded to an Austrian savant, with a Netherlander in second place, and a Frenchman in third (see §4.4).[14]

More specialized state institutions such as astronomical observatories, natural history museums, and mining schools gave employment to other savants. The first of these was of little relevance to the sciences of the earth, but the other two categories were of great importance. Most natural history museums had no more than a single curator, who was responsible for all branches of the subject. But at the "Cabinet du Roi" or royal museum of Louis XVI in Paris—the finest of its kind in the world—there was a large and varied staff, hierarchically graded, with divided responsibility for the various kinds of collections.[15]

Schools of mining were of course of particular importance for the sciences of the earth. Three had been founded by central European states after the devastation of the Seven Years' War in the hope that reformed mining industries might help revive their economies. Saxony set up the Bergakademie at Freiberg in 1765, and in 1770 Prussia followed with a school in Berlin, and the Austrian empire with one at

Schemnitz in Hungary (now Banská Stiavnica in Slovakia). Russia founded a school in St Petersburg in 1773, Spain at Almadén in 1777, and France its École des Mines in Paris in 1783. Sweden already had a different system of mining education under its Bergskollegium or state mining board. (Among the major powers, Britain was the exception in having no mining school of any kind, since the industry was left entirely to private enterprise.) Like universities, however, mining schools were primarily teaching institutions, designed to train managers and administrators for the service of the state. Students learned not only about mineral veins and ores, but also about the practicalities of mine shafts and pumping technology, all within the political and economic context of the sciences of statecraft and public administration [*Kameralwissenschaften*] (Fig. 1.6).[16]

Fig. 1.6. A scene symbolizing the mining industry of central Europe: an engraving that complemented the verbal information on the title page of Christoph Delius's *Introduction to the art of mining* (1773), the textbook for students at Austria's mining school at Schemnitz. It shows two mines administrators (with cylindrical hats) who would have been trained at the school, in relation to miners (with conical caps). One official stands with an arm outstretched in a conventional gesture of demonstration or command, while beside him a foreman pays close attention, ready to pass on his instructions to the ordinary miners. Another official is watching the practical techniques of some miners; other miners work with their hammers in the background. Beyond them are the mine shafts, and in the distance the buildings in which the ore is being smelted. (By permission of the British Library)

14. In Britain the Royal Society, although a private institution without state funding, also functioned *in practice* as a governmental advisory body, at least by the period with which the present book is concerned; it was also much involved in practical and technical questions: see Gascoigne, "Royal Society" (1999), and Miller, "Usefulness of natural philosophy" (1999). See also Heilbron, "Göttingen around 1800" (2002), for its Societät der Wissenschaften, one of the most productive of its kind.

15. See Laissus, "Cabinets d'histoire naturelle" (1964), and Spary, *Utopia's garden* (2000); the museum buildings were (and are) in the grounds of the botanic garden or Jardin des Plantes, as the museum was (and is) commonly known.

16. Fig 1.6 is reproduced from Delius, *Anleitung in der Bergbaukunst* (1773) [London-BL: 445.d.5], part of title page; of twenty-four large fold-out plates, all but two illustrate mining technologies. As its full title

Nonetheless, mining schools provided livelihoods for many savants with interests in the sciences of the earth. Around the time that Saussure climbed Mont Blanc, for example, the Saxon mineralogist Abraham Gottlob Werner (1749–1817) was earning a good salary at Freiberg and a European reputation for the quality of his teaching there; and his contemporary Johann Wolfgang von Goethe (1749–1832) was deeply involved with practical mines administration for the Duchy of Weimar, in addition to his multifarious literary activities.[17]

Many of the leading savants of Saussure's time—like the senior scientists who work in modern research institutions and universities—had onerous duties of teaching and administration, which left them little time for original research and publication. If the duties were less than onerous, the salary might be modest too, and the savant would need to supplement it with other sources of income. It was in fact quite normal to try to accumulate several paid positions—the French called it *cumul*—which would jointly provide an adequate income for a gentlemanly style of life. A few savants without private wealth enjoyed generous patronage that allowed them to devote themselves almost full-time to scientific work: an example was Saussure's fellow Genevan Jean-André de Luc [or Deluc] (1727–1817), whose position in Windsor as "reader" or intellectual mentor to the German-born Queen Charlotte (the wife of King George III) enabled him to be a savant of prodigious productivity.

Many other savants were "amateurs" in the original and non-pejorative sense of that word. They did scientific work—often of the highest quality—for the sheer love of the subject, as a spare-time interest: in the same way, they and their friends might be highly competent "amateur" practitioners of painting, music, or literature. Often their scientific work was funded in effect by the income they earned as clergymen, lawyers, or physicians—the three traditional learned professions—or in commerce or a civil service. Only a few savants were wealthy enough, for example from the rents generated by a rural estate or from having married into money, to indulge their tastes as full-time gentlemanly amateurs, though again their work was often anything but amateurish in the pejorative modern sense. Saussure was in that happy position, thanks to his own and his wife's substantial investments; he was able to spend much of his time on his Alpine research, though for some years he also held an academic position in Geneva. Another example would be his admirer James Hutton (1726–97), whose income from the Scottish farms he had inherited, and from being a sleeping partner in a chemical firm, enabled him to live a gentlemanly life in Edinburgh as a member of the brilliant intellectual circle that included such famous savants as the economist Adam Smith and the chemist Joseph Black (see Fig. 1.17).[18]

There was no correlation, however, between the way that savants earned their living—whether in the modern sense they were professionals or amateurs—and their scientific standing and esteem among others. As in the modern world, those who practiced the sciences two centuries ago recognized among themselves a subtle tacit gradient of competence and achievement. This ranged from an acknowledged elite of those with international reputations, down through those of more modest achievements and more local recognition, to the *dilettanti* who merely dabbled in the sciences and to those who as beginners had yet to prove their worth. Scientific work at any point on this gradient was sustained and made possible (as it still is) by many other people, whose vital role was rarely recorded in formal scientific publications, and who for that reason have been termed "invisible technicians". And finally, in a less direct sense, scientific work was of course also sustained by a wider public, who were in effect the patrons or consumers of what others produced. The rest of the present section reviews this wide range of historical actors.[19]

The Republic of Letters

The primary emphasis here, and throughout this book, will be on the scientific elite, comprising the leading practitioners of the relevant sciences. It was, above all, the interactions of these leading savants—their arguments and debates, their collaborations and controversies—that created the scientific knowledge that has since been found valid in far wider contexts than they could ever have imagined. Saussure himself can stand as an example of this almost wholly male and largely upper-class social group (Fig. 1.7).[20]

The intellectual elite around the end of the Old Regime was unified to a much greater degree than its counterparts today. All savants regarded themselves as citizens of an informal, invisible, and international "Republic of Letters", whatever the kind of civil government they were subject to in their everyday lives. They submitted themselves to a moral economy of obligations and responsibilities that transcended the boundaries of nation, language, and even to some extent social class. Needless to say, this scattered community of scholars did not always live up to the ideals it proclaimed, and savants were sometimes as selfish, dishonest, and

indicates, the book combined "Theoric" with "Ausübung", and dealt with the "Grundsätzen der Berg-Kammeralwissenschaft". On such "sciences of state", see Lindenfeld, *Practical imagination* (1997). The political economy of the early mining schools, particularly Werner's Freiberg, is analyzed by Wakefield, "Cameralist tradition" (2002), and the relation between mining schools and universities, particularly Freiberg and Göttingen, in his "Fiscal logic" (forthcoming); see also Roberts (G. K.), "Establishing science" (1991), 375–85, and Brianta, "Training in the mining industry" (2000). Schemnitz in Hungary should not be confused with Chemnitz in Saxony; Freiberg in Saxony should not be confused with Freiburg in Baden-Württemberg or with Fribourg in Switzerland.

17. In Goethe's house (Weimar-GW) the serried ranks of storage cabinets for his vast mineral collection are as conspicuous as the bookshelves of his huge library, but the attention of modern visitors is directed almost exclusively to his literary and artistic achievements. On his work in the sciences of the earth, see Hölder, "Goethe als Geologe" (1985); Hamm, *Goethe on granite* (1990) and "Goethe's collections" (2001).

18. See for example Wood, "Science in the Scottish Enlightenment" and other essays in Broadie, *Scottish Enlightenment* (2003).

19. The notion of a gradient of competence was introduced to interpret examples from geology: Rudwick, "Darwin in London" (1982), and *Devonian controversy* (1985), 418–26. Shapin, "Invisible technicians" (1989), introduced that term for those who operated seventeenth-century laboratories, and explored their role and the reasons for their near-invisibility then and in the modern scientific world; see also his *Social history of truth* (1994), chap. 8. On the history of the sciences in popular culture, see for example Cooter and Pumphrey, "Separate spheres" (1994).

20. Fig. 1.7 is reproduced from an undated print; the original painting is reproduced in Buyssens, "Saussure mémorable" (2001), 8. For simplicity and brevity, elite savants will often be described in this book as if they formed a group that was sharply separated from lesser mortals. In fact, they were simply the upper end of the continuum just described: as in the modern world, any individual savant would move into that zone in the course of time as he made his reputation, and sink back if he failed to maintain it.

Fig. 1.7. "Horace Bénédict de Saussure": a print engraved after a painting by Jean-Pierre Saint-Ours. The savant is dressed appropriately for his social status as a member of a leading family in Geneva, but is shown sitting outdoors as if resting after fieldwork. In one hand is a miner's hammer; in the other, a rock specimen collected with its aid. By his side is a specimen of a crystalline mineral, and also a collecting bag and a clinometer for measuring the "dip" of tilted strata. Behind him is a hygrometer (of his own design) for meteorological observations, and a telescope for studying the topography of the snow-covered Alps (in the background) where he had made his scientific reputation. All such instruments served to indicate his status as a serious savant. In a conventional pose, he looks up to heaven (or at least to the mountains) for inspiration. The portrait is highly idealized, and represents Saussure in his prime, around the time of his ascent of Mont Blanc; by the time it was painted in 1796, he had in fact suffered a crippling stroke (§6.5). Copies of this engraving would have been seen far more widely than the original painting in Geneva.

disputatious as any other similar group of people, and as much concerned with priority and credit as any modern scientists: Saussure's sedulous self-promotion after his ascent of Mont Blanc, for example, hardly showed saintly modesty. Nonetheless, the general acceptance of the virtue of adhering to the ideals of the Republic of Letters was important in practice.[21]

For example, when Saussure traveled through other parts of Europe, as he did extensively in the years before he climbed Mont Blanc, he took it for granted that he could visit other savants with similar interests, and that he would be given generous practical assistance and a free exchange of information and ideas; and they would expect the same from him in return, if and when they visited Geneva. When he traveled to Naples, to see Vesuvius and the other volcanic features of Campania, it was as a matter of course that he met the outstanding scientific authority on that region, the English aristocrat and diplomat Sir William Hamilton (1730–1803). Later, when Hamilton was passing through Geneva on his way home on leave, Saussure returned his help by taking him to Chamonix to see Mont Blanc; later still, he in turn sponsored Saussure as one of a batch of savants who were elected to foreign membership of the Royal Society in London. Saussure and Hamilton differed in nationality and mother tongue, but what they had in common was of far greater importance (Fig. 1.8).[22]

As this example indicates, the informal network of savants extended across the political frontiers of Europe's complex patchwork of kingdoms, principalities, duchies, republics, and territories of all kinds. It also extended across the different

Fig. 1.8. A portrait of Sir William Hamilton, the British ambassador to the court of the king of the Two Sicilies in Naples, painted by Sir Joshua Reynolds in 1777. Hamilton was a savant renowned internationally for his work in two distinct fields, both of which figure in this portrait. By his side are some of his great collection of Greek vases found in Italy, decorated with scenes from the life of the ancient world, and on his lap is one of the lavishly illustrated volumes in which he published and disseminated those designs in proxy form. In the background, less conspicuously, is the volcano Vesuvius (with a plume of vapor rising from its summit crater), on which Hamilton also published extensively. This portrait, like Saussure's, was also published as a print, making it widely accessible throughout the world of culture. (By permission of the National Portrait Gallery, London)

language areas. When, for example, Saussure stood on the top of Mont Blanc, he looked down in different directions into French-, Italian-, and German-speaking regions, as reflected in the names of the peaks he mentioned (Mont Blanc itself, Monte Rosa, and the Schreckhorn). Even the English-speaking world, the fourth major language area in Europe, was represented by the British and Irish tourists who were in Chamonix at the time, and by the anglophone aristocrats who frequently passed through Geneva on their Grand Tour across the Alps to the sunlit culture of Italy. Of those four languages, Saussure's native French was by far the most important in the world of the sciences, both natural and human, just as it was in the arts, politics, and diplomacy. In terms of the importance and quality of published work, most savants would have regarded the other three languages as being roughly on the same level, but not in the same league as French. For the sciences of

21. On the functioning of the Republic of Letters, see for example Daston, "Ideal and reality" (1991), where its norms of rigorous mutual criticism are placed in the wider context of a history of objectivity itself; see also Johns, "Ideal of scientific collaboration" (1994). Roche, *Republicains des lettres* (1988), gives detailed examples of its operation; Levine, "Strife in the Republic of Letters" (1994), describes the tension between savants and "men of letters"; Goldgar, *Impolite learning* (1995), analyzes its social dynamics to around 1750. For the sciences of the earth, by far the best account is in Rappaport, *Geologists were historians* (1997), chap. 1, though this too stops at midcentury. Goodman, *Republic of Letters* (1994), deals only with France, and primarily with literary rather than natural-scientific culture.

22. Fig. 1.8 is reproduced from the portrait in London-NPG, 680. Constantine, *Fields of fire* (2001), gives biographical background; see also the important essays in Jenkins and Sloan, *Vases and volcanoes* (1996), and, for his relation with Saussure, Rowlinson, "Common room in Geneva" (1998), and Vaj, "Saussure à l'Italie" (2001). Goethe was another of Hamilton's visitors during his subsequently famous Italian tour in 1787.

the earth, however, German would have been ranked ahead of either Italian or English. The most useful polyglot dictionary for mineralogy, for example, was based unsurprisingly on German, but provided synonyms for Latin, French, and Dutch (but not for English); Saussure had realized around 1770 that he would need to learn German in order to master its massive scientific literature, which was the product of the most advanced mining industry in Europe.[23]

French, however, was the primary international language of all the sciences, just as English is today. Thanks to their education, most savants in the non-francophone areas of Europe could read French easily, speak it more or less fluently, and compose letters in it adequately. Latin, which had fulfilled the same function throughout Europe in earlier centuries, had long been in decline, though it continued to be used in some parts of Europe, notably in Scandinavia (for example by Linnaeus) and Russia, and elsewhere in formal academic contexts. When scientific institutions were founded in other language areas, the Académie Royale des Sciences in Paris was often taken as their model, and French savants might be imported to help start them up, as in Berlin: in 1784 the Prussian academy awarded a prize for an essay that justified the universality of French on linguistic as well as historical and cultural grounds. At St Petersburg French was adopted as the language in which the Russian academy's work was publicized, reserving Latin for its more erudite activities (Russian was the language that the educated spoke to their servants). Even among the insular English, the Royal Society published scientific papers in French if they had been submitted in that language, while providing a translation in an appendix for the benefit of the linguistically challenged. Hutton, making no such concession, printed many lengthy quotations in French from Saussure's *Alpine Travels* in his own *Theory of the Earth* (1795), clearly taking it for granted that his intended readers would have no difficulty with them. De Luc would not have been thought inconsiderate for writing in French—even after living many years in England—to savants such as his neighbor the German-born astronomer William Herschel. The Dutch naturalist Petrus Camper and his son Adriaan, exchanging letters almost weekly while the younger man was living in Paris, both wrote as a matter of course in French, not in their native language.[24]

Any savant wanting his work to get full international attention would therefore try to have it published in French, or translated as soon as possible after its publication in another language. If the book was a costly one—which was always the case if it was generously illustrated—it might even be published with two texts in parallel, one of them French. For example, Hamilton published *Campi Phlegraei* (1776), his superbly illustrated work on the volcanic region around Naples, with its text in both French and English, the former being his own translation of the reports that he had earlier sent to the Royal Society in London. Given the subject matter, he might well have chosen to publish the work in French and German, had the text not been written originally in English, and had he not hoped to find purchasers among his wealthy compatriots who included Vesuvius, Herculaneum, and Pompeii in their Grand Tour.[25]

Despite the dominant position of French, however, other languages were also important in scientific work, particularly of course if the readership being sought for a publication was primarily a national or still more local one. Much valuable work, as already mentioned, was published in German, Italian, and English, and savants with any scientific ambition made sure they could read at least one of those languages in addition to French. Savants from the smaller language areas, such as the Dutch and the Scandinavians, were therefore obliged (as they still are today) to be even better linguists than the rest.

This sketch of the savants of Saussure's time has outlined a pan-European—but only European—scene. There were few scientific bodies outside "old Europe". The most prominent exceptions were the Academy of Sciences in St Petersburg, which was an important instrument of imperial cultural policy for making Russia more European; the American Philosophical Society in Philadelphia, the cultural center of the newly independent United States; and the Batavian Society [Bataviaasch Genootschap] in Batavia (now Jakarta) and the Asiatick Society in Calcutta. The two latter were composed not of Javanese or Bengalis but of Dutch and British expatriates, particularly the employees of the two nations' East India Companies. Apart from the members of such bodies, few savants worked outside Europe, except temporarily while on an overland expedition or voyage of exploration, or on a diplomatic or naval mission. Regions beyond Europe were of great scientific importance, particularly for the sciences of the earth, which frequently dealt with spatially extended features and spatially scattered materials. But distant regions were treated primarily as sources of factual information about such matters, not as arenas of debate about their interpretation and significance. The few savants who lived there were valued primarily as reliable sources of reports and specimens; they were indeed included in the informal network that constituted the Republic of Letters, but they were in all senses marginal. American savants such as Benjamin Franklin and Thomas Jefferson, although living on the civilized east coast of the new republic, were as much on the edge of the scientific world as the Dutch and British savants who worked in Asia. Except while they were living temporarily in Europe—Franklin in London, for example, or Jefferson in Paris—they were, and knew themselves to be, far indeed from the centers of scientific debate. The world of the scientific elite was a European world.

A variety of supporters

However, a focus on the scientific elite, as adopted in this book, is no more than a matter of relative emphasis. Also important for all the sciences were the local or provincial scholars, on whose limited expertise the internationally recognized

23. Schröter, *Lithologisches Reallexicon* (1779–88); Carozzi and Bouvier, *Library of Saussure* (1994), 49.

24. See for example Trabant, "New perspectives" (2001), 16, for Rivarol's prize essay at Berlin. Carozzi (M.), "Saussure: Hutton's obsession" (2000), estimates that about one-quarter of Hutton's full text is in French. De Luc to Herschel, letters dated 1784–1809 (London-RAS, Herschel MSS, W1/13, L60–67), of which only the last two are in (broken) English. Bots and Visser, *Correspondance de Camper et Camper* (2003), print all their letters written in 1785–87.

25. Hamilton, *Campi Phlegraei* (1776); a German edition followed a few years later, and a less expensive but unillustrated French one. Knorr, *Deliciae naturae* (1766–67), published at Nuremberg with lavish illustrations of objects of natural history, is an example of a comparable luxury work with text in German and French.

savants were often deeply dependent. Local experts were of particular importance in the sciences of the earth, for the reason just mentioned: the phenomena and physical features of scientific interest were intrinsically local in character, and those living in a particular region could often acquire an intimate knowledge of them, far beyond what more wide-ranging savants could hope to acquire without their help. Paccard, for example, as a resident of Chamonix and a man with a good Parisian medical education, may well have had a more detailed scientific knowledge of some aspects of the immediate environs of Mont Blanc than any occasional visitor such as Saussure. But unlike the cosmopolitan Genevan he could hardly generalize from it. Local expertise was (and still is) of much greater importance in the sciences of the earth, which dealt with spatially extended phenomena in the field, than in the sciences that captured or contrived natural processes within a laboratory. But even in the former, local research was guided by and modeled on the work of elite savants such as Saussure, Hamilton, and Werner. They were regarded as having transcended local limitations by reaching conclusions that were worth evaluating—applying, testing, extending, modifying—in every possible region. The work of local experts would have been literally inconceivable without such models, and it remained local in significance unless and until it was incorporated into the more comprehensive work of a savant with a broader viewpoint.

The inclusion of provincial savants and local experts does not exhaust the range of those who made scientific work possible in Saussure's time. Lower down the social scale were the "invisible technicians". They included, in particular, the skilled craftsmen who made the scientific instruments on which the work of leading scientific savants depended. Here some obvious examples are the barometers and thermometers, the hygrometer and electrometer, that Saussure took with him up Mont Blanc. His accurate measurements would have been impossible without the skills of the instrument makers: some of his instruments were made in Geneva to his own design, but in general the craftsmen in London were preeminent at this time.[26]

There were also the less obvious instruments composed of paper and ink: for example, the maps and landscape views that helped to give Saussure the comprehension of physical relationships that he needed, that aided his visual memory of important localities, and that would later give the readers of his *Alpine Travels* an understanding of the Alps beyond what any amount of text could convey (Figs. 1.2, 1.3). Such illustrations were the products of the skills of mapmakers, of artists such as Bourrit and L'Evêque, and of engravers and printers who specialized in making maps and plates for scientific publications. Then there were the printers who produced the texts that accompanied those visual materials; and finally the publishers who took commercial risks to put the books and periodicals on the market. However, while all these craftsmen and tradesmen (and some women) provided the means and the skills, it was the gentlemanly Saussures and Hamiltons who made their work significant. It was the savants who purchased the products of the instrument makers in order to record relevant measurements, who commissioned the artists to produce illustrations for specific scientific purposes, and who wrote the texts that the printers set in type and the publishers made generally available.[27]

The category of invisible technicians needs to be enlarged beyond those just mentioned: in the sciences of the earth, some such phrase must include many others whose role was likewise rarely acknowledged in formal publications. For example, neither Saussure, nor Paccard the previous year, could have reached the summit of Mont Blanc without the practical skills and experience of Jacques Balmat, their guide from Chamonix. Having got there, Saussure could not have made his detailed scientific observations without the aid of three more members of the Balmat family and fourteen other local men, acting at least as porters to carry his valuable and fragile instruments across the crevasses and up the slopes of rock, snow, and ice. None of the scientific work described in this book could have been done without the assistance of countless similar people. Those who helped Saussure to climb Mont Blanc are highly unusual—such was the importance that he attached to the expedition—in being recorded by name, although only in a footnote, by the gentleman who had paid them to risk their lives on the mountain. Most of the others are now anonymous.

In the sciences of the earth, those whose daily labor gave them opportunities to find and collect significant specimens, which might then be offered for sale to savants, were of obvious importance. They included farmers and peasants, who worked on the land and who, for example, might happen to find fossils in the fields. But there were also quarrymen and miners, whose work was more directly involved with rocks and minerals, and who often possessed substantial experience, tacit skills, and other forms of practical knowledge about such materials. The assistance of many such people in socially humble positions was certainly a necessary condition for the scientific achievements of those who paid them for their help or their specimens; but of course it was not a sufficient condition. Balmat and his relatives and neighbors in Chamonix could only contribute in limited ways to Saussure's scientific understanding of the Alps, and of mountains in general, simply because their opportunities, both educational and experiential, were far more limited. Similarly, a Saxon mining foreman, and even an ordinary miner, might possess a wealth of practical knowledge about the mineral veins in the Erzgebirge, or at least those in their own mines, but could hardly form any scientific generalization from it (Fig. 1.9).[28]

Right across this social spectrum, from elite savants to peasants and miners, one half of the human race was almost invisible in the scientific life of Saussure's time (and remained so until the twentieth century). Albertine Amélie de Saussure watched her husband through a telescope; he climbed the mountain, she stayed

26. Many of Saussure's instruments are now on display in Genève-MHS; they give a good impression of the high standard of precision engineering involved: see Archinard, *Collection de Saussure* (1979). His instruments are also illustrated in Turner (A. J.), *Early scientific instruments* (1987), 255–74, pl. XXVIII and pls. 291–330.

27. On the eighteenth-century scientific book trade, see Johns, "Print and public science" (2003).

28. Fig. 1.9 is reproduced from Charpentier, *Mineralogische Geographie* (1778) [London-BL: 457.a.16], pl. 2. The mines of the Erzgebirge are no longer worked, but one on the outskirts of Freiberg is now open as a tourist attraction; it gives a vivid impression of the mining methods used there, which changed little (apart from the replacement of water power by steam for pumping the mines dry) between Charpentier's time and the closure of the mines in the early twentieth century.

Fig. 1.9. Miners at work on narrow vertical veins of silver ore, deep within a mine at Ehrenfriedersdorf in the Erzgebirge [Ore Mountains] of Saxony: an engraving from Johann Charpentier's *Mineralogical Geography of Saxony* (1778). The plebeian miner's hammer was adopted—with great social incongruity—by savants who studied the earth (see the portrait of Saussure, Fig. 1.7). This picture, although probably much tidier than the reality, also gives a good impression of the world of three-dimensional rock *structures* in which miners spent their working lives, and which the science of "geognosy" (§2.3) took as basic for its practice. (By permission of the British Library)

in Chamonix. Women provided background support of many kinds, but it was almost always the men who did the visible scientific work: even safely down in Chamonix, it was Saussure's son, not his wife, who made the precision measurements with the barometer. The exceptions have been given generous attention by modern feminist historians, but in the sciences of the earth those exceptions were few and far between.[29]

Where women do appear in the historical record of the sciences of the earth, it is often for their possession of specific skills for which their limited opportunities were no handicap, or even a positive advantage. For example, some savants (of later generations than Saussure's) relied on their wives or daughters to sketch the scientifically significant landscapes about which they themselves would record only textual notes; or indoors at home the women might produce superbly skillful drawings of mineral and fossil specimens, ready for a professional engraver or lithographer—usually male—to turn into publishable form. In these cases women were often better qualified than men, because accurate drawing was a skill that was taught to young ladies as a standard accomplishment of their social class. Perhaps partly as a consequence of that visual training, such women were often highly skillful at finding fossils in the field, whenever and wherever social proprieties allowed them to venture there: this elusive visual skill had (and still has) no correlation with the scientific expertise needed to interpret the specimens once they were found. Some important private collections were formed and owned by women of high social rank, even if they themselves had not found the specimens but merely purchased them from dealers: for example Georgiana, duchess of Devonshire, amassed a superb mineral collection. Conversely, a woman of low social class and little education could become famous for her own collecting activities and make a modest living from such work: Mary Anning of Lyme Regis in southern England was celebrated among a later generation of savants for her skill in discovering the finest specimens of fossil reptiles, though she did not have the expertise to interpret them scientifically.[30]

The diversity of the social groups who contributed to the work of leading savants was matched by the diversity of those who read or heard about their conclusions. When savants published their work, they were usually concerned above all to be read by the relatively small number of other savants whose opinions they most wanted to influence. But they rarely failed to welcome the fame and respect that their publications might bring them in much wider circles of the educated classes of society: Hutton, who disdained any such popular appeal, was very much an exception. More specifically, many savants relied on the patronage of the most powerful members of society. De Luc published some of his most important scientific work in the form of discursive "letters" to his patron Queen Charlotte, and that highly intelligent woman is unlikely to have left them unread. At the very least, the work of savants such as de Luc and Hamilton was facilitated by the moral support and even the physical presence of their social superiors (Fig. 1.10).[31]

Scientific books such as Saussure's *Alpine Travels* and Hamilton's *Campi Phlegraei* were certainly accessible to, and probably read and viewed by, a wide range of people, including women; and many savants depended to some extent on royalties from their publications to support further scientific work. Even scientific periodicals had a range of subscribers and readers extending far beyond the ranks of those actively producing new scientific knowledge. For every provincial physician, clergyman, lawyer, merchant, or landowner who contributed in some local way to the stock of scientific knowledge, there were many more who were content to read what they had done, and to be instructed and entertained by their work and still more by that of more prominent savants. The interests and expectations of patrons and general readers, on whom the production and sales of many scientific books and periodicals depended, unquestionably affected the authors' style and the presentation of their work. However, as with the direct contribution of wider groups to the production of scientific knowledge, so likewise the indirect role of those who consumed it should not be exaggerated. The leading savants generally wrote and

29. Since women were completely absent from the ranks of the leading savants on whom this book is focused, no apology is necessary for the consistent use of masculine pronouns, etc. Any alternatives, however politically correct, would certainly be historically incorrect (and clumsy).

30. See Torrens, "Mary Anning of Lyme" (1995), and Kölbl-Ebert, "British geology: a female matrix" (2002). Georgiana Cavendish's collection is now on display at Chatsworth-CH (her husband was a kinsman of the famous savant Henry Cavendish). Knell, *Culture of English geology* (2000), describes collecting and collectors in the early nineteenth century, but much was unchanged from the later eighteenth.

31. Fig. 1.10 is reproduced from Hamilton, *Campi Phlegraei* (1776) [Cambridge-UL: LA.8.79], pl. 38. The tacit intentions here imputed to the king are those that Hamilton probably instructed Fabris to portray through his iconography; whether Ferdinand himself—notoriously a stupid and boorish young man—had such thoughts is quite another matter.

Fig. 1.10. Sir William Hamilton demonstrating an incandescent lava flow to the king and queen of the Two Sicilies on the night of 11 May 1771: a drawing made on the spot by Hamilton's artist Pietro Fabris (who depicted himself in the foreground), and published as a colored etching in *Campi Phlegraei* (1776). The king is tacitly acknowledging that an interest in Hamilton's research is appropriate for persons of the highest social rank, and thus is setting an example to his subjects by showing his fearless pursuit of an Enlightened knowledge of nature. The queen has reached the spot in the privileged comfort of a Sedan chair, but her presence and that of her female courtiers in such a potentially dangerous place indicates that women too are appropriate participants in Enlightenment. This particular lava flow originated not in the summit crater of Vesuvius (seen in the background) but lower on its flank. (By permission of the Syndics of Cambridge University Library)

published their work primarily for each other, as their correspondence often makes clear. Up-and-coming savants hoped that their publications would be noticed by the leaders in their field and help establish their reputation; or they wrote to attract the attention of a potential patron, who might advance their career by awarding them a lucrative appointment, and whom they were careful to flatter with a florid dedication. If the publications of savants also brought them fame among a wider public, that was an added bonus.

One specific point about the social composition of both the producers and the consumers of scientific work perhaps needs emphasis here. As already mentioned in passing, clergymen were represented in just the same way as the members of the other traditional learned professions of medicine and the law. Many modern historians, particularly Americans, have projected back into the past their own experience of the kind of fundamentalist religion that is currently so powerful, both politically and culturally, in the United States. That projection has led to a highly anachronistic concept of a perennial "conflict" between the natural sciences and religious practice and belief. In Saussure's time, by contrast, Protestant pastors, Anglican parsons, and Roman Catholic priests were all to be found in some abundance in the ranks of scientific authors; prelates were often as prominent as aristocrats in the lists of subscribers to scientific books, and their names were equally coveted by their savant authors (§1.3). Those who practiced and supported the sciences included many who were also—to put it at its lowest—professionally engaged in the practice of religion (their personal commitment and piety are of course far more difficult to gauge, and doubtless varied widely). Whatever reservations they had about specific knowledge claims by particular savants, they would have found ludicrous any suggestion that the learned world was, or ought to be, divided permanently into two warring camps.

Conclusion

This brief account of the wide range of social groups involved in scientific activity around the time of Saussure's ascent of Mont Blanc—the "golden spike" adopted for this synchronic survey—should justify, or at least explain, the focus adopted in this book. The emphasis will be on the rather small elite of prominent savants with international reputations, such as Saussure, Hamilton, Werner, and de Luc. Other categories of people also played various but lesser roles in the sciences: provincial savants and local experts; artists, printers, and publishers; "invisible technicians", miners, and peasants; women of all social classes; and wealthy patrons and educated readers (including women) with general scientific interests. Their significance must not be forgotten, but neither should it be exaggerated in the interests of an anachronistic egalitarianism or in the name of political correctness.

1.3 PLACES OF NATURAL KNOWLEDGE

Laboratories and museums

The world of the sciences in Saussure's time will next be sketched briefly in terms of the *where* of scientific work. This section outlines the places—spatial, social, and institutional—in which new claims to knowledge were constructed and tested. Again, there was much that was common to all the sciences, both natural and human, but the focus here will be on features that were specific to the sciences of the earth.

Scientific claims were generated (as they still are) in several distinct "places of knowledge". For savants of Saussure's generation, three such sites were of outstanding importance: the laboratory, the museum, and the field. In the modern world, the first enjoys by far the greatest prestige and as a result has received the greatest attention from historians. But in the age of revolution the laboratory was overshadowed, at least in public visibility, by the museum; and the glamour that is now associated with the laboratory was then attached to fieldwork, particularly if it involved travel to distant lands.[32]

In Saussure's time, the experimental laboratory was a site of increasing importance in some of the natural sciences: the famous chemical and physiological research of Antoine-Laurent Lavoisier (1743–94) in Paris was an outstanding example.

32. Ophir and Shapin, "Place of knowledge" (1991); Pickstone, "Museological science?" (1994) and *Ways of knowing* (2001); Livingstone, "Space, science, and hermeneutics" (2002).

In the sciences of the earth, by contrast, the role of the laboratory was marginal, at least as a site for gaining new understanding of the operation of terrestrial processes. Laboratory work was indeed important in the science of mineralogy, but it was used primarily as a diagnostic tool to determine the properties of minerals and thereby to identify and classify them (§2.1). In effect, it simply extended the range of characters used for identification beyond those that could be observed just by handling a specimen in a museum. On the other hand, laboratory *experiments* that were intended to replicate terrestrial processes, or at least to simulate them on a small scale, and thereby to test possible causal explanations of them, were quite limited. Moreover, they were often scorned—and not without reason—by those who had studied the products of those processes on a far larger scale in the field. Hutton, for example, famously dismissed the criticisms of the chemists by saying that "they judge of the great operations of the mineral kingdom, from having kindled a fire, and looked into the bottom of a little crucible." Although there were savants who rejected or ignored such criticism, the value, reliability, and relevance of indoor laboratory experiments were far from self-evident in the sciences of the earth.[33]

Museums, by contrast, had a central place in the practice of almost all the sciences. A museum gathered together collections of *samples* or "specimens" of natural and human products derived from spatially scattered localities: for example, exotic plants and animals from the far side of the world; masks, spears, and other artifacts from primitive tribes; examples of agricultural production, craftsmanship, and industrial manufacture from close at hand; rocks, minerals, and fossils, useful or spectacular, from near and far. The concentration of specimens in one place made possible their comparison, and hence their description, identification, and classification. The handling of materials in a museum was less obviously manipulative than in a laboratory experiment, but museum work did not leave its objects unchanged. Specimens of all kinds would be labeled; dried plants would be mounted on paper, and shells stored in boxes; animal bodies would be dissected and their perishable organs stored in jars of alcohol, their skeletons mounted and their skins stuffed in lifelike poses. Like laboratories, museums were primarily sites of *indoor* work: menageries (in modern terms, zoos) and botanic gardens were in effect their outdoor annexes, adding a modest selection of living—but therefore impermanent—specimens to the far wider range of dried plants and stuffed animals and skeletons preserved permanently indoors.

Museums were the primary places of work of many savants and many of their "invisible" supporters. Some museums were the private "cabinets" of wealthy individuals; others were owned by a royal or princely ruler; the British Museum in London was the first of those that were, in some still ill-defined sense, public property. In practice, all museums were expected to be open to any qualified savant, provided he arrived with a suitable letter of introduction to vouch for his scientific credentials and social respectability. Visits to museums, whether public or private, were prominent items on the agenda of savants when on their travels. For example, when Martinus van Marum (1750–1837), the director of Teyler's Museum in Haarlem, visited Paris for the first time in 1785, he made a point of recording in his diary

Fig. 1.11. A portrait of Georges Cuvier by Mathieu-Ignace van Bree, painted in 1798 after Cuvier's appointment to a junior position in the Muséum d'Histoire Naturelle in Paris (newly reformed from the pre-Revolutionary Cabinet du Roi, or royal museum). On his table, preserved in jars of alcohol, are specimens for his research in comparative anatomy. There is also a compound microscope for examining them in detail (its presence was largely symbolic, since simple lenses were usually more effective at this period). Behind him are the books and periodicals that form the basis for his work, and he is poised to write a text that will add to that published literature. His conventional pose, with head on hand, indicates profound thought, and like Saussure (Fig. 1.7) he looks up to heaven for inspiration.

the many public and private collections he saw day by day, with notes on their more important specimens. Though van Marum was not greatly impressed by it, one of the finest museums was, as already mentioned, the Cabinet du Roi in Paris. A few years after Saussure climbed Mont Blanc, and shortly after Louis XVI lost his head in the Revolution, this great institution would reform itself with a show of egalitarianism and become the Muséum d'Histoire Naturelle (§6.5). One of the aspiring young savants who would then join it was the zoologist Georges Cuvier (1769–1832); the setting he chose for his portrait epitomizes the museum as a place of knowledge (Fig. 1.11).[34]

Museums enlarged their holdings by the purchase of whole collections or outstanding single specimens, for example when a private collector died or went bankrupt and his—or sometimes her—collection was auctioned; by sending junior or subordinate staff into the field to collect specimens and to purchase them from local collectors or people of lower social class; and sometimes by donation. Also of great importance, however, was the practice of exchange, by which "duplicate" specimens were bartered between one museum and another, or one private collector and another. As Cuvier put it later, rather grandiloquently, "this reciprocal exchange of information is perhaps the most noble and interesting commerce that men can

33. Hutton, *Theory of the earth* (1795), 1: 251, responding to criticisms by the Irish chemist Richard Kirwan (see §6.4). Hutton's younger friend James Hall famously postponed publishing his experiments—designed in the hope of supporting Hutton's ideas—until after the latter's death: see Dean, *James Hutton* (1992), 88–89, 140–43. On the role of experiment, see also Newcomb, "British experimentalists" (1990) and "Laboratory variables" (1998).

34. Fig. 1.11 is reproduced from Bultingaire, "Iconographie de Cuvier" (1932), frontispiece. The pre-Revolutionary Jardin (and Cabinet) du Roi is described in Thiéry, *Guide des amateurs* (1787), 2: 172–84. Forbes *et al.*, *Martinus van Marum* (1969–76), 2: 31–52, 220–39, transcribes the relevant part of his diary. On the Jardin and its transformation into the Muséum, see Spary, *Utopia's garden* (2000), and Gillispie, *Science and polity* (1980), 143–84. Laissus, *Muséum d'Histoire Naturelle* (1995), is a superbly illustrated introduction.

have". Social sensibilities were respected by the avoidance of crude monetary transactions; but both parties would keep a sharp eye on the relative values of what was being bartered, whether the specimens were of strictly scientific importance or spectacular and costly items that would enhance a museum's public prestige. The tacit economy of the Republic of Letters was not as far from that of a street market as its outward forms might suggest.[35]

If particular specimens were rare or unique, exchanges might be made in proxy form. A highly realistic drawing or watercolor painting of the specimen would be commissioned from a professional artist and sent instead of the specimen itself; the practice was facilitated by the fact that still-life paintings in trompe l'oeil style were much in vogue in the art world (see, for example, Figs. 2.2, 2.3). In this way the real specimens in a museum could be supplemented with the proxy specimens of a much more extensive "virtual museum" or "museum on paper".[36]

By Saussure's time, museums were no longer regarded just as "cabinets of curiosities", collections of objects chosen for their rarity or oddity; they had become systematic "inventories of nature", in which common or frequent objects were appreciated almost as much as the rare and exceptional. The physical arrangement of museums therefore allowed for a distinction between casual visitors who merely wished to admire the most beautiful or spectacular items, and those who would

Fig. 1.12. A design for the cabinets in the natural history museum at Pavia, drawn around 1785. The more spectacular specimens (in this drawing, corals, sea urchins, and a large starfish) would be on permanent display, on shelves (B) protected behind glass; the bulk of the collection would be stored in the grid of compartments in the drawers (A) below, available for inspection by the more serious visitors or for research by the curator. The ornate design reflects the aesthetic dimension of all museums—including those of natural history—in the late eighteenth century. The specimens in Spallanzani's collection were (and still are) mounted on pedestals or held in dishes of elegantly turned and gilded wood. (By permission of the Archivio di Stato di Milano)

want to make a closer study of a wider range of specimens: as in modern museums, there were facilities both for display and for storage and research. The museum at Pavia, where Lazzaro Spallanzani (1729–99) was professor of natural history, is a good example (Fig. 1.12).[37]

Savants in the field

In contrast to the well-established indoor practice of museum work, it was a relatively novel idea that savants should also engage in outdoor fieldwork. As already mentioned, leading savants commonly dispatched their students, assistants, or other underlings into the field to collect specimens on their behalf. They themselves rarely did so, once their youthful apprenticeship was past, unless they accompanied some major expedition or voyage of exploration. For example, Linnaeus had undertaken arduous travels as a young man, but later became famous for sending his students to the ends of the earth to augment his collections in Uppsala. As a young man, the English naturalist Joseph Banks (1743–1820) had accompanied James Cook on his famous voyages to the Pacific, but in later life he remained in London, sponsoring expeditions that were undertaken by others, which would enrich the British Museum on their return. The reason for this focus on museum work was simple. It was the assembly of specimens in one central location—indoors, in a museum—that made possible their identification and classification, and that therefore rendered the specimens themselves truly scientific. So museum work had the highest priority and prestige; collecting in the field was treated in effect as a means to an end.[38]

This taken-for-granted scientific practice was beginning to be challenged in Saussure's time, not least by Saussure himself. He tirelessly insisted on the importance and value of firsthand outdoor fieldwork, as an activity that was appropriate

35. Cuvier, "Espèces de quadrupèdes" (1800 [1801]), 266; translation in Rudwick, *Georges Cuvier* (1997), 57. Cuvier's massive correspondence shows many traces of these negotiations, which were unchanged in character from the pre-Revolutionary era.

36. See Rudwick, "Cuvier's paper museum" (2000). The idea of a *museo cartaceo* dates from the seventeenth century, when one of the finest and largest sets of proxy images in natural history was assembled (though never published) by Cassiano dal Pozzo, a prominent Italian naturalist and one of Poussin's patrons: see Haskell, *Cassiano dal Pozzo* (1993).

37. Fig 1.12 is reproduced from Spallanzani (M. F.), *Collezione di Spallanzani* (1985), fig. 4 (after a MS drawing in Milano-AS). This gives a full description of the extant collection now preserved and displayed in Reggio-MC: see figs. 5, 6, showing a variety of specimens (including fossil fish) in their cabinets. Hamm, "Goethe's collections" (2001), 282, illustrates one of Goethe's cabinets for minerals in Weimar-GW: less ornate than Spallanzani's but with a similar division between display and storage spaces. Woodward's collection (superbly displayed in Cambridge-SM) is also similar, although several decades earlier: see Price, "John Woodward" (1989). On the earlier style, see Mauriès, *Cabinets of curiosities* (2002), the essays in Impey and MacGregor, *Origins of museums* (1985), and in Grote, *Macrocosmos in microcosmo* (1994), and, on Italy, Findlen, *Possessing nature* (1984). On the later transformation of that style into "inventories of nature", see Olmi, *L'Inventario del mondo* (1992), 165–209, and "From the marvellous to the commonplace" (1997); also Pomian, *Collectionneurs, amateurs et curieux* (1987), for Parisian and Venetian collecting, and, on Spallanzani, Spallanzani (M. F.), "Vom 'Studiolo' zum Laboratorium" (1994). Schaer, *L'Invention des musées* (1993), gives a well-illustrated general introduction.

38. See for example Miller, "Banks, empire" (1996), and Gascoigne, *Science in the service of empire* (1998); Koerner, "Purposes of Linnaean travel" and "Carl Linnaeus" (both 1996); and, more generally, Bourguet, "Collecte du monde" (1997), and Iliffe, "Science and voyages" (2003).

even for a savant as distinguished—both socially and scientifically—as himself (hence his choice of an outdoor setting for his portrait: Fig. 1.7). He championed fieldwork not primarily for Rousseauist or Romantic or gendered reasons: not for the moral virtue of exchanging effete urban luxury for the simple rural life, nor for the spiritual uplift generated by the sublimity of wild nature, nor for the supposed manliness of strenuous physical exercise (though Alpine exploration was later construed by others in all those ways). For Saussure and those inspired by his example, such motives were trivial by comparison with their main reason: fieldwork was demanded above all by the epistemic goal of understanding the *large-scale* features of the terrestrial globe, objects far too large to be collected and taken indoors into a museum (§2.2).

Given this new valuation of fieldwork, a savant had to show that he had indeed seen these features *with his own eyes*, that he had been there and studied them for himself, before he could establish any credibility or authority to pronounce on their scientific explanation or significance. Otherwise he risked being dismissed as indulging in indoor speculation, merely spinning a hypothetical spider's web or

Fig. 1.13. Lazzaro Spallanzani inspecting the interior of the crater of Vulcano (the eponymous volcano) in the Aeolian or Lipari Islands off the north coast of Sicily, during his fieldwork on active volcanoes in 1788. He noted that in this and other pictures he himself was depicted far too large in relation to the volcanic cones, "but the artist believed he ought to be allowed that liberty" because otherwise the savant would have been hardly perceptible. The cone is shown cut away (as in an anatomical drawing) to make him visible exploring the floor of the crater; the slope FG is carefully described as his route into the crater, and the fumaroles LL as among the hazards he encountered on the way. Such details emphasized his physical presence there as a real witness of the volcanic activity. The engraving was published in 1792 in his *Travels to the Two Sicilies* (i.e., southern Italy and Sicily itself). (By permission of the Syndics of Cambridge University Library)

theoretical "system" (§3.1). For example, when Spallanzani made an extensive tour of active volcanoes, he made sure that his university's official artist would record unambiguously his own presence on the spot, although this detracted from the impact of his published illustrations, making them—in effect though not in style—more like diagrams than proxies (Fig. 1.13).[39]

Likewise, Hamilton made sure that he himself was included in many of the pictures for *Campi Phlegraei*, although in his case his authority was not in doubt: he was famous for his extensive fieldwork observations on Vesuvius during his long residence in Naples, and for the frequency of the ascents he made from his villa at its foot. In some pictures Hamilton even got his artist Pietro Fabris to record *himself* recording the volcanic landscape—in Hamilton's presence and to his instructions—thus making doubly explicit the act of creating a proxy that deserved to taken as authoritative by those who had not themselves been there (Fig. 1.14).[40]

Fig. 1.14. Sir William Hamilton directing his artist Pietro Fabris how to record the landscape on the island of Nisida near Naples: a painting published as a colored etching in Hamilton's *Campi Phlegraei* (1776). As in many of its illustrations, the established artistic genre of an Arcadian rural landscape with figures in the foreground was adapted for scientific purposes: Hamilton interpreted the circular Porto Pavone as an ancient volcanic crater that had been breached and flooded by the sea. (By permission of the Wellcome Library, London)

39. Fig. 1.13 is reproduced from Spallanzani, *Voyages dans les Deux Siciles* (1795–97), 3/4 [Cambridge-UL: Gg.29.48], pl. 5, first published in *Viaggi alle Due Sicilie* (1792–97), 1, pl. 2 (explained on 2: 186), engraved from a drawing by Francesco Giuseppe Lanfranchi. The quoted comment (*Voyages* 1: 225n) is not in *Viaggi*, and was probably added in response to criticism that the pictures gave a misleading impression of the size of the volcanoes. The French edition had an introductory essay (1: 1–74) by Saussure's Genevan colleague Senebier; an English edition followed in 1798. The essays in Montalenti and Rossi, *Lazzaro Spallanzani* (1982), deal mainly with his biological work; but Vaccari, "Spallanzani and his geological travel" (1998), describes his volcanic research and includes fine reproductions of Spallanzani's other plates.

40. Fig. 1.14 is reproduced from Hamilton, *Campi Phlegraei* (1776) [London-WL: V25268], pl. 22. Nisida lies just offshore, about 10km southwest of the center of Naples. Fabris also depicted himself in the scene reproduced here as Fig. 1.10. Seta, *L'Italia del Grand Tour* (1992), reproduces fine examples of the contemporary genre of topographical landscape art in which Fabris was working.

Fig. 1.15. Ladies and gentlemen exploring the natural wonders of the Giant's Causeway on the coast of Antrim (in what is now Northern Ireland), with lower-class figures in the foreground: a detail from one of a pair of large engravings based on prizewinning paintings by the Irish artist Susanna Drury. These prints, published in 1744, had first made this locality famous throughout Europe for its superb display of regular hexagonal jointing in basalt: the dark fine-grained rock was highly controversial, being attributed either to a volcanic or to a sedimentary origin; the jointing was equally puzzling, being regarded as analogous either to the forms of crystals or to shrinkage cracks in mud (see §2.4).

However, it was not only savants who increasingly ventured into the field to see for themselves the large-scale features of scientific significance. For example, as already mentioned, aristocrats and gentlemen from all the European nations included Vesuvius on their Grand Tour, as well as the nearby Classical sites of Pompeii and Herculaneum and the famous Greek temples at Paestum; Hamilton complained at having his time wasted by visiting grandees who expected to be given conducted tours of the volcanic sights by the great savant himself. Many other natural sites and sights, scattered around Europe, were included in the nascent tourist itineraries, provided they could be reached without inordinately strenuous exertion; and some at least were even considered suitable for ladies (Fig. 1.15).[41]

The social life of savants

New claims to scientific knowledge were generated, then, in "places of knowledge" such as laboratories, museums, and the field, though for the sciences of the earth the first was of little importance. However, such claims remained no more than ideas inside the head of a savant until they were expressed at least in private to his friends, colleagues, and associates, or more openly at a scientific meeting, or in fully public form in an article or book. Such places constituted a gradient of relative privacy, on which mere claims to knowledge might become consolidated into established consensual "facts" about the world, or might disintegrate as their plausibility was eroded under the impact of expert criticism or further observation.[42]

Knowledge claims could be evaluated privately, of course, by trying them out on well-informed friends, for example in a coffeehouse or over a meal at home. In the main urban centers, the regular salons and soirées held in the homes of prominent savants—and, importantly, their wives and other female *salonnières*—were a more organized way in which scientific observations and ideas could be discussed informally in a congenial atmosphere. Such assemblies graded into the meetings of the informal private scientific clubs that flourished in urban settings in several countries, notably the Netherlands and Britain. Such bodies graded in turn into more formal—but still privately funded—societies, ranging from those based in a particular town or city to those on the national level. Here the Royal Society in London was preeminent, both as the first of its kind (founded 1660) and as the largest; but it had many imitators, such as the Hollandsche Maatschappij in Haarlem and the Royal Society in Edinburgh.[43]

Most such societies were very general in their scope, embracing *all* the sciences, both natural and human; some, such as the Royal Society of London, confined themselves more or less to the natural sciences, mathematics, and some of the technologies; only a few were still more specialized. All over Europe, prominent savants would belong to at least one such scientific body, if not to several; they might also find themselves elected to honorary membership of those in other countries, a custom that conferred honor and credit in both directions (the example of Saussure's election to the Royal Society in London has been mentioned already). These societies were important in scientific life, not only as places where work could be presented and discussed, but also because their libraries commonly provided resources far beyond the reach of their individual members. Scholarly books were often very costly, particularly if they were illustrated, and many could be owned only by the wealthiest savants. For those with more limited means, membership in a society with a good library was a practical necessity, unless they had access to the private library of an affluent patron with scientific interests. Saussure had to spend huge sums on scientific books, ordering them from booksellers all over Europe; but he was wealthy enough to do so, and he had no alternative, since there was no good institutional library in Geneva. By contrast, the professors at Göttingen, the most

41. Fig. 1.15 is reproduced from Drury, *Giant's Causeway* (1744), "West prospect", one of two large prints by the leading landscape engraver François Vivarès. Drury's original paintings had been submitted anonymously, apparently to avoid anticipated prejudice against a woman artist: see Anglesea and Preston, "Susanna Drury and the Giant's Causeway" (1980). Her primary purpose in depicting human figures was to indicate the scale of the columns that she described in detail in the accompanying text, but the inclusion of gentlewomen did also suggest the practicality and social propriety of their visiting the site, which a male artist might not have considered worth showing. How often women in fact visited the Causeway cannot of course be inferred from her iconography alone.

42. The notion of a gradient of relative privacy was introduced to help make sense of a geological example: see Rudwick, "Darwin in London" (1982).

43. McClellan, *Science reorganized* (1985), surveys such societies throughout Europe; Uglow, *Lunar men* (2002), describes the small and private Lunar Society, which had a crucial role in the early Industrial Revolution in England; on the scientific culture of London coffeehouses, see Stewart, "Other centres of calculation" (1999), and Levere and Turner, *Discussing chemistry and steam* (2002). Mijnhardt, *Tot heil van't menschdom* (1987), describes societies in the Netherlands, which are also reviewed in Snelders, "Professors, amateurs, and learned societies" (1992), and analyzed in Roberts (L.), "Going Dutch" (1999). Vaccari, "Accademia dei Fisiocritici" (2000), describes the institution in Siena, which paid particular attention to minerals and fossils; another, in Berlin, is described in Heesen, "Natural historical investment" (2004).

innovative university in Europe in terms of research, had access to what was probably the largest academic library in the world.[44]

Savants on their travels took every opportunity to attend meetings of the societies in the cities they visited and thereby to meet others with similar interests; as in the case of museums, suitable letters of introduction acted as passports throughout the Republic of Letters. But traveling—usually by public coach on badly surfaced roads, or more circuitously by sailing ship—was expensive, uncomfortable, and even hazardous, and was not undertaken lightly or frequently. Face-to-face contacts were therefore supplemented, and for many savants largely replaced, by the proxy meetings and conversations made possible by correspondence. Postal services were well developed and quite reliable; the costs were high, and were borne not by the writer but by the recipient, but this was treated as an acceptable expense among savants. In an age without telephones or electronic mail, the art of fluent letter writing was a standard accomplishment in the educated classes, and it was one that was exploited regularly by savants throughout Europe, and indeed as far as Asia and across the Atlantic. The exchange of letters between savants was quite as important in scientific debate as the meetings of scientific societies or the formal publication of articles and books. Individual savants took great care to build up and maintain their own networks of correspondents, as human investments on which they could expect a lifetime's return.[45]

Scientific publication

Sooner or later, however, a savant would seek a wider audience for his work by having it published. Scientific books were still the primary form of publication, and some publishers specialized in scientific work. Unless the book seemed likely to attract a large sale, it was common practice to print and distribute a brief prospectus for it and invite subscriptions in advance. In this way the publisher could be sure of covering his costs before having to invest in paper or commission an engraver to make the illustrations—usually the two largest single costs—or to start the printing and binding: Paccard's plans for an account of the very first ascent of Mont Blanc were probably aborted when he failed to attract sufficient subscribers (§1.1). Conversely, by paying for the book before they saw it, subscribers were expressing confidence in the author's work, or taking a calculated risk about it; in return, they would at least have their names recorded in print in the book itself and could be publicly seen to have been patrons of scholarship. Authors would hope to give their work social prestige and respectability by putting the names of royal or princely personages, aristocrats and prelates, at the head of their subscription lists (which therefore provide a revealing sample of the kind of readership that a work attracted).[46]

Illustrations were vitally important in many of the sciences, not least those concerned with the earth; but they added greatly to the costs of book production, and for that reason were used sparingly (and therefore give a valuable indication of what the author regarded as most important about his work). The four volumes of Saussure's *Alpine Travels*, for example, contained only twenty-three plates scattered among their 2,379 pages, and Hamilton's lavishly illustrated *Campi Phlegraei* was very much an exception. Engraving on copper was virtually the only medium used in scientific books; it was expensive both in copper itself and in the highly paid skills of the engraver (cheaper media such as lithography and wood engraving began to relieve this situation in the early nineteenth century). Illustrations were even more expensive if they were colored, since the only method available was to have the colors added by hand, usually by female labor, to each copy in turn. So unless a work was aimed at the luxury end of the market—as, for example, Hamilton's was—colored illustrations were used only where they were really essential, for example in books on birds and flowers, and for some maps.[47]

By Saussure's and Hamilton's time, the publication of scientific work in the form of relatively short articles rather than lengthy books was rapidly increasing in importance; many scientific periodicals, in a burgeoning number and variety, were being published throughout Europe and even beyond it. Some were linked to specific scientific bodies, as their *Transactions* or *Memoirs* (or their equivalents in other languages), and only contained papers that had been read out formally at a meeting: outstanding examples were the *Philosophical Transactions* of the Royal Society in London and the *Mémoires* of the Académie Royale des Sciences in Paris. Other periodicals were commercial ventures, owned either by the editor himself or by the publisher. Most of these independent periodicals were very general in scope and confined themselves to brief descriptions of recent publications: the long established *Journal des Savans*, for example, reported monthly on work by savants of all kinds, as its name implies, throughout the Republic of Letters; and the weekly *Göttingische gelehrte Anzeigen* [Göttingen erudite reports] similarly covered the whole range of learned or *wissenschaftlich* studies for the German-speaking world, summarizing new publications received from all the main centers of learning in Europe.[48]

Some independent periodicals, however, focused specifically on the natural sciences and related fields—though that still left them with a wide scope—and published original research papers in almost modern style, as well as reporting on recent publications elsewhere. The most important was the *Observations sur la Physique* edited in Paris, the full title of which added "sur l'histoire naturelle et sur les arts": "physics" still covered chemistry and much else, "natural history" included the sciences of the earth, and "arts" covered crafts and technologies (see §1.4). In 1786 one of its editors, Jean-Claude de La Métherie [or Delamétherie] (1743–1817), began to include in it an annual review of the scientific research published in the preceding year. This was not only a service to his fellow savants throughout Europe,

44. Saussure's book-buying activities are reconstructed and analyzed in detail in Carozzi and Bouvier, *Library of Saussure* (1994); on Göttingen, see Clark, "Bureaucratic plots" (2000).

45. Outram, "Cosmopolitan correspondence" (1983), makes this point for the case of Cuvier at a somewhat later period, but the pattern was unchanged from the pre-Revolutionary era.

46. Darnton, "History of reading" (1991); also Johns, *Nature of the book* (1998), a magisterial case study of comparable issues in early-modern England.

47. On Saussure's engravings, see Ripoll, "Iconographie des *Voyages*" (2001).

48. On the *Anzeigen*, see Clark, "Bureaucratic plots" (2000).

but also a significant indicator of the accelerating pace of scientific activity. Only a few periodicals specialized still further, in particular sciences: for example, the *Bergmännische Journal* was founded (in 1788) by the Bergakademie in Freiberg in order to service the sciences relevant to mining. Savants who found foreign languages hard going, or who had limited access to the full range of newly published work, were well served by the translated excerpts, summaries, and reviews that appeared in all these periodicals.[49]

Conclusion

In Saussure's time, claims to new scientific knowledge were generated in laboratories, in museums, and in the field. For the sciences of the earth, laboratories were as yet almost negligible, museums had long been of great importance, and fieldwork was rapidly growing in prestige and significance. Such claims were then evaluated by the social processes of debate among savants. The places where this took place were ranged along a gradient of relative privacy, from conversations and correspondence, through meetings of informal or formal societies, to full publication as scientific books and—increasingly—as articles in scientific periodicals. By these means, claims to new knowledge were in effect mobilized by being disseminated outwards from an individual savant and his immediate colleagues to ever wider circles of others, and thereby tested and evaluated. In the course of time they were either consolidated into consensually accepted and established "facts", or undermined and rejected as spurious or invalid.

1.4 MAPS OF NATURAL KNOWLEDGE

The literary and the philosophical

The third and last part of this brief survey of the scientific world around the time of Saussure's ascent of Mont Blanc deals with the tacit mental "map of knowledge" on which the various sciences were situated and related to each other. In other words, having outlined *who* was practicing the sciences (§1.2) and *where* the scientific knowledge was being generated and evaluated (§1.3), it remains to sketch in this section *what* it was that was being investigated. Specifically, this will serve to locate the sciences of the earth in relation to other kinds of knowledge.

The word "savants" has been used in this chapter as if they were an undifferentiated social group; in reality, of course, savants had diverse interests, skills, and competences. Specialization was far less pronounced than in the modern world; but most savants, however wide their interests, built their reputations in one specific field, or perhaps in two. Hamilton, for example, was renowned for his volcanic and antiquarian studies, Saussure for his work on the geography of the Alps and for his meteorology. Polymaths such as Goethe, who ranged with distinction even more widely, were almost as rare as they are today, and as much admired.

The mental map of knowledge in the late Enlightenment—the way in which the sciences were classified into various kinds of study—was strikingly different from its modern equivalent. The contrast is only incidentally a result of the vast growth of research in the past two centuries, and the consequent specialization of knowledge into ever-smaller fields and subfields. Primarily it is due to major differences in the way that the various kinds of knowledge were conceived in relation to each other. The modern classification of the sciences is not simply a direct and unmediated reflection of how the world is, but rather a contingent product of specific historical developments. The classification that it replaced, which by Saussure's time had long been taken for granted, owed its main outlines to the ancient concept of the varied human "faculties", as elaborated in the early seventeenth century by Francis Bacon. It had been made explicit, for example, in the tabular chart that Denis Diderot (1713–84), the editor of the famous multivolume *Encyclopédie*, had prefixed to this supreme manifestation of the Enlightenment (Fig. 1.16).[50]

As Diderot's chart shows, three faculties were relevant to the classification of the sciences, because they generated three major kinds of human knowledge: "memory", "reason", and "imagination" generated respectively "history", "philosophy", and "poetry". These words differed, however, from their modern usage. "History" embraced all those sciences that aimed to *describe* the diversity of the world; "philosophy" incorporated those that sought to *explain* how the world works. "Poetry"—to which the practical-minded Diderot gave short shrift—comprised all kinds of imaginative products, including drama, music, and the visual arts as well as poetry and novels.

In Saussure's time, a map of knowledge broadly of this kind was embodied in the academies, societies, and universities in which research and teaching were carried out. Many societies had separate "Literary" and "Philosophical" sections (or their equivalents in other languages): for example, in 1785 Hutton read a paper on his "theory of the earth" (§3.4) to the newly founded Royal Society of Edinburgh, which was later published in the "physical" part of its *Transactions*, while in the following year he contributed a paper on linguistics to its "literary" section. "Literary" savants included not only those who created works of literature, or studied it in either its ancient or contemporary forms, but also, for example, philologists, historians, and theologians, and those who were concerned with legal theory, moral philosophy, and metaphysics. Savants who regarded themselves as "philosophers" included those who studied any of the natural and medical sciences and mathematics, and those who were concerned with at least the more fundamental aspects of practical crafts and technologies. In order to denote that latter range of studies,

49. It is therefore realistic to assume, as in the narrative in Part Two, that any important scientific claims made in print in one part of Europe would generally become known to savants with similar interests elsewhere, at least in outline, within a year or two. Broman, "Periodical literature" (2000), reviews those of the later eighteenth century; there was an almost exponential increase in their number during the decades covered by the present book. La Métherie, "Discours préliminaire" (1786), was the first of its kind, and prompted by the controversies about the chemistry of gases; by this time he was de facto the sole editor of *Observations*. On its earlier years under Rozier, see McClellan, "Scientific press in transition" (1979).

50. Fig. 1.16 is based on the "Systême figuré des connaissances humaines" in Diderot, *Encyclopédie* 1 (1749), folding plate at end of "Discours préliminaire"; the original typographical design used a system of brackets that makes the hierarchy of categories difficult to discern. See Darnton, *Great cat massacre* (1984), chap. 5, and "Epistemological angst" (2001); also Yeo, *Encyclopedic visions* (2001), 27–32, 120–44, and "Classifying the sciences" (2003).

HISTORY (MEMORY)

- Sacred History (history of prophecies)
- Ecclesiastical History
- Civil History, ancient and modern
- **Natural History**
  - **Uniformity of Nature** [regular, law-like]
    - Celestial natural history [astronomy]
    - **Natural history of weather** [*météores*], **of earth and sea, of minerals,** of plants, of animals, of [chemical] elements
  - **Lapses** [*écarts*] **of Nature** [irregular, "accidents"]
    - Celestial prodigies
    - **Prodigies of weather** [*météores*], **of the earth and sea; monstrous minerals;** monstrous plants; monstrous animals; prodigies of the [chemical] elements
  - Uses of Nature (arts and crafts, manufactures)
    - [crafts and trades, listed as "works and uses" in gold and silver; precious stones; iron; glass; skins; stone, plaster and slate; silk; wool; etc.]

PHILOSOPHY (REASON)

- General metaphysics or ontology (science of being, [etc.])
- Science of God
  - Religion (or, by abuse, Superstition)
    - Natural theology; Revealed theology
  - Divination, black magic
- Science of Man
  - Pneumatology (science of the "soul" [psychology])
  - Logic
    - Arts of thinking (ideas, induction etc.); of retaining (memory, writing, etc.); of communicating (grammar, rhetoric etc.)
  - Ethics [*morale*]
    - General (good and evil, duty, virtue etc.)
    - Particular (science of law; jurisprudence etc.)

Science of nature [**Natural philosophy**]

- Metaphysics of bodies, or general physics (extension, motion, the void etc.)
- Mathematics
  - Pure mathematics
    - Arithmetic (numbers, algebra); Geometry (including theory of curves)
  - Mixed [applied] mathematics
    - Mechanics (statics, dynamics); Geometrical astronomy (cosmography etc.); Optics; Acoustics; Pneumatics [gases];
    - Art of conjecture, analysis of risks [statistics]
  - Physico-mathematics
- **Particular physics**
  - Zoology (anatomy, physiology, medicine etc.)
  - Physical astronomy, astrology
  - **Meteorology**
  - **Cosmology**
    - Uranology ["physical" astronomy]
    - **Aerology, Geology, Hydrology** ["physics" of atmosphere, lithosphere and hydrosphere]
  - Botany (including agriculture, horticulture)
  - **Mineralogy** ["physics" of minerals]
  - Chemistry (including metallurgy, alchemy etc.)

POETRY (IMAGINATION)

- Sacred and Profane
  - Narrative Poetry (epic, epigram, novel, etc.); Dramatic Poetry (tragedy, comedy, opera etc.); Parabolic Poetry (allegories);
  - [also, not clearly assigned to these categories:] music, painting, sculpture, architecture, engraving

Fig. 1.16. Denis Diderot's "Diagrammatic system of human knowledge", illustrating the introductory essay in the first volume (1749) of the *Encyclopédie*: here rearranged and abridged to clarify the hierarchical relations of the various branches of knowledge or sciences (editorial explanations, and some significant words in the original French, are given in *square* brackets). The categories (in **bold** type) that contain the subject matters of the modern earth sciences are scattered across the classification, distributed between "natural history" and the part of the "science of nature" that was termed "physics". ("Geology" features as a subordinate category, but not in its modern sense). The category of "lapses" or "prodigies" was used here to cover instances that seemed anomalous because they lay outside the normal lawlike regularities of nature.

Fig. 1.17. James Hutton (left) and Joseph Black portrayed as "philosophers": an etching made in 1787 by John Kay and sold at his print shop in Edinburgh. Although they seem to have been drawn separately, their juxtaposition in the published print—as if in conversation—expressed appropriately their participation in the primarily face-to-face and usually indoor activity of intellectual debate: it was as if they were discussing, say, Hutton's "theory of the earth", which he had expounded at the Royal Society of Edinburgh (and part of which Black had read for him there) just two years earlier. (By permission of the Syndics of Cambridge University Library)

the Royal Society in London entitled its periodical—the first of its kind in the world (1665)—the *Philosophical Transactions*; its junior transatlantic counterpart was called the American Philosophical Society. So when an Edinburgh artist surveyed the social and cultural elite of the new Athens of the North, and gently caricatured all its principal luminaries (by coincidence, in the year that Saussure climbed Mont Blanc), he featured Hutton and his friend the chemist Joseph Black (1728–99) as "philosophers" (Fig. 1.17).[51]

In effect, the distinction between the "literary" and the "philosophical" was between sciences that dealt primarily with the study of words and texts of all kinds (and with artifacts such as Greek vases that might illuminate them) and those concerned with natural objects and physical phenomena of all kinds (including those of the human body), and the conceptual, causal, and mathematical relations between them. The distinction corresponded roughly to the modern division between the humanities and the natural sciences. However, it was no sharp dichotomy, either in subject matter or in terms of the individuals involved; many savants did important work in both kinds of *Wissenschaft*, as the cases of Hutton and Hamilton illustrate.

Natural history and natural philosophy

As Diderot's chart shows, there was a major distinction between two complementary ways of studying the natural world. "*Natural history*" dealt with the description and classification of natural phenomena and natural objects of all kinds. "*Natural philosophy*"—or what Diderot called the "science of nature"—included the causal and mathematical relations between natural phenomena, as well as mathematics itself. Those who engaged in these two kinds of work were generally known as "*naturalists*" and "*natural philosophers*" respectively (like the word "savant", these terms will be used throughout this book, in order to avoid the anachronism of referring to "scientists").

Just as "philosophy" had a meaning quite different from its modern usage, so likewise the word "history" retained its original meaning of a description. If a book was entitled a "history" it did not necessarily—or even usually—imply that it contained any kind of temporal narrative. William Marsden's *History of Sumatra* (1783), for example, was primarily a description of the *present* "government, laws, customs, and manners of the native inhabitants" of that exotic island, and of its "natural productions"; only the final section dealt with the "ancient political state of the kingdom of Acheen" [now Aceh], and even that was little more than a chronicle of early visits by European travelers. In general, the modern meaning of "history" was no more than a secondary sense, denoting those descriptions of things that were organized on a basis of chronology rather than by some other criterion. The phrase "natural history" therefore denoted the description of the natural world, and the orderly classification of its diversity, without any temporal connotations whatever: a "natural history of mammals" did not imply any perception that mammals might have had a history in the modern temporal sense of a biohistory, let alone an evolutionary one.[52]

The description and classification of natural things were regarded as distinct from the study of their causal and mathematical relations, but the two kinds of work were treated as being of equal importance and difficulty. The title of naturalist was one that any savant would wear with pride, and certainly without any sense of inferiority to those who were natural philosophers. Some savants were both—Saussure is an example—but they were clearly aware of which role they were acting in while engaged on any given piece of scientific work. Natural history enjoyed the same high prestige as natural philosophy: the former had none of the pejorative or dismissive overtones that it often has in the modern world, particularly among scientists. Isaac Newton's famous book on the workings of the solar system under the laws of universal gravitation—his *Principia* or "Mathematical principles of natural philosophy" (1687)—was regarded as the supreme exemplar of its genre. But its prestige among Saussure's contemporaries was equaled, and complemented, by that of the *Systema Naturae* (in ever-expanding editions from 1738) by Linnaeus and the *Histoire Naturelle* (in many volumes from 1749) by the director of the Jardin du Roi in Paris, Georges Louis Leclerc, count Buffon (1707–88). Both these great works were celebrated as masterly surveys of the diversity of the natural world, even though their contrasted approaches split naturalists into rival camps.

The major division between natural history and natural philosophy gave the implicit map of knowledge of the late Enlightenment a character quite distinct from our modern classification of the natural sciences. Specifically, it cut right across the modern dichotomy between the physical and the biological sciences.

51. Fig. 1.17 is reproduced from Kay, *Original portraits* (1838), 1 [Cambridge-UL: Lib.4.83.3], no. 25. This is a collection of his etchings published in book form half a century after they were drawn and first sold (see also Fig. 3.2). Another etching of Hutton, entitled "Demonstration" (no. 99), showed him conversing outdoors with another Edinburgh savant, Lord Monboddo. Hutton, "Theory of the earth" (1788) and "Written language" (1790). The meaning of "philosopher" remained unchanged throughout the period with which the present book is concerned.

52. Marsden, *History of Sumatra* (1783). The descriptive meaning of "history" reached back to Antiquity, as for example in the Latin title (*Historia animalium*) of Aristotle's classic work.

Natural history embraced the description and classification of *all* kinds of natural entity, including even the celestial: Herschel treated the diversity of the puzzling objects that he called "*nebulae*" as a form of cosmological natural history. Hume's famous essay "On the natural history of religion" (1757) had been shocking precisely because, provocatively, it extended the category to the ultimate degree. For terrestrial entities, nothing would be left out if one's first question—as in the traditional guessing game—was "animal, vegetable, or mineral?". The mineral "kingdom" had parity with the other two, and "mineralogy" was often used in a very broad sense to include everything in it. Both Linnaeus and Buffon had ambitions to include all three kingdoms in their surveys of natural diversity, though in the event they got little further than botany and zoology respectively. Minerals and rocks, no less than animals and plants, were classified into orders, families, genera, and species (§2.1): such terms had no necessary association with living things, still less any evolutionary connotations.[53]

Natural history included the description of much of the inorganic world; conversely, important parts of natural philosophy dealt with organic phenomena. Physiology, for example, investigated functional and therefore causal relations within the living body, including most importantly the human body. In fact, as Diderot's chart showed, much of natural philosophy (including physiology) was often referred to as "*physics*"—in a sense that went right back to Aristotle—because it investigated the *causal* relations between natural entities of all kinds. Saussure's work on meteorology, and his aspiration to discover the causes of Alpine weather patterns, qualified him to be described—for example on the prints that recorded his ascent of Mont Blanc (Figs. 1.1, 1.4)—as Geneva's "célèbre physicien".

Precise measurement was coming to be seen as vital for such causal enquiries, since it might provide the data for formulating quantitative and even mathematical "natural laws" to match those of the great Newton himself. Those who practiced any kind of "physics" within natural philosophy therefore prided themselves on their efforts to quantify the phenomena they studied. They were of course dependent for this on the craftsmen who constructed instruments of ever-increasing accuracy. The many instruments that Saussure's porters carried up Mont Blanc were typical of an even wider range that was becoming available to his generation. His own almost obsessive concern to measure every imaginable phenomenon on the top of the mountain (§1.1) was characteristic of what modern historians have seen as the "quantifying spirit" of the late Enlightenment. Nor were the sciences of natural history left behind in this drive for exactitude: even if it was not usually expressed in quantitative terms, the accent—as seen again quite typically in Saussure's work—was on the utmost precision in description and classification.[54]

However, the sciences classified respectively as natural history and natural philosophy (or "physics") were not fixed for all time. For example, much of chemistry dealt with the description of substances and their classification, as part of natural history (chemicals too were arranged in genera and species); but the more recent kind of experimentation was shifting certain parts of chemistry in the direction of causal "physics". Lavoisier's self-styled "revolution" in chemistry affected both sides: it entailed both a reform of the classification of substances and new causal interpretations of familiar reactions between them. The next chapter shows how the sciences of the earth likewise straddled the boundary between natural history and natural philosophy, and the rest of the book traces their subsequent changes in relation to that boundary.[55]

Philosophy and theology

As in the sketch of the scientific community of Saussure's time (§1.2), here too the historical relation between "science" and religion needs to be rescued from the anachronism (and the ideology) of intrinsic and perennial conflict. Diderot, for example, like other leading French *philosophes*, was often highly critical of religion, seeing in his own Catholic world—as he put it in his chart—mainly its corruption by "abuse", leading to "superstition"; yet it is significant that he did nonetheless include it in his classification. Indeed, the religious realm figured in each of his three great divisions of human knowledge: "history" included "sacred" history; "poetry" embraced artistic and imaginative products both "sacred and profane"; and, above all, "philosophy" included the "science of God" or theology (Fig. 1.16). He thus conceded implicitly that theology generated *claims* about knowledge of God, to put it no higher, just as other sciences made claims about knowledge of nature and humanity.

Elsewhere in Europe, outside France, Enlightenment savants were often more sympathetic to theological claims. Theology was a respected field of study at almost all universities, and in some—notably those in the German lands—it was a field marked by distinguished scholarly research. Specifically, philological work on the biblical texts, carried out in a context of research on oriental languages and ancient history in general, was leading to a new appreciation of the cultural setting of the biblical documents. That in turn was generating a burgeoning field of textual criticism, which made the older style of biblical literalism clearly obsolete, not least in terms of religious practice itself. Some earlier critics of the Bible, such as Spinoza, had hoped and expected that textual analysis would undermine the foundations of traditional religion and make it intellectually untenable. But in Saussure's time many biblical critics believed that their work could well be in the service of religion. For example, those in the faculties of Protestant theology in German universities such as Halle and Göttingen were responsible for the training of pastors; they believed they were equipping their students with the intellectual tools they would

53. Schaffer, "Herschel in Bedlam" (1980); Hume, *Four dissertations* (1757), 1–117. Rudwick, "Minerals, strata, and fossils", and other essays in Jardine et al., *Cultures of natural history* (1996), review "natural history" as a historical category that is *not* synonymous with "biology"; see also Spary, "The 'nature' of Enlightenment" (1999), and, more generally, Pickstone, "Museological science?" (1994). In Lepenies, *Ende der Naturgeschichte* (1976), the title of the section "Von der Naturgeschichte zur Geschichte der Natur" (52–77) neatly summarizes the change of meaning with which the present book is also concerned.

54. See the essays in Frängsmyr *et al.*, *Quantifying spirit* (1990); Bourguet and Licoppe, "Voyages, mesures et instruments" (1997), sets Saussure's measurements on Mont Blanc in their context; Golinski, "Barometers of change" (1999), uses the barometer to analyze the culture of instrumentation in general.

55. On Lavoisier and his contemporaries, see for example Holmes, "Revolution in chemistry and physics" (2000), and Golinski, "Chemistry" (2003).

need to exercise a Christian ministry that would at the same time embody the ideals of Enlightenment scholarship.[56]

The importance of biblical criticism deserves emphasis. As suggested already, many historians now project the literalism of modern fundamentalist religion back into the intellectual world of the eighteenth century, with gross anachronism. In fact, attitudes to biblical interpretation—among those to whom such questions were matters of any concern—varied widely according to time, place, religious tradition, and above all social location. There were of course some writers and preachers, both Protestant and Catholic, who claimed that the meaning of specific biblical texts was obvious and unambiguously literal; and their readers and hearers often agreed with them. But there were other scholars who, following much older traditions, argued that those texts might have many layers of meaning, poetic and symbolic, allegorical and typological, which in religious terms might be far more significant. For them, the new biblical criticism could have a further liberating effect: it could clarify what the original writers might have intended and what their original readers might have understood, and hence it could facilitate the necessary translation of religious meaning from those ancient cultures into their own time.

Diderot's treatment of the "science of God" expressed what others too took for granted, in being divided into two distinct kinds of theology. Critical methods of biblical interpretation mainly affected the historical claims of "*revealed*" theology, that is, the claims to knowledge of God arising from what believers regarded as divine self-disclosure in and through the course of the historical events recorded in the biblical documents. This, however, was thought to be supplemented, complemented, or even supplanted—the relation was itself a matter of intense controversy—by the claims of "*natural*" theology, that is, the evidence for the divine that was believed to be discernible in the natural world, including human nature itself. More specifically there was a long tradition, among those who practiced the natural sciences, of searching for evidence of God's providential "wisdom" in the "design" of the natural world. This "*argument from design*" was set out most persuasively by those who claimed that they perceived it in the correlation between structure and function in living organisms (in modern terms, in the phenomena of adaptation); searching for design in nature had long been an important motivation for pious savants who did that kind of scientific research.[57]

In the practice of the sciences of the earth, however, the argument from design and other forms of natural theology were of relatively little importance, at least by Saussure's time. In contrast, the historical claims made by those sciences did have potentially important points of contact with revealed theology. For example, scientific claims about traces of a physical "deluge" or watery catastrophe in the distant past—or, conversely, the rejection of such claims—could obviously impinge on questions of the historicity and religious significance of Noah's Flood as recorded in Genesis (§2.4, §2.5). But that point of contact was not necessarily treated as material for conflict: it might just as well be interpreted in terms of a corroboration between natural and human forms of evidence. Conflict was not the only kind of relation between the science of theology and the sciences of nature, and certainly it was not widely regarded as intrinsic or inevitable.

The relation between theology and the natural sciences, like other aspects of the map of knowledge, was not just an abstract issue, but was embodied in concrete social and institutional forms. At specific times and in specific places, there was indeed ecclesiastical suspicion or even hostility towards certain claims to natural knowledge. But this was more on the grounds of their heterodox theological implications than because they contradicted a literalistic reading of biblical texts. For example, recurrent speculations about the natural origin of life, and more particularly of the human species, were often suspect. The suspicion arose, however, not just because such ideas seemed to contradict the literal sense of the creation narratives in Genesis; anyway, biblical scholars—and through them, the educated laity too—were well aware, for example, of the multiple meanings of the Hebrew words translated as "create" and "day". What often made this kind of speculation unacceptable was that it was seen as undermining the theistic concept of creation itself, with its radical distinction between God and *all* kinds of "creature"—from atoms to humans—and all the personal, social, and political implications that flowed from that emphasis on divine transcendence. In other words, certain claims in natural philosophy were objectionable to the religious because they were also tacit claims about the proper conduct and ultimate meaning of human lives: it was a case of one religion pitted against another, even if one of them was often called antireligious or even atheistic (see also §2.5).

Most importantly, however, the relations between scientific savants and ecclesiastical authorities varied greatly from one cultural environment to another. The highly centralized and absolutist French state, with Catholicism as its established form of religion, tended to equate any philosophical speculation with potential sedition. Yet even here the official censorship was generally either lax or relaxed: those responsible for checking the credentials of new books were often themselves savants, and might be sympathetic to their authors' scholarly goals; and anyway, in case of difficulty, books could be published in Amsterdam or Geneva and then shipped back to France, or else be printed in France but issued under the false imprints of such safely foreign cities. In England, by contrast, the political and ecclesiastical establishments were intertwined in such a way that it was usually taken for granted that sound scientific work was more likely to support religious practice than to subvert it; the doctrinally liberal "latitudinarian" party in the established Anglican church was generally able to call the tune. In the German and Italian lands the sheer multiplicity of sovereign states generated a political and cultural pluralism that ensured that most kinds of scholarly work could find a congenial home somewhere: even within the Catholic world, for example, what was unacceptable in the Papal States might be published not far away in the Republic of Venice.[58]

56. See for example Löwenbrück, "Michaelis et la critique biblique" (1986), on the famous Göttingen savant Johann David Michaelis (1717–91), whose less famous son Christian Friedrich worked on fossil bones (§5.3); more generally, Rogerson, *Old Testament criticism* (1984), chap. 1.

57. Brooke, "Science and religion" (2003), reviews these issues, and particularly *natural* theology, for the eighteenth century.

58. For France, see for example Darnton, *Business of enlightenment* (1979), and *Great cat massacre* (1984), chap. 4; and Darnton and Roche, *Revolution in print* (1989).

In general, therefore, discussion of even potentially disruptive ideas was usually tolerated in practice, provided it was safely confined to the conversations of gentlemen. Savants expected few problems with the authorities, provided their publications were deemed unlikely to "frighten the horses or excite the servants". And the specific topics associated with the sciences of the earth—even the earth's timescale (§2.5)—were still less likely to cause trouble in any of the centers of enlightened intellectual life.

Conclusion

Savants of Saussure's generation conceived the relations between different kinds of scientific work in ways that differ sharply from those of the modern world. Their distinction between literary and philosophical studies corresponded at least approximately to the modern division between the humanities and the natural sciences. But within the latter, their distinction between natural philosophy (or physics) and natural history cut right across modern divisions between physical and biological sciences because it separated mathematical and causal investigations—of living and nonliving phenomena alike—from the systematic description and classification of their diversity. These two kinds of study were regarded as being of equal importance, profundity, and difficulty: prominent naturalists enjoyed the same high prestige as leading natural philosophers. And theology was a science like others on the map of knowledge; it was taken to be related in specific ways to other sciences, human and natural. Revealed theology and natural theology might both have specific points of contact with the natural sciences, but they were just as likely to be treated as points of corroboration as of conflict. In practice, savants could usually get on with their studies without opposition from political and ecclesiastical authorities and often with their blessing; sometimes the representatives of those authorities were themselves among the relevant savants. And anyway, the sciences concerned with the earth were far less prone to such problems than those that impinged more directly and profoundly on human life and values.

This chapter has outlined—in very brief and general terms—who was involved in the sciences in the late Enlightenment, where they worked, and what they were investigating. The way is now clear for a more specific survey (in the next chapter) of those sciences that took the earth and its natural features as their subject. I shall claim that in Saussure's time the earth, like all other parts of the natural world, was usually the subject either of description and classification or of causal explanation; in either case it was *not* being treated systematically as the record of any *history* of nature in the modern temporal sense of that word. Subsequent chapters will evaluate a scientific genre that might seem to contradict that claim (Chap. 3) and consider the first modest signs of genuine exceptions to it (Chap. 4). The synchronic survey in Part One will conclude with a review of the role of fossil evidence in this nascent geohistory (Chap. 5).

CHAPTER 2

Sciences of the earth

2.1 MINERALOGY AS A SCIENCE OF SPECIMENS

Minerals and other fossils

In Saussure's time, four fairly distinct sciences were concerned with the material objects and phenomena of the earth; or they could be described as four distinct sets of practical activities, each with its own characteristic genres of texts and pictures. The differences between them were well recognized, even though one and the same savant might use more than one approach on different occasions. Three of the four were concerned primarily with the description and classification of the diversity of terrestrial things and were therefore treated as branches of natural history. The fourth, by contrast, dealt with the causal explanation of the materials presented by any of the other three, or at least with the regularities or "natural laws" displayed by their occurrence; it was therefore classed as a branch of "physics" or natural philosophy. None of these sciences individually, nor all of them collectively, can be called "geology" without serious anachronism: the word itself was just beginning to be used regularly, but for yet another kind of scientific project (§3.1). Since the four sciences of the earth straddled the major boundary between natural history and natural philosophy (§1.4), there was at the time no obvious need for any single term to denote them collectively. Here, whenever some such term would be helpful, they will simply be described as the "sciences of the earth", which covers much the same wide range as the modern terms "earth sciences" or "geosciences".[1]

1. Rappaport, "Earth sciences" (2003), is a fine survey for the first half of the eighteenth century; see also Rudwick, "Minerals, strata, and fossils" (1996), and, more generally, Laudan, "Tensions in the concept of geology" (1982). The four sciences, or bodies of practice, described in the present chapter could be referred to as four distinct "discourses" about the earth; but that fashionable term will be used only sparingly in this book, because it implicitly privileges the verbal above all other aspects of practice, or else—perniciously—turns even the material and the visual into yet more "text".

In this chapter these four sciences will be outlined in turn. The intention is not primarily to summarize their contents or their conclusions, but rather to sketch—with just a few examples—the kinds of problems that they tried to tackle and the methods and assumptions that were used in their practice. They can also be characterized by the genres through which their claims were disseminated, including the kinds of visual representation that were deployed. The conclusion will be that, although some incorporated a temporal dimension, none involved more than a minor element of true historicity.

This section describes the science of "mineralogy", in Saussure's time the core and foundation of the sciences of the earth. Mineralogy was the analogue of botany and zoology, the sciences of the other two "kingdoms" of nature; like them, it was primarily a science of description and classification, and therefore a branch of natural history (§1.4). Again like botany and zoology, mineralogy was primarily a science of *specimens*, of samples of the natural world that had been gathered from dispersed localities and concentrated in one place, in a museum or cabinet; and so, like them, it was a science practiced mainly indoors. The initial collection of the specimens, outdoors in the field, was not an end in itself; as a mere means to an end, it was often delegated by leading savants to their students or subordinates, or left to the contingencies of donation and purchase (§1.3). What made the study of mineral specimens—and the specimens themselves—truly scientific was their description and classification; their arrangement, display, and storage in a museum was the embodiment of the results of that indoor work (Fig. 2.1).[2]

Fig. 2.1. Rock specimens displayed as if in a "cabinet" or museum: an illustration from Hamilton's *Campi Phlegraei* (1776). This colored etching, based on a painting by Pietro Fabris, shows rocks from the volcanic region around Vesuvius. Hamilton sent a large collection of real specimens to London, for the museum of the Royal Society; these illustrations made selected specimens available in "proxy" or paper form wherever copies of the book were available. This picture is unusual only in that the trompe l'oeil style extends even to the shelving; it therefore gives a good impression of what such specimens might have looked like when displayed in a museum (see Fig. 1.12). The snuffbox, although an artifact, was not considered inappropriate for inclusion here; it was carved from travertine, and was the product of a local craft. (By permission of the Wellcome Library, London)

The word "mineralogy" had a wider meaning than it has today, because it covered the study of *all* kinds of objects within the mineral kingdom: not only minerals in the modern sense, but also rocks and even fossils. All such objects were "fossils" in the original sense of the word (which survives in the term "fossil fuels"), because they had all been "dug up" from the earth. Mineralogy in the sense of a study of specimens was often called *oryctognosy* or *oryctology*, denoting—with impressively Classical learning—the knowledge or science of "fossils" in this broad sense; English authors sometimes used the simpler terms *fossilology* or *fossilogy* to the same effect. Werner's first book, which made his reputation and secured his position at the Freiberg mining school, was entitled *On the External Characters of Fossils* (1774), although in modern terms it dealt only with minerals. Linnaeus's classification of the whole mineral kingdom included not only minerals and rocks but also fossils (in the modern sense); and when Johann Friedrich Gmelin (1748–1804), the professor of chemistry at Göttingen, published an "improved" version of Linnaeus in four volumes, its many illustrations included not only the crystal forms of minerals but also various fossil shells. Likewise, when van Marum was appointed director of the new and richly endowed Teyler's Foundation in Haarlem, he at once proposed buying a collection of "Fossilia" at a forthcoming auction in Amsterdam, and he spent a huge sum on "the finest ores, crystals, petrifactions, and other fossils" to form the core of a new museum.[3]

Identification and classification

In the practice of mineralogy, specimens were described and identified—the necessary prerequisite for classification—in terms of their diagnostic characters, the physical and chemical properties that established them for what they were. Here a laboratory was often required, as a place not of experiments but of diagnostic tests; its importance underscores the close affinities between mineralogy and the science of chemistry. Laboratory tests followed either the "wet way" or the "dry way", using either aqueous reagents or the intense heat of a blowpipe. Each method had its advocates, but the intention was the same: to develop reliable methods that would enable the mineralogist to determine unequivocally which known "species" of mineral was under investigation, or, of course, to discover that the species was new and undescribed. A further batch of tests had the advantage that they could be applied even in the field or down a mine, without recourse to laboratory facilities, because they were based on what Werner called "external characters". These were the tests—learned by generations of students well into the twentieth century—that required little more than a hand lens and a pocket knife: "characters" such as crystal form and hardness, color and cleavage. Werner became famous for the precision

2. Fig. 2.1 is reproduced from Hamilton, *Campi Phlegraei* (1776) [London-WL: V25291], pl. 48. Some contemporary paintings of shelved specimens of other kinds are reproduced in Pinault, *Le peintre et l'histoire naturelle* (1990), 230, 244. See also, more generally, Bourguet, "Collecte du monde" (1997).

3. Werner, *Kennzeichen der Fossilien* (1774); Engelhardt, "Linné und das Reich der Steine" (1980); Gmelin, *Natursystem des Mineralreichs* (1777–79); Wiechmann and Touret, "Van Marum als verzamelaar" (1987), 115.

of his classification of such features: the inaugural issue of the *Bergmännische Journal*, for example, began with his article "On the different degrees of solidity of rocks" and how they should be defined.[4]

In all this, the parallels between mineralogical practice and that of the other natural history sciences will be obvious. Botany, for example, was similarly focused on the study of diagnostic characters, and there was much argument about the validity of Linnaeus's insistence on the taxonomic primacy of the reproductive organs of plants, or conversely the feasibility of taking all characters equally into account, as Lamarck's Parisian colleague Michel Adanson advocated. In all branches of natural history, the determination of diagnostic characters was the first stage in both description and identification, which in turn were the foundations for classification.

The classification of minerals, like that of animals and plants, was hierarchical, descending from classes and orders down through families and genera to species and varieties. As already emphasized, these levels had no biological connotations, let alone any hint of evolutionary interpretation: they simply denoted degrees of decreasing generality or increasing specificity. The whole classification displayed the structured diversity of the natural world and the relations of similarity or "affinity" between natural kinds, without any temporal dimension or implications of temporal origin. In the second of his annual reviews of recent research, La Métherie noted that only some 500 mineral species had been described and named, compared with about 16,000 species of animals and 20,000 species of plants; but he assumed that the disparity was due simply to the less advanced state of mineralogy, and that "species" were comparable in all three branches of natural history. Like sulfur, iron, common salt, and other chemical substances, minerals were simply natural kinds, part of the atemporal diversity of nature. It would have made no sense to talk of the *history* (in the temporal sense of the word)—still less of the *origin*—of mineral "species" such as hornblende, feldspar, or quartz.[5]

However, the classification of entities within mineralogy was plagued by complications without exact parallel in the other two branches of natural history. For example, eighteenth-century mineralogists gradually became aware that they needed *two* distinct systems of classification, one for "simple minerals" and one for "compound minerals" or rocks. A well-known rock such as granite, for example, was visibly a mixture of several distinct mineral species, such as quartz, feldspar, and mica. But other distinctive rocks, such as marble and quartzite, were confusing in this respect, in that each was composed of a single mineral species; so it was only gradually, in the course of practice, that the distinction between minerals (in the narrower modern sense) and rocks became clearly established.

The problems of description and classification were further exacerbated in the case of the many rocks that are very fine-grained in texture. Bulk chemical analyses of such rocks were of little help, and they all looked much the same under a hand lens (polarizing microscopes and techniques of thin sectioning were not developed until a century later). For example, the common rock *basalt* (see Fig. 1.15) often looked much like the rocks that German mineralogists called "*Trapp*" or "*Wacke*" (one variety of the latter survives in modern terminology as "greywacke"). In this case the ambiguity affected a classification that was partly based on inferred modes of origin. Was basalt a hardened sediment or a solidified lava? Was it to be classed with sandstones and mudstones or with other volcanic rocks? Yet these remained primarily questions of classification, and therefore problems for natural history. Only if they were considered to impinge on causal questions—of, say, the role of volcanoes in the economy of the earth—were they regarded as belonging instead to the realm of natural philosophy.[6]

Fossils of organic origin

As naturalists came to recognize and agree on the organic origin of certain kinds of "fossil", those kinds were distinguished as "*extraneous*" or "*accidental*" fossils: they were distinct from the material in which they were embedded; they were not "essential" or defining components of the rocks, and they clearly had a different mode of origin (such adjectives gradually dropped out of use during the late eighteenth and early nineteenth centuries, leaving the word with its primary modern meaning). However, fossils of organic origin continued to be included with rocks and minerals in the science of mineralogy, not only because it was traditional to do so but also for reasons of sheer convenience.

By Saussure's time, earlier uncertainties about the organic character of most kinds of fossil (in the modern sense) had long been resolved. There had never been any serious doubt, for example, about the shells found almost unaltered in loose sediments near sea level. On the other hand, shell-like objects such as ammonites (see Fig. 5.2), which were of unfamiliar form, apparently composed of solid stone, and often found far from the sea, had been much more of a problem to understand. By the late eighteenth century, however, naturalists had come to have a better grasp of the varied ways in which organic remains could have been altered in appearance and in substance after being embedded in sediments. Familiarity with the *cire-perdu* [lost-wax] technique used in jewelry and sculpture enabled them to understand the otherwise puzzling appearances of natural casts and molds (see Figs. 4.7, 5.6). Likewise, the techniques by which anatomists impregnated organic tissues with wax in order to preserve them suggested how fossil wood and bone might have been "petrified" in rather the same way, by natural impregnation with mineral matter. And the botanists' practice of preserving plants by squashing them flat between sheets of paper, to form a *hortus siccus* [dry garden] seemed a plausible analogy for a process by which fossil plants might have been flattened naturally between layers of

4. Werner, "Graden der Festigkeit" (1788). On some aspects of the above, see Porter (T. M.), "Promotion of mining" (1981).

5. La Métherie, "Discours préliminaire" (1787), 11–12. On Linnaeus's "classes" of minerals (one of which included "*Petrifacta*" or fossils in the modern sense), see Engelhardt, "Linné und das Reich der Steine" (1980); for later classifications of minerals see for example Karsten, *Tabellarische Übersicht* (1791), and Eddy, "Geology in John Walker's lectures" (2001). The Germanic miners' names used (then and now) for some of the commonest minerals are evidence of what had long been the preeminence of the German lands in both mining and mineralogy.

6. The English industrialist and savant Josiah Wedgwood (1730–95) developed his "black basalt" pottery (still in production by the company he founded) as a passable imitation of the natural material that was the subject of such intense debate at the time. For anyone unfamiliar with the rock, a glance at a dish or vase made from this famous ceramic should be enough to show why the nature of basalt was so obscure.

Fig. 2.2. Impressions apparently similar to the fronds of a fern, found on the surfaces of black shales and illustrated in the great "paper museum" of fossils and other mineral objects (1755–75) published by Georg Knorr and Immanuel Walch. This specimen came from the private collection of Casimir Christoph Schmidel, the professor of medicine at Erlangen. Such fossils were accepted by eighteenth-century naturalists—with growing confidence—as genuine plant remains, because well-preserved specimens made it seem highly unlikely that their detailed resemblance to living plants was merely fortuitous. (By permission of the British Library)

shale. So, for example, the great *Collection of Remarkable Natural Objects* (1755–75), begun by Georg Wolfgang Knorr (1705–61) of Nuremberg and continued after his death by Johann Ernst Immanuel Walch (1725–78) of Jena, offered in four folio volumes a vast "paper museum" of proxy fossil specimens of all kinds. It described and illustrated the "moss agates" that had only a trivial resemblance to anything organic, and at the other extreme it included fossil leaves that were unmistakably lifelike in every detail (see Fig. 5.9). In between, it also showed, for example, the more problematic impressions of ferns on the surfaces of some shales, particularly those close to seams of coal, which by this time had come to be accepted as truly organic in origin although no plant "substance" remained (Fig. 2.2).[7]

The character of many other puzzling fossils was resolved gradually in the course of closer study, particularly by the cumulative effect of the discovery of ever better specimens. Some, such as belemnites and crinoid ossicles, were accepted as unambiguously organic in origin as soon as more completely preserved specimens were found (§5.2). Others, such as moss agates, dendritic markings on joint planes, and nodules shaped like kidneys and other body parts, were recognized as having only a trivial and accidental resemblance to plants or animals. So by Saussure's time the whole range of mineral objects with some apparent organic resemblance was clearly divided into "extraneous fossils"—the remains of plants and animals that had once been truly alive—and objects with a merely incidental resemblance to anything organic. Only a small and shrinking category of uncertain cases remained provisionally in between (where it persists even today as the aptly named "*Problematica*").[8]

However, recognition of the unambiguously organic origin of "extraneous" fossils did not alter the fact that they were still "fossils" in the original sense: objects as much "dug up" as, for example, distinctive minerals and rocks. So for reasons of convenience they were often stored and displayed in museums along with those other mineral specimens: for purposes of conservation, ammonites and coal plants had more in common with quartz crystals and chunks of granite than with fish preserved in jars of alcohol or bound volumes of pressed plants. Consequently the savants who made a special study of extraneous fossils were often called mineralogists rather than botanists or zoologists. Likewise the fossils themselves continued to be given informal descriptive names that reflected their status as mineral objects—for example, *ichthyolites* [fish stones] for fossil fish—long after their organic origin was generally accepted. Yet at the same time the status of extraneous fossils as zoological or botanical objects was not neglected or ignored, and by the late eighteenth century they were routinely assigned formal names—increasingly, the binomials advocated by Linnaeus—analogous to those of living animals and plants. This had the effect of highlighting the question of the similarities and contrasts between living and fossil forms, which became a major issue of interpretative debate (see Chap. 5).

Fossil localities

Fossils—using the word henceforth in its modern sense—were described and classified in the indoor settings of public museums and private cabinets. They were labeled and catalogued just like other objects of natural history; and collections were enlarged in the same way, not only by finding fresh specimens in the field but also by purchase and exchange. Major collections were then rendered mobile by being published; Knorr and Walch's paper museum of fossil proxies, published in Nuremberg, could be compared with real stony specimens anywhere that the volumes were available, perhaps by Saussure in Geneva, Gmelin in Göttingen, Spallanzani in Pavia, or Hamilton in Naples. As this suggests, pictures of specimens were vitally important in published works on fossils, just as they were in those dealing with animals and plants. Rather than being illustrative of textual descriptions, let alone merely decorative, the engraved plates were usually the raison d'être of the

7. Fig. 2.2 is reproduced from Knorr and Walch, *Merkwürdigkeiten der Natur* (1755–75), 4 [London-BL: 457.e.15], pl. 157, drawn "from nature", i.e., direct from the specimen, by Georg Carl Leinberger.

8. This early history of the interpretation of "fossils" is summarized in Rudwick, *Meaning of fossils* (1972), chap. 1. Early in the eighteenth century Johann Beringer of Würzburg had famously been the victim of a spiteful hoax, when he was duped into publishing a collection of spurious "fossils" that had been carved by rivals and "planted" for him to find: see Jahn and Woolf, *Lying stones* (1963), retold in Gould, *Lying stones* (2000), 19–26. By Saussure's time such an incident would have been almost inconceivable.

work, and the texts were sometimes little more than their verbal explanations. The function of the engravings was to supply virtual specimens that could be compared with the real specimens in the reader's—or rather, the viewer's—own collection, and then used to give them authoritative identifications. Of course, this was quite compatible with providing enlightened recreation and social prestige, in the form of a lavishly illustrated book that could be displayed prominently on a gentleman's library table. For example, Knorr had published a selection from his collection under the title *Deliciae Naturae* [Delights of Nature] (1766–67), clearly aimed at just such a dilettante readership; like the larger work completed by Walch, the original—in German, despite its title—was combined with a text in French, making it marketable throughout Europe.

Although fossil specimens had been collected in the field, all that they usually brought with them, as it were, from the outdoor world was a bare record of the locality from which they came, and even that information often got lost or forgotten in the course of exchange or sale. Specific kinds of fossil, like specific kinds of plant and animal, were treated as "characteristic" of some specific locality or region, not of anything else. Particular localities became famous for yielding a profusion of fine specimens of particular kinds, and published descriptions of fossils often took the form of monographs on all those found in one locality (Fig. 2.3).[9]

Fig. 2.3. Fossil shells from a famous locality near Hordwell (now Hordle) on the Hampshire coast in southern England: one of the plates published in Gustav Brander's monograph on them (1766). The original specimens were kept in the then newly founded British Museum in London, but these superb engravings served as accurate proxies or mobile substitutes for them; the text in Latin—by this time becoming less common than French—made the accompanying verbal descriptions accessible throughout the scientific world. The shells were named—on Linnaeus's then new classification—as species of *Murex* (62, 70), *Buccinum* (63, 71), and *Strombus* (64–69). Some of Brander's fossils were considered similar, if not identical, to species known to be living in *tropical* seas, and were therefore used as evidence for the reality of past "revolutions" or major changes—not necessarily sudden or violent—in the condition of the earth's surface (see §2.4, §5.2). (By permission of the Syndics of Cambridge University Library)

Several other localities, such as the underground limestone quarries just outside Maastricht in the Netherlands, were equally famous for their fossil shells. Superbly preserved fossil fish came from Monte Bolca in the Alpine foothills behind Verona (see Fig. 5.7); others were bizarrely preserved in copper in the *Kupferschiefer* [copper shale] near Eisleben in Halle (see Fig. 5.8); and others again were found, together with finely preserved fossil plants (see Fig. 5.9), at Oeningen (now Öhningen) near Konstanz in Switzerland. The quarries at Solnhofen near Eichstätt in Bavaria were celebrated for yielding a wide variety of fossils, exceptionally well preserved apart from being flattened on the surface of the rock (see Fig. 2.20). As a result of purchase and exchange, fossil specimens from these and other famous localities were dispersed in collections throughout Europe (Fig. 2.4).[10]

Fig. 2.4. A map of Europe showing some of the famous localities that were visited, described, and discussed in the late eighteenth century by savants with interests in the sciences of the earth (and which are mentioned in this volume).

9. Fig. 2.3 is reproduced from Brander, *Fossilia Hantoniensia* (1766) [Cambridge-UL: 7340.b.1], pl. 5, drawn and engraved by Benjamin Green.

10. Some of these fossil localities are still popular with fossil collectors; others, such as Maastricht and Monte Bolca, are accessible as tourist sites (and are well worth visiting) although collecting is no longer possible or permitted. The "lithographic stone" of Solnhofen (of Jurassic age) was the first major *Lagerstätte* to be exploited (the modern term denotes exceptional deposits in which the "soft parts" of animals are preserved), and its fame has only been overshadowed more recently by the Burgess Shale in British Columbia, of much greater (Cambrian) age.

From a modern perspective, what is striking about this way of treating fossils is the lack of any clear sense of relative geological ages. These famous localities were of course recognized as being in different kinds of rock, even as being in different "formations" (see §2.3). Yet the way the fossils were treated in the ordinary practice of "mineral" natural history removed them in effect from any such field context, leaving them with little more than the *spatial* attribute of coming from specific geographical localities. This will seem less strange, however, when it is recalled that the study of fossils was indeed a part of natural history; naturalists therefore regarded it as appropriate to treat fossil specimens in just the same way as specimens of animals and plants.[11]

The only exception to this—and it was slight and marginal—was that descriptions of fossil specimens sometimes included one further item of field information that would now be regarded as geological. Yet even here the modernity is deceptive. What was recorded was that a particular specimen had been found not only at a specific locality but also at a specific depth below the surface of the ground, or at a specific altitude. For example, in 1781 the young French naturalist Robert de Paul de Lamanon (see §4.3) described in *Observations sur la physique* how a huge fossil bone had been found in clay behind the wall of a wine merchant's cellar in Paris; and he carefully recorded that it was buried under 11 feet of clay, 14 feet above the level of the nearby river Seine, and 127 feet above sea level. Such records did reflect a slight sense of relative age, because the tacit assumption was that the deeper a fossil was buried, or the higher it was above sea level, the older it might be. But causal questions about the mode of emplacement of fossil specimens—whether shells, fish, plants, bones, or anything else—belonged not in the science of mineralogy (broadly defined) but in the study of the "physics" of terrestrial phenomena of all kinds (see §2.4, §3.5).[12]

Prize specimens

Just as mineral natural history differed from its botanical and zoological branches by requiring more complex systems of classification, so also its procedures for handling specimens accentuated the greater value—in all senses—that was attached to *particular* specimens. In botany and zoology, the exchange of "duplicate" specimens between museums or private collectors was generally unproblematic: another orchid or another butterfly of a specific kind could often be collected without difficulty from the same localities as before, and sent off—if necessary, to the far side of Europe or even across the Atlantic—in exchange for other desirable species. Only if the species was extremely rare, or if it lived in a very remote part of the world, might particular specimens have great value. In the flourishing eighteenth-century culture of natural history collecting, such value might be commercial as well as scientific: for example, specimens of exceptionally rare and exotic shells commanded astronomical prices among collectors (see §5.2), almost as absurd as the prices of rare tulip bulbs during the famous Dutch "tulipomania" of the previous century.[13]

For collectors of fossils, however, whether savants or amateurs, a further variable was added to that of mere rarity, by the exigencies of fossil preservation. Far

Fig. 2.5. The collection of an exceptionally important specimen, the gigantic toothed jaws of the "Maastricht animal" (later named the mosasaur), discovered in 1780 in the underground quarries outside the Dutch town. This imaginative reconstruction of the scene was drawn much later to illustrate a monograph on the Maastricht fossils (1799) by the French naturalist Barthélemy Faujas de Saint Fond, who had removed this famous specimen to Paris as cultural loot during the Revolutionary wars (his picture of the specimen itself is reproduced in Fig. 7.7). The traditional (pre-Revolutionary) social distinction is drawn between the workmen handling the huge mass of rock and the gentlemen observing the object of their manual labor. The strangeness of the find is accentuated by the dramatic quality of the scene. (By permission of the Syndics of Cambridge University Library)

more obviously than in the case of sea shells, stuffed birds, or dried plants, specimens of even quite common fossils varied vastly in their quality, from the poorly preserved or highly fragmentary to those preserved in superb and complete detail. This put a high premium—again, both scientifically and commercially—on particular specimens of exceptional quality. This effect was compounded if the species was also rare and of particular scientific interest. For example, the limestone quarries outside Maastricht yielded not only fine specimens of a profusion of fossil shells, but also—very rarely—the bones, teeth, and jaws of a huge unknown vertebrate. The discovery of a new specimen of the "Maastricht animal" was therefore a major event (Fig. 2.5).[14]

11. In modern terms, the localities listed above are highly diverse in geological age: Eisleben is Permian; Solnhofen is Jurassic; Maastricht is Cretaceous; Hordwell and Monte Bolca are Eocene; Oeningen is Miocene. Historically, however, the important point is that in the late eighteenth century all these rocks were classed as "Secondary" (see §2.3); in practice they were regarded as being of roughly the same age, and the differences between their respective fossils were tacitly attributed to different environments, just like regional differences in animals and plants.

12. Lamanon, "Os d'une grosseur énorme" (1781).

13. On the scientific and cultural valuation of particularity in museum specimens, as it first developed in the early modern period, see Daston, "Factual sensibility" (1988); on conchology, see Dance, *History of shell collecting* (1986).

14. Fig. 2.5 is reproduced from Faujas, *Montagne de Saint-Pierre* (1799) [Cambridge-UL: MA.38.59], 37 [in quarto edition; in folio edition, it is on the title page]. This famous specimen is still on display at Paris-MHN, where it has been ever since the Revolution; it was recently the subject of an attempt at politically correct "cultural" repatriation.

Such exceptional specimens could become, as it were, the objects of scientific pilgrimage. After van Marum had purchased a major collection of "Fossilia" or mineral objects of all kinds for the new Teyler's Foundation in Haarlem, he set out—literally, and on his honeymoon—to acquire some still more spectacular specimens to make its museum an irresistible attraction to savants and dilettanti alike. At Maastricht he purchased a large collection of the local fossil shells at auction; he also bought, from another local collector, the very first specimen of the Maastricht animal to have been discovered. He paid 400 ducats (2,100 guilders) for the fossil jaw alone, an enormous sum comparable to the 2,500 guilders that he had recently persuaded the managers of Teyler's to spend on building the world's most powerful electrostatic generator. These two acquisitions put Haarlem, and the museum itself, firmly on the European map, in both natural history and natural philosophy. The expenditures indicate how fossils and electricity were both perceived as "hot" topics for research, belonging equally to the eighteenth-century equivalent of modern Big Science (Fig. 2.6).[15]

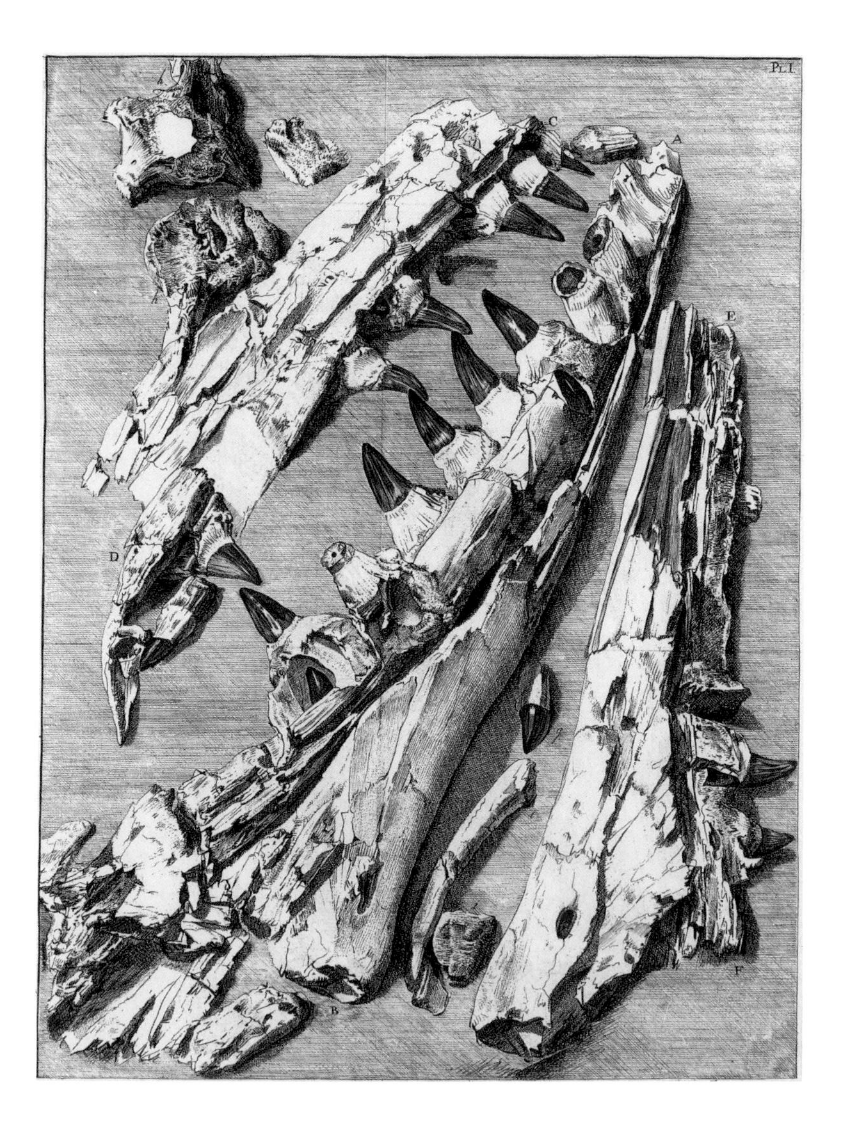

Fig. 2.6. The huge fossil jaws of the "Maastricht animal": the first specimen found (in 1766), which was purchased in 1784 by Martinus van Marum for the new museum at Teyler's Foundation in Haarlem. This engraving illustrated his published description of the specimen (1790), which had become one of the most famous exhibits in the museum; the proxy gave it further publicity and may have attracted further visitors to Haarlem. The jaw is about 1.2m (4ft) long. (By permission of the Syndics of Cambridge University Library)

Conclusion

"Mineralogy", which included the study of fossils, was flourishing in Saussure's time as a science of specimens, practiced indoors in museums and private cabinets. Collections were made for both scientific and more broadly cultural reasons, though it is rather anachronistic to draw that distinction. Specimens were rendered mobile and more widely available by being published in proxy form, in densely illustrated books and articles; this enabled specimens in one place to be compared with those in another, allowing them to be identified and classified. As in the other branches of natural history, classification was the ultimate goal; mineralogy was not concerned with causal explanation, and in the standard treatment of minerals, rocks, and fossils there was little if any sense of relative ages, still less of geohistory.

Although fossils were usually conserved and studied alongside rocks and minerals, they were well understood to be the remains of animals and plants that had once been truly alive. Exceptionally well-preserved fossils were highly prized, the more so if they were rare or of unusual kinds. But like living animals and plants they were treated simply as the products of specific localities or regions, not as characteristic of specific rocks, still less as the traces of life at different times in the past. Like minerals and rocks, fossils were arranged and classified primarily according to their different natural kinds—in families, genera, and species—and only secondarily according to the places in which they were found. The latter, however, linked the indoor museum study of mineral objects of all kinds to an outdoor science in which issues of geographical distribution were paramount. The latter is the subject of the next section.

2.2 PHYSICAL GEOGRAPHY AS A SPATIAL SCIENCE

Huge solid facts

The indoor science of "mineralogy", which included the study of fossils, was for most naturalists the foundation for any scientific understanding of the earth. But its primacy was being challenged in Saussure's time—not least by Saussure himself—through the forceful advocacy of another science, in which outdoor *fieldwork* became the very center of scientific practice and no longer merely a means to an end. Saussure and other francophone naturalists usually called this science "*géographie physique*".

Physical geographers studied the major features of the earth's surface, such as mountain ranges and volcanoes, rivers and their drainage basins, even continents and oceans: these were the "huge solid facts" on which the science was founded.

15. Fig. 2.6 is reproduced from van Marum, "Kop van eenen fisch" (1790) [Cambridge-UL: 911:01.b.6.4], pl. 1. See Wiechmann and Touret, "Van Marum als verzamelaar" (1987); his journey is described in Palm, "Van Marums contacten" (1987), and the relevant part of his journal is printed in Forbes, *Martinus van Marum* (1969–76), 2: 19, 208; Regteren Altena, "Verzamelaars te Maastricht" (1956), describes the collectors there. On the electrical research of the period, see Heilbron, *Electricity* (1979); on the huge electrical machine built for Teyler's, Hackmann, *John and Jonathan Cuthbertson* (1973), 29. Both the fossil and the electrical machine are still prominently on display at Haarlem-TM.

They also studied the smaller items of which those features were composed, such as valleys and hot springs, sand dunes and estuaries; and they considered them all in their distributional or *spatial* dimension. None of these features, not even the smaller ones, was mobile; none could be shipped into the arenas of scientific debate. Any specimens that were transported in that way were indeed just small-scale samples of parts of the object, not the real thing. In this respect there was no parallel with botany and zoology: those sciences had no comparable large-scale features, or at least none as materially immobile as mountains. Physical geography was unmistakably a branch of natural history: studying terrestrial features in spatial terms was an integral part of the project to describe and classify the large-scale diversity of the earth. As in the study of specimens small enough to store in museums, so here too the goals of description and classification were regarded as worthy ends in themselves. Any attempt to explain the terrestrial features in causal terms, or even in terms of regular natural laws, belonged to a different kind of science (§2.4).[16]

Physical geography covered a much wider range of large-scale features than the phrase might now suggest. In modern terms it included the study not only of the lithosphere but also of the hydrosphere and the atmosphere. Oceans and hurricanes were grist to its mill, as much as mountains and volcanoes: all were features that could be described in spatial terms and be classified according to what were perceived to be their natural kinds. Saussure's design for an accurate hygrometer, his precise measurements of many atmospheric variables on the top of Mont Blanc, and his careful observations on the thunderstorms that sometimes raged around the Alpine peaks were all tokens of his keen interest in meteorology. This in turn reflected his broad conception of physical geography, which was fully shared by his contemporaries. In the present context, however, the main focus will be on the features of the lithosphere or solid earth. Much work in physical geography was in fact focused on still more specific classes of terrestrial objects and phenomena. Many publications were not only regional in their coverage but also devoted to just one of these favorite topics: Saussure's own mountainous *Alpine Travels* and Hamilton's volcanic *Campi Phlegraei* are obvious examples.[17]

Physical geography was not of course a rival or alternative to mineralogy: all naturalists would have agreed in principle that a full understanding of the earth required both. To Saussure, however, it seemed clear that the large-scale features had been neglected in favor of small-scale samples of them. In his opinion, fieldwork was not an optional extra, fit to be delegated by a busy savant—or one overconscious of his social status—to his students or assistants. Only the naturalist's own *firsthand* observation was adequate to comprehend the large-scale features of the earth: the savant, like Mahomet, had to come to the mountain. As already mentioned (§1.3), this was a quite novel idea, and it put fieldwork at the heart of physical geography. Saussure made the point eloquently in the "preliminary discourse" with which he introduced his volumes on the Alps, using an apt analogy that exploited the contemporary vogue for the Grand Tour and for antiquarian studies of the material remains of the Classical world (§4.1). To focus attention exclusively on portable specimens was as foolish as to collect pretty scraps from Antiquity while ignoring the great buildings of which they had been a part:

> The only goal of most of the travelers who call themselves naturalists is to collect curiosities; they walk—or rather, they crawl—with their eyes fixed on the ground, picking up little pieces here and there, without aiming at general observations. They are like an antiquarian who scratches the ground in Rome, in the middle of the Pantheon or the Colosseum, looking for fragments of colored glass, without glancing at the architecture of those superb buildings. I do not at all advise the neglect of detailed observations; on the contrary, I regard them as the only basis for solid knowledge. But in observing these details I would wish that one should never lose from view the large masses and their ensemble, and that a knowledge of these great objects and their relationships should always be the purpose that one sets oneself when studying their small parts.[18]

The primacy of fieldwork

When Saussure reached the summit of Mont Blanc, his greatest thrill was to see, laid out below him as if on a map, the complex topography of the mountains and valleys for miles around (§1.1). He duly noted that the highest outcrop was of granite, but the specimens he collected there were important to him primarily as proof of the composition of the peak itself. When later he sat for his portrait (Fig. 1.7), he pointedly chose an outdoor setting, in which his use of mineral specimens was clearly placed in a context of fieldwork. What the portrait included were not specimens neatly labeled and stored indoors in the drawers of a museum cabinet: they were chunks of raw outdoor nature, recently detached from the solid rock by his use of a plebeian miner's hammer. All such details indicate Saussure's chosen identity as a physical geographer, and only secondarily as a mineralogist.

Several other prominent naturalists were following Saussure's example by supplementing, complementing, or even largely replacing work in the museum with work in the field. For example, Spallanzani planned his first major fieldwork tour, during the summer of 1788, primarily to collect specimens of volcanic rocks to fill some glaring gaps in his museum at the university of Pavia: the fieldwork was to contribute to the science of mineralogy. But as he traveled south to Campania and ultimately to Sicily, he became increasingly fascinated by the large-scale features of the volcanoes themselves, and on his return he wrote his six-volume *Travels to the Two Sicilies* (1792–97), the fruit of his extensive firsthand observations in the field and clearly a contribution to physical geography (Fig. 1.13). Such voluminous accounts of scientific travels—Saussure's *Alpine Travels* was of course another example—

16. The phrase "huge solid facts" (for which I am indebted to Mark Hineline) comes from a journalist's report of an early twentieth-century geological field excursion in the western United States, but is extremely apt as an expression of eighteenth-century attitudes to, say, the Alps or Etna. In botany and zoology the nearest equivalent to physical geography would have been the systematic study of the distribution patterns of individual species or whole faunas and floras (in modern terms, biogeography). But that kind of study had barely begun with the work of Linnaeus's students, and only got under way with Alexander von Humboldt's work at the turn of the century: see Browne, *Secular ark* (1983), chaps. 1, 2.

17. Carozzi and Newman, *Lectures on physical geography* (2003), reconstructs Saussure's course of 1775; see also Sigrist, "Géographie de Saussure" (2001).

18. Saussure, *Voyages dans les Alpes*, 1 (1779): ii-iii. He himself collected and conserved his specimens with great care: see Decrouez and Lanterno, "Collection géologique de Saussure" (1998).

were one of the standard forms of publication in physical geography. The format was adapted easily from the well-established genre of more general accounts of travels to distant or exotic places.[19]

Fieldwork in physical geography was not of course a complete novelty in Saussure's time. There had been a long tradition of "*chorographies*" (literally, descriptions of places), scholarly publications in which all the natural and human features of a specific region would be described in detail. Much of the material for such works might be compiled indoors by a scholar searching through archives or distilling information from local reports of all kinds. But if—as was generally the case—an account of the region's natural history was included, it might be based on the compiler's own observations, which in turn might well include a description of the physical geography as well as the local fauna and flora. Gilbert White's famous *Natural History and Antiquities of Selborne* (1789) is an example of a chorography on the modest scale of a single English parish (of which he was the Anglican parson); others, such as William Marsden's *History of Sumatra* (1783), more ambitiously covered much larger areas. However, these works remained compilations; the aim was to give as complete a description as possible of all the noteworthy features of a specific region. So even if they incorporated the results of field observations on physical geography, it was fieldwork carried out without any special goal beyond that of being accurate and comprehensive.[20]

The kind of fieldwork pioneered and advocated by Saussure, by contrast, was done with a far more precise sense of purpose. It is a sign of its novelty that he felt it necessary, or at least desirable, to spell out in detail exactly what his practical procedures in the field had been, and what he advised others to do in turn. He compiled a practical guide to the conduct of fieldwork in physical geography, which was eventually published not long before his death, at the conclusion of his wide-ranging "Agenda" for the future of the sciences of the earth (§6.5). Specifically, he explained how he had developed the habit of listing "*agenda*" in advance of each field trip, to remind himself exactly what he wanted to look out for, observe, and record. This was no trivial innovation, because it indicated his transcendence of mere compilation and his conception of field observation as an activity guided by goals of comparison and classification, interpretation and explanation. Only with such goals already in mind did it make sense to list in advance what specific observations needed to be made on the trip that was being planned, in order to *test* alternative interpretations and avoid having to visit the area again on some future occasion.[21]

The spatial character of physical geography, tackled through fieldwork, generated a major problem for its practice. The more the importance of firsthand field observation was emphasized, the more pressing was the problem of making that experience real and convincing to those who had *not* in fact witnessed what was being described. Unless that sense of veracity could be communicated, any conclusions based on fieldwork would be convincing only to the savant who had been in the field, and no wider consensus about the matter might ever be formed. The real witness of certain features had to recount his experience in such a way as to make others "*virtual witnesses*" of it: he had to convince them that it was *as if* they too had seen it, so that they might also be persuaded to accept the interpretation or explanation offered for it.[22]

One well established way of creating virtual witnesses was to use carefully constructed and persuasive prose: to describe what the writer had seen, with such vivid eloquence that the reader felt that it was *as if* he—and here it might also be she—had been present too. For example, de Luc's six volumes of *Lettres Physiques et Morales* (1779) published the 150 lengthy letters he had originally sent to his patron Queen Charlotte, describing the physical and human geography of wide tracts of western Europe (§3.3); and she and other readers may well have felt it was as if they had accompanied him everywhere and seen all that he had seen. Saussure's volumes on the Alps and Spallanzani's on the Italian volcanoes would not have been so successful if their texts had not been persuasive in the same way. But for a science such as physical geography, as much as for mineralogy, words were not enough; or at least, a work such as de Luc's that contained nothing but text was seriously handicapped in the task of converting its readers into virtual witnesses. That task was greatly facilitated if the text was supplemented, complemented, or even—as in Hamilton's *Campi Phlegraei*—almost supplanted by visual illustrations. As any modern geologist will appreciate, it was intrinsic to the subject matter that one good illustration was worth not just a thousand words, but ten or even a hundred thousand.[23]

Two contrasting visual methods were used to extend the plausibility of ideas based on fieldwork, from the primary observer to the other savants who might evaluate them; or, say, from Saussure to physical geographers elsewhere in Europe. Both methods helped to generate an indefinite number of virtual witnesses to what only one or a few individuals had witnessed in reality. Both were therefore in effect scientific instruments; their instrumental function has only been obscured by the uncritical assumption that all instruments at this period were made of glass and brass.

Proxy pictures

The first instrumental method was the use of *proxy pictures* of the large-scale features under discussion. As already pointed out (§2.1), proxy pictures were also

19. Spallanzani, *Viaggi alle Due Sicilie* (1792–97); see Vaccari, "Spallanzani and his geological travel" (1998).

20. See for example Jankovic, "Place of nature" (2000). The word "chorography" appears only rarely after the seventeenth century, although the genre of local descriptions continued and indeed flourished into the nineteenth. Many later editions of *Selborne* omitted the "antiquities" of the original title, thereby obscuring its true genre.

21. Saussure, "Agenda" (1796), also printed in his *Voyages* 4 (1796): 467–538. Drafts date from the 1770s: e.g. within "Alpes No. 1 du 14 juin 1774", a manuscript clearly added to at various later times (Genève-BPU, Saussure MSS, 81/4). See Carozzi, *Saussure on basalt* (2000), for abundant material on his methods in the field; on published agendas for fieldwork, see Vaccari, "Géologues voyageurs" (1998).

22. Shapin, "Pump and circumstance" (1984), introduced the notion of "virtual witnessing", and the "literary technologies" that made it convincing, to make sense of the experimental practice of Robert Boyle and others in the early modern period; see also his *Social history of truth* (1994). The techniques can also be related to those of traditional *rhetoric*, in the proper and non-pejorative sense of presenting a good case as persuasively as possible.

23. As already suggested, this important point has been missed by many historians, as a result of their literary training with almost exclusive attention to texts.

important in the science of mineralogy, because they made specimens conserved in one place accessible in paper form in many others; but they were never strictly indispensable, because the specimen itself, or at least a very similar "duplicate", could be sent to another museum or another savant. In physical geography, on the other hand, the importance of proxy pictures was heightened by the intrinsic immobility of the originals: a proxy could be made redundant only if and when a savant could travel to see the features with his own eyes.

In physical geography, pictures served effectively as proxies—standing in for the real thing—to the extent that they faithfully reproduced what the primary observer had seen in the field, making that experience convincing to others and thereby converting them into virtual witnesses. Of course, the primary observation itself was not an unmediated witnessing of nature; simple retinal impressions were (and are) processed in complex ways by the observer's previous experience and cultural circumstances. More specifically, an effective proxy experience was necessarily mediated by the social and artistic conventions that tacitly underlay *any* pictorial representations in a given historical and cultural context. For mineralogists of Saussure's generation, that meant—as already pointed out—that proxies were depicted with the visual conventions of still-life painting in trompe l'oeil style, effectively creating the illusion that one was viewing the three-dimensional specimen itself rather than an image on paper (Fig. 2.1). Likewise, for physical geographers it meant that mountains and volcanoes, glaciers and coastal cliffs, were depicted with all the contemporary conventions of topographical landscape art. By painting in that style, Fabris, for example, could make the topography of Campania and Hamilton's

Fig. 2.7. The Rhône glacier, with the meltwater in the foreground forming the source of the river: an engraving from the lavishly illustrated *Pictures of Switzerland* (1780–86). The large boulders on the surface of the ice are on their way to join those forming the "*moraine*" around the snout of the glacier. (The human figures standing behind the huts, like the figure of Spallanzani in Fig. 1.13, are exaggerated in size, and therefore give a misleadingly inadequate impression of the scale of the glacier.) (By permission of the Wellcome Library, London)

interpretation of it as volcanic, intelligible and convincing to those who had never been nearer to a volcano than a salon in Paris, Berlin, or London (Fig. 1.14). In the same way, a picture of the great Rhône glacier could serve to make that feature—and glaciers in general—vividly real to those who had never had the opportunity to visit the Alps in person (Fig. 2.7).[24]

As these examples suggest, the value of proxy pictures was proportional to the inaccessibility of the original features. For example, instances of prismatic basalt were avidly collected by savants taking sides in one of the great arguments of the time. As already mentioned (§2.1), basalt was a problem for mineralogy, because the study of specimens of that rock, either in museums or in laboratories, failed to determine decisively whether it should be classed with volcanic lavas or with sedimentary rocks such as greywacke. The same ambiguity extended to its appearance on a large scale in the field. Indeed, here the puzzle was heightened in many localities by the astonishingly regular hexagonal jointing of the rock (see Fig. 1.15), a feature that could not be shipped to a museum (though collectors tried their best to do so, and museums might display a single hexagonal block as one of their largest specimens). So the discovery of a new locality showing this striking feature of physical geography was as newsworthy as the finding of a remarkable fossil. If in addition the locality was as inaccessible as the Isle of Staffa off the remote west coast of Scotland, a proxy picture of it became indispensable; it might be years before another savant—accompanied by a competent artist—would have a chance to visit it again (Fig. 2.8).[25]

Fig. 2.8. A view of Fingal's Cave, eroded in the basalt of the Isle of Staffa: a drawing made for the naturalist Joseph Banks—the first savant to visit the island, in 1772—and published as an engraving in Thomas Pennant's *Tour of Scotland* (1774–76). Perhaps unintentionally, the artist made the cave architectural and even Classical in form, but the regularity of the prismatic jointing is hardly exaggerated. Until the arrival of steamships in the early nineteenth century, visits by savants to this remote and often stormy spot were rare and difficult, and pictures such as this were therefore invaluable proxies for the experience of seeing prismatic basalt at first hand. (By permission of the Syndics of Cambridge University Library)

24. Fig. 2.7 is reproduced from La Borde and Zurlauben, *Tableaux de la Suisse* (1780–88), pl. 181 [London-WL: V25130], explained on xcix–cii; a similar view of a glacier above Grindelwald is reproduced in Broc, *Montagnes* (1991), fig. 21. See also Wagner, "Gletschererlebnis" (1983), and, more generally, Klonk, "Science, art" (2003).

25. Fig. 2.8 is reproduced from Pennant, *Tour of Scotland* 1 (1774) [Cambridge-UL: Yorke.c.160], pl. opp. 263, engraved by Thomas Major, and illustrating the "Account of Staffa, communicated by Joseph Banks Esq." (261–69). On successive depictions of Fingal's Cave, see Klonk, *Science and perception of nature* (1996), 74–94. Turner's famous rendering of the scene, like Mendelsohn's *Fingal's Cave* overture (1830), dates from a much later period of Romantic fascination with the place, and of the tourist trade.

Most major features of physical geography were of course stable and permanent on a human timescale, and remained much the same in appearance whenever they were visited. A savant touring the Alps in Saussure's footsteps might be disappointed to find Mont Blanc concealed in cloud or still covered with winter snow, but the mountain itself and the rocks of which it was composed remained the same. Saussure's proxy pictures of it (Fig. 1.2) could therefore be checked for veracity quite easily by any savant with the opportunity to go there and wait for a spell of fine weather. For some features, on the other hand, proxy pictures were of heightened importance because they recorded the contingent details of unrepeated individual events: they purported to convey an accurate impression of what their viewers would have seen, had they been present in person at that particular unique moment.

Prominent among such specific features were scenes of major volcanic eruptions. When Vesuvius erupted in an exceptionally violent manner, only three years after Hamilton had published his general description of it in *Campi Phlegraei*, he considered it scientifically important—and commercially worthwhile—to issue a supplement devoted just to that latest eruption. Many of the pictures in the original volume had likewise been portraits of particular moments in other specific eruptions. Such scenes recorded important information about the character of the eruption, which could then be used to build up a fuller description of Vesuvius and its place in the natural history of volcanoes in general (Fig. 2.9).[26]

Fig. 2.9. The 1767 eruption of Vesuvius, as seen on the night of 20 October from the far side of the Bay of Naples: a gouache by Fabris published as a colored etching in Hamilton's *Campi Phlegraei* (1776). As in Fabris's other paintings of the volcanic region, the established conventions of topographical landscape art were used primarily to complement Hamilton's earlier report to the Royal Society (reprinted in the volume) by making an accurate pictorial record of the eruption; this was of course compatible with producing an aesthetically pleasing composition that would also be admired by general readers (or rather, viewers) of the book. (By permission of the Syndics of Cambridge University Library)

Major earthquakes were likewise the objects of special study by physical geographers. After a violent earthquake struck Calabria in 1783, Hamilton was among those who traveled to the devastated area to study it at first hand. Other savants were sent by the Royal Academy of Sciences in Naples, with artists attached, to record in detail what had happened. Most of the pictures in their report were of ruined buildings and other human effects of the catastrophe, but some portrayed the natural traces of the event, such as landslides, streams with altered courses, and open cracks in the earth (Fig. 2.10).[27]

This detailed attention by savants made the Calabrian earthquake of 1783 the best described event of its kind in the eighteenth century, although it was less devastating than the more famous Lisbon earthquake of 1755. However, the title of the official report, as a "[natural] history" of the earthquake, made it clear that it was

Fig. 2.10. Open fissures caused by the catastrophic earthquake in Calabria in February 1783, with the town of Monteleone in the background: an engraving published by the Royal Academy of Sciences in Naples in its *History of the Phenomena of the Earthquake* (1784). As in some of Fabris's scenes of Campania for Hamilton (Figs. 1.10, 1.14), credibility was added to the picture by the inclusion of the artist Pompeo Schiantarelli in the act of drawing it, as well as two of the gentlemanly savants who described it in words. (By permission of the British Library)

26. Fig. 2.9 is reproduced from Hamilton, *Campi Phlegraei* (1776) [Cambridge-UL: LA.8.79], pl. 6; the eruption was described in letter 2, written to the Royal Society on 29 December 1767 after it had subsided. The 1779 eruption was described and illustrated in Hamilton, *Supplement* (1779).

27. Fig. 2.10 is reproduced from [Reale Accademia di Napoli], *Istoria de'fenomeni del tremuto* (1784) [London-BL: 649.c.7], part of pl. 5; this and other plates are reproduced in Keller, "Sections and views" (1998). See also Hamilton, "Account of the earthquakes" (1783).

not primarily concerned to speculate on the *causes* of the event, any more than Hamilton's accounts of the eruptions of Vesuvius. Description and classification were the proper goals of physical geography, as of any other branch of natural history; causal explanations belonged in a different science (§2.4).

Maps as instruments

All these proxy pictures were or could be complemented by the second visual method for creating virtual witnesses of the large-scale features and phenomena of physical geography. This too was in effect an instrument, again composed not of glass and brass but of paper and ink (and often water colors too). Just as a microscope enabled botanists and zoologists to see what was too small for the naked eye, and a telescope made visible to astronomers what was too far away, so maps enabled physical geographers to see what was otherwise too large in scale to be comprehended (aerial views and those now derived from satellites being of course unavailable).

The improved instrument making of the eighteenth century, specifically in surveying instruments such as theodolites, led to greatly increased cartographical accuracy. Mapmaking had long been driven by political and strategic interests and by those of land tenure, commerce, and navigation; but the natural sciences were often its beneficiaries. Even if maps were made primarily for other purposes, they could give more precise understanding of the form of mountain ranges and coastlines, for example, and of the drainage basins of rivers, as well as of the distribution of volcanoes and other localized features; they provided material for classifying them all and for discerning their natural regularities. Conversely, the lack of a reliable map to act as a topographical base could greatly hamper the work of a physical geographer. Saussure complained that there were no good base maps for his observations in the Alps, and he conceded that his own were inadequate (Fig. 1.3). Physical geographers in some other regions were more fortunate. In particular, those in France benefited from an outstandingly fine and accurate set of maps, which was being made at great expense by that powerful nation, primarily for strategic and economic purposes. By the eve of the Revolution, sheets covering most of France on a large scale had been published under the supervision of César-François Cassini, the third in that famous dynasty of astronomers.[28]

Cassini's maps were used as a topographical base for the most ambitious thematic map of the time, the *Atlas Minéralogique* of France. Although never completed, this was a major attempt by the French state to survey the whole range of its mineral resources and to record them in cartographical form. It was commissioned in 1766 by Henri Bertin, Louis XV's minister responsible for mines, and entrusted initially to the naturalist Jean-Étienne Guettard (1715–86), assisted by the young Lavoisier, and later to the mineralogist Antoine-Grimoald Monnet (1734–1817). The project was plagued by conflicts among the surveyors, and by budgetary crises that threatened its continuation; but even in its highly fragmentary form it was one of the supreme achievements of Enlightenment physical geography.[29]

Just as proxy specimens were portrayed in the style of still life painting, and proxy scenes in that of topographical landscape art, so likewise the maps that served

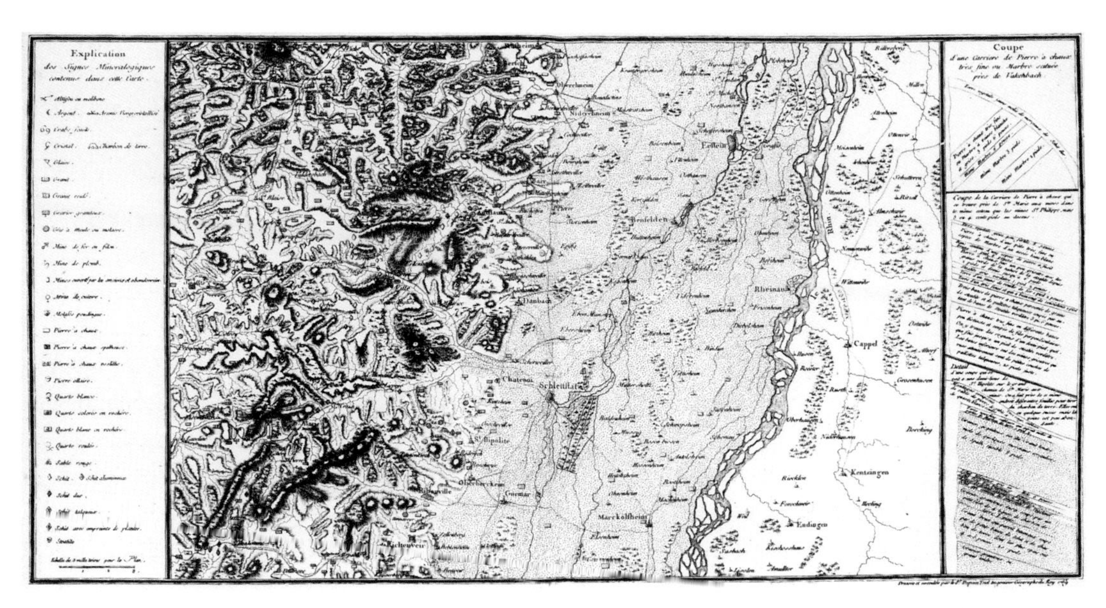

Fig. 2.11. A sheet of the *Atlas Minéralogique*, dated 1769, showing the hills of the Vosges in Alsace, and the braided river Rhine forming the eastern frontier of France; the town of Schlettstat (now Sélestat) south of Strasbourg is near the center of the map. The topography is plotted with much detail and accuracy, although the hills are depicted only by relatively crude hachuring; the alluvial plain of the Rhine is stippled. The map was based on the earlier topographical survey made for strategic and economic reasons of state; what made the atlas "mineralogical" was the inclusion of many spot symbols (see the key in the left margin) denoting the localities of useful or interesting minerals and rocks. The third dimension (see §2.3) was represented only by small sections of strata at specific localities (in the right margin). This sheet was based on some of the first joint fieldwork by Guettard and Lavoisier. This reproduction shows the general format of the maps in the atlas; Fig. 2.12 shows part of another sheet in greater detail.

physical geography, such as those of the *Atlas*, depicted topography in the cartographical style of the period. In particular, topographical relief was usually shown by hachuring, not contours. Even in the hands of a skilled engraver working from good sketches by the field surveyor, the resultant impression of the form of hills and valleys usually left much to be desired from the point of view of a physical geographer; in other respects, however, the sheets of the *Atlas* were impressive (Fig. 2.11).[30]

The way that the "mineral" component of the survey was recorded on the sheets of the *Atlas* reflects accurately the goals of the science of physical geography. The convention of spot symbols was long established in cartography and used on ordinary maps to show churches, castles, towns of varied status, and so on; here it was simply extended to show noteworthy and distinctive minerals, rocks, and even

28. This was the second and more ambitious Cassini survey (1747–88), with 180 sheets at a scale of 1:86,400: see Konvitz, *Cartography in France* (1987), chap. 1; his fig. 6 charts its progress; Broc, *Montagnes* (1991), figs. 12, 17, reproduces samples. Maps of a comparable standard did not begin to be produced in Britain, for example, until the threat of Revolutionary war on the Continent caused the military Board of Ordnance to set up an Ordnance Survey for that purpose.

29. Rappaport, "Geological atlas" and "Lavoisier and Monnet" (both 1969); Konvitz, *Cartography in France* (1987), chap. 4. By 1780, 31 sheets out of a projected 214 had appeared, and the final count was only 45. The engraver was Dupain-Triel (see Fig. 4.12). Monnet's MS records of some of his fieldwork for the *Atlas* are in Paris-EM, MS 9, 10.

30. Fig. 2.11 is reproduced from Monnet, *Atlas minéralogique* (1780), unnumbered sheet.

fossils of economic importance or scientific interest. They were shown as isolated spot symbols, because the evidence for them—for example in quarries, riverbanks, and road cuttings—was usually localized in the same way. Nonetheless, the effect was to make visible, as it could not be to the eye of the traveler on the ground, the distributions of all the distinctive mineral products of the area covered by the map, against the similarly spatial background of the hills and valleys, streams and rivers, and human reference points such as towns, villages, and main roads. It was a *spatial* representation of mineral and topographical diversity, and as such an appropriate portrayal of the physical geography of a region (Fig. 2.12).[31]

Guettard made this spatial conception even more explicit when in 1784 he published a small-scale map of the whole of France, showing the broad distribution of three "*bandes*" of distinctive kinds of terrain: the sandy, the marly, and the slaty. It would be tempting to a modern geologist to interpret these in stratigraphical or geohistorical terms (very roughly, as Cenozoic, Mesozoic, and Paleozoic formations respectively), but Guettard's conception was wholly lacking in any such sense of relative age. Nor was he unusual in this respect. The famous lectures by the Parisian chemist Guillaume-François Rouelle (1703–70)—which had given Lavoisier, among many others, his first taste of the sciences of the earth—had described France as

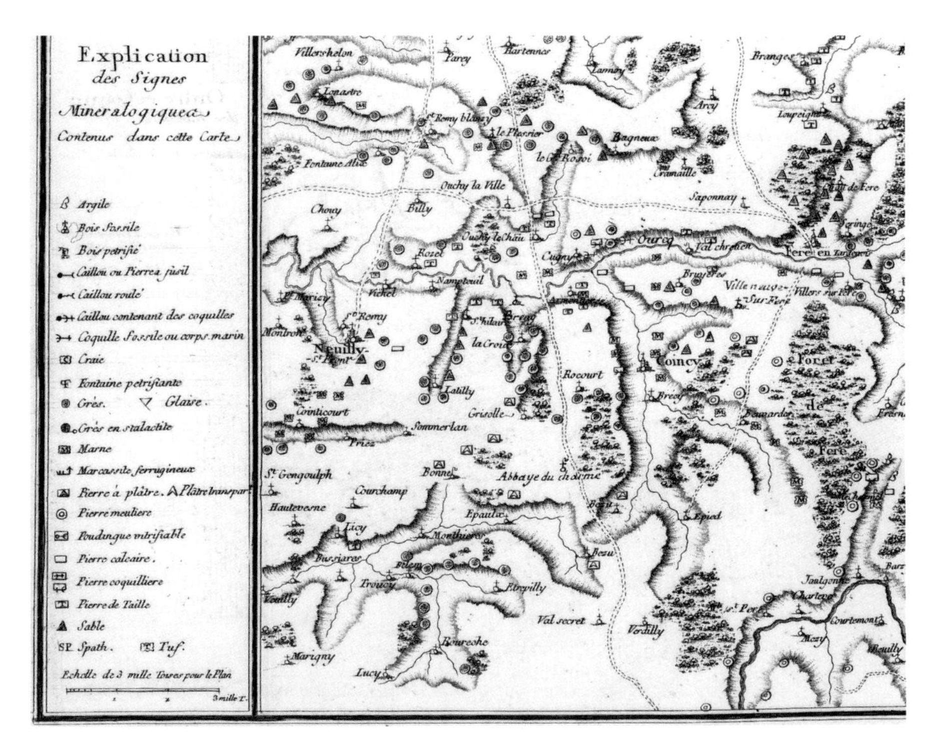

Fig. 2.12. Part of a sheet of the *Atlas Minéralogique*, showing an area (about 25km square) between the rivers Aisne and Marne about 80km east of Paris. The key identifies the many spot symbols that denote the localities of distinctive "mineral" materials. These include rocks such as chalk, sandstone, sand, and clay [*craie, grès, sable, argile*]; minerals such as gypsum and marcasite [*plâtre transparente, marcassite ferrugineux*]; and fossils such as petrified wood and marine shells [*bois petrifié, coquille fossile ou corps marin*]. Other materials of practical value include millstone and gunflint [*pierre meulière, pierre à fusil*], and there are curiosities such as a petrifying spring [*fontaine petrifiante*]. (By permission of the British Library)

divided into a series of regions, each a "*tractus*" characterized by specific kinds of fossil, grading at its edges into other regions with other fossils. Again, there was no sense of three-dimensional structure or stratigraphical sequence, let alone of geological time: the analogy, if any, would have been with the tracts of land characterized by specific assemblages of animals and plants.[32]

In this way, the science of physical geography lacked any strong sense of the third dimension that would have made it a science of the solid structure of the earth's crust, let alone any sense that structure could be translated into geohistory. In the *Atlas Minéralogique*, any graphical representation of the third dimension was literally marginalized and confined to a few small and local sections. Saussure was well aware of the problem of understanding the Alps in three-dimensional terms and gave much thought and attention to the extremely puzzling appearance of the large-scale folding of the rocks that was visible in certain mountainsides (see Fig. 2.25). Yet throughout his massive work on physical geography he concentrated on describing what could be directly observed *at the surface* of the earth, and the spatial or two-dimensional patterns of distribution of those surface features. Most tellingly of all, the great French physical geographer Nicolas Desmarest (1725–1815), who as a young man had written the article on "géographie physique" (1757) for the original *Encyclopédie* and who in his old age compiled four volumes with the same title (1796–1811) for its bloated successor the *Encyclopédie Méthodique* (§6.5), in effect *defined* physical geography as the study of those natural features and phenomena that could be represented on the two-dimensional surface of a map (see §4.3).[33]

Conclusion

Physical geography was built on the basis of outdoor fieldwork; it relegated to an ancillary role the subsequent indoor study of the specimens that were collected there. Firsthand observational experience of the large-scale features of the earth's surface was made mobile, and was transposed into the arenas of scientific debate, by being rendered not only into persuasive descriptive prose, but more importantly into proxy pictures and maps. Jointly, these media enabled other savants to become the virtual witnesses of what the physical geographer in the field had witnessed at first hand, and thereby put them in a position to accept—or to criticize or reject—whatever inferences or conclusions the original observer saw fit to offer. Throughout its practice, physical geography was a science of two-dimensional or spatial

31. Fig. 2.12 is reproduced from Monnet, *Atlas minéralogique* (1780) [London-BL: Maps C.25.c.7], part of sheet 27. A small section of the strata near Fère-en-Tardenois was engraved in the margin above the key (but is not shown here).

32. See Rappaport, "Geological atlas" (1969), which includes a small reproduction of Guettard's map (276); and "Lavoisier's theory" (1973), with a revealing quotation from the MSS of Rouelle's lectures (251).

33. The sections on the sheets of the *Atlas* were largely due to Lavoisier, and their marginal position indicates how his three-dimensional conception (§2.4) was overridden by the more traditional ideas of Guettard and Monnet. Desmarest, *Géographie physique* 1 (1796), 1, 792; although this was not published, and perhaps not written, until several years after Saussure climbed Mont Blanc, it was an idea that had pervaded his earlier work. Ellenberger, "Cartographie géologique" (1985), an invaluable study, has insightful comments on Guettard (18–21) and Desmarest (26–29).

distributions and, like other branches of natural history, a science of description and classification. Physical geographers did not aspire to offer causal explanations of the spatial features that they described and classified; or, if they did, they regarded that work as belonging to a different science (§2.4). Still less did they attempt, except occasionally and in passing, to reconstruct the *history*—in the modern temporal sense—of the features they described (some rare but important exceptions will be described in §4.3). Physical geography was not a geohistorical science.

2.3 GEOGNOSY AS A STRUCTURAL SCIENCE

The mining context

A third science of the earth, also flourishing in Saussure's time, was closely related to physical geography, but it treated the third dimension not as a marginal feature but as the focus of attention. German speakers, who were best at it, called this science "*Geognosie*" (literally, earth knowledge), and the word was adapted into other languages. Alternatively, it was called "*mineralogische Geographie*" or "*géographie souterraine*", both adjectives suggesting its goal of penetrating down into the mineral kingdom. Italian naturalists sometimes referred to it as "*anatomia della terra*", expressing even more clearly its involvement with structures hidden below the surface. Whatever the name given to the science, its practitioners sought to extend the two-dimensional spatial methods of physical geography into a three-dimensional or *structural* knowledge of the earth's crust. But geognosy, like physical geography, was still a branch of natural history, a science of description and classification. Any attempt to find causal explanations for the structures that were observed or inferred was considered to belong not to geognosy but to another science (§2.4).[34]

Attention to the dimension of depth did not of course entail any neglect of the spatial dimensions of surface distributions; in principle there was every reason why geognosy and physical geography should have formed a single unified science. That their practice was largely distinct in the late eighteenth century was a contingent result of differing social and cultural contexts. Physical geography belonged to the world of cultured travels and regional surveys of natural and human resources, the world of savants such as Saussure and Hamilton. Geognosy had a much more specific home, in the world of mining. Mining provided geognosy not only with empirical data on the dimension of depth in the earth's crust, but also—far more importantly—a distinctive way of thinking and even of seeing. Anyone involved in the mining industry, from ordinary miners right up the social scale to those who managed and administered mines, worked in a three-dimensional world of rock *structures* (Fig. 1.9). Geognosts, like physical geographers, put great emphasis on the importance of fieldwork; but in geognosy that term was extended in practice to include work in the confined and often dangerous underground world of mines.

The science of geognosy, like the practical mining from which it drew much of its inspiration, was of course concerned with the distribution of rocks at the surface as well as at depth. Even without benefit of the evidence of mine workings, exposures of rocks at the surface could be used as a basis for inferring three-dimensional

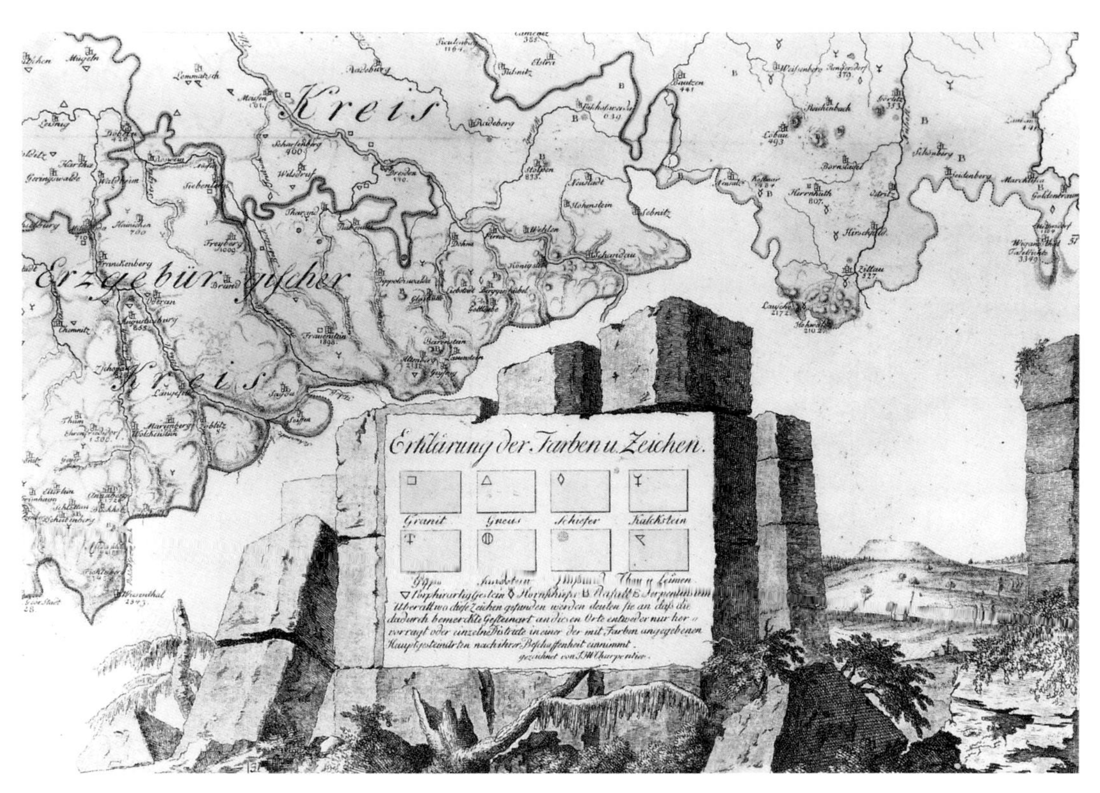

Fig. 2.13. A corner of the large map illustrating Johann von Charpentier's *Mineralogical Geography of Saxony* (1778), with the key to the (faint) colors and the spot symbols that denoted the surface distributions of eight major rock masses [*Gebirge*] such as granite, gneiss, sandstone, and limestone [*Kalckstein*]. The use of color washes implied that the direct evidence of isolated rock exposures (as in Fig. 2.12) could be extrapolated into claims about the broader areas composed of specific kinds of rock. The design of the key implies that the intention was simply to show the distributions of the rocks at the surface, not their structural relations, let alone their relative ages. The great mining area of the Erzgebirge is on the left; the town of Freiberg, its center and the site of the mining school where Charpentier (and Werner) taught, is just above the word "Erzgebürgischer". The mapped area ends at the frontier with Bohemia (now the Czech Republic). (By permission of the Syndics of Cambridge University Library)

structures. So it is not surprising that geognosts shared with physical geographers an appreciation of the value of maps as working tools. For example, almost the first illustration offered by Johann von Charpentier (1738–1805), Werner's older colleague at the mining school in Freiberg, in his book on the geognosy and mining industry of Saxony, was a fine colored map showing the surface distributions of the main classes of rock in that kingdom (Fig. 2.13).[35]

34. See Ellenberger, "Géographie souterraine" (1983); on *anatomia*, see Vaccari, "Mining and knowledge of the earth" (2000), and Morello, "Spallanzani geopaleontologo" (1982); Ciancio, *Autopsie della terra* (1995), uses an equally appropriate term for his study of Alberto Fortis. By a misunderstanding that would have surprised and offended those who practiced geognosy, their science has often been identified by historians of geology and modern geologists as an example of the genre of "theory of the earth" (see Chap. 3). In fact, geognosts made a sharp distinction between such highly speculative theorizing and their own soberly empirical science.

35. Fig. 2.13 is reproduced from Charpentier, *Mineralogische Geographie* (1778) [Cambridge-UL: Hh.20.28], part of pl. 1. The whole map is reproduced, though on a very small scale, in Laudan, *Mineralogy to geology* (1987), 103, and Klonk, "Science, art" (2003), 611. See also Ellenberger, "Cartographie géologique" (1985), 24–25. Charpentier's very first illustration was an *allegorical* design, a late example (among scientific books) of an older style of frontispiece: it is reproduced in Hamm, "Knowledge from underground" (1997), 94.

However, such maps, showing only the surface distributions of rocks, were merely the first stage in geognosy. They acted as a framework for studies of structures at depth, just as plain topographical maps served as a base for plotting the surface features of physical geography. The units that Charpentier mapped, and that all geognosts worked with as they tried to clarify the structure of the earth's crust, were the "*Gebirge*" (literally, the "mountains"). The preeminence of the mining industry in the German lands was such that equivalent words, such as "montagnes" and "monti", were used throughout Europe. However, whatever the language, the term referred to the solid *rock mass* of which a mountain or hill might be composed, not to the topographical feature itself. In other words, miners and geognosts found it natural to take a word ordinarily used for a surface feature and

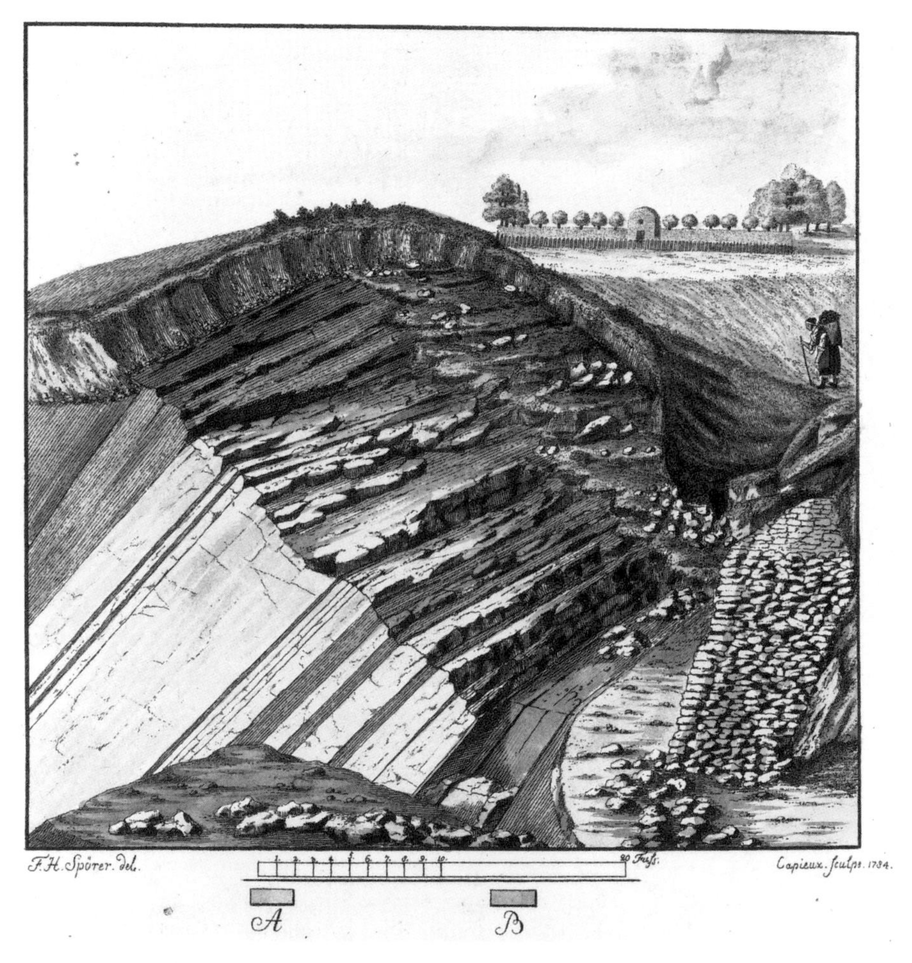

Fig. 2.14. A view of strata or bedded rocks in a quarry near Clausthal in the Harz mountains, with an imagined *section* extrapolated back from the observable quarry face to show the inferred three-dimensional structure of the rocks. This engraving served to explain to general readers what would already have been familiar to those who managed mines or quarries, or who worked in them; it was one of the first illustrations in the account of the mining industry in the Harz, *Observations on the Interior of Rock Masses* (1785), by Friedrich von Trebra, who had been the first student to graduate from Freiberg. The section shows the beds of sandstone (A) and shale (B)—elsewhere there were coal seams too—that characterized the "Secondary" formation flanking the massive "Primary" rock masses at the core of the Harz (see below). The scale is in feet. (By permission of the British Library)

extend it to the underlying mass of rock, a mass that could be penetrated mentally—and, in mines, materially—making it knowable in three dimensions.[36]

It was the responsibility of those who operated mines to understand as accurately as possible the three-dimensional structure of the rocks into which the mine was sunk, and of course in particular the veins of mineral ore or other economically valuable rocks that were the reason for the mine's existence. As in mineralogy and physical geography, visual representations were central to the practice of both mining and geognosy. Maps were of course invaluable, but they were complemented by *sections*, which were of paramount importance. Sections came to be the most characteristic graphical tool in geognosy. They allowed solid structures to be depicted; they helped to make those structures convincing to others; and above all they facilitated the process of thinking in three dimensions (a talent that remains indispensable in the earth sciences and is far from being evenly distributed, as every teacher of undergraduate geology knows from experience).

Sections depicted what it was thought would be visible if it were possible to slice the ground open along a specific vertical plane: they were in effect "virtual" cliffs or quarry faces. If the rocks were "*bedded*", or in the form of layers or "*strata*", surface exposures were quite easy to extrapolate into three-dimensional structures, at least on a small and local scale (Fig. 2.14).[37]

In any but the most primitive mines, the sinking of vertical shafts and their extension into horizontal adits and galleries needed to be recorded accurately. Sections were the most effective graphical means of doing so. In themselves, such sections were no more than the equivalents, in a vertical plane, of large-scale topographical plans. But they could also be used as a base on which to record the rocks encountered in the workings, and, more inferentially, the largely unseen course of the rock masses and their boundaries, thereby representing a clearly three-dimensional structure (Fig. 2.15).[38]

Like maps in physical geography, sections were thus an instrument—again, of paper and ink, and often of water-colors too—that made visible what could not be observed directly. However, sections necessarily embodied a greater element of inference: they extrapolated from observations to depict what could *not* in fact be seen. Yet those inferences were predictive, since they could be confirmed, modified, or invalidated by future mine workings at depth or by future exposures of the rocks at the surface (on Charpentier's map of Saxony, Fig. 2.13, the extrapolation from isolated exposures to broad areal distributions had a similar inferential character).

36. See Ospovat, "Reflections on Werner" (1969), and *Werner: Short classification* (1971), 97. The word *Gebirge* will be translated here as "rock masses" rather than "formations", because it was applied not only to bedded rocks but also to massive bodies of granite, basalt, gneiss, etc.; in other words, to rocks now interpreted as being intrusive, extrusive, and metamorphic as well as sedimentary. *Gebirge* had the advantage of being a purely descriptive term that was neutral in relation to questions of origins (§2.4).

37. Fig. 2.14 is reproduced from Trebra, *Innern der Gebirge* (1785) [London-BL: 457.e.19], pl. 1, fig. 2, based on a drawing by F. H. Spörer. The text does not state explicitly that the smooth rock surface on the left is a section inferred from the strata exposed in the center, but if it was a visible quarry face it was curiously idealized compared to the rest of the drawing.

38. Fig. 2.15 is reproduced from Trebra, *Innern der Gebirge* (1785) [London-BL: 457.e.19], part of pl. 6, a very long section extending from Goslar south into the Harz; pale color washes distinguish the ore body and the rocks above and below it.

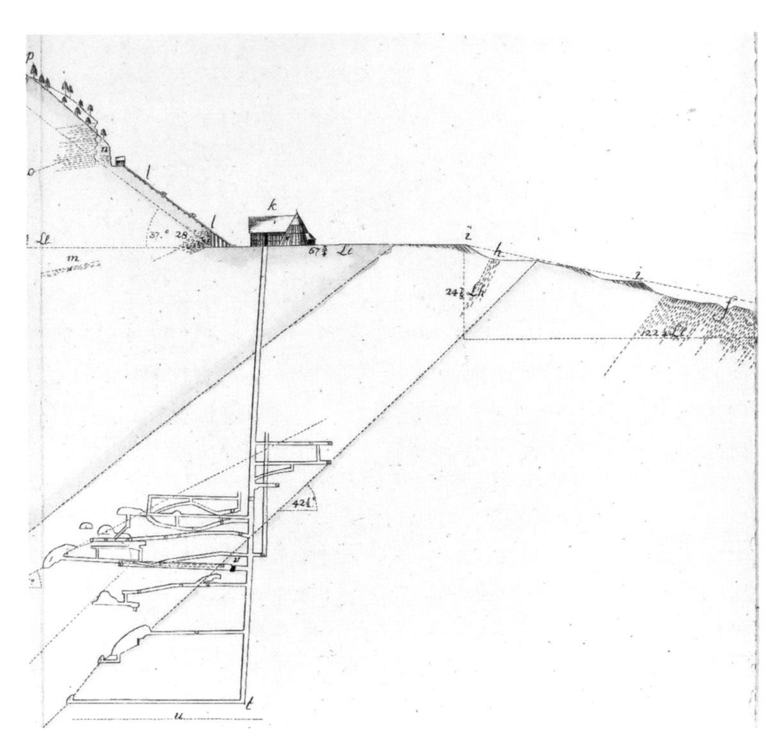

Fig. 2.15. A section through the Kannekuhl mine in the Harz mountains, one of the great mining areas in eighteenth-century Europe: part of a much longer engraved section published in von Trebra's *Observations on the Interior of Rock Masses* (1785). The mine is working a sloping ore body, the lower boundary of which is marked with its precise "*dip*" or angle from the horizontal, as measured from its intersections with the vertical mine shaft and horizontal adits, and from its exposures at the surface. Other rocks, known only from such exposures, are shown more tentatively, penetrating only a short way from the surface. Detailed sections of mines were rarely published, for reasons of state and industrial secrecy. (By permission of the British Library)

Structures and sequences

The "beds", strata, or layers of rock that might each be familiar to miners and quarrymen—and even be named individually—were grouped by geognosts into the larger units of the *Gebirge.* For example, the alternating beds of sandstone and shale portrayed in Fig. 2.14, along with the coal seams found elsewhere, were all grouped together as parts of a single rock mass. *Gebirge* might thus include several different kinds of rock in their subordinate parts, yet the whole rock mass would have some kind of perceived uniformity or coherence. Like the beds exposed in a quarry, major rock masses could be the subject of three-dimensional or structural inference. However, this entailed far greater extrapolation from what could be observed directly. Such a degree of speculation seems to have been considered inappropriate, except in a context of explicitly global theorizing (see Chap. 3); certainly, sections showing the inferred structure of any large region were rarely published. Nonetheless, many geognosts may have kept this kind of structural interpretation in mind as a desirable goal of their research, even if it could not often be attained (Fig. 2.16).[39]

As a branch of natural history, the primary goal of geognosy was the accurate description and classification of the *Gebirge* of the region under study. Their description included a record not only of the kinds of rock of which they were composed, but also of their structural relations to one another, and the kinds of topographical situation in which they were found. Their classification was based on the same disparate criteria; it was distinct from the classification of rocks, which in turn was different from that of their constituent minerals (§2.1).

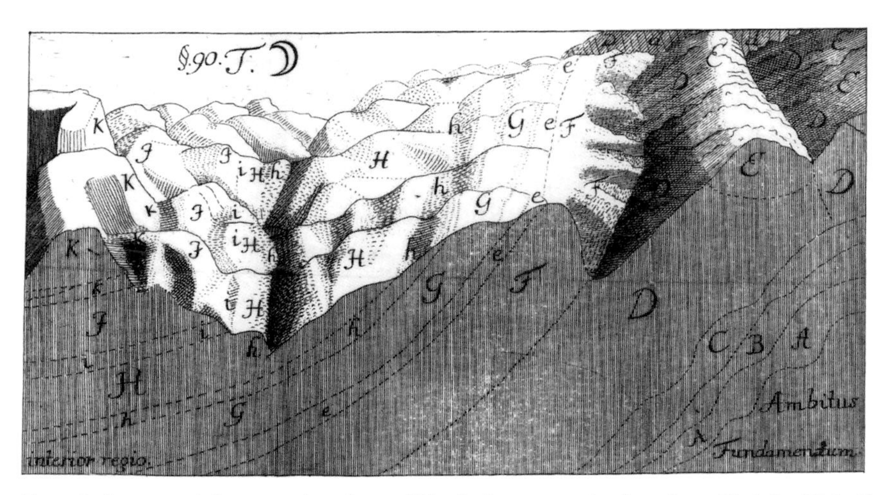

Fig. 2.16. A geognostic interpretation of part of Thuringia: an engraving from Georg Christian Füchsel's "History of Land and Sea" (1761). This drawing depicts the three-dimensional structure by means of a quasi-aerial perspective view of the surface outcrops of the *Gebirge*, combined with an inferred section through them at depth. Fourteen successive "Secondary" rock masses (see below) are shown overlying a basement rock mass [*ambitus fundamentum*], which is not exposed at the surface within the area shown. The vertical scale, and hence the dip of the rocks, is exaggerated to clarify the structure. This kind of visual representation of large-scale structure—in modern terms, a *block diagram*—was very rare in the eighteenth century. (By permission of the British Library)

It cannot be emphasized too strongly that the goal of geognosy, no less than the other branches of natural history, was to classify the diversity of nature; it was not to reconstruct geohistory. Werner made this clear in the article—soon republished as an influential booklet—in which he summarized part of his famous course on geognosy at Freiberg: he treated the subject explicitly as natural history [*Naturgeschichte*]. His article was not particularly original: it was a codification of what was becoming a consensus among geognosts throughout Europe. Indeed, he claimed that it was needed precisely because so much was currently being published in geognosy or "mineral geography" that a better classification was urgently required. He entitled his work unambiguously a "classification and description", and its subject was to be the "species" [*Arten*] of *Gebirge*, the basic units of classification in geognosy (Fig. 2.17).[40]

39. Fig. 2.16 is reproduced from Füchsel, "Historia terrae et maris" (1761) [London-BL: 963.a.35], part of pl. 5. The rest of this famous plate shows a less stylized quasi-aerial view of the same region with more specific detail of the outcrops over an area of about 30km by 50km, including such towns as Ilmenau, Weimar, and Jena: it is well reproduced in Ellenberger, *Histoire de la géologie* 2 (1994), 254; Guntau, "Natural history of the earth" (1996), 224; and Hamm, "Knowledge from underground" (1997), 90. See also Ellenberger, "Cartographie géologique" (1985), 31–34. Although based on fieldwork in Thuringia, Füchsel's volume was designed primarily to expound his "theory of the earth".

40. Fig. 2.17 is reproduced from Werner, *Kurze Klassifikation* (1787), title page. The original text (1786), almost identical, is reproduced in facsimile, with a parallel translation and valuable editorial notes, in Ospovat, *Werner: Short classification* (1971); see also his "Reflections on Werner" (1969) and "Importance of regional geology" (1980), and Wagenbreth, "Werners System der Geologie" (1967). Guntau, "Klassifikation natürlicher Objekte" (2002), rightly interprets Werner's work as taxonomy; other articles in Albrecht and Ladwig, *Abraham Werner* (2002), also review recent historical research on Werner and his contemporaries. Like the rest of this chapter, this analysis of geognosy, and particularly of Werner's practice, is based on evidence from around the 1780s and earlier. Many historical accounts have made substantial use of much later reports of Werner's teaching (up to the time of his death in 1817), which in part reflect what geognosy became under the impact of developments to be described later in this book (§8.1). Werner, like any competent modern scientist, moved with the times.

Kurze
Klassifikation und Beschreibung
der
verschiedenen Gebirgsarten,
von
A. G. Werner,
Bergakademie-Inspektor und Lehrer der Bergbaukunst und Mineralogie
zu Freyberg.

Dresden, 1787.
In der Waltherischen Hofbuchhandlung.

Fig. 2.17. The title-page of Werner's *Brief Classification* (1787). The title itself made it clear that the primary goal of his work was to set out a "brief classification and description of the various species of the rock masses" of which the earth's crust was composed, not to reconstruct its history (it was "brief" because it was a sketch for a longer treatise, which in fact never appeared). He himself is described as a mines inspector, and a teacher of mining technology and mineralogy, at the Freiberg mining school. The use of a stock decorative motif was common in publications of all kinds, and the "Gothic" typeface was usual in those written in German.

As in other branches of natural history, classification in geognosy was hierarchical in structure. Having established the "species" or basic natural kinds of rock mass, putatively universal in validity, the taxonomy could be extended in either direction: upwards in the hierarchy to more comprehensive and fundamental groupings, or downwards to local variants. Geognosts agreed that most kinds of rock mass fell into one of two fairly distinct high-level categories, just as animals either did or did not have backbones, and plants either did or did not have flowers. Together with two other categories of lesser importance but on the same level of generality, all kinds of rock mass anywhere in the world would then be covered; or at least, that was the goal towards which geognosts were working, and which Werner believed he had effectively reached.[41]

Primaries and Secondaries

The two main categories of rock mass were called *Primary* and *Secondary*. These came to be the standard terms in English (and will be used here); they were derived from what the great Italian geognost Giovanni Arduino (1714–95) had described as "*monti primari*" and "*monti secondari*", in work that became known in translation throughout Europe (see Fig. 2.32). German writers usually called them respectively the "fundamental rock masses" [*Urgebirge*] and the "layered rock masses" [*Flötzge-*

birge]. This expressed their structural relation, namely that the Primaries everywhere underlay the Secondaries and seemed to be the foundation of the earth's crust; and also a general contrast between them, namely that the Secondaries were usually bedded rocks lying in a structural sequence that was distinct, at least in any given region, whereas the Primaries rarely had any clear sequence. Furthermore, the word *Urgebirge* implied a belief that the Primaries were "primitive" or fundamental in a temporal sense; Werner cautiously called them "apparently primitive" [*uranfänglich*]. In French this was expressed less ambiguously: following Rouelle's lead, the rock masses were usually termed the "*ancienne terre*" and the "*nouvelle terre*", terrains old and new.[42]

The French terms, and more ambiguously the others, indicate that relative age was indeed one of the criteria by which these two great categories were to be distinguished. Yet relative age was inferred from structural position, and in practice that directly observable criterion was treated as more important: geognosts usually referred to rock masses as being "above" or "below" others, and only rarely as "younger" or "older". When, for example, the Swedish geognost Johann Jakob Ferber (1743–90) reported on Arduino's classification—and thereby made it more widely known to savants throughout Europe—he explained how the major groups of *Gebirge* were distinguished "by the arrangement of their beds below or above one another, and the different age and origin inferred therefrom". It was the structural position of the Primaries, as the apparent physical foundation of the earth's crust, that entitled them to be treated as primary; and it was the structural position of the Secondaries, clearly overlying the Primaries and sometimes manifestly composed of materials derived from their erosion, that made them secondary, rather than the lesser age that those observed features implied. Anyway, "Primary" and "Secondary" remained terms of classification, not geohistory: the inferred difference of age was just one taxonomic criterion among many, and for most purposes the others were more important.[43]

The Primaries were generally hard rocks, usually found in upland or mountain regions and lacking any trace of fossils; the Secondaries were often softer, they were

41. Laudan, *Mineralogy to geology* (1987), chap. 5, draws a sharp distinction between Werner's classification and others in natural history, on the grounds that Werner's posited gradations between species and used geological time as a major taxonomic criterion. But the nature and reality of the divisions between species was also controversial among botanists and zoologists (hence their lively debates about hybrids and "sports"); and the supposed introduction of time did not alter the fundamental goal of Werner's work, which was to *classify* the varied rock masses (see below).

42. The *Ur-* prefix was ambiguous, because it denoted some kind of fundamental status, but not necessarily a temporal one: in the case of botany, Goethe suggested a hypothetical "original plant" [*Urpflanze*] to represent the archetype or common structural ground plan on which the diversity of plants was constructed, rather than their temporal origin or ancestral form. Werner later proposed a "Transition" category [*Übergangsgebirge*] that bridged the gap between the Primaries and Secondaries (§8.1), but this did not alter the basic character of his classification, any more than the gradations that he allowed between many of the "species" within them. "Primary", "Transition", "Secondary", etc., are given initial capitals throughout this account, to indicate that they were technical terms, like their indirect modern equivalents such as "Carboniferous" and "Jurassic".

43. Ferber, *Briefen aus Wälschland* (1773), 38. French and English editions followed in 1776; however, the latter (*Tour through Italy*) missed the crucial structural point by translating Ferber's phrase only as "according to the difference of their beds and their presumptive antiquity and origin" (36). On Arduino himself, see Vaccari, "La 'classificazione' delle montagne" (1999).

usually found in lower hills or in flat country, and some of them contained abundant fossils. Their constituent rocks were also of contrasting kinds: the Primaries included masses of granite, gneiss, schist, and marble, among others; the Secondaries included masses of sandstone, shale, limestone, and "pudding-stone" (conglomerate), among others. (The distinction between Primary and Secondary rock masses was similar to the informal one often used by modern geologists, between "hard-rock" and "soft-rock" terrains.) The recognition of the structural relation between Primaries and Secondaries was not always straightforward. Primary rock masses were said to be "below" the Secondary ones, yet they usually rose to higher altitudes. However, what was verbally paradoxical was easy to comprehend visually: the relation was one of structural overlap. Where the junction was found in the field, Primary rock masses could be seen to emerge from below the Secondaries; or, put the other way round, the Secondaries overlapped onto the Primaries, which rose above them to form higher hills and even mountains (Fig. 2.18).[44]

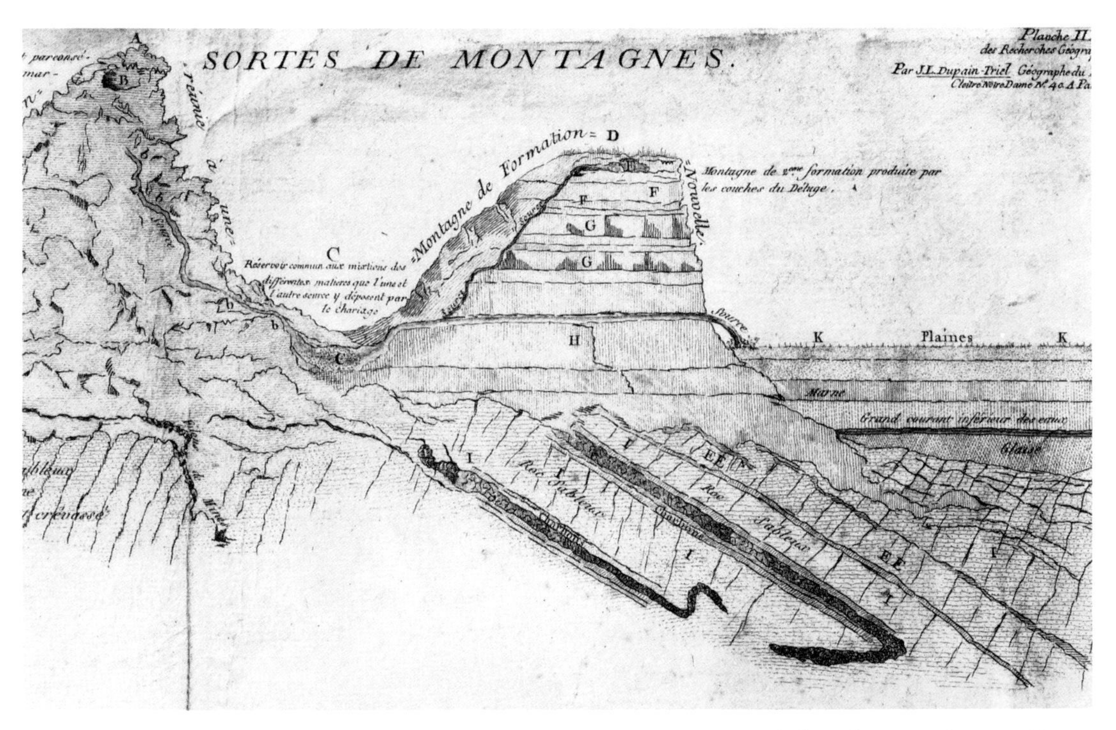

Fig. 2.18. Part of an idealized section drawn to explain the structural relation between Primary and Secondary rock masses [*montagnes*]. The Primaries are shown as rising to the highest altitude (left), with an internal structure that is "irregularly jointed" [*crévassé*] below and even more chaotic above. Resting on them are two sets of Secondary rocks, an underground sequence of dipping layers of sandstone and coal [*roc sableux, charbons*], and above them a horizontal sequence of other layers forming lower hills (center). Still lower in altitude, but still higher in structural position, are the "plains" (right), underlain by a sequence of alluvial or Superficial layers, among them marl and clay [*marne, glaise*]. This engraving was designed by Jean-Louis Dupain-Triel the younger, the French royal geographer at the time of the Revolution, for a booklet (1791) explaining the sciences of the earth to general readers; it is unusual in showing two distinct sets of Secondary rock masses, with (in modern terms) an unconformity between them, and in claiming that the upper one was produced by the "Deluge" (see §2.4). Its design also suggests the practical value of geognosy, in helping to clarify the situation of coal mines and the puzzling phenomena of springs and underground aquifers.

A further distinction between Primaries and Secondaries is well illustrated by this example. As already mentioned, the Secondary rock masses were usually composed of more or less distinct and parallel-surfaced beds or strata (see also Fig. 2.16), apparently deposited in succession (just how they might have been deposited, and from what medium, were separate problems). The Primary rock masses, on the other hand, had more diverse and often obscure structural relations with one another. Some, such as gneiss and schist, seemed to be layered or bedded—Saussure thought the granite near the summit of Mont Blanc was another case in point (§1.1)—but others were "massive" and without any trace of bedding. Some seemed to lie side by side, rather than being either above or below one another; and mineral veins usually cut right through rock masses of some contrasting kind. These confusing structures made the study of the Primaries more difficult than that of the Secondaries. However, it was also literally more rewarding, for most of the economically important minerals—particularly the ores of the precious and other nonferrous metals—were found in veins in the "hard-rock" terrains of the Primaries, which for that reason were sometimes called "veined rock masses" [*Ganggebirge*]. Coal seams and some iron ores were almost the only valuable materials found in the bedded rock masses of the Secondaries (which were also important, however, as sources of workaday materials such as building stone, brick-earth, etc.). Consequently, geognosts paid far more attention to the Primaries than to the Secondaries; the latter were often tacitly dismissed as being of little interest, either to the practical manager of mines or to the *wissenschaftlich* savant.

By comparison with the Primary and Secondary *Gebirge*, the other two high-level categories in geognostic classification were of relatively minor importance. Most of the *alluvial* or *Superficial* deposits were not rock masses at all, at least in the colloquial meaning of "rocks": they were the loose gravels, sands, silts, and muds usually found in river valleys and estuaries and on low-lying plains (see Fig. 2.18), though sometimes they formed low hills. Their origin was inferred to be very recent, relative even to the Secondary rock masses. More important, however, was the observable criterion that they were clearly derivative: Werner called them appropriately the "washed-out" [*Ausgeschwemmte*] deposits. River gravels, for example, might include easily identifiable pebbles of any of the Primary and Secondary rocks in the region drained by the river and its tributaries. They were therefore sometimes called *Tertiary* deposits.[45]

The fourth and last high-level category in geognostic classification was also treated as relatively minor in significance. The main diagnostic feature of the

44. Fig. 2.18 is reproduced from Dupain-Triel, *Recherches géographiques* (1791), pl. 3, fig. [1]. The modern distinction between "hard-rock" and "soft-rock" terrains (and hard-rock and soft-rock geologists) has no simple correlation with age, because it depends on the contingent geohistory of specific regions. Pudding stone or *poudingue* was so called because its embedded pebbles resembled, for example, the fruit and nuts in traditional Christmas puddings.

45. The word Tertiary was not being used here in its modern stratigraphical sense. However, Arduino had used "*monti terziari*" in a way that did approximate to that modern meaning, namely for what others regarded as upper and flat-lying beds of Secondary formations (as shown, though without that label, in Fig. 2.32), containing distinctive fossils such as nummulites. He then distinguished the alluvial deposits on the valley floors and plains—not "monti" at all—as a *fourth* category or "*quatro ordine*" (the indirect forerunner of modern "Quaternary"): see Vaccari, *Giovanni Arduino* (1993), chap. 3.

Volcanic rock masses, apart from the nature of the rocks themselves, was their mode of origin, not their relative position, still less their inferred relative age. On the question of origin Werner was in a minority: he rejected the volcanic origin of the widespread masses of the enigmatic rock basalt, assigning them instead to an aqueous origin of some kind and classifying them among the Primaries. He therefore concluded, "perhaps to the not inconsiderable displeasure of many fire-addicted [*feuersüchtigen*] mineralogists and geognosts", that *true* volcanic rocks—those associated with active volcanoes and of clearly "fiery" origin—were unimportant components of the earth's crust.[46]

By contrast, as Werner's wry comment indicates, most other geognosts were convinced that basalt was identical to some of the rocks known to be produced from the cooling of volcanic lavas (see §2.4). After Werner made his views public, the argument between "*Neptunists*" and "*Vulcanists*", between champions of water and fire as causal agents, raged for several years; a Classical education put all savants on familiar terms with the ancient gods. Even at Freiberg, Werner did not have it all his own way, for his senior colleague Charpentier was one of several Vulcanists there. But it was primarily an argument about the correct classification of basalt, which was just one puzzling kind of rock and rock mass: some Swiss naturalists summarized it bluntly, when in 1788 they set as a prize question "What is basalt? Is it volcanic, or is it not?" (see §2.4).[47]

On either interpretation, however, the treatment of basalt underscored how geognostic classification was only incidentally temporal, still less geohistorical. Werner classed basalt by its mineral character as a Primary rock, although the bodies of basalt that he knew at first hand were lying on top of both Primary and Secondary rock masses (see Fig. 2.22), and he thought the material had been precipitated relatively recently. By the same token, however, his critics could not treat volcanic rocks as characteristic only of recent times, since they regarded some basalts as the traces of volcanoes that had been active while the Secondary rock masses were being formed. More generally, Werner himself maintained that three of his four major categories were still being formed in the present world: not only the Volcanics and Superficials, but also—in the depths of the sea—the Secondaries. So he conceded that the latter were "earlier" only in the limited sense that they had *begun* to accumulate at an earlier time.[48]

Sequences of *Gebirge*

Geognosts, then, classified the huge diversity of rock masses into the four major groups: the Primaries, Secondaries, Superficials (or alluvials), and Volcanics. These in turn were classified into more specific kinds of *Gebirge*. Werner thought that the two dozen "species" described in his *Brief Classification* probably included all those reported so far, or likely to be reported in the future, from anywhere in the world. In that sense they were "universal", and he expected that his classification—with due refinement—would prove to be valid everywhere. This was a conclusion no more unrealistic, or conceited, than the confidence of mineralogists that they had described the broad outlines of mineral diversity on a global scale.

Gebirge of granite and gneiss, coal and gypsum, were "universal" in the same sense as minerals such as quartz and feldspar, hornblende and mica.

Werner followed normal geognostic practice in listing his species in an order based mainly on convenience of description. Among the Primaries, however, granite was generally listed first, because, as he noted with characteristic caution, it "seems to be the fundamental rock mass [*Grundgebirge*]". Granite often seemed to underlie *all* the other Primary rock masses such as gneiss and schist, and it rose from beneath them to form the central tracts and highest peaks of mountain ranges: it was no surprise to geognosts when Saussure reported that the highest rock on Mont Blanc was granite. Apart from granite, however, neither Werner nor other geognosts thought there was any invariable order to the structural sequence of the Primary *Gebirge*. There was often a crude ordering, and even a gradation, from granite through gneiss and schist to slate; but those rock masses might also recur at different points and be intercalated with "primary limestone" or marble, quartzite, and other rock masses. Such recurrences and inconsistencies eliminated any strict correlation between specific kinds of rock mass and inferred relative age. Anyway, the structural relations of the Primaries were often so confused and confusing that it was far from clear that they had any "sequence" at all.

With the Secondaries, on the other hand, a sequence or structural order was often clear and unambiguous, at least within a particular region. For example, Füchsel's block diagram of Thuringia (Fig. 2.16) showed a sequence of many distinct and distinctive bedded rock masses, inferred from surface outcrops. Part of the same region had in fact already been depicted in another section, one of the earliest of its kind in published form (Fig. 2.19).[49]

As already mentioned, a bedded rock mass [*Gebirge*] might be composed of more than one kind of rock. What made it a true "species" [*Gebirgsart*] was a distinctive *assemblage* of rocks, such as the alternating beds of sandstone and shale (and, elsewhere, of coal seams) depicted in Fig. 2.14; or, for example, a distinctive assemblage of beds of limestone and shale, or of beds of sandstone and pudding-stone. In all such cases the constituent beds had obviously been formed at the same period of time, so Secondary rock masses were often called "*formations*". But they could not be defined simply by their inferred time of origin, because many kinds of rock mass recurred at different points in the sequence. As Werner, for example, was well

46. Werner, *Kurze Klassifikation* (1787); see also Ospovat, "Importance of regional geology" (1980).

47. An adequate account of the debate between Neptunists and Vulcanists, free from the ignorant stereotyping of an earlier generation of geologist-historians, has yet to be written; however, Fritscher, *Vulkanismusstreit* (1991), makes a fine start with a careful analysis of its chemical and experimental aspects. Charpentier's cautious assessment of Saxon basalts, in the light of Hamilton's volcanic studies, is in his *Mineralogische Geographie* (1778), 408–9. The prize question set by the Naturforschenden Privatgesellschaft in Bern was the first of its kind to be reported in the new *Bergmännische Journal* (1: 378) edited in Freiberg by one of Werner's and Charpentier's colleagues.

48. Werner, "Entstehung des Basalts" (1788); in *Kurze Klassifikation* (1787), basalt is listed as one of the twelve kinds of Primary rock mass, *not* among the seven Secondaries. See also Ospovat, "Importance of regional geology" (1980).

49. Fig. 2.19 is reproduced from Lehmann, *Flötz-Gebürgen* (1756) [London-BL: 990.c.15], pl. 7. In modern terms, the section shows the basement rocks of the Harz, and overlying Carboniferous and Permian formations: see Ellenberger, *Histoire de la géologie* 2 (1994), 252–53. See also Rappaport, "Holbach's campaign" (1994).

Fig. 2.19. A general section from the Harz into Thuringia, showing many distinctive Secondary rock masses. They are numbered in sequence downwards (that is, in the *reverse* of any inferred temporal or geohistorical order), from the upper and more easily defined layers towards the more obscure lower units. The thin but highly distinctive *Kupferschiefer* or copper shale (no. 13), for example, was famous for its fossils (see *Fig.* 5.8). At the bottom of the pile the rocks with coal seams (nos. 19–30) rest on Primary rocks with mineral veins [*Ganggebirge*] (no. 31). The compass symbol indicates the north-south [*SE-ME*] orientation of the section, which was in fact highly diagrammatic, depicting a far longer tract of country than the trees, clouds, and landscape style might suggest. This engraving was published in Johann Lehmann's *Layered Rock Masses* (1756); a French translation by the *philosophe* Paul Henri, baron d'Holbach, had made Lehmann's work well-known throughout Europe. (By permission of the British Library)

aware, Secondary limestone [*Flötzkalk*] formations were known to lie both above and below the distinctive and important coal formations in several parts of Europe, and there were at least two separate formations of distinctive red sandstones.

Nonetheless, within a given region such as Thuringia, several formations were found to maintain the same structural order across quite wide tracts of country: they could be said to lie consistently above or below one another. A few unusually distinctive formations could be recognized even more widely, implying that they must have been formed over very large areas: the *Muschelkalk* [mussel-limestone], for example, was found right across central Europe. The outstanding case, however, was the brilliant white limestone of the Chalk, often with contrasting bands of black flint nodules: this was widespread in England and northern France (forming the famous white cliffs on both sides of the Channel), but also extended to Ireland and through the Netherlands into Denmark. Such an extensive distribution encouraged the hope among geognosts that eventually, after much further fieldwork, it might be possible to discover a structural order in the Secondary formations that would transcend any specific region and perhaps even be of global validity.

Geognosts were well aware that this structural order was likely to reflect a corresponding *temporal* order of deposition: as Füchsel and others recognized, "above" and "below" could be translated with confidence into "younger" and "older". The almost axiomatic principle of *superposition*, as modern geologists call it, had been familiar ever since the Danish naturalist Nils Stensen (writing under his scholarly Latin name of Steno) had formulated it in his classic essay on the strata of Tuscany a century earlier; that work had been republished in 1763, and Werner, for example, owned copies of both editions. Yet temporal terms were rarely used, as already mentioned, except when geognostic descriptions became raw material for the quite different sciences of causal origins and geohistory (see §2.4, §4.4). Usually the language was that of atemporal order and static arrangement: as the English clock maker and geognost John Whitehurst (1713–88) put it, "The arrangement of the strata in general is such, that they invariably follow each other, as it were, in alphabetical order, or as a series of numbers". Geognostic practice was primarily concerned with structural order, not temporal sequence, let alone geohistorical reconstruction.[50]

Fossils in geognosy

From a modern perspective, what may seem surprising about this ordinary practice is that geognosts paid little attention to fossils. Fossils were objects of great interest in "mineralogy"; they were avidly collected by both savants and dilettanti and were widely described and illustrated in publications (§2.1). Yet in geognosy their place was marginal. One reason for this is perhaps so obvious that it has often been overlooked. Geognosy was a science rooted in the practice of mining. Most mines exploited the valuable materials found in mineral veins in regions of Primary rock masses. The Primaries were found to contain no trace of fossils and were defined (in part) by that fact. Hence fossils were of marginal significance for the mainstream of geognostic practice; for this was directed towards providing the mining industry with scientific foundations that would facilitate the exploitation of known mineral resources and lead to the discovery of new ones. As already mentioned, only a few minerals of economic importance were found in Secondary rock masses, notably coal and some iron ores; these were the only mineral resources that might have been exploited more effectively, had geognosts given the Secondaries and their fossils more attention.

Geognosts did not neglect fossils altogether in their descriptive work on the Secondary formations, but they treated fossils as just one diagnostic criterion among others. There was no obvious reason to give fossils a privileged position above other criteria such as the exact type of rock or rocks that comprised the formation in question. Fossils were often noted in geognostic description, but only in general terms. If they were abundant and distinctive they might even figure in the name given to a formation: for example, the *Muschelkalk* contained abundant and distinctive "mussels" or bivalve mollusk shells that helped—along with the kind of

50. Whitehurst, *Inquiry into the original state*, 2nd ed. (1786), 178. Laudan, *Mineralogy to geology* (1987), chap. 5, in effect equates Werner's *Gebirge* with his use of *Formation*, and thereby interprets his geognosy as *geohistorical*, rather than (as argued here) fundamentally *structural*; see Rudwick, "Emergence of a new science" (1990).

limestone in which they were embedded—to distinguish that formation from others. Ferber reported how Arduino had found different kinds of fossils in the various formations in the foothills of the Italian Alps, and he named them in general terms; he even noted how, within one particular Secondary formation, there were "various marine fossils [*Seekörper*], which, however, are different in each bed." In a general sense and to a limited extent, some fossils were thus treated in geognostic practice as being "characteristic" of particular formations: but they were just one of the many diagnostic criteria used in recognizing the "species" of a given formation when a new exposure of it was being examined.[51]

A further reason for the relative neglect of fossils was more subtle in its impact and therefore more difficult to assess. Once it became apparent that many "fossils" were indeed the remains of animals and plants that had once been alive, they were distinguished from other mineral objects by being qualified as "extraneous" or "accidental" (§2.1). But that distinction also separated them conceptually from the rocks in which they were found; it encouraged naturalists to regard them as distinct from rocks and therefore appropriate for a different kind of study. Specifically, it tended to shift them from being part of the subject matter of mineralogy (in the broadest sense) to being part of botany or zoology. So, paradoxically from a modern viewpoint, the fruitful research by eighteenth-century naturalists—resolving many of the earlier puzzles about the nature of fossils—led them *away* from any conception of fossils as being "essential" characters of specific formations, let alone any conception of them as potential markers of geohistory.[52]

Conclusion

Geognosy was a flourishing science in Saussure's time (though his own work remained mainly on the plane of physical geography). Geognosts explored and made sense of the three-dimensional or *structural* relations of rock masses, as found by outdoor fieldwork, including the underground exploration that was possible in mines. In fact, as has been emphasized repeatedly here, geognosy was rooted in the world of mining. Even when it was extended to regions without mines, where the structure of the rocks at depth had to be inferred wholly from what could be observed in exposures at the surface, its conceptual models and its modes of visual representation—above all, sections—were derived from the practice of mining.

Like other branches of natural history, the goals of geognosy were those of description and classification. The structural position of a rock mass in relation to others was one of the diagnostic criteria for classifying it; others included the nature of its constituent rock or rocks, its topographical expression or characteristic altitude, and the nature of any fossils it might contain. The structural order of rock masses, as being above or below others, could often be translated confidently into a temporal order of younger and older; but priority was given to the structural, because it could be observed directly. Rock masses were classified first as belonging to one or other of the four major groups of Primaries, Secondaries, Superficials (or alluvials), and Volcanics; and then, within each major category, according to their natural kinds or "species", such as schist and sandstone. Secondary rock masses were generally easier to interpret than the Primaries, and bedded Secondary "formations" could often be detected in a consistent sequence, at least within a given region. But they were still treated primarily in terms of an observed structural order, rather than in the sense of an inferred temporal sequence, let alone a geohistory. Geognosy was a structural science; like mineralogy and physical geography, it was not a causal science, still less a science of geohistory.

2.4 EARTH PHYSICS AS A CAUSAL SCIENCE

The "physics" of specimens

Mineralogy was a science of specimens, practiced primarily indoors in museums; physical geography was a science of spatial distributions, based on outdoor fieldwork; and geognosy was a science of three-dimensional structures, also based on fieldwork but exploiting additionally the dimension of depth that was revealed by the practice of mining. All three were branches of natural history, in that their goals were those of description, identification, and classification. Jointly, they accounted for most scientific studies of the earth around the time that Saussure—a leading exponent of physical geography—climbed to the highest point in Europe.

Another set of problems was often broached, sometimes by the same savants and sometimes by others. However, in doing so they were clearly aware of embarking on a different kind of science; they usually presented their ideas in separate sections of a published work, and often even in separate articles or books. The fourth and last of the sciences of the earth being practiced in Saussure's time was recognized as belonging not to natural history but to natural philosophy or "physics". Rather than describing and classifying, it used the natural-history sciences as raw material for detecting the regularities or "natural laws" underlying the observable occurrence of terrestrial features and processes, with the ultimate goal of determining their physical *causes.*

Saussure's francophone contemporaries called this science "*physique de la terre*", which will be translated here as *earth physics* (it cannot be called "geophysics" without serious confusion, because it was quite different in scope from that modern science). Its name incorporated the contemporary meaning of "physics", denoting the study of the causes of anything in the natural world. Just as the causal science of physiology complemented the descriptive science of anatomy, so likewise the causal science of earth physics was regarded as complementing the descriptive sciences of mineralogy, physical geography, and geognosy. The character of earth physics around the time that Saussure climbed Mont Blanc will be illustrated here by a few examples, dealing in turn with the causal explanation of the subject matters of each of the three descriptive sciences of the earth.[53]

51. Ferber, *Briefen aus Wälschland* (1773), 45. The English edition (*Travels through Italy*) was again misleading, the subordinate clause being translated as "each stratum being filled with a [*sic*] peculiar species" (41).

52. See Laudan, *Mineralogy to geology* (1987), 146.

53. Savants working in the sciences of the earth generally used the notion of "cause" with little or no reference to the intense contemporary philosophical discussion of the concept of causality, for example by Immanuel

First, then, the specimens handled by mineralogists (in the broad sense that included those who studied fossils) had many features that called for causal explanation. One of the most pervasive problems was to explain the mode of origin of many of the commonest rocks. Shales, sandstones, and pudding-stones, for example, were relatively easy to understand. Apart from being solid or "stony", they looked similar to ordinary muds, sands, and gravels respectively and had evidently been deposited likewise from water or some similar liquid. Limestones could also be assigned an aqueous origin with some confidence, since many contained fossils of obviously marine animals or were visibly composed of the limey debris of broken shells and similar material. This accounted for the bulk of the Secondary formations. A few rocks such as basalt—often found within or on top of a pile of Secondaries—remained enigmatic; but even the most ardent Vulcanists agreed with their Neptunist opponents that *most* Secondary rocks had been deposited or precipitated from water or some similar liquid.

The rocks that constituted the Primary rock masses were much more puzzling. Gneiss and schist, for example, had strongly marked layering that made them look like bedded Secondary rocks; nothing similar was known to be forming in the present world, but it was plausible to suppose that they too had been deposited layer by layer, perhaps as chemical precipitates, and from a liquid medium, but perhaps one quite different in composition from sea water (see §3.5). Granite might have had a similar origin, since it was often layered in structure—as Saussure reported from Mont Blanc—though in other places it was completely unbedded or massive. Marble too could easily be imagined as a precipitate. Alternatively, some of these rocks recalled the slags and other products of the furnaces of the mining and metalworking industries, so it seemed possible that they might have been formed by crystallization from hot melts in the depths of the earth. But mineralogists recognized that their causal conjectures on such matters were highly speculative, since the chemical processes involved were so obscure.

However, the rocks that were most common among the Secondary formations also had their own difficulties for causal explanation. One that may now seem relatively unproblematic was their very "stoniness". For example, sandstones and pudding-stones were puzzling because, although their constituents were so similar to ordinary sand and pebbles, the material cementing the sand grains together was often composed of the same mineral, quartz, and it was not obvious how it had got there. Quartz was known to be almost insoluble in water, so any explanation in terms of precipitation from percolating water seemed highly implausible, particularly since the end product was a rock that was notably impermeable and resistant to weathering. The solidity of limestones was slightly easier to explain along these lines, since its material (in modern terms, calcium carbonate) was at least slightly soluble; "petrifying springs" and the stalactites in limestone caves also suggested how percolating water might be the causal agent of that particular case of stoniness. As a general problem, however, consolidation was difficult to understand, and it made the rocks of the Secondary terrains almost as puzzling as those of the Primary regions.[54]

Fossil specimens raised another set of problems for causal explanation. Quite simply, many fossils had long been recognized as the remains of *marine* animals, so

Fig. 2.20. A fossil lobster from the limestone of Solnhofen in Bavaria: one of the many exceptionally well-preserved fossils found at this famous locality. This engraving was published in Knorr and Walch's great "paper museum" of fossil specimens (1755–75). Finding the fossil remains of marine animals so far inland indicated that there had been major physical changes or "revolutions" at the earth's surface, but finding them so well preserved suggested that the changes had been slow and tranquil, not the result of any violent catastrophe. (By permission of the British Library)

their present position on dry land, often far from the sea, posed the question of how they had got there. Marine mollusks found near the Hampshire coast (see Fig. 2.3) were an easy case; by contrast, it was far more difficult to explain the marine fish found at Monte Bolca, high in the hills of Lombardy (see Fig. 5.7), or the wide variety of obviously marine animals spectacularly well preserved in the limestone of Solnhofen, far inland in Bavaria. In such cases, the museum study of fossils, supplemented of course by the information about location that normally accompanied the specimens, raised the causal question of emplacement (Fig. 2.20).[55]

That question was answered either by postulating some kind of transient violent event, by which marine animals had been thrown up on land, or by inferring that the distribution of land and sea had changed radically since whenever it was that the animals had been alive. The first kind of explanation found a small-scale analogue in the action of winter storms on low-lying coasts, which often hurled sea shells and other marine debris some way inland; or, on a larger scale, the rare but catastrophic "tidal waves" (in modern terms, *tsunamis*) that were known to follow some earthquakes. But the perfect preservation of fossils such as those at Solnhofen made that kind of event highly implausible, at least as a general explanation of marine fossils. More attractive was the inference that the present land areas, and perhaps whole continents, might once have been at the bottom of the ocean (and perhaps oceans had been continents). These two options had been discussed at

Kant (1724–1804) of Königsberg; the same latitude (or vagueness) will be adopted here. The discovery of phenomenal regularities or "laws" was treated in practice either as a worthy end in itself or at least as a preliminary stage towards the discovery of Newtonian "true causes" [*verae causae*].

54. See Laudan, "Problem of consolidation" (1977–78).

55. Fig. 2.20 is reproduced from Knorr and Walch, *Merkwürdigkeiten der Natur* 1 (1755) [London-BL: 457.e.12], part of pl. 13b.

length by savants for a century or more, as part of the earlier debate about the origin of "fossils" in general.

By Saussure's time, such changes were referred to routinely as the earth's "*revolutions*"; whatever their character and cause, the past occurrence of major changes was considered to be so obvious that this term was treated as unproblematic. In other contexts the same word was used to denote *any* major change, whether slow or rapid, smooth or violent. For example, in human history it was used to denote the rise and fall of empires, often slow, sometimes dramatically sudden. In astronomy, it denoted the perfectly smooth orbiting of the planets, just as it had done two centuries earlier, when Copernicus gave his famous cosmological work the title "On the *Revolutions* of the Celestial Spheres" (*De Revolutionibus Orbium Coelestium*, 1543). In the science of earth physics the word "revolution" was likewise quite neutral: it simply denoted *any* major change in physical geography: it did *not* necessarily imply that the putative event had been sudden, still less that it had been violent. Some savants did indeed argue that major changes in the positions of continents and oceans must have been both sudden and violent; but others thought they had been slow and gradual, even perhaps imperceptibly so from a human viewpoint (see Chap. 3).[56]

In practice, however, invoking revolutions simply put the causal problem one stage further back. For example, the former existence of the sea over areas that are now land might be explained either in terms of a global fall in sea level or of an upheaval of the earth's crust in that region. In a kind of causal regress, either of those changes might in turn be attributed to various physical agencies.

The "physics" of physical geography

Causal interpretations were also generated to explain specific features and phenomena of physical geography. The revolutions that had converted sea into land, and perhaps land into sea, were a case that was shared with "mineralogy": the evidence came both from a museum study of fossil specimens and from the field observation of the rocks from which the fossils were collected. A case that belonged more specifically to physical geography was the vexed question of the causal origin of valleys.

Valleys were observed to be of many forms. A few could plausibly be attributed to erosion by the streams that flowed in them, but most could not. For example, the "dry valleys" of the Chalk hills of northern France and southern England were just like some other valleys in form, but they contained no streams at all. Elsewhere, a huge valley might be drained by a puny little stream, which seemed utterly inadequate to explain its excavation, no matter how much time was invoked (see §2.5). In some such cases, the valley was broad and shaped in profile like a vast U (see Fig. 1.2), while the stream on its floor might flow within a smaller valley of contrasting V-shaped form. If the latter were attributed to erosion by the stream, the same agency could hardly be invoked to explain the former: by the principles of natural philosophy enunciated by the great Newton himself, like causes should have like

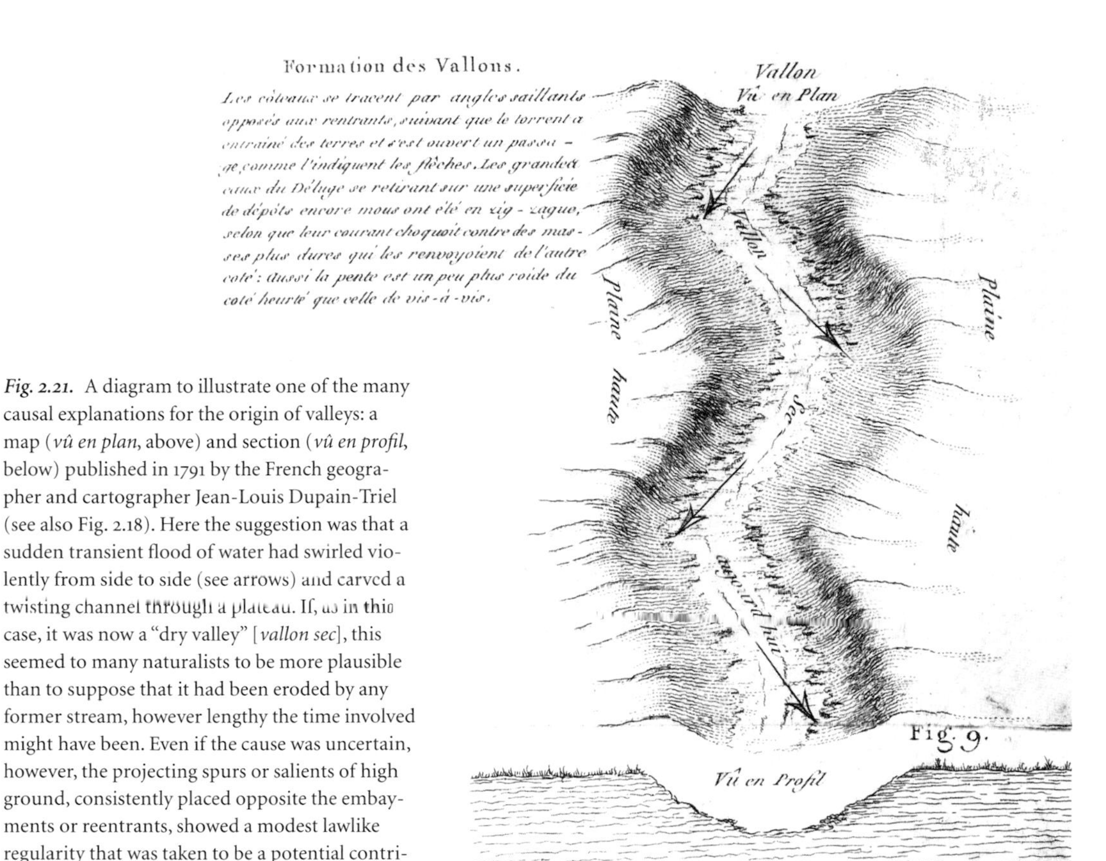

Fig. 2.21. A diagram to illustrate one of the many causal explanations for the origin of valleys: a map (*vû en plan*, above) and section (*vû en profil*, below) published in 1791 by the French geographer and cartographer Jean-Louis Dupain-Triel (see also Fig. 2.18). Here the suggestion was that a sudden transient flood of water had swirled violently from side to side (see arrows) and carved a twisting channel through a plateau. If, as in this case, it was now a "dry valley" [*vallon sec*], this seemed to many naturalists to be more plausible than to suppose that it had been eroded by any former stream, however lengthy the time involved might have been. Even if the cause was uncertain, however, the projecting spurs or salients of high ground, consistently placed opposite the embayments or reentrants, showed a modest lawlike regularity that was taken to be a potential contribution to a future causal explanation.

effects. So erosion by a sudden and violent rush of water, at some remote time, often seemed the most likely causal explanation of many valleys: the well-attested effects of violent flash floods were a persuasive small-scale analogue. In other cases, where the rocks on one side of a valley were quite unlike those on the other, it seemed possible that it might have opened up along a crack or "*fault*" in the earth's crust, a crack perhaps subsequently enlarged by erosion of some kind. All in all, the origin of valleys was puzzling and controversial; but clearly it fell within the scope of earth physics to try to resolve it (Fig. 2.21).[57]

This example illustrates how many naturalists sought to find phenomenal "natural laws" or "dispositional regularities" within the bewildering diversity of terrestrial

56. See for example Baker, "Revolution" (1988), and, for the earth sciences, Rappaport, "Borrowed words" (1982), and Ellenberger, "Terme révolution" (1989). Much gross misinterpretation in historical writing about the earth sciences has been due simply to a failure to attend to the contemporary meaning of the word "revolution": the history has been presented as one of relentless "catastrophism", when in fact the character of the earth's "revolutions" was a topic of fruitful and legitimate debate, as later sections of this book will suggest. Both meanings of the word "revolution" remain, of course, in ordinary modern usage: it is applied to the smooth running of automobile engines as well as to the seizure of power by dictators or their violent overthrow.

57. Fig. 2.21 is reproduced from Dupain-Triel, *Recherches géographiques* (1791), pl. 1, fig. 9. The caption referred to the eroding agent as "les grands eaux du Déluge", but the explanation offered did not depend on the identification of that event with the biblical Flood.

features. This was regarded as a worthy goal for earth physics, even if the underlying causal processes remained elusive. Newton, as so often, provided the model to emulate, for he had formulated his laws of universal gravitation as phenomenal regularities, as an essential step towards finding its ultimate cause. In earth physics, there was likewise much debate about the lawlike regularities to be detected within the diverse forms of valleys: for example, the alleged regularity with which spurs and reentrants were juxtaposed, or with which tributaries joined a main valley at precisely the same level. Even if such regularities failed to resolve any causal puzzles, they could at least serve as valuable criteria for a classification of the diversity, and thereby tended to blur in practice the sharp distinction in principle between natural history and natural philosophy.[58]

An equally puzzling phenomenon was that of "*erratic blocks*". These were large boulders that were found strewn over the surface of the ground in certain regions, and might even be perched on hilltops. They were "erratic" because they were quite distinct from the rocks forming the ground on which they lay and had evidently strayed from elsewhere. Many erratic blocks of granite, for example, had evidently been shifted somehow from Finland (where granite was exposed in situ) into the forests of northern Russia, from which one of them had famously been moved with enormous difficulty into the center of St Petersburg, to form the huge plinth for Falconet's celebrated equestrian statue (1782) of the city's founder, Peter the Great.

Erratic blocks such as this were far too large to have been shifted by the present streams or rivers (see §10.2). Some savants, such as Saussure, attributed erratics to the same kind of sudden and violent event that might have excavated many valleys: a *mega-tsunami* (as it will be termed here) might have swept violently across a region, dragging even large boulders in the turbulent water and leaving them far from their original sites. Other savants, such as de Luc, thought it more likely that they had been thrown up violently from the depths of the earth, when the crust collapsed suddenly into vast subterranean cavities. Both these hypotheses could explain why erratics consisted of quite different rocks from those on which they now lay: they came from sources far removed from their present sites, either horizontally or vertically. Anyway, erratics were extremely puzzling, particularly to those who had seen them at first hand rather than merely reading about them, because it was obvious that they could not be attributed to any of the ordinary physical processes to be seen in action around them.[59]

Valleys and erratics looked as if they were of rather recent origin. So it is not surprising that they were widely attributed to the most drastic physical event of which there was some *human* record, namely Noah's Flood or the "Deluge" recorded in Genesis. A century earlier, this kind of "*diluvial*" explanation had often been used, for example by Steno, and later by the London naturalist John Woodward, to account for *all* the Secondary rock masses; but by Saussure's time its application was far more specific, and confined to what seemed to be this relatively recent event. Although diluvial theories invoked a biblical source, they demanded a far from literal interpretation of the text: the story in Genesis, taken at face value, did not suggest anything as violent as a mega-tsunami. Moreover, diluvial theories were commonly reinforced by equating the biblical record with similar reports of huge floods (for example those named after Ogyges and Deucalion) in the literature of other ancient cultures; or by using the latter as further examples of a repeated kind of event. So although diluvial explanations were put forward by some writers in the hope of strengthening the authority of the Bible, they could also be treated as a quite naturalistic kind of explanation of some otherwise extremely puzzling phenomena. To attribute these features to some kind of natural "deluge", usually in the form of a mega-tsunami, was a generally acceptable feature of the practice of earth physics, and was not necessarily linked to any religious agenda.[60]

Other puzzling problems concerned the causes of earthquakes and volcanoes. Since the two were often associated, in both time and space, some kind of causal connection seemed plausible. One such suggestion, much discussed at the time and for long afterwards, was presented to the Royal Society in London by the astronomer and mathematician (and Anglican parson) John Michell (1724–93), who was subsequently appointed to the Cambridge chair that had been endowed by Woodward. The very title of Michell's paper made it clear that his "conjectures concerning the cause" of earthquakes were distinct from his "observations upon the phaenomena" of their occurrence, particularly those of the terrible Lisbon earthquake of 1755. As usual, natural philosophy was being based on natural history, but was distinct from it.[61]

But while it was widely agreed that earthquakes might be related causally to volcanic activity—at depth, if not at the surface—there was little reason to connect either with the origin of mountains. The violent shaking of the ground during an earthquake might be seen to open up fissures (see Fig. 2.10); but it was not clear that the result was any permanent elevation of the land, which, on a larger scale or with prolonged repetition, might produce anything like a range of mountains. Volcanic activity, being "fiery" in appearance and often sulfurous in smell, was sometimes attributed to the subterranean combustion of unseen deposits of coal or pyrite: but most savants, such as Hamilton, who had witnessed the scale and violence of many eruptions at first hand, dismissed this as a quite inadequate explanation. Once again, the problems were puzzling and far from being resolved, but there was no doubt that they were legitimate questions for earth physics.

58. The phrase "dispositional regularities" is proposed in Taylor, "Two ways of imagining the earth" (2002); see also his "Lois naturelles" (1988), and Carozzi (M.), "Salient and re-entrant angles" (1986).

59. Saussure, *Voyages* 1 (1779), 166–71; see also Chrysologue, "Franche-Comté" (1787), 280–82, and Razumovsky, *Histoire naturelle de Jorat* (1789), 2: 25–28, on erratics on the Jura and near Lausanne respectively.

60. On earlier diluvial theorizing, see Rappaport, "Geology and orthodoxy" (1978) and *Geologists were historians* (1997), chap. 5. On this issue, the modern geological reader must imaginatively suspend the advantage of hindsight: it was more than half a century before the *glacial theory* was proposed, and still longer before it resolved satisfactorily the phenomena of erratic blocks, U-shaped valleys, and many other recalcitrant puzzles in physical geography. The term "mega-tsunami" was first proposed in this historical context in Rudwick, "Glacial theory" (1970), 140; it has been vindicated unexpectedly (and of course coincidentally) by its more recent adoption by geologists to denote putative past catastrophes of this kind and anticipated future ones (thirty years ago most of them would have rejected such conjectures as outmoded and utterly "unscientific").

61. Michell, "Conjectures upon earthquakes" (1760); see Hardin, "Scientific work of John Michell" (1966). Hamilton, "Account of the earthquakes" (1783), likewise suggested a causal explanation—a submarine volcanic eruption—but kept it separate from his description of its effects.

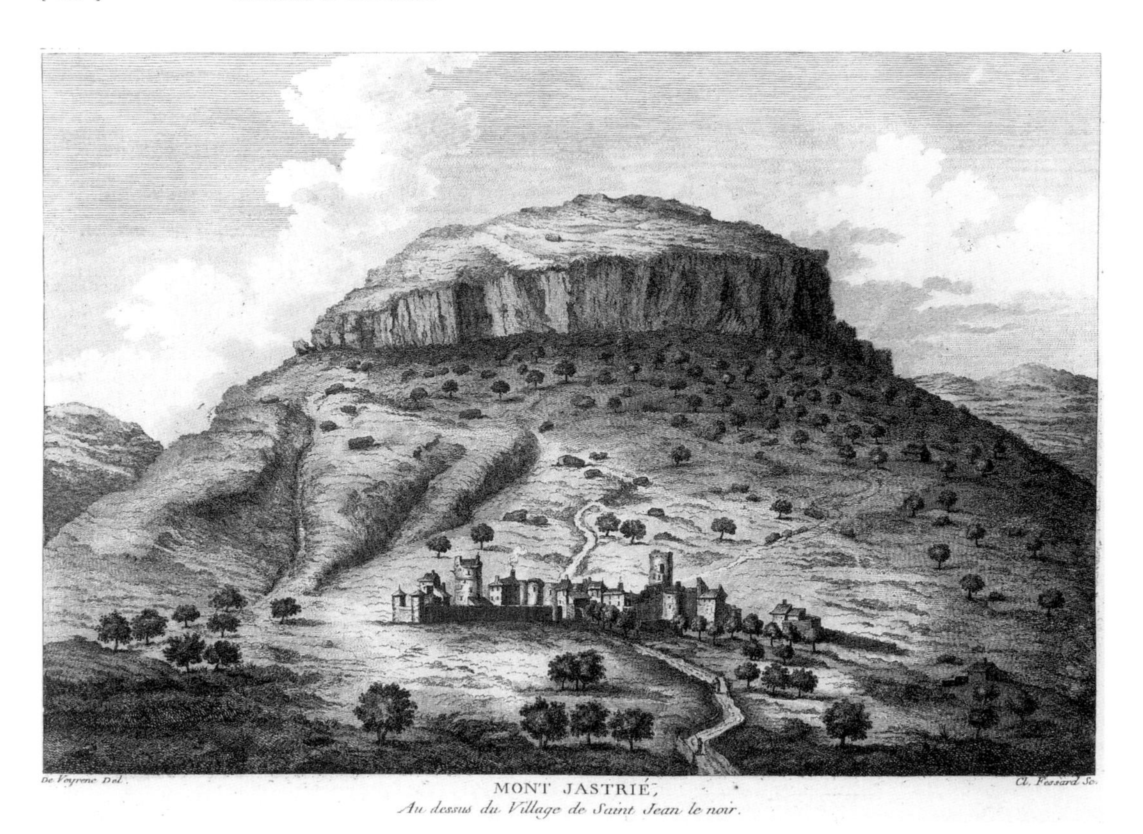

Fig. 2.22. A view of Mont Jastrié above the village of Saint-Jean-le-Noir in Vivarais: an engraving published by the French naturalist Barthélemy Faujas de Saint Fond (see below) in his *Researches on Extinct Volcanoes* (1778). Many naturalists claimed that basalt was a volcanic lava, even if it was unconnected to any obviously volcanic cone or crater; Faujas interpreted this particular hill as (in modern terms) a volcanic plug intruded from below. But some savants, notably Werner, argued that hills like this, capped with basalt, were isolated remnants of a widespread horizontal formation that had been some kind of precipitate of aqueous origin. (By permission of the Syndics of Cambridge University Library)

One specific puzzle concerned the causal origin of basalt. As already mentioned, this was a major problem within mineralogy, because it was unclear whether these common rocks should be classed as volcanic or sedimentary. Like the emplacement of marine fossils on dry land, it was a problem that spilled over into physical geography. After the discovery of spectacular extinct volcanoes in central France (§4.3), with obvious lava flows composed of a basaltlike rock, many naturalists had concluded that the basalts found elsewhere must also be the traces of ancient volcanic activity, however far they might be from any active volcanoes. But the appearance of basalt *in the field* suggested to others that basalt was just a massive Secondary formation, and nothing to do with any volcano. Werner became convinced of this—against the opinion of most other German geognosts—after finding in the field that there was an apparent gradation from some basalts into the bedded rocks immediately underlying them, and that the latter showed no signs of having been altered or "baked" by what would have been the high temperature of a molten lava. Particularly relevant to this argument, but also disconcertingly ambiguous, were the many cases in which the basalt formed an isolated hill or plateau (Fig. 2.22).[62]

De Veyrenc Del. CRATERE DE LA MONTAGNE DE LA COUPE, AU COLET D'AISA, Cl. Fessard Sculp.
Avec un Courant de Lave qui donne naissance à un pavé de basalte prismatique.

Fig. 2.23. The cratered volcanic cone of the Coupe d'Aizac, near Vals in Vivarais, "with a lava current that gives birth to a pavement of prismatic basalt" at the edge of the stream: an engraving in Faujas's *Extinct Volcanoes* (1778) that strikingly supported his claim for the volcanic origin of this particular prismatic basalt, and also by implication for others elsewhere. The scene is given scale and human significance by gentlemanly figures, with the carriage that has brought them as real witnesses to this remote spot. (By permission of the Syndics of Cambridge University Library)

Werner's Neptunist case was supported by contrasts between basalts on the one hand and what were agreed to be lavas on the other. For example, the spectacular phenomenon of regular prismatic jointing in many basalts, as at the Giant's Causeway and the Isle of Staffa (Figs. 1.15, 2.8), was said to be unknown in any true lava produced by an active volcano; and its similarity—however crude—to the roughly hexagonal cracks that were commonly seen when a muddy pool dried out reinforced the argument that basalt had originally been some kind of precipitate that had likewise dried out after it was formed.

However, this was a case in which fieldwork in physical geography yielded what was claimed to be decisive evidence to refute this Neptunist interpretation. For example, Barthélemy Faujas de Saint Fond (1741–1819), one of the naturalists attached to the royal museum in Paris, traveled extensively in central France, in the remote uplands of the provinces of Vivarais and Velay, and examined in detail some of its recently discovered extinct volcanoes.

62. Fig 2.22 is reproduced from Faujas, *Volcans éteints* (1778) [Cambridge-UL: MA.40.1], pl. 5, engraved by de Veyrenc; explained on 277–80 as the "volcan de Maillas". The hill, above what is now Saint-Jean-le-Centenier (Ardèche), between Aubenas and Montélimar, is not in fact the isolated mass that it appears to be in this view, but a spur projecting from the broad basaltic Plateau de Coiron in the background. The Neptunist case was put forward in, for example, Werner, "Entstehung des Basalts" (1788); particularly striking, in his opinion, was the Scheibenberg, a prominent basalt-capped hill in the Erzgebirge.

The fine engravings in his handsome volume *Researches on Extinct Volcanoes* (1778) made accessible to others what he claimed was clear field evidence for the Vulcanist interpretation of basalt: his proxy pictures demonstrated that there was a clear connection between prismatic basalts and what were unmistakably volcanoes, albeit ones apparently long extinct (Fig. 2.23).[63]

As always in scientific controversies, what Vulcanists like Faujas claimed as unanswerably decisive was found less than persuasive by Neptunists such as Werner: it was still possible for the latter to argue that while some basaltlike rocks might indeed be volcanic in origin, true basalts were not. Given the obscurity of the rocks themselves (without benefit of modern microscopic techniques), such a claim was no mere rearguard action or obscurantist quibble on the part of the Neptunists, but a scientifically legitimate conclusion. Anyway, as already emphasized, this debate between Neptunists and Vulcanists was primarily a quarrel about the correct classification and causal origin of just one specific kind of rock, albeit an important and widespread one: it was not a clash of "worldviews" or anything like it. Like the other problems mentioned earlier, it was controversial and unresolved; but it was accepted as an appropriate topic for the science of earth physics.

The "physics" of geognostic structures

The science of geognosy presented still more features that called for causal explanation. The mineral veins that criss-crossed the Primary rocks in mining regions such as the Erzgebirge, the Harz, and Cornwall, were particularly important to understand, for they contained ores of great economic value. Neptunists such as Werner regarded them as cracks in the earth's crust that had been filled from above by materials precipitated from whatever kind of proto-ocean had also produced the other Primary rocks (§3.5). Vulcanists thought they had been squirted into cracks from below, like the lavas brought up from the depths along whatever conduits underlay volcanoes. But once again, all such conjectures were ineluctably speculative, since in either case the putative causal processes were unobservable.[64]

The observed disturbances to the Secondary rocks were equally difficult to explain. In some regions these bedded formations were horizontal, as they might have been when first deposited on the floor of some vanished sea, but more often they were tilted (see Figs. 2.14, 2.19). If the dip was slight, it was plausible to regard it as original: the formations might have been deposited in turn on the sloping flanks of a preexisting massif of Primary rocks. But often the dip was too great for that causal explanation to seem credible. As an extreme and decisive example, Saussure discovered—above the village of Vallorcine, not far from Chamonix and Mont Blanc—some beds of pudding-stone in a *vertical* position, and he considered it inconceivable that any pebbly gravel could have been deposited originally like that. But the cause of the phenomenon remained obscure. Some savants, notably Hutton (§3.4), claimed that tilted strata must have been heaved up from below during the earth's revolutions; but others found it more plausible to suppose that they had collapsed under gravity. As in the case of basalts, the field evidence was ambiguous (Fig. 2.24).[65]

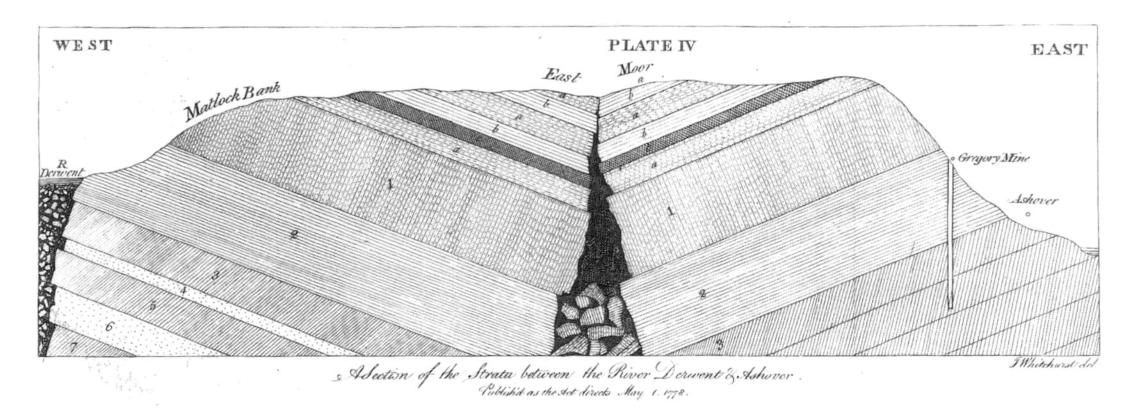

Fig. 2.24. An engraved section through part of Derbyshire in northern England, illustrating the geognostic study of the region (1778) by the Lunar Society member John Whitehurst, which was intended to support his "theory of the earth" (see Chap. 3). The causal interpretation of such structures was controversial. The change in the direction of dip, observable in surface outcrops and in a mine (right), could suggest that the rocks had *collapsed* along an unseen central crack (it would now be depicted as a smoothly curved synclinal fold); the analogy with a drawing of a collapsed building was heightened by the way the formations were shown with their boundaries ruled in perfectly parallel lines. But Whitehurst himself referred to the "subterraneous convulsions" that had produced this "apparent confusion and disorder in the strata" and claimed there had been some kind of explosive *upheaval* in the distant past. (By permission of the Syndics of Cambridge University Library)

The puzzle of tilted strata was in fact subsumed within a more general problem, that of accounting for mountains themselves. This was epitomized by another specific locality in the Alps—to which Saussure repeatedly returned—where the beds of solid rock were contorted on a huge scale. Initially he thought such structures might have been formed by successive precipitations on the sides of vast subterranean cavities, since exposed by erosion (the analogy was with a concentrically banded agate, though the scale was far larger). Later, accepting that the strata were hardened sediments, he concluded that the folding must be due to the collapse of originally horizontal layers, the mechanism invoked by de Luc and many others. In the end, however, he became convinced that such large-scale folding must be the

63. Fig. 2.23 is reproduced from Faujas, *Volcans éteints* (1778) [Cambridge-UL: MA.40.1], pl. 10; it was probably engraved from a drawing made professionally in Paris on the basis of mere field sketches by Faujas or his companions. A comparison with what can now be seen in the field indicates that although the scene was presented in the style of a true proxy for the direct fieldwork experience, it was in fact more like a diagram, with the visual evidence much enhanced to strengthen its rhetorical power (compare Fig. 1.13). The cone—now thickly forested and quite difficult to detect—is opposite the village of Antraigues, 6km north of the little spa town of Vals-les-Bains (Ardèche).

64. Werner, *Entstehung der Gänge* (1791), published the interpretation of veins that he had probably formulated much earlier; a French edition followed in 1802, an English one not until 1809.

65. Fig. 2.24 is reproduced from Whitehurst, *Original state* (1778) [Cambridge-UL: Hh.3.72], pl. 4; see 72–73, 165; his argument for crustal upheaval was supported by the account (and paintings) of Vesuvius in eruption, by his friend the famous artist Joseph Wright "of Derby": see Kemp, *Visualizations* (2000), 54–55. Saussure's discovery is described in *Voyages* 2 (1786), 99–104: Vallorcine is near the Franco-Swiss frontier, between Chamonix (Haute-Savoie) and Martigny (Valais), but the rock is now more easily seen, again in a spectacular vertical position, at another of his localities, near Dorénaz, 5km north of Martigny. Those without time for fieldwork can see a fine sample of this famous conglomerate—a striking rock with varied pebbles set in a contrasting dark matrix—in the form of a large erratic, unearthed during the construction of an autoroute near Geneva and now exhibited outside the entrance of Genève-MHN.

Tom. I.
Pl. IV.
Vüe de la Montagne du Nant d'Arpenaz entre Maglan et Salanche en Faucigny.

Fig. 2.25. A spectacular case of strongly curved strata, exposed at Nant d'Arpenaz in the valley of the Arve downstream from Chamonix: an engraving, based on a painting by Bourrit, published by Saussure in his *Alpine Travels* (1779). He claimed that the rocks in the cliff were just the lower part of an even larger S-shaped fold, the top of which was continued (here in foreshortened view) in the distant cliff on the right. To give his readers a sense of the vast scale of the phenomenon, he noted that the waterfall alone was about 800 feet (250m) high. This striking instance of hard rocks apparently folded like putty became a constant empirical reference point in Saussure's long search for an adequate causal explanation for the origin of mountains. (By permission of the Syndics of Cambridge University Library)

result of huge lateral compressive movements of the earth's crust, like the crumpling of a tablecloth, operating over vast spans of time (Fig. 2.25).[66]

It therefore became necessary to infer—as Saussure did—that the earth's crust had somehow been buckled, not just in the earth's infancy, but since the deposition of at least some of the Secondary formations. That implied that not all mountains were "primitive" or original features of the earth's crust; at least some must have been formed during the earth's later "revolutions". However, that inference did little to solve the causal problem. What deep-seated agency could have been capable of causing whole mountains made of Secondary formations to be heaved up on end? Or what could have caused the deep foundations of those mountains to be so undermined that they had collapsed? Or what huge lateral force could have buckled solid formations like a gigantic crumpled tablecloth? These were serious problems; worse, they were problems for which any proposals seemed unavoidably speculative and untestable, since they concerned putative processes or agencies hidden in the unobservable depths of the earth.

The "physics" of rock formations

A final example of the interpretation of geognostic structures in terms of physical causes dealt with the original deposition of the Secondary formations rather than with the disturbance that some of them had undergone subsequently. As a young man, Lavoisier had worked on the great *Atlas Minéralogique* of France (§2.2). That project lay firmly within the tradition of physical geography, but he had tried to expand its parameters to include—literally in its margins—some sense of the third dimension (see Fig. 2.11). Although the word "*géognosie*" was as yet rarely used in French, that in effect was the science he was trying to practice. Unlike most other geognosts, however, he wanted to interpret structural piles of rock formations in terms of causal sequences of past events. In the 1780s, after a long interval devoted more to his chemical research, he returned to this project and wrote a major paper that presented a causal interpretation of the varied Secondary formations that he had helped to map in the region around Paris (see Fig. 2.12). This was read at the Académie des Sciences shortly after the start of the Revolution (see §6.1). It was published not long before the Terror, during which the Jacobins abolished the Académie and executed Lavoisier himself, not for his scientific work but on account of his former tax-collecting function (§6.3).[67]

Lavoisier developed a distinction, already sketched by his teacher Rouelle among others, between "littoral" and "pelagic" formations [*bancs littoraux* and *pélagiens*]. Littoral formations were gravels, sands, and clays—often consolidated into pudding-stones, sandstones, and shales—that were taken to have accumulated in shallow coastal waters. Pelagic formations were finely bedded limestones, often containing delicate fossil shells, that looked as if they must have been deposited in deep water, beyond the reach of any surface turbulence due to tides and storms. In other words, an observed contrast in terms of types of rock was interpreted causally in terms of contrasting environments of deposition. Lavoisier took this distinction much further, however, by suggesting how both kinds of formation could have been formed *simultaneously* and side-by-side on the same sea floor, according to the depth and the distance from the shoreline of the time. This interpretation was based in the first instance on what he claimed was taking place at the present day, offshore from the Chalk cliffs on the coast of northern France (Fig. 2.26).[68]

66. Fig. 2.25 is reproduced from Saussure, *Voyages* 1 (1779) [Cambridge-UL: Mm.49.1], pl. 4. The Cascade de l'Arpenaz is 5km north of Sallanches, near the main road (and now also the autoroute) from Geneva to Chamonix: Saussure would have passed it every time he visited the Mont Blanc region. Carozzi, "Géologie" (1988), "Grande découverte" (1998), and "Refoulements horizontaux" (2001), all trace Saussure's successive interpretations of the fold, and include photographs of the locality (as striking now as it was then). In modern work on the extremely complex tectonics of the Alps, this huge S-shaped fold structure has been reduced—relatively—to a minor crinkle within the even more enormous Helvetic nappes, overthrust from the south, i.e., from the right.

67. The timing alone seems more than adequate to explain why Lavoisier's work had no immediate sequel, but it must surely have been known in published form to Brongniart and Cuvier when they surveyed the same region years later (§9.1).

68. Fig. 2.26 is reproduced from Lavoisier, "Observations générales" (1793) [Cambridge-UL: CP340:2.b.48.107], pl. 1, explained on 357–59. The present account is based on Rappaport, "Lavoisier's geologic activities" (1968) and "Lavoisier's theory" (1973), also retold in Gould, "Lavoisier's plates" (2000).

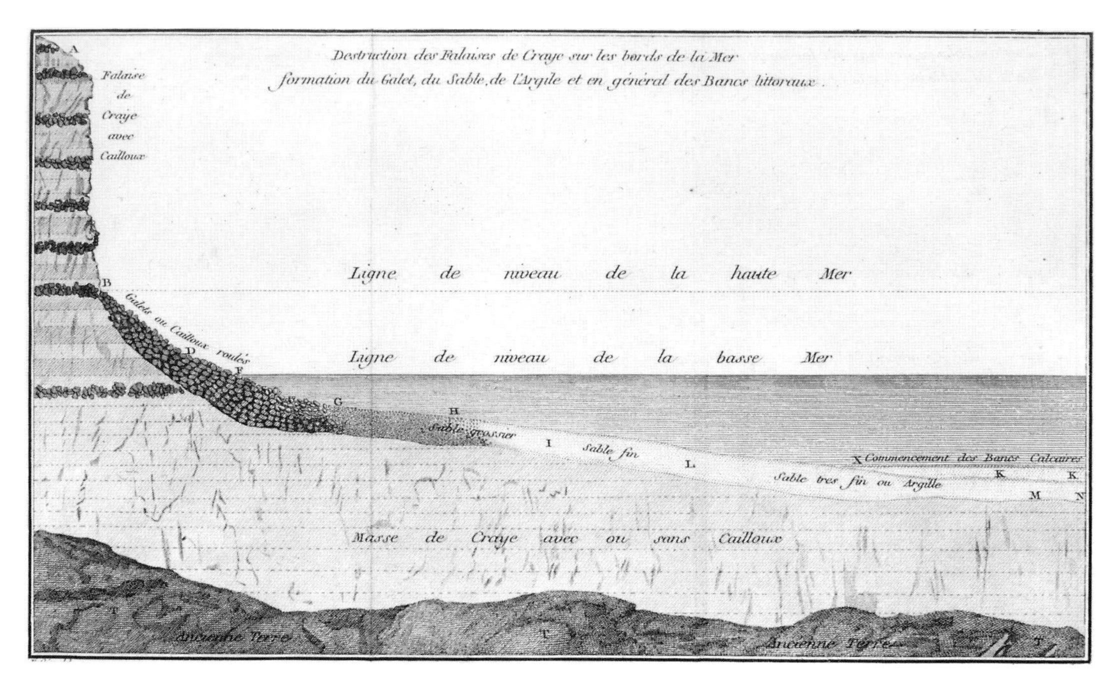

Fig. 2.26. Lavoisier's interpretation of the simultaneous deposition of different kinds of "littoral" materials—from gravel through coarse and fine sand to clay [*galets, sable grossier, sable fin, argile*]—derived from the erosion of coastal cliffs of the Chalk formation with its flints [*craie avec cailloux*]. Far offshore are the limey deposits [*bancs calcaires*] forming in deep water as a future limestone of "pelagic" origin. (In order to clarify the visual argument, the scale of the cliff, beach, and span between high- and low-tide levels [*haute mer, basse mer*] is greatly exaggerated relative to the depth and extent of the sea.) An unseen basement of Primary rocks [*ancienne terre*] was assumed to underlie the Secondary formation of the Chalk. The section was based on observations on the coast of Normandy, but was intended to be of universal application: it was the first in the series that illustrated Lavoisier's "general observations" on "modern horizontal beds" of marine origin, read in Paris in 1789. (By permission of the Syndics of Cambridge University Library)

Having established what could plausibly be supposed to be taking place in his own time on the floor of the Channel, Lavoisier then used it to infer what could have caused the Secondary formations that he had mapped around Paris. He postulated a far greater oscillation in sea level, like the tidal ebb and flow [*flux et reflux*] but operating over a vastly longer timescale. As the sea gradually rose, the pelagic deposits would encroach further inshore, covering the littoral sediments deposited at an earlier time; but when the vast cycle turned and the sea level began to fall again, the littorals would start to spread further out to sea and would cover the pelagics (in modern terms, a marine transgression would be followed by a regression). This cyclic process would thus generate an alternation, at any one locality, between littoral and pelagic formations (Fig. 2.27).[69]

Lavoisier's highly original interpretation was, as the title of his paper made clear, a *general* explanation for the alternation of littoral and pelagic formations, wherever they might be found in the regions of Secondary formations. It was a hypothesis—or, better, a scientific model—for the physical operation of certain causal processes, showing how they would necessarily produce certain observable results: his theory was explicitly what would now be called hypothetico-deductive

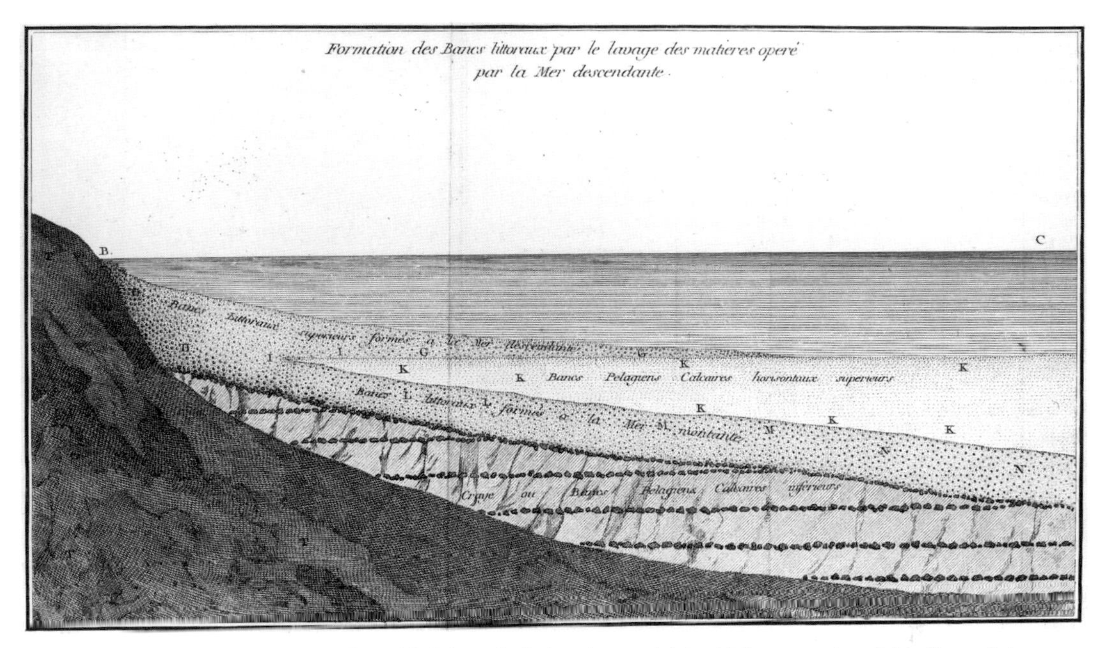

Fig. 2.27. One of Lavoisier's hypothetical sections explaining his interpretation of the effects of a long-term oscillation in sea level, analogous to the tides but on a far larger scale. Lying on a basement of Primary rock masses [*ancienne terre*] is the "pelagic" or deep-water formation of the Chalk [*craye*]. Above that in turn is a series of "littoral" and "pelagic" deposits [*bancs littoraux, bancs pelagiens*], formed simultaneously; during a long previous period of rising sea level, the latter encroached inshore, covering the former. Now, as the sea level falls again, new littoral deposits are spreading back out to sea, covering the earlier pelagic deposits and thereby continuing the *alternation* of littoral and pelagic formations building up on the sea floor. (By permission of the Syndics of Cambridge University Library)

in structure. As for the cause of the long-term oscillation in sea level that generated the cyclic changes, he rightly treated it as a separate issue and wisely left it to the physical astronomers, implying that it was probably extraterrestrial.[70]

In the last of his remarkable illustrations, Lavoisier presented in visual form the kind of concrete evidence that substantiated his hypothesis. The little sections he had had to squeeze into the margins of the maps in the *Atlas* (Fig. 2.11) were here brought center stage. Columnar sections through the formations he had actually observed around Paris showed how there was indeed the kind of alternation between littoral and pelagic deposits that his hypothesis predicted and explained in causal terms (Fig. 2.28).[71]

Lavoisier exemplified his causal interpretation by means of these particular formations, simply because he knew them at first hand and they—or at least the localities—were familiar to his Parisian audience. He would have been well aware, from his years of mapping further afield, that other regions had sequences of other rocks (for example, in modern terms, the varied Mesozoic formations beyond the Paris

69. Fig. 2.27 is reproduced from Lavoisier, "Observations générales" (1793) [Cambridge-UL: CP340:2.b.48.107], pl. 5, explained on 363–64.

70. Lavoisier, "Observations générales" (1793), 353, 369–70.

71. Fig. 2.28 is reproduced from Lavoisier, "Observations générales" (1793) [Cambridge-UL: CP340:2.b.48.107], pl. 7, explained on 365–68.

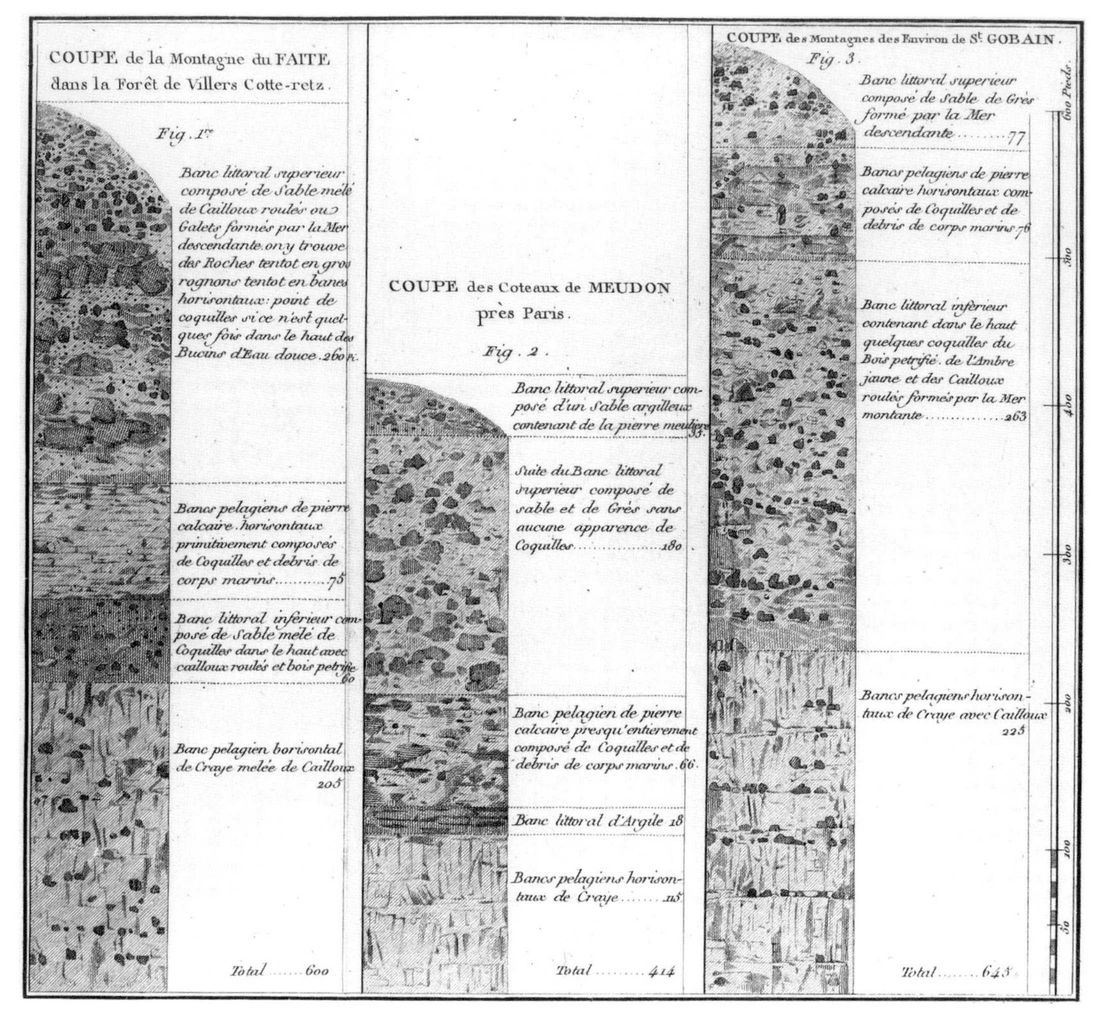

Fig. 2.28. Lavoisier's sections of the actual rock formations that his cyclic theory of sedimentation was designed to explain. The measured columnar sections (the vertical scale is of 600 feet or nearly 200m) were of the strata at three specific localities in the Paris region, and showed the alternation between what he interpreted as littoral and pelagic formations [*bancs littoraux, bancs pelagiens*], the lowest being in each case the pelagic Chalk [*craye*]. (By permission of the Syndics of Cambridge University Library)

basin) that could be interpreted in much the same way as the products of alternations between shallow-water and deep-water conditions. His interpretation was a general causal theory in earth physics, illustrated by the example of the formations of the Paris region. It was not—or not primarily—a reconstruction of a specific region at a specific period of geohistory.[72]

Conclusion

The limitations of what was being achieved in earth physics have been emphasized throughout this section. The problems were obvious but in many cases the solutions were highly ambiguous or just plain speculative. It is therefore not surprising that many of the savants who studied the physical features of the solid earth devoted more of their time and energy to description and classification: often they relegated their causal conjectures to an appendix, sometimes with an apology for straying into the field of speculation.

All the interpretations summarized here, explaining whatever physical features or phenomena the three natural-history sciences of the earth might present, were designed to give causal explanations that would apply quite generally. They were often formulated in the first instance as observable natural regularities, which were taken to reflect ahistorical natural "laws" like those of chemical combination and universal gravitation; but the hope was that they might lead in due course to truly causal explanations. There could, for example, be theories of the formation and consolidation of rocks, of the emplacement of marine fossils and erratic boulders, of earthquakes and volcanoes, of the formation of mountains by crustal upheaval or collapse, of the filling of mineral veins in the regions of Primary rocks and the deposition of formations in the Secondary terrains, and so on. All were causal theories in earth physics; they were not essays in geohistory. Of course they all invoked processes that operated in time; but those processes were ahistorical, in the sense that the explanations were put forward as being equally valid yesterday, today, and forever.

Earth physics was a causal science, not a geohistorical one. Nonetheless, it embodied the element of time in a way that the three natural-history sciences of the earth did not. The final section of the present chapter considers the role of time in the study of the earth; this will prepare the ground for a review (in Chap. 3) of a distinctive genre in which—in contrast to what has just been described—causal explanation took center stage.

2.5 THE QUESTION OF TIME

The short timescale versus eternalism

Those who practiced the science of earth physics offered causal explanations of many of the specific features described and classified by the sciences of mineralogy, physical geography, and geognosy. These features were attributed to a wide variety of physical processes, all of which necessarily had a temporal dimension. This final section of the chapter assesses the role of time in causal explanation, and specifically the magnitude of the timescale on which the processes were thought to be operating.

The question of time is important, because it is often assumed that the emergence of the modern earth sciences needed above all an adequate sense of the vastness of geological time; and that this is just what was lacking until at least the turn of the nineteenth century. More specifically, it is often claimed that on this issue

72. Gould, "Lavoisier's plates" (2000), rightly points out that Lavoisier did speculate briefly on how some shales below the Chalk might date from an earlier phase in geohistory with plants but no animals, and how the Primaries or "*ancienne terre*" might represent still earlier cycles that predated the origin of life itself. The point being made here, however, is that such hints towards a possible history of life were subordinate to the goal of explaining the diversity of formations in the causal terms of a cyclic theory in earth physics.

"the Progress of Science" was retarded by the "repressive" influence of "the Church". In fact the historical situation was more complex than that stereotype allows, and far more interesting. Contrary to the historical myth that persists today both among historians (who ought to know better) and among scientists, the savants of Saussure's generation were not constrained in their theorizing by having to squeeze the whole story of the earth into a few thousand years.

That traditional *short timescale* (as it will be termed here) had been derived from calculations made a century or two earlier by practitioners of the historical science of *chronology*. This was based on the analysis of textual records of all kinds, including but not confined to the Bible (see §4.1). James Ussher's figure of 4004 B.C. for the Creation has since become the most famous, but at the time it was just one among dozens of rival calculations that ranged—according to one tabulation—from 3616 to 6984 B.C.: the huge variation reflected the uncertainties of the complex textual analyses that the science of chronology required. It was a notoriously disputatious science; there was certainly no rigidly orthodox line about the date of Creation or of anything else. But anyway, a figure of a few thousand years had seemed quite adequate, even to naturalists, to account for most of the features they were trying to understand: matched against brief human lives, several millennia seemed quite long enough for a lot to have happened without any sense of rush.[73]

The short timescale was still taken for granted, throughout the eighteenth century and into the nineteenth, among less educated groups in society and in conservative religious circles (and it was of course revived in the twentieth century among American fundamentalists). But by the later eighteenth century most savants who took any interest in such matters recognized that it had become incompatible with a wide range of natural evidence (some of which is reviewed below). Those who were religious believers assumed that Nature, "the book of God's works", could not ultimately contradict Scripture, "the book of God's word"; so if the natural evidence seemed sound and persuasive, they simply inferred that the short timescale, in its application *to the age of the world*, must be based on mistaken assumptions.[74]

Anyway, for most people—educated and uneducated alike—the kind of analysis undertaken by chronologers was far less important than the imaginative impact of the Creation story and the religious perspective it sustained. When, for example, Josef Haydn was composing his great oratorio *The Creation* [*Die Schöpfung*] (1798), with a text based mainly on Genesis, he worked by his own account with a greater sense of religious commitment than at any other time in his career. Yet when he wrote the music to express how "God said, 'Let there be light', and there was light"—the startling fortissimo on the last word is surely one of the most dramatic moments in music of any age—it is inconceivable that he would have been worrying about exactly how many years ago that initial divine action had taken place, or indeed what it might have meant in scientific terms. Nor is it likely that such questions were in the minds of the aristocratic and bourgeois Viennese who gave his work such a wildly enthusiastic reception at its first performances. Yet those audiences could well have included savants who were aware of, and convinced by, the growing scientific evidence for an immensely long terrestrial timescale. Haydn himself probably knew about this debate: among his acquaintances, when he was living in London a few years earlier, had been the surgeon and anatomist John Hunter (1728–93), who was a contributor to research on fossil bones and certainly thought the earth's timescale was immense (see below, and §5.3). But the magnitude of the timescale, however intriguing scientifically, was simply irrelevant to the imaginative and religious impact of the Creation story.[75]

Even if the text of Genesis were taken to be authoritative and divinely inspired, it had been widely recognized among scholars—ever since Patristic times—that the seven "days" of creation were not necessarily to be understood as ordinary days: for example, the first three of them were said to have preceded the creation of the sun itself, without which ordinary days were literally impossible; and in prophetic language "the day of the Lord" clearly did not denote a period of twenty-four hours but a time of decisive significance. So when, in the course of the eighteenth century, it seemed to savants to be increasingly likely that the earth had existed long before the few millennia of recorded human history, the "days" of creation were simply reinterpreted in line with that philological scholarship, as periods of distinctive character but indefinite extension. Alternatively, the initial act of creation out of "chaos" was assumed to have been followed by an unrecorded period of vast but indefinite duration, before the humanly more important events of the rest of the narrative. If, at certain times and places, some guardians of orthodoxy grew alarmed at the new scientific claims about the vast timescale of the world, it was not always because those claims contradicted the literal sense of Genesis; religious authorities were, quite properly, more concerned with theology and its practical implications than with literalism of the crude kind adopted by modern fundamentalists.

The traditional short timescale was not challenged by "the Rise of Science", because it had been challenged far earlier by a much more radical alternative, that of the *eternalism* associated with Aristotelian philosophy. The spatial aspect of Aristotle's conception of the cosmos, with the earth fixed at the center of a vast but finite

73. The quoted figures are from Sale, *Universal history* 1 (1747): lviii–lx. Chronology had its roots in Antiquity, but its *wissenschaftlich* character was first established around 1600 when scholars such as Joseph Scaliger used newly rigorous historical methods on ancient sources both sacred and secular: see, for example, Grafton, "Scaliger and chronology" (1975), *Joseph Scaliger* (1983), and more generally *Defenders of the text* (1991); also Barr, "Ussher and biblical chronology" (1985), McCarthy, "Chronology of James Ussher" (1997), and Wilcox, *Measure of times past* (1987). Rossi, *Segni del tempo* (1979), and Rappaport, *Geologists were historians* (1997), 189–99, give excellent summaries of the issue as it concerned naturalists in the decades around 1700; see also Fuller, "Geology of time in England" (2001), and the broader surveys in Haber, *Age of the world* (1959), and Dean, "Age of the earth controversy" (1981). Chronology is not an extinct science, nor merely a feature of modern fundamentalist religion: it remains an indispensable aid to mainstream historical research, and its results are seen, for example, whenever the dynasties of China and ancient Egypt, and their events and artifacts, are given dates B.C. or A.D. (or B.C.E. or C.E.).

74. When it was applied to *human* history, from Adam to the present, the short timescale had a quite different historical trajectory, remaining much longer in the mainstream of intellectual life (see §5.4).

75. On the first performances of *Die Schöpfung*, see Landon, *Years of "The Creation"* (1977), 317–23, 448–58; Hunter's widow Ann was among the many English subscribers to Haydn's score, and made a new and better translation of its text. A modern analogue would be *Noyes Fludde* (1958), Benjamin Britten's church opera for children: the rather endearing literalism of the medieval miracle play on which it is based does not require modern audiences to adopt a fundamentalist attitude to the story of Noah, nor does it lessen in any way the musical and religious impact of the work. Both compositions were, in the proper sense of a much misused word, reworkings of *myths* of great imaginative and religious power.

universe, had been thoroughly absorbed into European culture in earlier centuries. But its temporal aspect, with the universe existing in uncreated eternity, had been emphatically rejected, on the grounds that it was radically inconsistent with the Christian (and Jewish) conception of the *created* status of the world and everything in it, from atoms to humans, in relation to a transcendent Creator. That theological objection to eternalism was at the root of the much later reluctance to extend the short timescale. The perceived threat to orthodox beliefs lay not so much in abandoning a literal interpretation of Genesis, but rather in undermining the foundations of human society by questioning the ultimate moral responsibility of human beings to their divine Creator.[76]

Given this profoundly religious objection, it is not surprising that eternalist ideas persisted in European culture largely as an "underground" alternative, visible more often when repudiated by the orthodox than in any direct advocacy. Right through to Saussure's time, discussions among savants about the timescale of the world continued to be colored by this clash of theologies. It was not a case of "Religion versus Science", but of one religious view of the world against another. If some Christian (and Jewish) theists believed they had a stake in the short timescale, because it helped guarantee the doctrine of creation, deists and atheists often felt an equal stake in the doctrine of the uncreated eternity of the world. For many deists, eternity seemed appropriate to an almost impersonal "Supreme Being", who (or which) presided over a perfect cosmos that ran forever in accordance with its own timeless natural laws. For atheists—a much more elusive breed in the eighteenth century—an eternal cosmos could seem the best guarantee of the absence of a creative deity of any kind. Anyway, both breeds of "skeptic" took the eternity of the cosmos to include the uncreated eternity of the human race, or at least of some such rational beings. The eternalist view assumed that there had never been a time when the world was without human or at least rational life; it allowed for no conception of a prehuman and therefore radically *nonhuman* world. In that sense, the eternalist option was as profoundly unmodern as the short timescale.

In the eighteenth century, claims that the age of the earth was inconceivably vast by any human standards were often linked to explicit commitments to the doctrine of an eternal universe. When, for example, a timescale of millions of years was suggested in the anonymous and notorious *Telliamed* (1748), on the basis of the rate at which the global sea level was supposedly continuing to fall, that unimaginably lengthy trend was interpreted as just one phase in an even longer cyclic movement from and to all eternity (see §3.5). Suggestions that the timescale might be far longer than that calculated by traditional chronology, *yet not infinite*, therefore ran the risk of being misinterpreted as signs of covert eternalism. Nonetheless, many eighteenth-century savants did in effect claim that this middle ground was more plausible than either extreme. Several lines of evidence suggested to them that the short timescale was no longer tenable, yet this did not oblige them to conclude that the earth might be eternal. Most of this empirical evidence related to the *rates* at which various natural processes could be seen to be operating; it was taken for granted that in this sense the present was the obvious key to the past, or at least the first point of reference to be consulted.[77]

Volcanoes, valleys, and strata

Volcanoes provided some of the best evidence for such natural rates, and the most intensely discussed. The regions around European volcanoes had long been densely populated, thanks to the rich soils produced by the eventual weathering of the lavas. The disastrous effects of some eruptions therefore ensured that historical records included a full chronicle of the volcanic activity. For example, the eruption of Vesuvius in A.D. 79 had buried the Roman towns of Pompeii and Herculaneum; but it had also left a textual record in the form of the younger Pliny's account of the catastrophe in which his uncle had perished. Later eruptions of Vesuvius, and those of Etna, had also been recorded in detail, particularly in the more recent centuries. Although the eruptions were irregular and notoriously unpredictable, the records did give savants a rough sense of the rate at which those great volcanic cones might have accumulated, and hence of their overall age.

Fig. 2.29. The excavation of the Temple of Isis at Pompeii, with several gentlemen (one of them probably Hamilton himself) and a lady (probably his first wife) watching the workmen remove the thick covering of volcanic ash: a colored etching published in Hamilton's *Campi Phlegraei* (1776), based on a painting by Pietro Fabris. Hamilton noted that the eruption that had buried it in A.D. 79 was "the first recorded in [human] history"; but the building itself stood on volcanic rock, proving that there must have been unrecorded eruptions at a still earlier time. An inscription mentioning the temple's dedication to Isis was the only *textual* element in this eloquent "monument" of the ancient world (see §4.1). (By permission of the Syndics of Cambridge University Library)

76. This crucial point may now be difficult for some intellectuals to grasp, from within a Western culture that is "post-Christian" and often theologically uninformed. It is much easier—but facile—to interpret the "conflict" in the light of modern American fundamentalism, and to regard any historical resistance to lengthening the traditional timescale as a sign of deplorable ecclesiastical repression or laughable obscurantism. Sometimes and in some places it was, but not always or everywhere.

77. It was on this middle ground, expanded through the cautious but progressive extension of the traditional short timescale, that the origins of modern geochronology can be found, rather than on the unlimited spaces of eternalism: for a very brief survey, see Rudwick, "Geologists' time" (1999).

For example, Hamilton became convinced that Vesuvius, and indeed the Campi Phlegraei as a whole, had been an active volcanic region long before recorded history. The flooded crater on the island of Nisida, for example, offered a peaceful scene that was surely far removed in time from the ancient eruption that had formed it (see Fig. 1.14). Closer to the present, and vividly linking the human timescale to Nature's, were the buildings of Pompeii, buried eighteen centuries earlier in the first recorded eruption of Vesuvius, and being excavated in Hamilton's time to the great excitement of savants and the wider educated public throughout Europe (§4.1). For Hamilton found that they were standing on volcanic rock, proving conclusively that Vesuvius must have had still earlier *prehistoric* eruptions (Fig. 2.29).[78]

The far larger mass of Etna in Sicily was still more instructive. Its flanks were dotted with dozens of minor cones, some of them hundreds of feet in height, thrown up by eruptions far below the summit crater. As Hamilton would have learnt from his guide Giuseppe Recupero, a canon of Catania who was writing a natural history of Etna, a few of these cones were known to date from recorded historic eruptions, but many others were covered in large trees and were evidently much older: they were, as Hamilton put it, "so very ancient, as to be out of the reach of [recorded] history" (Fig. 2.30).[79]

The implication was clear. If the volcano had been built up by a succession of eruptions similar to those recorded through the centuries of human history, its

Fig. 2.30. A portrait of Etna: a colored etching published in Hamilton's *Campi Phlegraei* (1776), based on a painting by Fabris; the volcano was quiescent when they visited Sicily in 1769. The picture helped convey the sheer scale of the main cone: Hamilton noted that Saussure, using de Luc's portable barometer, had measured its height as 10,036 feet above sea level. The lower slopes are dotted with "minor" cones: the prominent double cone of Monti Rossi (on the left) had been formed during the great eruption of 1669, together with a huge lava flow that had threatened to engulf the city of Catania (in the middle distance); but many of the other cones and their lava flows appeared to be ancient beyond human record. (By permission of the Syndics of Cambridge University Library)

total age must be vast beyond comprehension. Yet active volcanoes such as Etna were generally regarded as rather recent features of the earth's surface, compared with the underlying Secondary and Primary rocks (§2.3). Hamilton offered no quantitative figures, even for the volcanoes, let alone for the earth as a whole: he had no evidence on which to base such estimates, and purely speculative ones would have violated his own scientific principles. However, his proxy pictures were eloquent qualitative evidence that the terrestrial timescale must be literally beyond human imagination.

The year after Hamilton's visit, the English traveler Patrick Brydone had learned of evidence that made this inference about Etna more explicit. Recupero told him that a well dug recently at Jaci (now Aci Reale) on the lower slopes of Etna had penetrated no fewer than seven successive lavas, each with an upper surface weathered into a fossil soil. Anyone familiar with volcanoes knew how very slowly any lava surface became weathered enough even to begin to support vegetation: the flow of 1669, for example, was still almost completely barren after a century (and remains so today after a further two). With the benefit of that local knowledge, Recupero estimated that it would probably take more than two thousand years to generate a substantial soil on any surface of lava. So his well section alone—a minuscule fraction of the pile of lavas comprising Etna—implied an antiquity of at least 14,000 years, more than enough to knock the bottom out of the traditional short timescale for the whole world. As Brydone reported:

> Recupero tells me he is exceedingly embarrassed by these discoveries in writing the [natural] history of the mountain. That Moses hangs like a dead weight upon him, and blunts all his zeal for inquiry; for that really he has not the conscience to make his mountain so young as that prophet makes the world. What do you think of these sentiments from a Roman Catholic divine? The bishop, who is strenuously orthodox—for it is an excellent see—has already warned him to be on his guard, and not to pretend [i.e., claim] to be a better historian than Moses; nor to presume to urge any thing that may in the smallest degree be deemed contradictory to his sacred authority.[80]

Brydone might scoff—in typical Enlightenment style—at the bishop's venal obscurantism, but in any case the attempted censorship failed. Recupero's discovery, and his inference from it, became known throughout Europe when Brydone published this letter in his entertaining *Tour through Sicily and Malta* (1773), which was soon translated into French and German. But most savants would have taken its implications in their stride. Hamilton, for example, would have found it compatible with, and indeed confirmation of, his own sense of the immensity of time implied

78. Fig. 2.29 is reproduced from Hamilton, *Campi Phlegraei* (1776) [Cambridge-UL: LA.8.79], pl. 41, explained on 173–75.

79. Fig. 2.30 is reproduced from Hamilton, *Campi Phlegraei* (1776) [Cambridge-UL: LA.8.79], pl. 36; see his account of his ascent, 37–52.

80. Brydone to Beckford, 25 May 1770, printed in Brydone, *Sicily and Malta* (1773), 141–42. A French edition appeared in Amsterdam in 1776, a German one in Leipzig in 1777. Moses was traditionally assumed to have been the author of Genesis. Long after Recupero's death, when the political and cultural climate had changed radically, his two-volume *Storia naturale dell'Etna* (1815) was published by the university press in Catania.

by the volcanic features of Campania as well as those of Sicily. While Recupero's report was welcome to deists who were arguing for an eternal world, it was also accommodated without difficulty by theists, provided they were prepared to concede—as many were, including presumably Recupero himself—that the "days" of the Creation story in Genesis could not have been intended literally.

River valleys were a second feature that was likewise invoked as evidence to suggest that the traditional short timescale was inadequate. Valleys were difficult to explain in causal terms, not least because their sheer variety of form made it unlikely that any one explanation would cover them all (§2.4). But it seemed possible that at least some valleys could be attributed to erosion by the streams that still flowed in them. On a summer's day a stream might look too placid to do anything of the kind, but after a winter storm the swirling water might be seen to be scouring its banks and carrying away mud, pebbles, and even boulders. In principle, such erosion could have carved out a whole valley, though it would have had to be continued for an almost inconceivably long time. Nonetheless, such claims were made.

The young French naturalist Jean-Louis Giraud-Soulavie (1752–1813), for example, cited the case of the remote part of Vivarais where he had earlier served as a parish priest (§4.4). On the floors of some of the valleys there were unmistakable lava flows, which had been eroded into small gorges since their eruption (Fig. 4.12).

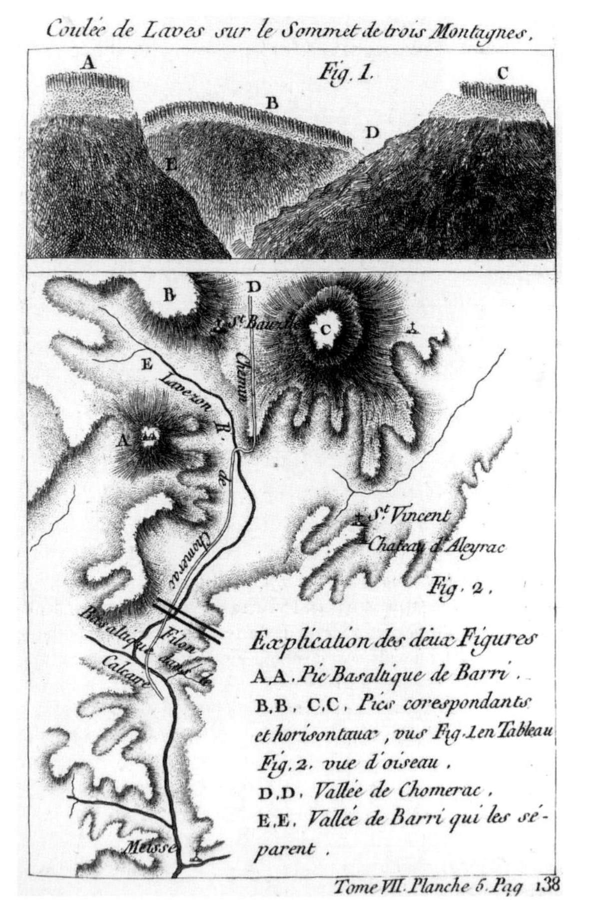

Fig. 2.31. A landscape sketch and map ("bird's eye view") of a hilly area in Vivarais, illustrating the claim by Jean-Louis Giraud-Soulavie that the valleys had been excavated over vast spans of time by the small streams that still flow in them: engravings published in 1784 in his *Southern France*. The hills A, B, and C are capped with basalt showing columnar jointing, underlain by a layer of gravel (stippled) and then by other rocks. Soulavie inferred that the basalt had originally been a single lava flow from an ancient volcano and that subsequently it had been fragmented by the erosion of the valleys D and E. As was usual in such landscape views, the topography was exaggerated vertically to clarify the point. (By permission of the Syndics of Cambridge University Library)

Soulavie claimed that he could "calculate the time" required for this erosion, and hence the age of the eruptions. He estimated that it would take "several centuries or thousands of years" just for angular fragments of the hard volcanic rock to become by attrition the smooth rounded pebbles found in the river beds further downstream; privately he estimated from this that some six million years must have elapsed since the lavas were erupted. Yet these were some of the most recent of the volcanic rocks in the area. He interpreted the basalt capping many of the hilltops as the remnants of far more ancient lava flows, and therefore inferred that the excavation of the valleys between them would have taken even more immense spans of time. Since he believed that the valley erosion was itself a relatively late feature in a far longer sequence of events (§4.4), the broader implication of his argument, for the earth's *total* timescale, would have been obvious to his readers (Fig. 2.31).[81]

Soulavie had to concede, however, that valleys did not in fact provide unequivocal evidence for estimating the timescale, since it was always possible that a far more powerful agency—such as a mega-tsunami or sudden and violent "deluge"—might have effected the same erosion in a much shorter time. The abbé Roux, a naturalist who had been Soulavie's neighbor (and fellow priest) in Vivarais, was one of those who doubted his explanation on just these grounds: he argued that it was more probable that the valleys had been excavated rapidly, perhaps during one or more of the "deluges"—those of Moses, Ogyges, and Deucalion—mentioned in early human records both sacred and secular. However, their disagreement was treated as a matter of civilized debate among savants, and Soulavie printed a long letter from Roux in one of his volumes, giving his critic's argument as much publicity as his own.[82]

Much more persuasive was a third class of evidence: the huge piles of Secondary strata that were being described in certain parts of Europe. A century earlier, when such rocks had yet to be studied closely, it had been quite plausible to suppose—with Steno, Woodward, and many others—that the entire pile of sediments could have been laid down all at once, perhaps in a violent Deluge, although even then this entailed taking great liberties with any literal reading of the story of Noah's Flood. However, once the sheer thickness of the Secondary formations was fully appreciated (§2.3), and detailed fieldwork suggested that many of them must have been deposited layer by layer under tranquil conditions, that kind of diluvial interpretation was quietly abandoned by most savants.[83]

81. Fig. 2.31 is reproduced from Soulavie, *France méridionale* (1780–84), 7 [Cambridge-UL: MA.47.33], pl. 5, explained on 126; see also Aufrère, *Soulavie et son secret* (1952), 45–47. The area shown is on the east flank of the broad basaltic Plateau de Coiron (Ardèche); the village of St Vincent de Barres (near the center of the map), from which the view was evidently sketched, is 12km north-northwest of Montélimar; the hilltops are about 300m above the valleys, and A and C are about 3km apart. His figure of 6Ma for the age of the most recent lavas was two or three orders of magnitude *in excess of* modern radiometric dates: he greatly underestimated the rate at which even a small stream can undermine and break up a flow of hard basalt and excavate a gorge along its edge.

82. Soulavie, *France méridionale* (1780–84), 1: 34, 128; Roux's letter is in ibid. 6: 304–91. Roux was *prieur* of Fraissenet (not far from Antraigues, where Soulavie had been *vicaire*) and one of many local clergy and landowners who had collections and libraries of natural history: see ibid. 2: 468–75.

83. On the earlier period, see Rappaport, *Geologists were historians* (1997), chaps. 5, 6. Astonishingly, the seventeenth-century form of diluvialism has been revived by some modern fundamentalists, who even apply

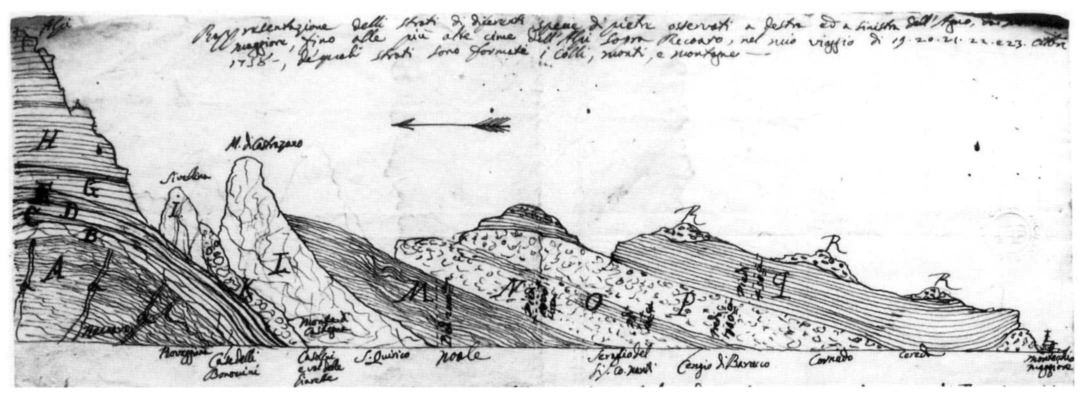

Fig. 2.32. Giovanni Arduino's manuscript section (1758) of the Secondary formations [*monti secondari*] in the foothills of the Alps, underlain as usual by Primary rock masses [*monti primari*]: the huge total thickness of the strata in sections such as this was widely taken as qualitative evidence of the vast scale of time over which the Secondary rocks had been deposited. The section shows the rocks exposed on the sides of the Val d'Agno, which leads from the north Italian plain towards the Alps (the arrow points north). As usual, the vertical scale is exaggerated, in order to clarify the structural relations of the rocks. The thicknesses of several formations are noted ("N", for example, is 1,000 feet thick), and each is described in an accompanying manuscript text. Although this section was never published, the extremely tattered state of the original suggests that Arduino demonstrated it repeatedly to the many savants who visited him, and perhaps even lent it to them to take into the field. (By permission of the Biblioteca Civica di Verona)

One of the most persuasive examples was offered by Arduino, who had explored the valleys leading from the north Italian plain towards the Alpine peaks, and had plotted the huge thickness of Secondary formations that he traversed in one of them (Fig. 2.32).[84]

Füchsel's published block diagram of the Secondary formations of Thuringia (Fig. 2.16), and Lehmann's earlier section through the same region (Fig. 2.19), had a similar effect. Lehmann had still attributed all his formations to a gigantic Deluge; but a generation later most savants who had seen such evidence with their own eyes in the field, or who had at least been turned into virtual witnesses by the persuasive accounts of others, concluded that the Secondary formations must have needed humanly inconceivable spans of time for their deposition. It seemed impossible to quantify the time that might have been involved; but the thought of thousands of feet of sediments, many of them fine-grained and finely layered, and some containing beautifully preserved fossils, was enough to make any savant's imagination reel at the likely immensity of time. The point was well summarized by La Métherie, the editor of *Observations sur la Physique*: "One feels that such enormous beds of limestone, gypsum, and shales, and such substantial masses of [fossil] shells, fish, and plants, could have been formed only in an innumerable sequence of ages [*siècles*] of which we have no conception, and perhaps at different epochs."[85]

Estimates of the timescale

There is much to suggest that it was indeed the human *imagination* that needed to be stretched, even among savants, before talk of vast amounts of time could begin to seem anything more than vacuous and scientifically irresponsible hand waving. For example, in the opening sentence of his *Alpine Travels* (1779), Saussure claimed that it was universally accepted—he meant, of course, among savants and other educated readers—that the earth's past revolutions or major changes had occupied "a long succession of ages [*siècles*]". The French word *siècle*, like its cognates in other Romance languages, was ambiguous in a way that was rather convenient at this time, since it could (and still can) mean either a literal century or an "age", a longer but indefinite span of time. Saussure could be read as suggesting a period that could just fit—at a pinch—within the traditional short timescale; but he himself certainly intended the other meaning of the word, for his vague phrasing was followed by further hints of a vast timescale. These were evoked, as so often, by an analogy with astronomy. Recalling the views that he had seen from the summit of Etna and from Alpine peaks (though not yet from Mont Blanc), Saussure commented on the "great epochs of nature" that they implied, which rendered the human world as insignificant in time as it was in space.[86]

Likewise, Werner commented in print—casually and just in passing—that the geognostic pile of rock masses must have accumulated "in the immense time span [*ungeheure Zeitraume*] of our earth's existence"; and in manuscript notes for his lectures on geognosy he estimated that the whole sequence might represent perhaps a million years. Lavoisier suggested that the "period" (in the sense of frequency) of his hypothetical oscillation of the sea level (§2.4) was perhaps "several hundreds of thousands of years"; and since he believed there had already been several such cycles, his conception of the earth's total timescale must certainly have run into millions. Figures of this order of magnitude seem to have been contemplated quite widely among savants who were familiar with the field evidence. And Kant's well-known earlier conjecture that "a series of millions of years and centuries have probably elapsed" in bringing the universe to its present state was almost a commonplace among cosmological theorists.[87]

Explicit figures were published only rarely, but it cannot be assumed that they were treated by savants as guilty secrets, to be whispered conspiratorially to

it, for example, to the huge thicknesses of sedimentary rocks exposed with spectacular clarity in the Grand Canyon in Arizona; the entire erosion of the canyon is then attributed to the draining away of the waters in the closing phase of the same event. This kind of exegesis has to take even greater liberties with the biblical story of the Flood than it did three centuries ago, yet it is invoked in order to support a literalistic interpretation!

84. Fig. 2.32 is reproduced from Verona-BC, Fondo Arduino, bs. 760, IV.c.11: see Vaccari, "Manoscritti di Arduino" (1993), 332. The section extends from Recoaro (left) to Montécchio Maggiore (right); the fieldwork is described in Vaccari, *Giovanni Arduino* (1993), 97–110. See also Ellenberger, *Histoire de la géologie* 2 (1994), 258–62, where the section is reproduced with its MS annotations, and the formations above the basement are dated in modern terms as ranging from Permian to Oligocene.

85. La Métherie, "Discours préliminaire" (1791), 34; on *siècles*, see below.

86. Saussure, *Voyages* 1 (1779): i, vi; the metaphor of the "époques de la nature" was borrowed from Buffon, whose famous work of that title had just been published (§3.2). The indefinite meaning of the words *siècle, saeculum*, etc., was familiar to both Catholics and Protestants, for example from liturgical use at the end of the Lord's Prayer: "*in saecula saeculorum*" and "*pour les siècles des siècles*" expressed what was translated less colorfully in English as "for ever and ever".

87. Werner, *Kurze Klassifikation* (1787), 5; see Guntau, "Natural history of the earth" (1996), 225. Lavoisier, "Observations générales" (1793), 364; Kant, *Theorie des Himmels* (1755), 1890 ed., 62.

colleagues, confined to the privacy of manuscript notes, or even just kept safely inside their own heads. Werner, for example, probably expressed his estimate openly in his lectures to students and in his conversations with colleagues. In the Enlightenment atmosphere of Freiberg he would have been very unlikely to be criticized for it, but he never wrote the kind of theoretical work in which it would have been thought appropriate to publish such an unavoidably speculative figure. In the cultural climate of the late Enlightenment, anywhere in Europe, savants were much more likely to be criticized by their peers for ill-founded speculation than they were to be pilloried by ecclesiastical authorities for impugning the reliability of Moses. Brydone evidently regarded Recupero's dilemma as an anecdote worth reporting precisely because it would not have arisen in parts of Europe more Enlightened than Sicily.

One case that might seem to be an exception to this general tolerance can better be regarded as illustrating the social dimension of the issue. Not long before he died (in 1793), John Hunter wrote an elementary essay on fossils, apparently as an introduction to a planned catalogue of his famous collection in London. Most of the essay was an unremarkable review of the whole range of "fossils" (in the old broad sense) and of the processes by which organic remains could have become "extraneous fossils" (§2.1). At one point, Hunter mentioned casually and in passing that he thought that fossils might "retain their form" as recognizably organic objects "for many thousands of centuries"; this was an estimate that, as just mentioned, was in line with the opinion of many other savants. He asked the geographer James Rennell—like himself a Fellow of the Royal Society—to read his essay and help him improve it. Among other comments, Rennell suggested that some of Hunter's anticipated readers, "very numerous and very respectable in every point but their pardonable superstitions", might object to "any mention of a specific period that ascends beyond 6000 years" and might find his conclusions more acceptable if "centuries" were reduced to "years".

There is no good evidence, however, that this was the reason that Hunter failed to publish his essay. A more plausible explanation is that he died before he could revise the whole text to his own satisfaction—it is a very muddled piece of prose—or complete the catalogue for which he had written it. In a later draft he did change the wording as Rennell had suggested; but the incident only reinforces the point being made repeatedly here, that everything depended on the social group within which the new scientific ideas were being discussed. What was acceptable and uncontroversial among Fellows of the Royal Society and other savants might be less so elsewhere, even among the socially "respectable" but highly conservative English public that might have read the essay.[88]

A million years (or even several million) will seem utterly inadequate to modern geologists; but in Saussure's time such estimates—the modern word "guesstimates" would be more appropriate—represented a significant stretching of the scientific imagination. Indeed, a million years was not strictly imaginable at all, any more than the vast cosmic distances that were being taken for granted by Herschel and other contemporary astronomers. Anyway, even Saussure's modest talk of "a long succession of ages" was enough to imply the transcendence of any traditional limitations set by the short cosmic timescale. However—to repeat the point—this was no problem, even for savants who regarded themselves as Christian believers, since it was widely recognized that the story of creation in Genesis should not be, or at least did not need to be, interpreted literally. What was far more important to them than sustaining literalism was to defend the created status of the world, and everything in it, by refuting any suggestion that the cosmos was eternal and therefore uncreated.

Despite, then, a widespread qualitative sense of the likely immensity of time, savants rarely offered quantitative figures for the ages of specific features, let alone for the earth as a whole. This was not so much to avoid criticism from religious conservatives, but rather because there was little good evidence to support such figures. By far the best known and most explicit quantitative dating was that proposed by Buffon, after he became convinced that the earth and other planets had all originated simultaneously from a plume of intensely hot material torn from the sun (§3.2). He claimed that this had subsequently condensed into discrete bodies—the planets and their satellites—which had then cooled slowly to their present temperatures. He tried to derive a timescale by replicating the cooling process experimentally; but it should be noted at once, as it was by his contemporaries, that the validity of his figures was necessarily dependent on a hypothesis that was in fact highly contentious, namely the allegedly incandescent origin of the earth.

Using the facilities of the industrial forge he had established on his country estate, Buffon conducted extensive experiments with small spheres, timing their rates of cooling from white heat to the stable ambiance of a cellar. Iron balls ranging in diameter from half an inch to five inches gave him some sense of how the rate varied with size, and one-inch balls of some thirty different kinds of material suggested how it might vary according to composition. Yet it still needed a very bold scaling up, from inches to thousands of miles, before he could derive a quantitative timescale for the successive phases of cooling of the earth itself. He calculated that 74,832 years had elapsed from the earth's first formation to the year of his experiments. However, he did not confine such calculations to the earth, for the same laws of cooling would have applied throughout the solar system, and indeed the universe. So he extrapolated his experimental results to the other planets, and even to their satellites: for each of them he calculated its time of consolidation and of three subsequent key moments in its gradual cooling. He set out a chronology for the whole solar system, past *and future*, expressed in years from its beginning (Fig. 2.33).[89]

88. Hunter (J.), *Observations on geology* (1859), iii; Rennell's undated letter is printed as an appendix, lvii–lxviii. Wood Jones, "Hunter as a geologist" (1953), analyzes the extant MSS and reconstructs the tortuous story of their posthumous publication. The claim that Hunter withdrew the essay on account of Rennell's comment is based on Richard Owen's third-hand recollection over sixty years later: see Hunter (J.), *Essays and observations* (1861), 1: 293–94. Such flimsy evidence would never have been taken seriously by historians had it not supported a conflict thesis that serves a powerful ideological agenda.

89. Fig. 2.33 is reproduced from Buffon, "Refroidissement de la terre" (1775) [Cambridge-UL: MG.7.17], 502–3 (the parts of the table printed on two successive pages have been combined); the validity of his figures depended of course on the assumptions involved in scaling them up. Leclaire, "Buffon géologue universaliste" (1992), reports on the "restaging" of Buffon's experiments, demonstrating the skill required in their performance.

TABLE plus exacte des temps du refroidissement des Planètes & de leurs Satellites.

CONSOLIDÉES jusqu'au centre.	REFROIDIES à pouvoir les toucher.	REFROIDIES à la température actuelle.	REFROIDIES à $\frac{1}{25}$ de la température actuelle.
	LA TERRE.		
En 2936 ans.	En 34270$\frac{1}{2}$ ans.	En 74832 ans.	En 168123 ans.
	LA LUNE.		
En 644 ans.	En 7515 ans.	En 16409 ans.	En 72514 ans.
	MERCURE.		
En 2127 ans.	En 24813 ans.	En 54192 ans.	En 187765 ans.
	VÉNUS.		
En 3596 ans.	En 41969 ans.	En 91643 ans.	En 228540 ans.
	MARS.		
En 1130 ans.	En 13034 ans.	En 28538 ans.	En 60326 ans.
	JUPITER.		
En 9433 ans.	En 110118 ans.	En 240451 ans.	En 483121 ans.
	SATELLITES DE JUPITER.		
1.er en 6238 ans.	En 71166 ans.	En 155986 ans.	En 311973 ans.
2.d en 5262 ans.	En 61425 ans.	En 135549 ans.	En 271098 ans.
3.e en 4788 ans.	En 56651$\frac{2}{3}$ ans.	En 123700$\frac{5}{6}$ ans.	En 247401$\frac{4}{6}$ ans.
4.e en 1936 ans.	En 22600$\frac{1}{3}$ ans.	En 49348 ans.	En 98696 ans.
	SATURNE.		
En 5140 ans.	En 59911 ans.	En 130821 ans.	En 262020 ans.
	ANNEAU DE SATURNE.		
En 4604 ans.	En 53711 ans.	En 88784 ans.	En 177568 ans.
	SATELLITES DE SATURNE.		
1.er en 3433 ans.	En 40021$\frac{9}{25}$ ans.	En 87392 ans.	En 174784 ans.
2.d en 3291 ans.	En 38451$\frac{1}{3}$ ans.	En 83964 ans.	En 167928 ans.
3.e en 3182 ans.	En 35878 ans.	En 78329 ans.	En 156658 ans.
4.e en 1502 ans.	En 17523$\frac{1}{3}$ ans.	En 38262$\frac{1}{2}$ ans.	En 76525 ans.
5.e en 421$\frac{1}{3}$ ans.	En 4916 ans.	En 10739 ans.	En 47558 ans.

Fig. 2.33. Buffon's table, published in 1775, showing the calculated lapse of time—since the formation of the solar system and the origin of all the planets and their satellites—for each body to cool enough to be solid throughout (column 1); and for the surface of each to become cool enough to be touched by the human hand (column 2), to be at the present temperature *of the earth* (column 3), and to cool to one twenty-fifth of that present temperature (column 4). No distinction was drawn between past and future states, since all were determined by the ahistorical physical "laws" of cooling bodies: the moment of the present was represented by the figure of 74,832 years (the figure for the earth [*la terre*] in column 3), which was therefore also the *age* of the earth. The great precision of the figures, without any "rounding off", was normal practice at this time in such calculations, and does not imply that Buffon was unaware of the likely margins of error; conversely, the qualitative definition of successive points in the cooling process reflects the uncertain state of thermometry at this time. (By permission of the Syndics of Cambridge University Library)

Even in print, however, Buffon expressed dissatisfaction with this modest timescale. In particular, like other savants, he was struck by the sheer thickness of the Secondary formations, many of them composed of finely layered sediments. He could not believe that some twenty thousand years—the share of the total time that he allotted to them—were adequate to account for their accumulation. He suspected that some "hidden cause" [*cause cachée*] had been overlooked in his calculation and that the true timescale was far longer (Newton's earlier use of *causae latentes* made this kind of suggestion scientifically respectable). In manuscript drafts of his later *Nature's Epochs* (§3.2), Buffon felt confident in putting a figure of about three million years on the whole sequence, and he toyed with figures up to ten million. Even his much lower published estimate was of course more than enough to breach the traditional short timescale, so he is unlikely to have kept his larger estimates to himself in order to avoid criticism from religious authorities. Rather, it was a matter of scientific caution: he could at least offer experimental evidence in favor of the lower figures, but nothing more concrete than a hunch to justify the higher ones.

Nor was Buffon unusual in this respect, even with his higher figures. As an order of magnitude, his three million years were in line with the guesstimates of others such as Werner and Lavoisier, and about as large as most savants would have regarded as justified by the evidence available. It is significant that even thirty years earlier the editor of *Telliamed* had silently reduced the few "billions" of years (for each vast cosmic cycle) in de Maillet's manuscript down to a few "millions" in the published text, evidently to bring that extremely fanciful theory somewhat more within imaginative reach of its readers.[90]

Encounters with theologians

After the publication of *Nature's Epochs*, Buffon did have a brush with the Sorbonne, the theological faculty in Paris, but it was no more than a little local difficulty (on an earlier encounter, see §3.2). The great public acclaim for his work obliged the theologians to react officially to his handling of the Genesis narrative, which expanded the biblical "days" into vast spans of time and treated the text as one intended for the understanding of common people rather than scholars. Although Buffon was not at all original in this, the sheer popularity of his work made it seem a threat to the Sorbonne's authority: in effect, Buffon the philosopher was trespassing on the intellectual turf that properly belonged, in their opinion, to theologians like themselves. He gave them a perfunctory statement of orthodoxy, but he was scornful of their action and openly derided it. He was far too powerful—and protected by the king—for the Sorbonne to dare censure him formally. The whole episode had no impact on Buffon's work or that of other savants, many of whom were themselves clergymen.[91]

Soulavie's corresponding little local difficulty had the same specific character. He became the target of malicious attacks by a fellow priest and native of Vivarais, the ex-Jesuit Augustin Barruel, but apparently more out of jealousy and resentment at Soulavie's worldly success than from any deep zeal for orthodoxy. In his pamphlet "Genesis according to Mr Soulavie" (promptly reprinted in his popular "Provincial philosophical letters" from Vivarais), Barruel parodied Soulavie as claiming that 356,913,750 years had elapsed (by 1780) since the formation of the

90. Buffon, "Époques de la nature" (1778), 67–70. See Roger, *Buffon philosophe* (1989), 537–43, and "Buffon: Époques" (1962), lx–lxvii. On de Maillet, see the modern edition by Carozzi, *De Maillet: Telliamed* (1968), 182, and 381n52, and Rappaport, *Geologists were historians* (1997), 229.

91. Stengers, "Buffon et la Sorbonne" (1974), describes the episode in detail, using inter alia the archival records of the Sorbonne's deliberations; see also Roger, *Buffon philosophe* (1989), 554–56.

earth, with another 57 million still to go. He was probably aping Buffon's precision in order to improve the malicious joke; but the figure (two orders of magnitude greater than Buffon's unpublished estimate) may be one that Soulavie had indeed suggested in private. Anyway, although his views were ridiculed, the reception of Soulavie's ideas among savants does not seem to have been much affected by Barruel's scurrilous attack.[92]

Far from being persecuted by "the church", it was Soulavie himself who in 1783 brought a court action against his critic, claiming that not only his orthodoxy but also—more importantly—his honor had been impugned. In context, it is clear that Soulavie was incensed that his scientific research was being ridiculed in a work widely read in his local community and among his extended family in Vivarais. Conversely, Barruel was indignant that a liberal interpretation of the Genesis story was being adopted by a fellow *priest*, who was thereby lending aid and comfort to skeptical lay philosophes such as Buffon. The court case dragged on, as Soulavie's patrons among the higher clergy tried to find some way to end an embarrassing quarrel between two obstinate subordinates, which was making the church itself an object of ridicule. Soulavie turned down a lucrative offer of a naturalist's position on La Pérouse's voyage to the Pacific, which would have taken him out of Paris and given him four years to cool off (Lamanon, who did serve in the naturalist's position that Soulavie rejected, died on the voyage: see §4.3). But his honor was eventually satisfied, and he dropped his charge against Barruel, when the archbishop of Narbonne—who had long been his supporter and was the current president of the Assembly of the Clergy in Paris—invited him to live in his household and obtained for him a savant's salary funded by the provincial government of Languedoc. As in the cases of Buffon and Hunter, the tangled story illustrates the localized and socially specific character of encounters between savants and ecclesiastical authorities or conservative publics in the late Enlightenment, over the question of "Genesis and geology". Anything further from the stereotype of a monolithic conflict between "Science" and "The Church" would be hard to imagine.[93]

Conclusion

The few examples mentioned here suggest how the savants of Saussure's generation had no need to be either alarmed or gleeful—depending on their other commitments—when they became aware of the increasingly strong evidence that the traditional short timescale was grossly inadequate. Their inability to put firm quantitative figures on the earth's true timescale did not lessen their qualitative conviction that it was literally unimaginable in magnitude. Further examples of this will be mentioned in other contexts, in the next two chapters. There is no good historical evidence that any of the leading savants, in any part of Europe, were constrained in their theorizing by a shortage of "deep time". They just took the new perspective in their stride and allowed for the possibility of vast spans of time—literally inconceivable in human terms—in the earth's remote past. Their sense of time may not yet have been very "deep" by modern standards, but it was quite deep enough for their immediate explanatory needs.

Above all, it should be noted that in stretching the timescale to even a million years they were transcending the stark alternatives available in earlier centuries, both of them profoundly unmodern in character. In contrast *both* to the short and finite timescale of traditional chronology *and* to the infinitely long perspective of traditional eternalism, they were beginning to open up the conceptual space for a third (and modern) option: the timescale might be unimaginably lengthy, *yet not infinite.* This novel option was a crucial precondition for the reconstruction of geohistory, as the rest of this book will suggest.

92. No copy of Barruel, *La Genèse selon M. Soulavie* (1783), survives, but it was reprinted in his *Helviennes* (1781–84 and later editions), letter 28; the title referred to the "Helvienne [i.e., female Vivarais] philosopher" whose letters were one side of the fictitious correspondence. The parody at Soulavie's expense came in the context of far wider criticisms of philosophes such as Voltaire and Rousseau: see McMahon, *Enemies of the Enlightenment* (2001), chap. 2. The figure of some 357Ma attributed to Soulavie for the age of the earth was compatible with his unpublished estimate (already mentioned) of 6Ma for the most recent eruptions in Vivarais; years later, he did in fact publish the latter figure, in his *Mémoires de Richelieu* (1790–93), 9: 448–49, using it to illustrate how his own Enlightened age had exploded the "fable" of a short timescale.

93. The reprinted version of Barruel's parody is itself a rarity because many copies of his *Helviennes* were seized by the authorities—acting in *Soulavie's* interest!—during the court case; in later editions (e.g., the fifth, 1812, 1: 337–90), Barruel excised Soulavie's name from the parody, which became *La Genèse moderne.* The story is summarized in Mazon, *Histoire de Soulavie* (1893), chap. 3, with copious excerpts from manuscript sources, including Barruel's correspondence; see also Aufrère, *Soulavie et son secret* (1952), 75–80, who, however, portrays the episode in somewhat Galileo-like terms. Gaudant, "Querelle des trois abbés" (1999), describes the later but equally instructive case of the fossil fish of Monte Bolca (see Fig. 5.7), in which the interpretations suggested by the naturalists Alberto Fortis and Giovanni Volta were contested by their fellow priest the bookish scholar Domenico Testa.

CHAPTER 3

The theory of the earth

3.1 GEOTHEORY AS A SCIENTIFIC GENRE

The meaning of "geology"

In the late eighteenth century, the science of earth physics [*physique de la terre*] offered causal explanations of the varied features described and classified by the sciences of mineralogy, physical geography, and geognosy; or at least it described lawlike "dispositional regularities" that might point towards causal explanations. However, such studies of specific features (as sketched in §2.4) were less prominent among Saussure's contemporaries than works of a far more ambitious kind, which were intended to take causal explanation on to a much higher level of generality. In fact, to describe the more restricted kind of study as if it were an end in itself is rather misleading, for such work was usually undertaken as a means towards a much more important end. The ultimate goal of many savants concerned with the sciences of the earth was to construct what they called a "*system*" or high-level theory about the earth. This would be not merely a theory to explain specific features, such as the elevation of mountains, the consolidation of rocks, or the emplacement of fossils, important though such problems were. On the contrary, a system would try in principle to include all such limited explanations within a single overarching causal theory.

These systems were treated as rival claims to the title of "*Theory of the Earth*". The aim was to emulate on a terrestrial scale the achievement of Newton in the realm of celestial mechanics. It was to discover the one and only true explanation of how the earth works, just as Newton was believed—quite generally, by Saussure's time—to have discovered the one and only true theory to explain the movements of the sun and its planets, and all other stars and their putative planets throughout the universe, under the laws of universal gravitation. In other words, "Theory of the Earth"—the initial capitals will distinguish it here from any more limited theory—was not just a human conjecture or "hypothesis", which might or might not be valid. It was Nature's (or God's) hidden construction, which another Newton might one day have the honor of discovering. If a savant gave his work the title "Theory of the Earth", as many did, it was as *his* attempt to formulate what he hoped would turn out to be the true terrestrial "system". Theory of the Earth was in effect a scientific *genre*, just as landscapes, operas, sonnets, and novels were artistic genres.[1]

The origin of the genre of Theory of the Earth reached back to Descartes in the seventeenth century. What first made it possible was that in the wake of Copernicus's cosmology the earth—no longer central in position—could be treated as a physical body that was not altogether unique and could be subjected to the kind of causal analysis characteristic of the then new "mechanical philosophy". Descartes had proposed a hypothesis, or scientific *model* (to use the appropriate modern term), of how *any* earthlike body—anywhere in the universe at any time—would have developed, reaching eventually the character familiar to humans on their particular planet, with atmosphere, oceans, continents, mountains, and so on. The specific phrase "Theory of the Earth" dates, however, from Thomas Burnet's work later in Descartes' century. Burnet had called his theory "sacred", but he was no biblical literalist. He had supplemented Cartesian natural philosophy with what he believed the Bible disclosed about the past and future physical states of the earth. Like Descartes, however, he had claimed to show how the earth has worked in the past and how it will work in the future, in terms of purely *natural* physical processes. His book generated lively and sometimes acrimonious argument among savants through the rest of the seventeenth century and into the early eighteenth. A variety of new causal agents, such as comets, and new empirical evidence, such as fossils, had been brought into play in a proliferating range of rival systems. Newton himself was among those who had taken an active interest in such matters: it was a debate not on the margins of intellectual life but at its very center.[2]

When Saussure climbed Mont Blanc, a full century after Burnet's work was published, the genre was flourishing as much as ever. De Luc called his own first contribution to it "the sketch of a treatise on cosmology", but immediately acknowledged that "cosmology" was not really the right word, because his work was to be limited to the earth. So he proposed, albeit obliquely, that as the terrestrial analogue of cosmology it should be termed "*geology*" (Fig. 3.1).[3]

Saussure at once adopted the word in de Luc's sense, on the very first page of his *Alpine Travels* (1779), while distinguishing it sharply from the observational sciences of the earth, such as his own field of physical geography. The latter, he claimed, was the evidential foundation on which any "geology" must be built: "the science that gathers the facts that alone can serve as the basis for the Theory of the Earth or *Geology*, is physical geography, or the description of our globe." Thus "geology" first consistently entered the language of savants as a synonym for Theory of the Earth. To avoid confusion with the modern meaning of the word, the term *geotheory* will be used here, to denote the genre of those scientific theories that aspired to offer a true causal account of the earth, its origin (if it had ever had one)

PREFACE,

SERVANT

D'INTRODUCTION.

JE deſirerois que le Lecteur voulût ſe donner la peine de parcourir cette Préface; elle pourroit l'aider à ſaiſir plus tôt un plan qui ne ſe développe que peu à peu dans le cours de l'Ouvrage; parce que dans le fait il ne s'eſt formé que peu à peu, & que ſa nature même l'exigeoit ainſi.

CES LETTRES ne ſont que le canevas d'un Traité de Coſmologie (a)

(a) Je n'entends ici par *Coſmologie* que la

* 4

viii PREFACE, SERVANT

que j'eſpérois de faire un jour, mais dont je n'ai pu recueillir les matériaux ſuivant mes deſirs. Ce devoit être un Ouvrage méthodique, où la partie des Faits, diviſée par claſſes diſtinctes, auroit été portée, après de longues recherches, à un certain degré de généralité & de préciſion dont je m'étois fait une idée. Un ſyſtème, né d'un grand nombre de premières obſervations, m'eût ſervi de motif pour en faire de nouvelles, qui l'auroient, ou détruit, ou développé & appuyé plus complettement.

connoiſſance de la *Terre*, & non celle de l'*Univers*. Dans ce ſens, *Geologie* eût été le mot propre; mais je n'oſe m'en ſervir, parce qu'il n'eſt pas uſité. J'employerai donc toujours ce mot *Coſmologie*, dans le ſens que je viens de définir, & par analogie à *Coſmographie*, & à *Coſmopolite* ſurtout, dont on ne ſe ſert que relativement à la *Terre*.

Fig. 3.1. The footnote on the opening pages of de Luc's *Letters on Mountains* (1778), in which he implied that the unusual term "*geology*" would best denote the kind of high-level theorizing about the whole earth to which he intended to contribute:

> I mean here by cosmology only the knowledge of the earth, and not that of the universe. In this sense, 'geology' would have been the correct word, but I dare not adopt it, because it is not in common use.

In spite of this initial hesitation, he did in fact begin to use the word for the established genre of "Theory of the Earth", and he was soon followed by other savants. (By permission of the Syndics of Cambridge University Library)

and its development or change through time. This will avoid the need for repeated reminders that "geology" was not being used in its modern sense.[4]

The goals of geotheory

Geotheory, then, was a flourishing genre in Saussure's time. Every savant with any ambition to make his mark in this area of natural knowledge aspired to construct

1. Ashworth, *Theories of the earth* (1984), was among the first studies to recognize it explicitly as a genre.

2. Descartes, *Principia philosophiae* (1644); Burnet, *Telluris theoria sacra* (1680–89). Burnet was considered far from orthodox, and his *Archaeologiae philosophicae* (1692) got him into ecclesiastical trouble. On seventeenth-century geotheories see, for example, Roger, "Théorie de la terre" (1973), and "Cartesian model" (1982); also Gould, *Time's arrow* (1987), chap. 2, and Rappaport, *Geologists were historians* (1997), chap. 5.

3. Fig. 3.1 is reproduced from de Luc, *Lettres sur les montagnes* (1778) [Cambridge-UL: CCC.28.25], vii–viii; see also *Lettres physiques et morales* (1779), 1: 4–6. Arduino had already used "*geologia*" in rather the same way, but it is de Luc's usage that is clearly continuous with—though not identical to—the modern meaning. Other earlier uses of "geology" or its cognates, such as Diderot's (see Fig. 1.16), had been sporadic and idiosyncratic; some of them are described in Dean, "The word geology" (1979).

4. Saussure, *Voyages* 1 (1779), i–ii; the term "geotheory" was proposed in Rudwick, "De Luc and nature's chronology" (2001), 53.

his own "system" or geotheory, or else—as in Saussure's case (§6.5)—to explain why he was not going to do so, or not yet. So there was a proliferating profusion of systems, often incompatible with one another, yet all claiming to be based on sound physical principles. Before illustrating this profusion with a few examples, it is worth outlining some general features of the genre as a whole. By Saussure's time these features were treated as normative; they defined what a geotheory should ideally be like, even if they could not be followed fully.[5]

First, a system or geotheory was expected to explain, at least in outline or in principle, *all* the major features of the earth, considered as a complex physical and biotic entity. Rather than accounting for a specific mountain range, for example, it would have to explain in general terms the origin of all mountains and relate them to oceans, volcanoes, fossils, and so on. Second, the proposed causes were expected to be clearly natural in character. If the biblical Flood, for example, were taken to have been a real past event, it would have to be assigned a natural cause and not just be treated as miraculous (its ultimate or "final" cause might still be divine, of course, and in some sense providential, even if the "secondary" or "efficient" cause were natural). Third, the causes were expected to be based as far as possible on known physical entities and observable processes, which meant in effect that the present world was to be the key to understanding how the world works. However, causes might be suggested even if they were not of everyday occurrence: for example, comets could properly be invoked, since they were known celestial objects, and past or future impacts or near-misses within the solar system were quite conceivable, even if unrecorded in human history and unwitnessed in the present. And fourth, a system was expected to offer a plausible explanation of the past, present *and future* development of the earth, since the underlying causal laws of nature were taken to be perennial features of the world. Ideally, therefore, a geotheory was expected to account both for the *origin* of the earth and for its ultimate *end*; or, alternatively, to explain how and why the earth was eternal and therefore without either origin or end.

With these four characteristics, the genre of geotheory was clearly related to the science of earth physics, but raised to a higher level of generality. As in earth physics, the causes proposed as explanatory agents were processes that necessarily operated in time; but they were assumed to do so in the same way, whether acting in the past, present, or future. The system that represented the true Theory of the Earth, like its constituent physical causes, would thus express the operation of ahistorical natural laws. In consequence, the sequence of changes—in past, present, and future—would be *predictable* (or retrodictable), at least in principle, once the true Theory was known. If the initial conditions could be specified correctly, all subsequent changes would be the predictable consequences of the operation of natural laws: in modern terms, the system was deterministic and the changes taking place were, in effect, *programmed* from the start.

This element of deterministic predictability was particularly clear if the earth were conceived as being in a steady state or as changing in a cyclic manner; for the continued operation of the same natural laws would ensure that the general features of the earth had always been broadly the same and would always remain so in the future. The obvious analogy here was with the predictable movements of the planets around the sun under the ahistorical laws of gravitation. Alternatively, if the earth had changed in a directional manner, an analogy was drawn between the sequence of physical changes and the "development" (in modern terms, ontogeny) of an embryo into an adult organism.[6]

A fifth feature of most geotheoretical systems was a consequence not of their status as causal explanations but of their character as scientific *models*. To use another modern formulation, they were hypothetico-deductive in structure. The strategy of argument was to outline the model as succinctly as possible, to show how it could in principle explain a wide range of terrestrial features or phenomena; and then to deduce its implications and to marshal detailed empirical evidence to justify the claim that this *in fact* was the way the world worked. First came the exposition of the system itself, then the display of the material in its support. Most geotheories were presented in just this way: first the model in outline, then the detailed evidence, the latter being in some cases published separately or at a later time. A savant might even postpone collecting most of his supporting evidence until *after* he had formulated and presented his system. Far from detracting from the epistemic status of the geotheory, this might be seen as enhancing it, since risky predictions would be offered in advance of the attempt to confirm them.

In consequence of this hypothetico-deductive character, the making of geotheoretical systems was primarily an indoor activity of thinking and talking, to be pursued by the savant alone in his study, or in dialogue with others in salons or meeting rooms. Only at the stage of searching for confirmatory evidence might it become desirable to move into the laboratory or the museum or to venture into the field; and not necessarily even then, since the evidence might be assembled quite adequately from reports already published by others. John Kay, the artist who portrayed Hutton in indoor dialogue with his fellow "philosopher" Black (Fig. 1.17), showed them appropriately as they might have been seen when discussing the geotheory that Hutton had recently expounded (§3.4). The same artist also struck precisely the right note when he portrayed Hutton in the field: the savant's encounter with the rocks was shown as being just as much an encounter with his fellow savants (Fig. 3.2).[7]

The sixth and last characteristic of geotheory stemmed from its ambition to provide in principle a "system" or *comprehensive* understanding of the terrestrial physical world. Most geotheorists claimed indeed to be "philosophers" in the broadest sense. Their goal was to explain not just the character of the earth itself

5. They should be regarded as defining a Weberian "ideal type" of geotheory, to which real examples merely approximated. On the genre that is here termed geotheory, and its relation to empirical observation, see especially Taylor, "Rehabilitation of theories" (1992), and "Two ways of imagining the earth" (2002).

6. Oldroyd, "Historical geology" (1979), describes the latter as "genetic" explanation; Gohau, "Géologie historique" (1988), deftly calls it "embryologie de la terre".

7. Fig. 3.2 is reproduced from Kay, *Original portraits* (1838), 1 [Cambridge-UL: Lib.4.83.3], no. 24; like the other etchings eventually published in this work (see Fig. 1.17), this was sold and circulated at the time as a separate print. Kay's similar portrait of Black (ibid., no. 22)—with a walking stick in place of a hammer—showed him confronting another "rock face" of savants; it is reproduced in Simpson (A. D. C.), *Joseph Black* (1982), 6.

Fig. 3.2. A caricature of James Hutton doing fieldwork, drawn and etched by John Kay in 1787. Like Saussure (Fig. 1.7), Hutton is dressed in elegant urban clothes appropriate to his social position, while wielding the usual plebeian miner's hammer. But the "rocks" he is hammering are in fact other savants (their faces shown in profile). His outdoor fieldwork was thus also a metaphor for his indoor activity of verbal debate, such as the argument about the Theory of the Earth that he had presented at the Royal Society of Edinburgh two years earlier. (By permission of the Syndics of Cambridge University Library)

but also its relation to the basic structures of nature and to its human and even divine significance. Any geotheory was therefore embedded in a dense intellectual matrix. It had to be related on the one hand to fundamental questions of physics and cosmology and on the other to basic concepts of human nature and human society, of morals and metaphysics, and indeed of theology (whether the theology was orthodox or heterodox; whether theistic, deistic, or atheistic). Such were the issues to which the practice of geotheory was connected: any modern distinction between "scientific" and "nonscientific" questions would have been regarded as inappropriate and indeed meaningless.

It was in consequence of its wide-ranging ambitions that the genre of geotheory had intrinsic links with other intellectual constructions that offered explanations of human meaning and significance. Specifically, of course, it was here that geotheory impinged on systems of religious thought and practice. Savants who practiced any of the four sciences outlined in the previous chapter could get on with their work—at almost all times and places—without being concerned, let alone troubled, by its religious implications, except insofar as they themselves chose to make such connections. But if they had ambitions to propose a geotheory, they were in effect obliged by the expectations of their audiences to consider those implications in full.

Conclusion

The genre of geotheory is illustrated in this chapter by describing four or five examples, all of which were matters of current debate among savants around the time that Saussure climbed Mont Blanc. The purpose of the summaries that follow is not to give full or detailed accounts of the systems, at least two of which (those of Buffon and Hutton) have been the subject of intense historical study. Nor is it to trace their antecedents or the stages of their formulation, or fully to assess their reception. They function here simply as examples of the range of geotheories publicly available towards the end of the eighteenth century as explanatory *resources* for further work.

The selection is not arbitrary, because all these geotheories were to remain highly influential *as exemplars* in subsequent debates, even if their original empirical details failed to stand the test of time. They illustrate quite effectively the wide range of the systems that were proposed within the genre; nonetheless, they are no more than a small selection from the bewildering profusion that was available to savants in the last years of the eighteenth century. But they provide an adequate basis for assessing how far, or in what sense, the genre of geotheory can be said to have been truly geohistorical rather than merely temporal.

3.2 BUFFON'S COOLING GLOBE

Buffon's first geotheory

Only a few months before the first volume of Saussure's *Alpine Travels* appeared, Buffon published an important geotheoretical "system" in one of the later volumes of his great *Natural History*. It was in effect a replacement of the equally influential one with which he had started that long-running series just thirty years earlier. Despite important differences between the two, both exemplify the general character of the genre of geotheory, and both deserve to be summarized here. Given his prominent position in the Republic of Letters, as the director of one of the leading institutions in the world center of the sciences, the profound impact of his ideas is hardly surprising.[8]

Buffon's first geotheory had been published in 1749 in the very first volume of his vast survey of natural history. Its position was appropriate, because a true knowledge of the earth—the physical environment on which life depended—was the proper foundation for understanding the animals that would be described in most of the subsequent volumes. After an introductory essay on how natural history in general should be practiced, his "second discourse" was entitled appropriately "History and Theory of the Earth". It offered a geotheory that was to be founded on the descriptive sciences of nature (as in the title of the whole series, "history" meant descriptive natural history). That relation between theory and evidence was reflected in the structure of the work. The quite brief discourse itself, setting out his system, was followed by a far longer compilation of the "proofs" [*preuves*] or evidence on which the theory was based and various ancillary topics.[9]

8. This section draws substantially on the work of the greatest modern Buffon scholar: see Roger, *Buffon philosophe* (1989), esp. chaps. 7, 13, and 23, and "Buffon et l'introduction de l'histoire" (1992). The vast range of recent Buffon research, most of it, however, on his work on living organisms, is surveyed in other essays in Gayon, *Buffon 88* (1992).

9. Buffon, "Histoire et théorie" (1749); see Roger, "La place de Buffon" (1988), Gohau, "La 'théorie de la terre' de 1749" (1992), and Ellenberger, "Sciences de la terre avant Buffon" (1992). Like its English equivalent, *preuves* was used in the legal sense of supporting evidence, rather than in the mathematical sense ("Q.E.D.") of irrefutable demonstration.

As expected from an Enlightenment philosophe of his generation, Buffon based his geotheory firmly on a repertoire of physical causes that could be seen to have the relevant effects: as Newton had put it, imputed causes had to be "true causes" [*verae causae*]. Specifically, this meant that Buffon's method for explaining the earth was to be rigorously *actualistic*, based on "actual causes" or processes observably in action in the present world:

> We should not be concerned with causes the effects of which are rare, violent, and sudden; they are not found in the ordinary course of Nature. But effects that happen every day, movements that are followed and renewed continuously, operations that are constant and repeated everywhere: these are our causes and our reasons.[10]

Deploying these ordinary causes, Buffon set out a model of continuous but directionless terrestrial change. Being well aware that fossil shells were widespread on land, far from the sea, he inferred that the oceans must have covered all parts of the globe in succession. Bedded formations were laid down on the ocean floors, of course, but Buffon claimed that hills and valleys were also sculpted underwater by tides and currents. At the same time, continents were continuously worn down and leveled by rain and rivers, until eventually they became the floors of new oceans:

> The waters of the heavens little by little destroy the work of the sea, continually lower the height of mountains, fill in valleys, estuaries and gulfs, and, by reducing everything to one level, will one day return this earth [i.e., continent] to the sea—which will seize it bit by bit, while laying bare new continents [already] broken up by valleys and mountains, just like those we inhabit today.[11]

So there was a continuous slow exchange of territory between continents and oceans: any particular point on the globe might have been both land and sea at different times. Since all this was the result of the ordinary physical laws of nature, it had happened throughout the past, it was taking place at present, and it would continue into the indefinite future. To borrow a term from modern cosmologists (though it has been rejected by them in favor of the Big Bang), Buffon's was a *steady-state* geotheory. The details of geography were in continuous flux, but overall the earth was in a kind of dynamic equilibrium: in past, present, and future it remained much the same kind of place.

There was little in all this that was particularly novel: Buffon was drawing on a rich legacy of still earlier geotheorizing, reaching back to Aristotle. There was also much about his geotheory that was not fully worked out: for example, the mechanics of the interchange between land and sea remained obscure, since he regarded volcanoes and earthquakes as relatively unimportant and discounted any major movements of the earth's crust. Nonetheless, it was a striking model of a dynamic earth that could provide living organisms with an ever-changing yet broadly stable environment over indefinite spans of time. Expressed in Buffon's famously eloquent prose, its persuasive naturalism had a profound impact on the way that savants thought about the earth in the middle decades of the century.

Buffon's earth was in a steady state of dynamic equilibrium, which might have been taken to be eternal; yet in fact he did not treat it as such. In a separate essay, the first in his "Proofs", he set out an even more ambitious theory, which de Luc and others would have regarded as "cosmological" in the proper sense of that word. It offered an explanation of the *origin* of the earth and all the other planets, and indeed—at least by implication—of all other planetary systems anywhere in the universe. Buffon suggested that at some time in the distant past a comet might have struck the sun obliquely and torn off a plume of matter that then condensed into a string of planets (comets were at this time believed to be very dense bodies, so the hypothesis seemed physically plausible). This could explain why—as had been long been known to astronomers—the planets all lie more or less in one plane, orbiting and rotating in the same direction. At first sight, Buffon's suggestion breached the principles he had set himself in his explanation of the earth: this putative event was sudden and violent, and of course unparalleled in human experience. Yet it was impeccably natural in character, and fully conceivable as a physical possibility: in short, a respectable "hypothesis". Still, there was a decidedly awkward disjunction between Buffon's general explanation of the earth, as a system that could well have been eternal, and his hypothesis to account for its origin, as an "accident" of cosmological chance at a specific moment in the remote past.

Buffon's volumes were an immediate success, and editions in German, English, and Dutch appeared in subsequent years; the *Natural History* was probably the most widely read scientific work of the century. But it was also criticized, by other naturalists, by the philosophes involved in the *Encyclopédie*, and by some theologians. The wider implications of Buffon's work could hardly be ignored by any of these groups, for he presented his ideas about the earth in a characteristic context of far wider theorizing, about the natural world as a whole, humankind, and—at least by default—God. The reactions of theologians were just one aspect of a much wider range of criticism in the Republic of Letters.

Without being openly atheistic, Buffon had simply redefined the scope of the natural sciences in such a way that divine action was marginalized. Theological reaction to his work was slow in coming and hesitant when it came: the theologians in Paris, particularly the Jesuits (who had not yet been expelled from France), were enthusiastic about Buffon's overall project in natural history, but were goaded into some response by the more conservative reaction of the Jansenist party. Leaving aside reactions to his broader theorizing about nature as a whole and about human beings, Buffon's geotheory was criticized mainly for its apparent eternalism; and he could refute that accusation without dissimulation, since he had also proposed a conjecture about the earth's temporal origin. Anyway, he was in far too powerful a position in Paris—as director of the royal museum and botanic garden—to be

10. Buffon, "Histoire et théorie" (1749), 99; the phrase "actual causes" [*causes actuelles*] was coined later by de Luc (§3.3). On the actualistic method, see the classic work by Hooykaas, *Natural law and divine miracle* (1959). The word "actual" was used in the sense still current in European languages other than English, meaning current or present-day: the news on French television, for example, is the day's "*actualités*".

11. Buffon, "Histoire et théorie" (1749), 124; this was the conclusion of Buffon's main text, immediately before the *Preuves*.

much affected by criticism from the Sorbonne: as was usual in such arguments, the matter was settled by careful diplomacy. Buffon was scornful of the whole affair, but treated the necessary verbal compromises as an acceptable price to pay for being left in peace. He issued an anodyne denial of any heterodox intentions—its wording was probably drafted for him by the theologians themselves—and he reprinted it in subsequent volumes of his work, using it in effect to protect himself against any further sniping of that kind. In sum, the criticism had little effect on Buffon's work or on its reception.[12]

Nature's epochs

Thirty years later, Buffon integrated the two components of his geotheorizing—and thereby remedied its major shortcoming—when he presented a quite different model of how the earth works. The change was provoked in part by specific new empirical evidence. Other savants had convinced him of the reality of the earth's internal heat, as implied by the warmth encountered in mines, increasing in proportion to their depth (in modern terms, the geothermal gradient). This could readily be regarded as "central heat", emanating from the earth's deepest interior, which in turn Buffon interpreted as evidence of *residual* heat. The implication, in his view, was that the earth had originated as a hot body, that it had been cooling ever since, and that it would cool still further in the future. Buffon had also been impressed by reports of the remains of apparently tropical mammals such as elephants and rhinoceros, found as fossils in high latitudes in Siberia as well as in Europe and North America (§5.3). This too suggested a formerly hotter earth. So he now formulated a new geotheory, in which a wide range of empirical material was marshaled in support of a model of the earth as a slowly cooling globe.[13]

Once again, there was little about the model in general that was original to Buffon. The idea of a cooling earth, for example, had been at the center of Leibniz's geotheory, devised in Burnet's time though not fully published until Buffon's; but Buffon elaborated it persuasively and presented it with his usual eloquence. It was constructed within the same genre of geotheory: as before, it was a system that reconstructed the past, interpreted the present, and predicted the future, with the whole sequence operating under the same ahistorical natural laws. Again as before, Buffon's work was divided between an exposition of the system itself and a long series of "*notes justificatives*" (the phrase recalled the "*pièces justificatives*" or documents and other exhibits produced to support a case in a court of law). And like his earlier geotheory, and indeed like most of its rivals, it was a work devised largely in Buffon's study, using published sources; it was not based on much fieldwork of his own.[14]

Buffon distinguished three classes of evidence, which he called "facts", "monuments", and "traditions". His facts were major observable features of the *present* earth. Prominent among them was the shape or geodetic "figure" of the earth itself. This had recently been proved to be an oblate spheroid, as predicted on Newton's laws of gravitation and as expected if the earth had originally been a spinning mass of hot fluid. There was also the fact, carried over from Buffon's first geotheory and generally agreed by naturalists, that marine fossils were found almost universally on the continents, far from the sea and high on mountain ranges, which indicated unequivocally the wide former extension of the oceans. Buffon's monuments were various *natural* vestiges or relics of the *past*, again including fossils as prominent examples. They served to show that the earth had passed through physical states of quite different kinds, and by implication it would also pass in future through yet other distinct states. Third, Buffon's traditions were the *human* textual records of events that provided evidence of the past condition of the earth; but in practice these were relegated to a minor position in his system, for the simple reason that he believed that human records only witnessed to the most recent phase in a far longer and largely *prehuman* sequence of changes.[15]

Buffon used his facts, monuments, and traditions to distinguish seven successive "*epochs*"; he used the word in its original sense, to mean a point in time rather than an extended period. These epochs were, as he put it on the very first page of his text, "a certain number of milestones on the eternal road of time" (see Fig. 4.14). Like any metaphor, this one had its limits; Buffon did not think his epochs were spaced like milestones at equal intervals. But they did mark important points along a continuum of temporal change, and they could be numbered in the correct sequence. By describing them in turn, or rather, by reconstructing what the earth must have been like at each successive milestone, Buffon claimed to show how it had *developed* continuously through an unrepeated sequence of distinct physical states. Unlike the steady-state picture offered in his first system, but like the Big Bang theorizing of modern cosmologists, it was a *directional* model of temporal change. Unlike his first geotheory, his second claimed that the earth had *not* always been much the same kind of place.[16]

At the first epoch, the earth had been a body composed of extremely hot fluid; in consequence of its rotation it had acquired the oblate spheroid shape that it has

12. See Stengers, "Buffon et la Sorbonne" (1974); Roger, *Buffon philosophe* (1989), 249–54. The episode illustrates the fallacy of lumping all theological responses together: even in a state with an absolutist monarchy and a supposedly monolithic Roman Catholic Church, there were in fact diverse reactions to Buffon in Paris, even among the clergy, just as there had been over a century earlier to Galileo in Rome.

13. Buffon, "Époques de la nature" (1778), reprinted in an outstanding critical edition, Roger, "Buffon: Époques" (1962); Gohau, *Buffon: Époques* (1998), reprints the text in a more compact form. There was no contemporary English translation; even more surprisingly, in view of the immense historical importance of the work, there is still no modern one either. For a summary interpretation, see Roger, *Buffon philosophe* (1989), chap. 23, and "Buffon et l'introduction de l'histoire" (1992); also the introduction to Gohau's edition, and Taylor, "Geology during Buffon's later years" (1992).

14. Leibniz, *Protogaea* (1749), had been published during his lifetime only in summary form; Cohen, "Leibniz's *Protogaea*" (1996), and Hamm, "Knowledge from underground" (1997), rightly emphasize how his geotheory was rooted in his experience of mines.

15. The global validity of Buffon's facts had become apparent from the great voyages and expeditions mounted by European powers in the course of the eighteenth century. The "figure of the earth" depended on exact geodetic measurements made in Lapland (in subarctic Scandinavia) and Peru. Fossils had been collected high up in many mountain regions around the globe, but not for example on the highest peaks in the Andes, a fact that Buffon duly built into his system. On his choice of the word "monuments", see §4.1.

16. On the origins and significance of the language of "epochs", see Taylor, "Buffon, Desmarest, and *époques*" (2001), and §4.1 below. The term "directional" was introduced in this context (primarily to define the view of Lyell's critics at a later period) in Rudwick, "Uniformity and progression" (1971).

retained ever since. As before, Buffon suggested that like the other planets it had originated as a result of a glancing blow by a comet against the sun (he was unaware of, or ignored, the more recent evidence that comets were far too insubstantial to have had any such effect). But this causal explanation of the earth's origin now formed an essential prelude to his account of its subsequent development: the two components were completely integrated.

By the second epoch the earth had cooled enough to be a solid body, or at least to have a solid crust. But it had not solidified into a perfectly smooth spheroid; its primitive surface had had irregularities—minor in relation to the size of the earth as a whole—the relics of which were the mountain ranges of "vitrifiable" Primary rocks such as granite. In effect, Buffon adopted the standard geognostic classification (§2.3), while siding with those who believed that most of the Primaries were relics of the earth's original or "primitive" state, and of "igneous" origin rather than being ancient precipitates or sediments.

Buffon's third epoch was marked by several major changes. Further cooling, he suggested, had caused the massive condensation of water from the original thick atmosphere. The resulting torrential rains had partly eroded the "primitive" mountains, depositing material in the newly formed oceans, thus accumulating the detrital Secondary rocks. In these early oceans, still warm with original heat, life was generated spontaneously, not only in small or primitive forms but also as large and complex organisms such as the ammonites that were abundant in some Secondary rocks (see Fig. 5.2). Indeed, the limestones that were so prominent among the Secondary formations were, in Buffon's view, all the products of living organisms; the lack of fossils on the highest mountains was taken as proof that the primitive oceans had never been quite universal.[17]

At the fourth epoch the oceans shrank more or less to their present bounds, leaving all these Secondary formations on the newly exposed dry land of the continents. The ocean waters drained away into subterranean cavities, sculpting the valleys as they did so; the Secondary formations remained in their original horizontal positions, unless subsequently tilted by local crustal collapse. At the same time, the first volcanoes were active; Buffon argued that the chemical reactions responsible for them required the proximity of water, so that active volcanoes were necessarily confined to the margins of the land. This explained why the known extinct volcanoes, for example those in central France, now lay inland, whereas those still active, such as Vesuvius and Etna, were near the sea. And with the appearance of dry land the first terrestrial organisms were generated, taking advantage of the new habitats.[18]

At the fifth epoch, despite further cooling, the continuing relative warmth of the whole earth was indicated by what Buffon took to be clear evidence that tropical terrestrial animals such as elephants and rhinoceros were living at high latitudes, in what are now subarctic climates. Their fossil remains were found close to the surface, obviously more recent in origin than the marine fossils embedded in the Secondary rocks. It was only the subsequent cooling of the globe that had caused ice sheets and a frigid climate to encroach fairly recently on the areas in which these animals had formerly lived (Fig. 3.3).[19]

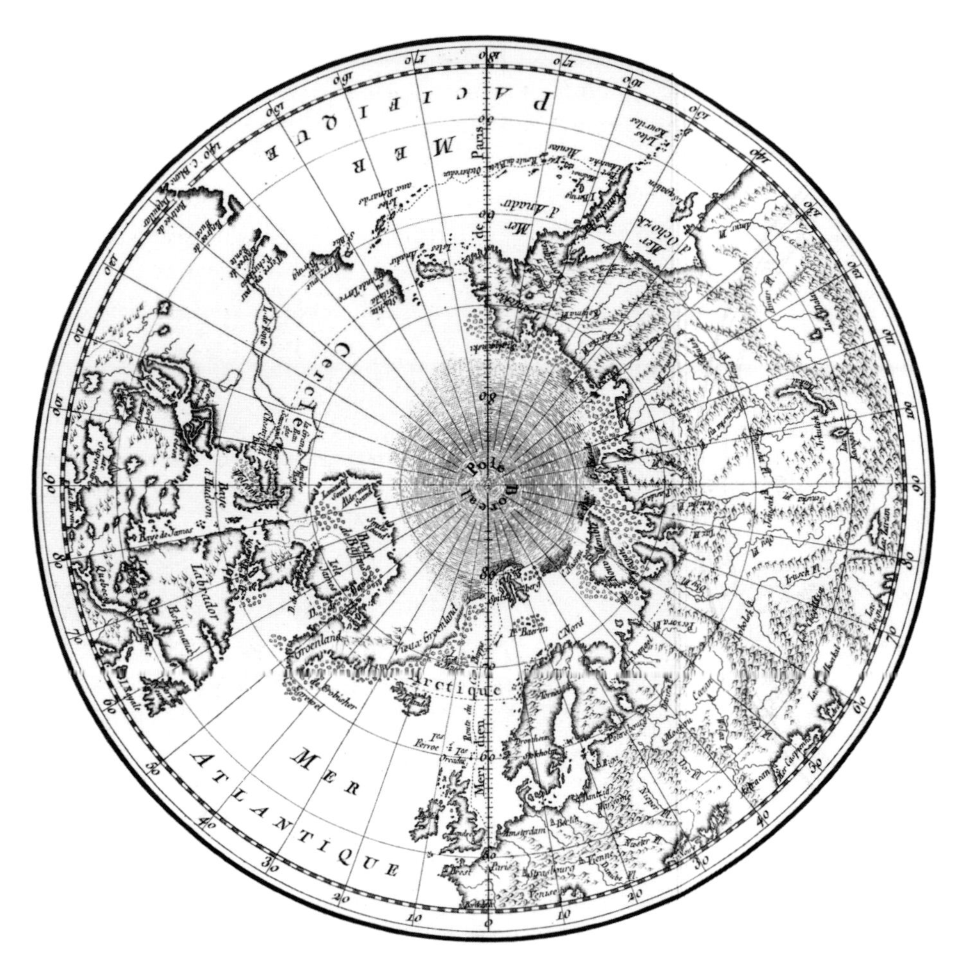

Fig. 3.3. Buffon's map of the earth's northern latitudes, in polar projection, from *Nature's Epochs* (1778), illustrating his theory of the progressive cooling of the globe. The vast areas of the polar ice-sheets indicated the encroaching arctic regime: "those that have already enveloped the vicinity of [Novaya] Zemlya", for example, were just one tongue of the huge polar ice cap, "throwing a shadow on this part of the earth for ever lost to us". The map also served to locate the finds of fossil remains of apparently tropical mammals at high latitudes, for example in eastern Siberia, near the mouth of the River Lena on the Arctic Ocean: they proved that those regions had formerly enjoyed far warmer climates than at present. As the map shows, the geography of Siberia was much better known at this time than that of North America, as a result of extensive explorations funded by the Russian imperial government. The map extends down to the latitude of 45°, which allows it to include Paris. (By permission of the Syndics of Cambridge University Library)

17. The "generation" of life from nonliving materials, by *some* kind of natural process, was a commonplace of Enlightenment theorizing. Buffon, with his well-developed theory of "*molécules organiques*" and "*moules intérieurs*", was a prominent contributor to these debates; but his ideas in this field have only tenuous links with geotheory, except analogically, and lie beyond the scope of this book. See the classic work of Roger, *Sciences de la vie* (1963), and more briefly in *Buffon philosophe* (1989), esp. chaps. 9, 10; also Sloan, "Organic molecules revisited" (1992), Hodge, "Two cosmogonies" (1992), and essays by other Buffon scholars in Gayon, *Buffon 88* (1992), part 5.

18. For Buffon, the heat of volcanoes had no connection with the earth's internal heat: see Taylor, "Geology during Buffon's later years" (1992).

19. Fig. 3.3 is reproduced from Buffon, "Époques de la nature" (1778) [Cambridge-UL: MG.7.20], pl. opp. 615; quotation on 603–4. The matching map (not reproduced here) showed the Antarctic, likewise apparently enveloped in a vast ice cap, but in fact known at this time from little more than its encircling zone of dangerous icebergs. Cook's famous second voyage (1772–75) had just proved conclusively that no *temperate* southern continent existed.

Buffon claimed that this explanation was far more satisfactory than to suppose that the animals' carcasses had been swept there from some tropical region, or that the earth's axis of rotation had changed. He envisaged new species being formed near the poles, migrating slowly and in successive waves to lower latitudes as the earth cooled further. Almost all these species, he argued, were still alive; the one possible exception, the "Ohio animal" (§5.3), might have been adapted to a climate warmer than any now found on earth. Since the relevant fossil bones were found in both the Old and New Worlds, Buffon inferred that the continents had still been joined at this time (Fig. 3.4).[20]

Buffon's sixth epoch marked the point at which further crustal collapse separated the Old and New Worlds, with the Atlantic appearing between them. Though his readers could not have known it, he had intended to place the appearance of human beings—also found on both sides of the Atlantic—*before* this event, back at the fifth epoch, making them the contemporaries of the probably extinct Ohio animal. But at the eleventh hour, as it were, while the volume was in press, he distinguished a final epoch, subsequent to the sixth, to mark the arrival of the human species. Whatever the reasons for this change of mind, or at least of strategy, Buffon's seventh epoch certainly had the effect of emphasizing unambiguously the *prehuman* character of the earth at *all* the preceding epochs. The whole of human history was confined to the most recent portion of a far longer temporal sequence. Buffon claimed that since this last metaphorical milestone there had been few further changes of any significance in the physical world: as an exception, the ancient story of the lost land of Atlantis might be a faint memory of a last episode of crustal

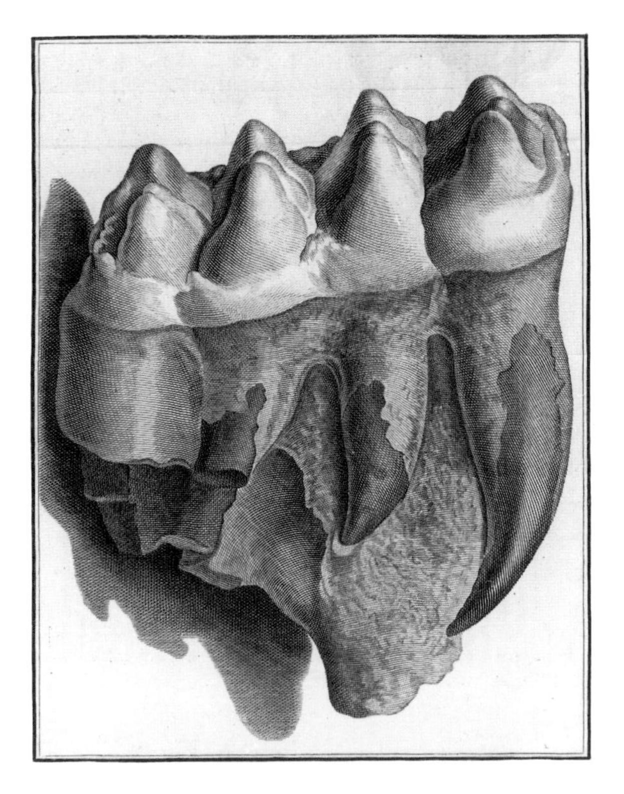

Fig. 3.4. Buffon's illustration of a large fossil molar tooth found in Little Tartary (the region west of the Urals); an engraving from *Nature's Epochs* (1778). He had become convinced—contrary to his own earlier opinion—that the similar teeth, bones and tusks found on the banks of the Ohio River in North America all belonged to a single species, unknown alive; this Russian specimen proved that the "Ohio animal" had lived in the Old World too. Buffon suggested that it might have been adapted to a hypertropical climate that no longer existed on the slowly cooling earth, and that it—perhaps alone among terrestrial species—was truly extinct (see §5.3): "everything leads us to believe that this ancient species, which must be regarded as the first and largest of all terrestrial animals, existed only in the earliest times and has not reached ours." (By permission of the Syndics of Cambridge University Library)

collapse. In general the earth had subsided into a state of relative repose, making it a suitable environment for human life.

But not forever: the physics of Buffon's system required that eventually, with further diminution of the earth's internal store of heat, the inexorable refrigeration shown by the present ice caps would extend over the whole globe and all life would be extinguished. This necessary extension of the system into the future was expressed most clearly in a brief essay in one of the final volumes of his *Natural History*. Published when Buffon was almost eighty (and only a year before Saussure climbed Mont Blanc), his comments were linked poignantly to his sense that his own life was coming to an end. He foresaw that his successors would draw out the significance of fossils, as relics of the distant past, in ways that he himself had barely glimpsed. But he believed that their conclusion about the future would necessarily be the same as his, namely that even "the [further] diminution of the waters, combined with the multiplication of organisms, will be able to retard by only a few thousand years the enveloping of the whole earth by ice and the death of [living] Nature from cold."[21]

The earth's timescale

In his earlier system, Buffon had avoided quantifying the timescale. For example, he gave no date for the formation of the earth; apart from that singular "accident", a timescale was almost irrelevant to an earth in dynamic equilibrium. In *Nature's Epochs*, by contrast, the unrepeated sequence of changes generated by the earth's continuous cooling demanded some kind of quantification, and Buffon obliged. As already mentioned, he extrapolated from experimental results to reach figures that dwarfed the traditional timescale, while privately suspecting that even these were quite inadequate (§2.5). However, Buffon did not confine such calculations to the earth. He extrapolated them to all the other planets, and even to their satellites, assigning to each the dates—in years since the formation of the solar system—at which they had reached or would reach specific temperatures during the inexorable cooling (see Fig. 2.33). Furthermore, he even gave figures for the time at which *life* had first appeared on each body (or would appear in due course), and for the time at which it would cease (or had already ceased) to exist. For he took it for granted that life would appear spontaneously on each and every body, as soon as its surface became cool enough, and would then continue there until it became too cold. Nothing could indicate more clearly the status of Buffon's geotheory as a *model* that would be equally applicable to an earth-like body wherever and whenever in the universe it might exist. Every such body would necessarily pass through the same sequence of stages in its development, at least in broad outline, just as every

20. Fig. 3.4 is reproduced from Buffon, "Époques de la nature" (1778) [Cambridge-UL: MG.7.20], pl. 1; quotation on 512. Surprisingly, this and other engravings of fossil specimens are not reproduced in Roger's otherwise comprehensive edition, "Buffon: Époques" (1962).

21. Buffon, "Pétrifications et fossiles" (1786), 173; see Roger, *Buffon philosophe* (1989), 565–69. Buffon thought that a reduced area of ocean and the heat generated by an expanding biosphere would both contribute to the earth's heat budget, but not enough to halt the net cooling.

COMMENCEMENT, FIN & DURÉE *de l'existence de la* NATURE ORGANISÉE *dans chaque* PLANÈTE.

	COMMENCEMENT.	FIN.	DURÉE absolue.	DURÉE à dater de ce jour.
Ve Satellite de Saturne.	5161 de la formation des Planètes.	47558 de la formation des Planètes.	42389 ans	0 ans
LA LUNE.....	7890.......	72514.......	64624...	0.
MARS........	13685.......	60326.......	56641.-.	0.
IVe Satellite de Saturne.	18399.......	76525.......	58126...	1693 ans
IVe Satellite de Jupiter.	23730.......	98696.......	74966...	23864.
MERCURE....	26053.......	187765.......	161712...	112933.
LA TERRE...	35983.......	168123.......	132140...	93291.
IIIe Satellite de Saturne.	37672.......	156658.......	118986...	81826.
IIe Satellite de Saturne.	40373.......	167928.......	127655...	93096.
Ier Satellite de Saturne.	42021	174784.......	132763...	99952.
VÉNUS.......	44067.......	228540.......	184473...	153708.
Anneau de Saturne.	56396.......	177568.......	121172...	102736.
IIIe Satellite de Jupiter.	59483.......	247401.......	187918...	172569.
SATURNE....	62906.......	262020.......	199114...	187188.
IIe Satellite de Jupiter.	64496.......	271098.......	206602...	196266.
Ier Satellite de Jupiter.	74724.......	311973.......	237249...	237141.
JUPITER.....	115623.......	483121.......	367498...	

Fig. 3.5. Buffon's table of figures for the dates (since the formation of the solar system, 74,832 years ago) at which life began and would end on each of the planets and their satellites (columns 1 and 2), its total duration (column 3), and the time still to run (column 4), all based on calculations of their inferred rates of cooling in relation to their size; the bodies are listed in the order of the putative commencement of life on each. No distinction was drawn between events in the past and the future, since all were governed by the ahistorical physical "laws" of cooling bodies. As in the table reproduced in Fig. 2.33, the precision of the figures does not imply that Buffon was unaware of the likely margins of error in his calculations. Both tables were published by Buffon in 1775 in a paper that prepared the ground for *Nature's Epochs.* (By permission of the Syndics of Cambridge University Library)

individual organism passes through the same sequence in its development from embryo to adult, from birth to death. That was the relevant analogy (Fig. 3.5).[22]

This bare summary of *Nature's Epochs* and other work related to it—which does scant justice to the subtlety of Buffon's argument or the diversity of his supporting evidence—does at least serve to illustrate its character as a piece of hypothetico-deductive geotheorizing. It was in effect a model of how the earth *must* have developed, if indeed it had originated as an intensely hot fragment of solar matter, cooling thereafter as it spun through space. It offered a plausible account, at least in broad outline, of all the successive phases through which the earth must have passed. It purported to explain the general features of the earth—continents and oceans, mountains and valleys, marine and terrestrial life—and their general place in the temporal sequence. More specific features, such as the former climate of Siberia or the separation of the New World from the Old, were brought in only as examples of more general processes. Even in relative details the method was unashamedly deductive. For example, Buffon argued that within the span of time represented by the Secondary formations, the deposition of shales must necessarily have preceded that of limestones, and should therefore underlie them; and he then reported observations that this was indeed the order in which such formations were found (he was apparently unaware of a conspicuous contrary case, not far from his country estate, that he could have found with a little fieldwork).[23]

Buffon's second geotheory, with its strongly directional picture of the earth's development, avoided the suspicion of eternalism that had hung about the first; but instead its explicitly vast spans of time invited comparison with the traditional short timescale of the world. Buffon simply adopted one of the standard solutions to this apparent problem: citing an earlier Benedictine scholar to support him, he claimed that the "days" of the Creation story in Genesis were not to be taken literally, because that ancient text had been adapted to the understanding of the ordinary people to whom it was originally addressed, not to savants in the age of Enlightenment. He maintained that his sequence of epochs was broadly compatible with the events of the successive "days" of Creation, and indeed his delineation of seven epochs was bound to suggest a concordance with the Genesis story, if not a sly parody of it. His eleventh-hour formulation of a seventh epoch for the arrival of human beings was probably intended to enhance the parallel, and to keep human origins clearly and safely separate from the animal realm; yet it cannot have escaped his readers' notice that its effect was to put Man into the climactic position, which in the biblical account marked the Sabbath completion of the creative activity of God.

As for Noah's Flood—which of course had to be placed still later than the seventh epoch—Buffon claimed disingenuously that since it was acknowledged to have been a miracle it was futile to expect it to have left any physical trace, and he consistently declined to attribute any observable features to its action: diluvial theorizing, at least in its classic form, was eliminated altogether. Despite all this, Buffon's own religious position remained ambiguous. Although he had marginalized the role of divine action in nature, he was—like most other leading philosophes—probably a deist rather than an atheist; yet in terms of religious practice he apparently regarded himself, to the end, and with whatever reservations, as a Catholic believer.[24]

Theological reaction to Buffon's work paralleled what it had been thirty years earlier (§2.5), but in an even lower key. There was again some criticism of his cosmology from the Sorbonne, and, this time, of his interpretation of the Genesis story. But he was a more powerful figure than ever, while the Paris theologians were becoming culturally marginal; their action had no impact on the savant world and was promptly forgotten. Even some of his clerical critics focused on the scientific flaws in his system rather than its religious implications. In fact, Buffon's geotheory was most widely faulted—not least by other naturalists—as a mere romance or "novel" [*roman*], or in modern terms as a piece of science fiction: entertaining as a speculation, but of little value when set beside the solid work of field naturalists such as Saussure. Saussure himself reported in just such unflattering terms the opinion of other savants in Paris:

> They do justice to the beauty of his style, but they think nothing of him as a man of science: they look on him neither as a physicist, nor a geometrician, nor a naturalist. His observations they account very inexact and his systems visionary.[25]

Conclusion

Buffon offered two distinct geotheoretical "systems" in the course of his long life. Both exemplified the genre in their broad explanatory scope and pervasive naturalism; both were based on the ahistorical "laws" of nature and therefore covered past,

22. Fig. 3.5 is reproduced from Buffon, "Refroidissement de la terre" (1775) [Cambridge-UL: MG.7.17], 513. For Buffon, it was in fact much more than an analogy: see Hodge, "Two cosmogonies" (1992).

23. Buffon, "Époques de la nature" (1778), 101–03; see Gohau, "Géologie historique" (1988), and Sloan, "L'Hypothétisme de Buffon" (1992).

24. Buffon, "Époques de la nature" (1778), 28–39; see Roger, *Buffon philosophe* (1989), 531–32, 559–60.

25. Quoted in translation from an undated letter to an unidentified correspondent, in Freshfield, *Life of Saussure* (1920), 93.

present, and future on equal terms; and both were immensely influential on Buffon's contemporaries. Yet they were also strongly contrasted in substance.

Buffon's first geotheory portrayed the earth as being in a steady state of dynamic equilibrium: processes of erosion and deposition were continuously altering the geography of continents and oceans, but without any overall directionality. All the processes involved were actualistic, the same as those observably at work in the present world; the earth had always been much the same kind of place and would continue to be so, indefinitely into the future. However, this potentially eternalistic model was combined somewhat awkwardly with one unique and unparalleled past event. The earth was not eternal, for it had originated at a finite point in past time, when it and the other planets had condensed from a plume of matter drawn from the sun by a passing comet.

In Buffon's second geotheory, that conjectural past event became the key to all that had followed. The earth had originated as a mass of incandescent matter in space; it had condensed into a spinning spheroid, which thereafter had cooled slowly to its present state and would continue to cool in the future. This process had taken the earth through a directional sequence of unique and unrepeated "epochs" or decisive moments, including the spontaneous origin of life and, in the relatively recent past, the first appearance of the human species; but in the distant future the earth would become too cold to support life in any form. This second geotheory was no less naturalistic than the first, but it portrayed an earth that had *not* always been the same kind of place. Buffon claimed that for most of the time since its origin the earth had been without human life. For the first time, this made explicit the possibility that the past, immensely lengthy and yet finite, had been largely *prehuman*. Superficially, this second geotheory might seem to anticipate modern reconstructions of geohistory. But in fact it was profoundly ahistorical, for it postulated a series of changes that had in effect been programmed into the system from the start, and that could be extended into the future with the same degree of confidence (see §4.5).

Buffon's models for the earth's temporal development were highly conjectural and could easily be dismissed as no better than a form of science fiction. Yet although most of their details were later abandoned, both of Buffon's geotheories were to remain powerful and fruitful exemplars for the future. The next section introduces another and equally influential geotheory formulated around the same time as Buffon's second one, but making much more direct connections to biblical history.

3.3 DE LUC'S WORLDS ANCIENT AND MODERN

The "Christian philosophe"

Around the time of Saussure's ascent of Mont Blanc, another geotheory, which had been published at almost the same time as Buffon's model of a cooling earth, was considered equally important, not least by savants who disagreed with it. Compared to either of Buffon's "systems"—and indeed to Hutton's (§3.4)—it has been grossly neglected by historians. The reason for this is no mystery. De Luc's system has been ridiculed and dismissed because he admitted, indeed emphasized, that his geotheory was an integral part of a Christian cosmology that he set against the deism or atheism of other Enlightenment philosophes. But he was not an intellectual lightweight, nor was he a biblical literalist; he deserves to be treated as seriously by modern historians—even if they do not share his religious beliefs—as he was by his contemporaries.[26]

Like Saussure, de Luc was a citizen of Geneva, and he proudly used that title to describe himself, even after he had settled in England and become Queen Charlotte's "reader" or intellectual mentor at Windsor and a Fellow of the Royal Society in London (§1.2). He had arrived in England in 1773 with a solid reputation as a savant, particularly in the field of meteorology. Coming from a Genevan clock-making family, he had a background in precision engineering; he was well known for his design of an accurate portable barometer, which was widely used to determine altitude by, for example, Saussure in the Alps and by Hamilton around Vesuvius and on Etna. But like his father and brother he was also known for his defense of religious belief against the corrosive influence of the "unbelievers" [*incrédules*] among the philosophes. The elderly Voltaire, one of whose homes was just outside Geneva, was perhaps the most influential of all and could not be ignored. Years earlier, when Rousseau had visited Geneva to be formally received back into the Reformed Church, de Luc had tried to persuade him that Buffon was an unreliable guide in matters concerning the earth, that Moses should not be dismissed as a mere spinner of fables, and that Genesis and the rest of the Bible should be treated as a genuinely divine revelation. After appointing de Luc, Charlotte had described him with satisfaction as a "proper philosopher" [*philosophe comme il faut*], on the grounds that he, unlike so many others, was not a skeptic in religion; he called himself simply a "Christian philosopher" [*philosophe Chrétien*].[27]

In 1778, shortly before Buffon published *Nature's Epochs* and Saussure the first volume of *Alpine Travels*, de Luc brought out the opening installment of his own multivolume work on geotheory; it was here that he hesitantly suggested "geology" as an appropriate word for what he was doing (§3.1). Published in the Netherlands, one of the main centers of the scholarly book trade, and of course in French, his work was ensured a wide distribution throughout Europe; an edition in German soon extended its readership still further. Like many other works by eighteenth-century savants, he presented his ideas in the form of discursive formal "letters", addressed in his case to his patron the queen. By the end of his seven volumes, he had

26. Gillispie, *Genesis and geology* (1951), 56–66, remains after half a century an influential example of the usual treatment of de Luc, patronizing and dismissive in tone. Ellenberger and Gohau, "Aurore de la stratigraphie" (1981), is an honorable exception, which Rudwick, "De Luc and nature's chronology" (2001), tries to follow. Rupke, "Gillispie's *Genesis and geology*" (1994), gives a balanced evaluation of a work that certainly had other more positive features.

27. De Luc, *Savans incrédules* (1762), is by the father, Jacques-François; de Luc, *Recherches sur l'atmosphere* (1772), and all further references below, are by or about the son Jean-André, whose brother Guillaume-François remained in Geneva and was a less important writer. Tunbridge, "Jean André de Luc" (1971), contains much biographical information (Charlotte's remark, in a letter of 1774, is quoted on 18); Hobson, "Causality in the *Inégalité*" (1992), 238–45, discusses his close relation to Rousseau. See also Archinard, "De Luc et la recherche barométrique" (1975), and Feldman, "Late Enlightenment meteorology" (1990).

printed no fewer than 150 of them, mostly written in the course of his travels in the Alps, the Low Countries, and Germany during the previous years. Unlike Buffon's "systems", de Luc's was not based primarily on printed sources but on extensive and prolonged fieldwork. However, the format he adopted did not encourage systematic or concise exposition. De Luc's readers had to wade through his voluminous and often verbose letters before they found, near the end, any summary of his ideas. Nor were they assisted by any illustrations: de Luc never learned that a picture might be worth thousands of words, and he provided none at all. Nonetheless, although he reversed the usual order and presented his supporting evidence before his explanatory model, much of de Luc's geotheory was as hypothetical as any other.[28]

De Luc entitled his volumes *Physical and Moral Letters*; they were to deal with sciences both natural [*physique*] and human [*morale*], and indeed with still wider metaphysical and religious issues. His first volume reported more on the social conditions and way of life of the Alpine peoples than on the "physics" of the mountain regions he had traversed (acting as companion to one of Charlotte's courtiers, who was traveling for the sake of her health). He did, however, describe how marine fossils were found over seven thousand feet up in the Bernese Oberland, telling the queen that this phenomenon was the main "apple of discord between savants", although the reason was left unexplained. In fact, after printing only fourteen letters de Luc suspended the work in its initial form, issued them in a slim volume, and resumed the following year with a modified title that omitted any mention of mountains. He claimed he had become convinced that certain lowland phenomena were even more significant for his purpose than any in the Alps. But although that was true there was probably a further reason for his change of tack. While he was in The Hague seeing his own work through the press he had read the prospectus for Saussure's *Alpine Travels*, and he may well have realized that his fellow Genevan's far more extensive research in the Alps would upstage his own brief and modest efforts there.[29]

Anyway, de Luc's remaining six volumes set out his geotheory in a substantial form that other savants could not well ignore. In fact, the first two volumes contained a classification and extensive review of the great variety of "systems" already put forward by others. This shows that de Luc was widely read and well informed about the current state of the genre; he later obtained Buffon's newly published *Nature's Epochs* just in time to evaluate it in some of the last of his letters.[30]

Like Buffon's ideas about the earth, de Luc's were set in a very broad context. Before resuming his letters, he printed fourteen introductory essays on a variety of topics, ranging from the basic properties of matter to final causes, human nature, and the grounds for tolerance in civil society; Helvetius, Hartley, and Priestley were among the prominent savants whom he tackled, particularly for what he saw as their materialism or their deism. Against such errors he juxtaposed his own explicit theism: human society and the moral values that underpinned it were in his opinion intimately related to the reality of divine revelation, mediated through human history. Nothing less than ultimate human happiness was at stake. Therefore, he explained, he was writing about the earth not only for savants but also for more general readers (of whom of course the queen was an eminent example): as experts, savants "have the first right to be the judges", he emphasized, using the language of the courtroom, "but their sentence on the matter concerns the whole of humanity so much that everyone ought to know the evidence in the case [*pièces du procès*]."[31]

De Luc claimed that the physical world confirmed the authenticity of certain historical claims, which in turn pointed to the reliable authority of revelation; so *physique* impinged necessarily on *morale*. Specifically, he set out to show that the biblical account of the relatively recent origin *of humanity* and the still more recent natural catastrophe of the Flood were supported by extensive physical evidence; the human species was not eternal, and human civilization in its present form was not extremely ancient, as many of his opponents claimed. The argument concerned the truth status of revealed theology; the usual form of natural theology, with its "argument from design" (§1.4), was not involved. Nor—to repeat the point—was biblical literalism at stake: de Luc was no fundamentalist in the modern mode.[32]

De Luc was well aware that to mention Genesis at all in a "philosophical" or scientific work was to invite a kneejerk reaction from many other savants. Far from expressing a view that was triumphantly dominant in his culture (as often portrayed by modern historical myth making), de Luc as a self-consciously Christian philosophe regarded himself as one of an embattled minority, indeed as part of a minority within a minority. He noted that even among his fellow theists—both Christian and Jewish—many now dismissed the early part of Genesis as unintelligible, while of course the skeptics treated it with open derision. He himself considered that it was often defended as weakly as it was attacked, and he promised he would criticize both camps equally. He claimed that "Genesis, the first of our sacred books, contains a true history of the world"; but he added immediately, "that is to say, the study of the earth shows us its broadest features, and does not contradict any of them." It remained to explain which were the "broadest features" of Genesis that the sciences of the earth confirmed, and which were the details that were not to be taken literally.[33]

28. The daunting bulk of de Luc's works cannot be discounted as a further reason for his neglect by modern historians, in addition to the unfashionable incorrectness of his apologetics. His only published illustrations were some purely topographical maps and a single plate of diagrams, published in one of his last volumes on the sciences of the earth, over thirty years after his first: de Luc, *Geological travels* (1810–11).

29. De Luc, *Lettres sur les montagnes* (1778); see letter 8 from Grindelwald (121–28) and "avertissement" (225). De Luc, *Lettres physiques et morales* (1779), 1: 53, explains the change of title. De Luc, *Geological travels* (1813), 1: 1, mentions Saussure's prospectus, the recollection serving retrospectively to establish his own priority. Pallas's recent and widely discussed work on the causal explanation of mountains (see §3.5) may also have deterred him from his original plan.

30. De Luc, *Lettres physiques et morales* (1779), 1 and 2: letters 15–53, on earlier geotheories; and 5 (2): letters 141–44, on Buffon's "Époques de la nature".

31. De Luc, *Lettres sur les montagnes* (1778), preface; like Buffon's "pièces justificatives", the last phrase was a *legal* term for the evidence produced in a trial in court. The *discours* are in de Luc, *Lettres physiques et morales* (1779), 1: 1–224, ccxxv–ccclxviii; see particularly discourse 2 (23–52); the Roman pagination is a huge late-stage insertion, mainly on Priestley's theological work.

32. De Luc's theological argument was carefully separated from his physical claims, in *Lettres physiques et morales* (1779), 5(2): letters 146–48. The conclusion set out the metaphysical and theological positions of "*Chrétien*" and "*Incrédule*" in parallel columns.

33. De Luc, *Lettres physiques et morales* (1779), 1: discourse 2 (23–52), quotation on 24.

De Luc's binary system

In fact, de Luc focused his scientific argument on one simple claim about the earth, namely "that *our continents are not ancient.*" Just three major phases were involved in his geotheory. There had been a time when the earth's present continents were on the sea floor; a "sudden Revolution" that saw their emergence; and a relatively short subsequent history of their human habitation: "All the phenomena of the earth, and also the history of Man, lead us to believe that the sea has changed its bed by a sudden Revolution; that the continents inhabited today are the bed that it formerly occupied; and that no great number of centuries [*siècles*] has elapsed since these new lands were abandoned by the waters."[34]

What de Luc described was in effect a reconstruction of the earth's temporal development, but the account was weighted quite differently from Buffon's *Epochs*. De Luc had nothing to say about the earth's origin, and almost nothing about the obscure times at which the Primary rocks had been formed: he explicitly refrained from indulging in what he regarded as "gratuitous hypotheses" or useless speculation about them. The "primordial rock masses [*montagnes*]" composed of puzzling rocks such as granite and gneiss were also "inexplicable rock masses", because they seemed to be "the effect of no known cause". It was only for the following phase, represented by the Secondary formations, that he felt there was enough evidence to construct some "reasonable hypotheses" about what the earth had then been like. The Secondary rocks and their fossils showed that many terrestrial features had been much the same as they are at present: the sea had had its tides and currents, and the fossil remains of plants and land animals proved there had been continents elsewhere at that time, with rivers sweeping such detritus out to sea. Like many other geotheorists at this time (§3.5), de Luc thought the global sea level might have subsided quite gradually, through a lengthy sequence of periods, uncovering ever wider land areas that would have been colonized progressively by plants and animals. He noted that "nothing indicates the duration of these distinct periods", but tacitly he treated the Secondaries as representing in total a vast span of time; his writing does not suggest that his timescale was any more constrained than, say, Buffon's.[35]

The phase represented by the Secondary rocks and their fossils had ended abruptly with the "sudden Revolution". The former continents had sunk out of sight, while the present continents had emerged from below sea level: in effect there had been an almost total interchange between continents and oceans. Like all his contemporaries, de Luc used the word "revolution" to mean simply a major change of any kind; his adjective defined this particular change as abrupt, but not necessarily as violent. He believed his Revolution was the same event as Noah's Flood, but it was far removed in character from what any literal reading of the story in Genesis suggested. It was no brief and transient inundation, but a permanent change in the earth's physical geography. For de Luc, it was enough to show that the physical evidence confirmed the basic historicity of the event: as he had put it to Charlotte at the start of the work, "we Christians have no need, in order to believe in the Flood, to know how it was caused; it is enough for us if it cannot be proved [physically] impossible, and that is very far from being the case." This was an important point: de Luc was claiming that, for any putative *past* event such as the Flood, establishing its historical reality was a legitimate and worthwhile task in its own right, and quite distinct from trying to find its natural cause.[36]

De Luc certainly did believe that his Revolution had had a natural cause of some kind: he explicitly repudiated those earlier savants who had treated the Flood as a miracle, or who resorted to supernatural causes to conjure its waters into existence and then get rid of them again. But he was content to leave the physical cause of his Revolution quite vague. Like Buffon and many others, he invoked vast subterranean caverns as a plausible possibility: "like the service rooms under palaces", as he put it, in a homely analogy that would have been readily understood by his patron at Windsor Castle. But he merely sketched the way in which a collapse of the crust into such caverns might first have caused the former continents to disappear below sea level, leaving a worldwide ocean, after which a second phase of collapse might have drained the water off the former ocean floors and left the new continents high and dry. Plants and terrestrial animals could have survived these catastrophic events if even a few small islands remained immune and above sea level: in effect a purely natural mechanism played the role of Noah's Ark. Throughout this discussion, de Luc's tone was openly hypothetical, and his explanation unashamedly ad hoc, as was customary in the genre of geotheory.[37]

Natural measures of time

Whatever its physical cause, de Luc's sudden Revolution had initiated the present world. He claimed that observation of the physical processes now active indicated the finite duration of the present state of the earth. Most of his detailed fieldwork was directed at assembling evidence that this most recent phase in the earth's development—the phase in which we live today—had been quite short, indeed coextensive with human history, or at least with reliably recorded history: "by studying the everyday effects that we see in progress," he promised, "I shall get back to the time when they must have started." It was the duration of this phase, not the age of the earth, that he tried hard to quantify, at least approximately. When he referred to the "age of the world" [*l'âge du monde*], he stated clearly that he did *not* mean the total age of the planet, but the age of the present state of things, with human civilizations on the present continents. De Luc's extensive fieldwork convinced him that the best evidence that our "world" (in this sense) was of very recent origin was to be found

34. De Luc, *Lettres physiques et morales* (1779), 1: discourse 1 (1–22); the whole quoted sentence (8–9) was italicized for emphasis. The context implies that the word *siècle* here denoted centuries, not indefinite ages. De Luc always referred to "our continents" in the plural, but his supporting evidence was confined to western Europe, the only continent he knew at first hand.

35. De Luc, *Lettres physiques et morales* (1779), 5(2): letters 137, 138.

36. De Luc, *Lettres physiques et morales* (1779), 1: letter 15, 241. A modern parallel would be the efforts of some geologists in the mid-twentieth century to establish the geohistorical reality of "continental drift", in the face of criticism from geophysicists that there was no adequate causal explanation for any such crustal displacement.

37. De Luc, *Lettres physiques et morales* (1779), 5 (2): letters 138, 139.

not among mountains, as he had perhaps expected when he started his fieldwork, but in the lowland plains. So his work was centered not in Saussure's Alps but—discreetly flattering the royal family—in George III's Hannover (and tacitly in Charlotte's native Mecklenburg further east).[38]

That the vast north German plain had formerly been on the ocean floor was proved adequately by the marine fossils found in some of the underlying rocks; this much was uncontroversial, and beyond it de Luc showed little interest in the Secondary formations. For his purposes, the heathlands [*bruyères*] of Hannover were crucial because they preserved the nearest there was to the pristine state of the continents immediately after their emergence from the ocean floor: "There are still many uncultivated lands there, which, like the teeth of a young horse, can give us some idea of the age of the world; I mean, of the date when the present surface took the form in which we know it today." The analogy—more generally familiar in de Luc's day than in ours—was apt and illuminating. A horse's teeth, progressively worn down by its grassy diet, were an infallible guide to its age (hence it was proverbially inappropriate to "look a gift horse in the mouth" to assess its age and its value before accepting it). In the same way, de Luc claimed, the thin peaty soil of the heathland was an unmistakable sign that the land had not long been exposed to the slow process [*tourbification*] by which peat accumulated from the decay of plants. The heathlands were in this respect "privileged places"; cultivated areas, in contrast, could not be used to date the emergence of the continents, because human agency had disrupted the natural process.

His conclusion was "a bold claim" but he thought it well founded. "I had been transported into the first ages of the world", he told the queen, as if he had been taken there on a time machine. It even seemed to him that the heathland itself was saying, "you have been transported very close to the start of the present world; here is the earth still completely untamed." On a later visit de Luc reinforced his conclusion with evidence drawn from the prehistoric burial mounds found in one part of the heathland. His own brief excavations showed that the peat covering the mounds was no thinner than that around them, so that these early relics of human activity could be little younger than the land itself; it suggested how the continent—as a *land* mass—was of about the same age as the totality of human history.[39]

The thickness of peat on uncultivated heathland provided only a rough measure of the time since the continents emerged. De Luc tried hard to quantify the time more precisely by studying other features of physical geography. Some of the best evidence related to the growth of new land at the mouths of rivers. Deltas such as that of the Rhine, with historical records of their growth through the centuries, indicated that the present physical regime could not be indefinitely ancient: deltas were of finite size, and had clearly begun to accumulate at some finite time in the past: "This is the true *clepsydra* of the centuries, for dating the Revolution: time's zero is fixed by the unchanging sea level; and its degrees are marked by the accumulation of the deposits of the rivers, just as they are by the piling up of sand in our ancient instruments of chronometry." Here was another apt analogy, although de Luc confused the dripping water of a primitive water clock or clepsydra with the trickling sand of an hourglass. The latter was as familiar in his day as horses' teeth, being used on board ship and in the parson's pulpit, as well as in the kitchen (where in the form of the humble egg timer it survived into the twentieth century, before being revived as an all-too-familiar icon on computer screens). The slowly accumulating deposits in a delta marked the passage of time since the continent emerged, just as reliably as the amount of sand in the lower half of an hourglass marked the time that had elapsed since it was last upended. In both cases the measure itself might be crude, but it indicated clearly that the process had started at some definite moment in the past.[40]

Many other features were cited in the course of de Luc's work, all having the same import: for example, the finite size of the screes below Alpine crags, and the fact that lakes have not yet been fully silted up and converted into alluvial plains, even if the rivers that flow into them are heavily charged with sediment (Lac Léman, at the outlet of which lay de Luc's native Geneva, was a striking example). All such features showed that the operation of present processes (or "actual causes", as de Luc later termed them) had started at a finite time, not very far back in the past. He inferred that the continents must have emerged from the ocean floor within the past few thousand years, at the dawn of human history.[41]

This was as far as de Luc took this final part of his reconstruction. It was enough to show that the continents in their present form, as landmasses and as the sites of human civilizations, were no older than was suggested by the analysis of the biblical record of early human history (§4.1). With this approximate concordance, it became clear that Moses, the putative author of Genesis (and of the rest of the Pentateuch), had not just concocted a fictional fable about the ancient Jews. The record of divine revelation that their history embodied, and above all the theology that it expressed, could therefore be relied upon for ultimate human happiness; in theological language, for salvation. To de Luc, that was what ultimately mattered most.

Conclusion

De Luc himself summarized his geotheory in some of his final letters to the queen. There had been "two very distinct Periods", separated by "the great Revolution". The first period was of unknown but probably vast duration; the Revolution had been sudden, and dramatic in causing a permanent change in the earth's geography; and the subsequent period had been relatively brief, and almost coeval with recorded human history. The earth's temporal development was thus "divided into ancient

38. De Luc, *Lettres physiques et morales* (1779), 1: 10. Herschel, who lived in Slough (on the other side of the Thames from Windsor) and was likewise an immigrant and a beneficiary of royal patronage, famously offered the king a similar tribute shortly afterwards, when he gave the name *Georgium Sidus* to the first new planet to be discovered since Antiquity (though *Uranus* was the name that stuck).

39. De Luc, *Lettres physiques et morales* (1779), 3: letter 52 (quotations on 11–12), and 5: letter 119; he did not in fact use the modern notion of a *machine* for his imaginative time travel.

40. De Luc, *Lettres physiques et morales* (1779), 5 (2): letter 139 (quotation on 497).

41. Readers who are geologists will appreciate the cogency of de Luc's reasoning. Many of the features he cited are now interpreted as products of the postglacial regime in the regions he knew at first hand, the present processes having in effect restarted there since the last retreat of the Pleistocene ice sheets some ten thousand years ago.

and modern," separated sharply by a unique event. De Luc insisted that the modern world was no older than the traditional short timescale *for humanity*; there is no evidence that he was trying to compress the overall age *of the earth* into that framework.[42]

De Luc's geotheory can be defined as strongly *binary* in character: a relatively undifferentiated "ancient" or "former world" was sharply separated by a brief and unique Revolution from a distinctly different "present world". And since the latter was distinct above all in being a *human* world, de Luc's system had the effect of sharply separating the human from the prehuman (or at least from the prehistoric). More clearly than other directional models such as Buffon's sequence of epochs, de Luc's binary system made the deep past radically different from the present: not because natural laws—or ordinary physical features such as continents and oceans, marine animals and terrestrial plants—had once been quite different, but because the former world was cut off from the present by the high barrier of a uniquely Revolutionary event, and because the present world was the *human* world.

Finally, de Luc's system has been summarized here as being a reconstruction of the earth's temporal development. But unlike Buffon's second geotheory it posited no underlying causal chain that would have necessitated the successive phases and made them in principle predictable (or retrodictable). Unlike Buffon's deterministic or programmed sequence of epochs, de Luc's geotheory was imbued with a sense of the contingency of what he himself referred to repeatedly as the earth's "*history*". What that meant will be analyzed more closely in §4.5. Meanwhile, the next section deals with a third major geotheory, devised around the same time as de Luc's and Buffon's, but not published until shortly after Saussure climbed Mont Blanc.

3.4 HUTTON'S ETERNAL EARTH MACHINE

A deistic geotheory

Around the time that Saussure climbed Mont Blanc, the attention given by savants throughout Europe to Buffon's second geotheory and to de Luc's first was matched by that given to one emanating from Scotland. Unlike de Luc's, James Hutton's geotheory has not suffered from historical neglect. On the contrary, it has received so much uncritical adulation that its place in the sciences of the earth of the late eighteenth century has been seriously distorted. Anglophone geologists have treated Hutton as their iconic "founder" or "father", with such pious veneration that his relation to his contemporaries has been obscured and misunderstood, despite a large body of fine research by modern historians. Hutton was no neglected or persecuted genius. Many of his ideas were commonplace among geotheorists, though he combined them in an unusual and original way. His system was well known at the time, and was discussed by other savants with the respect it deserved. And while most of them found it highly implausible, they rejected it on grounds quite other than those commonly supposed. Above all, however, Hutton's work has been misunderstood because it has not been treated, as it was by his contemporaries, as yet another "system" within the well-established genre of geotheory.[43]

Hutton first presented his geotheory in 1785 as a long paper to the "physical" section of the newly founded Royal Society of Edinburgh. A brief summary was printed almost at once and was probably distributed widely, certainly beyond Britain. The full paper formed the most substantial item in the inaugural volume of the society's *Transactions*, when it appeared at last in 1788 (the year after Saussure climbed Mont Blanc). The volume would have been sent to the society's sister institutions throughout the Republic of Letters, as part of the customary network of exchange; in addition, Hutton evidently sent offprints of his paper to favored colleagues and correspondents, among them de Luc. Anyway, his geotheory soon became well known and was summarized or abstracted in several periodicals, in Britain and on the Continent.[44]

The title of Hutton's paper, as recorded by the society, captured the essence of the work that was to grow eventually into three or four volumes: it was "The System of the habitable Earth with regard to its Duration and Stability". As a "system", it was placed squarely in the genre of geotheory: it was to offer a hypothetical model for how the whole earth works. It was to consider primarily the "habitable" earth, that is, those parts of the dry land that were capable of sustaining regular human life. And the linking of "duration" with "stability" hinted that Hutton would be concerned not with quantifying a timescale but rather with the earth as a body existing indefinitely in stable equilibrium.[45]

Three years later, the full printed paper had a title, "Theory of the Earth", that again proclaimed its genre unmistakably. And the subtitle stressed how it would aim to identify the ahistorical natural "laws" that produced a cyclic stability in the earth's land areas: it was to be "an investigation of the laws observable in the composition, dissolution and restoration of land upon the globe". The three linked nouns summarized what was described in the text as a set of processes that jointly ensured the enduring habitability of the earth. With its language of ends and purposes, of machines and their wise design, and of an unseen power behind the appearances,

42. De Luc, *Lettres physiques et morales* (1779), 5(2): 488, 506, and passim; throughout, de Luc used the word *ancien*, which can refer to either an "ancient" or a "former" state.

43. The tradition established by Geikie, *Founders of geology* (1897), chap. 9, and continued in Bailey, *Hutton the founder* (1967), and McIntyre and McKirdy, *Hutton the founder* (1997), still flourishes in the "historical" introductions to many modern works by geologists, and even, in attenuated form, in some works by historians, for example Dean, *James Hutton* (1992), 268–69. In addition to Dean's otherwise valuable survey, Hutton's work has been analyzed and set in historical context in Davies, *Earth in decay* (1969), chap. 6; Porter (R. S.), *Making of geology* (1977), chaps. 6–8; Laudan, *Mineralogy to geology* (1987), chap. 6; Gould, *Time's arrow* (1987), chap. 3; and in other work cited in subsequent footnotes. The following review is indebted to all these modern scholars.

44. For example, Blumenbach summarized it in 1790, with translated excerpts, in a leading German periodical (§6.1), and another brief account appeared in 1793 in *Observations sur la physique*. Desmarest, *Géographie physique* 1 (1794–95), 732–82, analyzed it at length, Hutton being one of only three recent authors whom he deemed important enough to add at a late stage to his roster of geotheorists (§6.5). Such reactions to Hutton on the Continent have been inadequately explored, owing perhaps to the linguistic limitations or chauvinistic inclinations of some anglophone historians.

45. [Hutton], *System of the earth* (1785). The relevant minutes are printed in Eyles (V. A.), "Original publication of Hutton's *Theory*" (1950), 378–79; see also Eyles's "Earliest printed version" (1955). In contemporary usage the word "habitable" referred to *human* life: Saussure's high Alpine peaks, for example, were *not* habitable, still less the vast expanses of Cook's Pacific Ocean, although the latter was of course inhabited by whales and much other nonhuman life. In Hutton's printed abstract the word was dropped from the title, but the text still took the human qualification for granted, as central to the argument.

this was no "scientific" theory in the narrow modern sense. Nor were these phrases a conventional gesture confined to the first and last pages, like the Marxist effusions that often disfigured Russian scientific works during the Soviet era: Hutton's teleological perspective pervades his writing throughout. Even his opening words referred eloquently and unambiguously to the deistic metaphysics and theology that underlay all his ideas about the earth and gave them human meaning:

> When we trace the parts of which this terrestrial system is composed, and when we view the general connection of those several parts, the whole presents a machine of a peculiar construction by which it is adapted to a certain end. We perceive a fabric, erected in wisdom, to obtain a purpose worthy of the power that is apparent in the production of it.[46]

Hutton's essay in geotheory was in fact just one part of a much more ambitious intellectual project. Soon after his paper was read he planned to enlarge it into a book, but in the event he shelved this in favor of other projects that he evidently regarded as having higher priority. By the time the first two volumes of *Theory of the Earth* appeared in 1795—no fewer than ten years after the paper had been read—Hutton had already published *Natural Philosophy* (1792), a large quarto volume of essays dealing with meteorology, the nature of "fire", and the fundamental theory of matter; a smaller but still substantial volume, *Light, Heat and Fire* (1794), criticizing Lavoisier's new "anti-phlogistic" chemistry; and above all a set of three massive quartos, *Principles of Knowledge* (1794). The length of texts and the order of their publication are not an infallible guide to their relative significance to an author; but when Hutton's books are placed side by side in chronological order there can be little doubt where his scholarly priorities lay.

Hutton's intellectual project was nothing less than to establish the grounds for rational human knowledge, following in the tradition of earlier savants such as Locke, Berkeley, and Hume. In the course of over 3,200 pages of dense prose, Hutton's *Principles of Knowledge* set out an idealist philosophy that dealt with such fundamentals as perception and conception, ideas and reason, time and space, cause and effect, matter and motion, piety and religion. His theology was openly and unmistakably deistic, and explicitly justified his pervasive teleology. He was concerned above all to demonstrate that the world showed "system" in the sense of orderly designful purpose, and that any appearance of "accident" or disorder was deceptive:

> It is thus that a system may be perceived in that which, to common observation, seems to be nothing but the disorderly accident of things; a system in which wisdom and benevolence conduct the endless order of a changing world. What a comfort to man, for whom that system was contrived, as the only living being on this earth who can perceive it.[47]

To discover "system" in this sense throughout the natural world was the underlying goal of all Hutton's writings in natural philosophy; his geotheory too is unintelligible except in the light of his deistic theology. In Hutton's view, the capacities of human thought and rationality alone gave meaning to nature; so a wisely designed world would necessarily make provision for the permanent existence of the human race, and hence for maintaining the habitability of the earth. But although Hutton believed that human existence "rises infinitely above that of the mere animal"—and he intended his work to lend credence to the immortality of the soul—the life of the mind still depended for its continuation in this world on a substrate of animal and plant life, utilized as food. And animal life too depended on plants, which in turn depended on a literal substrate of fertile soil. So the designful system of nature as a whole comprised a chain of subsystems, connecting the life of the mind to the earth itself. In this perspective, Hutton's "mineral system", the subject of his "Theory of the Earth", had a crucial though subordinate role in the wider project, and therefore justified its modest place in his life's work.[48]

Cyclic processes

More specifically, the crucial material link between human life and the earth itself was the soil. This was a theme on which Hutton had meditated long and hard, as the owner of farms near Edinburgh and as one deeply involved in agricultural improvement (at his death he left a massive work in manuscript, *Principles of Agriculture*). Unlike de Luc, with his notion of soil being gradually built up in the course of time, Hutton regarded the soil as a wasting asset, continually being washed away by rain and exhausted by the growth of plants and needing to be replenished as much by the slow breakdown of the underlying rocks as by the decay of plants. So the habitability of the land depended paradoxically on its slow disintegration. The landmasses on which plant, animal, and human life depended were slowly wasting away, although imperceptibly to human observation, and the soil and other products of erosion were being swept out to sea to be deposited on the ocean floor.[49]

The wise design of the whole system—the continuing habitability of the earth as a whole—could be ensured in the long run only if the continents thus wasting away were somehow replaced. Hutton argued that this could be achieved if the materials deposited on the ocean floor were eventually raised above sea level to form *new* continents, and that this in turn could be due to the expansive power of the massive heat that he claimed was present in the depths of the earth. But any such virgin continent would then be washed away too quickly to provide a long-term source

46. Hutton, "Theory of the earth" (1788), 209. The hoary legend of Hutton's unreadable prose has served various ideological purposes during the past two centuries. Soon after Hutton's death, Playfair, *Illustrations* (1802), used it as a reason for bowdlerising the work by detaching it from its theological framework and suppressing its teleology (see §8.4). He has been followed by countless other scientific commentators ever since. A distinguished recent example is Şengör, "Is the present the key to the past" (2001), 20–22, where Hutton's deistic metaphysics is interpreted as a mere "heuristic aid", or as "cant" that served to conceal his atheism; for a critique, see Oldroyd, "Manichaean history of geology" (2003). To recognize Hutton's unmodern metaphysics and theology does not, *pace* Şengör, detract from his scientific stature.

47. Hutton, *Principles of knowledge* (1794), 2: 239; similar sentiments are in Hutton, "Theory of the earth" (1788), 216–17. Jones (P.), "Philosophy of Hutton" (1984), is an excellent brief account of the *Principles*; Grant, "Hutton's theory" (1978), gives the best analysis of the relation between his epistemology and his geotheory; see also O'Rourke, "Hutton's *Principles* and *Theory*" (1978).

48. Hutton, *Principles* (1794), 2: 239.

49. See Jones (J.), "Hutton's agricultural research" (1985); Withers, "Georgics and geology" (1997).

for soil—and hence for plant, animal, and human life—unless the loose sediments had first been consolidated. That in turn could be achieved, he claimed, by the power of the same internal heat, melting the gravels, sands, and muds while they were buried below the sea bed and fusing them into hard resistant rocks. Then, after their elevation as a new continent, they would disintegrate slowly enough to release the essential mineral components for a fertile soil.

Hutton thus proposed a cyclic set of processes by which habitability could be ensured indefinitely. If there was indeed a wisely purposeful system to the earth—as he believed profoundly—some such cycle *must* be built into the earth's structure and function, or else its habitability would necessarily be limited to a finite period. However vast the timescale on which the present continents were wasting away, without "restoration" the land available for human life would be reduced eventually to zero. On the other hand, if there were a process for renewing the continents, enduring habitability could be guaranteed.

So Hutton claimed that the earth was a "beautiful machine", artfully designed and constructed—just like machines of human origin—in order to achieve an intended effect. Hutton did not live, as we do, surrounded by a bewildering variety of machines: he and his contemporaries understood by that word one specific device above all others, namely the steam engine. Steam engines dominated the new industrial scene, about which Hutton was well informed not only as a savant with wide interests but also through his personal stake in the chemical industry. The improved steam engine devised by James Watt was still a novelty (though Hutton knew both him and it); but the earlier, slower, and cruder Newcomen engine was in fact a more apt analogy for what Hutton had in mind (Fig. 3.6).[50]

Like the Newcomen engine, Hutton's earth machine showed in effect a slow oscillation, with new continents rising while others subsided. The analogy was not

Fig. 3.6. A Newcomen steam engine of Hutton's time, as illustrated in the then current edition of Chambers's *Cyclopedia* (1786). The alternating action of the expansion of steam in the cylinder (above the spherical boiler) and its induced condensation back into water powered a slow oscillation of the huge beam, which in turn worked a pump that extracted water from the shaft on the right (the brickwork, steps, and winch indicate the scale). The Newcomen engine, which was widely used for pumping water out of mines, was still the most noteworthy "machine" in the late eighteenth century, and the one that Hutton is likely to have had in mind as an analogue of his dynamic earth. (By permission of the Syndics of Cambridge University Library)

perfect, of course, since Hutton's continents were lowered by slow surface erosion rather than by crustal subsidence. On the other hand, the rise of the Newcomen engine's beam by the expansion of steam was a highly appropriate analogue for his notion of crustal elevation. The sheer irresistible power of steam was just what impressed all who witnessed a steam engine in operation, and it made the "machine" an equally powerful image to convey Hutton's argument for the dynamic equilibrium of the earth, based on huge unseen forces deep below the surface. Just how those forces worked in the earth was what he tried to elucidate through his physics of heat, a major topic in his other writings. But in any case he was clear that heat represented an expansive force that was in perpetual interaction with its opposite, the contractive force of gravitation. The oscillation of a Newcomen engine was an eloquent image of that dynamic equilibrium in nature.[51]

Hutton complemented his machine imagery with that of an organism: not one developing from an embryonic state, but an adult organism maintaining itself through time in a dynamic equilibrium. The elevation of new continents from the floor of the ocean, which was so essential to the design of the whole system, was its "reproductive power"; or, changing the metaphor somewhat, "this earth, like the body of an animal, is wasted at the same time as it is repaired". The circulation of the blood in the "microcosm" of the human body—the subject of Hutton's own medical dissertation at Leiden many years earlier—fitted perfectly into this metaphor of the organism, as an analogue no less appropriate than a steam engine. Likewise Hutton's meteorology, and in particular his theory of rain, was directed towards elucidating what was well recognized as another process of circulation (in modern terms, the hydrological cycle). Steady-state models of all kinds, based on the dynamic interaction of opposed entities or powers, were in fact commonplace in Enlightenment thinking about both the natural and the human world: the economic theorizing of Hutton's Edinburgh friend Adam Smith—for whom he later acted as literary executor—was just one example from the human world. So this aspect of Hutton's theorizing was no surprise to his readers. Natural theology had long emphasized the significance of systems that maintained themselves in dynamic equilibrium; Hutton was simply extending that kind of argument from the organic world to the inorganic, from the mechanisms of animals and plants to those of the earth itself.[52]

50. Fig. 3.6 is reproduced from Chambers, *Cyclopedia* (1786), 2 [Cambridge-UL: S900.bb.78.2]: Hydraulics and hydrostatics, pl. 6, fig. 71. See Jones (J.) *et al.*, "Correspondence between Hutton and Watt" (1994).

51. Lavoisier's self-styled "revolution" in chemistry—eliminating phlogiston altogether, among other changes—soon made Hutton's chemistry look obsolete, while the development of thermodynamics in the mid nineteenth century further undermined the intelligibility of his ideas, by making his physics seem to entail some kind of perpetual motion; both contributed to the later legend that his writing was obscure. On his physics of heat, light, and "fire" as manifestations of a single "solar substance" opposed to gravitation, see Heimann and McGuire, "Newtonian forces" (1971), 281–95, and Grant, "Hutton's theory" (1978); see also Gerstner, "Hutton's theory of the earth and of matter" (1968), Donovan, "Hutton, Black, and chemical theory" (1978), and Allchin, "Hutton and phlogiston" (1994).

52. Hutton, "Theory" (1788), 216; *Theory* (1795), 2: 562; see Ellenberger, "Thèse de doctorat" (1973), Roger, "Le feu et l'histoire" (1974), and Donovan and Prentiss, "Hutton's medical dissertation" (1980). The notion of circulation long predated Enlightenment thinking, but was highly congenial to it; Hutton, "Theory of rain" (1788), criticizing similar theorizing by de Luc, contributed to a long-running meteorological debate.

A theory confirmed by fieldwork

Hutton's geotheory was, if anything, even more purely deductive in structure than either of Buffon's. It proposed a highly abstract model of how the earth *must* work, if, as he insisted, it was "a thing formed by design". His system was not, of course, unsupported by evidence; but at the time he presented it, its empirical basis was largely limited to what was commonplace and uncontroversial among savants. For example, Hutton himself conceded explicitly that his knowledge of fossils—indeed his interest in them—was just that they were widespread on land, that they were mostly the remains of marine organisms, but that there was also fossil wood and other remains of land plants. That much was all he needed to know in order to claim—like many other savants—that the present continents had once been on the sea floor, but that even at that remote time there had also been continents in existence elsewhere. He had collected fossils and mineral specimens in the field and observed various features of physical geography, both around Edinburgh and in the course of earlier travels in England and the Low Countries. He was also well read: his paper cited recent articles in periodicals such as *Observations sur la Physique*, and the full version quoted pages on end—in French—from Saussure's *Alpine Travels*, for example, and criticized other geotheorists such as Burnet, Buffon, and de Luc. But Hutton evidently felt that he knew quite enough of the relevant "facts", from publications if not at first hand, to make his system public *before* undertaking any fieldwork specifically directed towards finding empirical support for it. Like Buffon with his *preuves*, Hutton clearly considered that his lengthy "proofs and illustrations"—the subtitle of his full *Theory of the Earth*—could best be presented separately from the model itself, and in his case long afterwards. Indeed, in the full publication he simply reprinted his paper as the opening chapter, unmodified by anything he had observed subsequently.[53]

In that original paper, the empirical material that Hutton discussed in detail was limited to the one crucial part of his argument that was *not* generally agreed. What most startled other savants (and ought to startle modern geologists too) was not his assumption of an indefinitely vast timescale for the earth, but his claim that stratified rocks—those that others called Secondaries as well as the Primaries—had all been more or less completely melted or fused while buried on the ocean floor. The *only* pictorial evidence that he chose to present, in illustration of his entire geotheory, consisted of specimens that he claimed would support this surprising interpretation. His two large and expensive plates of engravings—probably the allowance given to his paper by those in charge of the society's finances—portrayed "septarian" nodules and "graphic" granite, from the Secondary and Primary rock masses respectively, both interpreted as the products of fusion (Fig. 3.7).[54]

As soon as Hutton had launched his geotheory in Edinburgh, he set out on fieldwork designed explicitly to find evidence to support it: one of the chapters reporting his results was entitled unambiguously "The theory confirmed from observations made on purpose to elucidate the subject". Unlike physical geographers, he did not try to describe whatever he found worthy of record within a chosen region; instead

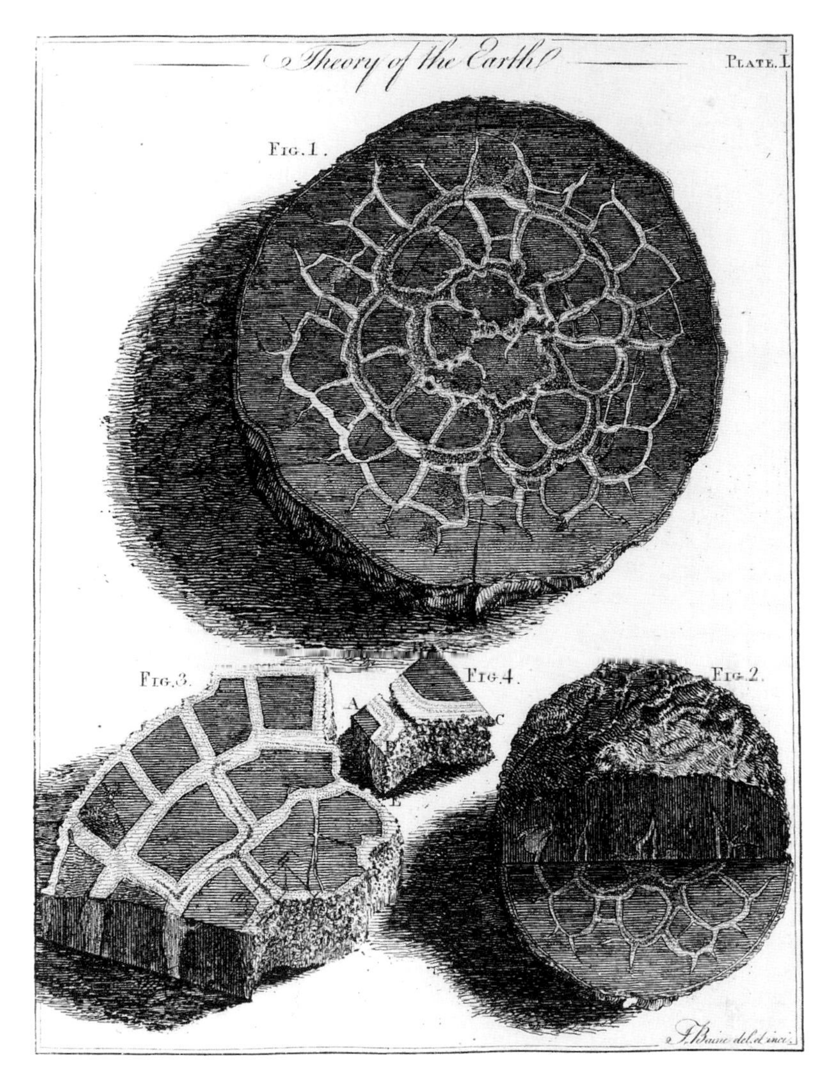

Fig. 3.7. Hutton's specimens of "iron-stone septaria" from the Coal formation near Edinburgh, as engraved for the original version of his "Theory of the earth" (1788). The flattened nodules have been sliced and polished along the plane in which they lay in the strata (above), and also at right angles (below right), in order to show the three-dimensional form of the characteristic cracks ("septa") filled with calcite ("spar"). Hutton argued that the cracks must have opened and been filled simultaneously, by a differentiation of the original mineral substance as it crystallized from a melted state in the depths of the earth, rather than being opened by shrinkage and filled by precipitation from percolating aqueous solutions. His only other pictures were of polished specimens of "graphic" granite showing interlocking crystals, which he claimed must have crystallized likewise from a melt, rather than being precipitated from any kind of fluid medium. (By permission of the Syndics of Cambridge University Library)

53. Hutton, *Theory of the earth* (1795), 1; in the subtitle, "illustrations" denoted examples expressed in prose, not pictures. Carozzi (M.), "Saussure: Hutton's obsession" (2000), translates the quotations and reprints Hutton's comments on them.

54. Fig. 3.7 is reproduced from Hutton, "Theory of the earth" (1788) [Cambridge-UL: P340:1.a.1.1], pl. 1, explained on 246–47; the granite specimens in pl. 2 are explained on 255–57. Both sets of images were reused (and this plate reengraved) in the two-volume *Theory of the earth* (1795), and were still prominent, there being only six plates in all. A septarian nodule also featured prominently in the famous portrait of Hutton by Henry Raeburn, painted about 1790 and now in Edinburgh-NGS: see Thomson, *Raeburn* (1997), no. 10. See Roger, "Le feu et l'histoire" (1974), and Laudan, "Problem of consolidation" (1977–78). To put Hutton's consolidation theory in modern terms, as nearly as it can be, he claimed in effect that *all* rocks, including for example the Chalk and its flints, are highly *metamorphosed* as a result of intense heating.

he searched all over Scotland for decisive features that his system led him to expect. In short, he was trying to verify predictions *deduced* from his hypothetical model.[55]

Two kinds of evidence were important to make his case persuasive. The first was to find more direct evidence of the fluid origin of rocks such as granite and basalt, as tokens of the elevating power of the earth's internal heat. Granite, as the most fundamental of all rock masses (§2.3), was crucial. He admitted that at the time he wrote his paper he had scarcely seen any in situ, so when he started his fieldwork the junctions between granites and other Primary rocks became his first priority. In three successive summers he duly found veins of granite penetrating what he called "schistus", first in the Highlands (in Glen Tilt near Blair Atholl), then in the hills of Galloway in the Southern Uplands, and then on the Isle of Arran off the west coast. Such veins had already been described by Saussure and others, but Hutton claimed that they were clear evidence that the granite had been squirted into the other rock from below, as a hot fluid that had crystallized as it cooled. This implied that the lowest rock mass in the geognostic pile was not in fact the oldest, so that there might be no truly "primitive" rocks at all. Furthermore, if granites were intruded from below, they became strong evidence of great heat, and hence of the required agent of elevation; they could be the analogue of the steam in a Newcomen engine, forcing the crust of the earth upwards to form a new landmass (Fig. 3.8).[56]

Similar evidence for the forcible elevation of landmasses came from Hutton's interpretation of the many layers of "whinstone" or basalt intercalated among the other Secondary formations in the region he knew best. Hutton adopted the

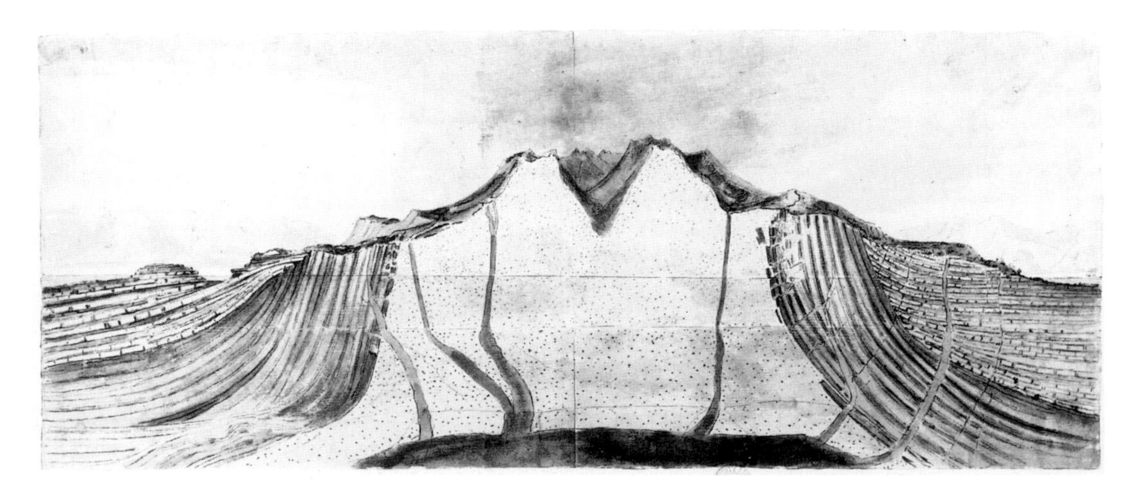

Fig. 3.8. Hutton's section through the Isle of Arran, based on fieldwork there in 1787, as drawn for him by the son of his friend John Clerk. The granite forming the island's central hills was of course observable only at the surface, but the way it was here depicted hypothetically to great depth gave persuasive visual expression to Hutton's claim that it had been intruded from below as an intensely hot liquid, tilting the overlying rocks as it forcibly elevated the landmass. Another variety of granite was shown as having been intruded later from an even deeper source, upwards through cracks in the solidified first granite and the overlying strata. Note also the angular "junction" (in modern terms, an unconformity) between the lower and upper sets of strata (see below). This section was probably intended to illustrate Hutton's description of Arran in the part of his *Theory* that was still unpublished at his death; it shows the vertical exaggeration usual in such sections. (By permission of the Syndics of Cambridge University Library and Sir Robert Clerk, Bt.)

Vulcanist conclusion that basalt was a rock of volcanic origin (§2.4), but for him it was material that had often been squirted forcibly into the pile of sediments deep within the earth (forming, in modern terms, an intrusive sill), thereby contributing to crustal elevation, without reaching the surface as a lava in a volcanic eruption. Indeed, he claimed that volcanoes were simply nature's safety valves, regulating and preventing excessive pressure below; hence they too were marks of designful order, not of destructive disorder (the governors being fitted to the latest steam engines would have provided him with an instructive analogy). He found field evidence, almost on his own doorstep, for this interpretation of basalt as an intrusive rock (Fig. 3.9).[57]

The second decisive feature for which Hutton searched specifically was evidence for the cyclicity that his system demanded. He had adopted from the geognosts the broad distinction between Primary and Secondary rocks. Even if—as he believed—the former did not strictly deserve their name, because they were not truly "primitive",

Fig. 3.9. Clerk's sketch of rocks on Salisbury Crags—the hill that rises dramatically above where Hutton lived at the lower end of Edinburgh's Old Town—exposing cliffs of "whinstone" or basalt, with stratified rocks above and below. This drawing shows the base of the massive basalt with irregular columnar jointing; the underlying horizontal strata were curved upwards at one point, with a tongue of basalt below. This suggested to Hutton that the whole mass of basalt had been intruded forcibly between preexisting layers of rock, rather than being a lava extruded subaerially, let alone a sediment or precipitate: it was persuasive evidence of the elevating power of the hot fluids emanating from the earth's deep interior. Like the section of Arran (Fig. 3.8), this sketch may have been destined for the part of Hutton's *Theory of the Earth* that remained unpublished at his death. (By permission of the Syndics of Cambridge University Library and Sir Robert Clerk, Bt.)

55. Hutton, *Theory* (1795), 1: 214. Gould, *Time's arrow* (1987), chap. 3, rightly emphasizes that Hutton's postponement of relevant fieldwork until *after* he had presented his geotheory in public does not detract in the least from his scientific stature, unless the latter is improperly hitched to a conception of good scientific work as necessarily inductive.

56. Fig. 3.8 is reproduced from the facsimile in Craig, *Hutton's lost drawings* (1978) [Cambridge-UL: Tab.a.59] (fig. 25 of accompanying text), of a MS drawing in Edinburgh-RS. Arran is described in Hutton, *Theory* 3 (first published by Geikie in 1899, with illustrations from Geikie's time, not Hutton's), chap. 9; a planned fourth volume was never written, or has not survived even in manuscript: see Dean, *Hutton in field and study* (1997), 11–19. Hutton, "Observations on granite" (1794), made his important new interpretation public in advance of his full *Theory* and gave it much wider circulation.

57. Fig. 3.9 is reproduced from part of the facsimile in Craig, *Hutton's lost drawings* (1978) [Cambridge-UL: Tab.a.59], fig. 4 in accompanying text; Craig dates it about 1785. His fig. 5 (and facsimile) reproduces Clerk's smaller-scale section and view of the whole of Arthur's Seat, showing the basalt of Salisbury Crags in its geognostic context among varied Secondary formations and other masses of "whinstone".

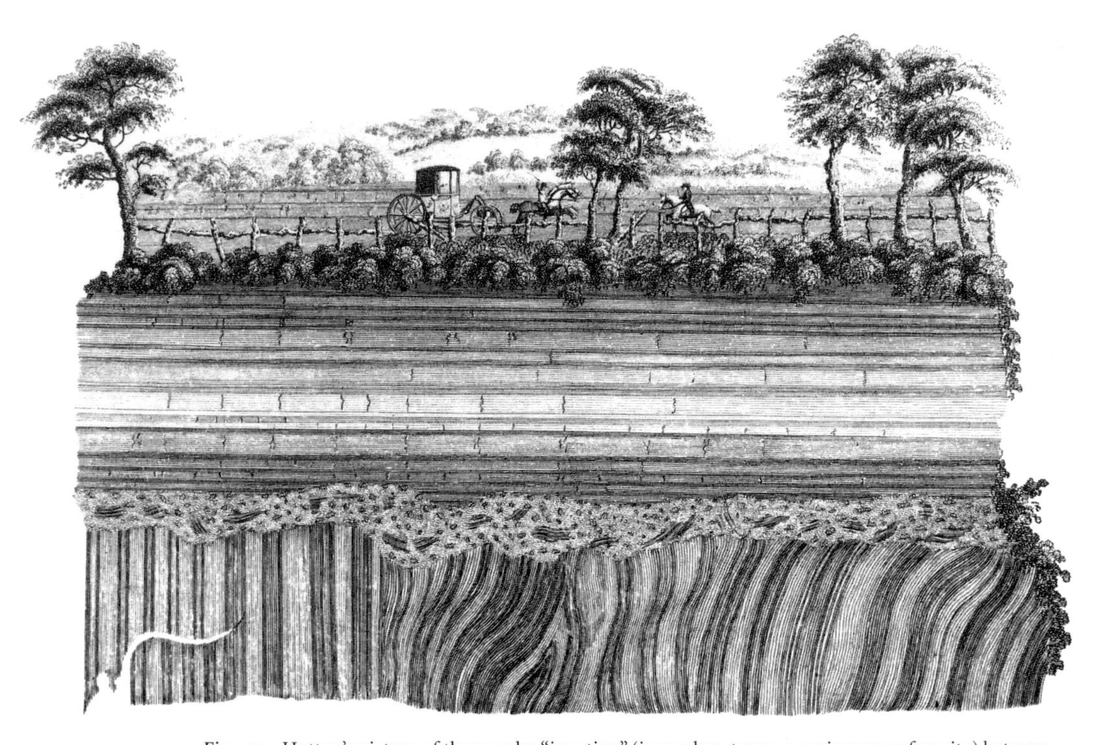

Fig. 3.10. Hutton's picture of the angular "junction" (in modern terms, a major unconformity) between two sets of stratified rocks, exposed in a river gorge at Jedburgh in the Southern Uplands of Scotland. Vertical layers of "schistus" have been planed off and are overlain by horizontal layers of sandstone and shale; at the base of the latter is a bed composed of debris clearly derived from the older rocks. Hutton interpreted such sections in terms of a cyclic "succession of worlds" preceding the present: each set of rocks had been deposited on the floor of a former ocean; each had later been elevated to form a landmass; each had been or was being worn down, providing soil on which life could flourish. The vegetation and the human figures with their horses represent the life of the present world, i.e., the most recent in the "succession", as well as denoting scale and giving the scene verisimilitude. A small vein of intrusive rock (bottom left) hints at the unseen power in the deep interior, responsible for the successive episodes of forcible elevation. This engraving, based on a drawing by Clerk, was published in Hutton's *Theory of the Earth* (1795). (By permission of the Syndics of Cambridge University Library)

most of them were clearly older than the overlying formations (the intrusive granites being exceptions). The two sets of rocks might therefore be relics of at least two "former worlds", both preceding the present set of continents. A *sequence* of "worlds" would go far to establish the cyclicity of the whole system. What he needed to find was evidence that the Primary rock masses were sediments that had been elevated to form an ancient set of continents; that they had then been worn down; and that the overlying Secondary formations represented a later set of sediments—the debris from yet another set of continents—that had eventually been elevated to form the present continents. Several earlier naturalists had described angular "junctions" (in modern terms, unconformities) of the kind required, but as in the case of the granite veins Hutton intended to interpret them quite differently. In the summer of 1787, two years after reading his paper in Edinburgh, he duly found an angular junction between the "schistus" and the overlying stratified rocks on Arran (Fig. 3.8) and again in the Southern Uplands near the border with England (Fig. 3.10).[58]

Time and eternity

The following summer, Hutton, in the company of his younger friends John Playfair and James Hall, found a third locality for this angular "junction" at Siccar Point on the coast not far from his farms in Berwickshire. Many years later, after Hutton's death, Playfair recalled how Hutton had expounded on the spot his interpretation of the long sequence of events that had produced what they saw before their eyes, and he recalled that "the mind seemed to grow giddy by looking so far into the abyss of time". The idea of time as an abyss was borrowed from Buffon, but it encapsulates what Playfair's generation (and others since) found most striking about Hutton's system.[59]

Yet Hutton's concept of time was in fact a commonplace among Enlightenment savants (§2.5). Like Buffon with his "eternal road of time" (§3.2), Hutton treated time as a dimension that necessarily stretched without limit into past and future: "Time, which measures every thing in our idea, and is often deficient to our schemes, is to nature endless and as nothing". So he claimed that the unimaginably vast time necessary for "the total destruction of the land" was no argument against the reality of that process. However, he did not infer a vast scale of time by extrapolating from a very slow observable rate of erosion. On the contrary, he flatly denied the validity of anything like de Luc's natural measures of time (§3.3); he claimed that *no* clear evidence of the rate of erosion of the continents could be detected, even within the whole of recorded human history back to the ancient Greeks: "It is in vain to attempt to measure a quantity which escapes our notice, and which [human] history cannot ascertain; and we might just as well attempt to measure the distance of the stars without a parallax, as to calculate the destruction of the solid land without a measure corresponding to the whole." The analogy with astronomy was revealing. Hutton argued that in time, as in space, the vastness was beyond direct observation and could only be inferred indirectly. The land *must* be wasting away, in order to maintain the whole "system" of habitability; it could not be observed by any detectible "measure"; therefore it must be happening too slowly to be humanly perceptible. Far from inferring a vast timescale from observation, Hutton deduced it from first principles and then explained away the awkward fact that its effects were unobservable.[60]

In Hutton's system, continents were successively wasted away by erosion and renewed by crustal elevation, all on an indefinitely vast timescale. Like many other savants, Hutton also assumed that the species of animals and plants continued

58. Fig. 3.10 is reproduced from Hutton, *Theory of the earth* (1795), 1 [Cambridge-UL: S365.c.79.1], pl. 3, explained on 430–32; see Tomkeieff, "Unconformity, an historical study" (1962). Clerk's original drawing is reproduced in Craig, *Hutton's lost drawings* (1978), facsimile and Fig. 38. On earlier reports of similar "junctions" by other naturalists, see for example Ellenberger, *Histoire de la géologie* 2 (1994), 312. The Jedburgh section is now overgrown.

59. Playfair, "Life of Hutton" (1805), 71–73; Hutton, *Theory of the earth* (1795), 1: 458, describes the spot but without reference to time; on Buffon, see Rossi, *I segni del tempo* (1979), chap. 15.

60. Hutton, "Theory of the earth" (1788), 215, 298–99. Astronomical parallax, the best *direct* evidence for the distances to the stars (vast even in relation to the size of the whole solar system), was not detected by telescopic observation until the mid nineteenth century.

through the same vast spans of time: nothing that he knew about fossils suggested any kind of progressive change in the living world. Even on the sensitive issue of the *human* species he was carefully ambiguous. He knew that "the written history of man", including of course the account in Genesis, suggested that humans were of quite recent origin, and he conceded that there was no natural evidence—of human fossils, for example—to contradict that traditional assumption (see §5.4). But natural history *as a whole* pointed to a quite different conclusion:

> There has not been found, in natural history, any document by which a high antiquity might be attributed to the human race. But this is not the case with regard to the inferior species of animals, particularly those which inhabit the ocean and its shores. We find in natural history monuments which prove that those animals had long existed; and we thus procure a measure for the computation of a period of time extremely remote, though far from being precisely ascertained.[61]

Taken in context, it is clear that Hutton regarded marine fossils as, in effect, *surrogates* for the missing evidence of the human presence in the remote past. Abundant fossil shells showed clearly and directly that the marine component of the system of organic nature extended indefinitely into the past. The sparser evidence of fossil wood and other plant fossils showed that terrestrial life also extended back in the same way. All these fossils therefore indicated indirectly that human life must also have existed, long before the recent period of extant written documents. For Hutton's system was meaningless without human life: "the globe of this earth is evidently made for man", as he insisted repeatedly, and its wisely maintained habitability would have been quite pointless if in fact it had supported human life only in the most recent times.[62]

So in every part of Hutton's system—all the way from the dynamic equilibrium of continents and oceans to the enduring human presence that constituted its ultimate purpose—an assumption of eternalism was implicit, and indeed crystal clear to any informed reader. Given Hutton's concern with establishing a sound basis for human understanding, it is not surprising that he phrased his eternalism with due regard to the limitations of human knowledge: as he put it in his earliest summary, "with respect to human observation, this world has neither a beginning nor an end". Likewise he concluded the full version of his paper by claiming only that this is what *we find*. Nonetheless, despite that properly careful wording, no reader at the time would have been left in any doubt that Hutton believed the terrestrial system was in fact eternal, even though it was beyond human capacities to demonstrate it conclusively. Any lesser conclusion would have negated the whole purpose of the wisely designed machine that Hutton had taken such pains to reveal beneath the apparent disorder of the natural world (Fig. 3.11).[63]

It is not surprising, then, that one of the two points on which Hutton was criticized most vigorously by his contemporaries was his eternalism. It was not his concept of the vastness of time that they rejected, but his scarcely concealed claim that the earth—and by implication the cosmos as a whole—had had no origin and would have no end. One general review, for example, summarized Hutton as

We have now got to the end of our reafoning; we have no data further to conclude immediately from that which actually is: But we have got enough; we have the fatisfaction to find, that in nature there is wifdom, fyftem, and confiftency. For having, in the natural hiftory of this earth, feen a fucceffion of worlds, we may from this conclude that there is a fyftem in nature; in like manner as, from feeing revolutions of the planets, it is concluded, that there is a fyftem by which they are intended to continue thofe revolutions. But if the fucceffion of worlds is eftablifhed in the fyftem of nature, it is in vain to look for any thing higher in the origin of the earth. The refult, therefore, of our prefent enquiry is, that we find no veftige of a beginning,—no profpect of an end.

Fig. 3.11. The final paragraph of Hutton's "Theory of the earth" (1788), with his famous concluding sentence. In context, its meaning lay not in the magnitude of the inferred timescale but in the enduring "system" of stable equilibrium by which the "wisdom" of the whole was both revealed and ensured. The explicit analogy was with the equally stable and designful Newtonian "system" of orbiting planets. (The "f"-like form of the letter "s", except at the end of a word, was common in printed texts at this period; Fig. 3.1 shows the same convention in French.) (By permission of the Syndics of Cambridge University Library)

claiming that there had been "a regular succession of earths from all eternity! and that the succession will be repeated for ever!!". The mineral surveyor John Williams complained that Hutton "warps and strains everything to support an unaccountable system, viz. the eternity of the world". Conversely, the radical writer George Toulmin (1754–1817), the author of the explicitly eternalistic work *The Antiquity and Duration of the World* (1780), must have been delighted to find he had such a prestigious ally in Hutton and duly noted this when he enlarged his claims to the ultimate degree in *The Eternity of the Universe* (1789). The English natural philosopher Erasmus Darwin (1731–1802), in the "philosophical notes" to his versified *Botanic Garden* (1791), noted that "the ingenious theory of Dr Hutton" clearly implied that "the terraqueous globe has been, and will be, eternal". And the *Encyclopaedia Britannica* (1797) likewise commented that "as the Doctor makes no mention of any thing prior to a world nearly similar to what we see just now, we must necessarily conclude that its eternity is part of his creed". So there can be no doubt about how Hutton's contemporaries, friendly and hostile alike, interpreted this aspect of his geotheory.[64]

The other and even weightier objection that featured in all but the most general reviews was directed against Hutton's claim that ordinary rocks were consolidated by intense heating and melting on or under the ocean floor, rather than by the precipitation of minerals from percolating aqueous solutions. His critics found this idea simply incredible, and they had little trouble in demolishing it on the grounds

61. Hutton, "Theory of the earth" (1788), 217.

62. Hutton, "Theory of the earth" (1788), 209–10.

63. Fig. 3.11 is reproduced from Hutton, "Theory of the earth" (1788) [Cambridge-UL: P340:1.a.1.1], 304. The earlier quotation is from [Hutton], *System of the earth* (1785), 28. At one point in his full *Theory* (1795), while defending himself against Kirwan's critique (§6.4), Hutton did state in passing that "here then is a world that is not eternal" (1: 223); but in context the meaning of the comment is obscure, and anyway it was embedded in a restatement of the wise design demonstrated by the world's indefinitely prolonged equilibrium, so that the disclaimer was hardly convincing.

64. Anonymous, "Theory of the earth" (1788), 37–38; Williams, *Mineral kingdom* (1789), lix; Darwin, *Botanic garden* (1791), part 1, notes, 65; *Encyclopaedia Britannica*, 3rd ed. (1797), 6: 255. See Hooykaas, "Ewigkeit der Welt" (1966), and Porter (R. S.), "Toulmin and Hutton" (1978). Toulmin made explicit what Hutton left implicit, namely that the human race was as eternal as the earth: Porter (R. S.), "Toulmin's theory of man" (1978). Toulmin's earlier book, and its second edition (1783), were published by the same prominent London firm (Cadell's) that marketed Hutton's *Theory*. Toulmin's work was in no way a samizdat or "underground" production.

of detailed empirical evidence. Most implausible of all was the case of the thick Secondary limestones (including the Chalk) that Hutton had to include in his argument. At this time many rural regions were dotted with limekilns, in which limestone was heated to produce lime for use in mortar and as a fertilizer. So it was common knowledge that heat did not melt limestone, but led to its chemical decomposition. Hutton had to explain away this awkward fact by appealing to the huge putative pressures on the ocean floor (his follower James Hall later tried to replicate this effect experimentally). But at the time it seemed an unconvincing explanation. So a point that Hutton himself regarded as crucial to his system was found deeply implausible by most of the savants among his contemporaries.[65]

Conclusion

Hutton's contribution to the genre of geotheory combined familiar elements in an unfamiliar way. Other savants deployed processes such as erosion and sedimentation; others took for granted an indefinitely vast scale of time; others even used deep-seated volcanic forces as a possible cause of elevation, though not on the huge scale of whole continents that Hutton proposed. Indeed, his conception of the earth as a highly *dynamic* body, at depth as well as at the surface, was almost without precedent. But Hutton combined all these processes in a novel way to form a tightly integrated *system*. In his view it deserved that epithet above all because it displayed pervasive wise design; it was a "machine" constructed by a benevolent if remote "Author of nature" to ensure the human habitability of the earth for all time, from and to eternity.

However vast—indeed infinite—its putative timescale, nothing could have been more profoundly ahistorical. Hutton showed no interest in plotting the particularities of geohistory; indeed, he explicitly rejected that kind of project. Even the successive "worlds" that he inferred, for example from the rocks at Jedburgh (Fig. 3.10), were of significance only as evidence for an ever-repeated cyclic process. The Primary "schistus" and the Secondary strata, like the ancient soil and the modern one, were important only as *instances* of passing phases in an ahistorical cyclic regularity. To use Hutton's own analogy, his successive "worlds" were as unspecific as the successive orbits of the planets around the sun, events with temporality but without history. A sense of the *history* of the earth, whatever its source may have been, certainly did not come from Hutton.

3.5 THE STANDARD MODEL OF FALLING SEA LEVELS

The multiplicity of geotheories

The preceding sections of this chapter have described three geotheories that were prominent, and much discussed by savants and the educated public, around the time that Saussure climbed Mont Blanc. But there were many other theories on offer. There were so many that de Luc felt it necessary to sort them out and classify them, applying on a metalevel, as it were, the procedures of natural history itself.

One important distinction was between those that postulated an earth in steady state or cyclic equilibrium and those that saw the earth's temporal development in directional terms. Hutton's geotheory was the purest example of the first kind, together with Buffon's original system (leaving aside the "accident" of the earth's origin); Buffon's later geotheory (in *Nature's Epochs*) was an outstanding example of the second kind, and de Luc's binary model was equally directional in character although different in other ways. Most other geotheories belonged in the latter group. They treated the earth's temporal development as clearly directional, from remote and obscure beginnings towards the familiar present. But they did not always make de Luc's sharp distinction between the present world and all that had preceded it, nor did they follow Buffon in attributing the overall directionality to the slow cooling of an originally incandescent globe.

The most prominent feature of these other geotheories was that they envisaged a gradual fall in the global sea level (in modern terms, purely eustatic change). Starting with an initially worldwide proto-ocean, mountain ranges would have been the first features to emerge as dry land, in the form of islands; further lowering of the sea level had progressively uncovered land at lower altitudes, until finally the continents were revealed to their full extent and the seas were confined to their present limits. As already mentioned, this kind of directional change in world geography was present, albeit in a subordinate role, in both Buffon's and de Luc's geotheories. Both savants invoked hypothetical subterranean cavities to explain where the surplus water had gone, using an ad hoc explanatory resource that had been common in geotheories ever since Descartes and Burnet. A century later, in Saussure's time, a progressively falling sea level—whatever its causal explanation—remained a standard feature of a wide range of geotheoretical systems.

In fact, geotheories based on a falling global sea level were so general that they will be grouped together here and termed the *standard model* of the earth's temporal development. The standard model provided a starting point for many savants to develop their own variant systems, which they hoped would account more adequately for the observable evidence. Conversely, it provided the standard against which others could distinguish their own systems as original by bringing in some more or less novel causal factor; in Buffon's case, for example, this was the idea of a hot origin and a subsequent steady cooling. It makes little sense to attribute the standard model to any single savant; at least by Saussure's time it was just a part of a theoretical repertoire that was taken for granted by all savants with interests in geotheory, whether they adopted it themselves in some variant form or rejected its central tenets in favor of some alternative.

All variants of the standard model were strongly directional in character, but there was no consensus about other features, such as the tempo and mode of the changing sea level. In some systems the fall in sea level was taken to have been slow and steady; in others, long static periods were thought to have alternated with episodes when the level dropped suddenly and even with catastrophic effects; in yet

65. Hall, "Series of experiments" (1812). See Gerstner, "Hutton's use of heat" (1971); Newcomb, "British experimentalists" (1990) and "Laboratory variables" (1998); and Laudan, *Mineralogy to geology* (1987), chap. 6.

others, some fluctuation was envisaged within the generally downward trend, so that the sea level might occasionally have risen somewhat, temporarily reflooding areas that had earlier been exposed as dry land. Then there was variation over the causal means invoked to explain the loss of water from the earth's surface. As already mentioned, subterranean cavities, necessarily vast in dimensions, were one of the favorites; but alternatively an extraterrestrial cause might be invoked, such as a slow evaporation into space or, more catastrophically, with comets commonly the culprits. Above all, there was no consensus about how the proto-ocean itself was conceived, in terms of its chemical composition and physical state, or about how the "liquid"—as it was usually called with convenient vagueness—had changed in the course of time before becoming the ordinary salt water of the present oceans.[66]

The earliest earth was generally taken to have been covered with a universal fluid of complex chemical composition, out of which different materials were precipitated in succession. With the global level falling, the earth was eventually left with its present restricted oceans composed of merely salty water, and this was usually taken to be the earth's final and stable form. Some such process of falling sea level, combined with chemical differentiation and precipitation, seemed adequate, at least in outline, to account for the broadly uniform sequence of rock masses that geognosts described in many different regions, and for what mineralogists discovered about their chemical composition (§2.1, §2.3). It is not surprising that the standard model was particularly attractive to geognosts and mineralogists, who often adopted it as a working hypothesis or as the explanatory background to their everyday descriptive work, even if they did not aspire to elaborate any geotheory of their own.

Primary rocks such as granite and gneiss could be conceived as crystalline precipitates from a complex primal chemical soup, perhaps hot and certainly quite different in composition from present seawater. Minerals such as quartz and feldspar (the main components of many Primary rocks) were known to be almost insoluble in ordinary water, but this did not detract from the plausibility of the model: the natural laws of chemical reaction were assumed to be stable and unchanging, but the circumstances in which they had acted might have varied greatly. The precipitation of the Primary rocks would necessarily have changed the composition of the remaining fluid, and this might in principle account for the quite different kinds of material found among the Secondary rocks; the second generation of precipitates, as it were, would necessarily differ in composition from the first. The prevalence of thick limestones among the Secondaries, for example, could be a product of this specific phase in the progressive modification of the chemical composition of the proto-oceans. At the same time, the lowered sea level would have exposed the mountain ranges of Primary rocks to subaerial erosion, contributing detrital materials to form other Secondary formations such as pudding-stones. Finally, the oceans would have drained off the continents altogether, and the remaining fluid—necessarily changed still further in composition—would have been reduced to the merely salty water that we know today. Since this kind of geotheory assigned a primary role to aqueous fluids, it was sometimes called "Neptunist", by extension from the controversy about the origin of basalt (§2.4).

Neptunist geotheory

The standard model has been sketched up to this point while deliberately refraining from naming any names, in order to underline its character as a *standard* kind of causal explanation of the earth's development, so much taken for granted that it was in effect anonymous. Historians and historically minded geologists have commonly ascribed the Neptunist system to Werner (and many of them, until recently, routinely castigated him for it). But Werner was merely giving his own expression to a widely held kind of geotheory, and he would have been the first to disclaim any originality in the matter. Only in his claim that basalts had originated as aqueous precipitates was he going against the general opinion of his geognostic contemporaries; and he clearly saw this as improving on a theory that had been prevalent long before he himself came on the scene. In any case, he regarded himself primarily as a sober descriptive geognost, not as a speculative geotheorist: his famous summary of geognosy was carefully entitled "A Brief Classification and Description . . .", *not* "Theory of the Earth" (§2.3). Like many other geognosts, mineralogists, and physical geographers, Werner never elaborated a geotheory of his own, or at least he never published one; and his lecture notes suggest that he simply fitted his descriptive work—when it seemed desirable to do so—into a causal framework of high-level theorizing already formulated by others.[67]

Long before Werner rose to prominence at Freiberg, the seeds of Neptunist geotheory can be detected in ideas that emerged out of much earlier debates. For example, around the turn of the eighteenth century the English physician and naturalist John Woodward, hoping to improve on Burnet, had proposed a system in which all the bedded rocks (later to be termed Secondary) had settled out of a kind of primitive soup, in order of their specific gravities, as an initially global ocean subsided to its present level. Woodward himself had equated this global ocean with the Flood and saw no problem in setting the whole sequence within the traditional short timescale. In the early eighteenth century his system was adopted by a wide range of savants throughout Europe, but with significant and growing modifications. Geognostic fieldwork soon put paid to the idea that the Secondary formations were arranged in order of specific gravity; a dense sandstone, for example, might be found overlying a much lighter limestone. And a growing awareness of the sheer thickness of the formations, many of them composed of finely layered sediments, made the short timescale increasingly implausible (§2.5). So by midcentury the equation with the biblical Flood had been generally abandoned. However, Woodward's basic idea of a sequential deposition or precipitation of the Secondary formations, from a proto-ocean that had subsided gradually to its present level,

66. The directional standard model could, exceptionally, be turned into a steady-state and eternalistic system by making the falling sea level just one phase in an even longer cycle. As already mentioned, this was the strategy that had been used in *Telliamed* (1748): see Carozzi, *De Maillet: Telliamed* (1968) and "De Maillet's Telliamed" (1992).

67. See Wagenbreth, "Werners System der Geologie" (1967); more recent research is represented in Albrecht and Ladwig, *Abraham Gottlob Werner* (2002); Şengör, "Is the present the key to the past" (2001), is almost alone among modern studies in reverting to a highly critical view of Werner.

remained intact; and that idea could readily be extended to include the Primary rock masses too.[68]

The widespread adoption of this basically Neptunist model generated two important empirical predictions for the science of geognosy. Both concerned the interpretation of the geognostic or structural order of rock masses in terms of their *temporal* sequence of deposition. First, if these rock masses had been deposited out of a proto-ocean that was steadily falling in level, the oldest rock masses should also be found at the highest altitudes, and successively younger ones only at lower elevations. This *altitude criterion* of relative age, as it will be termed here, seemed to be confirmed by fieldwork that found Primary rocks such as granite on the highest peaks of mountain ranges (as Saussure confirmed when he climbed Mont Blanc), Secondary formations on somewhat lower hills, and the alluvial or Superficial deposits on still lower ground, particularly valley floors and plains (see §2.3). Second, if all the rock masses, of whatever age, had been deposited or precipitated out of the complex soup of a primal proto-ocean, each specific type of rock would have characterized a specific phase in this process of chemical differentiation and would therefore be confined to a specific part of the whole pile of formations. This *lithic criterion* of relative age, like the altitude criterion, seemed broadly to be confirmed by geognostic fieldwork: granites and gneisses, for example, were found only among the Primary rock masses, deepest in the pile; thick limestones only among the Secondary formations, much further up the pile (although usually lower in altitude). Minor exceptions could easily be accommodated within this kind of interpretation; for example, repetitions of the same rock type at different levels in the pile could be explained by invoking occasional minor reversals of the general downward trend in sea level.

One influential example of this kind of geotheory was sketched by the Prussian savant Peter Simon Pallas (1741–1811) in a paper read to the academy at Saint Petersburg in 1777 and published in its *Acta* the year before Saussure's *Alpine Travels*, Buffon's *Nature's Epochs*, and de Luc's *Physical and Moral Letters*. Pallas had been commissioned by the Empress Catherine to lead a major expedition to survey the natural and human resources of her increasingly vast domains. Even before he finally returned to Saint Petersburg, he had begun to publish three massive volumes of *Travels in the Russian Empire* (1771–76), recording observations ranging from the Caspian Sea to the Arctic Ocean and as far away as eastern Siberia. Pallas's approach was primarily that of a physical geographer; but since he was expected to look out for mining opportunities he also had an eye for the third dimension of geognosy. He was particularly impressed by what he saw of the structure of the Urals and by his discovery of a carcass of a rhinoceros in frozen ground in a riverbank near the shore of the Arctic Ocean. Such features were included in his *Travels*, but only as descriptive natural history: their interpretation was reserved for a separate occasion, and became the topic of his later paper "On the Formation of Mountains" (1778). Although it was based on what he had seen in the Urals, Pallas's ambition was to construct a causal explanation of mountain ranges in general, and indeed to explain mountains in relation to other major physical features of the globe. In effect, an analysis of one mountain range became the basis for a wide-ranging geotheory.[69]

Pallas described how the structure of the Urals conformed in broad outline to what had been reported from other mountain ranges such as Saussure's Alps. The core was composed of granite flanked by "schistose" Primary rocks, with no trace of fossils and therefore apparently older than life itself. Then came two sets of limestones and other Secondary rocks, both with fossils, the lower and older ones dipping steeply, the overlying and younger ones almost horizontal. Disclaiming any originality for his interpretation, Pallas took a falling sea level almost for granted as the most plausible explanation for this sequence, though he conceded that there had been some crustal elevation elsewhere, for example in the Alps.

Overlying all these rocks, though as usual on still lower ground, was a "third order" of deposits, a hundred fathoms (about 200m) thick, with tree trunks and the bones of exotic animals, which had subsequently been cut through by deep valleys. Pallas attributed these deposits to a violent mass of water, sweeping suddenly and briefly over the continental landmass, and carrying not only sediment but also plant and animal debris from quite different regions. The evidence—and most vividly the frozen rhino in Siberia—had convinced him, explicitly against his own preconceptions, of "the reality of a deluge over our land". He demonstrated his Enlightenment credentials, however, by equating the biblical Flood with the records of similar events in other cultures and attributing "our globe's most recent catastrophe" to a mega-tsunami caused perhaps by the sudden elevation of the volcanic islands of the East Indies (a suitable source of rhinos). Finally, he suggested, the waters had retreated as suddenly as they came, excavating valleys in their own deposits as they did so, and disappearing into the usual convenient subterranean "abysses".

All this was presented explicitly as a hypothesis, but Pallas claimed that it was a plausible explanation of what he had seen on his travels in the Russian empire and of what others had observed elsewhere. The "catastrophe" that had formed the "third order" of deposits might have been the most violent for which there was clear physical evidence, but it was certainly not unique: earlier ones had perhaps been responsible for disrupting the older rock masses, and others were likely in the future. In Pallas's view, the largest and most recent event of its kind had been recorded—in however garbled a manner—by human beings early in their history. As usual in such theorizing, Pallas left the still earlier timescale vague and indefinite, but he was certainly not forced to invoke catastrophes from any failure to concede or imagine the magnitude of deep time.

68. Woodward, *Natural history of the earth* (1695); the enlarged edition of 1726 was translated into French in 1735. Woodward is used here simply as an example: it is beyond the scope of the present book to analyze the filiation of these ideas, but see for example Porter (R. S.), *Making of geology* (1977), chaps. 2, 3, and Rappaport, *Geologists were historians* (1997), chaps. 5–7. Woodward endowed a lectureship at Cambridge to promote his geotheory and conserve his collection of the specimens that supported it, thus in the event founding both the chair and the museum of geology there.

69. Pallas, *Reisen durch der Russischen Reichs* (1771–76), and "Formation des montagnes" (1778). Carozzi (A. V. and M.), "Pallas' theory" (1991), includes a translation from the original German text of the latter.

The explanatory scope of the standard model of geotheory can be further illustrated by just two brief examples. Linnaeus had adopted the standard model as a way of explaining what was known about plant and animal distributions. He speculated that all known forms of terrestrial life might have originated on a single primal island—the first to emerge above the early proto-ocean—and might have spread subsequently to all the varied landmasses that later appeared from beneath the steadily falling sea level. This provided him with a plausible way of explaining many of the otherwise puzzling features of what would now be called biogeography.[70]

Like Buffon, most savants who adopted some variant of the standard model wanted if possible to confine their explanatory repertoire to what de Luc later called "actual causes", those that could be observed directly at work in the present world. The standard model had therefore gained some plausibility from the careful observations reported by one of Linnaeus's compatriots. Anders Celsius (1701–44) had claimed that the level of the Baltic Sea was sinking steadily, even within the timespan of human records. Harbors once accessible to ships of a certain draft were now barred by solid rock, hunters reported that seals had shifted their habitual sites onto lower rocks, and the marks that Celsius had made to record the sea level were already being left high and dry.[71]

In this case the advantage of hindsight—the change in the Baltic is now attributed with good reason to a postglacial rise of the land on a merely regional scale—only accentuates a historical puzzle of more general importance. Most geotheorists took it for granted that any change of relative sea level must have been due to a fall in the sea, not to a rise in the land. The reasons for this generally unexamined assumption are clearly complex and multiple. But prominent among them must be the fact that in most of Europe, where most savants lived, the land was experienced as fixed and immovable, and there was little direct evidence for crustal movements. Major earthquakes shook the ground, of course, and might even crack it (see Fig. 2.10), but there was little unambiguous evidence that they permanently elevated the land. And visible evidence of the much earlier derangement of layered rock masses was even more ambiguous and could often be ascribed to crustal collapse just as plausibly as to crustal uplift or buckling (§2.4). As the example of Buffon shows, even a geotheory constructed around the earth's putative internal heat did not necessarily treat its interior as a place of dynamic activity or give any significant role to movements of its surface crust.

Conclusion

The standard model for geotheory provided a moderately satisfactory causal explanation of many of the most prominent features of the earth, particularly those disclosed by the descriptive work of geognosts and mineralogists. Many specific points might be left unexplained, or filled in with unashamedly ad hoc explanations, such as the subterranean cavities that were often invoked to receive the waters of the putatively subsiding ocean. And these systems usually left the origin of the earth obscure, or else set it aside as being beyond the proper scope of "geology", relegating it—or perhaps promoting it—to the realm of cosmology. But once the earth had become a going concern, as it were, the Neptunist geotheory in its various forms offered a coherent explanation of the larger geognostic features of the earth's crust, such as the structural relations between Primary and Secondary rock masses and their respective suites of distinctive rock types. All variants of the standard model were unmistakably directional in character. Like Buffon in his later years, proponents of the standard model insisted in effect that the earth had not always been the same kind of place. Indeed they claimed, like him, that it had passed through an unrepeated sequence of distinctively different phases; only in the course of a long physical development had the earth become the place where humans now make their home.

In this chapter, a few geotheoretical models have been summarized, out of the dozens that were on offer around the time that Saussure climbed Mont Blanc. Most savants who had ambitions to make their mark in the sciences of the earth seem to have felt obliged to put forward their own high-level explanations of how the earth worked: to every savant his own system. In effect, they all aspired to become the Newton of the terrestrial world. Yet this very multiplicity of rival systems generated a growing sense of unease about the whole genre of geotheory. For example, as already mentioned, Buffon's later system was dismissed as a mere fantasy or novel [*roman*], or in modern terms as little more than science fiction. Hutton's geotheory, hinging on highly implausible physico-chemical processes and on the deeply questionable assumption of eternalism, struck many of its readers even more clearly as a castle in the air. De Luc's arguments for a recent and radical change in world geography were equally questionable for a diametrically opposite reason, in that they were explicitly invoked to support the historicity and authority of the Bible. The problem was that almost any model was as good as any other, or as bad. Any savant could build a system on the basis of some bits of accepted empirical evidence and more or less convincingly explain away those that did not fit. To put it in modern terms, there were so few empirical constraints that the theorizing was grossly underdetermined: any hypothetical model was about as plausible as any other.

Saussure, as one of the most highly respected savants dealing with the sciences of the earth, repeatedly postponed offering his own geotheory because he was too well aware of the complexity of the evidence and the lack of relevant observations (§6.5). Others, particularly those of a younger generation, were therefore likely to follow his implicit example and adopt a moratorium on geotheory. A more fruitful way forward might be to focus on more limited problems: to search for empirical evidence that might first enable these to be resolved before aspiring to build a

70. See Frängsmyr, *Geologi och skapelsetro* (1969), chap. 4, and *Linnaeus* (1983), 110–55; also, more generally, Browne, *Secular ark* (1983), chap. 1.

71. Wegmann, "Moving shorelines" (1969). Such changes were reported only from Sweden; but this was no argument against the global reality of a falling sea level, because the almost enclosed and almost tideless Baltic offered optimal conditions for making such precise observations. But the Baltic case was not uncontested, since the historical experience of Venice, for example, appeared to indicate a steadily *rising* sea level. (Celsius also devised the temperature scale now named after him: another example of the eighteenth-century passion for precision in measurement.)

comprehensive geotheory. One particular class of evidence, and one particular set of unsolved but potentially soluble problems, came to the fore—and with good reason—in the last years of the century and of Saussure's life. But before the context of those investigations can be reviewed (in Chapter 5), it is essential to deal with an issue that has been hovering in the background throughout this review of the genre of geotheory, and indeed in the earlier survey of the sciences of the earth (Chapter 2). In what sense and to what extent is it right to claim—as has been claimed here repeatedly—that neither the genre of geotheory nor the sciences on which it was based were significantly *historical* in character around the time that Saussure climbed Mont Blanc? Since the main claim being made in this book is that the sciences of the earth first became truly "geohistorical" during the subsequent years, it is essential to establish what the situation was at the start of that period. This is the subject of the next chapter.

CHAPTER 4

Transposing history into the earth

4.1 THE VARIETIES OF HISTORY

The diversification of history

The central thesis of this book is that the sciences of the earth became historical by borrowing ideas, concepts, and methods from human historiography. The suggestion is hardly novel; but many such claims have either been pitched much too late or hitched to the wrong kind of history. They are linked either to the supposed emergence of modern historical methods in the work of nineteenth-century German historians such as von Ranke, or to the "philosophical" or "conjectural" histories of eighteenth-century writers such as Voltaire, Montesquieu, and Herder.

The first proposal sets the action much too late in the day. Leopold von Ranke's famous definition of truly *wissenschaftlich* history—as the accurate determination of "how things really happened" [*wie es eigentlich gewesen*]—could have been endorsed by "erudite" historians well back in the eighteenth century. If history based on the detailed critical use of documentary evidence was indeed an example and inspiration to naturalists, it was available to them long before the supposed births of modern history and modern geology in the age of von Ranke and Lyell. The second proposal is equally misconceived, though for different reasons. The genres of philosophical and conjectural history constructed in the Enlightenment shared with the genre of geotheory (Chap. 3) the goal of providing an overarching explanation of all the main relevant features: respectively of human nature and society, and of the physical earth. But if—as suggested in the previous chapter—geotheory was *not* in fact a major source for the historicization of the earth, nor a fruitful one, then the relevance of "philosophical" history as an example or inspiration also becomes questionable.

My claim in this book is that both proposals look for historicity in the wrong place. What was appropriated by naturalists, and transposed by them into the natural world, was not grandiose philosophical or conjectural history but rather the Cinderella that was scorned by the philosophes: erudite histories of more limited scope, based on detailed critical study of massive documentary evidence. Put another way, the historicization of the earth came from the transposition of those historical studies that were analogous to the empirically oriented sciences of the earth (Chap. 2), and not from those analogous to the genre of speculative geotheory (Chap. 3). Later sections of this chapter will show how that transfer of ideas and methods, from the human to the natural world, was initiated in the late eighteenth century (and in some respects even earlier), having no need to wait until well into the nineteenth.

However, the contrast between philosophical and conjectural histories on the one hand and erudite histories on the other should not be drawn too sharply, because some of the most innovative historiography involved their interpenetration; and anyway it does not begin to portray the full diversity of historical studies during the Enlightenment. The primacy of narrative political history—"history" tout court, in traditional usage—had long been taken for granted; studies based on the erudite description of local "antiquities" were treated in practice as merely ancillary. But during the eighteenth century these erudite or antiquarian histories became more prominent and more varied and began in effect to claim parity with the political. Traditional political narratives, focused on the public lives and actions of major historical figures, continued to be a highly valued genre, but they were increasingly supplemented by histories of other kinds: histories of trade and commerce, of "manners" and social relations, of religious and legal practices, of arts and sciences, and so on.[1]

How these diverse genres were to be related to each other was the subject of much experiment. For example, a main thread of political narrative might be interspersed with "digressions" on other topics; or the narrative might have its evidential foundations laid out separately, in one instance as "proofs and illustrations" (a phrase later adopted in Hutton's strategy for his geotheory; see §3.4); or several separate narratives might be offered in parallel, to be read in whatever way the reader chose. The ultimate ambition behind all such experiments was to find a way to integrate political narrative with erudite attention to documented detail on many other aspects of history. Perhaps the first major work in which this was achieved with some success was the acclaimed—but also controversial—*Decline and Fall of the Roman Empire* (1776–88), which Edward Gibbon completed in Lausanne in 1787 (coincidentally just as Saussure was climbing Mont Blanc some fifty miles away). All such historical works gave heightened value to the use of erudite or "antiquarian" sources of many kinds, not only documents and archives but also artifacts such as ancient buildings, coins, and inscriptions.[2]

By contrast, the style of "philosophical" history favored by many of the leading philosophes was relatively detached from any detailed documentary evidence; and it aspired to account for all the main features of human history, and in particular for the history of civilization, in terms of relatively simple and overarching concepts of human nature and progress. Voltaire's "Essay on General History and on the Manners and Spirit of Nations" [*Essai sur les Moeurs*] (1756) had set the tone in its very title, and his scorn for dry-as-dust erudition became a standard refrain. The related genre of "conjectural" history, which offered general explanations of how human societies "ought" to have developed in the past in order to have arrived at their present state, had equally tenuous links with any specific items of historical evidence: here Montesquieu's work "On the Spirit of Laws" [*De L'Esprit des Lois*] (1748), with its constant appeal to the "*esprit général*" of human societies, had set the tone. Hutton's Edinburgh contemporary Adam Ferguson had been just one of many other distinguished contributors to the genre, with his *Essay on the History of Civil Society* (1766): the first part dealt with "the general characteristics of human nature", and was followed by "the history of rude [uncivilised] nations", as the baseline for tracing "the history of policy [politics] and arts".[3]

This outline of historical genres in the late Enlightenment, although very brief and even crude, is enough to suggest the wide range that was in principle available to naturalists to transpose from the human to the natural world. Many historical models were almost irrelevant, however, simply because the perennial problems of morals and motives, conscience and decision, free will and determinism—all that made history a distinctively *human* story, and the very stuff of "philosophy of history"—were not obviously applicable to the natural world. Nonetheless, that still left naturalists with a wide variety of other historical models to draw on, if they were so inclined. The rest of this section will review briefly those that were in fact transposed into the sciences of the earth in the late eighteenth century.

Chronology and biblical history

One of the most important historical genres was also one of the most traditional. The science of *chronology* has already been mentioned in the context of the brief timescale that it assigned to human and even cosmic history (§2.5). It continued to

1. In this section I am much indebted to discussions with Mark Phillips: see his "History and antiquarianism" (1996) and *Society and sentiment* (2000), 3–30. This work in turn is inspired by, and reevaluates, Momigliano's seminal research on the history of historiography: see the latter's classic essay "Ancient history and the antiquarians" (1950) and others in his *Studies in historiography* (1966) and later collections. The present brief treatment of the topic cannot begin to do justice to a substantial literature; but see also for example Hay, *Annalists and historians* (1977), chap. 8; Levine, *Humanism and history* (1987); Haskell, *History and its images* (1993), chap. 6; and Iggers, "German Enlightenment historiography" and other essays in Bödecker et al., *Aufklärung und Geschichte* (1986).

2. See the examples of historical experiment cited by Phillips: the "digressions" in Hume, *History of England* (1754–62); the "proofs and illustrations" in Robertson, *Reign of Charles V* (1769), 1: 193–394; and the seven parallel narratives in Henry, *History of Great Britain* (1771–93). A crucial synthesizing role for Gibbon's work was claimed in Momigliano, "Gibbon's contribution" (1954), and many later papers. Phillips, "History and antiquarianism" (1996), argues persuasively that that thesis can no longer be sustained in its original form; but Gibbon remains at least an important *example* of how late eighteenth-century historians were striving to write narrative histories that would incorporate erudite "antiquarian" detail as never before.

3. See for example Spadafora, *Idea of progress* (1990), chap. 7, for Hume, Smith, Ferguson, and other Scottish writers.

be practiced throughout the eighteenth century; even the *Encyclopédie*, despite its subtly subversive goals, had a fair and balanced entry on "chronology". On a larger scale, the great collaborative *Universal History* synthesized in twenty volumes all that could be gleaned from textual scholarship on the chronology of the ancient world and was later continued in its *Modern Part* in another forty-four volumes. Chronologers emphasized one point that was a basic prerequisite for any other kind of detailed history, namely the determination of *temporal sequence*. To them it mattered supremely what happened when: if possible to get the dates precisely right, but at least to get events in exactly the right order. As the editors of *Universal History* put it, "an exact distribution of time is, as it were, the light of history: without this it would be only a chaos of facts heaped together". And since their goal was to compile a chronicle of "universal" or global history, it was necessarily based on, or derived from, the histories of specific cultures and civilizations, which in turn were based on the concrete particularities of local events. Since the acceptance of such events as authentically historical entailed a judgment of the reliability of the sources in which they were recorded, sources required interpretation and critical evaluation. Chronology entailed textual criticism.[4]

The ambition of chronologers to compile accurate "annals" of world history was a conceptual resource that could well be translated into similar ambitions for understanding the natural world in terms of its history. It would matter little, from this point of view, whether the natural world could yield quantitative dates; it would be a good start just to get events in the right order as precisely as the evidence allowed. It would also matter little that many chronologers still aspired—as the subtitle of *Universal History* put it—to date events accurately "from the earliest account of time" in the Creation story. That ambition could be soft-pedaled, ignored, or abandoned without affecting the broader emphasis on temporal sequence, even in whatever periods, short or long, might have preceded human history. In other words, the chronologers' passion for temporal precision in history could well be absorbed by naturalists without their having to adopt the short timescale as part of the package.

The science of chronology had been born from religious controversy, and it continued to be closely linked to issues of biblical interpretation. The time line of "sacred" history—ancient Jewish, with a Christian sequel—was usually treated as the main line with which other "profane" chronologies needed to be matched and correlated; where there were discrepancies between them, the biblical one was usually and unsurprisingly assigned greatest authority and authenticity. The crucial moments in this biblical narrative marked the "epochs" that gave qualitative shape and religious meaning to the story. The editors of *Universal History*, for example, stated at the outset that "the creation of the world, the deluge, and the birth of Christ [are] our three epochs"; the last, as the point of divine Incarnation, provided of course the reference point for years B.C. and A.D., which had become the standard timescale for everyday use (though it had long been recognized that the watershed date was not quite right). Leaving aside the Creation story itself—which, being mostly pre-Adamic, had to be treated as a product of direct revelation—the rest of the long biblical narrative was regarded in practice as *history* recorded by *human* authors, however much their writing was held to have been divinely inspired. Moses, to whom Genesis and the rest of the Pentateuch was traditionally attributed, was often referred to as "the historian", as for example in Brydone's letter about Etna (§2.5). The editors of *Universal History* claimed—though with some exaggeration of scholarly agreement—that Moses "is by universal consent allowed to be [i.e., accepted as] the most antient historian now extant". However privileged it might be—or, alternatively, treated with skepticism—the biblical record counted as a variety of history.[5]

Most events in biblical history concerned divine and human actions in the changing "covenant" relation between God and Man. They were set in a context of the material world but were primarily concerned with the moral and religious framework of human lives. Only a few events had obvious correlates in the natural world; some of these, such as the sun supposedly standing still for Joshua, had long been the subject of hermeneutic debate. In the physical history of the earth, however, just one event was of outstanding importance. The "Flood" or "Deluge" in the time of Noah was recorded as having been so drastic that it would surely have left physical traces in the present world, even if it had not been literally global in extent (a question that had long been the subject of scholarly argument). Throughout the eighteenth century there was much discussion among savants—including both naturalists and biblical scholars—about how it should be interpreted; there was certainly no rigid line of orthodoxy in the matter. But it did constitute a supremely important point at which geohistory might be tied into human history: not just analogically or metaphorically, but substantively, as an event marked in both natural and human records. So it is hardly surprising that "*diluvial*" theories were prominent in debates about the earth: savants such as Buffon, who virtually denied that the Deluge had had any physical role at all, were very much in a minority.[6]

This substantive conjunction between textual and natural records was not confined to biblical sources. Back in Classical times there had been a scholarly tradition—known as *euhemerism*, after its pioneer Euhemerus—of demythologizing traditional stories of gods and heroes into the more prosaic but historical realities of ancient kings and patriarchs. This kind of textual criticism had been extended in the seventeenth century, when savants such as Hooke had naturalized the stories still further and read them as garbled accounts of ancient physical events such as floods and earthquakes. In effect, the natural world became a resource to be read historically *in parallel* with textual sources: in that sense, it was indeed a time "when

4. Anonymous, "Chronologie" (1753); Sale, *Universal history* (1747–48), continued as the *Modern part of universal history* (1759–66); quotation from 2nd ed., 1 (1747), original preface, xxiii. Chronology in the eighteenth century has not been much explored by historians, but for erudite scholarship and textual criticism generally, see for example Grafton, *Defenders of the text* (1991), chap. 9, and essays in Grell and Volpilhac-Auger, *Nicolas Fréret* (1994).

5. Sale, *Universal history* 1 (1747): quotations from preface to 2nd ed., xi, and original preface, xi. That the Bible was treated as pervasively *historical* is particularly clear from attempts to interpret it pictorially, as in Scheuchzer, *Physica sacra* (1731): see Rudwick, *Scenes from deep time* (1992), chap. 1.

6. See Rappaport, "Geology and orthodoxy" (1978) and *Geologists were historians* (1997), chaps. 5–7.

geologists were historians". But the two classes of evidence could only corroborate one another because the history was taken to cover more or less the *same* span of time, the limited span of the traditional timescale.[7]

By the middle of the eighteenth century, however, that assumption was unraveling, and the yawning "abyss" of a far longer history was opening up, largely pre-historic and prehuman (§2.5). Yet the demythologizing of ancient human traditions could still be useful for making sense of the murky "heroic age" preceding reliably documented human history, as was shown for example in Nicolas Boulanger's "Antiquity Unveiled" [*L'Antiquité Dévoilée*] (1766). Buffon later drew heavily on Boulanger's unpublished but well-known *Nature's Anecdotes*, with its systematic use of natural evidence, for his imaginative reconstruction of his seventh and last "epoch", the time of the earliest human beings. Like the use of the biblical story of the Flood by other naturalists, this created a substantive link between human history and the history of nature; it involved treating human and natural evidence as complementary "witnesses" to the same distant past. Although it was now applied only to the most recent portion of a far longer total history, it did demand that at least *that* part of the natural evidence be read in geohistorical terms. It therefore created a powerful precedent for treating still earlier periods in the same way, even if in that deeper time the human records were not merely mythologized but totally absent.[8]

Chorographers and antiquarians

The science of chronology, with its emphasis on temporal precision, and the related scholarly traditions of "sacred" and "profane" narrative histories, were complemented by the genre of chorography. As already mentioned, this comprised the detailed scholarly description of specific localities or regions; in effect, local geography (§2.2). Although there was nothing intrinsically historical about the genre, in practice chorographies usually included descriptions of local "antiquities" of all kinds: not only material artifacts such as ancient buildings, megaliths, and earthworks, but also local history derived from documentary evidence or even from oral tradition and folklore. In other words, chorographies incorporated inventories of diverse historical materials, even if there was no attempt to synthesize them into any unified historical narrative of the region being described. Chorographers did much of their work on documents assembled in archives, preferring "public" documents such as charters and legal codes to the "private" records of possibly biased individuals. Like chronologers, they used textual criticism to detect anachronisms and hence to distinguish authentic records from fakes and forgeries. In doing so, they developed a keen sense of "period", a sense of which cultural traits belonged together and which were alien to a particular time. These were conceptual resources that could readily be transposed into the natural world, even though nature's documents and archives owed nothing to human agency. Deliberate forgeries might have no precise parallel, but nature's historical records might still be deceptive in other ways and would certainly require the same kind of critical evaluation.[9]

Closely related to chorography, indeed largely overlapping that genre, was the work of those who called themselves "antiquarians". They were concerned less with the textual records of the past than with its material artifacts. However, some of the latter were in fact textual. Greek or Latin inscriptions on stone, for example, often threw important new light on ancient history, beyond what could be gleaned from the surviving texts of Classical authors. They were therefore readily integrated into the exclusively textual studies of traditional scholarship, and *epigraphy* became a highly esteemed branch of erudite antiquarianism. But some inscriptions could not be read because their language was unknown: in particular, the Etruscan ones found in the heartland of ancient Roman territory highlighted the need for the relics of the past to be *deciphered* before they could be used as historical evidence (the hieroglyphic inscriptions from ancient Egypt were equally enigmatic, but were less well-known in the eighteenth century). Other artifacts allowed at least a tenuous link with the traditions of textual scholarship: for example, antiquarians could extract valuable historical information from the combined images and lettering on ancient coins, and *numismatics* was another well-developed and highly respected branch of antiquarian studies.

However, many other artifacts were almost totally nonverbal and *nontextual*, and the intense study of them by antiquarians represented a major enlargement of the scope of erudite research. They included "monuments" or ancient buildings such as the Colosseum in Rome, but also much smaller artifacts such as the elaborately decorated Greek vases that Hamilton collected (Fig. 1.8), and even more mundane objects from the ancient world (see Fig. 4.3). Nontextual antiquities showed that the distant past had many "witnesses" beyond the verbal reports and records of contemporaries; objects without words could still be made to tell a story. This extension of erudite history beyond the evidence of texts rendered the work of antiquarians ripe for analogical application to the natural world: fossils and other mineral objects might likewise be made to speak as "witnesses" to nature's past, even a deep past in which there had been no human presence whatever.

The erudite work of antiquarians enormously enriched the contents of the ancient world as it was apprehended by the educated classes throughout Europe. The world of the Romans, and beyond it the more remote world of the Greeks, had of course been central to European culture ever since the Renaissance. However, in the age of the Grand Tour an appreciation of the ancient world could extend beyond familiar Classical texts to the more immediate visual experience of seeing at first hand the Colosseum and the Pantheon in Rome, the Greek temples at Paestum beyond Naples and at Agrigento far away in Sicily, and Roman amphitheaters,

7. Euhemerism, like chronology, is not an extinct science and remains a valid and valuable approach to ancient history: see for example Greene, *Pre-Classical antiquity* (1992).

8. Boulanger, a civil engineer and orientalist, had died in 1759. On *Anecdotes de la nature*, see Roger, "Un manuscrit inédit" (1953); Hampton, *Boulanger et la science* (1955); and Sadrin, "Boulanger: avant nous le déluge" (1986).

9. On chorography in the early-modern period, see for example Cormack, *Charting an empire* (1997), chap. 5, and, more generally, Mendyk, *Speculum Britanniae* (1989).

Fig. 4.1. Sir William Hamilton and his mistress Emma Hart (soon to become his second wife) watching workmen recovering decorated "Etruscan" vases—which Hamilton claimed were in fact Greek—from a tomb at Nola, inland from Naples: an engraving dedicated by Hamilton to the Society of Antiquaries in London and used as the frontispiece of his great *Collection of Engravings from Ancient Vases* (1791–95), published in Naples. He relied on selling his vases to wealthy connoisseurs to cover his expenses. This tomb at Nola was deeply buried in layers of volcanic ash from eruptions of Vesuvius: this accentuated the analogy with finds of fossil bones, which likewise were often buried by later sediments well below the present surface (§5.3). (By permission of the British Museum)

aqueducts, and other remains right across western Europe (Greece itself, as part of the Ottoman Empire, remained in effect off-limits to all but a few intrepid travelers). And an appreciation of immobile monuments such as these could be supplemented—for those with money to spend—by the contemplation of smaller and mobile artifacts that could be taken home to adorn a fashionable town house or country mansion (Fig. 4.1).[10]

A new breed of art historians, among whom the German scholar Johann Winckelmann (1717–68) was a prominent pioneer, interpreted all these artifacts in ways that gave educated Europeans a new depth of understanding of the ancient world. While highlighting its "otherness"—its historical distance from their own time—this erudite scholarship also paradoxically made the ancient world more accessible than ever before, as a cultural resource that could be imported into the present, generating the pervasively Classical taste, fashion, and design that now seem such defining features of the eighteenth century. Likewise, in northern Europe, medieval churches and cathedrals, and even enigmatic megaliths such as Stonehenge and prehistoric burial mounds everywhere, attracted other antiquarians, who studied and described them with equal devotion to detail, inspiring "Gothick" and other trends in taste and fashion. Nonetheless, the historical consciousness of most educated Europeans remained strongly periodized, and indeed polarized, into Ancient and Modern, with the "Middle" Ages being treated as a more obscure and less

Enlightened time in between, and the "pre-historic" periods before the Classical as an even more shadowy presence.[11]

Herculaneum and Pompeii

The centrality of the ancient Classical world in European culture was heightened, then, by the greater accessibility of its major "monuments". It was further emphasized by the discovery of important new material evidence, the product of a new kind of excavation. Ancient artifacts of stone, pottery, or metal—prehistoric, Classical, or medieval—had long been retrieved by excavating graves, burial mounds, and other sites; excavation remained the usual way of obtaining such treasures. But during the eighteenth century, excavations at two specific and related sites transformed that practice, almost inadvertently, from traditional treasure hunting into what would be recognized in retrospect as the first truly scientific archaeology.[12]

In 1736 the sinking of a well at Resina, not far from Naples, had yielded some striking statuary from what turned out to be a deeply buried Roman theater. From that time onwards the site was in effect mined for its antiquities. As the underground excavations were extended it became apparent that a whole Roman town was buried there, though the hardness of the rock made its exploration slow and difficult. A few miles away and a few years later, another and more extensive site, buried in loose volcanic ash, had proved much easier to excavate, though the ruined buildings were preserved to a lesser height. In due course, inscriptions had been found that identified the towns as Herculaneum and Pompeii respectively (Fig. 4.2).[13]

These discoveries excited intense interest among savants and connoisseurs throughout Europe. "O what a great adventure of our times that we discover not just another monument, but a city", the Veronese scholar Scipione Maffei exclaimed; "by excavating, and leaving everything in its place, the City [of Herculaneum] would become an unequalled museum." But such hopes were soon dashed. The excavations were jealously guarded by Charles III, the then king of the Two Sicilies, who wanted to adorn his palace at Portici and enhance his cultural prestige with the spectacular objets d'art that they were yielding. Visiting savants such as Winckelmann were affronted to find the norms of the Republic of Letters flouted, for they were denied access to the sites, or allowed to see them only under strictly controlled conditions, being forbidden to take notes or make sketches. Two decades later, Hamilton, as the hunting companion of the new king, Ferdinand IV, was able to have drawings made (Fig. 4.2), but he too was critical in private:

10. Fig. 4.1 is reproduced from a loose copy of the engraving (London-BM, neg. PS 126155), which is based on a drawing by Christoph Heinrich Kniep. Hamilton's *Collection*, like his *Campi Phlegraei*, had its text in both English and French; on the exceptional depth of the Nola tomb, see 1: 23–24. See also Jenkins and Sloan, *Vases and volcanoes* (1996), 144; and Burn, "Hamilton and the Greekness of Greek vases" (1997).

11. Hamilton's folio volumes were explicitly intended to give contemporary artists an accessible collection of Classical examples to inspire their own designs.

12. See, for example, Schnapp, *Conquête du passé* (1993), chap. 4.

13. Fig. 4.2 is reproduced from Hamilton, "Discoveries at Pompeii" (1777) [Cambridge-UL: T468.b.36.4], pl. 6. The original painting by Fabris is reproduced in Jenkins, "Hamilton's affair with antiquity" (1996), 43.

Fig. 4.2. Roman buildings newly excavated at Pompeii: an engraving, after one of a series of drawings sent by Hamilton to the Society of Antiquaries in London, published in 1777 in its periodical *Archaeologia*. Uncovered beneath thick volcanic ash (B), the building (C) to the right of the colonnade (A) was identified as barracks; the ruins contained skeletons with bronze armor and helmets. Some skulls are shown, shelved on the far right for "the curious" to inspect. Hamilton noted that the teeth were in excellent condition, and suggested that this was the result of a diet without sugar; the comment exemplified the new ambition of antiquarians, inspired by the discoveries at Herculaneum and Pompeii, to reconstruct even small features of the everyday life of the ancient world. (By permission of the Syndics of Cambridge University Library)

> The arts here are at the lowest ebb, and the little progress made in the searches at Herculaneum and Pompeii proceeds solely from vanity. . . . they have been digging here and there in search of antiquities and by that means have destroyed many curious monuments and clogged up others with the rubbish. . . . Glorious discoveries might still be made, if they would pursue the excavations with vigour.[14]

Over the years, however, Hamilton's hopes were in fact fulfilled: the antiquarians appointed to supervise the digging did manage to transform the royal ambitions. From merely mining for artistic treasures, the excavations gradually became a systematic attempt to uncover the ruins, to plot the street plans of the two towns, and to study in detail not only the major public buildings but also shops and private houses. Particularly striking were the extensive frescoes decorating the interiors of many of the houses, which gave vivid new insights, including the titillatingly erotic, into the everyday life of the Roman world. Even such artistically modest objects as terra-cotta lamps, which had been mass-produced in Roman times, were newly appreciated and were stored and displayed in the royal museum along with more spectacular pieces such as statues and bronzes (Fig. 4.3).[15]

Above all, the sites themselves did become something like the open-air museums that Maffei had imagined; as a later savant put it, "The most singular and

Fig. 4.3. Terra-cotta lamps found in the excavation of the Roman town of Herculaneum (the lamps burned vegetable oil, with a wick protruding from the spout): a typical plate of engravings from the massive eight-volume description of its antiquities published under the auspices of the king of the Two Sicilies. The whole of the final volume (1792) was devoted to lamps and candelabra, the originals of which remained in the royal museum near Naples. Such artifacts were published in a style matching that used for illustrating fossils and other natural objects (see Figs. 2.3, 3.7 etc.): in both cases it was a matter of producing mobile paper *proxies* for the original specimens. (By permission of the Syndics of Cambridge University Library)

interesting spectacle [at Pompeii] . . . has been that of a Roman town emerging from the tomb, almost with the same freshness and beauty that it had under the Caesars." The comment expresses the sense that these excavations, more than any before, were in effect *resurrecting* the Classical world. They were bringing Herculaneum and Pompeii back to life in all their everyday detail, and providing, as it were, an unprecedented window of time travel. Likewise Hamilton's many visitors from all over Europe could enjoy the celebrated Classical "attitudes" or mildly erotic tableaux vivants enacted by his beautiful young mistress Emma Hart (eventually his second wife, but also Horatio Nelson's mistress in a famous ménage à trois), which were literally an embodiment of Hamilton's vivid sense of the accessible reality of the life of the ancient world. To be fully historical, the antiquarian could aspire to reconstruct that world with all the immediacy of a firsthand witness and to imagine what it would have been like to be there in person. In one instance that ambition was even expressed pictorially, in strikingly matched representations of

14. Hamilton to Henry Temple, Lord Palmerston, 19 August 1766, quoted in Jenkins, "Hamilton's affair with antiquity" (1996), 42. Maffei, *Tre lettere* (1748), 33, 36, quoted (in translation) in Bologna, "Rediscovery of Herculaneum and Pompeii" (1990), 84–85.

15. Fig. 4.3 is reproduced from Bayardi, *L'Antichità di Ercolano* (1757–92), 8 [Cambridge-UL: S524.bb.75.9], pl. "Pag. 161".

Fig. 4.4. The Temple of Isis at Pompeii—fully excavated since Hamilton's earlier view of it (Fig. 2.9)—with savants and connoisseurs studying and discussing the ruins: an engraving published in 1782 in a lavishly illustrated account by the abbé Jean-Claude Richard de Saint-Non of his Grand Tour through the Kingdom of the Two Sicilies. (By permission of the Syndics of Cambridge University Library)

Fig. 4.5. The Temple of Isis at Pompeii, "as it must have been in the year 79, when it was destroyed by the eruption of Vesuvius", with Roman Pompeiians participating in the cult of the Egyptian god: an imagined scene, "reconstructed [*rétabli*] after, and following, what still remains of it today", published in 1782 to match the view of the extant ruins in Fig. 4.4. (By permission of the Syndics of Cambridge University Library)

the present and the distant past, showing what could now be observed at one specific place, and how—from that concrete evidence—an otherwise vanished scene could be reconstructed (Figs. 4.4, 4.5).[16]

Furthermore, apart from a few inscriptions, and the famous literary texts discovered in one building (the Villa dei Papyri), all these reconstructions of the life of the ancient world had been achieved entirely by the study of *nontextual* artifacts. This made the excavations near Naples a powerful precedent for any attempt to achieve analogous reconstructions based on relics of the past history of nature. Indeed the analogy was almost inescapable, since this spectacularly rich slice of ancient human history had been preserved by the accident of an equally spectacular *natural* event, the eruption of Vesuvius in A.D. 79. At Herculaneum and Pompeii, human history and the history of nature were entwined inextricably, and Hamilton's two areas of achievement as a savant were fused into one.

Conclusion

The varied genres of erudite or "antiquarian" history, far more than "philosophical" and "conjectural" histories, were based on detailed concrete evidence, whether texts or artifacts. They were histories compiled bottom-up, not deduced top-down. They claimed to describe the particularities of how human events had truly happened, where and precisely when, rather than the idealized generalities of how human civilization "ought" to have developed given some overall view of human nature and society. They described specific events, and sequences of events, that could not possibly have been predicted in advance, or even retrodicted afterwards. Their sheer *contingency* was due not only to the uncertainties of all human actions and decisions, but also to the unpredictable complexities of natural circumstances surrounding human lives: the first recorded eruption of Vesuvius, with its catastrophic human effects at Herculaneum and Pompeii, stood as a vivid reminder of that ineluctable contingency. So if erudite or antiquarian history were to be used as a model or analogy for recovering and reconstructing the history of *nature*, it would bring with it a strong sense of history as a contingent and unpredictable sequence of events, rather than as any kind of intrinsically predictable or "programmed" temporal development.

By the time Saussure climbed Mont Blanc, a few savants had begun to transpose concepts and images from human historiography into the natural world; they interpreted the earth's features in terms that were knowingly borrowed from some of the varied historical practices of their erudite or antiquarian contemporaries. They were beginning to think about the earth not merely in temporal terms—which was a commonplace for anyone who considered the possible natural causes of anything

16. Figs. 4.4 and 4.5 are reproduced from Richard, *Voyage pittoresque* (1781–86), 2 [Cambridge-UL: S578.bb.78.2]: pls. 74 and 75bis, based on drawings by the Parisian artist and architect Louis-Jean Desprez; the numbering suggests that the reconstructed scene was a late addition. It was certainly an innovative one: although Desprez could draw on a long tradition of historical painting, what was novel was the tight correlation between the imagined scene and the extant traces of a specific locality. Latapie, *Fouilles de Pompeii* (1766), is quoted in Bologna, "Rediscovery of Herculaneum and Pompeii" (1990), 85.

terrestrial—but in truly *geohistorical* terms. Even if the causal explanation of particular features remained obscure and uncertain, these savants tried to establish the historicity of the specific past events that those features represented and to place such events in their correct temporal sequence. Sometimes they went still further, and tried to set that sequence within some kind of narrative account leading towards the present. In the rest of this chapter some examples of this kind of geohistorical interpretation are reviewed. To illustrate how it was beginning to be applied to the whole range of the sciences of the earth, the examples will be drawn from each of those sciences in turn, in the order in which their more conventional practice was first introduced in Chapter 2.

4.2 FOSSILS AS NATURE'S DOCUMENTS

Human history and its natural records

The science of mineralogy, with its focus on small-scale specimens that could be studied indoors in a museum (§2.1), had already been a site of geohistorical interpretation for more than a century. However, that bald statement requires two major qualifications. First, the specimens that had been the subject of historical treatment were *fossils*, rather than minerals and rocks in general. It was the "extraneous" or "accidental" character of fossils—the very fact that they were distinct from the rocks in which they were embedded—that gave them historical potential. And second, the historical framework within which they were interpreted had been that of human history, not prehuman geohistory.

Over a century earlier, back in Burnet's time, Robert Hooke had lectured regularly to the then newly founded Royal Society in London; in discussing fossils he had famously suggested that it might one day be possible to "raise a Chronology" from them. The history that Hooke imagined being constructed—with at least some of the rigor of the chronologers of his time—would have been a history of nature, but only as it had run parallel to the relatively brief history of the human race. Like his contemporaries, Hooke scarcely envisaged the possibility that there had been vast spans of totally *prehuman* geohistory. Fossils simply had the potential to supplement more conventional sources of historical information, particularly in the very earliest or "fabulous" periods of human history, for which the textual sources—if any—were sparse, unreliable, and difficult to interpret.[17]

Hooke suggested how fossils could be treated as nature's "medals" or coins, but the context shows that he meant this in the sense that they could supplement the kind of information supplied by ordinary coins of human origin. In northern Europe, for example, antiquarians found Roman coins, which supplemented the textual records and confirmed that the Roman Empire—or at least Roman commerce—had once extended to many regions far from Rome. Likewise, naturalists found fossils of shellfish such as clams and scallops far from the coast, which implied that the sea had once extended well beyond its present limits. The period at which it had done so could then be set within a broader narrative history; clearly it had been even more remote than Roman times. But it was taken for granted by

Hooke's generation that even such a remote period had already been within the span of early *human* history and that the fossils were simply supplementing what could be gleaned more obscurely from Genesis and ancient secular sources, enigmatic or garbled though they might all be.

However, as already mentioned, there had been a growing suspicion among savants, during the decades after Hooke's death, that much of the earth's history might have preceded the human presence altogether. This was the hunch that Buffon made explicit when he defined the first appearance of human beings as the very last of "nature's epochs" (§3.2). Fossils continued to be treated routinely as "nature's coins", but the meaning of that metaphor was slowly transformed. Having been regarded merely as supplementary to textual evidence, fossils came to be treated as historical evidence in their own right; they were evidence of events for which there could never be any human records because the periods had apparently been prehuman. Along with other kinds of natural evidence, fossils could then, in principle, be used to construct a geohistory for those long spans of prehuman time, which—it was hoped—could be linked to the more recent and familiar history of the human race. From being merely supplementary to human records, fossils became complementary, providing evidence for the long periods *before* human history. This new use of fossils will be illustrated here from the work of one particular savant from Saussure's time.

The natural history of fossils

François-Xavier Burtin (1743–1818), a francophone Netherlander and—as his name suggests—a Catholic, had received a medical training at Louvain in the Austrian Netherlands (now Belgium) and set up in practice in Brussels, its political and cultural capital. Like those of many other physicians, his interests ranged far beyond his professional work. One of his earliest scientific studies was of the fossil wood associated with the rich coal seams of the province; but he first became widely known in the Republic of Letters when he published an illustrated account of his collection of fossils from the region around his home. His *Oryctography of Brussels* (1784) was dedicated to the royal couple who were currently the Habsburg governors of the province, and its 142 subscribers included some of its prominent aristocrats and prelates. But there were also among the subscribers many savants and booksellers, from London to Turin, from Paris to Vienna and St Petersburg. Copies must have been seen, if not owned, by a wide range of naturalists all over Europe. It was a luxury work, with handsome print and a profusion of plates depicting his specimens, though the draftsman was less than expert in the difficult art of depicting the forms of shells (Fig. 4.6).[18]

17. Hooke, *Posthumous works* (1705), 317–28; the lectures had been given from the 1660s onwards, but their content did not become widely known until they were published after his death in 1703. Rappaport, *Geologists were historians* (1997), puts Hooke in the context of his contemporaries; Drake, *Restless genius* (1996), chap. 6, almost turns him into a modern geologist.

18. Fig. 4.6 is reproduced from Burtin, *Oryctographie de Bruxelles* (1784) [London-BL: 33.i.9], pl. 8, engraved by J. A. Balconi; the subscribers are listed on 137–41. Only some much more extensive studies of his contemporaries will show how far Burtin's work was representative, but it is at least exemplary.

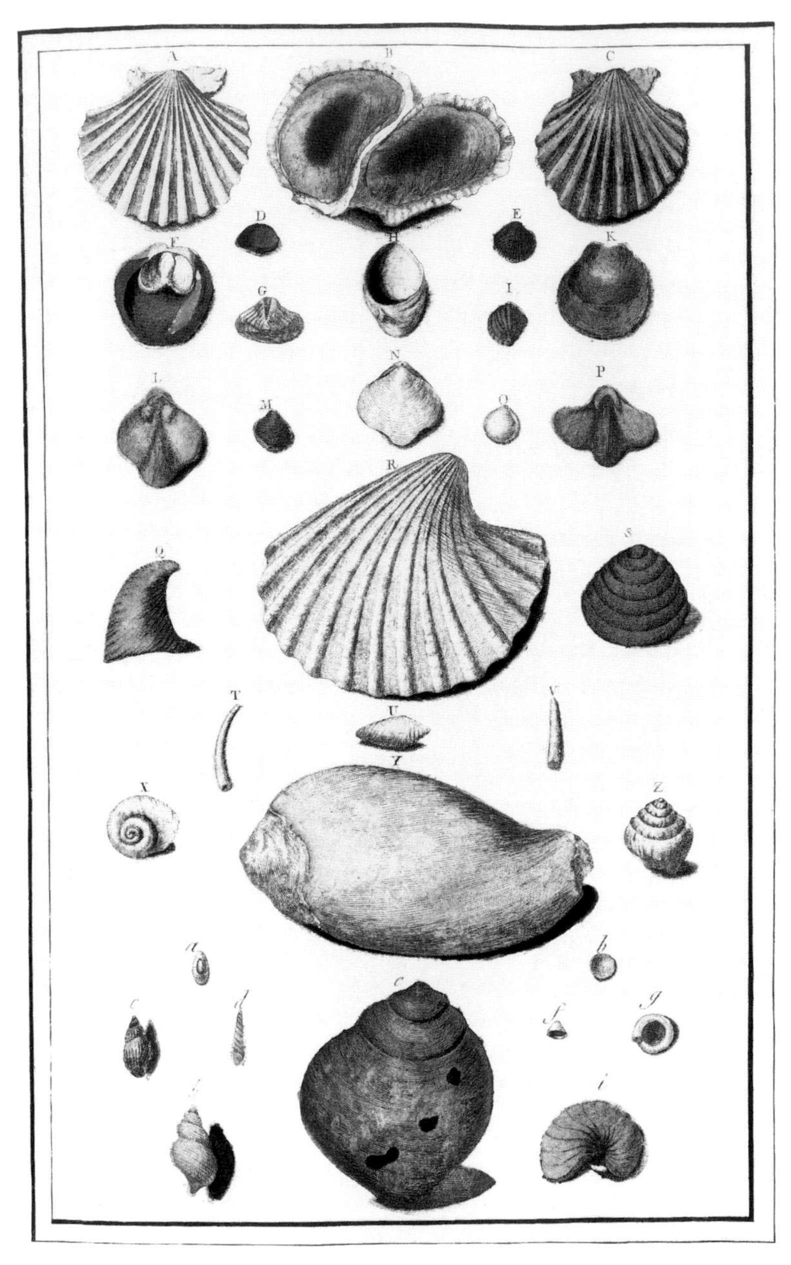

Fig. 4.6. Fossil shells from the region around Brussels: a plate of engravings, typical of those published in François-Xavier Burtin's *Oryctography of Brussels* (1784). Among the shells were those of scallops (A, C), oysters (B, F), cockles (R), and other genera abundant in present-day seas; but also others such the "*térébratule*" (L, N, P), known only rarely and from deep water (see Fig. 5.5). For Burtin and most other naturalists, such fossils implied clearly that the region had been on the sea floor at some remote period, before whatever "revolution" had radically changed the geography (§5.2). Burtin also claimed that the ancient sea must have been tropical, because for example the fossil nautilus (i, bottom right) closely resembled the living "pearly nautilus" of the East Indies (see the more spectacular specimen in Fig. 4.7). (By permission of the British Library)

Burtin's book was explicitly a work of traditional natural history, and it offered a systematic description of all the "fossils" found around Brussels. Burtin still used that word in its old broad sense (§2.1): "natural fossils" (rocks and minerals) were classified according to the kinds of material of which they were composed, and "accidental fossils" according to the kinds of organism of which they were the remains. It was an "oryctography", that is, a description of *specimens* of fossils, not of the formations from which they had been collected. Burtin described the sands, clays, and limestones around Brussels, insofar as they were represented by the specimens in his collection. But he did not describe them as rock *masses*, still less attempt to place such bodies of material in any geognostic order; the book contained no maps or sections and no illustrations from the field. The locations of his fossils (in the modern sense) were duly recorded, in some cases with a note that they been found at a certain depth below the surface of the ground, but they were not located geognostically as coming from specific formations. Still less did any of his specimens—"fossils" of either class—have any sense of *time* attached to them on the basis of where they were found.

In another way, however, Burtin's work went significantly beyond the level of description and classification. He claimed that fossils were potentially the most important kind of evidence for "*géologie*" or "theory of the earth" (§3.1). But he shared the increasing skepticism about geotheory, contrasting "the bold and impatient spirit of the system-maker [*Systematique*]" with his own status as a "modest observer". He would not even call himself a "*physicien*" or causal theorist, because to fulfill that role would first require many more local observations (some of which his book was of course designed to supply).[19]

Instead of embroiling himself in controversial matters of causation, let alone proposing an overarching causal geotheory, Burtin confined himself to some modest *geohistorical* inferences. He rejected emphatically what he regarded as the old and outworn use of Noah's Flood to explain the emplacement of fossils. Like other naturalists he took it for granted that his fossils were of marine origin, but their frequently perfect preservation made it inconceivable that they had been swept to Brussels from far away in any violent and transient deluge. On the contrary, they had evidently lived where they were now found, so the Brussels region had been on the sea floor at some distant period. Burtin also claimed that the ancient sea must have been tropical in climate, for those fossils that could be matched closely with living animals all had their "analogues" among the faunas of the tropics. The most notable examples were his fossil specimens of shells that were clearly similar to the famous "pearly nautilus", a shell of striking beauty known only from the waters around the tropical East Indies and a highly prized item in any shell collection. Burtin's finest fossil specimen was so important for his argument that it deserved a plate to itself (Fig. 4.7).[20]

This was as far as Burtin would go, even with geohistory: at some time in the extremely distant past, the region around his city had been on the floor of a vanished tropical sea. This inference was of course far from original, but he felt he had given it further empirical support. He mentioned briefly the various causal explanations that had been offered to account for such a major geographical and climatic

19. Burtin, *Oryctographie* (1784), 129.

20. Fig. 4.7 is reproduced from Burtin, *Oryctographie de Bruxelles* (1784) [London-BL: 459.e.20], pl. 14; a much smaller specimen was illustrated on the plate reproduced here as Fig. 4.6.

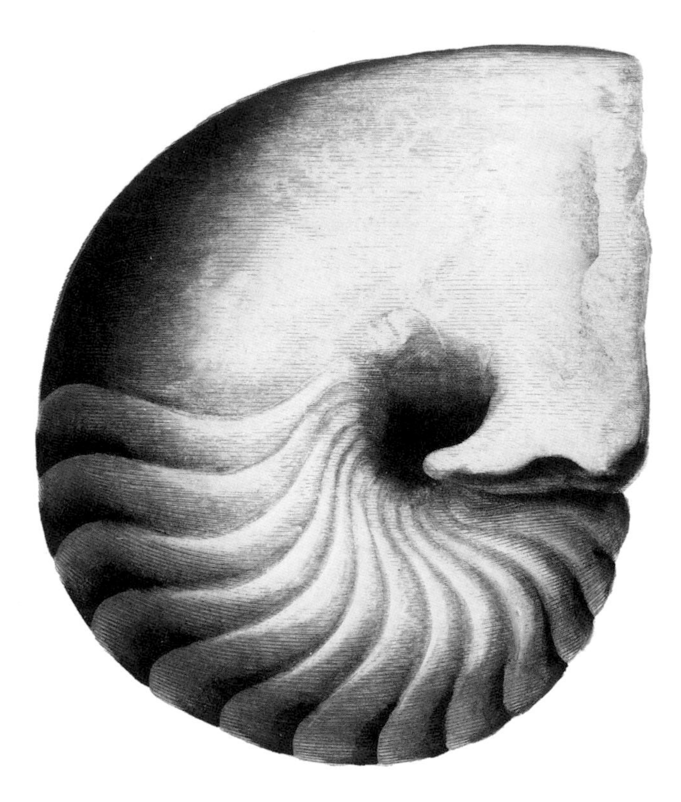

Fig. 4.7. A large fossil nautilus shell (about 40cm in diameter), as depicted in one of the engraved plates of Burtin's description of the Secondary fossils from the area around Brussels (1784). As he recognized, the shell itself had been dissolved away naturally, leaving a "cast" [*noyau*] composed of the material that had filled its interior after the death of the animal. The repeated wavy lines marking the junctions between the outer shell and the internal partitions were therefore visible on the surface; in the shell of the living pearly nautilus from the East Indies, the partitions and the chambers between them could be seen only when the shell was sliced open (see Fig. 5.6). But the similarity of original form was so great that Burtin identified his fossil as the "nautilus of the Indies". (By permission of the British Library)

change, but he ended inconclusively: the "physics" of the matter was beyond the scope of a work on natural history.[21]

Fossils and the earth's revolutions

Among the subscribers to Burtin's book was Martinus van Marum, who, as already mentioned, had set out on his working honeymoon to acquire important fossils for his new museum at Teyler's Foundation (§2.1). In Brussels he met Burtin, and it is inconceivable that they did not discuss their common interest. Later and back in Haarlem, the directors of Teyler's agreed that their next prize question should be on the wider significance of fossils; van Marum may well have proposed it in the hope that it would spur Burtin into making public his ideas on that topic. It was set as a question in the domain of "natural science" [*natuurkunde*]; but the wording referred not to "physics" or causal explanation but to geohistory (Fig. 4.8).[22]

The strongly empirical orientation of the wording probably reflects van Marum's experience a few years earlier, when adjudicating an analogous prize competition at the Hollandsche Maatschappij (housed on the other side of the same canal in Haarlem). In 1781 this Dutch society, which had modeled itself on London's Royal Society (§1.2), had set a question on the validity of the "scale of beings" [*trapswijze opklimminge*] in natural history. But it had received, as the better of only two entries, an essay on the metaphysical and theological implications of that venerable concept. Since its initially anonymous author was then revealed as being no less a savant than de Luc, the society could hardly reject the essay outright, though

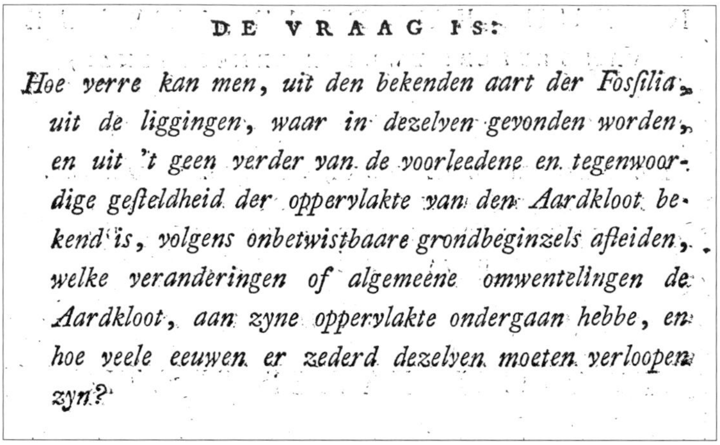

DE VRAAG IS:

Hoe verre kan men, uit den bekenden aart der Foſſilia, uit de liggingen, waar in dezelven gevonden worden, en uit 't geen verder van de voorleedene en tegenwoordige geſteldheid der oppervlakte van den Aardkloot bekend is, volgens onbetwiſtbaare grondbeginzels afleiden, welke veranderingen of algemeene omwentelingen de Aardkloot, aan zyne oppervlakte ondergaan hebbe, en hoe veele eeuwen er zederd dezelven moeten verloopen zyn?

Fig. 4.8. The prize question set in 1784 by Teyler's Foundation in Haarlem.

The question is: How far can one infer—on indisputable principles, [and] from the known character of fossils, from the beds in which they are found, and from what is known of the past and present condition of the earth's surface—what changes or general revolutions the surface of the globe has undergone; and how many ages must have since elapsed?

The prize was awarded to Burtin in 1787 for his essay "On the general revolutions of the earth's surface, and on the age of our globe", which was published in French and Dutch in 1789. The Dutch text of the question is shown here as a reminder that the "minor" languages, and their native speakers, were not unimportant in the life of the Republic of Letters; the prize was also advertised in French, however, to ensure that it became known internationally. (By permission of the Syndics of Cambridge University Library)

it was not what it had hoped to get. As a compromise, de Luc was awarded a silver medal, not the gold one on offer, and the prize was readvertised with the rider that what was required was an answer firmly within the realm of natural history. After that experience, van Marum—who had been one of those making the decision—might well have wanted to ensure that the question set by Teyler's should not attract an essay on the interpretation of Genesis, or one that proposed yet another speculative geotheory, rather than concrete reasoning about specific fossils. Anyway, his adroit steering of the patronage offered by Teyler's induced Burtin to complement his earlier work on the natural history of fossils with a new one on their geohistorical interpretation. In 1787 Burtin was declared the winner for his book-length "Answer to the Question" that had been set.[23]

Burtin began his work with renewed criticism of the whole genre of geotheory: this time the modest observer was contrasted scornfully with "the fertile mind who

21. Burtin, *Oryctographie* (1784), 130–33.

22. Fig. 4.8 is reproduced from Burtin, "Algemeene omkeeringen" (1789) [Cambridge-UL: 911:01.b.6.4], 244, the Dutch translation of his prizewinning essay. This was the second prize for *natuurkunde* to be offered by Teyler's, the first (on phlogiston) having been won by van Marum: see Himbergen, "Prijsvragen" (1978). The word *eeuw*, like the French *siècle*, could mean either a literal century or a lengthy and indeterminate "age"; the latter translation was used in an anonymous English "[Review of] Burtin" (1790), 539, and seems appropriate here. However, Burtin did take the question to refer also to quantitative chronology (see below).

23. Burtin, "Révolutions générales" (1789), published in a volume with a Dutch translation [*Algemeene omkeeringen*]. Both texts were reprinted by Teyler's in 1790 in its periodical, together with a short paper by van Marum on the museum's great specimen of the "Maastricht animal" (Fig. 2.6), intended as an illustration of Burtin's ideas; it is not clear whether there were any other candidates. On the other prize, see de Luc, "Mémoire sur la question" (1788), 457n, and Bruijn, *Prijsvragen* (1977), no. 42; no gold medal was ever awarded for the amended question.

from within his study comfortably builds everything as creator of the world". He praised those who had set the question for sharing his hostility to speculation, since their wording had in effect ruled geotheory out of bounds. However, like many geotheorists he himself was still almost exclusively an indoor savant. Apart from references to the coal seams and associated strata that he had studied much earlier, his essay showed little trace of any field experience. He may well have collected some of his fossils in person rather than buying them from quarrymen or dealers; but his thoughts about them were based primarily on an indoor study of the specimens, supplemented by a reading of the main periodicals and the works of leading savants such as Saussure, de Luc, and Buffon.[24]

"Revolutions" was the key word in Burtin's essay, as it was in the question set. But as usual its meaning was very broad: it denoted *any* major changes in the earth's surface, whether they were thought to have been sudden or gradual, violent or calm. Without offering causal explanations of those he inferred, except in general terms, Burtin tried to define what kinds of events they had been and to determine their correct temporal sequence. His main claim referred to the "great revolution" that had turned the floors of ancient seas into dry land (as for instance around Brussels): he reckoned that his own work and that of other naturalists had clearly established the historical reality of this event. Beyond this point, however, Burtin distanced himself from de Luc, whose work of a few years earlier (§3.3) he criticized relentlessly. For he inferred that the event had been "a *general revolution of the globe*, on a much larger scale and much older than that of the Flood that occurred in Noah's time". Burtin argued that de Luc had in fact conceded that Noah's Flood could *not* account for the features that needed explanation; yet by "continuing to apply physical evidence to the sacred cataclysm he confounds two epochs that I reckon to prove were entirely different". De Luc's singular revolution had been real enough, in Burtin's opinion, but it could not have been the biblical Flood: it had not been a brief and transient event, but had effected a permanent major change in physical geography. And it appeared to have been "general"—that is, more or less global—since the fossils around Brussels were matched by similar reports from around the world. In other words, Burtin adopted much of de Luc's argument for the reality of this great event, but decisively rejected de Luc's assignment of its place in geohistory. By claiming that it was a real event that had long preceded even the oldest of textual records—as Genesis was widely believed to be—Burtin highlighted the capacity of fossils to act in their own right as *nonhuman* sources of historical evidence.[25]

In fact, Burtin looked forward to a time when fossils would become a far richer source for geohistory. For the present, they could yield only a mere sketch [*ébauche*], but eventually they would show that "the surface of the globe is but a series of documents that demonstrate a series of revolutions on this planet". For example, he rejected the common view that the fossils of the older Secondary formations, such as ammonites, were still living somewhere in the world's oceans; he claimed that they were truly extinct, having been wiped out by some revolution still earlier than the "great" one that had turned sea floors into dry land (see §5.2). Indeed, he thought that the distinction between fossils of different formations "gives rise to a well-founded suspicion that the ancient seas must have occupied and abandoned the land more than just once", implying a complex *sequence* of revolutions during those remote periods. Likewise, he suggested that the large fossil bones found widely on land (§5.3) were relics of yet another revolution, more recent than the "great" one, but still much older than Noah's Flood or any other event recorded in human history. And all these specific and distinct revolutions were just the most notable events that had occurred against a background of slow and continual change: "Apart from large revolutions, due to major causes, the surface of the globe suffers a continuous one by the hand of time—slow, it is true, but certain and general—which never ceases to destroy on the one hand in order to generate on the other; which erodes the heights and excavates the valleys in order to extend the plains and fill up the seas."[26]

Burtin did not reconstruct his revolutions in any detail; doubtless he would have said that the observational evidence was too incomplete. Still less did he offer any detailed causal analysis, though he evidently thought that at least the gradual background "revolution" was caused by the processes observably at work in the present world (de Luc's "actual causes"). However, he did in effect assemble a *sequence* of events, a rudimentary geohistory. He regarded all of it as far older than any human records. He had little to say about Noah's Flood—though he clearly assumed that some kind of real event lay behind the ancient story—because it was far more recent than *any* of the revolutions attested by fossils. For Burtin, as for Buffon, the whole of human history, Noah's Flood and all, was simply tacked on to the end of a vastly longer geohistory.

Burtin argued that human history and geohistory were analogous, yet also in sharp contrast. The analogy was apparent in his use of historical metaphors to describe his fossil evidence. Fossils were the "coins", the "documents", and the "monuments" of the natural world; their meaning needed to the "deciphered" and "read". Such terms were used pervasively, indeed routinely and almost casually, throughout his work. Burtin was not unique or even exceptional in this, but his work is a fine example of the interpretative power of these historical metaphors. Fossils were not merely the remains of animals and plants that had once been alive and that ought to be described and classified like living organisms. They were also records of specific conditions at specific moments in the past, allowing the reconstruction of successive events in geohistory. Hard interpretative work was needed before they would yield that historical meaning; but with such work the naturalist could take on a new role as a *historian* of nature. The contrast between human history and geohistory became apparent, however, in the final chapter of Burtin's essay, where he tackled the final part of the prize question and considered the timescale of this eventful geohistory:

24. Burtin, "Révolutions générales" (1789), quotation on 3.

25. Burtin, "Révolutions générales" (1789), 5, 159. One striking find made him uncertain whether the distant period at which his fossil animals had been alive had really been prehuman, though he was sure it was far older than any *textual* records (§5.4).

26. Burtin, "Révolutions générales" (1789), 150; other quoted phrases on 5, 235.

> The history of the earth is quite other than that of nations. The one is devoted only to the hand of men, and—as petty [*mesquine*] and limited as they are—is calculated minutely in dates. The other, written in a majestic language, though obscurely for our feeble means, is found engraved in permanent letters in nature's great code, of which we have scarcely been able to decipher a few pages; but it teaches us that this history, the origin of which is lost in the immensity of time, admits neither dates nor rigorous calculation, but [does disclose] epochs and a perceptible direction.[27]

In other words, geohistory was far grander in scope than human history, but could not be described with the quantitative precision to which chronologers aspired (and which those who set the prize question may have hoped to be given). In rejecting de Luc's identification of the "great revolution" with the biblical Flood, Burtin also explicitly rejected his rival's "*calcul*". He argued that far more time had elapsed since that physical event than the few millennia allowed by de Luc, enough indeed for erosion by rain, rivers, and seas to modify greatly the surface of the land and its coastal outlines. Even the still more recent revolution that had left the fossil bones of large animals scattered across the continents was beyond the reach of any quantitative dating. Nonetheless, the "epochs" or crucial events in geohistory could be placed in their correct temporal order, and that sequence disclosed an intelligible directionality.

Burtin insisted that the vast timescale—literally incalculable—that he outlined for his geohistory was just as compatible with the Bible as the traditional timescale of some fifty-eight centuries. Everything here depended on biblical hermeneutics, and he had devout biblical critics on his side. He cited an earlier scholarly work in favor of a reading of Genesis that allowed for an indefinite lapse of time before the Creation story as preferable to Buffon's stretching of its "days" (§3.2). But either way, he claimed, the principle agreed upon by the best scholars was that the text had been written to be intelligible to ordinary understanding, and not only to savants, since its purpose was religious, not scientific. The bishops and other prelates who had subscribed to his earlier book, some of whom may well have read his prize essay too, are unlikely to have found any problem with his timescale. Alluding clearly to his own reaction to the vast geohistory he had sketched, Burtin commented that "the mind is astounded when it follows step by step the antiquity of the earth and the epochs of its different revolutions"; confronted with that vast panorama, he concluded, "in silence I admire the Creator in all his works". The sentiment was conventional, but that is no reason to dismiss it as insincere.[28]

Conclusion

Burtin's work was a fine example of how "mineralogy"—or more precisely "oryctography", the science based on fossil specimens—was beginning to be used as a basis for a newly *historical* interpretation of the earth. Fossils still needed to be described and classified, as Burtin himself showed in his earlier book. Causal analysis was also needed, to explain their mode of emplacement, for example, and the "revolutions" in which they had been involved. But distinct from either natural history or "physics", as traditionally understood, there was now the possibility of using them to reconstruct geohistory. This new task required that they be treated as nature's "coins", "documents", or "monuments": they needed to be "deciphered" and "read"—or in effect *interpreted*—in much the same way as the texts and artifacts that recorded human history. However, although Burtin answered the prize question by detecting several distinct "general revolutions" of different relative ages, he did not use them to reconstruct the history of the earth—or even the history of the region he knew best—as a connected story. To do so would perhaps have been to come too close to the format of Buffon's work, too close for comfort to the genre of geotheory that Burtin rejected so decisively.

4.3 VOLCANOES AND NATURE'S EPOCHS

The making of a physical geographer

Burtin treated fossils as nature's documents, and used them to distinguish several revolutions in the earth's past history; the greatest such event had turned former sea floors into the present continents, but it was not to be confused with the far more recent event recorded as the biblical Flood. This was a study based almost entirely on the specimens in Burtin's collection and the publications in his own or others' libraries; apart from perhaps collecting some of his fossils in person, it was not based on any work in the field. By contrast, any comparable use of physical geography as a basis for geohistory would demand intensive fieldwork, just as much as the purely descriptive practice of that science. This can be illustrated by the example of Nicolas Desmarest (§2.2), who was one of the most distinguished physical geographers of his generation, fully worthy of comparison with Saussure.[29]

Desmarest, a man of modest social origins in provincial France, had first come to the notice of savants when in 1751 he won a prize for an essay on the possible former physical link between France and England. He had described the topographical and mineral similarities between the opposite coasts of the Channel; but it was physical geography based entirely on written sources, and he himself had never yet seen the sea, there or anywhere else. On the other hand, the topic was also historical, for it had required him to consider the possible relation between human historical records of the Channel and the physical evidence for a former land bridge between the two countries. Having made his debut in this way, Desmarest's entry on "Physical geography" (1757) for Diderot's *Encyclopédie* had then set out his conception of the science as one that ought to be based on disciplined and systematic observation, and hence implicitly on the geographer's own firsthand fieldwork.[30]

27. Burtin, "Révolutions générales" (1789), 216. "*Code*" usually denoted a *legal* code—in this case, of *nature's* laws—yet here it is also associated with decipherment.

28. Burtin, "Révolutions générales" (1789), 240. The textual suggestion (226) was that the Hebrew *bara* would be rendered more accurately by "had created" rather than the Vulgate's "created".

29. Taylor, *Nicolas Desmarest* (1968), is an indispensable source, not only on Desmarest's volcanic research but also on his many other activities. Some of Taylor's many important later articles are cited below.

30. Desmarest, "Géographie physique" (1757).

The opportunity to follow that prescription for himself had arisen around the same time, when he was appointed to a position in the civil service as an inspector of manufactures. In the years that followed, Desmarest had reported first on the French textile industry and later on manufactures as disparate as cheese and paper: in modern terms he had become an industrial scientist. But although these technical analyses were in fields far from the study of physical geography, they had entailed extensive travel, and the work had transformed him from a relatively bookish scholar into a disciplined observer of local detail. This was an orientation that could readily be transposed into the study of physical geography. Desmarest's travels as an inspector gave him firsthand experience of the diverse topography of France and the opportunity to think about its meaning; above all, however, travel convinced him of the value of fieldwork, indeed its necessity. While in Paris he had followed Rouelle's famous lectures, and on some of his earliest travels he applied Rouelle's distinction between Primary and Secondary terrains [*ancienne* and *nouvelle terre*] (§2.3) to the countryside of Champagne and Burgundy. But his decisive moment had come in 1763, when, having been appointed to a position based in Limoges, his duties as an inspector took him for the first time to the province of Auvergne.[31]

Guettard, as the naturalist in charge of the great mineralogical survey of France (§2.2), had startled savants a few years earlier, when he reported that there were volcanoes in Auvergne so fresh that they looked as if they were merely dormant and might still menace the region; but he had not studied the lava flows and cratered cones in any detail. Desmarest had shown no previous interest in volcanoes, but on seeing those in Auvergne with his own eyes he made them the focus of his research in physical geography. He convinced the governor [*intendant*] of Auvergne that it would be in the public interest to have the area surveyed thoroughly. The skilled surveyor François Pasumot, released from military duties by the end of the Seven Years' War, was assigned to make a map; together they undertook a detailed study of the extinct volcanoes and volcanic rocks of the province; and two years later Desmarest presented a preliminary report to the Académie des Sciences in Paris. Soon afterwards, acting as guide and tutor to Alexandre, duke de La Rochefoucauld, a young nobleman making the Grand Tour, Desmarest had embarked on travels that greatly enlarged his experience of physical geography. Like many others, he and his companion were astonished at their first sight of Mont Blanc and other high mountains and glaciers, while they were crossing the Alps. But an even more decisive experience, when they reached Naples, was that Desmarest saw Vesuvius for himself. It was the only active volcano he ever visited, but it gave him a crucial point of reference for all his later studies of extinct volcanoes.[32]

The volcanoes of Auvergne

Back in France, Desmarest had resumed his research in Auvergne. His work was focused not so much on the obvious extinct volcanoes and their lava flows, but rather on the wider problem of basalt, to which he believed they were the key (§2.4). He already knew Susanna Drury's pictures of the Giant's Causeway in Ireland (see Fig. 1.15), and he made the phenomenon much more widely known when in 1768 he published illustrations of this and a similar French locality (at La Tour d'Auvergne) in the *Encyclopédie*. The accompanying explanation of the engravings had first made public his claim that prismatic basalt was a *volcanic* rock. If valid, this implied that volcanic action had been far more widespread in the distant past than it was in the present world.[33]

In 1771 Desmarest presented a full report on the basalts of Auvergne to the Académie (and was elected a member); it was published three years later. The full title of his paper made its place among the sciences quite explicit: it was "On the Origin and Nature of Basalt with Large Polygonal Columns". This identified it as a project in earth physics. But the solution to the causal puzzle had been "determined by the natural history of that rock, observed in Auvergne"; it was based on a descriptive study of basalt, the distribution of which Desmarest and Pasumot had plotted across the region with the usual methods of physical geography (§2.2), and displayed on Pasumot's fine though still incomplete map (Fig. 4.9).[34]

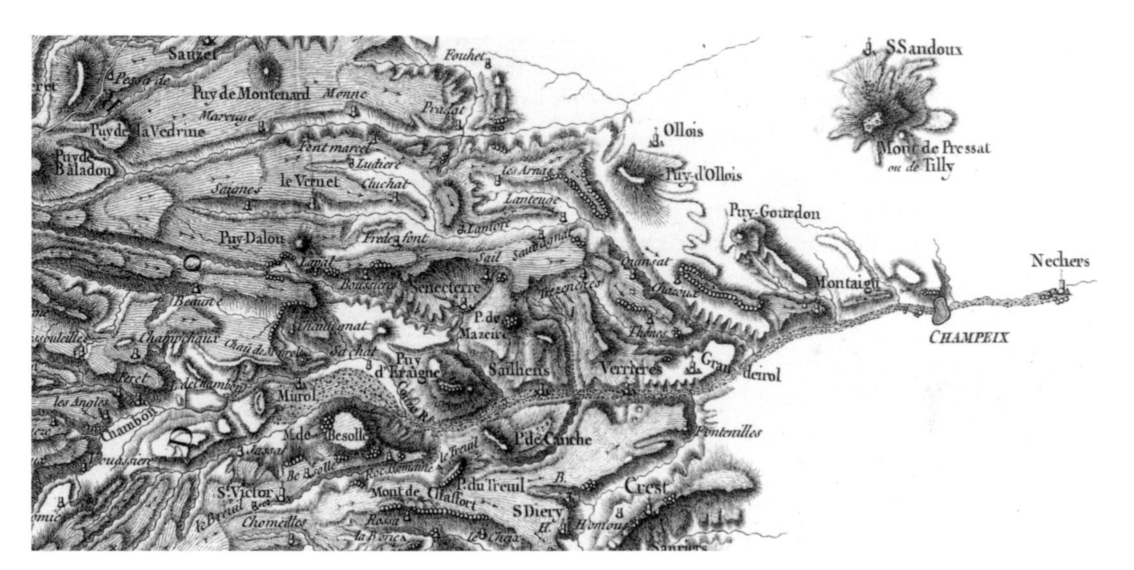

Fig. 4.9. A small portion of François Pasumot's map of Auvergne, recording his and Nicolas Desmarest's study of the basalts and extinct volcanoes, as published in 1774 with Desmarest's first major paper to the Académie des Sciences in Paris. The map showed the sheets of basalt (marked with fine parallel lines) capping the plateaus and hilltops, which Desmarest interpreted as ancient lava flows, left isolated by the subsequent erosion of the valleys; small arrows indicate the slopes and inferred directions of flow. Also mapped were the obviously volcanic lavas (stippled), flowing from cratered cones of volcanic "cinders" down the present valleys, which Desmarest interpreted as much more recent in origin. One of them is shown here, flowing from west to east; its cone (just west of Murol) has blocked the valley and ponded back a small lake (Lac de Chambon, between Chambon and Murol) on the upstream side. On the steep edges of the basalt plateaus, the many places where large polygonal "*prismes*" were exposed are marked with beadlike symbols; since they were also visible here and there in the "modern" lavas (e.g., at Nechers on the eastern edge of the map), they supported Desmarest's claim that *all* basalts were volcanic in origin. The areas not yet surveyed were left blank; the map was eventually completed on a larger scale but not fully published until long after Desmarest's death. The lava flow is about 20km in length. (By permission of the Syndics of Cambridge University Library)

31. See Taylor, "Beginnings of a geological naturalist" (2001).
32. Taylor, "Desmarest and Italian geology" (1995).
33. *Encyclopédie*, Recueil des planches 6 (1768): *Règne minéral*, 6e collection, pls. 6, 7, explained on 3–4.
34. Fig. 4.9 is reproduced from Desmarest, "L'Origine et la nature du basalte" (1774) [Cambridge-UL: CP340:2.b.48.88], part of pl. 15; the final part of the paper, "Basalte des anciens" (1777), was an erudite study

Pasumot's map showed the sites at which "large polygonal columns" could be seen, just as the sheets of the *Atlas Minéralogique* marked the localities at which various distinctive rocks, minerals, and fossils could be collected or exploited (Fig. 2.12). However, unlike those in the atlas, the map of Auvergne also linked the sites of prismatic basalt with the areal distribution of that rock. This showed that the basalts fell into two main "classes"; the use of that word underlined the classificatory or natural-history basis of the work. Some basalts lay along the bottoms of valleys, others on the hilltops or plateaus between. Basalts of the first class were unambiguously of volcanic origin, since many of them were visibly connected to cratered cones of loose volcanic "cinders" [*cendres* or *scories*], just like the minor cones around Etna (Fig. 2.30). Desmarest termed these basalts *courants modernes*. Basalts of the second class were not associated with any cones or craters, but Desmarest interpreted them nonetheless as lava flows, naming them *anciens courants*. He ascribed a volcanic origin to them primarily on grounds of their analogy to the "modern" flows: the rock itself was similar, and so was the distinctive prismatic jointing that could be seen in many places. In other words, the more recent flows acted as a key to the understanding of the more ancient. In fact, each of the two classes just summarized was subdivided, forming a series of four—in modern terms a morphological series—ranging from very well preserved "modern" flows, complete with cratered cones, all the way to isolated hills or buttes topped with "ancient" basalt, not at all flowlike in form and with no trace of any cones (see Fig. 2.22).

Desmarest's main claim was that basalt, wherever it might be found and in whatever topographical form, represented former lava flows; in particular, its most conspicuous form, with regular "polygonal columns", was an infallible mark of the former existence of volcanoes. Prismatic jointing was rare and indistinct in the lavas flowing from active volcanoes such as Vesuvius and Etna, but its occurrence in some of those flowing from extinct but "modern" volcanoes in Auvergne clinched the proof of its volcanic origin. As Desmarest remarked to Saussure, after the Genevan had visited Auvergne to see for himself what Desmarest had described in print, "you will have been able to see that Auvergne is infinitely more favorable for studying the operations of heat [*feu*] than the environs of Naples and Rome". Desmarest claimed that in mineralogical classification basalt belonged clearly among the volcanic rocks. In terms of earth physics, equally clearly, it found its cause in volcanic heat or "fire"; it had crystallized into a solid rock on cooling from incandescent liquid lavas, just like those that could be seen flowing during an eruption of Vesuvius (Fig. 1.10). The profuse local details in the text, combined with the map itself, were designed explicitly to guide others in the field to the localities that would persuade them of the truth of these conclusions; and in the event they duly did, as for example in Saussure's case.[35]

Desmarest's massive memoir on the origin of basalt was an essay in earth physics based on physical geography; it scarcely dealt with the temporal dimension, still less with any geohistory. The morphological series of four classes of basalt was, at least implicitly, a series representing temporal changes in the appearance of any lava flow, under the progressive impact of erosion, from its pristine "primitive state" to one in which its volcanic character was hardly detectable except by analogy.

Significantly, however, Desmarest numbered those four classes in just that heuristic order, from the clearest (his first class) to the most obscure (his fourth); or, in other words, in the very *reverse* of their geohistorical order from "ancient" to "modern". Indeed, those latter epithets were the only clear indication that he recognized that his explanation of the causal origin of basalt also embodied an interpretation of the geohistory of the region.

Epochs of volcanic activity

Desmarest was not unaware of that geohistorical dimension, but it did not belong, except in passing, in a paper devoted to a problem in earth physics. He made it the focus of a separate paper, which he read at the Académie in 1775, the year after the first had been published. Like Burtin a decade later (§4.2), Desmarest treated geohistory as a genre distinct from either natural history or natural philosophy, or, in his case, from either physical geography or earth physics. The *historical* orientation of the second paper was clear from its full title, for its key word was borrowed from the historical science of chronology (§4.1). It was "On the determination of some of Nature's epochs from the products of volcanoes, and on the application of these epochs to the study of volcanoes".

Desmarest had been using the word "epochs" informally in this context for several years, in letters and doubtless more generally in conversation. Even before he went on his Grand Tour he mentioned how his and Pasumot's mapping had "fixed the epochs" of all the lava flows in Auvergne; and later, after his return, he referred to "the distinction of Epochs, which has not yet been thought of being introduced into the study of natural history".[36]

The latter remark indicates clearly how Desmarest regarded it as an innovation on his part to apply the chronologers' concept of epochs to the natural world, and it is almost inconceivable that he did not use it when he read his paper in 1775. But that work had still not been published by the Académie when, early in 1779, Buffon's *Nature's Epochs* made its appearance to a great fanfare of publicity (§3.2). Desmarest immediately wrote a short summary of his memoir for *Observations sur la Physique*, implicitly staking his claim to the idea, but a fuller version was not published until many years later, by the Académie's Revolutionary successor and long after Buffon's death. Whether Buffon was responsible for blocking it earlier is

---

of the rocks named "basalt" in Classical texts. On his fieldwork, see also Taylor, "Pasumot-Desmarest collaboration" (1994). The whole area is now protected within the Parc Naturel Regional des Volcans d'Auvergne; Champeix is about 20km south of Clermont-Ferrand.

35. The quoted comment is in Desmarest to Saussure, 26 December 1776 (Genève-BPU, Saussure MSS, dossier 9, 85–86); Carozzi, "Saussure on the origin of basalt" (2000), charts Saussure's later vacillation on this difficult problem. Desmarest disagreed strongly with some of Hamilton's inferences about active volcanoes, to judge from a manuscript (anonymous but almost certainly by him) transcribed and analyzed in Taylor, "Commentaire anonyme inédit" (2002); naturalists did not divide into "Vulcanists" (goodies) and "Neptunists" (baddies) as neatly as some modern commentators might like.

36. Desmarest to [Pierre-Jean Grosley?], 11 December 1764; Desmarest to La Rochefoucauld, 15 May 1769, quoted in Taylor, "Geology during Buffon's later years" (1992), 374n; see also Taylor, "Buffon, Desmarest" (2001). Arduino was using similar language ("*epoca prima*", etc.) around the same time, but in a memoir that was never published: see Vaccari, *Giovanni Arduino* (1993), 267–81.

unclear; he was powerful enough, and it would not have been out of character. But in any case Desmarest's argument was weakened by the fact that his maps of the extinct volcanoes—more extensive and detailed than Pasumot's earlier one, and far more persuasive than any amount of text—were not generally accessible for another quarter century or more. Nonetheless, since Desmarest's argument was published in summary form in one of the leading periodicals, it became widely known to savants throughout Europe.[37]

In fact, Buffon and Desmarest used the notion of nature's "epochs" in distinctly different ways. Buffon adopted the primary meaning of the word and defined seven decisive *moments* or "milestones" [*pierres numéraires*] along the earth's programmed developmental pathway (§3.2; see Fig. 4.14). Desmarest used the secondary meaning, and described three extended *periods* in the history of Auvergne, each of which included distinct and successive "ages". A more important difference, however, was that whereas Buffon's epochs were boldly described in chronological order from earliest to most recent, Desmarest again treated his epochs in heuristic order from the clearest to the most obscure. In effect, Desmarest interpreted the two main morphological classes of basalt described in his earlier paper as the products of two distinct "epochs" in the history of Auvergne, and then added a third and still earlier one. But he retained the same kind of numbering, from the youngest to the oldest, because he was deliberately applying what he claimed was a rigorous and reliable *method* for inferring the deep past. His "methodical" or "analytical route" [*marche méthodique* or *analytique*] was to start from the present and penetrate step by step into the more remote past; in this case, to start with the well-preserved lava flows and unmistakable volcanic cones of the relatively recent first epoch and then to penetrate to the earlier second epoch of the more problematic basalts on hilltops, and beyond that to the even more obscure traces of the oldest and third epoch.[38]

The first epoch, then, was the time during which the "modern" lava flows (the *courans modernes* of the earlier paper) had been extruded from points that were still marked by cones of loose "cinders"; the flows were preserved throughout their length, uninterrupted by later erosion, and their upper surfaces were still as rough as the lavas from any active volcano. Yet among them were variants that showed that they had not all been erupted at the same time; they dated from different "ages" within this "epoch". The most recent of all were those with the best preserved cones and craters, with lava flows along the very floors of their valleys. Others showed abraded and less distinct cones, with no clear craters, and lavas that had been left somewhat above the streams flowing down the valleys, as a result of subsequent erosion by those streams.

In Desmarest's view, that same gradual erosion by streams and rivers, carried much further in the course of time, accounted for the far greater fragmentation of the basalts of the older second epoch (the *anciens courans* of the earlier paper). They were those now left high and dry on hilltops or plateaus between the present valleys, with their original length more or less interrupted and broken up, even to the extent that only an isolated butte of basalt might remain. All trace of the cratered cones from which they had originated had long since disappeared, and their upper surfaces had been worn relatively smooth (Fig. 4.10).[39]

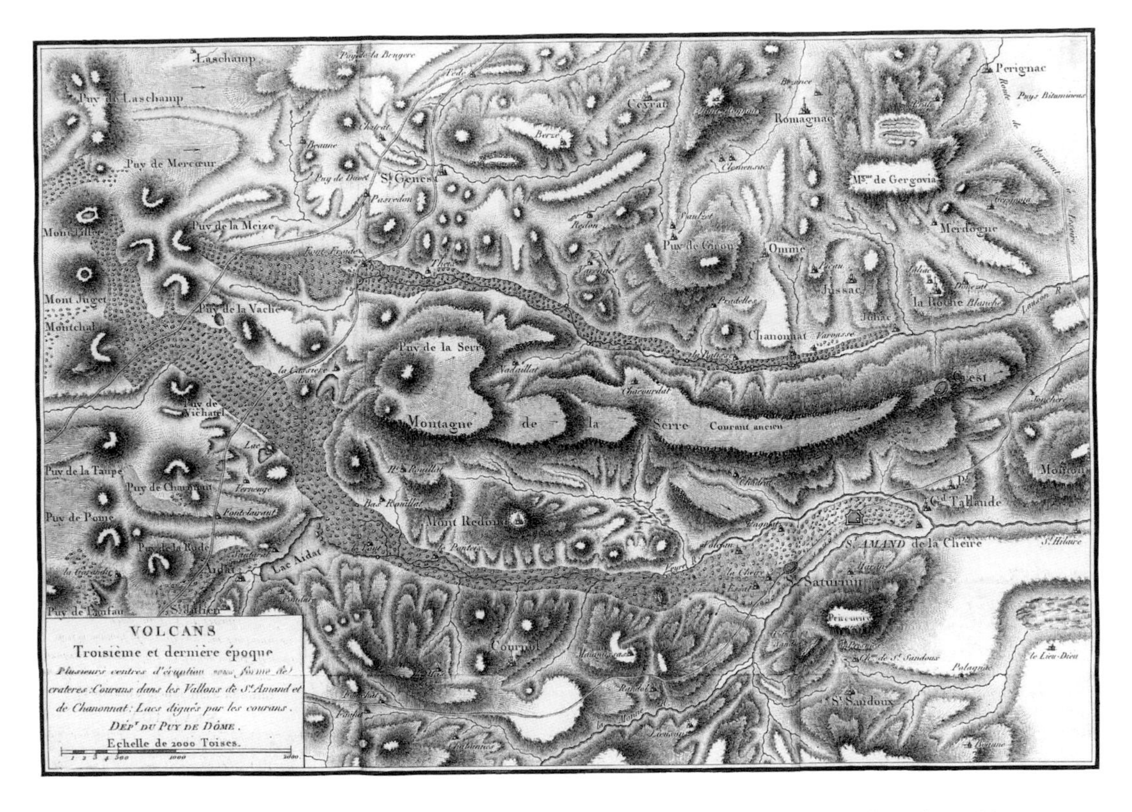

Fig. 4.10. Desmarest's detailed map of an area in Auvergne showing two "modern" lava flows of the most recent "epoch", originating from cratered volcanic cones in the west and flowing eastwards down two parallel valleys; the southern flow has blocked a side valley and ponded back a small lake (Lac Aidat) on the upstream side. Between them is a long narrow plateau (Montagne de la Serre), capped with basalt and sloping gently to the east (the conventional hachuring gives an unavoidably false impression of a stepped profile). Desmarest interpreted this basalt as an "ancient" lava that had also flowed eastwards—small arrows show the slope and inferred direction of flow—and that had become isolated by the later erosion of the valleys on either side; erosion had also destroyed its original volcanic cone. The isolated hill or butte of Gergovia (top right), also capped with basalt, was interpreted as the more fragmentary relic of another "ancient" flow—all the rest of which had been destroyed by erosion—dating from an earlier "age" within the same "epoch". This map was published with the full version (1806) of Desmarest's paper on the "epochs" shown by the volcanic rocks of Auvergne, but was redrawn from the far more extensive map that he and Pasumot had already made when he read the paper to the Académie des Sciences in Paris in 1775. However, by the time this small map was engraved he had changed the original numbering of his epochs, and the "first epoch" in his heuristic or "analytical" order from recent to ancient had been renamed the "third and last epoch" [*Troisième et dernière époque*] in his geohistorical order from ancient to recent. The scale is of 2000 *toises* or about 4km. (By permission of the Syndics of Cambridge University Library)

37. Desmarest, "Détermination de quelques époques" (1779), revised and expanded—but still not as extensively as promised—in "Détermination de trois époques" (1806), with maps of four small but significant areas (see Fig. 4.10). These maps were extracted from a far larger map in six sheets, which by 1794 was already sufficiently near completion for one thousand copies to be ordered by the government; but in the event it was not published until long after his death, as *Carte de Puy-de-Dôme* (1823) (see Fig. 6.2). Some fortunate savants may have had access to it at a much earlier time.

38. Desmarest's method and system of numbering may have been inspired by Steno's similar analysis of the rocks of Tuscany over a century earlier; Steno's famous *Prodromus* (1669) was reprinted in 1763 and was quite well known in Desmarest's time.

39. Fig. 4.10 is reproduced from Desmarest, "Détermination de trois époques" (1806) [Cambridge-UL: CP340:2.b.48.114], pl. 7, showing an area south of the Puy de Dôme and Clermont-Ferrand. Pasumot's earlier map had already shown the plateau of La Serre, but at that time he and Desmarest had not yet mapped the "modern" flows on either side of it.

Finally, Desmarest's fieldwork showed that these older basalts lay—in different places—either on a bedrock of granite or on top of a thick pile of horizontal sediments. Intercalated among the latter were layers of basalt, but they also included beds of consolidated gravel that contained, among other kinds of rock, rounded pebbles of basalt. Desmarest interpreted all these as the products of the "third epoch", a period of extremely ancient volcanic activity of which no other traces remained. Once again, more than one "age" could be distinguished within it, since the pebbles of basalt must have come from volcanic eruptions even older than the layers of basalt included among the sedimentary strata.

Desmarest's "methodical" or "analytical route" proceeded from the most recent and clearest "epoch" back to the oldest and most obscure; but he was well aware that he was also describing the traces of a continuous *geohistory* of the region, from an extremely remote past all the way to the present. The proper method, he argued, was first to follow an "analytical route" from the present back into the past, and then to use those results to *reverse* the movement and construct a geohistory from the past forwards to the present: "Having fixed the circumstances in which the igneous products [*produits de feu*] are found in each epoch, following the analytical route he has adopted in his researches, Mr Desmarest then reverses that order and goes back over those epochs, considering them according to the natural sequence of time."[40]

The final section of Desmarest's paper therefore offered a brief geohistorical reconstruction of Auvergne. The oldest basalts of all were traces of volcanic activity even before the "invasion of the sea", which in turn had led to the deposition of the sedimentary strata. Here were traces of two distinct "ages" within the oldest epoch; and the epoch itself had lasted long enough—implicitly a very long time indeed—for some 150 *toises* (300m) of strata to accumulate. During the following epoch, both the granitic and sedimentary areas had emerged to form dry land, and lava flows from a new series of eruptions had spread widely across the resultant plains, not yet marked by any valleys. This epoch was clearly later than that of the sediments, by the criterion of superposition; equally clearly, it antedated the excavation of the present deep valleys, which had left the older basalts isolated on plateaus and hilltops. Implicitly again, that erosion represented another extremely long period of time, since the observably slow action of ordinary streams and rivers had cut through and exposed those hundreds of feet of sediments. Finally, the most recent epoch comprised the time since most of that erosion took place; during this period still more volcanic eruptions had produced lavas that had flowed down the valleys that by then had been carved out.[41]

Desmarest's reason for insisting on the methodological priority of his "analytical route" over any such geohistorical reconstruction is not difficult to infer. He wanted to distinguish his own epochs—"the result of the analysis of facts" based on fieldwork—from the tissue of conjectures represented by the epochs of Buffon's geotheory, conceived not in the field but in Buffon's study. The rivalry between the two savants was more than a race for priority in the use of a valuable word; it was an argument about the best way to reconstruct a reliable geohistory. Desmarest did not enlarge his geohistory into geotheory; it remained a story about the region he had studied so thoroughly at first hand. He did hint that he thought the same sequence of events had occurred far beyond Auvergne, but even then the reference was to other regions, such as parts of Italy, that he had at least seen briefly for himself. His geohistory was not a comprehensive geotheory, or even a theory of everything in Auvergne; he had little to say about the granite, for example, although other savants would have thought it of the greatest interest because it was nearest to the earth's origin. Above all, Desmarest's geohistory was not deducible from any supposed laws of nature; there was no sense of inevitability about it, no unfolding of a predetermined program of development. On the contrary, Desmarest, like many other naturalists, regarded volcanoes and their eruptions as "accidental" features of nature; in modern terms they were *contingent* events.[42]

It was by the contingent "accident" of sporadic eruptions—first here, then much later there, and so on—that successive moments in the long history of the region had happened to be preserved for scientific study. It was in this sense that what Desmarest was doing made him a *historian* of nature. Throughout his work, he described his material features—rocks, valleys, and so on—as the "documents" and "monuments" of nature, metaphors that underlined their meaning as traces of a distant past that could be recovered and interpreted. And the concluding sentence of his summary paper made this antiquarian analogy explicit and specific: "The diverse witnesses to these changes that the earth's surface has undergone, preserved by the lava, are thus as precious for the naturalist as those artifacts preserved at Herculaneum by an envelope of similar material can be [precious] for the connoisseurs [*amateurs*] of a more modern antiquity."[43]

Like the historians who toiled in dusty archives and the antiquarians who carefully excavated the ruins at the foot of Vesuvius, Desmarest was using the exceptionally favorable circumstances of Auvergne to piece together a fragmentary but true *history*. Borrowing from the chronologers, he could organize this history around a sequence of epochs; Desmarest referred in passing to the "dates" of his natural events, but he made it clear that this alluded only to their sequence, not to their position on a quantifiable timescale.

The historical sciences of his time gave Desmarest powerful analogical resources for reconstructing a reliable geohistory. But his history referred to times far earlier than even the oldest human records. He stressed that his epochs had "nothing or

40. Desmarest, "Détermination de quelques époques" (1779), 123; see also 220, 258 (the text was written formally, in the third person, like the impersonal style of modern scientific papers). Like his method of retrospective analysis, Desmarest's procedure for then reversing the order, to generate geohistory, may have been derived from Steno's example. In "Détermination des trois époques" (1806) his numbering of the epochs was inconsistent; but on its maps he adopted a geohistorical order unambiguously, so that the *last* epoch became the "third" (Fig. 4.10).

41. Desmarest, "Détermination de quelques époques" (1779), 123–26. He made no mention of any fossils in the strata from the oldest epoch and tacitly assumed that any such obviously aqueous deposits must have accumulated on the floor of a vanished sea.

42. The implicit contrast with Buffon is in Desmarest, "Détermination des quelques époques" (1779), 117. See also Taylor, "Volcanoes as accidents" (1998).

43. Desmarest, "Détermination de quelques époques" (1779), 126. Like any other savant, he would have been well aware of current antiquarian research; he must surely have seen Herculaneum for himself while he was in Naples; and he had met Winckelmann while in Rome and later corresponded with him.

almost nothing in common" with those of historians, and that he would not be dealing at all with the "known or suspected times" of human history. Even the most recent of the volcanoes in Auvergne had, he believed, become extinct long before the earliest human records in the region; human history could be tacked on at the end of his geohistory, but there was no overlap between them (except in the sense that the slow erosion of the valleys was still continuing as it had done in the distant past).[44]

A lake on the site of Paris

Desmarest's work was a fine model for turning physical geography into geohistory. But the notion of doing so was still novel, and at first there were few other cases of a similar kind of work. One of the best was published, like Desmarest's article and only three years later, in *Observations sur la Physique*. It showed that geohistorical reconstruction need not be confined to areas with extinct volcanoes or volcanic rocks. Although it did not cite Desmarest's work and made no explicit use of analogies with human history, its affinities are unmistakable. Its author, the naturalist Robert de Paul de Lamanon (1752–87), had already traveled widely in the Alps and elsewhere. His paper was primarily on the fossils found in the Gypsum formation around Paris, which was worked for building stone, and the gypsum itself as raw material for plaster of Paris. There were quarries, for example, on the hill of Mont-

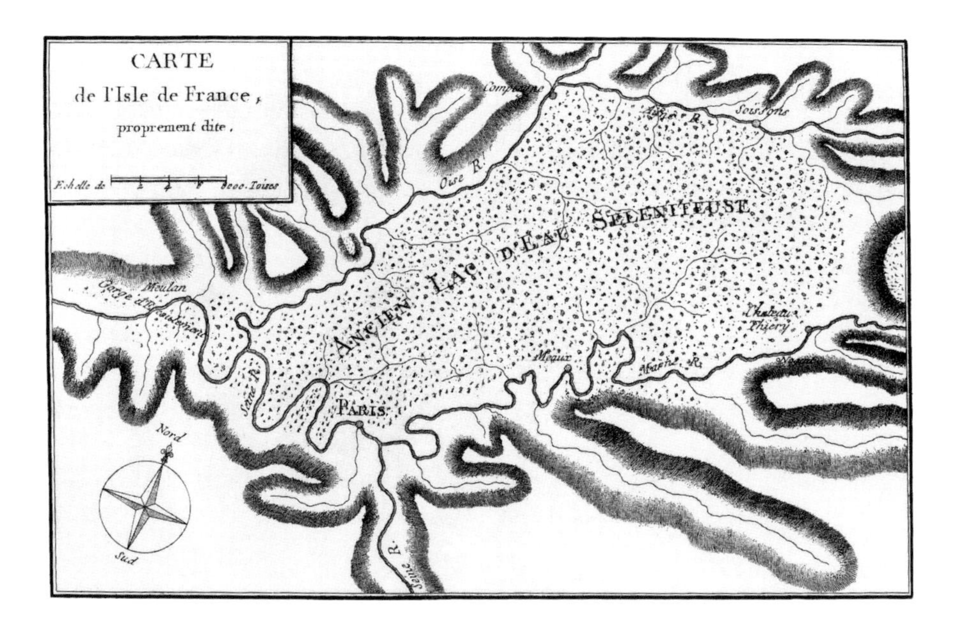

Fig. 4.11. Robert de Lamanon's map (1782) of the Île de France (the region around Paris), with the stippled area representing a vast former lake [*ancien lac*] inferred from the distribution of the Parisian gypsum formation, bounded by the present valleys of the Seine, Marne, Oise, and Aisne and with an inferred outlet at the narrow "gorge" of the Seine (left), downstream from Paris (the scale is of 8000 *toises* or about 15km). Lamanon argued that the gypsum must have been precipitated from water with "sulfate of lime" in solution [*eau séléniteuse*]; later the lake had sustained shellfish of freshwater species, and after it had drained away those same species had continued to live in the present rivers. This kind of reconstruction—in modern terms a *paleogeographical* map—was almost without precedent. (By permission of the Syndics of Cambridge University Library)

martre just outside the city (but long since absorbed within it) and many others further afield; Lamanon used the sheets of the *Atlas Minéralogique* (§2.2) to plot its full distribution.

Lamanon claimed that the fossil bones found in the gypsum at Montmartre belonged to an animal with no living "analogue", of which "the species is lost" (§5.3). More immediately relevant, however, was his further claim that in the shales directly overlying the gypsum were fossil shells typical of *freshwater* mollusks, "shells the analogues of which it would be useless to search for in the sea, but similar to those that are still found in the River Marne" and elsewhere around Paris. This led him to reconstruct a vanished lake—even larger, as he pointed out, than the present Lac Léman (Lake of Geneva)—corresponding in extent to the known geographical range of the Gypsum formation itself (Fig. 4.11).[45]

In effect, this reconstruction transformed the gypsum around Paris and the strata immediately adjacent to it into an episode in *geohistory*, and a surprising one at that. Lamanon inferred that this vast lake had at first borne in solution the mineral selenite (in modern chemical terms, calcium sulfate), the gypseous material itself. After that mineral had been precipitated to form the thick deposits of gypsum—the details of Lamanon's chemical explanation of this are not important here—the lake had sustained freshwater mollusks. At some later time it had drained away, exposing its former bed as dry land, though the shellfish continued to live in the rivers that now flowed around the former margins of the lake. Like Burtin with his "revolutions" and Desmarest with his "epochs", Lamanon left it to be inferred that his former lake dated from long before any of the history known from human records.[46]

Lamanon's reconstruction of this episode in the remote past history of the Paris region alluded in passing to its place in his own geotheory, which postulated a sequence of vast lakes on the sites of the present continents, rather than the former seas of the standard model (§3.5). But Lamanon's paper was not in itself geotheory; this aspect of it, at least, was geohistory. This particular lake—its areal extent, the composition of its water as "selenitic" and later fresh, and its shellfish, extinct mammals, and other fauna—could not have been predicted from any geotheoretical model: these were *contingent* features that could be inferred only from the detailed field evidence of the gypsum and its associated fossils. Lamanon's brief essay was a suggestive sample of the kind of analysis that could also be applied to other kinds of sediments, to reconstruct the geohistorical circumstances of their formation. But Lamanon himself did not live to extend it in that way: a few years later he

44. Desmarest, "Détermination de quelques époques" (1779), 117. Here and elsewhere, the parallels between Desmarest's work and what Burtin did a decade later (§4.2) are probably not coincidental: Burtin cited other authors from *Observations sur la physique*, and had almost certainly read Desmarest's well-known and widely discussed article.

45. Fig. 4.11 is reproduced from Lamanon, "Fossiles de Montmartre" (1782) [Cambridge-UL: T340:2.b.16.22], pl. 3; explanation on 186–88, 192 (the stippling on a hilltop on the right margin of the map was noted as an engraver's error). The portion of the *Atlas* reproduced in Fig. 2.12 lies at the eastern end of Lamanon's former lake. In the present context an "analogue" was a living species *identical* to a fossil one (§5.2).

46. He did suspect, however, that human beings had already been living around the shores of his lake: the period had been prehistoric, but not prehuman (§5.4).

died on the far side of the world, at the tragically early age of thirty-five, while serving as a naturalist—the position that Soulavie had turned down (§2.5)—on La Pérouse's ill-fated expedition to the Pacific.[47]

Conclusion

Desmarest's research on the basalts and volcanoes of Auvergne, even without the full texts and maps that would have made it more persuasive, was a powerful exemplar of how the established science of physical geography could be made the basis for reconstructing geohistory. His work was not, like Buffon's *Nature's Epochs*, a global geotheory concocted in a study, but a history grounded in detailed fieldwork and guided by a careful method of inference. It was a geohistory of one specific region. The contingent "accidents" of sporadic volcanic eruptions in Auvergne had preserved the crucial "monuments" and "documents" that enabled him to reconstruct a sequence of "epochs" based solidly on factual evidence. The antiquarians, chronologers, and other erudite historians provided him with the fertile metaphors and analogies that enabled him to become, at least in relation to this exceptionally favorable region, a true *historian* of nature.

Apart from Desmarest's important paper, there was little other work at this time that used physical geography as a basis for the new genre of geohistory. Nonetheless, he had provided a seminal example of how it could be done; and Lamanon did at least show—perhaps inadvertently—how it might become a more general approach, still linked to specific local "accidents", but not necessarily to those of volcanoes.

4.4 ROCK FORMATIONS AS NATURE'S ARCHIVES

The volcanoes of Vivarais

Desmarest distinguished three "epochs" in the geohistory of Auvergne, on the basis of careful fieldwork carried out according to the standard procedures of physical geography. He was aware of the three-dimensional structure of the rocks and used the criterion of superposition to prove the temporal relation between the pile of sediments and the basalts that overlay them. But his approach remained primarily two-dimensional: he was concerned above all to map and then to interpret the *spatial* distribution of his basalts and lava flows. Lamanon likewise offered an interpretation of the spatial distribution of the gypsum deposits around Paris, but did not allude to any of the other sedimentary formations in the same region. Examples of geohistorical work based on the three-dimensional sequences of rock masses or formations must be sought elsewhere, among those who practiced the science of geognosy (§2.3).

Yet such examples are hard to find in the work that was being done in the heartland of geognosy, the German-speaking regions of central Europe and the Scandinavian lands to the north. There the emphasis remained almost exclusively on the description of the three-dimensional structure of the Primary rock masses and their classification. This is not surprising: these were the rocks of greatest economic importance, as the sites of most of the valuable minerals; and their structural relations were often complex and obscure, and not readily translated into temporal terms, let alone into terms of geohistory. Expressions of temporal or geohistorical meaning were usually confined to generalities about the sequential accumulation of the rocks in the putatively global proto-ocean; they were usually little more than passing remarks that assumed the broad validity of the standard model of geotheory (§3.5). More specific interpretations of the origins of successive rock formations, and more concrete reconstructions of their place in geohistory, are hard to find anywhere in the late eighteenth century. In view of the subsequent use of this kind of evidence—in its modern incarnation as "stratigraphy" (§9.5)—for reconstructing a detailed history of the earth, it is important to avoid anachronism and to stress that such a use of rock masses was quite exceptional in Saussure's time.

The use of geognosy as a basis for geohistory will be illustrated here with just one such exceptional case; but it is an instructive one both for what it achieved and for how it failed. Jean-Louis Giraud-Soulavie, Lamanon's almost exact contemporary,

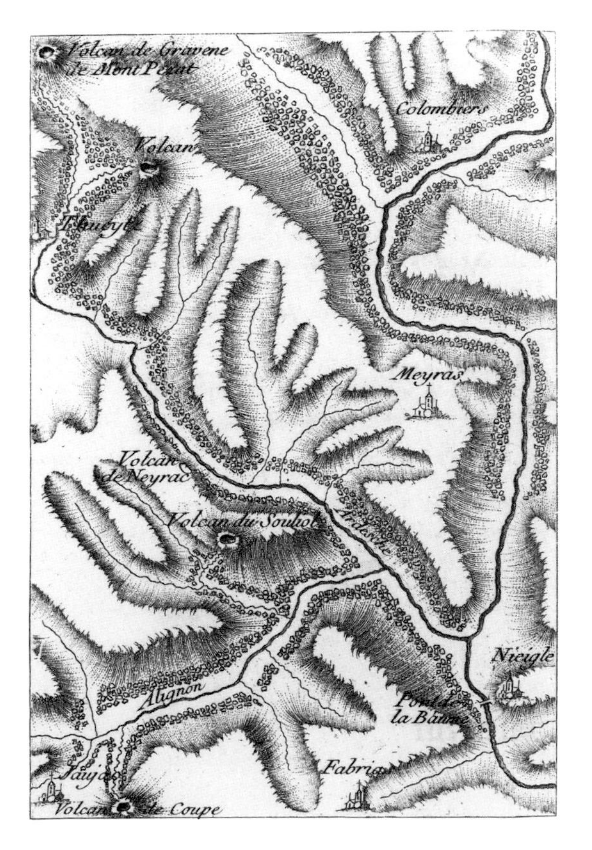

Fig. 4.12. A map of some of the extinct volcanoes in the province of Vivarais, published by Soulavie in 1781 in his *Southern France*, based on his own fieldwork, but surveyed and engraved by the royal cartographer Jean-Louis Dupain-Triel. The map shows four small cratered volcanic cones—others are beyond the area depicted—each with a former lava (shown by the beadlike pattern) flowing down its valley; the rivers had subsequently cut through the lavas, revealing them to be prismatic basalt. On the basis of their topographical relations, Soulavie defined these volcanoes as dating from the "fifth epoch" in his geohistory of the region (see below). The area shown (about 5km by 8km) is to the west of Antraigues, the village in which Soulavie had served as a young parish priest and the site of a similar extinct volcano (Fig. 2.23). (By permission of the Syndics of Cambridge University Library)

47. Pichard, "Robert de Paul de Lamanon" (1992), gives a detailed account of his work based on manuscript sources and reproduces (56) his map of another inferred former lake in the upper valley of the Drac, north of Gap (Hautes Alpes). Cartwright, "Lamanon an unlucky naturalist" (1997), describes his later oceanographic and geomagnetic work.

has been mentioned already for his implicitly huge timescale (§2.5). The son of a provincial lawyer in Vivarais, he had trained for the priesthood and then served as curate [*vicaire*] in the village of Antraigues. For a young man who already had a taste and talents for natural history, this remote spot in the hilly interior of his native province had one precious asset. From his church, perched on a spur above a deep valley, Soulavie could look across to the nearby cratered cone of Coupe d'Aizac, one of the finest of the extinct volcanoes that Faujas and Guettard had discovered a few years earlier and that the former had just made famous by illustrating it in his book on the region (Fig. 2.23). In other valleys within a few miles of Soulavie's village were several similar cones, with lava flows likewise revealed by subsequent erosion to be composed of prismatic basalt (Fig. 4.12).[48]

These volcanoes and basalts were easily recognizable as similar to Desmarest's "modern" ones in Auvergne; Soulavie was familiar with the older naturalist's work and acknowledged its affinities with his own. Desmarest's kind of "ancient" basalt was also found in Vivarais: Soulavie knew, for example, that extensive plateaus to the north and east of his village were capped with prismatic basalt, which had been deeply eroded. Here then was a basis for constructing his own analogous sequence of "epochs" (Fig. 4.13).[49]

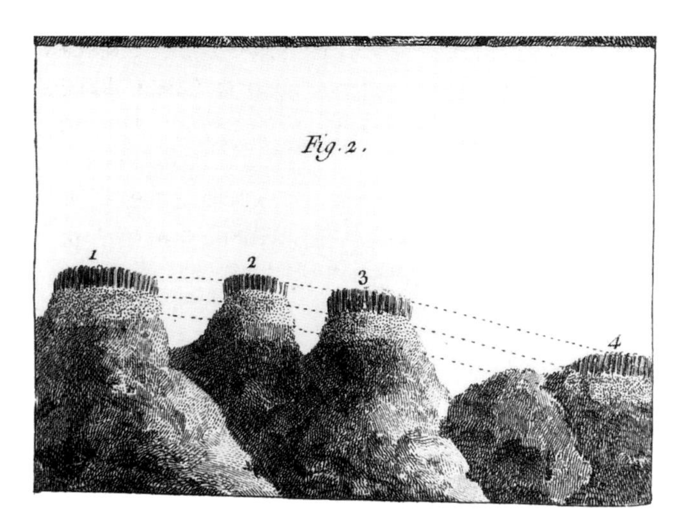

Fig. 4.13. Soulavie's sketch of the volcanic rocks of his "second and third epoch" in Vivarais (see below), as published in 1781 in his *Southern France*. Here he traced with dotted lines the inferred former extent of a sheet of prismatic basalt capping a plateau (the product of a "third epoch" of eruptions), and the underlying bed of ancient gravel or conglomerate (containing basalt pebbles derived from an earlier "second epoch" of eruptions elsewhere), before their erosion to form the present topography (see also the similar locality shown in Fig. 2.31). Soulavie argued that the erosion had been wholly due to rain and streams, which "have cut these solid rocks almost sheer, and the same waters are still destroying these ancient structures every day, and above all during heavy rains". Like the rest of his geohistorical analysis, this kind of eroded topography was compared explicitly to the evidence of *human* history, and more specifically to the antiquarian studies of his time: "a former contiguity of the horizontal formation [*terrain*] should be assumed with as much confidence as when one assumes that a stone has been broken, if it bears only one half of an inscription". (By permission of the Syndics of Cambridge University Library)

Soulavie did not stay long in his rural parish: ambitious and talented, he had powerful ecclesiastical support when, just as Buffon's *Nature's Epochs* was published, he moved to Paris to build a career as a savant. He outlined his research to the Académie des Sciences and then returned to southern France for more extensive fieldwork. Back in Paris in 1780, he published a paper on "nature's geography" in *Observations sur la Physique*; in effect it was a prospectus for a comprehensive work on the region's natural history and especially its "mineralogy". He offered his subscribers a free copy of his new map of Vivarais, but he also described how it was based on another construction that reveals more clearly the kind of work he was doing. He had made a *three-dimensional* map of Vivarais out of seventy pounds of clay, sculpted to model the topography of the province and subdivided into separate blocks. The surface was painted in colors to show the outcrops of the various rock masses he had distinguished, while the sides of each block showed their structural sequence; the model as a whole thus displayed the inferred solid geometry of the rocks (see the similar block diagram in Fig. 2.16). Soulavie described his work as "subterranean geography" rather than "geognosy"—a term that did not become common in French until later—but geognosy was clearly what he was doing.[50]

Nature's erudite historian

Equally clearly, however, Soulavie planned to turn his three-dimensional and structural study into geohistory, for he claimed that the pile of "six superposed formations [*couches*]" that he had mapped were "the products of six separate and distinct epochs". He claimed that this "ancient history of the terrestrial globe", unlike speculative geotheoretical systems, could be founded on "an unquestionable principle, amenable to the most rigorous mathematical demonstration." This referred to the solid geometry of rock masses and to the principle of superposition. Evidently he thought this principle would still be unfamiliar to his readers, for he explained it by using a simple but vivid analogy with the well-known church of Sainte Geneviève near the Sorbonne in Paris. He pointed out that the hill on which the church was built obviously dated from before the towers, and the foundations of the towers from before their upper parts; in the same way the lower formations in Vivarais must unquestionably date from before those that overlay them and the underlying bedrock from some even earlier time. Superposition offered a reliable way to turn a pile of formations into geohistory.[51]

48. Fig. 4.12 is reproduced from Soulavie, *France méridionale* 4 (1781) [Cambridge-UL: MA.47.30], pl. 5. More thorough surveys later showed that Soulavie had idealized the basalts: the gorges are always to one side, carved not through the solid basalt itself but along the weaker line between it and the basement rock; the flows are also less extensive than shown here. Antraigues is in another side valley, east of this area and upstream from Vals-les-Bains. The River Ardèche, which runs diagonally across the map, later gave its name to the *département* in which the volcanic area lies; the town of Aubenas is further downstream to the southeast. Ellenberger, "Sources de la géologie française" (1979), is an invaluable guide for "re-treading" the fieldwork of Soulavie (and other savants) in the Massif Central.

49. Fig. 4.13 is reproduced from Soulavie, *France méridionale*, 4 (1781) [Cambridge-UL: MA.47.30], pl. 1, fig. 2, explained on 122–23. His own map of the area (pl. 3) shows that the apparently isolated buttes 1–4 were in fact spurs projecting from the edge of the plateau, seen in foreshortened view; as usual the vertical was greatly exaggerated to clarify the point being made. The locality was near Dornas, on the east flank of the dissected plateau of Mont Mézenc, north of Soulavie's village of Antraigues. The second quotation in the caption (referring to the similar plateau illustrated in Fig. 2.31) is from ibid., 2 (1780), 393.

50. Soulavie, "Géographie de la nature" (1780), 65; see also *France méridionale* 1 (1780), 144–49. On "géographie souterraine", see Ellenberger, "Vocabulaire de la géologie" (1983). Soulavie's model also marked the vegetational zones, as they related to altitude, but botany was subordinate to mineralogy. Later he offered copies of this "carte en relief" for sale, noting that it would fit conveniently into a drawer of the "cabinet de physique" kept by many savants to house their other scientific models, instruments, etc.: see *France méridionale* 3 (1781), "Avis". No example of the model—perhaps the very first of a kind familiar to modern geologists—appears to have survived, and possibly none (apart from the original) was ever made. His published map, and his profile of vegetational zones, are finely reproduced in Broc, *Montagnes* (1991), figs. 14, 15.

51. Soulavie, "Géographie de la nature" (1780), 64–65. One tower (Tour de Clovis) from the old church is still extant; no expert knowledge of architectural history is needed to see the contrast between its Romanesque lower part and its sixteenth-century upper part. In Soulavie's time, Louis XVI was replacing this church with a new one in Classical style on an adjacent site, now known by its secularized Revolutionary name as the Panthéon.

Soulavie's massive work, the first two volumes of which appeared later the same year, was entitled *Natural History of Southern France* (1780–84). Its intended subject matter was as wide as Buffon's own great *Histoire naturelle*, although it was to be regional rather than global in scope. The long subtitle began by describing it as "Researches in Mineralogy", but Soulavie made it clear that these volumes were merely the "mineralogical" portion of a work that would also cover the animal and plant kingdoms (in the event, only one botanical volume appeared, and none on zoology). He also construed "natural history" in a characteristically Enlightenment sense that included humanity itself; human life would be related to its natural environment. And the subtitle placed the work squarely in the tradition of *local* histories or chorographies (§2.2), for Vivarais was just the first in a list of no fewer than twelve provinces and dioceses, and Soulavie's volumes consisted largely of a series of monographs on the natural history of these specific regions.[52]

However, in his introductory essay Soulavie stressed that all this local descriptive work was to serve as the basis for reconstructing *geohistory*. As with superposition, he could not assume that the idea of geohistory was familiar, and indeed he stressed its novelty: he explained how, after making "a large collection of facts" about the *present* state of things, "one can unravel nature's past epochs, and it is only in our own time that naturalists have conceived the idea of doing so." Like Desmarest (though without mentioning him by name), he proposed investigating these traces of the past in order of increasing age, "from the best known to the least known", back along a "chain of causes" to nature's most ancient features, before reversing direction to construct "the chronological history of the earth". But unlike Desmarest he boldly numbered the resultant "epochs" in their true historical order, from the most ancient to the most recent. Tactfully, he praised Buffon's similar work (§3.2) on "the history of nature's past ages", though he rejected the use of global cooling as the "clock [for] calculating the very ages of the world". As a more reliable criterion he proposed the steady retreat of the seas—"now recognized for a fact", as he put it, alluding to the standard model (§3.5)—though he recognized that it was hardly possible to derive any quantitative dates from this process. Above all—and not merely in his use of "epochs"—Soulavie made abundantly clear his intention of tracing this geohistory by working analogically with *human* history. For example, the regions of granites, limestones, and volcanic rocks each offered "monuments" of the world's epochs, and local observations allowed each to be "assigned its natural place in the *annals of the physical world*", constituting "the earth's chronological history". The natural monuments observed in the hills of southern France could be used "with even more ardour than the literary scholar [*littérateur*] or genealogist who digs in libraries and archives to unravel the civil events [*faits moraux*] of the human species, to assign dates and to give them a chronological order" in the history of nature. Soulavie defined himself as *nature's* erudite historian.[53]

This work of geohistorical reconstruction was set out more fully when, after three volumes, Soulavie suspended his regional descriptions "in order to meditate on the vestiges" of the extinct volcanoes. Like Burtin and Desmarest, but more explicitly than either, Soulavie recognized that geohistory was distinct from descriptive natural history. His fourth volume—which was also sold separately, and probably reached an even wider circle of readers—was "The Physical Chronology of the Eruptions of the Extinct Volcanoes of Southern France" (1781). The key word "chronology" indicated that the historical metaphors that pervaded all his volumes would here become the focus of the work. As before, Soulavie claimed the novelty of what he was doing: "in natural history as in human [*morale*] history, there exists a geography and a chronology, coins and monuments, and an unknown art—even more sublime—*the art of verifying the physical dates and epochs of nature*". Soulavie—not for nothing a lawyer's son—recalled that a century earlier Louis XIV had decreed that notaries must file dated documents for all civil contracts, to prevent forgery and allow for the later verification of their authenticity; "in nature, in the same way, there are checked registers [*registres de contrôle*] that place the successive events of the physical world in their natural order, and which demonstrate the reality of the various periods and ages of nature". The meaning of rocks, valleys, and volcanoes could not be discovered, he claimed, "unless the observer knows how to construct their chronology, to assign them to their respective epochs of formation, to decipher these old charters [*chartes*] of the world and the succession of physical events, just as the erudite [historian] forms a chronology of the deeds [*gestes*] of a people by comparing the documents in the nation's archives". Although the physical laws of nature did not change, nature, like the human world, had had its revolutions: "The craggy hills alone offer inscriptions of these events, just as our ancient buildings offer reliable signs of human revolutions, battles, and conquests: the naturalist who admires and describes, but goes no further, is like the slavish artist who copies the ruins of ancient monuments without telling us what they were". In sum, nature's archivist had a viewpoint distinct from, and higher than, that of the traditional and merely descriptive naturalist.[54]

This rich repertoire of historical metaphors was put to work in Soulavie's reconstruction of the geohistory of his native region. Notwithstanding his repeated invocation of "dates", and indeed of "annals" and "chronology" itself, in practice what concerned him was the *relative* ages of the events that had left their traces in the rocks; like Burtin a few years later, he emphasized that the precise dates sought by chronologers for human history could not be matched in the natural world. His method was to be based on three criteria of relative age, which he listed in order of their reliability. Superposition was the most trustworthy, and the preferred criterion wherever it could be applied; the degree of subsequent erosion was a useful guide, though more equivocal; relative altitude—which depended of course on the validity of the standard model's assumption of a steadily falling sea level (§3.5)—was now treated as the most fallible and subject to important exceptions. However, all three required arduous fieldwork for their application, and Soulavie lost no

52. Soulavie, *France méridionale* 1 (1780), title page; it was to include a treatment of "l'homme & la femme de ces contrées", or in modern terms their ethnography. Lengthy subtitles served to advertise the scope of a work (acting like the publisher's blurb on a modern book), so the string of names also had a marketing function. On the broad intended scope of the work, see also the outline, ibid. 1: 1–2; on the three parallel series of volumes, see ibid., 7 (1784), "Avis au relieur".

53. Soulavie, *France méridionale* 1 (1780): 1, 10–11, 30–33.

54. Soulavie, *France méridionale* 4 (1781): 7–8.

opportunity to castigate those who studied the earth only from the comfort of their indoor cabinets: mere classifiers or *nomenclateurs* were repeatedly contrasted with true *observateurs* such as himself.[55]

Soulavie distinguished six "epochs" of volcanic activity in Vivarais, amplified from the three that Desmarest had defined in Auvergne. His "physical chronology" of the eruptions was to run all the way "from those that were close to the earth's formation to those that are described in [human] history". The first was related to the most ancient granitic rocks; the second was contemporary with an overlying series of limestone formations; and the third was represented by the sheets of basalt on plateaus (Fig. 4.13). The fourth was contemporary with the long-continued erosion of all these rocks, forming the present valleys, and the fifth saw the subsequent eruption of volcanoes within those valleys (Fig. 4.12). The sixth and last was represented by similar volcanoes that he suspected had erupted within the human period; the Coupe d'Aizac at Antraigues (Fig. 2.23) was, he thought, an example. Like Buffon, Soulavie explicitly hitched human history onto the tail end of a far longer and largely prehuman story, though his human epoch—unlike Buffon's but like that of Genesis—was the sixth.

As this bare summary implies, Soulavie's "epochs" of volcanic eruptions punctuated a still fuller geohistory that was also recorded by a variety of other rocks. Even the granites were of two distinct ages, the first being the "foundation" of the region. The overlying limestones comprised three successive formations, distinguished by their fossils (§5.2); together they recorded an immensely long period of three successive "ages" during which the region was still wholly submerged beneath the sea. Above the limestones but below the basalts on the plateaus—and therefore intermediate in age—was a thick bed of "fluviatile gravel" or conglomerate, an "astonishing bed [that] even scares the imagination" with its mass of varied pebbles, interpreted as the waterworn debris of the oldest mountains; pebbles of basalt witnessed to earlier volcanic eruptions (Fig. 4.13). This bed also contained freshwater shells and fossil wood and bones, indicating that with the steadily falling sea level the region had now emerged to form a continent, with land animals and vegetation as well as fresh water. Then came the sheets of basalt that had preserved these older rocks, "just as the lavas of Vesuvius have preserved the antiquities of Herculaneum for us". Still later was the slow erosion of the whole region to form the present deep valleys. Finally came the eruption of small volcanoes within those valleys: some ancient enough to have had gorges eroded subsequently through their lava flows (Fig. 4.12); others recent enough for a faint memory of their fiery origin to have been preserved—so Soulavie claimed—in textual records from late Roman times and even in their vernacular names.

Censors and critics

Soulavie's timescale for all this geohistory was implicitly vast, principally because like Desmarest he was convinced that the deep erosion of the region had been due mainly if not solely to the ordinary action of rain, and of the streams and rivers that still flow down the valleys. But he offered no quantitative figures, at least in public, less for fear of criticism from other clergy than because he had no good evidence on which to base any such dates (§2.5). Unlike the personal attacks by his clerical critic Barruel, the official disapproval that Soulavie's work aroused in Paris was not on account of his implicit timescale.

His first two volumes were given the coveted endorsement [*privilège*] of the Académie des Sciences, but only on condition that two short chapters be excised (a few copies had already been printed, and they show the original text). This was censorship by the Académie, not by Soulavie's ecclesiastical superiors; the objections were not to the religious implications of what Soulavie had written, but to its scientific status. In the first of the excised chapters, Soulavie offered an explanation of the fossils in his three successive limestones in terms of the "dégénération" (in modern terms, evolution) of one set of species into another, attributing the change to nature's own inherent powers (§5.2). In the other chapter, he rejected the biblical Flood as an explanation for all the formations filled with fossils—as was usual by this time among savants, Burtin being just one (§4.2)—yet also claimed that other physical evidence did confirm the reality of that event at a still later time. The Académie recorded that it "neither approved nor condemned" these chapters, but that as a body devoted to physical researches it could not extend its formal approval to this "theological part" of Soulavie's work: in effect, he had ignored the tacit and tactful boundary that the savants at the Académie had worked hard to define and to have respected by others. Ironically, therefore, Soulavie's work was censored in part because he *had* tried briefly to equate "Genesis and geology".[56]

Soulavie did not meekly accept this and other criticisms. In his volume on natural chronology, he concluded his reconstruction of "epochs" with a spirited defense of his work. He argued that his critics had wrongly equated geohistory with geotheory: "it has been assumed that the *chronology of events* was a product of the imagination, a systematic [i.e., geotheoretical] work offered as recreation for the mind, as a novel [*roman*] that was more or less agreeable to read". He claimed that determining a sequence of events in geohistory was, on the contrary, quite different from explaining them causally, and that the former merited far greater confidence. He insisted that his chronology of past events was firmly based on irrefutable field evidence and quite distinct from any causal suggestions he had made in passing to explain those events: "I stand not at all by the physics [i.e., causal explanations] that I have offered for these ancient events, but by the comparative chronology of the processes". His suggestions about a possibly warmer climate in the past or the

55. Soulavie, *France méridionale* 4 (1781): 179–82; the degrees of reliability were termed respectively "évident", "probable", and "paroît vraisemblable". His earlier ambition to derive quantitative estimates of elapsed time (§2.5) seems to have been abandoned, or at least shelved, in the face of the courteous criticism he had received from his former neighbor and priestly colleague Roux.

56. The censored chapters (9 and 10) are preserved in a few extant copies of Soulavie, *France méridionale* 1 (1780): 348–64; I have used the copy originally in the library of the abbey of Saint-Victor, which was transferred to Paris-MHN at the Revolution. In most copies, the excised pages are replaced with new material added to the previous chapter; but the excision was hardly concealed, since the censored chapters were still listed and summarized in the table of contents. Soulavie claimed that calcite in cracks in the granite on one of the highest hills in Vivarais could have been precipitated only during a Flood that covered the whole region, *after* all the other events he had reconstructed (360–62 in uncensored text). See Aufrère, *Soulavie et son secret* (1952), 73–75, which quotes from the archival record of the Académie's deliberations.

transformation of one set of species into another were comparable to the conjectures of other savants about the possible central heat of the earth or changes to the ecliptic. All were in the realm of hypothesis [*l'hypothétique*]: they were conjectures that might be useful for guiding further research, but they were not to be confused with reality [*le réel*]. Finally, Soulavie tried to recruit the powerful Buffon to his own side, or at least to neutralize his opposition, by claiming that *Nature's Epochs* had established its author as "the first annalist of nature", and that the purpose of that work, just like his own, was "to prove the succession of events in nature rather than to describe its physics" (a reading that Buffon would surely have repudiated). So Soulavie claimed that geohistory was a respectably scientific project, and a new one at that, distinct from either descriptive natural history or speculative geotheory.[57]

Soulavie's massive work—eight volumes were published over four years before the series petered out—had the highest credentials. As already mentioned, the first volumes appeared with the coveted approval of the Académie des Sciences, and later ones had that of Louis XVI himself. The list of nearly two hundred subscribers was headed by the royal family and included most of the relevant leading savants around Europe, among them Buffon, Faujas, Guettard, Lamarck, and Lavoisier in Paris, Saussure in Geneva, Hamilton in Naples, and the Royal Society in London; also included were no fewer than twenty-one bishops and archbishops, and twenty other clergy, among them one of the professors of theology at the Sorbonne. Soulavie's ideas—including prominently his ideas about how geohistory could be reconstructed—were certainly not obscure or neglected. Buffon thought he had made the elementary mistake of identifying a coarse sandstone as a granite, "but this is the least of this young curate's blunders; he's only a student and writes with the tone of a master." Yet despite the sneer—Soulavie was still in his twenties at the time—Buffon took him seriously enough to pay good money for *two* sets of his volumes (perhaps one for Paris and one for his country retreat).[58]

Exporting geohistory to Russia

There was an important ambiguity about Soulavie's geohistory. It was based almost exclusively on what he had personally observed during his extensive fieldwork in southern France, so it could claim to be a reliable reconstruction of the history *of that region*. Yet Soulavie himself referred repeatedly to his epochs and his history as being those *of the world*. For example, since the beds with fossil plants were underlain in Vivarais by the limestones full of fossil shells, he inferred that marine life had appeared on earth before land plants (a sequence that appeared to contradict the Genesis story). Had he treated this geohistory as a purely local narrative and conceded that other regions might have had a different sequence of events, he would have avoided Barruel's criticism that he was rewriting Genesis, or at least he would have blunted its edge. Although Soulavie had insisted on the distinction between geohistory and geotheory, it therefore seems likely that he had ambitions to establish a geotheoretical "system" that would be founded on more adequate geohistorical grounds than, say, Buffon's. This interpretation is compatible with what turned out to be his last major work on natural history.

While Soulavie's volumes on southern France were being published, the academy at St Petersburg offered a valuable prize for an improved classification and nomenclature of rock masses. The question was almost certainly formulated by Johann Ferber (1743–90), the Swedish mineralogist who had just been appointed the academy's professor of natural history: he had already traveled widely around Europe, and had published accounts of the mineralogy and geognosy of several countries; it was his report of meeting Arduino in Venice, for example, that had made that savant's geognosy widely known outside Italy (§2.3). Anyway, the prize question for 1785 was worded mainly in conventional geognostic terms; but it did briefly suggest that other features might also be useful for classifying rock masses: "the diversity of their origin and their antiquity could also be determined, by making known by what natural process they have been formed in the course of the successive revolutions of our globe, and to arrange [them] in classes relative to those epochs".[59]

Soulavie seized this opportunity to extend and generalize what he had been doing in southern France, and duly sent an essay to St Petersburg. He praised the savants there for asking for "a history of the first ages of the world" rather than a more conventional kind of work on minerals and their uses. Yet in fact the question had called for a classification that would be of practical use and had referred only briefly to the earth's history. In effect, therefore, Soulavie bent the question to his own interests. He offered a novel classification in which geohistory became the *primary* criterion. It would be founded not on museum specimens but on the rocks in mountain regions; not on chemical tests in a laboratory but on fieldwork, "the study of nature in its own laboratory". Soulavie claimed that "in each of the distinctive epochs of the physical world, nature has formed a particular class of minerals . . . each of these classes thus represents a global process in nature, a major epoch, a large event, an age, a revolution". This simply made a standard feature of the standard model of geotheory more explicit, namely that the main types of rock had been formed in succession, each in its own period of the earth's development (§3.5). But Soulavie elaborated that idea into a hierarchy of classification: "in sum, each age corresponds to a class, each order to a revolution, each phenomenon to species, each accident [i.e., particular instance] to varieties".[60]

In the body of his essay Soulavie used these categories to turn the conventional framework of geognosy into a schematic geohistory. He defined nine successive "classes" or "ages", beginning as usual with the granites as the products of the first and oldest, and extending through the Primary and Secondary rock masses to

57. Soulavie, *France méridionale* 4 (1781): 183–85.

58. Buffon to Faujas, 3 October 1781, printed in Nadault de Buffon, *Correspondence inédite de Buffon* (1860), 2: 109–10. The subscribers are listed in Soulavie, *France méridionale* 6 (1782): 5–16; that he printed the list here, rather than in the customary place in the *first* volume, was probably a move in his ongoing battle with his critics. The tangled micropolitics of the publication of Soulavie's work apparently involved tension between the Académie and Louis XVI's court.

59. The prize question, dated 10 October 1783, was later printed in *Acta Petropolitana* for 1783 (1787), Histoire 151–52, and in the introduction to [Académie de Saint Petersbourg], *Mémoires présentés à l'Académie* (1786), in which two of the essays submitted were published (see below).

60. Soulavie, "Classes naturelles" (1786), 5.

reach the "volcanoes ancient and modern" as the products of the seventh. Soulavie also defined four successive major "revolutions", not clearly correlated with his "ages" but likewise contributing to an outline of geohistory. The first, once again, was that of the granites; the second, that of the earliest volcanoes; the third, "the first reign of living beings" as recorded by the fossils in the Secondary formations; and the fourth and last, the production of the present topography and the appearance of quadruped animals. The whole temporal sequence was summarized quite conventionally as "this passage from primitive chaos towards the repose of the elements and the reign of man".[61]

All this amounted indeed to a schematic geohistory—a sequence of distinct and unrepeated periods—but compared with Soulavie's volumes on southern France it was sketchy and muddled. Even the analogies with human history were somewhat perfunctory, though the obscure Primary rocks were described vividly as the "fleeting scraps [*lambeaux fugitifs*] [of] nature's heroic ages", like the mythical periods before reliably recorded human history. There was nothing on the timescale of geohistory. All in all, Soulavie's essay must have been a disappointment even to his admirers. His geohistorical approach had been richly suggestive when applied to a specific region; but when extended explicitly to the global scale, it added little to the already conventional features of the standard model of geotheory.

Soulavie did not win the competition. The academicians at St Petersburg gave the prize to Karl Haidinger (1756–97), a former teacher at the Schemnitz mining school and a mines administrator and mineralogist at the natural history museum in Vienna. His essay outlined a traditional geognostic classification, arranging the rock masses [*Gebirge*] hierarchically under classes, orders, genera, and species, primarily by their mineral composition. There was almost no reference either to causal issues or to relative ages, except for the standard comment that ages could be determined only by superposition. Even the terms "primary", "secondary", and "tertiary" were not used in Arduino's sense, but to characterize successive "orders" within the "class" of what others called Primary rocks; and Haidinger showed little interest in the rocks of more recent periods, describing only breccia and sandstone among those that others called Secondary.[62]

The runner-up (first "*accessit*") in the competition was Louis de Launay, who like Burtin was a civil servant and amateur naturalist in Brussels. Launay had already given the academy of sciences in that city a "discourse" on the "Theory of the Earth", to which in the usual way (§3.1) he had appended some empirical material in the form of an account of the "accidental fossils" of his province. In passing, he had adopted—explicitly from Buffon's *Nature's Epochs*—the idea of fossils as the "precious monuments [of] these ages [*siècles*] buried for us in the night of time", but otherwise his approach had hardly been geohistorical at all. Launay's essay for St Petersburg was equally conventional, being mainly a catalogue of "earths and stones" classified into genera, species, and varieties, and defined briefly by their constituent minerals. Prominent localities were noted for each, just as they would have been in a botanical or zoological classification, but there was no suggestion that the rocks were in any structural order, let alone a chronological one. Like Haidinger, Launay treated the Primary rocks in much more detail than the Secondaries; and his geohistorical remarks were confined to the usual inference that the Primaries were truly "*primitives*" in that all of them "appear to have been earlier than organic Nature" and most of them "seem to be as ancient as our globe".[63]

Soulavie was put in third place (second "*accessit*"), though his and Launay's essays were printed in full by the academy (Haidinger's having been published separately). Most hurtfully, the judges rejected Soulavie's main thesis—his use of geohistory as a basis for classification—as a mere "hypothesis" of the kind that the academy had repudiated, and even his third place was awarded only because some unspecified "ingenious ideas and very interesting details" merited recognition.[64]

This criticism of Soulavie's work is rather puzzling. In the same year that the Russian academy published his prize essay, its *Acta* contained the first part of a lengthy paper by Ferber, entitled "Reflections on the Relative Antiquity of the Rocks and Earthy Beds that Comprise the Crust of the Terrestrial Globe". The wording suggests that this was in effect his own response to the academy's question. Most of his paper, and perhaps all of it, was read to the academy *after* the deadline for submission of essays for the prize, so that Ferber would have had the benefit of a preview of Soulavie's work; certainly he knew of, and cited, the earlier article that Soulavie had published as a prospectus for his volumes on central France. Anyway, Ferber's paper shows that he for one was thinking seriously about the feasibility of geohistory, despite the academy's curt dismissal of Soulavie's ideas as mere "hypothesis". Possibly Ferber was overruled by his colleagues; or he wanted to steal Soulavie's thunder; or the Frenchman's essay changed his mind and convinced him that reliable geohistory was possible after all.[65]

Ferber began by coming close to claiming that no true *history* of the prehuman earth was possible, by definition, because no humans had been present to witness and record what had happened. Had there been good observers present, he claimed, we would by now have "the complete history of the earth, the physical Chronology of its revolutions"; but despite his use of Soulavie's distinctive phrase, Ferber made that chronology dependent on human witnesses. While accepting that we can find traces of past events only by studying the present state of the earth's crust, Ferber rejected the efforts of those who had tried to "dig in nature's archives" on the grounds that they had produced little more than an amusing tale [*roman*]:

61. Soulavie, "Classes naturelles" (1786), 99–161. Rather surprisingly, his last two "ages"—presumably even more recent than the "modern" volcanoes—were those of coal and rock salt respectively, choices that hardly accorded with the fieldwork of other geognosts.

62. Haidinger, *Eintheilung der Gebirgsarten* (1787): the untitled "erste Klasse" was subdivided into three "Ordnungen", identified as "monti primarii, secondarii [and] tertiarii", but all these were the Primary rocks of other geognosts (10–73); the "zweite Klasse" of "zusammenküttete Gebirgsarten" was described much more briefly (74–82). The essay was first published in St Petersburg (1786); it was one of those cited by Werner (§2.3) as a reason for publishing his own *Kurze Klassifikation* (1787). See Flügel, "Haidingers und Werners 'Klassifikation'" (2003).

63. Launay, "Histoire naturelle des roches" (1786), 31, 87; the same year he published in Brussels an enlarged version in book form. On his earlier work, see "L'Origine des fossiles" (1780), 528–29, 566; a lack of illustrations made it far less useful than the later work of his compatriot (and likely rival) Burtin.

64. Anonymous, *Question minéralogique* (1786), introduction to essays by Launay and Soulavie.

65. Ferber, "L'Ancienneté relative" (1786–88). The first part—most unusually—has no date of reading attached; the second and third parts were read in January and February 1786; the deadline for the prize essays had been 1 July 1785.

the allusion was clearly to Buffon, but perhaps to Soulavie too. However, Ferber then developed a less negative view of the feasibility of geohistory. He attributed the geognostic pile of formations [*couches*] to a temporal sequence of "revolutions" and inferred that the different kinds of formation "differ greatly in date and origin [*naissance*] and in range of antiquity". This was all quite conventional; but Ferber then applied the idea to geognostic classification, in just the way that the prize question had suggested. The two or three major classes (i.e., Primary, Secondary, and perhaps Tertiary), defined in part by their relative age, could be increased in number, he suggested, because "rock masses [*montagnes*] or rocks of which the epochs of origin were far removed from each other have [hitherto] been mixed up and confused in the same class". Ferber took this as self-evident for the Secondary formations: "all sandstones or limestones are certainly not of the same age". But he suggested that it could also be applied to the Primaries, and he anticipated that further research would allow for an increased number of classes, as the rock masses were assigned to their correct places "in the physical Chronology of the globe". So Soulavie's concept was accepted after all, however grudgingly, as an analogy that was *not* restricted to the period of human history: Ferber concluded that the classification of rocks could properly aspire to be based on their "relative antiquity". Yet his ambivalence about extracting geohistory from the structure of rock masses indicates how the new approach was still far from becoming part of the standard practice of geognosy.[66]

For Soulavie the rebuff from Russia may have been the last straw. His great work on southern France had been published with even its "mineralogical" volumes incomplete, and the botanical series hardly begun. The scurrilous personal attacks by his relentless critic Barruel (§2.5) had long been undermining his will to continue, and it is significant that around this time he offered to sell his whole mineral collection to van Marum, who was visiting Paris. However, even if his natural history had received greater acclaim, Soulavie might still have left it unfinished, for he had already begun to develop another area of scholarly interest. Having proclaimed himself *nature's* erudite historian, it is hardly surprising that he was now becoming a prolific historian of *human* affairs. Soon after the outbreak of the Revolution, his nine-volume account of the politics of the reign of Louis XV began to appear, signaling the start of his second scholarly career. Based on intensive study of the papers of the duke de Richelieu, and those of other courtiers, this was erudite history of just the kind that he had tried to transpose into the natural world.[67]

Conclusion

Soulavie's work has been worth describing in some detail because it was perhaps the supreme example—around the time of Saussure's ascent of Mont Blanc—of a consistent attempt to apply the methods and concepts of human history to the three-dimensional field science of geognosy, in order to generate a thoroughly *geohistorical* interpretation, not only of a striking landscape but also of the sequence of rock masses that underlay it. Yet it is a sign of the novelty of this approach that Soulavie's reconstruction of the geohistory of southern France had so little immediate impact on those studying other regions; and that Ferber, a leading representative of the mainstream Germanic tradition of geognosy, was at best ambivalent about Soulavie's extension of his historical interpretation to the whole earth, and at worst inclined to dismiss it as a speculative exercise of little scientific value.

4.5 GLOBAL GEOHISTORY

Causal processes and geotheories

Around the time that Saussure climbed Mont Blanc, a few savants were beginning to explore how the descriptive and classificatory work of natural history might be used as the basis for what they recognized as a *novel* kind of scientific practice—the detailed reconstruction of the *history* of the earth—which modern scientists would describe as geohistory. In doing so, these naturalists made explicit use of powerful analogies with the work of contemporary antiquarians and other erudite historians. This transposition of metaphorical resources has now been illustrated with examples from the three natural-history sciences of the earth. However, it must be emphasized that the practice of detailed empirical geohistory was not only novel but also quite rare. Most work on the natural history of the earth continued to follow traditional lines of description and classification, and attempts at geohistory might be met with skepticism or even rejection: in the prize competition at the St Petersburg academy, Soulavie's placing behind Haidinger and Launay was a case in point. It remains to be seen whether the same was true of the natural-philosophical science of earth physics, or its more ambitious extension, the genre of global geotheory.

Earth physics has been defined here as the body of *causal* explanations that were proposed for specific terrestrial features such as valleys and volcanoes (§2.4). By suggesting antecedent causes for observed effects, or continuing processes by which those effects had eventually been produced, all such explanations necessarily embodied a *temporal* dimension. But they were also intrinsically *ahistorical*, because a satisfactory causal explanation for a specific kind of feature would be one that was valid at all times and in all places, just as in any other branch of "physics". Any causal explanation would of course cite concrete examples, just as an explanation in other parts of "physics" might cite experimental results; but in both cases those examples would be valued insofar as they were instances of a general phenomenon, not as specific and perhaps unrepeated configurations.

If, for example, the tilting and folding of strata were attributed to "revolutions" of upheaval in the earth's crust, or alternatively to episodes of crustal collapse, those

66. Ferber, "L'Ancienneté relative" (1786–88), 185–89, 198–209.

67. Soulavie, *Mémoires de Richelieu* (1790–93); he claimed (9: 480) that his historical research, with its portrayal of the corruption of the monarchical system, had been politically influential during the early Revolutionary period. He himself served as the Republic's ambassador in Geneva in 1793–94. After Thermidor and the fall of the Jacobins (§6.5), he was imprisoned briefly and then withdrew from public life to pursue his scholarly interests (he had earlier abandoned the priesthood and married). And so, having been the self-styled "archivist of nature" he eventually became, under Napoleon, a leading archivist of the political history of the Old Regime and of the Revolution: see Mazon, *Histoire de Soulavie* (1893), chaps. 4, 24.

explanations gained plausibility the more generally they could be applied. A convincing causal explanation of, say, Saussure's huge Alpine fold structure at Nant d'Arpenaz (Fig. 2.25) would be one that would also cover less spectacular cases of folded strata, not one that was unique to that specific locality. Likewise, if it was claimed, as it was by de Luc, that at the dawn of human history there had been a major "revolution" in the earth's geography (§3.3), a satisfactory causal explanation of that event would be one that made it an instance of a wider class of similar events, not one that focused on its unique place in time. In other words, in the practice of earth physics the more successful and persuasive a causal explanation became, the more it would divert attention away from what was specific about a particular instance of the phenomenon, as a distinctive episode at a particular point in geohistory, and on to what it shared with other instances as products of the same physical cause. Causal explanation and geohistorical reconstruction were not incompatible, but they were decidedly different ways of understanding the earth, and they were not easy to combine. The *intrinsically* ahistorical character of earth physics must surely account for the lack of any clear examples of geohistorical inferences being based on the practice of that science, comparable to what, say, Burtin (§4.2), Desmarest (§4.3), and Soulavie (§4.4) were doing with the three natural-history sciences of the earth.

The genre of geotheory, or "theory of the earth", was the ambitious extension of the science of earth physics on to the global level; it integrated causal explanations of specific features into more general "systems" explaining how the earth works and how it has come to be as it is (§3.1). In effect, geotheories offered accounts of how the earth was *programmed* to change in the course of time. At least in broad outline, if not in detail, they left no room for contingency; at least in principle, if not in practice, they were just as ahistorical as descriptive accounts in the natural-history sciences and causal accounts in earth physics.

Two distinct kinds of geotheory have been outlined, the steady-state and the directional. Steady-state or cyclic systems made a virtue out of being ahistorical. Their authors claimed in effect that there was *nothing* distinctive or unique about the configuration of features on earth at any particular time, past, present, or future: the details might change continuously, but the overall character of the earth remained the same at all times, because it was always governed by the same unchanging laws of nature. This argument can be seen in its purest form in Hutton's geotheory, with its radically ahistorical interpretation of the earth as a "machine"—eternal at least to human apprehension—in which one "world" would succeed another in an indefinite succession, with nothing specific to distinguish any of them (§3.4). In such a system, a unique event or distinctive configuration could be invoked only as a kind of deus ex machina, something introduced from outside the ordinary regular course of nature, or, in eighteenth-century parlance, an "accident". Such, for example, was Buffon's explanation for the origin of the earth, which in his early geotheory was almost detached from his steady-state explanation of its subsequent operation (§3.2). Hutton rejected a priori the possibility of any such unusual "catastrophe" at any point in time, precisely because it would not be part of the law-bound design of nature: "accidents" were anathema. Cyclic or steady-state systems

were thus the very antithesis of any geohistory, and they could not be made geohistorical without in effect betraying their own foundations.

Directional geotheories, in contrast, postulated a linear sequence of events or configurations on earth. This might seem to make them historical, just as intrinsically as steady-state or cyclic systems were ahistorical. Yet in fact many of those who proposed directional geotheories aspired to make them no less deterministic. The relevant analogue here was not a machine but a developing organism. Just as an embryo is (in modern terms) programmed genetically to develop along a determined pathway of ontogeny towards a specific kind of adult organism, so the earth was taken to be programmed to develop through an analogous sequence of stages, fully determined by its putative starting point and the unchanging laws of nature. If, as Buffon claimed in his later system (*Nature's Epochs*), the earth had begun as an incandescent globe in orbit, and had then continuously lost its heat into space, the main "epochs" or metaphorical milestones along that path of gradual cooling were determined ineluctably by the laws of cooling bodies: first a solidifying crust, and then the condensation of water to form oceans, with hypertropical waters slowly cooling to their present temperature, polar ice caps encroaching further in the future, and so on (§3.2). A narrative of the earth's inevitable sequence of changes, such as Buffon's later geotheory embodied, can of course be regarded as historical, but only if "history" be defined to include this kind of determinate development, in which case the epithet would also have to be applied to every instance of ontogeny in organisms.[68]

The true parallel with human historiography will now be apparent: as already suggested, geotheory was the analogue, in the natural world, of the genres of philosophical and conjectural history in the human world. Like them, it aspired to discover some relatively simple key to the complexity of the actual course of events; some simple causal explanation of how things "must" or at least "ought to" have happened in the past in order to reach the world we know today, and how they "ought to" continue into the future. In both the natural world and the human, this was history without contingency, history that at least in principle was fully determined and fully predictable (or retrodictable).

The place of contingency

In practice, however, the genre of geotheory was rarely pure; it has been treated up to this point as if it was, in order to clarify the analysis. Even if the broad outlines of the earth's temporal development were predictable in advance of any empirical investigation, the details might still need to be filled in and would at least provide concrete evidence for the broader features, and hence for the validity of the whole model. In describing such details, analogies drawn from human history might well be valuable, not least rhetorically. Anyway, many geotheories allowed in practice for some degree of contingent particularity in the earth's development. In effect, they

68. See Roger, "Buffon et l'introduction de l'histoire" (1992). The distinction being made here is similar to that between "historical" and "genetic" explanations in Oldroyd, "Rise of historical geology" (1979).

can be placed at various points on a continuum, from the most rigorously deterministic to those in which the narrative conceded substantial contingency. This can be illustrated from the small selection of geotheories that was outlined in chapter 3.

If Hutton's system be taken to exemplify the purest and most deterministic end of the spectrum (§3.4), Buffon's later one belongs a little way along it. In *Nature's Epochs*, Buffon amplified the bare outline of his deterministic model of a cooling earth with a mass of detail, some of which he would not have claimed was predictable from the model itself (§3.2). Significantly, however, the most obviously contingent features were those of organisms, as reconstructed from their fossil remains; and fossils were precisely the features that many of Buffon's contemporaries would have classed as "accidental", or outside the regular and lawlike course of physical nature. Only empirical research on fossil bones, for example, could have established that—so Buffon claimed—the first land animals were elephantlike (Fig. 3.4). But that they had been *huge* animals was something he treated as predictable (or rather, retrodictable) from the tropical or even hypertropical climate of the "epoch" at which they had flourished; the large size of some of the ammonites that characterized the Secondary formations seemed to confirm this deductive inference. The exact form the organisms had taken might be a contingent matter, but their general character was necessarily determined by their temporal place in the long process of a cooling earth. Unpredictable contingent detail thus had a place in

HISTOIRE
NATURELLE.

DES
ÉPOQUES DE LA NATURE.

COMME dans l'Hiſtoire civile, on conſulte les titres, on recherche les médailles, on déchiffre les inſcriptions antiques, pour déterminer les époques des révolutions humaines, & conſtater les dates des évènemens moraux; de même, dans l'Hiſtoire Naturelle, il faut fouiller les archives du monde, tirer des entrailles de la terre les vieux monumens, recueillir leurs débris, & raſſembler en un corps de preuves tous les indices des changemens phyſiques qui peuvent nous faire remonter aux différens âges de la Nature. C'eſt le ſeul moyen de fixer quelques points dans l'immenſité de l'eſpace, & de placer un certain nombre de pierres numéraires ſur la route éternelle du temps. Le paſſé eſt comme la diſtance; notre vue y décroît, & s'y perdroit de même, ſi l'Hiſtoire &

Supplément. Tome V. A

Fig. 4.14. The first page of *Nature's Epochs* (1778), in which Buffon introduced his directional "system" or geotheory with an analogy drawn from the "erudite" study of human history:

> As in civil history title deeds are consulted, coins are studied, and ancient inscriptions are deciphered in order to determine the epochs of human revolutions and to fix the dates of human events; so also in natural history it is necessary to excavate the world's archives, to extract ancient monuments from the earth's entrails, to collect their remains, and to assemble in a body of evidence all the marks of physical changes that are able to take us back to the different ages of nature. This is the only way to fix some points in the immensity of space, and to place a certain number of milestones on the eternal road of time.

(By permission of the Syndics of Cambridge University Library)

Buffon's geotheory, although only a subordinate one. It was enough, however, to make him appreciate the rhetorical potential of metaphors and analogies drawn from human history, as the purple prose of his opening page makes clear (Fig. 4.14).[69]

The ambiguity of the phrase "natural history" was crucial. Buffon normally used it in the traditional sense of the *description* of nature (as in the title of his multivolume work), but here he made it express the still quite novel sense of a *history* of nature that would be analogous to human (or "civil") history. The temporal dimension provided by the human science of chronology was applied directly to the natural world, if not in its quantitative use of precise dates then at least in its qualitative use of "epochs" to delineate the main features of the story. But in addition, erudite and antiquarian histories provided an even richer source of analogies. They suggested how the deep past could be known from its material relics, but that these relics needed to be "deciphered" or *interpreted* before they would yield their full historical meaning.

Buffon's eloquence and prestige ensured that the heuristic power of this analogy with human history was a lesson well learned by his readers. Conceived within the genre of geotheory, his system was akin to the "philosophical" and "conjectural" histories of some of his fellow philosophes. But his use of metaphors drawn instead from erudite and antiquarian histories was a resource that others could then appropriate and apply to more empirical and contingent conceptions of geohistory; indeed, Desmarest was already doing just that, before Buffon upstaged him (§4.3). So it is only an apparent paradox that one of the most influential of all geotheories contained—on its opening page—one of the most eloquent statements of the parallel between human history and the history of nature.

The kind of geotheory that has here been termed the "standard model" (§3.5) often showed a similarly inconsistent mixture of the deterministic and the contingent, and some variants should clearly be placed still further along the continuum, away from any programmed purity. The causal model of a gradual chemical differentiation—with successive precipitations from an initially global proto-ocean—was at least in principle as deterministic as Buffon's model of a cooling earth. Given a certain initial composition of the chemical soup of the proto-ocean, it was tacitly assumed that the various kinds of Primary rocks would have been precipitated in turn, in an order that would be predictable (or retrodictable) if only the relevant and obviously complex physics and chemistry were fully understood. Given the changed composition resulting from those first precipitates, it was inevitable that some of the Secondary formations would be composed of different kinds of rock; and given the steadily falling level of the proto-ocean (by whatever means it had been caused), it was inevitable that in due course the Primary rocks would have emerged and become subject to erosion, thus providing materials for the detrital sediments that became Secondary rocks of other kinds. Finally, given the eventual retreat of the waters to their present bounds, it was predictable that all the earlier

69. Fig. 4.14 is reproduced from Buffon, "Époques de la nature" (1778) [Cambridge-UL: MG.7.20], 1.

rocks would be eroded still further, to form the alluvial or Superficial deposits on the lowest ground, in the river valleys and on the plains. Even volcanic action could be subsumed within this schema—and be treated as regular and not "accidental"—if eruptions were attributed to the subterranean combustion of materials in some of the rocks of earlier origin.

However, the deterministic purity of an ideal version of the standard model of geotheory had to be modified in practice by the brute facts revealed by the fieldwork on which it was based. The sequence of rock types was found to be far from invariable, and some of them were found to recur at different points in the geognostic pile of rock masses. So the elegant simplicity of the idealized model had to be qualified with various ad hoc adjustments, invoking, for example, recurrent phases in the chemical differentiation or repeated fluctuations in the level of the ocean. Simplicity might appeal to an indoor savant such as Buffon, snugly ensconced in his study; but the more the research was taken outdoors into the field the more elaborate the necessary adjustments seemed to become. The standard model, if it were to account for all the details revealed by fieldwork, had to lose some of its deterministic character and concede some place for contingency.

Saussure as a geotheorist

This is best illustrated in the case of Saussure. Perhaps the supreme exemplar of a fieldworking savant, he also became the one most bewildered by the complexity of what he observed, and the one least able to formulate a system that satisfied him. To the end of his life he planned to produce his own geotheory, and he left behind a research "agenda" that he hoped would ultimately yield—to those who followed him—a satisfactory explanation of the earth (§6.5). Yet his chronic difficulties with geotheory mask a more profound transformation. The complexities that were disclosed by his fieldwork seem to have pushed him inexorably towards a more contingent conception of the earth than geotheory allowed. The details he observed became not so much illustrations of a simple underlying causal model, but rather clues to an altogether more complex narrative of geohistory; its components could be pieced together only by interpreting the specific features—often unexpected and even startling to him—that his fieldwork revealed. For example, the vertical conglomerate at Vallorcine (§2.4) led him to accept that the elevation of the Alps had included not just one but repeated "revolutions", each with distinctive effects; the strange "erratic" blocks of rock scattered on the lower ground towards Geneva led him to postulate another distinctive episode in the much more recent past. These were events or processes that had not been predicted on any geotheoretical system; they were features of a geohistory that could be constructed only bottom-up from the detailed empirical evidence of the deep past. So it is perhaps not surprising that it was Saussure who wrote and published one of the most eloquent passages expressing this sense of geohistory—the sense of an unrepeated and unpredictable sequence of distinctive events and processes—in its most vivid form. This was an imaginative *reconstruction* of geohistory as he himself might have seen it at the time, albeit from a lofty perspective more divine than human; it was the narrative analogue of an antiquarian's pictorial reconstruction of a scene at Pompeii, based on its ruins (Figs. 4.4, 4.5).

At one point in his *Alpine Travels*, Saussure described the "vision" of geohistory that he had experienced years before, when he climbed alone to the peak of Crammont, overlooking the huge south face of Mont Blanc, and for three hours contemplated the mountains all around him. This and a later ascent with two friends were, he recalled, the most pleasurable times of studying nature that he had ever enjoyed. He had already confirmed that the great massif of Mont Blanc was composed of Primary rocks, and he was beginning to be convinced that they were precipitates or sediments like the Secondaries. From the peak of Crammont he saw those bedded Secondary rocks apparently resting against the Primaries and tilting away from them, first at a high angle and then, further away, more gently. This was a standard structural arrangement that he interpreted as supporting the standard model of geotheory. But in his mind's eye Saussure transformed this and other Alpine features into a vivid narrative of imaginatively witnessed geohistorical events:

> So then, retracing in my head the sequence of the great revolutions that our globe has undergone, I saw the sea—which then covered the whole surface of the globe—form first the primitive rock masses [*montagnes*] and then the secondary ones, by successive deposits and crystallizations; I saw these materials being arranged horizontally in concentric beds; and then [I saw] fire—or other elastic fluids contained in the earth's interior—elevate and disrupt this crust, and thus push up the internal and primitive part of the crust, while the external and secondary parts remained leaning against the internal beds. Then I saw the waters pouring into the abysses that were burst open by the explosion of the elastic fluids; and these waters, in flowing to these abysses, swept to a distance those enormous blocks that we find scattered on our plains.[70]

Saussure's causal explanations are not the most important aspect of this passage; in fact, having recorded his "vision", he went on to explain that he had subsequently changed his mind on some important points. What is far more significant is that it shows that he was beginning to transcend the framework of natural history and to use his descriptive work as a basis for *geohistorical* reconstruction, however tentative it had to be. The literary device of imaginative time travel is not a necessary condition for thinking historically (or geohistorically), but it is certainly a striking indication that a writer was indeed thinking that way. Yet the *novelty* of doing so in the sciences of the earth is suggested not only by the rarity of such flights of fancy, at least in published form, but also, in this case, by Saussure's own description of it as a "vision" and by its stylistic affinity to biblical prophecy. In a Bible-reading age,

70. Saussure, *Voyages* 2 (1786), 339–40. The beds were "horizontal" in any one locality, but "concentric" on a global scale; in modern terms, most "elastic fluids" were gases, but "fire" [*feu*] or heat was generally regarded as a similarly rarified substance, capable of powerful expansion (as, for example, its role in Hutton's system illustrates). Saussure's interpretation of the structure of the Alps was of course far simpler than the one offered by modern tectonic geology. The peak of Crammont (2,737m) gives superb views not only of the Mont Blanc massif to the north but also over the lower ranges of the Val d'Aosta down towards the plains of Piedmont to the south (see Fig. 1.3). Saussure's similar though less striking "vision" of the formation of a ancient conglomerate (*Voyages* 2: 191) shows that Crammont was not just a flash in the pan.

his repeated "Then I saw . . ." would have recalled, for example, Isaiah's vision of the Lord surrounded by six-winged seraphim or John the Divine's vision of the new heaven and the new earth. At least for Saussure, geohistory was not yet sufficiently established as a scientific discourse to be readily expressed in more mundane terms.

De Luc as a geohistorian

Among the many variants of the standard model of geotheory, de Luc's has been singled out as particularly influential in Saussure's time. De Luc proposed a strongly *binary* system, in which a very lengthy but ill-defined "ancient" or "former world" was sharply separated, by a uniquely radical "revolution" in geography, from a familiar "modern world" that had so far lasted only a few millennia (§3.3). De Luc's system can now be located on the continuum between the strictly programmed and the contingently geohistorical: it lay even further from any deterministic purity. Although he considered himself to be writing within the genre of geotheory—for which he had proposed the name "geology" (§3.1)—what de Luc produced has a greater claim than any other "system" of its time to be regarded as geohistory. Although he defined a long sequence of "epochs", from near the earth's origin to the present, he did not suggest any kind of inevitability about it; he assumed that the events had had natural causes of some kind, but he did not suggest any causal linkage between them, comparable, for example, to Buffon's use of progressive cooling. De Luc suggested that his great "revolution" had been due in part to the waters of the ancient oceans draining away into vast caverns—an explanation also used by Saussure and many others—but he treated it as an event that could hardly have been predicted (or retrodicted), either in its timing or in the form of the new "world" that emerged from it. Above all, de Luc repeatedly described what he was doing as "history", and he deployed the metaphors of erudite and antiquarian human history more substantively and pervasively than perhaps any other savant of the time, apart from Soulavie.

The contingent historicity of de Luc's system was rooted explicitly in his theistic apologetics, just as—at the opposite end of the continuum—the determinism of Hutton's system was rooted in his deistic metaphysics. De Luc wanted to identify, in scientific terms, the distinctive events in the deep past that corresponded to the sketchy outline—clearly *not* scientific in its primary purpose—that he believed had been revealed to Moses. In taking the Creation story in Genesis as his model, he committed himself knowingly to an understanding of history that was radically contingent, because it was perceived as being dependent on divine "sovereignty", or God's "voluntaristic" freedom of action in the world. Although this fundamental concept was acknowledged by all the Christian churches, it was one that resonated particularly in de Luc's native city of Geneva, Calvin's city two centuries earlier, and still in de Luc's and Saussure's time the theological center of the Reformed tradition.

De Luc did not use—and might not even have understood—the modern concept of contingency, but his explicit belief in God's sovereignty led him in practice to treat historical events as contingent, and unpredictable to mere human beings.

This would be as true of the history of the natural world as it was of human history, as the Creation story itself implied. Each "day" was framed with the formula, "And God said, 'Let there be . . .' and it was so". No literalism—and de Luc was no literalist—could obscure the obvious point about the story, that God was sovereign and that the will, the act, and the outcome might all have been otherwise. Transposed into the scientific realm, as de Luc wanted it to be, this implied that deterministic models of geotheory were radically misconceived. The course of events in the deep past could not possibly be predicted (or retrodicted) on the basis of any simple set of natural laws, because things might always have happened otherwise: God—usually acting of course through ordinary natural or "secondary" causes—might have chosen to organize events in another way, or with different timing, and so on. This was no abstract piece of theology: it had consequences for scientific practice, because it implied that geohistory, like human history, would have to be compiled bottom-up from the empirical evidence of how things had *in fact* happened, rather than being deduced top-down from some simple physical principles that stated how they "must" or "ought to" have happened.

In this way, the theistic metaphysics of a self-styled "Christian philosophe" such as de Luc favored the same approach to geohistory as that practiced by a savant such as Soulavie, inspired more by the newly flourishing erudite and antiquarian style of human history. Both wanted to construct geohistory from the often surprising and unexpected empirical relics of the past. Both were therefore opposed to those more typical Enlightenment philosophes, such as Hutton and Buffon, who sought to find the key to the complex past, present, and future of the earth in some set of simple causal principles, analogous to those favored by the "philosophical" and "conjectural" styles of human history.

Conclusion

In principle, then, the genre of geotheory was as antithetical to geohistory as the science of earth physics on which it was based. In practice, however, some geotheories did allow for some degree of contingency, and the further they deviated from deterministic purity the more the genre could become a vehicle for geohistory. Since geotheory aspired to be global in application, this meant that some systems might become the basis for *global* accounts of geohistory: sketches of a worldwide sequence of contingent events and distinctive periods, all the way from the remotest past to the present, but of course no longer into the unknowable future. Saussure's "vision" on Crammont was a sign of that aspiration, but it was a fleeting one; Soulavie's prize essay was another, but it failed to convince his peers. Much more substantial was de Luc's system, which in particular offered detailed arguments for the global historicity of one distinctive major event—his "great revolution"—at a specific moment in the past. Its geohistorical character was underlined by his claim that it corresponded to and confirmed—though in a far from literal way—what he believed was a reliable human record, the story of Noah's Flood. At this point geohistory and human history were not merely laid end to end but showed a significant overlap or fusion. The irony can hardly be missed. Far from "retarding the Progress of Science", a lively concern to understand Genesis in scientific terms, and more particularly an interest in identifying the physical traces of the Flood, facilitated just the kind of thinking that was needed in order to develop a distinctively *geohistorical* practice within the sciences of the earth.

This chapter has outlined how in Saussure's time the varied sciences of the earth were just beginning to be affected by the equally varied sciences of human historiography. Of those varieties of history, the erudite and antiquarian traditions were flourishing as never before, claiming greater prestige and gaining wider public attention, particularly in the wake of the famous excavations at Herculaneum and Pompeii. They provided powerful conceptual resources for understanding how the past could be reliably known in the present, on the basis of specific relics both textual and artifactual. At least a few naturalists in the late eighteenth century transposed these resources from the human world into the world of nature and thereby explored their potential for reconstructing geohistory. This is shown most clearly in their use of historical terms (which are italicized in the following paragraphs), qualified as being "*nature's*. . . ."

From the science of *chronology* came the basic idea of a temporal sequence, quantifiable at least in principle by specific events with precise *dates* assembled into *annals*, but in any case showing how history could be divided qualitatively into a sequence of *epochs*, that is, by decisive moments or periods. Transposed into nature, this suggested that an accurate sequence of real and distinctive geohistorical events could be reconstructed in the same way, even if they could not be dated.

From chorography and erudite local histories came the practice of a critical evaluation of the varied *documents* preserved in *archives*, which, if judged authentic, directly witnessed to the historical reality of specific events. Transposed into nature, this implied that geohistory would have to be constructed likewise from specific past local events, reliably authenticated by critical assessment of their material relics.

From numismatics and epigraphy came the idea that *coins* and *inscriptions* might valuably supplement more conventional documents and give further direct evidence of the past, provided the *language* in which they were written could be *deciphered*. Transposed into nature, this emphasized that the meaning of the relics of geohistory was likewise not self-evident, that the relics themselves needed to be interpreted, and that nature's own "language" would have to be decoded and learned in order to do so.

Finally, from antiquarian studies, and particularly from the new techniques of careful excavation, came the recognition that the preservation and discovery of *monuments* and other artifacts might owe much to chance or to such "accidents" as the eruption of a volcano or the sinking of a well. The very riches of Herculaneum and Pompeii indicated how much had been lost elsewhere and how fragmentary was the evidence of the human past. Conversely, however, those riches might allow the everyday life of certain vanished cities to be reconstructed in detail and their inhabitants to be brought vividly back to life, at least in the mind's eye. Transposed into nature, all this reminded naturalists of the similarly fragmentary character of the evidence of geohistory, and the chanciness of its preservation and discovery; but also, more encouragingly, that they too could aspire to reconstruct at least some episodes in geohistory in all their unpredicted particularity.

These were some of the resources that in Saussure's time were beginning to be transposed into the natural world. The most effective of these early essays in geohistory, examples of which have been summarized in this chapter, were based on the analogy with erudite and antiquarian history, of history constructed bottom-up from a mass of detailed evidence of what had *in fact* happened in the course of time. In contrast, the intellectually more prestigious genres of philosophical and conjectural history favored by the most prominent philosophes in the Enlightenment were *not* conducive to this process of constructing geohistory. They found their analogue in the equally fashionable genre of geotheory; for this, at least in its purer forms, was constructed similarly by deductive inference top-down from putative principles or natural laws, which in effect dictated what "must" or "ought to" have happened in the past and what "must" or "ought to" happen in the future.

However, in contrast again, there was one overarching historical interpretation that did by analogy facilitate and foster the new style of geohistorical thinking. This was, ironically, the biblical narrative of human history that underlay both Protestant and Catholic traditions (and their common Judaic foundations). Its pervasive theme—that the course of events might have been otherwise—molded the religious imagination profoundly, and hence also the scientific imagination. In the Creation narrative itself, God might have chosen to build the present world by a different sequence of creative actions; some other means than a Flood might have been used to give human society a fresh start; Abraham might have been unwilling to sacrifice Isaac; Jesus might have chosen to avoid being arrested in Gethsemane. Only in retrospect could any kind of inevitability be ascribed to the course of events recorded in biblical history, and then only as a sense that God's intentions had ultimately been fulfilled through the choices of human beings and despite their frequent failures.

This theological point deserves emphasis here, in order to explain why the sense of history as contingent and intrinsically unpredictable, and therefore as based necessarily on detailed empirical evidence of what had *in fact* happened, was particularly congenial to savants such as de Luc the Protestant and Burtin the Catholic, and decidedly uncongenial to Hutton and Buffon the deists. Given this sense of the sheer contingency of human history, a newly *historical* science of nature, stressing a similar contingency in the deep prehuman past, could well be built on foundations borrowed not only from erudite antiquarianism but also from the radically historical Judeo-Christian religion that was so scorned by most of the philosophes of the Enlightenment (and still is by many, though not all, modern scientists and historians).

CHAPTER 5

Problems with fossils

5.1 THE ANCIENT WORLD OF NATURE

The deep past as a foreign country?

Around the time that Saussure climbed Mont Blanc, a few naturalists were consciously applying the methods and concepts of "erudite" or antiquarian history to the natural world and suggesting how *nature's* history could likewise be reconstructed bottom-up from detailed evidence, even in the absence of human witnesses and human records. Some examples of this kind of research were reviewed in the previous chapter, but it needs to be emphasized again that they were few and far between. In the late eighteenth century, most savants who studied the earth continued to follow the traditions of either natural history or natural philosophy: they produced works either of description and classification or of causal explanation and geotheory. Those who explored the possibility of using such research as a basis for reconstructing geohistory—which they saw as a distinct and *novel* project—were a very small minority.

The work of that small minority has been singled out and highlighted in the previous chapter because this book aims to explore how the initially marginal genre of geohistory eventually became a constitutive feature of a distinctive new science, which came to be known—with a significant change of meaning—by de Luc's term "*geology*". This goal entails treating cursorily, or even ignoring, most of what continued to be done within the already well-established sciences of mineralogy, physical geography, geognosy, and earth physics. Focusing instead on geohistory, the present chapter reviews the particular problems that surrounded the use of fossils as one of the main sources of evidence for reconstructing a history of the earth analogous to—and contiguous with—human history. This will conclude the synchronic survey of the sciences of the earth around the time of Saussure's ascent of Mont Blanc, which constitutes the first part of this book, and which prepares the ground for the diachronic narrative that will follow in Part Two.

One of the fundamental issues underlying the scattered essays in geohistory reviewed in the previous chapter was the character of the deep past that was disclosed by rocks, fossils, volcanoes, valleys, and so on. Was the deep past of geohistory familiar territory or, as it were, a foreign country? Had the earth, at its unimaginably remote "epochs", been a place much like it is at present, or had it been significantly different? Hutton and de Luc represented the alternatives in their starkest form. For Hutton, a "succession of worlds" extended indefinitely into the past, all of them much the same in general character and no more distinctive than successive orbits of the planets around the sun (§3.4). For de Luc, the "present world" was sharply distinct from the "ancient" or "former world" that it had replaced at the dawn of human history, and was separated from it by a major "revolution" (§3.3).

De Luc's binary geotheory was in fact an elaboration—backed by a mass of new field evidence—of a much more pervasive assumption among savants. De Luc adopted and gave empirical substance to a dichotomy that was already widely taken for granted: the sharp contrast between the familiar world of the present and recent past on the one hand, and on the other the unfamiliar world of the deeper past that had left its enigmatic traces in the form of fossils and other relics. Significantly, the key phrase that expressed that dichotomy was almost invariably used in the singular and with the definite article: *the* "ancient" or "former world" [*l'ancien monde, die Vorwelt*, etc.] was treated implicitly as an *undifferentiated* period occupying the entire history of the earth before the establishment of the "present order of things". This strange territory—the foreign country of the deep past—had earlier been treated in effect not merely as unknown but almost as unknowable: the sheer proliferation of geotheoretical "systems" reflected a sense that any one speculation about it might be as good as any other. By the late eighteenth century, however, the body of research on specific problems in earth physics, and even more the scattered essays in geohistory, were beginning to make the former world seem knowable, at least in principle, in that some of its relics could be interpreted in terms of events and processes of kinds familiar in the present world.

In fact, as has been emphasized repeatedly in previous chapters, all savants concerned with understanding the earth took it for granted that the present was the obvious key to the past, at least in the sense that the events and processes observable in the present world were clearly the best source of clues to the former world, however strange the latter might have been. Even de Luc, who emphasized so strongly the contrast between ancient and modern, used what he called "*actual causes*", or processes demonstrably at work in the present, as the primary means of penetrating back into the past. For him it was simply a matter of empirical discovery that these processes had operated in their present form only for the past few millennia, that is, since the time of the great "revolution". But in practice, as mentioned earlier, he took it for granted that similar processes had already existed long before that time: for example, tides and currents had swept the plant debris of vanished continents out to sea—there to be embedded in marine sediments as future plant fossils—just as they still do in the present world.

This use of actual causes was of course much more pervasive in the work of those who proposed what was—in contrast to de Luc's binary model—a *unitary* conception of geohistory. For example, Desmarest used the slow erosive action of the present streams and rivers in Auvergne as a key to the formation of even the deepest valleys, punctuated by the contingent "accidents" of occasional volcanic eruptions (§4.3). For him there had been no radical revolution in the recent past, and there was no fundamental contrast between the present world and what had preceded it; erosion had been much the same all along, and so had volcanic action. His proof of the volcanic origin of basalt depended precisely on demonstrating the *identity* between ancient and modern instances of that previously enigmatic rock. In other words, the more that actual causes were used successfully to decipher the deep past, the less strange and unknowable it seemed and the more familiar it became.

Yet Desmarest insisted that his occasional volcanic episodes could be used to define the successive epochs of a real *history* of the region: the erosive action of the streams and the material of the lava flows might have remained unchanged throughout, but the eruptions had affected a topography that was changing continuously as the valleys were eroded. And this process was not just one phase in a repeated cycle, as Hutton would claim. For Desmarest it was part of an even longer and directional geohistory: the thick sediments dating from a still earlier period pointed to a time when Auvergne had been under water, and the underlying granite might represent the original foundations of the earth. Desmarest, like those such as Soulavie who followed his lead, claimed that geohistory could be reconstructed by tracing the action of familiar processes back into an increasingly unfamiliar past.

It remained, however, to determine just how familiar or unfamiliar the deep past of the former world had in fact been, how like or unlike the present world that could be observed and experienced directly. Yet the detection of extinct volcanoes, for example, depended obviously on the recognition of their similarity to those active in the present world: the method used to decipher the deep past necessarily highlighted the ways in which it had been *like* the present. Indeed, such features were incapable in principle of revealing any way in which the deep past might have been quite *unlike* the present, or of defining with any clarity how each of the successive epochs of geohistory might have been uniquely *distinctive*. In principle it remained possible to claim—as in effect Hutton did—that the period at which Auvergne had been under water might also have been a time when other regions were experiencing subaerial erosion and occasional volcanic eruptions: on a global scale there might have been nothing distinctive to characterize Desmarest's early period of geohistory.

This dilemma was recognized by Soulavie, when he defined the three criteria by which nature's chronology could be constructed: superposition, degree of erosion, and relative altitude (§4.4). But superposition was not always available, and anyway was difficult to extend from one region to another, let alone on to a global level. The degree of erosion was not an unequivocal criterion, since its rate might not have been constant or uniform. And altitude was still less useful, because it depended on the validity of the standard model in its purest form, and it seemed increasingly questionable whether global sea level had in fact subsided regularly and without oscillation. There was, however, one further criterion, which Soulavie did not list with the others, but which he used to a limited extent elsewhere in his work. As already mentioned, he described three successive limestone formations, clearly of different ages by the criterion of superposition, which contained distinctive assemblages of fossils. Many other naturalists used fossils in the same way, if more crudely, as indices of geohistory, when they noted that ammonites and suchlike fossils were confined to the lower and therefore older Secondary formations, and that fossil shells of kinds known alive were most common in the upper and therefore younger ones. That very broad distinction was general enough to be, in effect, a standard component of the standard model of geotheory: it looked as if living things might have changed directionally in parallel with the materials being precipitated or deposited to produce the successive formations.

Fossils and geohistory

As potential clues to geohistory, fossils were different in an important way from, say, volcanoes, in that they were recognized to be *both* like *and* unlike their counterparts in the present world, both familiar and unfamiliar. By Saussure's time, fossils—in the newly restricted (and modern) sense of organic remains—were regarded as "accidental" or "extraneous", because their origin was clearly distinct from that of the rocks in which they were found (§2.1). Like volcanoes, which were also widely considered to be "accidental", fossils were therefore possible markers of geohistory. But their potential was much greater than that of volcanoes, because they were not only analogous to living organisms but also distinct from them. They could be, and were, assigned their places within the classifications constructed for living organisms: they were recognized to be the remains of ancient elephants or oysters or ferns or whatever. But many of them—perhaps all of them—proved on closer study to differ in detail from any animals or plants known alive; they could not readily be assigned to the same species, or in many cases even to the same genera or families. They seemed to prove that the former world had been *distinct* from the present.

However, the significance of the difference was far from clear. De Luc defined the problem when he referred to three alternative explanations of the substantial—and perhaps even complete—contrast between the kinds of animals and plants found as fossils and those known alive. The older organisms might have been wiped out in one or other of the earth's many revolutions, whether those changes had been sudden or gradual. Alternatively, all the species found as fossils might still be alive in the present world, but they might have migrated in the course of time to such remote or inaccessible places that they had yet to be discovered. Or third, they might all have been transmuted into the species known alive, perhaps as a result of the changed environmental circumstances brought about by the earth's revolutions. It cannot be emphasized too strongly that in de Luc's and Saussure's time these three explanations were treated as alternatives, as it were on a par with one another. Put succinctly, the difference might be due to *extinction*, or to *migration*, or to *transmutation* (in modern terms, evolution).[1]

None of these alternative explanations was obviously more plausible than the others. Each entailed grave difficulties and further problems. Most naturalists felt an almost gut revulsion against the idea that extinction might be an ordinary feature of the natural world. For it implied that some of the wonderfully diverse species, the fossil remains of which they took such delight in collecting and describing, might have been lost forever. That extinction could happen in the present world was not in doubt, however. For example, it was well known that animals released by sailors landing on Mauritius—on the main trade route between Europe and the East—had found the flightless dodo a prey that was all too easy to hunt down, and that the bird had apparently become extinct within about a century of being discovered (Fig. 5.1).[2]

However, the dodo and other recently extinct or threatened species had been or were likely to be wiped out by *human* action; the gaps they had left or would leave

Fig. 5.1. The flightless dodo of the island of Mauritius in the Indian Ocean, which was thought to have become extinct about a century earlier through human interference with its habitat. This version of an often copied seventeenth-century picture of the famous bird was published in 1799 by Johann Friedrich Blumenbach, the professor of natural history at Göttingen, as one of a series of colored engravings of notable subjects of natural history. (By permission of the Syndics of Cambridge University Library)

1. De Luc, *Lettres physiques et morales* (1779) 5(2): 613. De Luc listed them in this order, which probably reflected his opinion of their decreasing plausibility. Modern Darwinian theory incorporates all three, thereby transcending the apparent need to choose between them, but it would be grossly anachronistic to project such insights back into the eighteenth century.

2. Fig. 5.1 is reproduced from Blumenbach, *Abbildungen* (1796–1810) [Cambridge-UL: MB.46.75], pt. 4 (1799), pl. 35. The dodo had been found unpalatable for human consumption, but that had not saved it. Within the period covered by the present book, it remained in fact somewhat uncertain whether it was completely extinct: see Geus, "Animals and plants extinct in historical times" (1997).

in the diversity of the living world were due to human beings, greedy, shortsighted, or just careless. To postulate analogous gaps that were wholly *natural* in cause and that owed nothing to sinful humans was quite another matter. For many theists it seemed almost inconceivable that a caring and personal God's providential oversight of the created world could ever have lapsed in this way. For deists, equally, it seemed seriously to mar the perfect design of nature that was to be expected from the Supreme Being. For savants of both kinds, it also negated the traditional and widespread assumption of "*plenitude*": all forms of existence—living and nonliving—that were possible in this world should surely always exist, so that the universe would be permanently "full" in its diversity. To claim that extinction was part of the regular course of nature could therefore seem tantamount to supporting an atheistic view of the world, in which there was no providence, no design, and no plenitude. Yet even without these powerful metaphysical and theological arguments, there would have been good reasons for naturalists to be skeptical about extinction in nature, because there were at least two other plausible ways of explaining what was observed.[3]

Migration and transmutation

One of those alternatives was, for most naturalists, much more plausible than extinction. They were acutely aware how little was yet known of the faunas and floras of distant lands, and even more of the world's seas and oceans. The interiors of the continents were barely known at all (apart from little Europe, which on a global scale hardly deserved to be treated separately from Asia); and Australia, although still barely known, had already begun to disclose striking new forms such as kangaroos and eucalypts. Likewise, the deeper waters of even the seas nearest to Europe, let alone those on the far side of the world, were almost totally unexplored (and remained so until the era of the *Challenger* expedition in the mid-nineteenth century): the longest lines were limited to a few hundred fathoms and could not touch the sea floor except near land. Almost every expedition and voyage of exploration—even if its primary purpose was strategic or commercial—brought previously unknown animals and plants back to Europe. Every naturalist with any ambition could hope for renown as the discoverer or describer of new species or to be immortalized by having such species named after him. Every newly described living species enlarged the pool of those available for comparison with fossils and increased the chance that the latter would be identified in due course as belonging to species still alive in the present world. Again and again that expectation was realized, as fossil animals and plants that had been assumed to be "lost" or extinct were found to be still alive, if not the identical species then at least members of the same genus or family.

Such discoveries—of what are now popularly known as "living fossils"—made many naturalists hesitant to conclude that *any* fossil animal or plant could be pronounced extinct: such a claim was all too liable to be resoundingly disproved by the next expedition to return from distant parts of the globe. This *living-fossil effect*, as it will be called here, made it highly plausible—to a degree that is now difficult to recapture—to infer that the difference between fossils and living organisms should be explained in terms of their migration in the course of time: any kind of fossil animal or plant might still be alive and well and living in some remote part of a distant continent or somewhere in the depths of the ocean.[4]

The other alternative explanation was much less widely discussed. This was that earlier forms of life had not become extinct, nor merely migrated to obscure places, but had instead been transformed or transmuted into present forms in the course of time. The possibility of a process of *transmutation* or of a theory of *transformism*—in modern terms, the idea of evolution—was canvassed widely in the eighteenth century. It was on occasion played down on account of its perceived overtones of materialism, and even of atheism, though its risqué character made it all the more attractive to some savants. But the idea was of dubious status not just because of its heterodox implications, but much more because it seemed to be irredeemably speculative. There was simply no good evidence for the kind of lability in organisms that transmutation implied and required; on the contrary, the practice of natural history seemed to confirm the stability and constancy of species, the quite narrow limits of variability, and the rarity and marginality of hybrids, at least under natural conditions. So although many naturalists treated transmutation as a possibility, at least on a limited scale between related species, transformism was hardly attractive as a *general* explanation of the diversity of the whole of the living world.[5]

Conclusion

None of the three alternative explanations—extinction, migration, and transmutation—was obviously superior to the others: each had its advocates, but none was overwhelmingly persuasive, and the issue remained controversial and unresolved. Although it was obviously important in terms of what would later be called "biology"—the alternatives implied quite distinct ideas about the origin and history of

3. Many modern historians writing about the question of extinction at this period have treated the metaphysical and theological arguments as primary and the empirical evidence as epiphenomenal or mere window dressing. This imbalance needs to be redressed, and the evidential reasons are therefore given greater emphasis here; but it is beyond the scope of this book to attempt a full evaluation of this complex issue.

4. Perhaps the best modern examples of living fossils are the coelacanth fish (*Latimeria*) and the dawn redwood tree (*Metasequoia*). The former was for many years known only from a single specimen caught in deep water off South Africa, though it has since been found to be quite common—and well-known to local fishermen—off the Comoro Islands. The latter was first found in the grounds of an ancient temple in a remote part of China, but has since become a gardeners' favorite worldwide. Both genera belong to groups that were previously well-known as fossils from formations of Mesozoic age, but were assumed to have been extinct for seventy million years or more.

5. The terminology for what would now be called evolution was highly variable at this time; that one of the options (in French) was *dégénération* signals the important point that the direction of change was *not* always assumed to be onwards and upwards. In this book, rather arbitrarily, the term "transmutation" is used for the process and "transformism" for the theory. Confusingly for modern readers, the word "evolution" (literally, unrolling) was used to denote the *ontogenetic* transformation of an *individual* organism from embryo to adult.

the diversity of living organisms—it was equally significant in the context of geohistory. The highly plausible option of migration implied that the former world might have been more or less the same in character as the present, in its organisms as much as in its volcanoes, its valleys, and so on, apart from changes in their spatial distribution. In contrast, the option of extinction (and, less plausibly, that of transmutation) implied a more or less radical disjunction between past and present. So what remained unresolved was not only biological, but also a fundamental matter of geohistory. It remained unclear whether the life of the deep past had been like or unlike the present, familiar or unfamiliar. Only dimly was a third possibility envisaged, which might make it unnecessary to choose between those alternatives. This was that organisms might have become *by degrees* more familiar in the course of time (though not necessarily by any kind of transmutation), so that a highly unfamiliar living world might have become progressively more familiar.

In the rest of this chapter, these vast uncertainties about the past world of life will be analyzed in more specific terms by referring in turn to three distinct kinds of fossil evidence: first, the fossils that were taken to be the relics of marine life; then, those that were identified as the remains of land animals (and subordinately of land plants); and finally—rarest and most controversial of all, but obviously of outstanding interest—the possible traces of human beings.

5.2 RELICS OF FORMER SEAS

Vanished shellfish

The most abundant fossils to be found in the field, and those most commonly displayed and stored in museums, were (as they still are today) the remains of animals that were taken to have been the inhabitants of former seas. Mollusk shells in great variety were the most common fossils of all (see Figs 2.3, 4.6); corals, sea urchins, and other invertebrates (to use a term that had yet to be coined) were also prominent; fossil fish were much rarer, but found abundantly in particular places (§2.1). That these fossil animals had all lived in the sea was generally taken for granted, and for several good reasons. Many were identified as belonging to the same genera as well-known living marine shellfish; corals and sea urchins were known to live only in the sea, never in freshwater; and the standard model of geotheory (§3.5) encouraged the assumption that the oceans had been more widespread in earlier periods than they are today. The large fossil bones and teeth that were taken to be the relics of land animals always excited great interest and were often displayed prominently in museums as their prize exhibits (§5.3); but they were very rare by comparison with the prolific abundance of fossil shells. The question of the relation between fossils and living organisms therefore had to be tackled first in relation to these relatively unspectacular specimens.

Fossil shells might be less spectacular than fossil bones, but many were highly prized nonetheless, by amateur collectors as much as by savants. Well-preserved specimens could be almost as attractive to collectors as their counterparts from the present world: what they lacked in beauty, for example in original coloring, they gained in interest from the very fact of being relics of the former world. Comparisons between fossil shells and their living counterparts were greatly facilitated at just this time by the vogue for *conchology* as a fashionable branch of natural history. Many illustrated handbooks of conchology were published to satisfy the market. The more lavish compilations, with colored engravings, formed in effect virtual museums on paper: one of the finest, begun by the physician Friedrich Martini in Berlin and continued by the pastor Johann Chemnitz in Copenhagen, was actually entitled a "shell museum" [*Conchylien-Cabinet*]. Leading conchologists such as these were able to exploit the riches of their own and many other private collections. Collectors put a high premium on rarity, and they often paid astonishing prices for shells of the rarest species. What was brought back from remote parts of the world therefore received immediate attention from savants as well as collectors; whatever could be known of the marine faunas of distant seas was known without delay. At another period of history, without the commercial incentives of the eighteenth-century fashion for conchology, similar specimens might well have languished unexamined in museum basements.[6]

Naturalists were generally agreed that, on the face of it, many fossil shells were quite distinct from living shellfish, and likewise with other marine fossils; the problem was to discover the extent of the difference and then to explain it. In fact, the difference varied greatly from one group of fossils to another. In some there seemed to be no close similarity at all, but in others it was arguable that at least some of the fossils were identical to living species. This spectrum between contrast and identity supported a corresponding range of explanations, from claims of widespread extinction to the denial of any extinction at all. The diversity of the problem can be illustrated first by reviewing the groups of fossils involved, and then by summarizing the ways in which they were interpreted by some representative savants.

Some of the most common and distinctive fossils in the Secondary formations had no obvious counterparts at all in the present world. The most striking examples, and certainly the most frequently cited in this context, were the *ammonites*. These beautiful fossils were often rather confusingly preserved; but by the late eighteenth century they were well understood to have been delicate shells coiled in a plane spiral, their interiors divided into a series of chambers. In detail, however, they displayed an astonishing diversity of form, and they varied in size from a coin to a cartwheel; clearly they represented a profusion of different species. Yet not a single ammonite shell had ever been found in the present world. The most closely similar living mollusk shell was the famous pearly nautilus from the East Indies, highly prized as an elegant and exotic "natural curiosity": also a chambered spiral shell, but with partitions of far simpler form than any ammonite (see Fig. 5.6 below; also Fig. 4.7). To many naturalists, ammonites therefore seemed to be strong

6. Martini and Chemnitz, *Conchylien-Cabinet* (1768–95), extended to eleven volumes and over two hundred plates with superb colored engravings of shells (pl. 213 depicted a few *fossil* shells). Dezallier, *Conchyliologie* (1780), was a representative handbook: it was a new edition of a classic work first published in 1742, but revised extensively after the author's death; eighty plates were crowded with illustrations of shells, including a few fossils such as ammonites. On conchology at this period, see Dance, *History of shell collecting* (1986), chaps. 4, 5.

Fig. 5.2. A variety of ammonites, as portrayed in one of the plates published in 1768 in the great "paper museum" *Remarkable Natural Objects* compiled by Georg Knorr and continued by Immanuel Walch. Ammonites were common fossils in many Secondary formations, but totally unknown alive; they posed in an acute form the problem of accounting for the difference between fossils and living organisms. The repeated looped lines visible on the two lower specimens marked the junctions between the shell itself, coiled in a plane spiral, and the regularly spaced shelly partitions that had made it a "chambered" shell. This gave ammonites a general resemblance to the pearly nautilus, but they were also clearly distinct from that living mollusk. As was frequently the practice at this time, the localities of the fossils were not given, let alone any information about the specific rock formations from which they had been collected. (By permission of the British Library)

evidence—perhaps the best evidence of all—for extinction on a large scale (Fig. 5.2).[7]

Often closely associated with ammonites, and almost equally abundant in some Secondary formations, were the puzzling fossils known as *belemnites*, solid bullet-shaped objects with a strongly crystalline internal structure. Even the organic origin of belemnites had earlier been in doubt; but as progressively better specimens were collected in the course of the eighteenth century it became clear to naturalists that the belemnite itself had grown around the apex of a delicate conical shell, partitioned internally into a series of chambers recalling those of the nautilus. Belemnites therefore seemed to be mollusks, perhaps distantly related to the nautilus and even to the ammonites; but as with the latter, there was no trace of them in the present world.[8]

A third group of common fossils, likewise completely unknown alive, were the objects that, from their coinlike size and form, were given the name of *nummulites* (Fig. 5.6). When sliced open they were found to be tightly coiled plane spirals, closely divided into chambers, giving them a slight resemblance to both ammonites and nautilus shells; but nothing like them had ever been found alive, and their affinities were utterly obscure (in modern terms they were giant foraminiferans or shelled protozoans). They were so abundant in certain Secondary formations—though not in the same ones as ammonites or belemnites—that the rock was composed of little else: a striking example, well known from reports by travelers, was the limestone of which the ancient Pyramids in Egypt were composed.

Living fossils

Another class of fossils common in many Secondary rocks had, like belemnites, been subject earlier to doubts whether they were organic at all. These were small pill-shaped or star-shaped objects, with an internal structure identical to that of

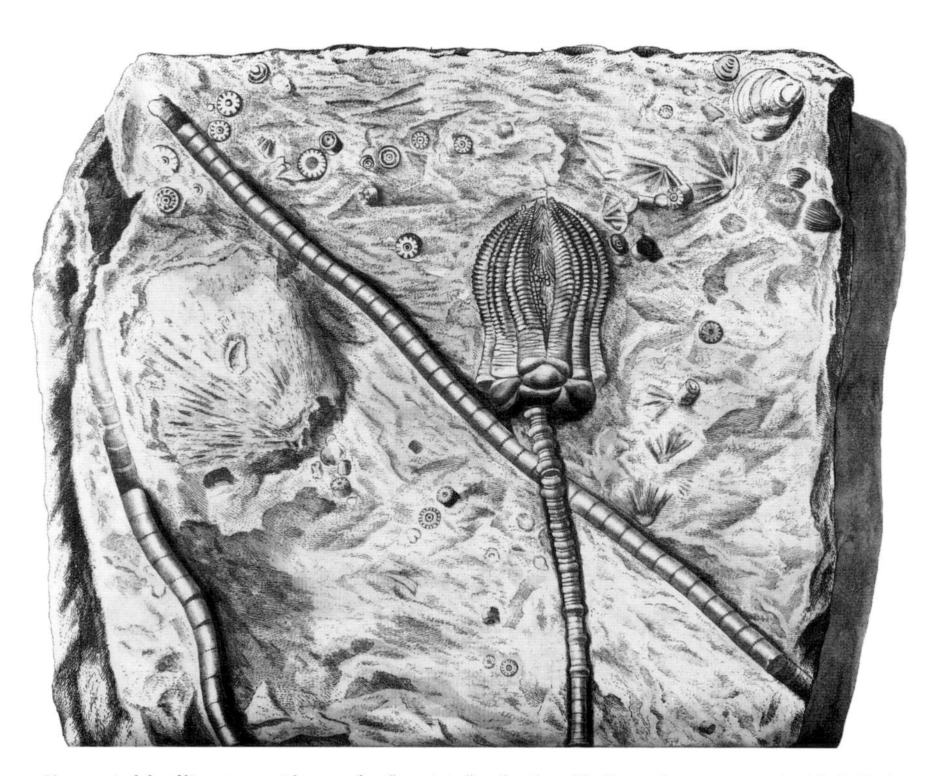

Fig. 5.3. A slab of limestone with part of an "encrinite" or fossil sea lily (in modern terms, a crinoid): half of a double-page engraved plate published in 1755 in Knorr and Walch's paper museum of fossils. The body and stem of one individual, with tapering arms, lie across part of the stem of another, belonging to a different species. Isolated "trochites" or cylindrical segments of stem are scattered across the surface of the rock, and there are also several different kinds of fossil shell (mostly "anomias" or, in modern terms, brachiopods). The specimen had been in the private collection of a merchant in Halle, but no information about its locality was given, and perhaps none was available. Unlike the case of the ammonites, sea lilies had been found alive, although extremely rarely (Fig. 5.4). (By permission of the British Library)

7. Fig. 5.2 is reproduced from Knorr and Walch, *Merkwürdigkeiten der Natur* 2 (1768) [London-BL: 457.e.13], pl. 1a. In modern terms the specimens came from quite different Mesozoic formations: the lower ones Triassic in age, the others Jurassic.

8. Guettard, "Sur les bélemnites" (1783), in his *Mémoires* (1768–86) 6: 215–96, reviewed no fewer than thirteen interpretations; this article, and the later historical review in Blainville, *Bélemnites* (1827), are still two of the best sources for this instructive case of the relation between fossil material and zoological interpretation. In modern terms, belemnites are indeed related, as *cephalopods*, to both ammonoids and nautiloids and, more closely, to living cuttlefish and octopuses.

crystals of inorganic "spar" (calcite). They were of varied form, and were known by a corresponding variety of names (e.g., *trochite, asterite*); in places they were so abundant that, like the nummulites, they comprised the bulk of thick beds of limestone. Whatever organism they represented had evidently flourished profusely in the former world. Often several of these puzzling objects were preserved stacked together (forming what was known as an *entrochite*); and as still better specimens were found it had become apparent that these were segments of the long and originally flexible stems of some larger organism, more or less broken up after death. Then, in rare and highly valued specimens, such stems were found with a kind of root at one end and a kind of body at the other; the body was itself composed of small shelly plates, and extended into five segmented arms, often highly branched. These strange fossils were known by various names such as *encrinite.* Just as ammonites were slightly similar to the nautilus, so the plated structure and striking fivefold symmetry of encrinites suggested an affinity to the brittle stars of modern seas, and more distantly to starfish and sea urchins. Although rather plantlike in general form, encrinites or "*sea lilies*" seemed to have been animals rooted permanently to the sea floor (Fig. 5.3).[9]

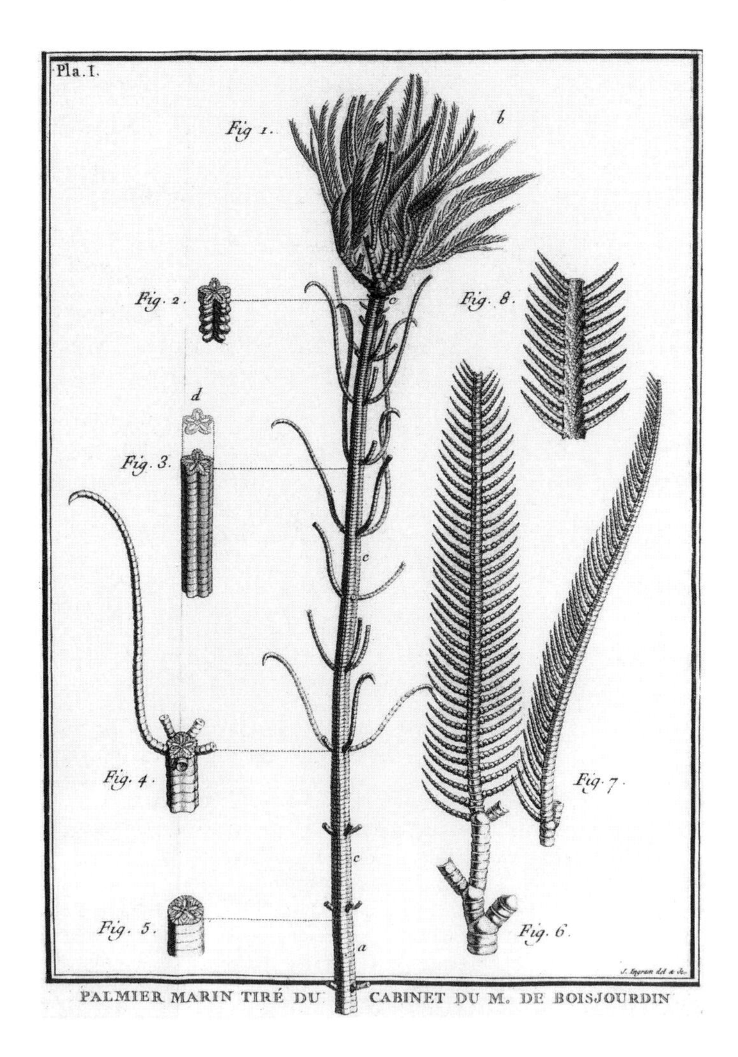

Fig. 5.4. A living sea lily or encrinite—here called a "sea palm" [*palmier marin*]—that had been dredged by chance from the sea floor in the West Indies. This extremely rare specimen, from a private collection, was described by Guettard in 1755 to the Académie des Sciences in Paris; he noted that another had been found at great depth off Greenland. The drawings on the left show the structure of the stem, with its star-shaped segments revealing the fivefold symmetry of the whole animal; on the right, the structure of one of the five arms. To many savants, living sea lilies suggested by analogy that ammonites, belemnites, and other fossils unknown in the present world might likewise still be living in the ocean depths, but as yet undiscovered. (By permission of the Syndics of Cambridge University Library)

Unlike the ammonites, however, sea lilies were certainly *not* extinct. Indeed, at the time they were the best example of what has here been called the "living fossil effect" (§5.1), and they offered strong evidence in favor of the migration option for explaining the difference between fossils and living organisms. Just occasionally, living sea lilies were brought up from great depths in the ocean by mariners who had let down long lines for soundings or anchorage. Such specimens were extremely rare, however, and highly prized in consequence (Fig. 5.4).[10]

The possibility that the contrast between living and fossil organisms might be due to their migration in the course of time was reinforced by another important group of fossils. Unlike the ammonites, belemnites, and nummulites, with absolutely no trace in the present world, and unlike the sea lilies, known only with extreme rarity, many other common fossils were well known as living organisms, at least of the same genus if not exactly of the same species. Such living species were referred to as the "*analogues*" of the fossil species. Yet in many cases these analogues were found living in places far removed from where their fossil counterparts were collected. For example, in many Secondary formations the fossil shells known generally as "*anomias*" were as abundant and diverse as the ammonites. Like the shells of clams, mussels, and oysters, anomia shells consisted of two parts hinged together, the body of the animal being housed and protected between them. But unlike the more familiar shellfish, all anomia shells had a perfect bilateral symmetry that gave them a beautiful sculptural quality and made them highly attractive to collectors (in modern terms they are not mollusks at all, but members of an unrelated group of shelled animals, the brachiopods). Their living analogues were usually found in waters deeper than those normally exploited by fishermen, and this seemed an adequate explanation for their relative rarity in the present world. Like the sea lilies, anomia shellfish might still be just as abundant as they had been in the former world, but living at such depths in the sea that their shells were not often dredged up (Fig. 5.5).[11]

The rarity of the living analogues of other common fossils seemed to be due to their having changed their position in a different way. Eighteenth-century naturalists noticed that many fossils that were common in the Secondary formations of Europe found their living analogues among the faunas of the *tropics*, not those

9. Fig. 5.3 is reproduced from Knorr and Walch, *Merkwürdigkeiten der Natur* 1 (1755) [London-BL: 457.e.12], pl. 11a. The specimen was drawn in Halle in 1750 by G. W. Gründler; this kind of encrinite was found in the *Muschelkalk* formation (in modern terms, of Triassic age). Again in modern terms, all encrinites were *crinoids*, and were indeed related to the other echinoderms.

10. Fig. 5.4 is reproduced from Guettard, "Sur les encrinites" (1761) [Cambridge-UL: CP340:2.b.48.72], pl. 8. Another Caribbean specimen was described and illustrated by Ellis, "Account of an *Encrinus*" (1762), and was given the same Latin name as the fossil "sea lilies"; in 1780 it became part of William Hunter's collection, and is now on display in Glasgow-HM: see Durant and Rolfe, "William Hunter" (1984), 18–19. Such specimens are an important example of how, in the natural-history sciences, the scientific importance of specimens was not (and is still not) dependent on their abundance: "replication" does not have the same significance as in the experimental sciences. In this case, even a single specimen, or half a dozen, was enough to make the point decisively.

11. Fig. 5.5 is reproduced from Bruguière, *Vers testacées* 4 (1797) [Cambridge-UL: XXVI.1.123], pl. 239. Bruguière's incomplete text—covering only genera A–C, and therefore not the "*terebratule*"—was published in 1789–92, but this volume of plates not until 1797. See also the *Anomien* in Martini and Chemnitz, *Conchylien-Cabinet* (1768–95) 8: 65–118, pls. 76–79.

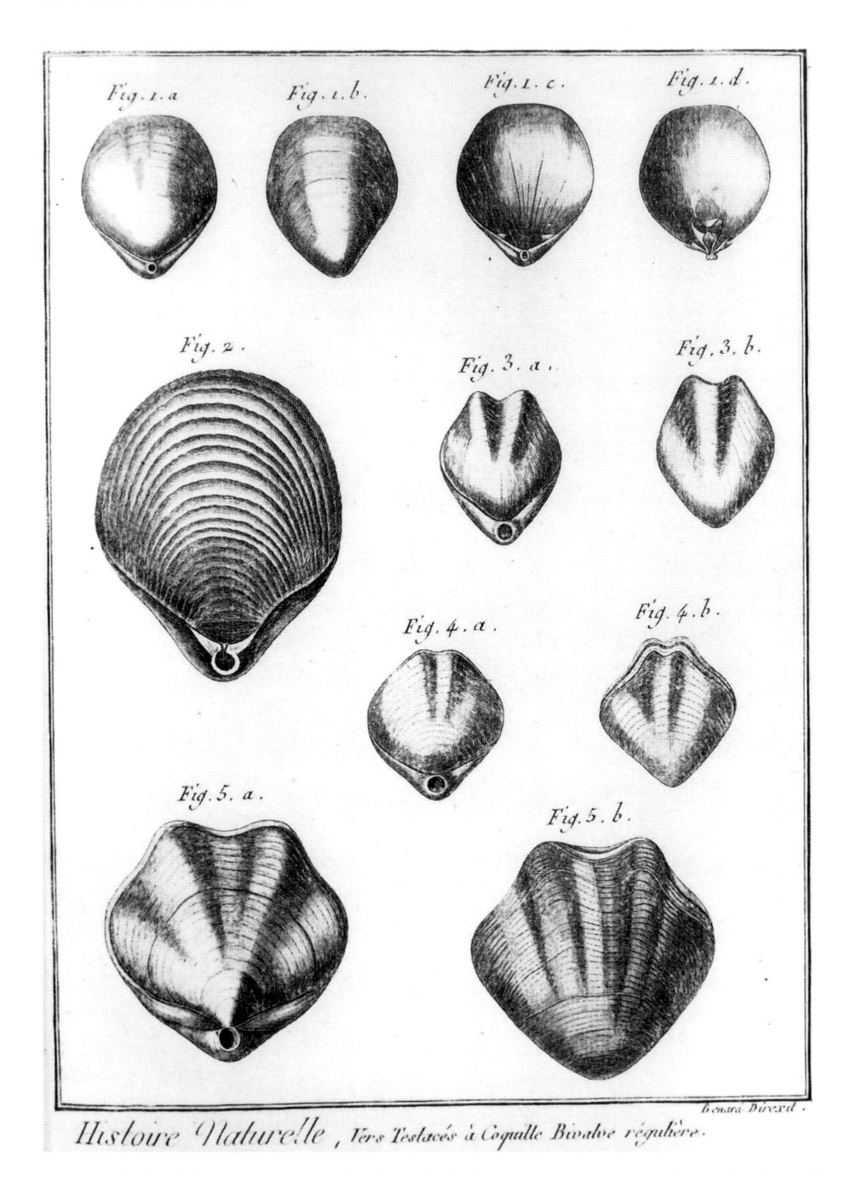

Fig. 5.5. Shells of five living species of *térébratule*, the least uncommon genus of anomias (in modern terms, brachiopods) to be found in present seas: as illustrated in one of the hundreds of engraved plates of shells published (this one in 1797) in the natural history volumes of the vast *Encyclopédie Méthodique*. One specimen (fig. 1) is shown with its two hinged parts separated and the interior visible (figs. 1c, 1d). Such shells were usually found in deep water, suggesting that their relative rarity—compared to their abundant fossil counterparts—might be due simply to migration, as in the case of the sea lilies. (By permission of the Syndics of Cambridge University Library)

of the nearby North Atlantic, North Sea, or even the warmer Mediterranean. For example, many of the shells found in abundance at famous European fossil localities, such as Hordwell on the Hampshire coast of England (Fig. 2.3), were assigned to genera—if not to species—that were known alive only from warm waters such as those of the Caribbean or the East Indies. The most notable case of all was the nautilus, which Burtin, for example, found as a fossil in his Secondary rocks near Brussels (Fig. 4.7). As already mentioned, this was unmistakably similar, if not identical, to the beautiful pearly nautilus, a rare and highly valued shell that was brought to

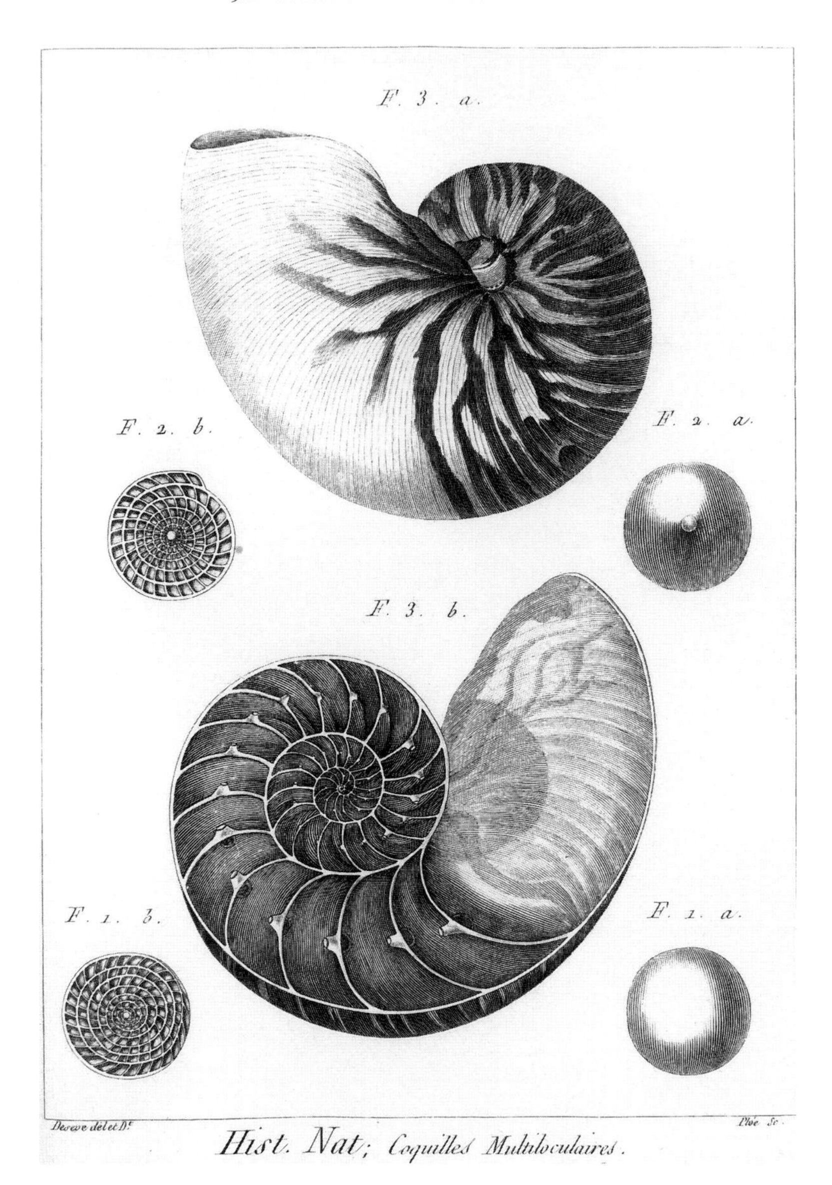

Fig. 5.6. The pearly nautilus shell from the seas of the East Indies: engravings published in 1797 in the *Encyclopédie Méthodique*. The upper drawing is of the shell with its external color markings. The lower shows it sliced open along the plane of symmetry, with its chambered interior and regular shelly partitions; the open space at the end was known to house the body of the animal (somewhat similar to a cuttlefish) and the chambers were thought to be filled with gas to give it buoyancy. The smaller drawings are of the puzzling fossils, extremely abundant in some Secondary formations, known from their coinlike size and shape as "*nummulites*": on the right their featureless external form, and on the left sliced open to show their complex internal structure of a tightly coiled plane spiral with close partitions. (By permission of the Syndics of Cambridge University Library)

European collectors from the East Indies, but which was quite unknown in temperate waters (Fig. 5.6).[12]

Fossil fish and possible whales

The remains of marine vertebrates, although far rarer and more fragmentary, seemed to be as ambiguous as fossil shells and other invertebrates. Fossil fish were generally rare, apart from isolated sharks' teeth; but in a few famous localities they were well-preserved and—given intensive collecting—quite abundant (§2.1). Like

12. Fig. 5.6 is reproduced from Bruguière, *Vers testacées* 4 (1797) [Cambridge-UL: XXVI.1.123], pl. 471; his unfinished text (1792) described neither the nautilus nor the nummulite. See also the nautili in Martini and Chemnitz, *Conchylien-Cabinet* 1 (1768): 222–6, 241–53, pl. 19; engraving of "soft" body, 222.

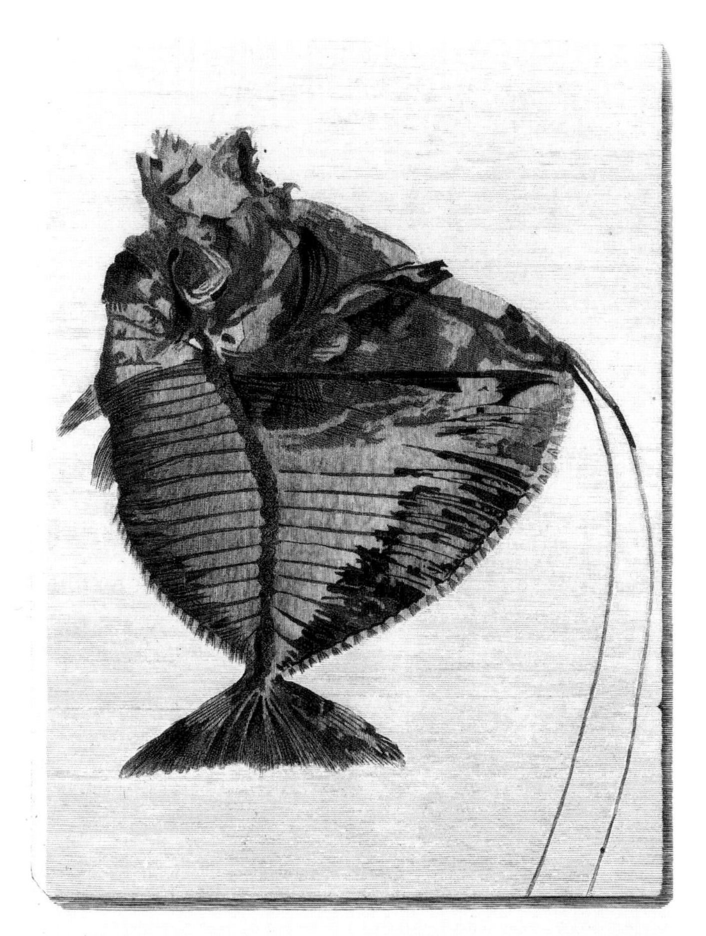

Il Rombo indiano ——— *Scomber Rhombeus.*

Fig. 5.7. The Indian turbot [*rombo indiano*] or *Scomber rhombeus*, one of the many superbly preserved fossil fish found at the famous locality of Monte Bolca, as depicted in one of the large engraved plates of Giovanni Volta's *Fossil Fish of Verona* (1796[–1809]). As the name implies, this fish was identified as a species known alive in the East Indies. Like the fossil nautilus, it suggested a major change of climate in Europe since the time when the limestone of this Secondary formation had been deposited. (By permission of the British Library)

the nautilus, some were identified as closely similar, if not identical, to species now living in tropical seas. These could be grouped with the nautilus and other shells of tropical genera, and treated as evidence that the apparent contrast between fossil and living marine animals was due simply to a change of climate and a consequent shift in habitats (Fig. 5.7).[13]

Other fossil fish, however, were unlike any known alive, and seemed as likely as ammonites to be truly extinct. For example, those preserved in the famous *Kupferschiefer* [copper shale] formation in Germany had no obvious close analogues in the present world (Fig. 5.8).[14]

Marine fossil vertebrates other than fish were even rarer, and no less ambiguous. The most celebrated were the huge toothed jaws and other scattered bones of the "Maastricht animal", which had been found in the underground quarries just outside the Dutch town (Figs. 2.5, 2.6). The local naturalists had identified their animal as a giant crocodile, which might have made it a strong candidate for extinct status. But in 1782 the great Dutch anatomist Petrus Camper (1722–89) purchased some of the best of these "antiquities of the old world" for his private collection. He was

Fig. 5.8. A fossil fish from the *Kupferschiefer* [copper shale] formation of Eisleben, near Mannsfeld, as depicted in 1755 in Knorr and Walch's paper museum of fossils. Its mosaic of solid diamond-shaped scales was quite unlike the thin overlapping scales of most living fish. This and other fossil fish from this famous locality were widely assumed to have become extinct in the "revolutions" of the deep past. (By permission of the British Library)

convinced it was a marine animal, since it was associated with so many other clearly marine fossils. While visiting London in 1785, he was able to compare the fossil specimens with a crocodile skeleton at the Royal College of Surgeons. As he reported to the Royal Society (to which he had been elected), it confirmed his suspicion that the "fossil monster" was significantly different from any crocodile, and he identified it as a cachelot, or toothed sperm whale. It was much larger than any known toothed whale; but even with the intensive whaling industry of the time no naturalist would have been so imprudent as to claim that all living species had yet been sighted. So the Maastricht animal, like many less spectacular marine fossils, was ambiguous: it might be extinct, as Camper probably suspected, but it might be a species as yet undiscovered in the present world.[15]

13. Fig. 5.7 is reproduced from [Volta], *Ittiolitologia Veronese* (1796[–1809]) [London-BL: 459.g.6], pl. 18, drawn by the local painter and architect Leonardo Manzatti (Volta's authorship is given on cccxiii). The 1793 prospectus [copy in London-BL: 111.g.65/5] for this great paper museum states that printing and engraving had started three years earlier, so it must represent research begun in the 1780s; its completion was delayed by the subsequent French occupation of Verona. Fortis, "Différentes petrifications" (1786), argued for the tropical character of the fauna in the light of Cook's voyages in the Pacific; Gaudant, "Querelle des trois abbés" (1999), describes the pamphlet war of 1793–95 on this and related points of interpretation. The main private collection on which Volta's work was based was expropriated by Bonaparte in 1797: see Frigo and Sorbini, *600 fossili per Napoleone* (1997). Some of the purloined specimens are on display in Paris-MHN; fine collections of similar specimens found subsequently can be seen in Bolca-MF, near the site, and Verona-MSN.

14. Fig. 5.8 is reproduced from Knorr and Walch, *Merkwürdigkeiten der Natur* 1 (1755) [London-BL: 457.e.12], pl. 18, fig. 1.

15. Camper, "Petrifactions near Maestricht" (1786); see also Visser, *Petrus Camper* (1985), 127–32.

Explaining the former world

Putting all these examples together, it is hardly surprising that many naturalists considered it plausible to suppose that the apparent contrast between fossils and living marine animals might be due simply to the fossil species having changed their habitats in the course of time, moving either to deeper waters or to warmer parts of the globe. Or, as a still simpler explanation, they might not have had to change their position on the globe at all, if the Secondary formations in which they were so common had been deposited in much deeper or warmer waters than those of present European seas. In either case the contrast would only be apparent, and overall the world of life might not have changed significantly in the course of time. On the other hand, the *total* absence of any trace of living ammonites and belemnites, for example, seemed to other naturalists to be a strong argument in favor of the reality of extinction as a result of natural events—the earth's revolutions—in the deep past. Less plausibly than either of those alternatives, it was also at least conceivable that the fossil species might have been transmuted into the living ones, without either migration or extinction. Anyway, the issue was clearly inconclusive. Having outlined some of the evidence that weighed one way or the other, the diversity of opinion can now be illustrated by referring to the work of a few representative savants.

Johann Friedrich Blumenbach (1752–1840), the young professor of medicine at Göttingen, faithfully reflected the consensus—or the lack of it—among savants, in the successive editions of his elementary *Handbook of Natural History*, which he wrote for the use of his students. In the first edition (1779) he alluded only briefly to the problem, in the context of reviewing the number of animal species known. "Since we know so many animals only as petrifactions, and not yet in [living] nature," he noted, "some distinguished men have concluded that many genera and even whole families must be *extinct* [*ausgestorben*]." On the other hand, he also noted that the seabed was almost unexplored, so they might still be living there. But yet again, there were so many of these fossil genera unknown alive, notably the ammonites, that they might have been lost in the "very great catastrophes" that the earth had undergone. A decade later, in the third and "greatly improved" edition of his textbook (1788), Blumenbach included a much fuller section on the study of fossils [*Petrefacten-kunde*], which he now treated as a branch of "mineralogy" that promised to give "most important clarification to cosmogeny", that is, to geotheory or to what de Luc had called "geology" (§3.1). Reviewing the major groups of animals in turn, Blumenbach listed the fossil species of each under two headings, the known [*bekannte*] and the unknown [*incognita*]. Ammonites, for example, were listed as *incognita*, with the implication that they might be extinct. However, Blumenbach prudently sat on the fence, leaving the problem unresolved, as indeed it was among his fellow naturalists.[16]

In fact, some naturalists changed their opinion one way over these years, and some the other. For some, an explanation in terms of migration was strengthened every time a new "analogue" was discovered in distant waters or at great depths. Their argument was necessarily probabilistic: it seemed increasingly *probable* that all fossil species might still be living somewhere in the world's oceans. Jean-Guillaume Bruguière (1750–98), one of the naturalists who argued this most forcefully, was also one of those best qualified to judge the matter. After a medical training at Montpellier, he had served as surgeon and naturalist on Kerguelen's voyage, gaining valuable field experience of natural history—including doubtless the collecting of exotic shells—in southern Africa and around in the Indian Ocean on Mauritius and Madagascar, not to mention firsthand experience of the sheer scale of the world's oceans. Back in France he became excited by the fossils he collected around Montpellier and began writing about them as the relics of the earth's "revolutions". In 1781 he moved from the provinces to Paris—like Soulavie at around the same time—to build a career as a naturalist. Buffon's colleague Daubenton, who was in charge of the volumes on natural history for the *Encyclopédie Méthodique* (planned as the successor to Diderot's famous earlier *Encyclopédie*), commissioned Bruguière to deal with the "worms" [*vers*] (which in modern terms included the mollusks and many other invertebrates, as well as wormlike animals).

Bruguière soon made himself a leading expert, particularly on the conchology of both living and fossil mollusks. Although his work was published among the sporadic installments of a vast and sprawling encyclopedia, his reputation would have ensured that his descriptions and conclusions were keenly read by shell collectors as well as by naturalists and other savants. In his introduction, he looked to the fossil remains of "worms" as the empirical key to geotheory, because they gave reliable evidence of a geohistory analogous to human history:

> By comparing their fossil remains from remote times with those of species that inhabit the vast expanses of the seas, exact notions of the true theory [of the earth] can now be reached. The alterations that are continually produced there [in the sea] imprint on nature historical monuments more effective and durable than those that man, aided by the arts, tries in vain to perpetuate.[17]

However, when Bruguière discussed the fossil mollusks in more detail, he almost negated this highly geohistorical approach, for he interpreted the contrast between fossil and living species solely in terms of environmental change. For example, early in the order of topics for publication (under the letter "A") he had to deal with the ammonites. He described twenty-three species, all known only as fossils and all quite distinct from the living nautilus. However, he rejected what he implied was the usual assumption that they must be extinct, arguing instead "that the races of ammonites still survive, and that they live at the greatest depths of the sea". He claimed that ammonites, belemnites, and many other fossils as yet unknown alive were all "*pelagic*" or deepwater in habitat, and that "the vast floor of the sea could still be paved with them", yet we might have no direct evidence of their existence. This was supported, he argued, not only by the sea lilies and anomias that were

16. Blumenbach, *Naturgeschichte* (1779), 43–44; ibid., 3rd ed. (1788), 673–74.

17. Bruguière, *Histoire naturelle des vers* (1792), Introduction (1789), iii; see Laurent, "Jean-Guillaume Bruguière" (2002).

known to survive in deep water, but also by the analogues of well-known fossils that were found among "the marine species that arrive daily, [brought back] from the most distant seas".[18]

Bruguière also supported this inference with a well-established observation from the field. Ammonites and other putatively "pelagic" fossils were confined to the "ancient beds" of the older Secondary formations and were overlain—even thousands of feet higher in the pile of formations—by beds with more familiar "*littoral*" species. This was to be expected on the standard model of geotheory that Bruguière, like so many other naturalists, took tacitly for granted. All the apparent contrast between ancient fossils such as ammonites and the more familiar living mollusks could be due to the progressive retreat of the oceans from what had now become the continents. This radical change in physical geography had shifted the habitats of mollusks in the course of time, without any true faunal change, let alone extinction. So the former world had not been fundamentally different from the present. The conclusion was not original to Bruguière, but being based on such an expert evaluation of detailed evidence it was one that other naturalists could not lightly dismiss.[19]

Bruguière's argument can be contrasted with the views expressed by Burtin (§4.2) around the same time. In his earlier book on the fossils found around Brussels, Burtin had noted that all the fossil species distinctive enough to be identified accurately—such as his "nautilus of the Indies" (Fig. 4.7)—had their analogues among species now living in the tropics. Here he opted for an explanation in terms of a faunal migration, whatever the physical cause of the climatic change might have been. Yet he also denied that ammonites, belemnites, and "anomias" might still be living in deep water in northern oceans, so in their case he rejected the option of migration in favor of extinction. In fact, Burtin's subsequent prize essay on the earth's "general revolutions" showed him to be an unusually strong advocate of widespread if not universal extinction: "I dare to claim that *nonextinct* [*non perdus*] fossil species are a true rarity", he wrote, implying that this was now a minority opinion. He insisted on the importance of accurate identifications; mere generic resemblances were not enough. He claimed that, with the possible exception of some anomia shells, there were no "true analogues" or *exact* matches among living species, only approximate analogues at the generic level. But Burtin's claim about the generality of extinction depended on his forthright rejection of the supposition that the fossils without known analogues might still be living in the deep oceans: "It is thus wrong to claim the sea floor as the habitat [*séjour*] of the [living] analogues of the fossil shells that are alleged to be *pelagic*, of which not the least specimen has yet been found either in museums, or by the researches of naturalists [in the field], or by divers, or by soundings, or [thrown up] by the most violent movements of the sea."[20]

This assertion was of course supported by the total failure to find the shells of any living ammonites or belemnites; but Burtin ignored completely the living fossils of other kinds, such as the sea lilies, which did indeed seem to be exclusively deepwater in habitat. His conclusion, that there had been widespread natural extinctions in the course of the earth's successive revolutions, was therefore not as firmly based as it appeared. However, he does seem to have had a vague sense of how the issue might be resolved, if the general tacit assumption of a single *undifferentiated* "former world" were to be abandoned. As a local fossil collector he had little firsthand field experience of the varied formations in which fossils were found. But he was evidently aware at second hand that it was only the older Secondary formations, those closest to the Primary rocks, that contained ammonites and belemnites; his own collection from around Brussels included none at all. "Does not all this give rise to a well-founded suspicion," he suggested, "that they have been interred by a revolution different from that to which we owe so many other accidental fossils?" That rudimentary sense of a *plurality* of revolutions (§4.2) was supported by what he had read—as reported by Ferber—of Arduino's fieldwork many years earlier (see Fig. 2.32): in his own words, "these Alps are composed of beds [*couches*], and each bed contains a species of petrifaction that is proper to it, which always differs from the species contained in the other beds."[21]

Burtin was not alone in being aware that several Secondary formations contained distinctive kinds of fossils: some, such as the *Muschelkalk* of the German lands, were even named after their fossils. There was also a rather vague sense of a *sequence* of Secondary formations, each perhaps with its own distinctive suite of fossils, at least within a given region (§2.3). However, in Soulavie's analysis of the province of Vivarais this sense became highly specific, and was supported by fieldwork as careful as Arduino's a generation earlier. Soulavie described three distinct limestone formations, clearly sequenced by the direct evidence of superposition, with fossil faunas that indicated a directional trend towards those of the present world. The lowest and therefore oldest limestone contained no forms known alive: among them were ammonites, belemnites, "*térébratule*" anomias, and "*entroques*" or fragments of the stems of sea lilies (he was apparently unaware that the latter had been found as "living fossils"). The middle limestone contained fossils that he considered to be a mixture of unknown and living forms, such as cockles, scallops, and nautilus shells. The uppermost and therefore youngest formation contained shells of forms well known alive in the nearest sea (the Mediterranean), such as oysters, whelks, and cowries, as well as sea urchins.[22]

How were these differences to be interpreted? Soulavie's readers were left on no doubt that he was claiming a true sequence of faunal change: the three limestones

18. Bruguière's striking verb [*pavé*] indicates clearly that at this time the word "pelagic" denoted what would now be called the *benthic* or bottom-living fauna of the ocean floors, not the organisms now termed "pelagic" in the open water above. Each of his ammonite species would now be regarded as at least a genus, if not a taxonomic family.

19. Bruguière, *Histoire naturelle des vers* (1792), 1: 28–43.

20. Burtin, "Révolutions générales" (1789), 81, also 7n, 27; he is unlikely to have read Bruguière's article on ammonites before revising this for publication the same year. See also his *Oryctographie de Bruxelles* (1784), 13–14, 132–33.

21. Burtin, "Révolutions générales" (1789), 199; and 198n, citing the French edition (1776) of Ferber, *Briefen aus Wälschland* (1773), letter 5. The French translation (and the English) implied incorrectly that Arduino thought a single species, rather than a distinctive assemblage of many, characterized each "bed" or formation (§2.3).

22. Soulavie, *France méridionale* 1 (1780): 317–32. He described the fossils only by their vernacular names (the French equivalents of the words used here), not in terms of precise species. The limestones are now dated respectively as Jurassic, Cretaceous, and Tertiary (Miocene): see Ellenberger, *Histoire de la géologie* 2 (1994): 278.

represented three successive periods of time, each the "reign" of a distinct fauna; and "as the ages [*siècles*] multiplied, new families thus appeared in the bosom of the sea." How this had happened remained unclear, though Soulavie suggested rather vaguely that "life's principle has seemed to develop and extend itself in the more recent Secondary formations [*carrières*]". Only a few readers would have seen in addition his more explicit reference to the possibility of transmutation, in the brief chapter on the "metamorphosis of several families of animals" that was censored by the Académie des Sciences (§4.4). There he promised he would "prove that the recent families that are not found in the ancient limestones are nonetheless species that descend from the primordial families". He explained that "as the passing ages [*âges*] gave living matter the time necessary to vary its forms, new species inhabited the empire of the seas".[23]

Soulavie's censored speculations about the possible transmutation of species were in fact less important than the sheer fact of faunal change itself, however it had been caused. He himself acknowledged this, when the following year he declared himself ready to leave the controversial causes of "the transmigration of shellfish, the transmutation [*dégénération*] of their species, etc., etc., in the class of more or less proven hypotheses [i.e., speculations]", because what mattered far more was "the comparative chronology of the processes" (§4.4). Geohistory was paramount; what mattered was to determine—*not* as mere speculation—the order of events, which included defining a sequence of faunal "ages". While abstaining from speculation about causes, Soulavie was convinced that the geognostic pile of formations in Vivarais bore witness to real faunal changes in the course of time. The contrasts between fossil and living species were not just a product of changes in physical geography, as Bruguière would claim a few years later. Soulavie insisted instead that the changes in faunal composition that led step by step from the age of the ammonites to the shellfish of the present world proved that there had been a real *history* of marine life. In other words, Soulavie's careful example from Vivarais suggested that the relation between the world of fossils and the present world might be neither one of radical contrast nor one of virtual identity (apart from changed distributions), but instead a matter of stepwise geohistorical change, shifting in the course of time from contrast to similarity.

Conclusion

The handful of savants whose opinions on this issue have just been summarized are enough to suggest how the matter was still highly inconclusive around the time that Saussure climbed Mont Blanc. The contrast between the shellfish and other marine animals preserved as fossils in the older Secondary formations and the species now alive in the present world (which were similar to those preserved in the younger formations) was acknowledged and well-understood by naturalists. How to account for the contrast remained, however, unclear and controversial. The case of the ammonites seemed to many to be strong evidence for widespread extinction, but "living fossils" such as sea lilies made that assumption debatable, and made it plausible to suppose that even the most ancient fossil species might still survive somewhere in the world's oceans, either at great depths or in remote parts of the world. Lurking in the background as a third option, disreputable mainly because it was so speculative, was the idea that the ancient species might have been transmuted into the present ones, without any extinction at all. Finally, and not yet clearly formulated, there was the possibility that the contrast seen in the oldest fossils had changed step by step in the course of time into the similarity or near-identity seen in the youngest, as a result of true faunal change. That *geohistorical* option would allow the search for the causes of change—as a problem in "physics"—to be shelved in favor of first plotting its course.

5.3 WITNESSES OF FORMER CONTINENTS

Fossil plants

Most of the fossils found in the field and stored in museums were of animals that had evidently lived in the seas that once covered the present continents. However, a few fossils proved that there had also been land areas in those remote times. This was anticipated on the standard model of geotheory (except for the most distant past), because the global sea level was thought to have subsided gradually, uncovering ever larger tracts of the earth's surface (§3.5). It was also built into the idiosyncratic geotheory proposed by Hutton, since his cyclic model required that at every stage there be habitable continents undergoing slow erosion (§3.4). On almost every other "system", including Buffon's (§3.2) and de Luc's (§3.3), land areas in the deep past were either to be expected, or were at least not incompatible with the theory.

The best evidence that continents had in fact existed came from fossil plants. These were not seaweeds, which—had they been preserved—would have been nothing more than further evidence of former seas. They belonged unmistakably to the more complex plants that now cover most land areas with vegetation. They were taken to be the remains of plant material washed out to sea by the rivers that must have drained the ancient continents, falling eventually to the bottom and being buried there in marine sediments. This assumption was supported by the knowledge that driftwood and other plant debris often floated far out to sea: a sight that was common to mariners and naturalists alike, on board sailing ships that were often becalmed and allowed ample time to notice such things. However, in evaluating the similarities or differences between fossils and living organisms, fossil plants were just as ambiguous as the remains of marine animals, if not more so. Fossil plants were generally uncommon, but like fossil fish they were abundant at particular localities. There they might be superbly preserved, for example as leaves looking as fresh as if they had just been blown from the trees by autumn winds (Fig. 5.9).[24]

23. Soulavie, *France méridionale* 1 (1780): 327, 365; uncensored text, 349, 354.

24. Fig. 5.9 is reproduced from Knorr and Walch, *Merkwürdigkeiten der Natur* 1 (1755) [London-BL: 457.e.12], pl. 9. The similar specimens in the following plate (9a) were said to be probably from Oeningen.

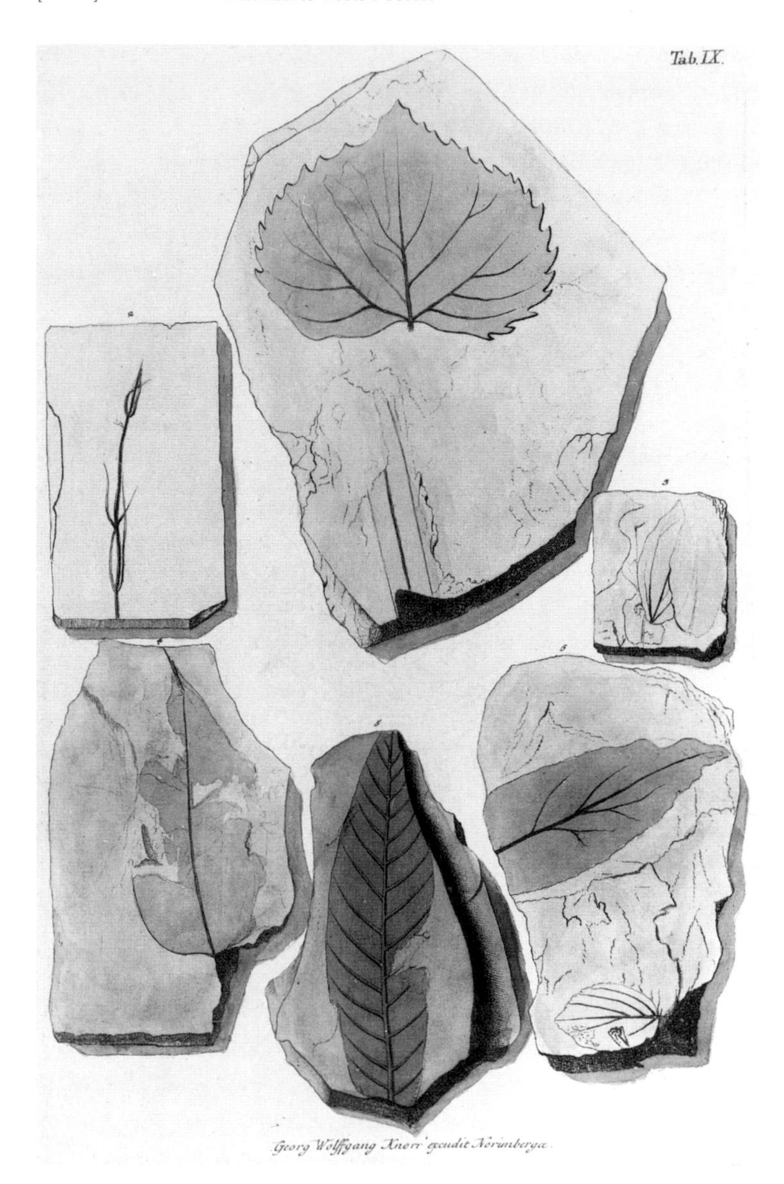

Fig. 5.9. Fossil leaves, as depicted in Knorr and Walch's great paper museum of fossils. This plate (published in 1755) showed specimens from two private collections; their localities were not given, but they probably came from the famous locality of Oeningen near Konstanz. Nor were they identified, but they had an obvious general resemblance to the leaves of living species. (By permission of the British Library)

Such fossil leaves, and the fragments of fossil wood that were found in some Secondary formations, were in effect equivalent to many of the commoner fossil shells. They suggested a world not much different from the present one: the leaves looked like those of familiar living species, and so did the wood. Yet in the case of plants that general impression was much more difficult to confirm with any precision. Most mollusks were identified largely on the basis of their shells, which were often fully preserved as fossils (apart from the loss of their color). But plants were classified primarily—indeed almost exclusively, by Linnaeus and his many followers—on the basis of their flowers, fruits, and seeds, which were extremely rare as fossils; and some plants that were classified in quite different families had almost identical leaves. Fossil wood was rarely preserved with any trace of its internal structure and was even harder to identify. So the degree of contrast or similarity between living and fossil plant species was difficult to assess.

Where the fossil plants could be identified with some confidence, at least at the level of taxonomic "family" if not more precisely, they seemed to conform to the same pattern as the fossils of marine animals. For example, certain Secondary formations were found to include not only beds of coal—locally of great economic importance—but also shales containing well-preserved fronds that were identified as those of ferns (Fig. 2.2). But the best specimens, and some of the fossil trunks found in the same beds, suggested that these ferns had been much larger than those now living in European habitats; they were identified as belonging to the *tree ferns* known to be living in the rain forests of the present tropics. So they were in effect equivalent to the fossil shells of the nautilus and other tropical mollusks: they too suggested that the climate in the part of the globe that now forms Europe might once have been much warmer. They were in fact routinely cited by those naturalists, such as Buffon, who argued that the earth had indeed been hotter in the distant past (§3.2).[25]

However, there was another possible explanation, which was much more plausible here than in the case of marine organisms. This was that the debris of tropical plants might have been swept northwards into Europe by the sudden and violent "deluge" (or mega-tsunami) that was widely held responsible for the excavation of valleys, the displacement of erratic blocks, and many other puzzling features (§2.4). Anyway, whether they had lived where they are found or far away, fossil plants remained highly ambiguous, apart from confirming the existence of dry land in the deep past. They could hardly help resolve the question of the degree of similarity or contrast between the former world and the present.

Large fossil bones

Apart from plants, the most important fossils that witnessed to former land areas were the bones and teeth that were taken to be the remains of terrestrial vertebrates. Bones and teeth were occasionally found in the solid rocks of the Secondary formations. The Maastricht animal was the most striking example (see Fig. 2.5); but once Camper had identified that "monster" as a toothed whale, it took its place as a relic of marine life, just like the mollusk shells and sea urchins with which it was preserved (§5.2). It then seemed likely that other cases of bones and teeth in Secondary formations, such as the supposed crocodile fossils found at Honfleur in northern France and Whitby in northern England, could likewise be interpreted as *marine* fossils. However, most fossil bones and teeth were found not in solid beds of limestone or shale but in Superficial deposits: they were buried in loose gravels and silts close to present rivers or strewn across the surface of the continents. Clearly these fossils were of relatively recent origin, compared to those in the underlying

25. Coal itself was recognized as being composed of plant debris, but it was generally assumed that this was material that had drifted out to sea; the standard model of geotheory encouraged the assumption that virtually all Secondary formations must be of marine, not freshwater, origin. The modern view, that coal accumulated in situ in a continental environment in freshwater swampy forests of some kind, was not widely adopted until the early nineteenth century, when the economic interests of the burgeoning Industrial Revolution stimulated more intensive study of the coal strata.

Secondary formations; they might even date from de Luc's recent "sudden Revolution" in physical geography (§3.3). Since they included, for example, teeth and tusks that looked like those of elephants, it was taken for granted that they must be the remains of the *terrestrial* fauna of the ancient world.[26]

The apparent former presence of tropical animals such as elephants, in what are now much cooler northern climates, generated among naturalists a familiar set of alternative explanations. The animals might have migrated southwards as the climate changed in the course of time. But since they were terrestrial, it was also possible that—like exotic tree ferns—they had been swept northwards out of their unchanged tropical habitat by some kind of transient mega-tsunami in the relatively recent past. Or they might be species distinct from any living forms; if so, they might be extinct, or they might still be flourishing undiscovered in unexplored areas. These alternatives were discussed in relation to several distinct kinds of fossil bones and teeth, found in several different parts of the world.

Huge bones and teeth had been dug up from time to time in Europe for centuries past and were highly prized items in early "cabinets of curiosities" (§1.3). Right from the start they had been integrated into *human* history. They were interpreted first as the remains of the giants who were believed to have lived in the early centuries of the human race, a view that long persisted among the uneducated. When closer study revealed their contrast to any human bones and their similarity to those of elephants, savants took them to be the remains of the elephants that Hannibal had famously imported from Carthage to use in his wars against the Romans. But the textual sources indicated that these military vehicles had been quite limited in number, and anyway they were likely to have been of the small north African variety. So the much larger size of the fossils and, above all, their sheer abundance and wide distribution across Europe had made this explanation implausible. By the later eighteenth century their date had been pushed back into the mists of "antediluvian" times, and they were widely taken to have been swept northwards into Europe during the biblical Flood or in some other violent revolution in the earth's early history. However, although that inference assigned them to the former world, they still did not seem to differ from the elephants of the present world, except perhaps in their larger size.

The problem of accounting for these large fossil bones and teeth became much more acute in the course of the eighteenth century, as an indirect result of the commercial and strategic expansion of the European powers. Important new finds were made both in the vast spaces of Siberia and in the even less explored interior of North America (see Fig. 3.3). They were incidental products of the expansion of Russia to the east and of France and England—in competition with each other—to the west, a situation not significantly affected by the later establishment of the independent United States in one part of North America.[27]

In Siberia there had long been an important trade in what was called "fossil ivory", and the wide though localized distribution of large bones, teeth, and especially tusks was well-known. Several expeditions setting out from St Petersburg, the imperial capital, had brought back specimens of the remains of what the indigenous Siberian people called the "*mammut*" (whence "mammoth"); and in 1741 an illustrated report published in London by the Royal Society had made its identification as an elephant convincing to naturalists and well-known throughout the Republic of Letters. The sheer abundance of fossil elephants in Siberia made it clear that they were a problem for natural history: it was almost inconceivable that their emplacement had anything to do with human activities. Since the nearest known living elephants were in India, many savants interpreted the Siberian fossils as the remains of animals swept northwards from south Asia during some kind of "deluge", just as the European bones might have been swept out of Africa.[28]

This kind of explanation was reinforced by one of the most ambitious and successful expeditions to Siberia. Pallas spent several years exploring vast tracts of the Russian empire under orders from the Empress Catherine (§3.5). His massive published *Travels* (1771–76) made his observations and conclusions widely known, but one discovery was so important that it merited a report to the Academy of Sciences in St Petersburg even before the expedition finally returned. Pallas described how the partial remains of a *frozen* carcass had been found deep within a sandy deposit, exposed in the banks of a tributary of the River Lena, flowing towards the Arctic Ocean in the far east of Siberia (see Buffon's map, Fig. 3.3). That the remains were preserved in the permanently frozen ground, with parts of the skin and even the flesh, seemed to Pallas to be clear evidence that the animal had been swept *suddenly* "out of its native country in the south into the frozen lands of the north." The carcass was not that of an elephant, however, but apparently a rhinoceros; he later recruited Camper to check his identification authoritatively against a living rhinoceros (Fig. 5.10).[29]

More than one large species was now involved in the puzzling phenomenon. Pallas argued that the carcasses must have been swept northwards in the biblical Flood [*Sündfluth*] itself, on the grounds that human historical records mentioned no *later* event massive enough to have effected the transport; it is significant that at this point he did not consider the possibility that it might have been unrecorded because it was too *early* for any human records. A few years later, however, in his great essay on the formation of mountains, he recognized a multiplicity of successive revolutions (§3.5). He attributed the formations near the Urals, with their large fossil bones, to "our globe's most modern catastrophes", and he cited the

26. The Whitby fossils were later identified as extinct marine reptiles, most commonly ichthyosaurs. Lamanon seems to have regarded the Paris gypsum as a Superficial lake deposit of quite recent origin (§4.3), rather than as part of a sequence of Secondary formations; so the fossil terrestrial vertebrate that he reported from the gypsum might not create an exception to the inference that all Secondary vertebrates were marine.

27. The close parallel between the Russian and American sides of this issue has been obscured by the assumption of "exceptionalism" among United States historians of the colonial and early republican periods. For naturalists, there was no practical difference between a fossil bone collected by a servant of the absolutist Russian empire and one acquired from a citizen of the newly independent and relatively democratic United States; the sharply contrasted political cultures of the respective territories were totally irrelevant to the interpretation of the fossil specimens.

28. Cohen, *Destin du mammouth* (1994), chap. 4, summarizes this early history and reproduces the decisive engravings from Breynius, "Mammoth's bones in Siberia" (1741).

29. Fig. 5.10 is reproduced from Pallas, "De ossibus rhinocerotum" (1769) [Cambridge-UL: T340:9.a.1.27], pl. 9; quoted phrase from *Russischen Reichs* (1771–76), 3: 99. Camper, "De cranio rhinocerotis" (1780), read in St Petersburg in 1777, described and illustrated the osteology of the living African rhinoceros and confirmed Pallas's conclusion: see also Visser, *Petrus Camper* (1985), chap. 5.

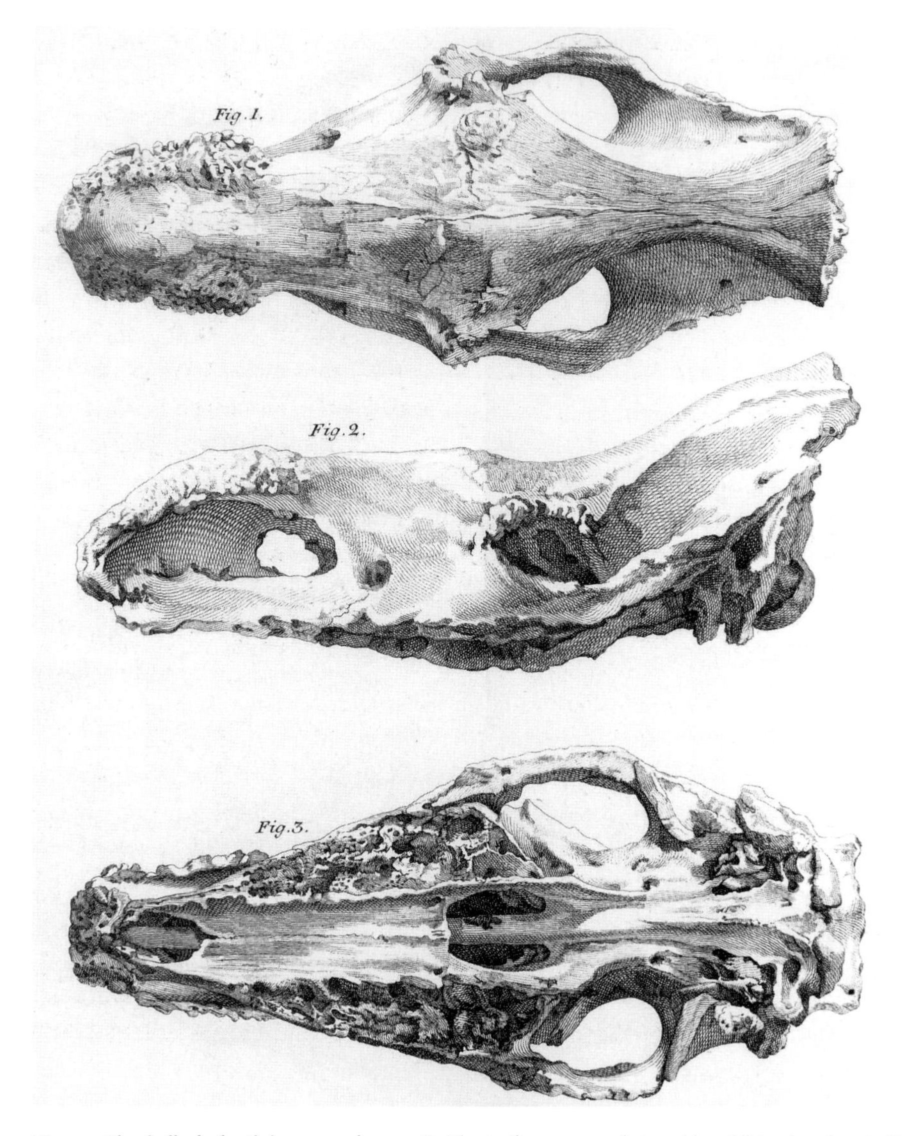

Fig. 5.10. The skull of a fossil rhinoceros from arctic Siberia: three engraved views (dorsal, lateral, and ventral) of a specimen brought back by Pallas, illustrating his report to the Academy of Sciences in St Petersburg in 1768. Finding this large tropical mammal in frozen ground in the far north, added to the already well-known remains of fossil elephants, suggested strongly that they had all been swept from the south in some kind of transient "deluge" or mega-tsunami. (By permission of the Syndics of Cambridge University Library)

frozen rhinoceros of Siberia as "undeniable evidence" that the relevant event had been sudden.[30]

The "Ohio animal"

Meanwhile the puzzle had been extended by the discovery of apparent "mammoth" bones in North America, still further from any known living elephants. Back in 1739, a French military party had traveled from Niagara to the Ohio and then down the Mississippi to strengthen French forces in one of the Indian wars in that contested but barely explored region. On the way, at a salt lick near the banks of the

Ohio, the soldiers had come across masses of fossil bones, teeth, and tusks. Their commanding officer had shipped some specimens down the Mississippi and thence to Paris. The locality became famous as Big Bone Lick. It was later visited by many other travelers, and further specimens found their way not only to coastal cities such as Philadelphia, the cultural center of the British colonies that were to become the United States, but also to all the scientific centers in Europe.[31]

Buffon's initial reaction to the American bones had been, not unnaturally, to equate them with the Siberian bones and to treat the mammoth as one of the animals common to both Old and New Worlds. Claiming that it had been far larger than any living elephants, he had inferred that it was an extinct species of elephant. On the other hand, his colleague Louis Jean Marie Daubenton (1716–99) had considered that the difference in size between living elephants and the fossils from both America and Siberia was within the range of variation to be expected, allowing for age, sex, and environment. The implication was that no new species was involved, let alone one that might be extinct (Fig. 5.11).[32]

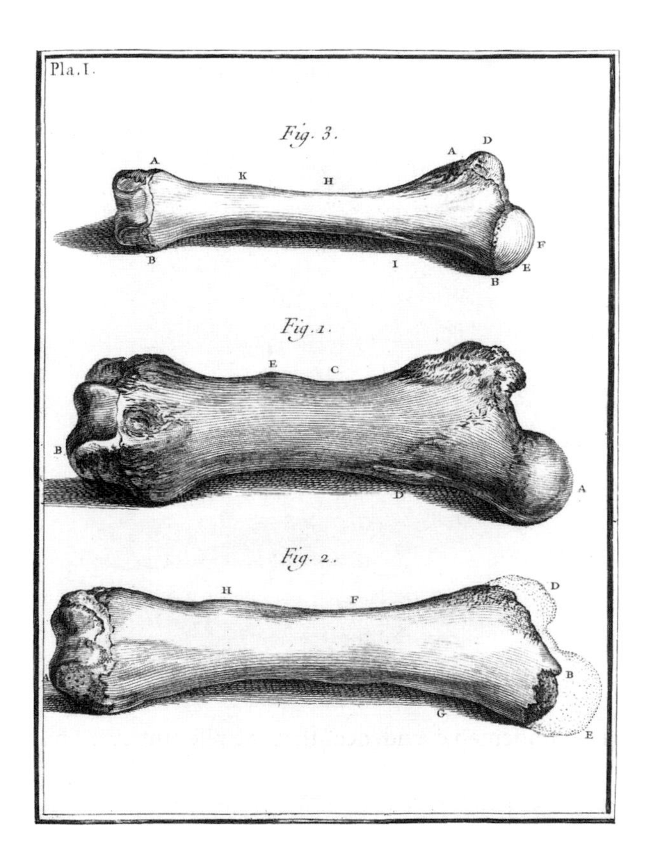

Fig. 5.11. The femurs of living and fossil elephants, as engraved to illustrate Louis Daubenton's paper, read to the Académie des Sciences in Paris in 1762. The huge bone sent back from the Ohio (fig. 1, middle) and a similar one from Siberia (fig. 2, bottom) were much larger than that of a living—or rather, deceased—elephant formerly in the royal menagerie at Versailles (fig. 3, top). But Daubenton argued that this was within the normal limits of intraspecific variation. He noted the length of the Ohio specimen as over 3 feet 4 inches (about one meter). This kind of illustration, portraying equivalent (or, in modern terms, homologous) bones at a uniform scale, became the standard way of representing the *comparative* anatomy of fossils in pictorial form. (By permission of the Syndics of Cambridge University Library)

30. Pallas, *Russischen Reichs* (1771–76) 3: 99; "Formation des montagnes (1778), 53–55.

31. One of the best accounts is still Simpson, "Beginnings of vertebrate paleontology" (1942), 135–51, a piece of detailed historical scholarship all the more remarkable for having been written in his spare time by an already distinguished professional paleontologist. See also Bell, "Box of old bones" (1949), and Semonin, *American monster* (2000), chap. 4. Big Bone Lick was about 25km southwest of the future site of Cincinnati, on the left (now Kentucky) side of the river.

32. Fig. 5.11 is reproduced from Daubenton, "Sur des os et des dents" (1764) [Cambridge-UL: CP340:2.b.48.79], pl. 13 (engraved in error as pl. "3"; it was pl. 1 of his paper). The Ohio bone was from

There was one major problem with Daubenton's neat solution. Unlike the bones and the tusks, the teeth from Big Bone Lick were not at all similar to the distinctive "grinders" or molars of living elephants: the American teeth had a knobbly surface (see Fig. 3.4) very different from the convoluted but flat surface of elephant teeth. Daubenton inferred, quite reasonably, that the consignment from the Ohio had included a mixture of specimens from two distinct species, both of which might have died and been buried while visiting the salt lick; he identified the teeth as similar to those of the living hippopotamus. If this were the case, *two* tropical species might have lived in North America, and their presence there would be more puzzling than ever. But given the vast tracts of that continent that were still unexplored, there was no reason to conclude that either species was extinct.

Three years after Daubenton's paper was published, however, a larger collection from the Ohio had reached London, made this time by an Irishman who had been in the region to negotiate with the Indians in the English interest. One of the Ohio specimens that had been donated to the British Museum was a jawbone with its molars still in place. The leading anatomist and fashionable surgeon William

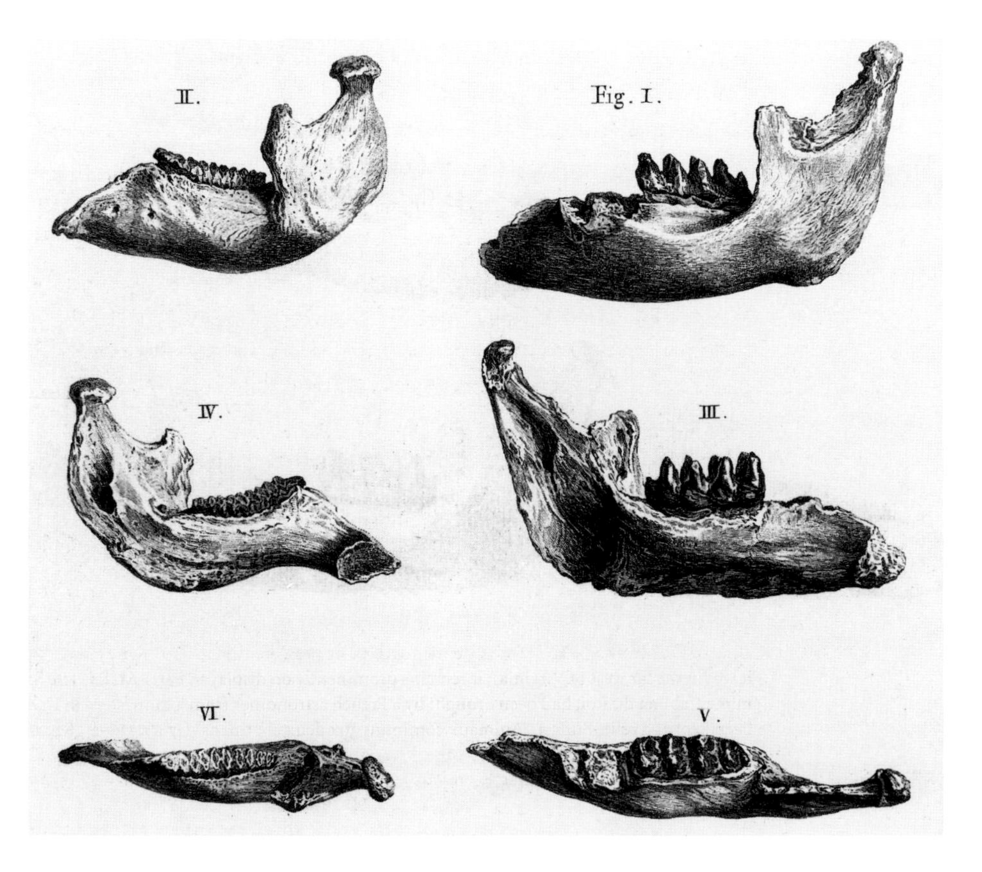

Fig. 5.12. William Hunter's illustrations of a specimen of the lower jaw of "the American *incognitum*" from the Ohio (right), with its characteristic molars in place, compared to that of "a full-grown elephant" from his younger brother John's collection (left), as engraved for the paper that William read to the Royal Society in 1768. He claimed that the American animal had been a distinct species unknown alive and concluded that it was probably a true case of natural extinction. (By permission of the Syndics of Cambridge University Library)

Hunter (1718–83) treated this as decisive evidence that the putative hippopotamus teeth belonged in the jaws of the putative elephant and that only one large species was represented at Big Bone Lick. He inferred that it had been an animal related to, but quite distinct from, the elephants of the present world. He thought the knobbly molars were those of a carnivore, but he knew of no report of any such fearsome beast being sighted alive or even rumored among the hunters and trappers in America or their Indian informants. He therefore concluded that "though we may as [natural] philosophers regret it, as men we cannot but thank Heaven that its whole generation is probably extinct". What he called the "American *incognitum*" or "pseud-elephant", and what others often called the American "mammoth", had thus become a likely case of true extinction (Fig. 5.12).[33]

Buffon had promptly adopted Hunter's conclusion and reverted to his earlier view, though now on this much stronger evidence. In *Nature's Epochs* (1778) the Ohio animal was the only example of a species that had definitely become extinct in the course of the earth's gradual cooling; the inference was that it had been adapted to a hypertropical climate that no longer existed anywhere (§3.2). It was so important in Buffon's geotheory that its distinctive teeth were the *only* fossils he illustrated. Since by this time they had been found in Russia (Fig. 3.4) as well as North America, the species must have dated from his fifth epoch, before the Atlantic separated the New World from the Old.[34]

However, Hunter's inference that the Ohio animal was extinct, although adopted by Buffon and several other naturalists, did not go unchallenged. It was strongly denied, for example, by Thomas Jefferson (1743–1824)—an amateur naturalist as well as a prominent political figure in the young United States—when in 1787 he reported on Virginia, in which Big Bone Lick then lay. Responding to a French request for information about his native state, Jefferson, who was then its governor, compiled an inventory of its human and natural resources, its climate, customs, and commerce: in short, a standard chorography. In the long section on its natural history ("Productions, mineral, vegetable and animal"), he began his inventory of its animals with a description of its "mammoth", which he assumed to be the same as the Siberian animal of the same name; it also had first place in his list of species that were "Aboriginals of both [continents]". He suggested that it might be, in effect, the circumpolar cold-climate cousin of the tropical elephant; anyway, a

"Canada", a term that the French extended southwards to cover the region where it was found (to the English, it was in the far west of Virginia); it remains prominently on display in Paris-MHN. The Siberian bone, also in the Cabinet du Roi, had been brought by a French astronomer from a monastery in Kazan, where it had been kept as a relic. Buffon, "Animaux communs aux deux continens" (1761), 126–27. See also Semonin, *American monster* (2000), chap. 5.

33. Fig. 5.12 is reproduced from Hunter (W.), "Bones near the river Ohio" (1769) [Cambridge-UL: T340:1.b.85.57], pl. 4; the bones are shown, top to bottom, in external, internal, and dorsal views; the engraving was by Jan van Rymsdyk, after Hunter's own drawings. See Rolfe, "William and John Hunter" (1985), and, more generally, Durant and Rolfe, "William Hunter" (1984). Part of the collection was sent to Benjamin Franklin (who was then in London, representing the interests of the American colonies), who came to the same conclusion: see Simpson, "Beginnings of vertebrate paleontology" (1942), 142, 146, and Semonin, *American monster* (2000), chap. 6.

34. Buffon, "Époques de la nature" (1778), 165–90, 504–16. Of his eight plates, six (at 512) were devoted to these specimens; one (pl. 5) was of a molar of a living hippopotamus, to show that it was not the same species as the fossils.

distinct species and one that might still be flourishing in its proper habitat. He argued that the vast tracts beyond the settled coastal areas of North America might well contain *living* "mammoths"; travelers, trappers, and hunters were bringing back to civilization a variety of equally novel if less spectacular animals and plants. He himself had not traveled in the deep interior, but as an American he would have had a much better sense of its unexplored immensity than any European (except the very few who had had that experience at first hand). He concluded that "he [the Ohio animal] may as well exist there now, as he did formerly where we find his bones."

Jefferson rejected extinction mainly on the traditional grounds that it would imply a break in the "chain of being". In a characteristically deistic manner, he made the point in terms of a personified Nature: "such is the economy of nature, that no instance can be produced of her having permitted any one race of her animals to become extinct; of her having formed any link in her great work so weak as to be broken". By comparison with this powerful argument from plenitude, he conceded that any other evidence "would be adding the light of a taper to that of the meridian sun". But he claimed that his inference was in fact also supported by the ethnographic evidence of Indian traditions that involved huge beasts: like the ancient written records routinely cited by European savants, these oral traditions might well preserve genuine historical evidence, albeit under the form of myth and legend.[35]

Although Jefferson may not have been aware of it, his argument was exactly parallel to the one being put forward by other naturalists to explain the apparent extinction of marine fossils. The Ohio animal, like the ammonites, might be alive and well and living in remote places: not on the ocean floor, but in the deep interior of the North American continent. At any moment a hunter or trapper might report seeing a live "mammoth", and even bring back bits to prove it, just as the supposedly extinct sea lilies had been proved to be still flourishing in the deep ocean. The future president of the American Philosophical Society (and of the United States) may well have hoped that his country would in fact be able to boast the world's largest living terrestrial animal. But his argument was not just a piece of chauvinistic boosterism. It was also a variant of an argument long used by other naturalists to explain more humble fossils: the Ohio animal, like the sea lily, might turn out to be a living fossil.

Even without Jefferson's forceful denial of its extinction, the American "mammoth" continued to be very puzzling; indeed it became more so. The young physician Christian Friedrich Michaelis—the son of the great orientalist and biblical scholar Johann David Michaelis of Göttingen—had been interested in the Ohio fossils even before he went to America to serve as surgeon-general to the Hessian forces fighting in the loyalist interest. After peace was restored and the United States established, Michaelis hoped to visit Big Bone Lick in person to collect his own specimens, but this proved impossible owing to unrest among the Indians. As the next best thing, he commissioned the American portraitist Charles Willson Peale (1741–1827) to make him drawings of the fine specimens in a private collection in Philadelphia (doing this gave Peale the idea of opening his own natural history museum there). On his way back to Germany with his proxies, Michaelis studied further real specimens in London; back home, he wrote a short paper on this animal of the "primeval world" [*Urwelt*] for a Göttingen periodical. Like several other naturalists he rejected Hunter's claim that it had been a carnivore; but more surprisingly he argued that it had had neither tusks nor trunk, an inference based on his interpretation of a specimen of the upper jaw (which, as it turned out later, he had got back to front). Michaelis sent copies of his proxy specimens to Camper, and the Dutch anatomist had sufficient confidence in his German colleague's competence that he adopted this new conception of the animal in a paper that he sent to be read at the Russian academy.[36]

The form and status of the Ohio animal was therefore almost as confused and uncertain around the time that Saussure climbed Mont Blanc as it had been some thirty years earlier when Buffon and Daubenton studied the first specimens to reach Europe. It was still unclear whether the American and Siberian animals—both of them usually called "mammoth"—were the same or different species, let alone their relation to living elephants. Unless and until such uncertainties were resolved, these fossils could hardly be used as decisive evidence about the former world.

Giant elks and bears

However, the "mammoths"—whether one species or two—were not the only fossil bones over which naturalists continued to argue. There was, for example, the "Irish elk". The bones and enormous antlers of this giant deer were found occasionally in many parts of Europe, but the best specimens were dug up from the peat bogs in Ireland. Like the Siberian mammoth, the Irish elk was well-known to naturalists, having been described and illustrated long before in a report published by the Royal Society. Hunter thought it was probably extinct, like the Ohio animal; he wrote a paper on the American moose—and commissioned the famous horse portraitist George Stubbs to paint the first living specimen to reach Britain—in order to show that it was *not* the same species as the Irish fossils. Yet he never published that paper, apparently because he came to suspect, as the eminent naturalist Thomas Pennant had suggested, that the huge elusive "waskesser" reported live in the wilds of Canada might in fact be the putatively extinct Irish elk. In other words, the case of the Irish elk remained unresolved for exactly the same reason as the Ohio "mammoth": it too might survive as a living fossil as yet undiscovered.[37]

35. Jefferson, *State of Virginia* (1787), 64–72, 77: this London edition was the first full publication of the original text in English (see its "advertisement"). Much of Jefferson's passage on the "mammoth" was devoted to refuting Buffon's suggestion that life was in some sense less vigorous in America than in the Old World, by claiming that American animals were if anything larger. The claim that the single species involved was probably still extant but undiscovered had been made earlier, for example, by Pennant, *Synopsis of quadrupeds* (1771), 92. See also Bedini, "Jefferson and American vertebrate paleontology" (1985), and Semonin, *American monster* (2000), chap. 9.

36. Michaelis, "Thiergeschlect der Urwelt" (1785); Camper, "Complementa varia" (1788); see Bell, "Box of old bones" (1949), and Visser, *Petrus Camper* (1985), chap. 5. On Peale's role, see Semonin, *American monster* (2000), chap. 11.

37. Rolfe, "Hunter on Irish elk" (1983), transcribes the unpublished paper; see also Rolfe, "William and John Hunter" (1985), 310–15.

Fig. 5.13. The "very beautiful and quite regular" entrance of the cave at Gailenreuth in Bavaria, as illustrated in Esper's account (1774) of the fossil bones found abundantly in the deposits on the floors of this cave and others nearby. Its position well above the valley floor made it difficult to imagine how the bones had got there, even during a violent Deluge, and its regular form suggested to Esper that it had not been torn open by that event. This was the only *field* illustration in his book. (By permission of the British Library)

Another well-known case was that of the bones found in great abundance in deposits on the floors of certain caves in Bavaria. They were described in a magnificent monograph (1774) by Johann Friedrich Esper (1732–81), a Lutheran pastor and amateur naturalist who lived and worked not far away; a simultaneous French edition, translated for him by the professor of anatomy at Erlangen, the nearest university, made the work accessible and well-known to naturalists throughout Europe. The entrances of most of the Bavarian caves were well up on the steep sides of the valleys, so that the bones could hardly have been swept in by the present rivers, even in times of flood. Esper assumed in the usual way that they dated from the earth's "general devastation" [*allgemeine Verwüstung*] during the Deluge, but he was baffled by the problem of their emplacement: "it remains inconceivable to me", he admitted, "how such an immense quantity of animal skeletons has got into the caves [I have] described" (Fig. 5.13).[38]

Esper was equally baffled by the identity of the animals. He thought the most abundant bones might be those of large bears, along with hyenas and even lions: not what was now found in the region, even in its dense Teutonic forests. But the bones did not seem to match exactly those of any species known to him; local naturalists had told him they agreed. He therefore inferred in the usual way that the species must either be undiscovered or extinct. He was not averse to the possibility of extinction; indeed, he countered one standard argument against it—as used later by Jefferson, for example—by recruiting natural theology on to the other side: "It seems to me to be a proof of the greatest wisdom and most precise providence of him who in the animal kingdom is also monarch and conserver of his works, that animals that are created for certain purposes, for certain times, should withdraw again from the theater of what is visible [*der Schaubühne des sichtbaren*], when those purposes are achieved."[39]

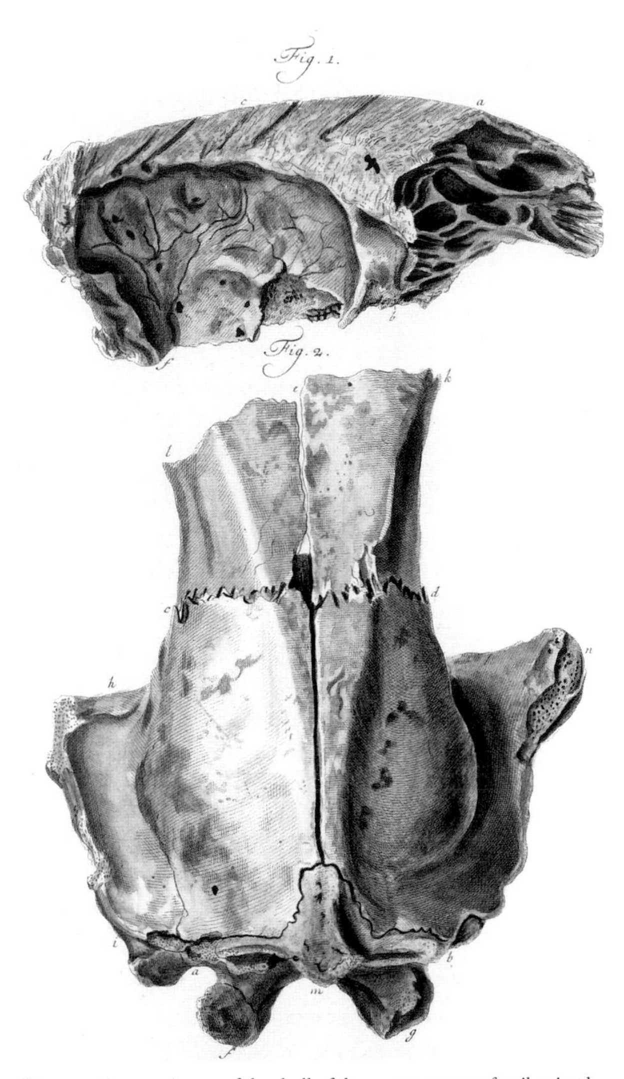

Fig. 5.14. Two specimens of the skull of the most common fossil animal found in the Bavarian caves: one of the plates engraved for Esper's monograph (1774). He thought it might be a large bear, but he knew of no living species like it, and therefore left it unidentified. The upper drawing is of a skull sliced along the median plane, viewed from within to show the brain cavity; the lower is of another skull in dorsal view, showing the sutures. (By permission of the British Library)

In the face of all this uncertainty, Esper evidently felt he should simply describe the bones in detail and illustrate them with a set of proxy pictures that might one day enable others to resolve the problem (in fact he also sent real specimens to naturalists such as Buffon and Daubenton in Paris, Camper in the Netherlands, and Hunter in London). His descriptions showed osteological competence, at the very least, and the engravings were superb. Even if the cave animal or animals remained unidentified, Esper had certainly established yet another case of a strikingly large fossil species in the fauna of the former world (Fig. 5.14).[40]

Esper's work added to a debate that rumbled on among a network of naturalists who, although they were scattered around Europe, were in contact by letter, exchanged publications, and took every opportunity to visit each others' collections when they could. For example, Johann Heinrich Merck (1741–91), a prominent chemist in Darmstadt, published "letters" to his colleagues elsewhere—in effect they were brief scientific papers—describing the bones found in his region, particularly those of elephants and rhinoceros. After visiting Camper, Merck

38. Fig. 5.13 is reproduced from Esper, *Description des zoolithes* (1774) [London-BL: 35.i.6], 1 (quoted phrase, *Neuentdeckten Zoolithen* (1774), 9). Like Faujas's pictures of extinct volcanoes (Fig. 2.23), this was probably drawn by a professional artist or engraver on the basis of the naturalist's rough field sketch; to modern eyes it appears highly idealized. See Heller, "Zoolithenhöhle bei Burggailenreuth" (1972). The cave is on a forested slope above the village of Gailenreuth, which is near Muggendorf, between Erlangen and Bayreuth (see the much later map, Fig. 10.9).

39. Esper, *Neuentdeckten Zoolithen* (1774), 90.

40. Fig. 5.14 is reproduced from Esper, *Neuentdeckten Zoolithen* (1774) [London-BL: 459.e.7], pl. 2, engraved by J. A. Eisenmann and explained on 40, 146. The gifts of specimens are mentioned in Rupke, "Caves, fossils" (1990), 242.

reported that the specimens in the Dutchman's fine collection—which he regarded as second only to John Hunter's in London—had enabled the two savants to make the crucial comparisons: Merck's fossil rhino teeth were distinctly different from Camper's specimens of the living species, either the African or the Asian, but also from Pallas's Siberian fossils. Above all, Merck insisted that only such detailed comparisons would ever lead to firmer results.[41]

Conclusion

This survey of the problem of fossil bones up to the time of Saussure's ascent of Mont Blanc, although far from complete, will at least have shown that it was a highly active and productive field of research. It engaged naturalists in a debate that straddled national and linguistic frontiers, throughout Europe and all the way from Russia in the east to what had recently become the United States in the west; the cosmopolitan ideals of the Republic of Letters were far from nugatory in the late Enlightenment. Yet the debate remained highly inconclusive. The explanations invoked to account for ordinary marine fossils were also deployed in relation to these more spectacular specimens: the fossil bones might belong to known living species or to species living but as yet undiscovered; or the species might be truly extinct. They might have lived where their fossil remains are found; or, like some fossil plants, they might have been swept there from alien climates in some kind of megatsunami, perhaps as recent as the biblical Flood itself. Almost any alternative was plausible, or at least defensible.

The major cause of all this uncertainty will be apparent. Even where fossil specimens were abundant, even when they were fully described and faithfully illustrated to make them mobile in proxy form (as they were by Esper, for example), no firm decisions about their identification could be reached without an extensive knowledge of the osteology of all known living species and a sense of the range of their intraspecific variation. What was required was not only a thorough understanding of comparative anatomy but also access to a comprehensive collection of the skeletons of living species to act as a database for comparison. Even a substantial paper museum of proxy bones was not available. There was nothing in the field of osteology to match Chemnitz's great *Shell Museum* for conchology (§5.2); it would of course have been a vast undertaking to compile one, with so many vertebrate species and so many bones in the skeleton of each. Even Hunter and Camper, for example, with access to large private collections, did not have adequate means to resolve the problem. As Burtin commented, those two "princes of modern anatomy" had been reduced to calling the Ohio animal an "incognitum", in effect a confession of failure: "*anatomie comparée*"—Burtin used what was still a quite novel phrase—was too undeveloped to resolve the chronic uncertainties surrounding fossil bones.[42]

However, although the problem remained unresolved, a gradual shift of opinion can be detected among naturalists during the years leading up to Saussure's ascent of Mont Blanc. Unlike the parallel debate about marine fossils, in the case of terrestrial animals the balance of plausibility seemed to shift progressively in favor of the option of extinction and away from the notion that the species might survive as living fossils. The chronology of Camper's work, for example, shows this trend quite clearly. For him and other naturalists, the shift was probably due to the *cumulative* impact of a growing number of examples of fossil bones for which extinction seemed the most plausible explanation. But unless these cases could be resolved more decisively, fossil bones could hardly provide firm evidence with which to gauge the degree of contrast or similarity between the former world and the present; they could scarcely be recruited as reliable indices of geohistory.[43]

5.4 THE ANTIQUITY OF MAN

Humans in geohistory

Around the time that Saussure climbed Mont Blanc the problem of reconstructing the former world and assessing its relation to the observable present world—the prerequisite for any true geohistory—remained unresolved. Fossils seemed to offer the best hope, since they were both familiar and unfamiliar, capable of being classified with living animals and plants but often quite distinct in detail (§5.1). Yet they too were inconclusive. Whether they were of marine organisms (§5.2) or of animals and plants that had lived on land (§5.3), their relation to living faunas and floras was uncertain and controversial. They were nature's unambiguous witnesses to former seas and continents, but the relation between that former world and the present was obscure. In particular, the discovery of even a few living fossils, and the reasonable anticipation of many more to come, threw doubt on the reality of the apparent contrast between the present and the deep past. It was conceivable that the world had always been much the same kind of place, with much the same kinds of animals and plants, at least back to some primal point of origin of all living things. Yet set against that view was the powerful fossil evidence—if it were taken, as it were, at face value—that the former world had been substantially different from the present, whatever might be the explanation of the change. In trying to resolve this uncertainty, plant and animal fossils were ambiguous.

The problem had one further very important aspect. The most puzzling form of life to understand geohistorically was also the one of greatest intrinsic interest: the human species itself. In Saussure's time it remained quite uncertain whether humans had lived alongside animals and plants more or less from the start, or whether they were relative newcomers. The latter would imply the reality of a lengthy *prehuman* world; it would turn the relation between the present and the former world into—most fundamentally—a contrast between a human world and one that was totally *nonhuman* because it was prehuman. This was a possibility that was hardly anticipated, however, and the evidence for it was once again inconclusive.

41. Merck, *Lettre à Monsieur de Cruse* (1782), *Seconde lettre* (1784), and *Troisième lettre* (1786), all published in Darmstadt but written in French to make them accessible internationally.

42. Burtin, "Révolutions générales" (1789), 29.

43. Camper, "Complementa varia" (1788); see Visser, *Petrus Camper* (1985), 135–38.

That there might have been a lengthy prehuman world was not anticipated on either of the rival conceptions of time and history inherited from earlier generations (§2.5). The picture of cosmic history derived from the Creation narratives in Genesis—which remained immensely powerful at the imaginative level, even though they were no longer interpreted literally, at least by savants—implied that the universe had had a human presence from the start, apart from a brief prelude to set the stage for its primarily human story. But the Aristotelian picture of an uncreated eternity—which was equally familiar among the educated—likewise assumed that humans had *always* been part of the cosmic scene: literally always, from all eternity. Both conceptions took it for granted that humans had an essential and permanent role in the universe. Neither facilitated thinking about what was in effect a third alternative. This was that cosmic history—or at least the more accessible history of the earth and life—might have been very lengthy but not eternal and that human life might have appeared only at a late stage in the relatively recent past. Those who supported (or took for granted) either of the traditional views found this third alternative startling and even difficult to conceive, because it implied that for most of its history the earth had lacked the human presence that would have given it ultimate meaning. Yet the third alternative became increasingly plausible in the course of the eighteenth century, at least to savants who had seen or knew about the relevant empirical observations.

Texts and bones

There were three distinct sources of evidence that might help in evaluating this surprising possibility: textual records, human skeletal remains, and artifacts. Human history with substantial textual documentation went back—fairly consensually among scholars—only as far as the Romans and more fragmentarily to the ancient Greeks. Beyond that was the more problematic ground of ancient Jewish history, often scorned and dismissed by savants hostile to religion, and the even more obscure histories of the Egyptians and other ancient peoples around the Mediterranean: obscure because, in particular, the hieroglyphic script had yet to be deciphered, and Egyptian history was known only at second hand from Greek sources. Much further away were China and India, with other and quite distinct ancient histories, still poorly known to European scholars. None of these, however, seemed to take human history further back than a few millenia without entering highly debatable territory. Even Hutton could concede—honestly and as it were with a straight face—that the known record of *human* history was compatible with the short timescale calculated by chronologers (§3.4).

To take human history back beyond those last few millennia entailed accepting as authentic the earliest putative records of Egyptian, Indian, and Chinese history; and most scholars dismissed these as mere myths and legends that had been spun in ancient times to boost the prestige or establish the legitimacy of long-vanished dynasties. On the other hand there were some who, following Boulanger's lead (§4.1), suspected that genuine records of historical events might survive under the guise of myths and legends, and that if properly interpreted these unpromising texts could push human history further back than conventional chronology allowed. But even if they did, it would only slightly extend human history, relative to the vastly greater spans of time that seemed to be required for geohistory. It needed a robust faith in the eternity of the human world to assume, as Hutton did implicitly and Toulmin explicitly, that humans must have existed indefinitely far back in the deep past, yet without any surviving textual traces except in the most recent millennia.

Where textual sources failed, it was reasonable to hope that other kinds of evidence might take human history still further back into the past. Obviously the most direct evidence would be the fossil remains of humans, their skeletons or at least some of their bones. But the continuing failure to find any such fossils made a long *prehuman* geohistory seem increasingly likely. The accumulating evidence in favor of a total timescale of unimaginable magnitude was not matched by any parallel evidence that these vast expanses of time had been filled with human history. The great piles of Secondary formations yielded masses of marine fossils, and at least the scattered relics of plants drifted from vanished continents, but no trace whatever of humans (or none that was not highly controversial). Skeletons were found, of course, in burial mounds in northern Europe and in supposedly Etruscan tombs in Italy (Fig. 4.1); but all these were taken to be, unproblematically, remains from within the span of recorded human history. Unambiguously *fossil* remains of human beings could not be found. Many were reported, as they had been in earlier centuries when almost every fossil bone was assumed to be human. Again and again, however, closer study by savants with the requisite anatomical knowledge showed that such reports were spurious: the bones always turned out to be those of animals.

A celebrated case was the skeleton of "a man a witness to the deluge" [*Homo diluvii testis*], which the Swiss naturalist Johann Jacob Scheuchzer (1672–1733) had found and claimed as such early in the eighteenth century. Decades later, van Marum purchased this famous specimen (Fig. 9.7) for Teyler's Museum in Haarlem, but by then any claims for its human status had long been abandoned: Burtin, for example, relegated it and similar cases to a mere footnote, identifying it as a fish [*silure*]. By that time, almost all such claims could be dismissed in the same way, as products of an earlier and less enlightened age or of plain anatomical ignorance.[44]

The only notable exception shows in fact why there was such general skepticism among naturalists about reports of fossil human bones. When Esper carefully described the animal bones found abundantly in some Bavarian caves (§5.3), he also mentioned finding shards of pottery on the cave floors, but he guessed that these human artifacts dated from historic times and were perhaps no more than a

44. Scheuchzer, *Homo diluvii testis* (1726), Burtin, "Révolutions générales" (1789), 17n; see Jahn, "Notes on Dr Scheuchzer" (1969). The specimen, which remains prominently on display in Haarlem-TM, was van Marum's most expensive purchase on his travels in 1802: see Forbes et al., *Martinus van Marum* (1969–76) 2: 199, 375. The animal—later identified as a giant amphibian (§9.3)—is well-preserved, and its nonhuman anatomy is so obvious that it is surprising that Scheuchzer, a physician by profession, ever claimed that it was human and that any of his contemporaries, on seeing the specimen or at least his quite accurate engraving of it, ever agreed with him.

thousand years old. Quite distinct from such signs of relatively recent human occupation were two human bones, a maxilla and a scapula, that had been found in the same deposit as the bones of his possibly extinct animal species. This was wholly unexpected, and gave Esper "quite awesome pleasure [*ganz schröckhaften Vergnügen*]", evidently because it would help to locate his animals within human history:

> Did both pieces [of bone] belong to a Druid, or to an antediluvian, or to an inhabitant of the earth of more recent times? Since they lay under the animal bones with which the Gailenreuth cave is filled, since in all probability they were found in the original bed [*Schichte*], I conjecture—not without adequate grounds—that these human structures are also of the same age as the other animal petrifactions. They must have got here with them, by the same accident [*Zufall*].[45]

As already noted, Esper inferred that the "accident" or exceptional event that had emplaced all the bones had been the Flood itself. This implied that the human bones must be those of an "antediluvian", not a postdiluvian, individual. His pleasure at the discovery was therefore understandable: the biblical account of Noah's Flood stressed the human casualties in that catastrophe, but no unambiguously human remains had yet been found in the Superficial deposits that were supposedly of "diluvial" origin. However, doubts can be detected between the lines of Esper's report. Did the human bones really come from the same deposit as the animal bones? That this was the case "in all probability" implies that it was not certain; or rather it suggests that Esper could not be sure because he had not seen them in place with his own eyes. The bone deposits were probably being excavated by workmen, not by Esper himself, and probably he watched them only intermittently. Since he mentioned the possibility that the human bones were postdiluvian he must have realized that they might be coeval with the pottery, or that they might come from a burial in the floor of the cave, dug later into the truly fossil bone deposits. And finally, it was always possible that he had been duped by the workmen, who might well have known what kind of bone the naturalist most hoped to find and would be willing to reward them for.

Above all, however, other naturalists who read Esper's report on these specimens—which in any case was buried in his dense descriptive text—would be bound to reserve judgment on it, since he provided no proxy pictures to match his fine engravings of animal bones (Fig. 5.14). Anyway, even if the crucial specimens were accepted as genuinely human and genuinely contemporary with possibly extinct animals, Esper's find would merely confirm that human antiquity extended a short way back into the former world; and this was no surprise, because some antediluvian humans—Noah's contemporaries—were already vouched for by traditional textual history. In any event, Esper's claim seems not to have caused any great stir among naturalists. Nor did a later and similar claim by a more distinguished naturalist. Spallanzani described a mass of bones [*montagna dell'ossa*] on the island of Cythera or Cerigo (now Kithira) off the west coast of Greece, mentioned that they were "petrified" and therefore truly fossil, and claimed almost in passing that most of them were human; but since he gave no detailed anatomical evidence his claim, like Esper's, was tacitly discounted. In effect, the consensus of naturalists was that no human bone had yet been found that could confidently be regarded as authentic and contemporary with the possibly extinct animal species of the former world.[46]

History from artifacts

With no textual or fossil evidence that humans had existed before the last few millennia of geohistory, only one possibility remained: that there might be traces of human activity in the form of artifacts more durable than records on parchment, papyrus or paper, or even perhaps than bones. The idea that human history might be pieced together from the nontextual evidence of artifacts such as buildings and pottery was relatively novel, though it was greatly strengthened by the sensational finds at Herculaneum and Pompeii (§4.1). However, if—unlike those at these famous sites—the artifacts were unaccompanied by inscriptions and coins and unmentioned in contemporary texts, they could not be dated with any certainty and their place in world history remained obscure.

Megalithic monuments in northern Europe, for example, and burial mounds containing pottery and other objects, were evidently pre-Roman; they were generally attributed to the shadowy "Druids" or "Celts" who were believed—on the textual authority of Tacitus and other Latin authors—to have been living in those regions before the Romans arrived. But there was no obvious reason to regard these artifacts as any older than, say, early Greek or Jewish history. They might be traces of less civilized people, even of illiterate people, but they were not obviously the traces of people *earlier* than those featured in the histories based on textual records. For example, de Luc used the peat cover on burial mounds in Hannover as one of his "natural chronometers" for dating the emergence of the present continents, concluding that the mounds were constructed soon after that decisive natural event (§3.3); but since in his opinion the event was none other than the "deluge" reported in biblical and other ancient records, which textual chronologers dated to just a few millennia ago, the same kind of date clearly applied to these traces of early human history.

More problematic were the chipped stone tools that were often picked up off the fields. Unlike the pottery found in burial mounds, these had no context to help in their interpretation. That they were of human origin was certain: earlier doubts on that score, like the similar doubts about the organic origin of some fossils, had long since been dispelled. Some of these tools looked crude in workmanship, and were therefore readily attributed to uncivilized people (no antiquarian had yet tried to

45. Esper, *Neuentdeckten Zoolithen* (1774), 25–26. "Druids" were a customary catch-all for northern European peoples at the dawn of the historic period (see below). Grayson, *Human antiquity* (1983), 89, notes that the first sentence and the phrase quoted before it were, rather strangely, omitted from the French edition.

46. Spallanzani, "Isola di Citera" (1786), 451–59; the only plate (at 464) depicts oysters and bone fragments, the latter not identified as human. The paper was summarized in French, but not until years later, in *Journal de physique* (1798).

replicate the flint knapping that produced them, which has given modern archaeologists a healthy respect for the skills of Palaeolithic people). But this did not necessarily imply that they had been made at an earlier time than the literate cultures that had left textual records in more civilized parts of the world. Found on the surface of the ground, as most of them were, they provided no unambiguous evidence about the dating of early humans.

Even if stone tools were to be found below the surface, embedded in Superficial deposits, their position in human history would merely be pushed back into antediluvian times, to join Esper's problematic human bones. It was not just provincial naturalists like Esper, but also internationally respected figures such as Pallas, who attributed the fossil animal bones found in these deposits to the mega-tsunami obscurely reported as the "deluge" in biblical and other ancient records (§5.3). Those records made it clear that this catastrophic event had happened *within* human history, not before it: it had almost totally destroyed some kind of antediluvian human society. The records of this time were extremely scanty and its character obscure, but it was taken for granted that there had been such a human world. So it would not be surprising if the deposits left by the most recent revolution in geohistory contained some traces of human life and activity. When, a decade after Saussure climbed Mont Blanc, stone tools were first reported from deep within Superficial deposits, the discovery caused no great stir (Fig. 5.15).[47]

Fig. 5.15. Two views of a chipped "flint weapon" found in a Superficial deposit of brick-earth at Hoxne in eastern England: an engraving to illustrate a report sent in 1797 to the Society of Antiquaries in London by John Frere, a member of Parliament and local landowner. Ancient stone implements of this kind were already well-known. But most of them were not demonstrably any older than the earliest textual records in more civilized parts of the ancient world, even if—as Frere suggested—they were made by people without knowledge of metals. However, his report was one of the first to claim that such flints might be much older, dating from "a very remote period indeed; even beyond that of the present world", on the grounds that at this particular locality they had been found *below* layers of sand with marine shells and large fossil bones. (By permission of the Syndics of Cambridge University Library)

De Luc, for example, who was among the many naturalists who made that equation between "sudden revolution" and biblical Flood, clearly assumed that human beings had been present in the former world, living on the ancient continents. That he knew of no traces of them did not surprise him, however, since what was preserved of that world was primarily the ancient sea floors that formed the present continents. So if he later heard of Frere's report of human artifacts being found in the Superficial deposits, he would surely have assimilated it into his geohistory without difficulty. The great revolution that divided the present from the former world in de Luc's binary geotheory could almost be equated with the divide between the human world and the prehuman, but not precisely. To concede that human history extended some short way back into the former world entailed no radical adjustment of its place in traditional chronology. Human history might still have been quite brief in total duration.

However, a few naturalists reported finding human artifacts apparently dating from much earlier than this. These reports constituted the slender empirical evidence for supposing that the human species might extend back beyond the last few millennia into the depths of geohistory. But they were highly dubious, and not just because—if genuine—they would entail breaching the short timescale of human history embedded in traditional chronology.

Soulavie, for example, while bringing out his multivolume work on southern France (§4.4), published a short paper describing some further volcanic rocks that he had found in Auvergne. By his own criteria this particular lava dated from "a very ancient eruption" because it had since been affected by the slow erosive action of rain and rivers. Underneath the lava, and therefore even older, was a deposit with fossil plants, which he regarded as terrestrial in origin. This much was uncontentious. But he also reported finding in this deposit a wooden plank that had been shaped by an adze [*hache*]. Its significance was clear: "It shows that this former world [*ancien monde*] was inhabited; for these worked planks indicate an intelligent being who fashioned wood for use." Soulavie thought the "arts" might still have been in a primitive state at that remote time—the adze might have been made of stone, not metal—but nonetheless he regarded the find as evidence of an extremely lengthy human history. It was further grist for his geohistorical mill: it was integrated into his outline reconstruction of the whole history of the region (§4.4). As the sea level subsided, the uplands of Auvergne had emerged as islands; and his new find proved that these islands had been inhabited by "intelligent beings", before the land was buried under volcanic lava. The material traces of early human life, together with all the other evidence, enabled the naturalist to "read nature's archives, and to give these successive events [*faits*] a chronological order".[48]

47. Fig. 5.15 is reproduced from Frere, "Flint weapons at Hoxne" (1800) [Cambridge-UL: T468.b.36.13], pl. 15, read in June 1797; Frere emphasized that it was only the position of the artifacts that made this case worth reporting. The village of Hoxne is near Eye, which is 30km north of Ipswich (Suffolk). It would be anachronistic to attribute the lack of impact of this report to any distinction between archaeologists and "scientists": Hamilton, for example, was only one of the many English savants who were fellows of both the Royal Society and the Society of Antiquaries and would have read the periodicals of both bodies.

48. Soulavie, "Volcan de Boutaresse" (1783), 291–93. Boutaresse is on the north flank of the Monts du Cézallier, about 30km southwest of Issoire (Puy-de-Dôme).

However, the antiquity of Soulavie's wooden plank would have seemed questionable to other naturalists. He had found it in a deposit that others would have regarded as Superficial and relatively recent in origin, had it not been covered by a lava flow; it was only by Soulavie's controversial criterion of extremely slow erosion that the eruption was estimated as being of very great antiquity (§4.4). In any case, on his own reconstruction of geohistory the artifact was less ancient than the youngest of his Secondary limestone formations, which was full of marine fossils of living species (§5.2). And above all, other naturalists could not judge the authenticity of the plank without seeing it for themselves; and Soulavie, like Esper with his human bones, provided no picture of it as a proxy for that firsthand experience.

A second example was more striking because it came from what was apparently a thick Secondary formation of solid rocks. When Lamanon interpreted the gypsum around Paris as a precipitate from a vanished lake (Fig. 4.11), he treated the bones and teeth found in that deposit as the remains of an animal that was probably "lost" [*perdu*] or extinct. In the same paper, however, he also reported evidence that human beings had been living around the shores of the lake. This would imply a human presence at an extremely remote time, but only if the workers in the gypsum quarries could be believed. The issues of trust and credibility, implicit in Esper's earlier report, here became quite explicit:

> A quarryman—a sensible [*sensé*] person—told me that two years ago he found a key almost eighty feet deep, in the heart of the gypsum stone at Montmartre. He recounted this—and he was not seeking to tell me—with such simplicity that I could not help believing him. He drew the form of the key for me on the sand, and my figure [Fig. 5.16] is based on his.[49]

Lamanon judged the report credible, because the worker seemed trustworthy and had not been trying to tell what he might have known the savant would want to hear and be prepared to reward him for. But of course there was no material object to back up the report: only the quarryman's recollection of its form, sketched by him on the floor of the quarry, copied by the savant, and later reproduced by an engraver. If it had not been an object of such significance, it would probably have been rejected by Lamanon, or at least by the editor of the respectable periodical, as utterly unreliable evidence. However, the author and the editor were of course well aware of the high significance of finding an iron key, of all things, at such a depth within solid rock. As Lamanon put it:

> I therefore consider it certain, not only that the existence of men preceded that of the present surface of the Île de France [the region around Paris], but also that the shores of this lake of selenitic waters were inhabited by men living together socially [*réunis en société*], and that in their time the art of working mines and forging iron was [already] known.
> I know of several other facts analogous to this, which prove incontestably that the crafts [*arts*] were cultivated in the times that preceded the great physical revolutions that have happened at the surface of the globe.[50]

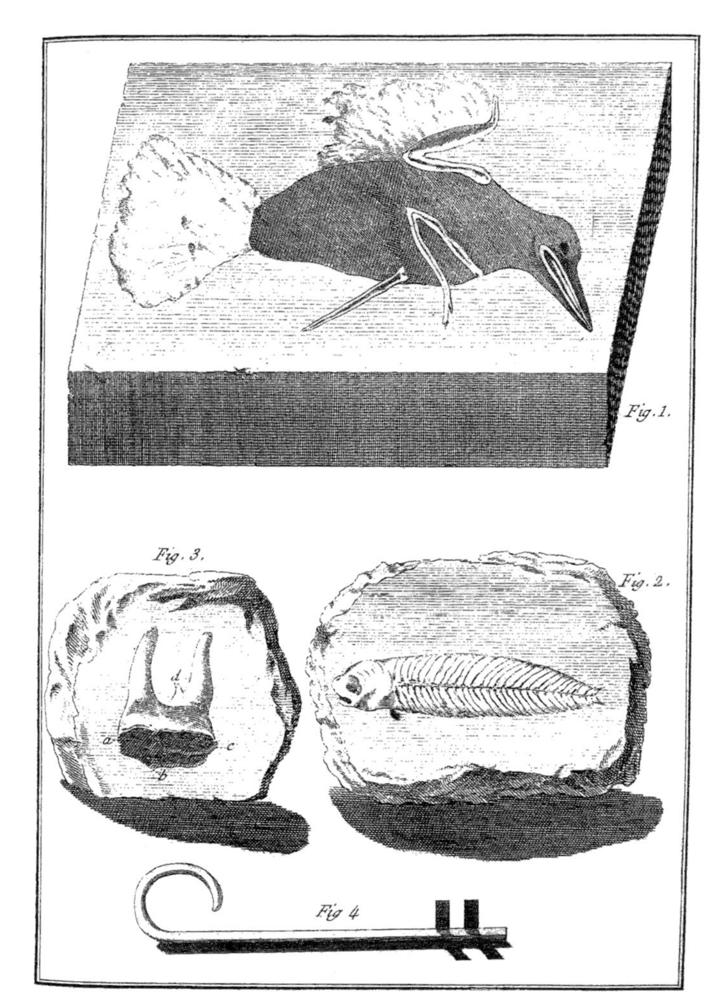

Fig. 5.16. The iron key that a quarryman claimed to have found preserved deep within the thick Gypsum formation outside Paris: a drawing based on the worker's sketch, illustrated along with a suspiciously well-preserved bird (§7.3), a fossil fish, and the tooth of a possibly extinct fossil mammal, as engraved for Lamanon's paper on the gypsum and its fossils, published in 1782 in *Observations sur la Physique.* If genuine, this striking find would be evidence for civilized human life at an extremely remote period. (By permission of the Syndics of Cambridge University Library)

In other words, Lamanon claimed that his key was good evidence that human history extended back before the relatively recent physical changes that had formed the present world, back into the far more remote ages of geohistory. And these extremely ancient humans had not been primitive savages, but civilized people as accomplished technically as eighteenth-century Frenchmen. This was a conclusion that was anticipated, and would have been welcomed, by those savants who rejected the biblical account of the recent origin of humanity, and who assumed that human history might have been coextensive with the history of nature, perhaps from all eternity. Yet although Lamanon knew of a few similar reports—for example, of copper nails allegedly found in another Secondary formation near

49. Fig. 5.16 is reproduced from Lamanon, "Fossiles de Montmartre" (1782) [Cambridge-UL: T340:2.b.16.22], pl. 1; quotation on 192. The rather crude engraving (fig. 4) was intended to depict a key with two tines throwing oblique shadows; the size of the key was not mentioned, and no scale was given.

50. Lamanon, "Fossiles de Montmartre" (1782), 192. It is not clear whether he regarded the gypsum as relatively recent and Superficial, since the lake's freshwater mollusks continued to live in the present rivers (§4.3), or as very ancient and Secondary, since the gypsum was in a thick and solid formation.

Nice—it was highly precarious evidence on which to support such a momentous conclusion.[51]

Only two years later, Burtin's book on the fossils around Brussels reported another specimen of the same kind. Unlike Lamanon's, this was a big one that had not got away; and unlike Soulavie's, the evidence was presented in the most persuasive form, as a pictorial proxy of the crucial object. It was a polished stone tool of a kind well-known to antiquarians, being found quite commonly, along with pottery, in burial mounds and similar sites. Such artifacts could reasonably be attributed to the indigenous peoples of northern Europe just before the Roman invasions. Burtin's precious specimen, by contrast, was the only one of its kind to have been found deep within the rocks of a Secondary formation: he thought it "an estimable monument of such a remote antiquity" that it deserved the closest attention. He emphasized that it had been found deep in a quarry, embedded in a layer that lay *under* one containing the nautilus (Fig. 4.7), turtle bones, and other exotic marine fossils. It was so important that he took care to verify its location by interrogating the workers who had found it. So the authenticity of the find, and its wider significance, depended on the trustworthiness of humble workmen, as it had in the cases

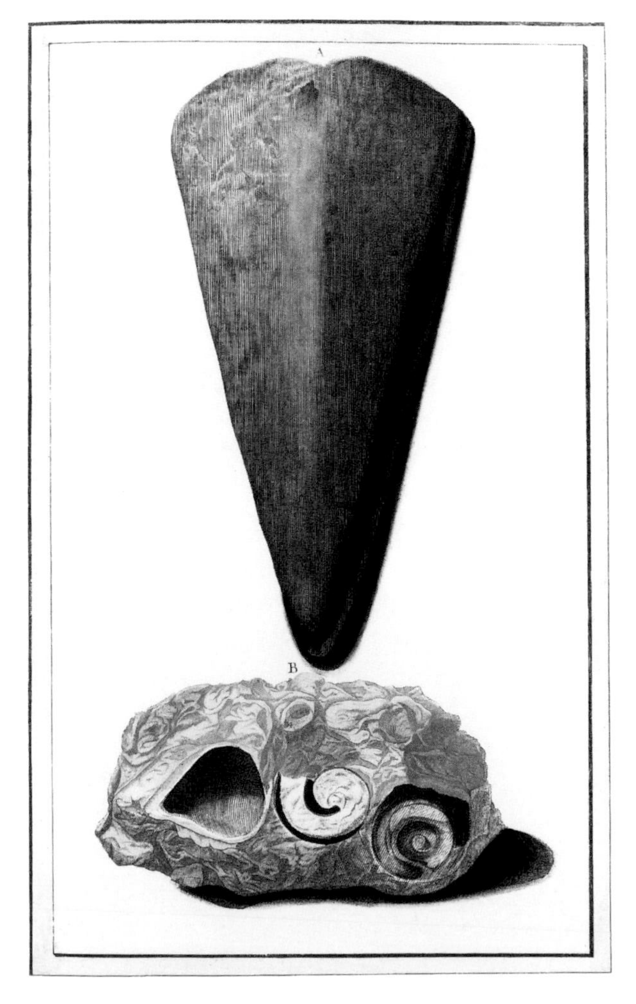

Fig. 5.17. A polished stone ax or adze [*hache*] that quarrymen claimed to have found deep within a Secondary formation near Brussels, under a bed (a specimen of which is shown below) containing the nautilus and other exotic fossils: an engraving published by Burtin in 1784 in his book on the fossils of that region. Like the iron key reported by Lamanon (Fig. 5.16), this find—if genuine—would be striking evidence of the extremely high antiquity of humans, or at least of "intelligent beings" of some kind. (By permission of the British Library)

of Esper's bones and Lamanon's key. Burtin declared himself satisfied, but other naturalists might be more skeptical (Fig. 5.17).[52]

The tool was said to be composed of jade, which Burtin found extremely puzzling, for he knew of that rare and precious material only from South America and the East Indies. Were the people who made the tool able to navigate around the globe to import it, or might jade be present somewhere in Europe as yet undiscovered? Faced with such uncertainties, Burtin simply offered his specimen—or rather, his proxy of it—to the "cosmologists" among his readers, to make of it what they would. Soon afterwards, however, his prize essay on the earth's "revolutions" released him from the limitations of descriptive natural history and enabled him to discuss more fully the origin and significance of his polished ax. He had been told that such objects were not uncommon in England, and that they were attributed to the Druids; but their significance was unclear, because they were collected by "people too inattentive to record the situations and the beds" in which they were found. In view of that uncertainty of context, Burtin in effect discounted them as objects of any great age. He claimed in fact that "at the moment of the great revolution men did not yet inhabit the earth": the ancient ocean floors were abruptly turned into the present land areas *before* the first appearance of humans on earth. This matched his insistence that that major change in physical geography was *not* the biblical Flood, as de Luc claimed, but a far earlier and prehuman event (§4.2).[53]

However, this conclusion left Burtin with the obvious problem of accounting for his precious stone implement: by his own account it was even older than his putatively prehuman revolution, since he claimed that it had been found in a Secondary formation with tropical marine fossils. The stone ax, he emphasized, "*says so much to the thinking man*" that it was highly regrettable that it was unique. But if in due course further specimens were found, it would support the inference—suggested to him by another savant—that the artifact was the work of *another* "intelligent being", a *prehuman* rational species, or in effect a "*pre-Adamite*". This was delicate ground, particularly for Burtin, for "pre-Adamitism" had been regarded as heretical by Catholics ever since it was suggested by the Protestant scholar Isaac La Peyrère a century and a half before. Burtin protested that he was not advocating pre-Adamitism, which may have been true in a strictly theological sense. He argued that even if his conjecture were correct, Adam would still be theologically the "father" of all truly *human* beings, so that Christ could still be regarded in orthodox manner as having redeemed all humankind. Nonetheless, Burtin's suggestion did broach publicly the possibility that *other* rational species might have supplied a

51. This and the other reports summarized here are too early to be among the claims reviewed in Cremo and Thompson, *Forbidden archeology* (1993). This work, inspired by Vedic sources, advocates what is now a decidedly unorthodox view of extreme human antiquity; it represents a modern revival of the kind of eternalism for humanity that was not uncommon in the Enlightenment, as expressed implicitly by Hutton and explicitly by Toulmin (§3.4).

52. Fig. 5.17 is reproduced from Burtin, *Oryctographie de Bruxelles* (1784) [London-BL: 459.e.20], pl. 13. The *hache* was said to have come from a quarry at Loo outside Brussels.

53. Burtin, "Révolutions générales" (1789), 17–24, 192; *Oryctographie de Bruxelles* (1784), 66–67. His "*cosmologues*" were those interested in the science that de Luc had proposed calling "geology"; that is, geotheory (§3.1).

humanlike presence much further back in geohistory. That he was not alone in speculating along these lines at this time is suggested by Soulavie's choice of the same carefully noncommittal phrase "intelligent being" to account for the humanlike traces that he too believed he had found in the deep past.[54]

Conclusion

Unless more artifacts (or better still, human bones) were found in Secondary formations, and found in much less ambiguous circumstances, Soulavie's plank, Lamanon's key, and Burtin's ax would remain mere anomalies of dubious status: far too slender evidence on which to base momentous conclusions about the place of humans (or of humanoid "intelligent beings") in geohistory. Apart from these extremely rare and questionable finds, all the signs were that humans were very recent newcomers on earth. Their traces seemed to be no older than the time of the Superficial deposits, which in turn might be no older than the few millennia of traditional chronology. So the chronologers might have been right to date Adam as quite recent, although mistaken in dating the Creation to around the same time (§2.5); and Buffon might have been right—at his eleventh hour—to define the very last "epoch" of his conjectural narrative as the quite recent moment of appearance of the human species (§3.2).

By the late eighteenth century the traditional picture of a brief cosmic history, almost coextensive with human history, looked utterly untenable. But so, equally, did the Aristotelian picture of an infinitely long cosmic history, also with a human presence throughout. In their place, as an increasingly plausible third option, was the novel picture of an ancient but not eternal earth, and of an unimaginably lengthy geohistory that, until the final few millennia, had probably been nonhuman because it was *prehuman*. However, this remained little more than an intriguing possibility. Although it was unavoidably based on negative evidence—the absence of any trace of human existence—it would need much more consistent support before it could be treated as a reliable and well-established conclusion.

However, even if this novel picture were adopted, at least as a working hypothesis, the relation between the present world and the deep or prehuman past, as summarized in earlier sections of this chapter, remained problematic and controversial. Fossils were increasingly recognized as a potentially decisive kind of evidence about the former world because they were both like and unlike the living animals and plants of the present world. They were similar enough to show that the deep past had had much in common with the present, in its organisms as much as in, say, its volcanoes; but they were different in detail in that many of the larger groups, and certainly most of the species, were not the same. But even that generalization was difficult to interpret: the apparent contrast between present and past might be merely a result of changes in the spatial distribution of animal and plant species, due in turn to a gradual climatic change or some other cause, or the result of a sudden "deluge" or mega-tsunami that had strewn their remains far beyond their natural habitats. At the same time, it remained utterly uncertain whether extinction was a genuine natural phenomenon and a frequent event; still more dubious was any suggestion that earlier forms had somehow transmuted (or evolved) into present ones without extinction.

Above all, the problem was generally formulated in terms of a dichotomy between the present world and a tacitly *undifferentiated* former world. Even when several successive assemblages of fossils could be distinguished within the latter, it was not clear whether they represented successive "ages" in a genuine *history* of life, as Soulavie claimed, or merely a local record of a general trend from "pelagic" to "littoral" conditions as the global sea level declined. All in all, the use of fossils as indices of geohistory remained more a matter of promise than achievement; only a much more rigorous and extensive study of them, and in particular a much closer comparison between living and fossil species, might turn them into clearer evidence of the character of the former world. Around the time that Saussure climbed Mont Blanc, fossils could not show unambiguously whether, or how far, the former world had differed from the present, and whether the deep past had indeed been a foreign country where nature did things differently.

54. Burtin, "Révolutions générales" (1789), 192–95. La Peyrère had suggested in effect that the early human history recounted in Genesis (from Adam onwards) had referred only to the Jews, not to the human race as a whole. This had solved the problem of explaining the dispersal of humans—not least to the Americas—in the short time since the Flood (which on a literal reading had wiped out all except Noah's family). But in its place it had raised the problem that Adam could not then be the ancestor of *all* humans, and that therefore on Pauline typology the "second Adam" could not have died for all ("As in Adam all die, so in Christ shall all be made alive"). On pre-Adamitism see Popkin, *History of scepticism* (2003), chap. 14, and *Isaac La Peyrère* (1987), chap. 9; the latter deals briefly with the eighteenth century and notes Burtin's claim. Livingstone, "Preadamites" (1986) and "Preadamite theory" (1992), deal mainly with its later nineteenth-century forms.

INTERLUDE

Chapter 5, just summarized, brings to a close the first part of this book, with its synchronic survey of the sciences of the earth as they were being practiced around the time that Saussure achieved his epoch-making ascent of the highest peak in Europe. That event, cited perhaps ad nauseam in the preceding chapters, has served as a convenient golden spike by which to mark the last decade of the Old Regime, before the whole of European civilization—including its scientific life—was thrown into turmoil by the upheaval of the Revolution in France and the long subsequent wars. What has been portrayed was a flourishing scientific culture, pan-European in extent and even reaching distant outliers such as Russia and North America; highly international and multilingual, at least among its elite; and constituting the natural-scientific wing of a wider Republic of Letters. This culture was as lively among the subset of savants who worked in the sciences of the earth as it was among those whose interests lay in other aspects of the natural world (and whose work has attracted far more attention from modern historians).

The sciences of the earth, like the natural sciences generally, were divided rather sharply into natural history and natural philosophy: sciences of description and classification and sciences of mathematical analysis and causal explanation. The museum science of mineralogy (which included the study of fossil specimens), the field science of physical geography, and the field and subterranean science of geognosy were all regarded as branches of natural history. The science of earth physics, in contrast, was equally clearly a branch of "physics" or natural philosophy, and so was the more ambitious genre of geotheory that was based on it. The scene was one of diverse and flourishing activity. Certainly it was not a scene of "pre-paradigmatic" confusion, for each of these sciences had its own well-defined tradition of methods, norms, and genres and its own set of "paradigms" or exemplary achievements (for example, Saussure's *Alpine Travels* for physical geography).[1]

Mineralogy, physical geography, and geognosy, as sciences of natural history, described and classified the diversity of the mineral world in atemporal terms. Earth physics, as a branch of natural philosophy, interpreted the features of that world in terms of natural causes, and hence necessarily in terms of processes operating in time. But these temporal processes were as ahistorical as those of gravitation or chemical reaction, in that they too were proposed as being valid for all time, past, present, and future. The related genre of geotheory integrated such processes into comprehensive "systems" that attributed to the earth either directional or cyclic sequences of events or states. Some geotheories might appear to have offered historical interpretations of the earth. But this was deceptive, because the putative history was more or less rigorously determined or programmed by the underlying ahistorical laws of nature, either in the sense that an embryo develops into an adult organism in every generation or in the sense that the planets orbit repeatedly around the sun. The course of change would in either case be predictable (or retrodictable), at least in principle, once the initial state and the relevant laws of nature were correctly identified; geotheory, like the earth physics on which it was based, aspired to account for the past, present, *and future* states of the earth on equal terms.

None of the sciences of the earth, therefore, nor the overarching genre of geotheory, aimed to construct a true *history* of the earth and its life, in all its unpredictable and contingent particularity. The few savants who did try to do so recognized that they were exploring a *new kind of science*, which was as distinct from the ahistorical causal explanation of phenomena as it was from the description and classification of atemporal natural diversity. Their inspiration came either from the sciences of human history, and particularly from the burgeoning practices of antiquarianism and erudite scholarship, or—ironically—from the profoundly historical perspective embodied in traditional Christian (and Jewish) theology. From the former, these savants derived a powerful set of metaphors and analogies, which clarified how the deep past could be recovered from its traces in the present, even if that past had not been witnessed or recorded by any human beings. From the latter, they (or at least some of them) derived a sense of the deeply contingent character of history—contingent because it was ultimately under the sovereignty of God—and a conviction that the history of nature must somehow overlap and be continuous with the history of humankind. In either case, these tentative essays in geohistory were based on the bottom-up reconstruction of specific past events from the observable traces of what *in fact* had happened, rather than the top-down formulation of what "must" or "ought to" have happened, given certain general principles or laws of nature.

However, this new practice of geohistory remained marginal: unusual within the scientific study of the earth, confined to a handful of savants, and often repudiated, misunderstood, or at best ignored by the others. It has been necessary to survey the activities of late eighteenth-century savants at length in the first part of this book, in order to demonstrate that the approach that is now called geohistorical was *not* a prominent part of the scene. The aim of the second part of this book is to show how this marginal genre came to be incorporated into the mainstream; and how, starting from a few scattered essays, geohistorical interpretation became a routine component of a practice that incorporated all the earlier sciences of the earth and integrated them into a new science. Transforming de Luc's definition, this new science came to be known as "geology".

Leaving behind the golden spike of Saussure's famous climb, using it simply as a starting point, this book now changes gear from the synchronic to the diachronic, embarking on a narrative that will cover the subsequent three or four decades of scientific study of the earth. However, three points about the narrative must be emphasized at once. First, as already mentioned in the introduction, the focus will be on the development of a geohistorical perspective; the narrative that follows makes no claim to be a comprehensive account, or even a representative survey, of all the sciences of the earth during that period. Second, there will not be space in which to describe in detail all the arguments and counterarguments among savants, which led some knowledge claims to be accepted consensually and others to be rejected: it should just be assumed that all but the most banal claims were in fact contested more or less vigorously in the "agonistic field" that characterizes all scientific activity, then and now. And third, the narrative does not attempt to stick rigidly to a chronology unfolding year by year, let alone month by month. At some points, different themes will need to be pursued in parallel for a few years, with copious use of flashbacks, pluperfect tenses, and other literary devices. Nonetheless, the narrative is designed to convey and replicate in broad outline the sense of development—indeed, of cumulative *progress*—experienced by the historical actors themselves, as they lived and worked through what can be seen in retrospect to have been the most formative period in the whole history of the earth sciences.[2]

1. Kuhn's earlier use of the term "paradigm", with its emphasis on exemplary cases, is more useful in this context than his later, better known but more abstract use: see *Scientific revolutions*, 2nd ed. (1970), 175.

2. Rudwick, *Devonian controversy* (1985), explored how a narrative that observed a rigorously precise chronology—reconstructing the debate week by week, and sometimes even day by day, without ever anticipating its future course—could replicate the actors' sense of an extremely complex puzzle that was gradually resolved. This kind of historiography would be neither feasible nor desirable here; but on a coarser level of temporal resolution, as it were, the narrative in the present book should convey something of the same experiential sense of movement and progress.

Part Two

RECONSTRUCTING GEOHISTORY

CHAPTER 6

A new science of "geology"?

6.1 REVOLUTIONS IN NATURE AND SOCIETY (1789–91)

Meanings of revolution

On 14 July 1789 a Parisian mob broke into the royal prison of the Bastille to seize the gunpowder and weapons that were kept there, also liberating a handful of criminals; lives were lost on both sides in the affray. The relative anticlimax of the event has not stood in the way of patriotic mythmaking, which much later turned the storming of the Bastille into France's national day and a "golden spike" for students of European history. However, as historians of the Revolution have long emphasized, political and economic squalls had been around for several years, and—to extend the metaphor—it was to be another year or two before the thunder and lightning began in earnest and it became clear that what was going on in France might well engulf the rest of Europe. In the short run, the pressure from reformist elements among the nobility and clergy, with their increasingly powerful allies among the professional and bourgeois classes of the "Third Estate", to refashion the constitution, to eliminate anachronistic privileges and to temper the absolutism of the monarchy, were widely welcomed inside and outside France. The granting of press freedom—or at least, much more than before—and the defining of civil rights were likewise approved as bringing the country into line with politically more Enlightened lands. It seemed that France might make a relatively smooth transition to the kind of constitutional monarchy that Britain had long enjoyed and that was the envy of many other Europeans.

For the savants who did original research in the sciences, and for the dilettanti who followed their work with interest, what was happening in France later became a matter of the gravest concern, because that country, and above all Paris, was so much the center of the scientific world. However, in 1789 it seemed almost inconceivable that France would be engulfed in violent revolution or that some of its leading scholarly institutions would be abolished and many of its savants dispersed into hiding or exile. For the time being, the place of the sciences in French society appeared secure, perhaps more so in the new atmosphere of Enlightened reform than under the absolutist Old Regime. The Académie des Sciences, for example, was assigned new and useful tasks for the public good and seemed set to consolidate under any new regime the dominant position it had enjoyed under the old. Still, even the events in the months following the incident at the Bastille were recognized at the time as being already a "revolution", in one of the well-established senses of that word: they constituted a major political change, although marked by only sporadic violence and not effected suddenly overnight.

As emphasized repeatedly in the first part of this book, the word "revolution" was used in that same sense when it was applied to the world of nature, and particularly to the past history of the earth. A constant theme was that of "nature's revolutions". Some savants believed they had been sudden, and even violent, but others thought they might have been—and might continue to be—as smoothly regular as the "revolutions" of the planets around the sun. What they all agreed, in the face of evidence as disparate as mountains, fossils, and volcanoes, was that the earth had somehow undergone massive changes; and that those changes had happened over a timescale that was certainly vast in relation to human lives, perhaps unimaginably so in relation to the whole of human history. It was agreed that the earth had had its revolutions; it remained to work out—and to argue over—just what these revolutions had been like, when they had taken place, and exactly how.

These questions had been regarded as belonging primarily to earth physics [*physique de la terre*] (§2.4). This was a science that sought to discover the "physics" or causal mechanisms—or at least the "laws of nature" or phenomenal regularities—that would explain all nature's revolutions, alike in the past, the present, and the future. But around the time that the political revolution in France began to unfold—just two years after that other golden spike, Saussure's climb to the summit of Mont Blanc—this ahistorical style of research was beginning to be complemented by another and contrasting kind. The new *geohistorical* approach imported into the natural world the methods and the imagery of historians and antiquarians. One of its effects was to allow questions of causal mechanisms to be shelved in favor of first discovering exactly what had happened, where, and when. For nature's revolutions were treated not as being "programmed" in a predictable way by ahistorical physical laws, but as being the result of causal webs as complex and unpredictable as the contingencies of human history. And so, like past changes in the human world, they would have to be reconstructed bottom-up from a detailed study of their surviving traces, not deduced top-down from the supposed laws of nature.

However, this geohistorical way of investigating nature's revolutions was hardly in the mainstream of research: most scientific studies of the earth continued to be done within the well-established traditions of natural history (§§2.1–2.3), or else in that of earth physics as a branch of natural philosophy (§2.4). The marginal position of geohistory will be reviewed in this section by recalling some examples that were described earlier, while introducing others that became available to savants throughout the Republic of Letters during the period corresponding to the earliest phase of the Revolution in France, before that country turned itself into a republic of another kind. This will help to tie the synchronic survey in the first part of this book into the diachronic narrative that follows.

Blumenbach's "total revolution"

Some of the most striking work was being done in the tradition of "mineralogy", by studying specimens, and particularly fossils, indoors in museums. It was in 1789, for example, that Burtin's prizewinning essay on nature's revolutions was published at Haarlem in its original French and in Dutch translation (§4.2). Burtin used fossil specimens, not least those in his own collection from around Brussels, as raw material for geohistory. He claimed that a major revolution had turned former seabeds into the present continents, at a time extremely remote by any human standards, indeed well before the first appearance of true human beings, though possibly not before there were humanoid pre-Adamites around (§5.4). However, he did not claim that this event had been unique: other revolutions might have been responsible, for example, for the disappearance of fossils such as ammonites at a still more remote time; and, at some more recent time, for the bones of tropical animals found near the surface in northern latitudes. Such events might have been sudden, even violent, but Burtin took it for granted that they had had natural causes of some kind. And this Catholic naturalist insisted that all of them had been prior to whatever local inundation had given rise to the story of Noah's Flood: here was no conflict between "geology and Genesis". Unlike human history, it was impossible to give dates for nature's revolutions, or to quantify nature's vast timescale, but nonetheless it was feasible in Burtin's view to reconstruct a sequence of events in the earth's *history*.

Whether or not he read Burtin's essay, Blumenbach at Göttingen was developing rather similar ideas about nature's revolutions. In the most recent edition of his popular textbook on natural history, he had already begun to give more prominence to fossils, treating them as either "*bekannte*" or "*incognita*", species either known or unknown to be still alive (§5.2). In 1790, however, he expounded his emerging ideas more clearly, both in an article in a leading German scientific periodical published in Gotha and in the introductory sections of a little book on variability in the natural world. The title of the first, "Contributions to the Natural History of the Former World [*Vorwelt*]", indicated that he would give his traditionally atemporal field of natural history a truly historical dimension. Blumenbach turned the spotlight on fossils, but now with powerful use of antiquarian metaphors. What he had previously treated as a subordinate branch of "mineralogy" now became crucial evidence for the past *history* of the earth and of life, provided that fossils were treated like historical documents:

> If one regards fossils from the great standpoint that they are the most infallible documents in nature's archive, from which the various revolutions that our planet has undergone—and even their kind and manner and to some extent their epochs—can be determined; and that consequently even the respective ages of the various important kinds of formation [*Gebirgsarten*] can be fixed: then it is self-evident that their history must be seen as one of the most important and instructive of all parts of natural history, but particularly of scientific mineralogy.[1]

Blumenbach argued, however, that fossils could not reveal their geohistorical significance unless they were given "the utmost meticulousness in observation", which they had not always received from mere collectors. It was essential to compare them accurately with their counterparts in the living world, distinguishing the truly identical [*wirklich gleich*] from the merely similar [*blos ähnlich*]. He criticized Hutton—whose new geotheory (§3.4) he summarized for German readers in the very next article—for implying that the organic world had not changed over time and that fossil species were identical to living ones. He rejected that assumption by citing some concrete cases. In Westphalia, for example, one of the commonest fossil anomia shells [*Terebratuliten*] (see Fig. 5.5) was similar to the *Anomia venosa* that Daniel Solander, the Swedish naturalist who had accompanied Banks on Cook's famous voyage, had described from waters off the Falkland Islands; but it was not exactly the same. However, Blumenbach's prime case, the encrinites or sea lilies, took the argument right into the opposing camp. Sea lilies did indeed survive in the modern world as perhaps the best example of living fossils (§5.2). Looked at in detail, however, the species were not identical: the fossil ones—of which he had fine specimens in his collection—had only a "generic similarity" to the living ones (see Figs. 5.3, 5.4). He therefore concluded that the fossil encrinite was "a true *incognitum* of the pre-Adamitic former world [*Präadamitischen Vorwelt*]"; and judging by the profusion of the little fossil "trochites" that were now recognized as the segments of its stem, it must have been one of the commonest creatures in those vanished oceans. Blumenbach's definition of the former world as "pre-Adamitic" made explicit its place in *prehuman* geohistory.[2]

This argument was weakened, however, by continuing doubts about the reality of extinction, at least in the case of marine mollusks. It had just been questioned by Bruguière, for example, and with well-earned authority. In the first installment (1789) of his survey of conchology for the *Encyclopédie Méthodique*, he argued that ammonites might still be living in abundance, yet lurking undiscovered on the floors of the present oceans (§5.2). In later installments of the same work, he extended this inference by pointing out that the coiled gastropod shell *cerite denticulé*, for example, had been known as a fossil at Courtagnon near Reims long before Cook's expedition found it alive off the Friendly Islands [Hawaii]; with that precedent, the fossil species *cône antediluvien* and *cône perdu* ("pre-Flood" and "extinct" cone shells) might well turn out to have been named prematurely.[3]

Blumenbach was not unaware of this kind of argument. In the opening sections of his little book on variability in nature, he conceded that *some* fossil species might indeed survive in remote places: he alluded to the progress already made in surveying the oceans, by the expeditions that had been supported by his sovereign the Elector of Hannover (better known as King George III of Great Britain). But in his opinion the scale of the disparity was such as to make the survival of *all* fossil species highly unlikely: he mentioned that some two hundred species of ammonites had already been described, not one of which was known alive (§5.2). So he concluded that "an entire pre-Adamitic organic creation [*Schöpfung*]" must have perished. Only after that "*Totalrevolution*" had the "present creation" taken its place, including the human species in all its diversity.[4]

Blumenbach's use of the word "creation" did not imply that either the former or the present set of organisms had been produced directly by divine action: like other naturalists at this time (and well into the nineteenth century), he assumed that *some* kind of natural process had acted as a "secondary cause". In fact he had already outlined his own ideas on this: the "formative drive" [*Bildungstrieb*] that guided the generation of organisms was a force just as material in its effects as gravitation, although equally mysterious in its ultimate nature. He argued that the change could not be explained in terms of "mere transmutation [*Degeneration*] during a long sequence of millennia": the contrast in specific forms was "not a result of degeneration [*Ausartung*] but of re-creation [*Umschaffung*] through a changed direction of the formative drive". The explanation might be somewhat obscure, but Blumenbach's readers were left in no doubt that he was claiming the reality of organic change at the time of the total revolution: not only the mass extinction of old species but also the formation—by some kind of natural process—of a set of new ones. Taking a concrete example, he described how one well-known mollusk from northern seas was exactly matched by a fossil species, except for one striking feature: the fossil shell was coiled in the opposite direction. At least in this exceptionally clear case, it was obvious that it could not have been happened gradually, however great the time allowed (Fig. 6.1).[5]

However, Blumenbach's brief writings on this subject, like Burtin's more ambitious essay, simply extended to the organic world the kind of binary model that de Luc had already sketched on the basis of physical geography: geohistory was divided into a former world and a present world, sharply separated by a major revolution

1. Blumenbach, "Naturgeschichte der Vorwelt" (1790), 1–2.

2. Blumenbach, "Naturgeschichte der Vorwelt" (1790), 2–6, and Hutton, "Theorie der Erde" (1790). In modern terms the anomia shells and encrinites were brachiopods and crinoids respectively (§5.2). That they were pre-Adamitic denoted only that they predated the Adamitic world of human beings, not that Blumenbach thought the former world had included humanoid "Pre-Adamites": in his "Bitumindsen-Mergelschiefer" (1791) he explicitly rejected the latest claims about human fossils.

3. Bruguière, *Histoire naturelle des vers* (1789–97), 1, s.v. "Cerite" [*Cerithium*], 472, and "Cônes" [*Conus*], 601.

4. Blumenbach, *Beyträge zur Naturgeschichte* (1790), 6–8; in context it is clear that he was using the word *Gattungen* to denote Linnean species, not genera. Most of the book dealt with organic variability in the wild and under domestication, preparing the ground for his later and more famous research on physical anthropology.

5. Fig. 6.1 is reproduced from Blumenbach, *Abbildungen* (1796–1810) [Cambridge-UL: MB.46.75], pt. 2 (1797), no. 20; quotation from *Beyträge zur Naturgeschichte* (1790), 25–27. The shells had been found near Harwich on the east coast of England and may have been collected and given to him by de Luc (§6.2); in modern terms, they were from the Red Crag of early Pleistocene (Quaternary) age. On his theory of organic change, see *Über den Bildungstrieb* (1781), and Richards, "Kant and Blumenbach" (2000); Lenoir, *Strategy of life* (1982), chap. 1, defines his position in the perennial debates on "generation" as that of "vital materialism".

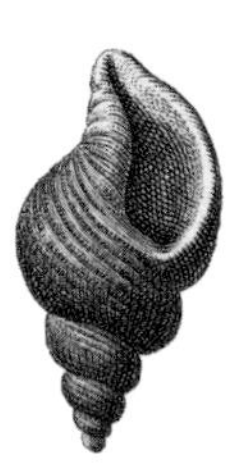

Fig. 6.1. Adult and juvenile shells of the fossil *Murex contrarius*, coiled in the opposite direction from the otherwise identical shells of the living *Murex despectus*. Blumenbach used this example in 1790 to illustrate how, in a "total revolution" in the distant past, the species of the prehuman former world had been replaced by those now alive, through some natural but nonevolutionary process. He published this engraving a few years later in a series illustrating some notable objects of natural history. In modern terms, the sinistral coiling of this species is a highly unusual exception to the dextral direction that is normal in gastropod shells: obviously, no *gradual* "transmutation" (evolution) could change one into the other. (By permission of the Syndics of Cambridge University Library)

before or soon after the first appearance of the human species (§3.3). In the face of the continuing argument over living fossils, as represented by Bruguière, it would need more than Blumenbach's sketchy speculations to establish convincingly the reality of faunal change, let alone his conception of the revolution itself as a *total* replacement of species. Nonetheless, given his growing reputation as a naturalist, Blumenbach's work was a significant example of the new emphasis being given to fossils as evidence for geohistory, within the tradition of working indoors on collections of specimens in museums.

Montlosier's continuous revolution

These museum studies continued to be complemented with outdoor fieldwork on physical geography. Using such evidence, Nicolas Desmarest had earlier reconstructed sequences of geohistorical events, not for the earth as a whole but just for Auvergne, where the spectacular extinct volcanoes provided the key to nature's successive "epochs" (§4.3). Although this research had been published only in preliminary form, it was well-known to savants in France and elsewhere. In 1789 it was significantly extended in a little book published by François-Dominique de Reynaud de Montlosier (1755–1838), an Auvergnat landowner and naturalist, and a moderate royalist who served as a member of the Constituent Assembly in the early phase of the Revolution. Rather than acknowledging his debt to Desmarest, however, Montlosier was more concerned with contrasting his own work with a recent travel book about Auvergne: unlike its author, a mere visitor, he himself lived there and his account was based on prolonged and detailed fieldwork.[6]

The title of Montlosier's book was significant: it was an "Essay on the theory of the volcanoes of Auvergne". It was indeed concerned with "theory", but in two respects Montlosier distanced it from the traditional genre of geotheory (§3.1). It was limited to a specific class of phenomena, the volcanoes, and to the specific region of Auvergne; and it offered no more than an "essay", implying that it would doubtless be improved in the future. In other words, Montlosier's work exemplified an emerging alternative to geotheory, which would break up the big global questions into more readily soluble parts and not claim prematurely to offer the final answer to everything everywhere. Yet he claimed the title of *géologie* even for this more modest project, indicating that de Luc's crucial word might already be changing its meaning. And right from the start, he associated it with a recently developed geohistorical approach characterized by its analogy with human history:

> From that moment [of Buffon's *Nature's Epochs*], the history of the earth has started to become interesting. Erudition has appropriated nature's archives; savants have come from all parts into the provinces to interrogate its [nature's] monuments and to search its memoirs; and so geology has become a major science, to which mineralogy, assaying, and chemistry have had the honor to be subordinate.[7]

Montlosier's book dealt with both the "physics" of the Auvergne eruptions and the effects of erosion on the volcanic cones and lava flows. Central to his analysis was another metaphor drawn from the human sphere: the various lava flows were "nature's witnesses [*témoins*]", offering evidence that was no less eloquent for being in reality nonhuman and mute. Montlosier followed Desmarest in distinguishing two classes of volcanoes in Auvergne, the well-preserved cones and lava flows in the valleys and the basalts capping the hilltops. The modern ones were the key to the ancient, the present to the past. Adopting a metaphor from Classical philosophy, he treated the ancient basalts as nature's esoteric teaching [*doctrine cachée*], which could only be understood after mastering the exoteric teaching [*doctrine publique*] offered by the more easily intelligible modern volcanoes. However, he was clear that this modernity was relative. Even the recent volcanoes were very ancient in human terms. For example, the fifth-century Roman author (and bishop of Auvergne) Appolinaris Sidonius had lived beside Lac Aidat—from which came fish that were, as Montlosier put it, "as much cherished by gourmands" as its volcanic features were by naturalists—yet this lake clearly owed its very existence to one of the "modern" eruptions (see Fig. 4.10).[8]

6. Montlosier, *Volcans d'Auvergne* (1789), v–ix, criticizing Le Grand, *Voyage d'Auvergne* (1788); Le Grand was a prolific literary figure but no naturalist.

7. Montlosier, *Volcans d'Auvergne* (1789), iv; the allusion was to the passage with which Buffon—who had died the previous year—had opened his work (Fig. 4.14). The museum- and laboratory-based sciences were "subordinate" in that although they provided evidential grounds on which geohistorical inferences could be based, the "monuments" studied in the field were in Montlosier's view paramount.

8. Montlosier, *Volcans d'Auvergne* (1789), 14–19, 32–34. Like Lac Chambon further south (Fig. 4.9), Lac Aidat was (and is) clearly dammed by a "modern" lava flow.

Even among those recent but prehistoric eruptions there were significant differences of age, to judge by the varying degrees to which the lavas had been eroded subsequently. Some were so recent that the streams had still had virtually no effect on the lavas that had flowed down their valleys. In contrast, the lava that had ponded back Lac Aidat had already been eroded into a deep gorge further downstream (at Saint-Saturnin). Another locality was described by Montlosier as "one of the most beautiful pieces of geology that the naturalist could desire". It had been mapped by Desmarest's collaborator, the surveyor Pasumot, but Montlosier gave it a characteristically *geohistorical* interpretation. At one point a valley had been completely blocked by a modern lava flowing into it from the side. Yet in the period since this eruption the stream had already excavated a new valley, bypassing the obstruction altogether and leaving a small lake ponded back by the lava in the abandoned old one. If the erosive action of streams and rivers was as slow as it appeared to be, this was an unmistakable sign of the magnitude of the time that had elapsed even since the modern eruptions in Auvergne (Fig. 6.2).[9]

Montlosier not only emphasized the different ages represented within each of his (and Desmarest's) two classes of volcanic rocks; he also regarded the classes themselves as no more than a convenient approximation, with many "*nuances intermédiaires*" between them. In other words, intermittent eruptions had left "witnesses" to many successive stages in the immensely long process of slow erosion, all

Fig. 6.2. A striking case of valley erosion during the time since one of the most recent volcanic eruptions in Auvergne: a detail from Desmarest's posthumously published map of the region. Downstream from the village of Olby (bottom right), the River Sioule turns sharply to the west before resuming its northward course. Montlosier inferred that this westward channel had been excavated by the river—through hard Primary rocks [*gran(ite) intact*]—*after* a lava (middle right, stippled), flowing westwards from the Puy de Côme volcano, had blocked the original valley north of Olby, forcing the river to change its course and ponding back a small lake (the silted Étang de Fung) in the abandoned valley. The area shown here is about 8km by 6km.

the way from the "primitive" topography uncovered when the seas first retreated from the continent. Yet another historical metaphor made the point: "one finds coins [*medailles*] struck by nature at every age, bearing witness to all the stages of its work and its progress". There had been major changes in the landscape of Auvergne, but these revolutions had been unimaginably slow and gradual, caused by nothing more extraordinary than the streams and rivers that still drain the region. Like other naturalists, Montlosier offered only a vague impression of the vast timescale of this transformation (§2.5). But by wording it extravagantly as "an infinity of ages [*siècles*]" he laid himself open to a suspicion of eternalism, although in fact the phrase was applied to a clearly finite process. Anyway, whatever the timescale, Montlosier's vision was no less geohistorical than Burtin's and Blumenbach's, even though his conception of nature's revolutions contrasted so sharply with theirs.[10]

Soulavie had extended Desmarest's work in another direction only a few years before the start of the Revolution, reconstructing a sequence of "epochs" based on the similar extinct volcanoes and basalts of Vivarais, south of Auvergne (§4.4). But he had transcended the two-dimensional methods of physical geography by incorporating into his geohistory the three-dimensional analysis of rock formations that his German contemporaries called "geognosy". However, Soulavie had abandoned his multivolume account of his native region, discouraged by persistent sniping from one malicious critic, and perhaps also by his failure to convince the savants at St Petersburg that geohistory could provide a workable basis for the geognostic classification of rocks. By the time the Revolution began, Soulavie was already lost to the natural sciences and was about to make his debut as a chronicler of *human* history instead.

However, Soulavie's claim about the reality of organic change during geohistory, based on the concrete evidence of the distinct faunas of successive limestone formations in Vivarais, was not lost from sight. It reinforced the much earlier and well-known report by Arduino, that the formations on the southern flanks of the Alps were similarly characterized by distinct assemblages of fossils (§5.2). Both must have helped to convince Ferber in St Petersburg—who had edited Arduino's work and judged Soulavie's—that it was indeed feasible to use the relative ages of formations as a criterion for classifying them (§4.4). The process has yet to be fully explored by historians, but it seems clear that by 1789 such opinions had become widespead among naturalists throughout Europe. For example, as mentioned already, Blumenbach noted in passing that if fossils were studied more closely, and treated as documents in nature's archive, "even the respective ages of the various important kinds of formation can be fixed". The validity of this claim had yet to be

9. Fig. 6.2 is reproduced from Desmarest, *Carte de Puy-de-Dôme* (1823); the area was also included on the earlier map (see Fig. 4.9) published in Desmarest, "L'Origine et la nature du basalte" (1774), but at that time the Puy de Côme lava had not yet been mapped. Montlosier may have seen the later map in unpublished form, but in any case he had evidently explored the area thoroughly for himself: see *Volcans d'Auvergne* (1789), 28–35. Olby is 8km west-southwest of the famous volcanic peak of Puy de Dôme (west of Clermont-Ferrand), after which the département is named. The Étang de Fung is now a patch of flat marshy meadow bounded to the north by the vast *cheire* or rough-surfaced lava (much of it now forested).

10. Montlosier, *Volcans d'Auvergne* (1789), 75, 93–94.

demonstrated *in detail* for any specific region or set of formations; but it did at least express what was clearly seen as a fruitful direction for future research.

In contrast, Lavoisier's analysis of the formations around Paris—presented to the Académie des Sciences a few months after the start of the Revolution—was primarily an essay in earth physics rather than geohistory (§2.4). It proposed a causal explanation for the alternation of putatively shallow-water and deep-water sediments, which were interpreted as such partly on the basis of the organisms fossilized with them. But this was an explanation that, if valid, would apply to all such formations anywhere, deposited at all times, past, present, and future. It was not—or not primarily—a reconstruction of any specific passage of geohistory.

Geotheory as a flourishing genre

Finally, the range of rival geotheories on offer continued to expand as it had in the preceding decades. For example, Ermenegildo Pini (1739–1825), the professor of natural history at Milan and a member of the Barnabite order, published in 1790 a "New theory of the earth", and then more fully a "*Memoria geologica*" on "the revolutions of the terrestrial globe". He was well-read in the relevant literature, and included, for example, a reference to Montlosier's newly published work. Pini's system was based on a variant of the standard model of a gradually falling global sea level (§3.5); but despite the plural "*rivoluzioni*" in his title, he focused in fact on one specific "general, extraordinary, and brief inundation" in the relatively recent past. Like de Luc and many other diluvial theorists (though not all), he identified this exceptional global event as the one recorded as Noah's Flood, thereby hitching human history onto geohistory. However, that biblical link in no way disqualified his system from serious attention by other savants: it was published in the prestigious periodical of the Società Italiana delle Scienze, and it became widely known outside Italy through de Luc's critical review of it in the Parisian *Observations sur la Physique*.[11]

In this context, Hutton's system, published in the Edinburgh *Transactions* two years earlier, was naturally being treated by other savants as yet another example of the same genre (§3.4). It is significant that when (as mentioned already) Blumenbach made that new work accessible to German readers, he objected not to Hutton's indefinitely vast timescale—a commonplace by this time—but to his ahistorical conception of an unchanging organic world. Anyway, the examples of Pini and Hutton—others from these years will be mentioned later in this chapter—are sufficient to show that the genre of geotheory was alive and well. Such systems were treated with respect, though not without criticism, by other savants. But as Desmarest commented a little later, the sheer multiplicity of systems, often incompatible with each other, threw increasing doubt on the very genre itself (§6.5).

Conclusion

This brief review of the sciences of the earth during the earliest phase of the Revolution in France has suggested how all the four traditions that had structured research in the previous decades were continuing to flourish, alongside a minor component of studies that can properly be called geohistorical. The work that has been mentioned, however briefly, has also been representative in being drawn from all parts of Europe, from France to Russia, Scotland to Italy: as the Revolution began to erupt in France, the sciences of the earth remained as cosmopolitan as the rest of the Republic of Letters, at least among the elite of leading savants. In addition, the earlier trend towards putting greater emphasis on fossils, as some of the best evidence for nature's history, was also continuing, as in Blumenbach's plea for giving them much closer and more accurate attention. However, such prescriptive suggestions had not been turned into concrete achievements; fossils had yet to be shown to be capable of yielding a reliable and detailed reconstruction of geohistory. Meanwhile, de Luc's renewed efforts in geotheory were so important in their implications, despite growing doubts about the genre, that they deserve to be described separately.

6.2 GEOTHEORY AS GEOHISTORY (1790–93)

De Luc's new system

A few months after the Revolution broke out in France, de Luc began sending an article almost every month to La Métherie, the Parisian editor of *Observations sur la Physique*, which continued to be one of the most substantial scientific periodicals anywhere in Europe. Thirty-one long "letters" were published in its monthly issues over the following three and a half years, as the Revolution lurched towards its most violent and radical phase. De Luc, safely ensconced at the court of George III at Windsor, was sheltered from these winds of change, and what he wrote made no allusion to the turbulent events in Paris. In parallel with this series, he also wrote four equally public letters to Hutton; they appeared in translation in the *Monthly Review*, a leading British periodical of general scope and liberal outlook, which published original articles alongside extensive reviews of recent British and foreign books. Both series of letters set out a geotheory that revised the one that de Luc had published over a decade earlier as letters to his patron Queen Charlotte (§3.3). Coming from such a prominent savant, they brought several important new features into the debate about geohistory.[12]

De Luc's wings were significantly clipped by his choice of the French periodical. There was no place in it for the metaphysical and theological essays with which he had introduced his letters to the queen; yet in his own view the *physique* of his system was linked indissolubly to its *morale*, the key terms in his earlier title (§3.3). In fact he added a postscript to one letter, after he had read Burtin's essay (§4.2), complaining that the Brussels naturalist had ignored his *morale*—his "principal goal"—

11. Pini, "Nuova teoria della terra" (1790) and "Rivoluzioni del globo terrestre" (1790–92); de Luc, "Lettres à La Métherie" (1790–93), letter 16 (1791).

12. Overlapping with the tail end of his letters to Paris, de Luc also sent a series of letters to Blumenbach for the benefit of German-speaking savants, setting out his theory yet again (§6.4).

and therefore misunderstood his *physique*. He emphasized that his current letters were confined to the latter, and he hoped they would clarify his ideas about the former world and the revolution that had brought it to an end. But he chafed under this restriction, and the following year he seized an opportunity to restate the "moral" dimension, when the Hollandsche Maatschappij offered a prize for an essay on the universal foundations of morality. The essay that de Luc sent to Haarlem explained that it was intended as an introduction to his "*géologie*". Yet he could not publish the two together. In effect, he traded the advantage of bringing his geotheory more widely to the attention of scientific savants at the cost of insulating it from his deeper concerns.[13]

Nonetheless, the restriction still left de Luc with a wide field, and he exploited it to the full. His usual prose style—verbose, rambling, and repetitious—must have blunted the impact of his work, but in part it was an unavoidable consequence of his chosen genre. Like other geotheorists, de Luc felt obliged to examine everything from first principles. If the causes of past events were to be understood correctly, "*physique terrestre*" had to be based on "*physique générale*". Any adequate geotheory had to identify the unchanging general physical causes at work in the world before tackling the origin of specific features on earth or the causation of specific events in its past history. Yet these questions of "general physics" (which in modern terms included chemistry) were being fiercely contested at just this time, as a result of Lavoisier's self-styled "revolution" in chemistry. So de Luc was necessarily—though not unwillingly—embroiled in arguments that ranged far beyond the sciences of the earth. In the tables of contents of successive issues, La Métherie classified de Luc's first few letters as chemistry, the next few as physics, most of the rest as natural history, returning finally to physics. Nothing could show more clearly the awkward position of geotheory in relation to the more disciplinary conception of the sciences that was emerging at just this time, as epitomized by *Observations* itself.

De Luc began by sending La Métherie seven letters on general problems of physical science—such as the nature of liquids and gases, acids and electricity, heat and light, and the puzzling substance phlogiston that Lavoisier had scrapped altogether—and on their implications for meteorology. Only then did he feel able to embark on an equally prolix discussion of "terrestrial physics". This was the area for which he now adapted his own earlier term *géologie* (§3.1): no longer for geotheory as a genre, but in effect for earth physics (§2.4) as the sum total of causal explanations of terrestrial phenomena. It was, he conceded, "a science in which we are still looking for the first rudiments"; but he was trying to find the relevant natural laws, analogous to those of chemistry, "by which geology will become a real science". It is no coincidence that La Métherie first adopted the term *géologie* in his influential annual surveys of the whole range of the natural sciences at just the time when he had begun publishing de Luc's letters. From that time onwards, "geology" began to be used with increasing frequency by some savants, and not only in France, though its earlier life as a synonym for "theory of the earth" left others skeptical about it.[14]

A differentiated "former world"

However, de Luc's geotheory was based not only on causal "geology" but also on his reconstruction of "the ancient history of the earth". Here he was clearly using the key word "history" in its modern temporal sense: his geotheory, unlike Hutton's and far more than Buffon's, was once more to be truly geohistorical (§4.5). It would trace a sequence of past events that could *not* be predicted (or rather retrodicted), even in principle, from the causal laws of physics, but would need to be pieced together from the natural "monuments"—the antiquarian metaphor was ubiquitous—that remained from those remote times. De Luc was explicit about the proper method for this kind of science: "a true *theory of the earth*" would have to be built on "the ensemble of the *events* on our globe, linked to their true *causes*". To establish the "connection between past and present phenomena" it was necessary to replace vague generalities with "determinate ideas of definite *events*, the *causes* of which, by their distinct nature, must have led to those that act on our globe today". Present causes were a useful key to the deep past, but only insofar as the concrete "facts" or monuments surviving from that time were consistent with them. It could not be assumed in advance that all the causes of the past were still acting today: some might have ceased to act, without of course abrogating the "general physics" on which they were based.[15]

De Luc's earlier geotheory had been strongly binary in character: he had focused his analysis on the recent major "revolution" that—so he claimed—had transformed a relatively undifferentiated "former world" into the familiar "present world" of recorded human history (§3.3). His new geotheory, while hardly downgrading the significance of that last great event at the earth's surface, recognized it as just the most recent in a long series of major natural changes. In effect, a multiplicity of revolutions now differentiated the former world into a sequence of distinct stages in geohistory. By his own account, the decisive factor in this deepening of his sense of geohistory had been his reading of Saussure's *Alpine Travels*. This had supplied him with a vast store of proxy field observations to amplify his own, as well as Saussure's own interpretations of them. He was, for example, convinced by the case of the now famous vertical pudding-stone at Vallorcine (§2.4), and equally impressed by the vision of the broad sweep of geohistory that his fellow Genevan had experienced on the summit of Crammont (§4.5). He rejected criticism from a Parisian archivist—a scholarly chronologer who advocated the traditional short

13. De Luc, "Lettres à La Métherie" (1790–93), letter 10 (1790), 350–51; see also letter 23 (1792), 455. The Haarlem prize question—chosen in the light of Kant's notoriously difficult work—was set in 1791 and de Luc wrote his essay in 1792, but it did not win him the prize: see Bruijn, *Prijsvragen* (1977), no. 66. It remained unpublished until de Luc used it as a "discours préliminaire" to his *Lettres à Blumenbach* (1798), iii–cxxviii (§6.4).

14. De Luc, "Lettres à La Métherie" (1790–93), letter 11 (1790), 442; letter 21 (1792), 290. See La Métherie, "Discours préliminaire" (1791), 28–40; in his review the previous year the same topics had still been classed as "*minéralogie*". Montlosier had already used *géologie* in much the same sense (§6.1).

15. De Luc, "Lettres à La Métherie" (1790–93), letter 13 (1791), 174; letter 14 (1791), 271–72. These passages are merely illustrative; de Luc's methodological remarks, like everything else, are scattered throughout his text.

timescale—on the grounds that his critic had failed to consult "the archives of nature, that is, the mountains, where one can read of so many past events"; even if the critic was unable to go into the field in person he could at least have read "extracts from these archives" in Saussure's volumes.[16]

Like Buffon's geotheory (§3.2), de Luc's new geohistory was divided by decisive "epochs" into seven distinctive "periods". The parallel with Genesis was not close, because he also followed Buffon in defining the seventh and final period not as the divine sabbath but as the world of man. Still, in a more general sense he clearly did intend his geohistory to be a scientific rendering of the biblical narrative, making due allowance for the poetic imagery of the latter and its primarily religious purpose, and hence the need to interpret it in a nonliteral way (see §6.4). Certainly he insisted on "the enormous antiquity of the earth", adding that "naturalists who have thought otherwise were not attentive observers". "What time must there have been for the formation of this pile of beds", he exclaimed, referring to the Secondary formations; and he reminded his readers how he had long rejected any attempt to compress the whole early history of the earth into six literal days, as being "in effect as much contrary to natural history as to the text that is to be explained". Above all, his objectives were unmistakably geohistorical. As he put it at the outset, in language that echoed Soulavie's (§4.4), "I shall trace a sequence of *events*, linked by distinct *causes*, and certified by our *beds*; I shall divide these *events* into different *periods*, not by any determinate *duration* but by the order of their *succession*."[17]

The core of de Luc's series of letters was devoted to expounding his reconstruction of the seven periods of geohistory. Like Buffon's system, this was a confusing mixture of the highly speculative and the soberly empirical, the latter being, in de Luc's case, mostly derived from his own observations in the field. He had emphasized that it became increasingly difficult to reconstruct what had happened the further back in deep time the savant tried to penetrate, owing to the effects of successive revolutions in effacing the traces of earlier ones. So it is hardly surprising that his earliest period was also the most speculative. It began with the first and most decisive "epoch", the addition of light to a previously inert world—the natural manifestation of the divine "*fiat lux*"—and hence the start of chemical reactions on earth. His treatment of light as a material substance indicates how his geohistory was intimately connected with his "general physics"; but the latter, though profoundly unmodern, was no more strange to his contemporaries than, say, Hutton's.

The following periods were familiar to de Luc's readers from many other systems, particularly those on the standard model of a gradually falling sea level (§3.5). Although still highly speculative, these periods were at least related to observable features. The second period had seen the formation of granites and other Primary rocks, as successive "precipitations" from an initially global body of chemically complex "liquid"; like Saussure, de Luc claimed that granite was sometimes clearly stratified. The third and fourth periods saw further precipitations of other kinds, increasingly local in extent, together with the first evidence of life in the form of fossils; and the fifth period was marked by the formation of the first limestones, and by the first volcanic action and "catastrophes" of local crustal collapse. De Luc pointed out that, in interpreting such Secondary formations, "superposition is our only guide, as regards the order of the times"; as Soulavie had recognized (§4.4), an understanding of this basic guide to geohistory could not yet be taken for granted and still needed to be stated clearly.[18]

Only after sketching this long sequence of periods, each marked by "monuments" in the form of distinctive kinds of rocks and fossils, did de Luc approach the specific revolution that had been the prime focus of his earlier geotheory. His sixth period was marked by the first appearance of terrestrial animals and by the Superficial sands and gravels in which their bones were interred. He interpreted these deposits as the final precipitations of the ancient sea. Then, in the most recent major revolution, oceans and continents had in effect changed places, not by any upheaval of the earth's crust but by its fracture and collapse. That great epoch was the sharp boundary between the present world and all that had gone before, and it marked the start of the seventh and last period. De Luc assumed that the physicochemical processes that had precipitated the huge pile of varied Primary and Secondary rocks were no longer operating. Once the Superficial deposits had been formed, "the liquid found itself reduced to the water of our sea", and the earth was now settling gently into a final state of "repose".[19]

As before, de Luc was more concerned with establishing the historical reality of the most recent revolution than with fixing its physical cause. Once again, vast unseen caverns in the earth's interior were invoked as an explanatory deus ex machina for the putative crustal collapse. Far more important to him was to determine the date at which the former seabeds had been left high and dry as the present continents. As in his earlier theory, the most recent period of geohistory—since that great event—differed from all the preceding ones in being "determinate" in duration: not as precisely as traditional chronologers would have liked, but at least to the correct order of magnitude. If, as he claimed, the present continents had appeared quite suddenly, by an abrupt drop in sea level, the various physical agents now visibly operating on the continents—such as fluvial erosion and deposition—must have started at that point in time. In terms of his earlier analogy, that event would be like turning an hourglass: the limited amount of sand that had since trickled through showed that the glass had been upended at a finite time in the past (§3.3). De Luc now developed that metaphor by transforming the humble hourglass into a *chronometer*: John Harrison's famously precise timekeeper—the supreme high-tech achievement of the century—was much in the public eye, since it at last

16. De Luc, "Lettres à La Métherie" (1790–93), letter 8 (1790), 206; letter 22 (1792), 369; Viallon, "Lettre à Delamétherie" (1792). Viallon was the librarian of Sainte Geneviève, the church whose building had been used by Soulavie as a vivid analogue for the geohistory that could be read from rock formations (§4.4).

17. De Luc, "Lettres à La Métherie" (1790–93), letter 10 (1790), 332; the emphases are original, and typical of de Luc's prose style. See also letters 17 (1791), 334; 21 (1792), 282; 23 (1792), 455; the quoted phrases are just a few of those scattered throughout this work.

18. De Luc, "Lettres à La Métherie" (1790–93), letters 10–13 (1790–91).

19. De Luc, "Lettres à La Métherie" (1790–93), letters 14 (1791), 26 (1792); the narrative was interrupted for almost a year and a half by letters responding to his critics and dealing with other topics. It is beyond the scope of this book to explicate his chemical concept of precipitation; he showed the customary disregard for what might be happening on the unobserved floors of present seas.

made possible a dramatic improvement in the determination of longitude and hence in global navigation. Of course de Luc did not claim any comparable accuracy for "nature's chronometers", but he did insist that they proved that only a few millennia could have elapsed since the last major revolution in the earth's geography.

De Luc's most extensive example of nature's chronometers was derived from his own earlier travels in the Netherlands. The vast delta of the Rhine had many features that could be dated reliably from recorded human history: Roman settlements, dykes of which the date of construction was known, and so on. Together, they indicated the rate at which the delta had grown in area, even since Roman times. Since its total size was finite, one could extrapolate back to the time when it must have begun to accumulate:

> Here then is a true *chronometer*: one finds the total operation since the birth of our *continents*; one sees there its *causes* and their *progress*, and one can distinguish the parts of the *whole* that have been produced in known *times*. Doubtless there are too many causes of irregularity in this progress to be able to count the *centuries*; but it is evident that their number would not be found to be considerable.[20]

De Luc concluded that his analysis of the continents had now shown "that the *epoch* of their *birth* is not extremely remote". This was in direct conflict with those such as Desmarest and Soulavie, and more recently Hutton and Montlosier, who claimed that the present land areas had been subject to subaerial erosion for incalculably long periods of time. The argument was not about the earth's total timescale but about the length of time that the present continents had been above sea level (and that the earlier ones had been drowned). He claimed that they had appeared so recently that a memory of these dramatic events, however faint and even garbled, had been preserved in human records, most importantly of course in the biblical story of the Flood. The claim was crucially important for his geotheory, because it linked geohistory to human history and thereby helped demonstrate the truly *historical* character of the earth itself. Conversely, in his opinion, those who rejected his evidence for the recent date of this revolution, or even for its reality, put the nascent science at risk: "there is perhaps no opinion that has done more to damage [natural] philosophy than that of an immense antiquity for our *continents*; for it based *geology* on an error, and *geology* covers our whole field of knowledge."[21]

The role of fossil evidence

De Luc's revised geotheory turned his earlier binary distinction between the present and former worlds into just the most recent change in a more subtly differentiated geohistory. His new exposition also reflected wider trends in that he gave much closer attention to the evidence provided by fossils. This affected his interpretation of both his recent revolution and the far earlier changes recorded in the Secondary rocks.

In defining the character of the recent event that had turned ancient seabeds into present continents, de Luc now supplemented the evidence of physical geography with much greater use of the fossil bones and shells found in the Superficial deposits. He argued that the famous rhinoceros carcass found by Pallas in frozen ground in Siberia (§5.3) invalidated Buffon's inference that such tropical species had migrated slowly to lower latitudes as the earth cooled (§3.2); de Luc agreed with Pallas that they were evidence for a sudden and quite recent event. He inferred that the fossil shells that he himself had collected in the foothills of the Apennines must likewise be quite recent, for they were preserved with some of their original color and with the ligaments on some of the bivalves. The bones that Esper had described from Bavarian caves (§5.3) were also relevant, for in that savant's later research—published posthumously, and translated for de Luc into French by a Dutch woman naturalist—Esper claimed that the cave bones were mainly those of the polar bear. De Luc argued that the region must therefore have been near sea level at the time and that the bears had used the caves as dens, just as seals now used coastal caves in Scotland. That the bones were preserved in the caves under a crust of stalagmite—which accumulated slowly but not imperceptibly on a human timescale—proved that the bears had been living there at a remote but not incalculable time in the past. Finally, de Luc reported that an English naturalist had recently found an elephant tusk, hippopotamus teeth, and a large bovine skull in sands that also contained common seashells such as scallops, whelks, limpets, and oysters. Yet none of these could be relics of the present world, because among the shells was the famous left-spiraled whelk, known only as a fossil (Fig. 6.1): "the *time* when *elephants* and *hippopotamuses* lived along with *cattle* in this part of the globe—a *time* when a distinct *shellfish*, unknown among those alive, was still abundant in this part of the *sea*—was not remote *by any very great number of centuries*".[22]

However, de Luc's revised geotheory showed still greater changes from its earlier version, in his use of the fossils in the Secondary formations. In the intervening years he had become far more aware of the huge pile that overlay the confusing Primary rocks. Like other naturalists, and above all the geognosts, he was trying to detect some kind of order in the formations—a specific sequence of sandstones, shales, limestones, and so on—that might be valid globally, or at least across Europe (§2.3). Beyond that, however, he too had become aware that in general terms these formations contained "characteristic" fossils. Doubtless he derived this idea from his wide reading, but it came also in part from his own fieldwork. For example, he mentioned how he had been struck by what he saw on the south coast of England, during a tour that apparently extended right round to the east coast. In the Dorset cliffs he traced three major formations piled on one another, with distinctive fossils.

20. De Luc, "Lettres à La Métherie" (1790–93), letter 27 (1792), 344; characteristically, he provided no map or other image to convey his argument graphically (in either sense of that word). He distanced himself from those who treated geohistory as deterministic by adding that the *future* development of the delta could not be predicted, owing to the complexity of the causal circumstances.

21. De Luc, "Lettres à La Métherie" (1790–93), letter 28 (1792), 414–15. The meaning of the final clause of the quotation became clearer in his later writing (§6.4).

22. De Luc, "Lettres à La Métherie" (1790–93), letter 14 (1791), 275–83; letter 28 (1792), 427–30; letter 18 (1791), 462–64. Esper, "Osteolithen-Höhlen" (1784), 100–106.

> Here is a really grand scene for the geologist [*géologue*]. The order of the beds is known by the position in which they succeed one another, plunging in the same direction below the same sea level; the beds of *clay* pass under those of *limestone*, and those under the beds of *chalk*. Moreover, we judge that the beds of *clay* and *limestone* are surely of a more ancient date than those of *chalk*, in that the former contain *ammonites* that are no longer found in the latter; and this proves that the change in the state of the sea, which produced the *precipitations* of the *chalk*, were fatal to that kind of animal.[23]

Here the fossils found—or not found—in the formations were integrated into de Luc's broader reconstruction of the history of life, in that the Chalk was taken to date from after the demise of the highly distinctive ammonites. But his comments also proposed a *causal* explanation for the faunal change: the ammonites had been wiped out by a change in the composition of the "liquid" from which all the formations had been precipitated in turn. De Luc had no truck with those such as Bruguière who denied any true faunal change; he insisted on the reality of extinction, citing not only ammonites but also belemnites, nummulites, and many anomias and other bivalve mollusks (§5.2). How then to account for those fossils that seemed to have survived these revolutions? In contrast to the mass extinction or total revolution suggested by Blumenbach (§6.1), de Luc proposed, in effect, three degrees of change, depending on relative proximity to the putatively localized events that had caused a sudden alteration in the "liquid":

> 1. All the species of *marine animals* ceased to exist in the places where this modification happened suddenly. 2. Some species of these animals were destroyed throughout the sea. 3. The species that were conserved on seabeds elsewhere underwent great changes, and thus came closer by degrees to those that we find in the present sea.[24]

Whatever the process that de Luc may have had in mind, he clearly envisaged some kind of organic change *by degrees*. But unlike most speculations involving transmutation (or evolution), this process was combined with extinction: only the two causes acting together could account for the observed contrasts—not total but partial—between the fossils of successive formations, and between all of them and the species still alive. This concept of a natural process of piecemeal organic change through time, rudimentary though it was, enabled de Luc to see how fossils might act as monuments not only to the history of life but also to that of the physical world that was life's environment. This remained little more than a passing suggestion, but it was one that other naturalists might well notice and appropriate.

A critique of Hutton

De Luc's published letters to Hutton can be summarized more briefly, because they covered much the same ground as those he was sending at the same time to Paris. However, they did serve to make his ideas more familiar among savants and amateurs in the anglophone world, and of course they dealt specifically with the points on which he and Hutton disagreed. The two had already clashed—in a civil manner, as befitted savants—over the theory of rain, so de Luc was naturally one of those to whom Hutton sent an offprint of the article setting out his geotheory (§3.4). However, de Luc expressed his surprise at finding in it no mention of his own work, although Hutton can hardly have been unaware of it. Like other critics, he was particularly scornful of Hutton's ideas on the consolidation of rocks by fusion, noting that Hutton admitted he had never seen in the field some of the decisive cases. De Luc also rejected Hutton's notion that the present continents would eventually disappear through being worn down by erosion, pointing out that on many coasts they are actually growing in area by new deposition. Rain and rivers, he argued, are merely smoothing the rough edges of what emerged quite recently from beneath the sea. The jagged peaks of the Alps proved that fluvial erosion was not involved, since they were in the realm of snow, ice, and glaciers, not running water. Again and again, de Luc pointed out the consequences of Hutton's limited experience in the field; most seriously, he had never seen any high mountains. Yet paintings and engravings of the Alps should have made "such striking monuments of revolution" familiar in proxy form, even to one who had never seen them at first hand. In short, de Luc suggested to Hutton that his kind of philosophy "may have induced you to speculate more than to observe": the unkind cut expressed the general view among naturalists that Hutton, like Buffon, was too much of an indoor savant.[25]

Echoing Montlosier's metaphor, de Luc criticized Hutton for neglecting nature's most ordinary and familiar witnesses: "not being consulted, they remain silent". He did not object to a vast timescale, but to Hutton's use of "the vague idea that time has no bounds" as an *explanatory* principle. For "time effects nothing" by itself. Here de Luc put his finger on the weakest point in Hutton's reasoning. Hutton himself conceded that he could not demonstrate that the continents had been perceptibly eroded during the period since Antiquity (§3.4), so no amount of hand waving about the vastness of deep time could possibly count as evidence for what had happened in the still longer run. De Luc claimed that the alleged "unbounded antiquity" of the continents was disproved by the "hour-glass-like chronometers" of real causal agents accurately observed: nature's monuments showed that "our continents have undergone but very small and well determined alterations" since they emerged. "Speculative geologists" such as Hutton offered "poetical descriptions of dreadful effects sometimes produced by swollen rivers and torrents", in place of precise descriptions of the changes effected in known spans of time. In sum, Hutton had *failed* to use the present world as a key to the past. As de Luc put it in his conclusion:

23. De Luc, "Lettres à La Métherie" (1790–93), letter 18 (1791), 458–60. His fossil-based stratigraphy was of course very crude in comparison with what was being done by William Smith in England a decade or two later (§8.2). In modern terms, he was probably referring to some of the Jurassic shales and limestones (such as those of Kimmeridge and Portland respectively) that underlie the Chalk of Cretaceous age; ammonites are in fact found in the latter, but are rare and local.

24. De Luc, "Lettres à La Métherie" (1790–93), letter 12 (1791), 101; see Gohau, *Sciences de la terre* (1990), 291–92.

25. De Luc, "Letters to Hutton" (1790–91): see for example letter 1 (1790), 206–7, 211–14; letter 2 (1790), 600–1; letter 3 (1791), 574–77. Hutton had not yet published his full *Theory* (1795), with its textual "illustrations" based in part on fieldwork (§3.4).

> Those real inquiries [such as his own!] into the history of our continents, when more generally attended to, will be the tomb-stone of every theory of the earth, the agents of which, and their agency, must be hidden under the veil of unbounded antiquity, for fancy [fantasy] to take the appearance of genius, and assertion that of knowledge.[26]

Conclusion

De Luc's revised geotheory, like its earlier version, portrayed the former world and present world—roughly, the prehuman and the human—as being sharply separated by the major revolution that had somehow turned former seabeds into present continents; and nature's chronometers of present causal processes proved that this had happened no more than a few millennia ago. For all preceding periods of geohistory, on the other hand, the timescale was immeasurable but clearly immense. But it was still possible to do geohistory, because all the earth's earlier revolutions could be placed in their correct order by attending to nature's monuments such as rocks and fossils. However, de Luc sketched an outline reconstruction of geohistory that was far more differentiated into distinct periods than his earlier writing had suggested. He was now giving much more attention to fossil evidence, having realized—in part from his own fieldwork—that fossils could be used to characterize successive Secondary formations. In short, he was now aware that fossils could help significantly in piecing together the history of the earth. Yet these geohistorical insights were scattered across two prolix accounts of a wide-ranging geotheory, and they could easily be missed. De Luc conceded that his theory might seem to be just another among the many "geological novels [*romans*]" already on offer—the metaphor had become a code word for geotheories like Buffon's (§3.2)—but he insisted that his had solid foundations, because it was built on careful field observations and sound physical principles.[27]

De Luc's new attention to fossils did indeed reinforce what had already been distinctive about the earlier version of his geotheory, namely its geohistorical character (§4.5). This in turn came in part from his adoption of antiquarian metaphors such as that of nature's "monuments": rocks and fossils, as relics of real past events, were a more reliable foundation for causal theorizing than any ahistorical laws of nature. But de Luc's sense of geohistory was also not unconnected—to put it at its lowest—with the profoundly historical interpretation of the world that this self-styled "Christian philosophe" derived from his underlying theistic beliefs. Avoiding misplaced literalism, the strongly historical character of the biblical narrative as a whole also applied, in de Luc's view, to the vast spans of prehuman time disclosed by geology. More specifically, his identification of the most recent physical revolution as the biblical Flood, although mentioned only in passing in his letters to La Métherie and Hutton, served to tie an immensely long and varied geohistory to a meaningful narrative of the few millennia of human history; it linked the natural realm of *physique* to the human realm of *morale* and hence to his own "ultimate goal".

6.3 THEORIZING IN A TIME OF TROUBLE (1790–94)

Geotheories and focal problems

De Luc's lengthy published letters, addressed nominally to La Métherie and to Hutton, made his revised geotheory widely known to savants on both sides of the Channel, over just the period when events in France were slipping from moderate reform towards violent political upheaval. When de Luc sent his first letter to Paris, from the quiet safety of Windsor, the Revolution was still in its earliest phase, of relatively orderly change. Around the time he was expounding the core of his ideas, the increasing radicalism of the new regime in France was driving many of the nobility and clergy into hiding or exile; the royal family had tried to flee Paris and had been brought back in disgrace, and the king put on trial; and France was at war with Austria, a major power that controlled the southern Netherlands as well as much of central Europe. By the time de Luc sent his last letter to Paris, the French monarchy had been abolished and a republic declared; all titles had been replaced by plain "Monsieur" and "Madame" (soon to be replaced in turn by the even more egalitarian "Citizen"); Louis XVI had been executed by the new "humane" method devised by the physician Joseph-Ignace Guillotin; and France was at war with most of the rest of Europe, including Britain. These events shocked the political classes throughout Europe, far more than the earlier and relatively decorous revolution in America. Yet de Luc's last letter from Windsor promised a sequel—which might have been sent, had war not broken out between France and Britain—and the whole series made no reference to the turbulence developing across the Channel.

Even in Paris, however, scientific life had continued at first without much disruption. In 1790 La Métherie's annual editorial review of the sciences began with an enthusiastic political endorsement of what the Revolution then stood for, but *Observations* went on publishing original articles—among them, de Luc's—just as it had in earlier years. The Académie des Sciences remained the most prestigious scientific forum; Lavoisier, for example, supplemented his earlier paper on the "physics" of sedimentary formations (§2.4) with one on the topography of those outcropping around Paris and on the barometric measurements that would be needed to plot their three-dimensional structure. The Académie also sponsored useful practical reforms: for example a new "metric system", based on the "natural" meter, was to replace the old regime's confusing multiplicity of local weights and measures. And foreign savants continued to make their pilgrimages to Paris to hear the great men lecture and to meet them socially in the *salons*.[28]

26. De Luc, "Letters to Hutton" (1790–91): see letter 3 (1791), 577–78; letter 4 (1791), 564–68, 585.

27. De Luc, "Lettres à La Métherie" (1790–93), letter 26 (1792), 228.

28. La Métherie, "Discours préliminaire" (1790); Rappaport, "Lavoisier's geologic activities" (1968), 383. Chaldecott, "Scientific activities in Paris in 1791" (1968), reconstructs the visit of Hutton's friend James Hall to meet Lavoisier and other savants. Among administrative reforms around this time, the old French provinces were replaced by a series of smaller and more manageable *départements* named after natural features such as rivers and mountains.

In contrast to the continuing vogue for comprehensive geotheories, many savants engaged in "geology" were now giving increasing attention to more circumscribed problems: Montlosier's "essay" on volcanic activity in Auvergne has been mentioned already (§6.1). One recalcitrant focal problem was that of accounting for the "stoniness" or consolidation of rocks of all kinds (§2.4). De Luc and others developed a concept of precipitation from some kind of aqueous medium; La Métherie and others formulated it in terms of crystallization; and Hutton's idea of fusion under conditions of intense heat, although generally considered far less plausible, was clearly a part of the same debate. The controversy over the origin of basalt was also at its height at just this time (§2.4), as La Métherie noted in his annual review for 1791, when he first adopted the term *géologie*; he gave space to both sides of the argument, printing translations of Werner's Neptunist work as well as articles favoring a Vulcanist interpretation. But all these were problems within earth physics, in that they concerned the causal origins of particular kinds of rock, whatever their age; they did not directly affect questions of geohistory.

The parallel problem of the origin of granite, on the other hand, had major implications for geohistory. In 1790, both Hutton and his younger friend Sir James Hall (1761–1832)—who had succeeded to his father's baronetcy while still a teenager—read papers in Edinburgh claiming new evidence that at least some granites, like basalt, had an igneous origin, and that they had crystallized from an intensely hot melt in the "Plutonic" depths of the earth. This struck, quite literally, at the foundations of the standard model of geotheory, at least in its commonest form, for it denied the "primitive" character of what appeared to be the lowest and therefore oldest rock of all (§3.5), and left the earth, in Hutton's famous phrase, "without vestige of a beginning" (§3.4). De Luc probably knew about the Scottish work, even if the Parisian savants did not, long before its eventual publication. Nonetheless, he could properly treat the Plutonic interpretation of granites as less than proven and therefore leave it on one side. For the debates on all these focal problems—volcanic activity, consolidation, basalt, and granite alike—were inconclusive for the same reason: the questions of "general physics" on which they depended were far too obscure, and currently controversial, for any consensus about them to emerge.[29]

By contrast, the age of the present continents—*not* the age of the earth itself—was a focal problem that was central to de Luc's and Hutton's geotheories and also more amenable to resolution by appropriate fieldwork. What de Luc's published letters hinted at, and what became more explicit in the work to be described in this and the following section, was an underlying polarization on this issue. De Luc argued that his own conclusion—that the continents had emerged quite recently as land areas—was crucial to the nascent science of "geology" and that any theory about their extreme antiquity undermined its very foundations (§6.2). What gave that argument wider resonance was his claim that this last great revolution had been none other than Noah's Flood. As already pointed out, the equation entailed a far from literal interpretation of the biblical story, and anyway diluvial theorists were quite prepared to recruit extrabiblical sources—the flood stories in the records of other ancient cultures—to reinforce its historicity. Nonetheless, all such theories were treated by their critics as covert attempts to use natural evidence to bolster revealed religion. Conversely, however, claims to the unimaginable antiquity of the continents were seen by diluvialists as equally covert attempts to deny the historicity of the Flood, and with it the credibility of the biblical texts as a whole. More specifically, if the continents had been dry land for "an infinity of ages", as Montlosier put it (§6.1), without any recent disruption, they could have been home to humans for far longer than biblical tradition maintained. Even if the universe itself was not treated as eternal and uncreated, the human presence on earth might still have been almost eternal in relation to the few millennia of recorded history, as Hutton had implied and his admirer Toulmin made explicit (§3.4).

Dolomieu's mega-tsunamis

Once again, this was not a conflict between "Science" and "Religion", but an argument in which alternative interpretations of natural evidence were used—symmetrically, as it were—to support different theologies. The issue of the age of the continents was deployed by both sides in a struggle between traditional theism and the deism espoused by many Enlightenment savants. As the Revolution in France turned increasingly radical and those in power sought to eliminate all traces of the culturally Christian past, this scientific issue became embroiled in the political struggle between old and new regimes.

This can be illustrated with the case of a savant whose geotheory was akin to de Luc's, but who was present at the center of the Revolution rather than on the sidelines beyond the Channel. It will also show that in such debates there was no confessional difference between a Protestant and a Catholic: both were on the same side in opposing a deism that increasingly tended to shade into atheism. Matching the Genevan Protestant de Luc was a Parisian Catholic savant, Dieudonné (or Déodat) de Gratet de Dolomieu (1750–1801). Like some other younger sons of Catholic provincial nobles, Dolomieu had been entered almost at birth into the international lay military order of the Knights of Malta. As a talented young man he had risen swiftly through its ranks. His scientific interests were stimulated by travels that took him, for example, to Vesuvius and the Alps, and his patron La Rochefoucauld—whom Desmarest had earlier accompanied on his Grand Tour (§4.3)—secured him a place as a "corresponding" (i.e., non-Parisian) member of the Académie des Sciences. In the years before the Revolution, after further extensive travels and fieldwork, Dolomieu had published a report on the great 1783 earthquake in Calabria, and books on the active volcanoes off the coast of Sicily and the extinct ones off the coast of Italy. All this was within the established

29. Hutton, "Observations on granite" (1794), and Hall, "Formation of granite" (1794), read to the Royal Society of Edinburgh in 1790–91, presented evidence based respectively on field and museum work. Hutton was careful to restrict his Plutonic interpretation to "massive" granite, accepting that Saussure's "foliated" [*feuilleté*] kind (in modern terms, gneiss and similar rocks) might be sedimentary in origin. There should be no surprise at the lack of agreement on these problems, for they involved the difficult physics and chemistry of complex silicates under conditions of high temperature and pressure.

Fig. 6.3. Déodat de Dolomieu at the age of thirty-nine, painted in Rome by the famous portraitist Angelica Kauffmann on the eve of the Revolution in France. His elegant clothes indicate his social status, but like Saussure (Fig. 1.7) he is seated outdoors, hinting at his work as a *field* naturalist. He is portrayed with a field notebook on his lap, and the volcano and palm tree in the background suggest the extensive travels on which his scientific reputation rested. (By permission of the École des Mines de Paris)

research tradition of physical geography (§2.2); editions in German and Italian made him well-known throughout Europe among other naturalists with similar interests. He also wrote an important paper for *Observations sur la Physique*, claiming to have found in Sicily ancient lava flows interbedded with Secondary limestones, proving that there had been submarine volcanic activity in the area, long before Etna even began to erupt; it was at the very least a gesture in the direction of geohistorical interpretation (Fig. 6.3).[30]

Like many other idealistic people, Dolomieu supported the Revolution enthusiastically when it began, to the consternation of his aristocratic family and his superiors in Malta. The latter exiled him from the island, and eventually he moved to Paris and became active in its scientific life. His reputation as a mineralogist, as well as a physical geographer, was enhanced by a fine paper on a common but puzzling variety of limestone; Saussure's son Nicholas-Théodore (who by now had a modest reputation of his own) named it "dolomite" [*dolomie*], an unusual accolade for a living savant. Dolomieu also published an agenda for the use of the naturalists (among them Lamanon: §4.3) going on La Pérouse's expedition, in which he noted that what they observed on the far side of the world would be important for "the ancient history of our globe".[31]

Such hints of a geohistorical perspective became more explicit later in 1791, when Dolomieu published in *Observations* the first installment of a lengthy monograph "On compound stones and rocks". As the title suggests, this was primarily an analysis of the relation between rocks and the "simple" minerals of which they are composed and of the ways in which rocks of all kinds might have been formed and consolidated. It was far from being a comprehensive geotheory, but Dolomieu clearly regarded it as contributing to one: he admitted being reluctant "to add my system to the ten thousand already formed", an exaggeration that captures the general view of an overcrowded genre. Anyway, his interpretation was based on fieldwork as well as the indoor study of specimens, and what he had seen in the field led him to consider how whole formations might have been emplaced. He knew of limestones in which the rock was homogeneous and the fossils fragmentary, with apparently tropical species mixed with temperate, marine species with freshwater. He inferred that such formations could not have been deposited quietly layer by layer, as other naturalists supposed; instead, "these floods of scarcely fluid mud" were formed "at one time, as it were at a single throw", by seawater "in a most violent state of agitation". The vast spreads of gravel that he had seen elsewhere, and the valleys filled with materials that could not be matched among local rocks, pointed to a comparable effect. Dolomieu, like others before him, made such features the basis for a theory of occasional violent events.[32]

By itself, Dolomieu's theory was a piece of earth physics, not geohistory, since it purported to account for the production of thick limestone formations, gravels of erratic pebbles, and so on, at any point in time. It differed from Lavoisier's theory (§2.4), as Dolomieu himself noted, in that "it is not time that I shall invoke, it is force". He argued that neither kind of explanation ought to be preferred a priori to the other, but that his was better fitted to what could be seen in the field (he was not in fact parsimonious with time itself, since he assumed there had been long intervals of calm between the violent episodes). It was a matter of following the Newtonian principle that inferred causes should be commensurate with the effects to be explained. In trying to imagine what the putative events had been like, Dolomieu relied mainly on the traces of the most recent one, which was also the best preserved. He agreed in part with de Luc, but his own concept was of *transient* events, not one that had effected a permanent change in physical geography; and his were events that had been repeated many times in still earlier geohistory, not one that might well have been unique.

Dolomieu suggested that the events had been huge tidal waves [*très-grandes marées*] or mega-tsunamis, which had swept occasionally over the earth's surface, depositing massive beds of rock and carrying exotic materials far from their source. Like an ordinary wave breaking on the shore, or the terrible tsunami that broke over Lisbon in 1755—but on a far larger scale—the effects of these mega-tsunamis would have reached to a much greater height than the original wave out at sea, depositing suspended material across the land surface and then eroding the continents in the undertow as the mass of water retreated. Since the traces of the most recent "diluvial" event were said to reach an altitude of some 2000 *toises* (about

30. Fig. 6.3 is reproduced from an oil painting (now in Paris-EM) enlarged from a miniature painted in June 1789; see Bourrouilh-LeJan, "Déodat de Dolomieu" (2000), 86. Dolomieu, *Voyage aux Iles de Lipari* (1783); *Tremblemens de terre de la Calabre* (1784); "Volcans éteints du Val di Noto" (1784); *Mémoire sur les Iles Ponces* (1788). Dolomieu himself used "Déodat" as his first name, being in either case content to be called God-given. Lacroix, *Déodat de Dolomieu* (1921), prints a substantial correspondence and includes a valuable biographical essay (1: xiii–lxx).

31. Dolomieu, "L'Origine du basalte" (1790); "Genre de pierres calcaires" (1791); "Note à Messieurs les naturalistes" (1791); Saussure (N. T.), "Analyse de la dolomie" (1792); see also Carozzi and Zenger, "Discovery of dolomite" (1981). The name was later extended from the rock itself to the spectacular range of mountains in the Italian Alps which is largely composed of it.

32. Dolomieu, "Pierres composées et roches" (1791–92), pt. 1, 400–405.

4000m), Dolomieu estimated that in this case the original wave must have had an amplitude of some 800 *toises* (about 1600m). He concluded that such mega-tsunamis "would have been able to produce all the phenomena, an explanation of which seems to me impossible by any other means". The idea was not novel—Pallas for example had used it to account for his rhino carcass in arctic Siberia (§5.3)—but Dolomieu certainly gave it a clearer articulation.

Dolomieu claimed that his theory was no more extraordinary than the idea that the sea had completely covered the continents "for thousands of ages [*siècles*]" (the phrase suggests again that he had Lavoisier's theory in mind). He noted that on both theories antecedent physical causes would have to be invoked, as much to explain Lavoisier's huge tidal cycle as his own mega-tsunamis. He thought that in either case the causes would have to be supplied by the astronomers: like Lavoisier, he was thinking of extraterrestrial causes of some kind, and he had already mentioned privately that he was attracted by much earlier ideas of possible cometary impacts.[33]

What made Dolomieu's theory geohistorical was his use of it to explain specific and unrepeated events in the deep past, such as the major structural discordance between the Primary and Secondary rocks. More important still—for Dolomieu as for de Luc—was the most recent event, not only because its effects were the best preserved, but also because it linked geohistory with recorded human history. For since there was no earlier trace of humans—he was unaware of, or discounted, the few that had been claimed (§5.4)—Dolomieu rejected "any great antiquity for the present order of things". Like de Luc's "present world", that phrase suggested the binary character of his underlying conception of geohistory. He estimated that the most recent mega-tsunami—which had ended the "former order of things" and ushered in the present one—had been about four millennia before the Christian era and about six before the present. This was sufficiently similar to the calculations of the chronologers (§2.5) to leave his readers in no doubt that he too equated it with the biblical Flood and its extrabiblical equivalents. "In this respect", he wrote, "the historical facts are in accord with those of nature, and the human race was surely quite recent six thousand years ago, or at least it was then renewed after an almost complete destruction."[34]

It should be obvious that for Dolomieu, as much as for de Luc, any such equation with the Flood entailed a far from literal interpretation of the story in Genesis. But the estimated date did clearly align the two savants, as Dolomieu acknowledged: like de Luc, he inferred that "*the present state of our continents is not ancient*" and that they had not for long been the "the empire of man". But he added a comment that reveals the tensions underlying this scientific claim:

> This truth would not perhaps have been attacked so fiercely and so strongly combatted, had it not been related to religious opinions that one wanted to destroy . . . It was believed to be an act of courage, showing oneself exempt from prejudice, to increase—by a kind of bidding up [*enchère*]—the number of centuries that had elapsed since our continents were given over to our [human] industry.[35]

In other words, Dolomieu argued that the prejudices of those who had their own agenda for opposing traditional theism were what led them to argue for a vast antiquity for the continents as land areas. He recognized that in claiming otherwise he was even laying himself open to ridicule, and to "the kind of disfavor that now surrounds those who do not give themselves up to exaggerations and leaps of imagination". Here was a striking reversal of stereotypical roles: Dolomieu claimed that it was the critics of religion who were blinkered by prejudice, not the believers; it was the skeptics who indulged in irresponsible speculation. With the regime in Paris outlawing the monastic orders and attacking those clergy who refused to abjure their loyalty to the papacy, this was not a line that was likely to advance his career.

Dolomieu on the Nile delta

Dolomieu made a sharp distinction between the brief recent period of geohistory and all the rest: "time cannot be measured in the earlier epochs, and the imagination can lavish thousands of centuries on them as easily as minutes." Like de Luc, he was certainly not restricting the earth's *total* timescale for reasons of religion or anything else. In the same pregnant footnote, he gave notice that he intended to publish a work that would "unite historical monuments with geological observations", in order to show that some ten thousand years was, if anything, probably an *over*-estimate for the habitability of the present continents. Before he could deliver on that promise, however, the political situation for even a constitutional monarchist worsened still further. The next issue of *Observations* contained a note that he was postponing further publication in order to give all his time to the defense of the rightful sovereign. Only a few weeks later, after Louis XVI was imprisoned, Dolomieu was with La Rochefoucauld when the latter was brutally assassinated. He then moved to his patron's rural chateau to protect his widow, foreboding the "terror" that seemed likely to follow. But his retreat from the capital did at least give him time for scientific work, and the first installment of his promised paper appeared early in 1793, shortly before the king was executed.[36]

Dolomieu's lengthy paper on "The physical constitution of Egypt" exemplified a still quite novel genre in the sciences of the earth: he himself called it "a new

33. Dolomieu, "Pierres composées et roches" (1791–92), pt. 1, 398–404. Dolomieu to Picot de La Peyrouse, 31 December 1788, printed in Lacroix, *Déodat de Dolomieu* (1921), 1: 211–14, mentions his positive reaction to William Whiston, Newton's academic successor at Cambridge and a contributor to the debate generated a century earlier by Burnet's *Sacred theory*. In modern terms, Dolomieu's putative events were something like a combination of a large-scale tsunami caused by a submarine earthquake and a massive submarine turbidity current flowing at high speed down a continental slope, dumping its suspended material as a single deposit on the ocean floor.

34. Dolomieu, "Pierres composées et roches" (1791–92), pt. 1, 404.

35. Dolomieu, "Pierres composées et roches" (1791–92), pt. 2, 42–43n; under the impersonal construction [*on vouloit* etc.] his targets remained discreetly anonymous. He expressed the same argument in Dolomieu to Saussure, 26 April 1792 (Genève-BPU, Saussure papers, dossier 8, 332–33), printed in Lacroix, *Déodat de Dolomieu* (1921), 2: 40–43, in which a quotation from Lucretius put it in the usual context of combatting eternalism.

36. Dolomieu, "Pierres composées et roches" (1791–92), pt. 2, 43n; Lacroix, *Déodat de Dolomieu* (1921), 1: xxvi–xxvii; 2: 53–54n. Dolomieu, "Distribution méthodique" (1794) and "Roches composées en général" (1794), in effect the sequels of his earlier paper, must have been sent to Paris around the same time.

method", uniting physical geography with history, the work of the "*naturaliste-géologue*" with that of the "*littérateur*". He extended de Luc's analysis of the Rhine delta to the even larger delta of the Nile; and for historical evidence he used not the merely Roman and later remains in northern Europe but the far richer "monuments" of the oldest civilization then known. In other words, he used the work of antiquarians and Classical scholars to throw light on the physical processes that had formed the Nile delta, to estimate the rates at which those processes had operated, and hence to reconstruct the geohistory of the region. Human history and its antiquarian investigation were linked to the natural world: not just transposed as metaphor and analogy—powerful and suggestive though that had already proved to be (Chap. 4)—but combined substantively, to throw light on the *same* period of history.[37]

Dolomieu had been sent rock specimens from Egypt but had not himself been there; he had to rely largely on accounts by others, but that did not seriously hamper his analysis (long footnotes drew parallels with the Po delta in Italy, of which he did have firsthand knowledge). He set himself two basic questions. Had the rates of the physical processes, such as the annual flooding of the river and the formation of new land at the edge of the delta, changed in the course of time? And could ancient authors, even back to Homer, be used as reliable witnesses to the state of things in their time, or should their writings be discounted as mere "fables" or "poetic fiction"? For the latter question, Dolomieu drew in effect on the newly refined practices of textual criticism (§2.5), which had been pioneered in the German universities for the analysis of the biblical texts and then applied to those of nonbiblical Antiquity. He concluded that, *when properly interpreted*, even Homer's record was compatible with other evidence about the growth of the delta (the parallel with the proper interpretation of biblical texts such as the Flood story would have been obvious to his readers). Ancient sources were reliable, at least for his purposes: although, for example, Homer's primary goals had been purely literary, indeed poetic, there was no good reason to doubt the accuracy of features that formed the backdrop to his story.[38]

Dolomieu illustrated his argument with a fine map of the valley and delta of the Nile, reconstructing their physical and human geography in Classical times. In itself, the map was not novel in form: ancient geography was a flourishing branch of Classical scholarship, and it was often illustrated with maps of this kind, identifying the present location of ancient sites and reconstructing the geography of ancient civilizations. What was novel, however, was the publication of such a map in a periodical devoted to the *natural* sciences. Dolomieu's use of this map in his paper on the geohistory of the Nile was a further sign of a quite new turn in the sciences of the earth, towards a detailed integration of human history with the final phase of geohistory (Fig. 6.4).[39]

Dolomieu claimed that the trenchlike valley of the Nile, passing through limestone country upstream from the delta, had not been formed by the river's own erosion, but by a cracking of the earth's crust at the time of the last great revolution. That same event had also produced the vast sandy deposits, evidently brought from elsewhere, that constituted most of Lower Egypt. Only the part of the delta

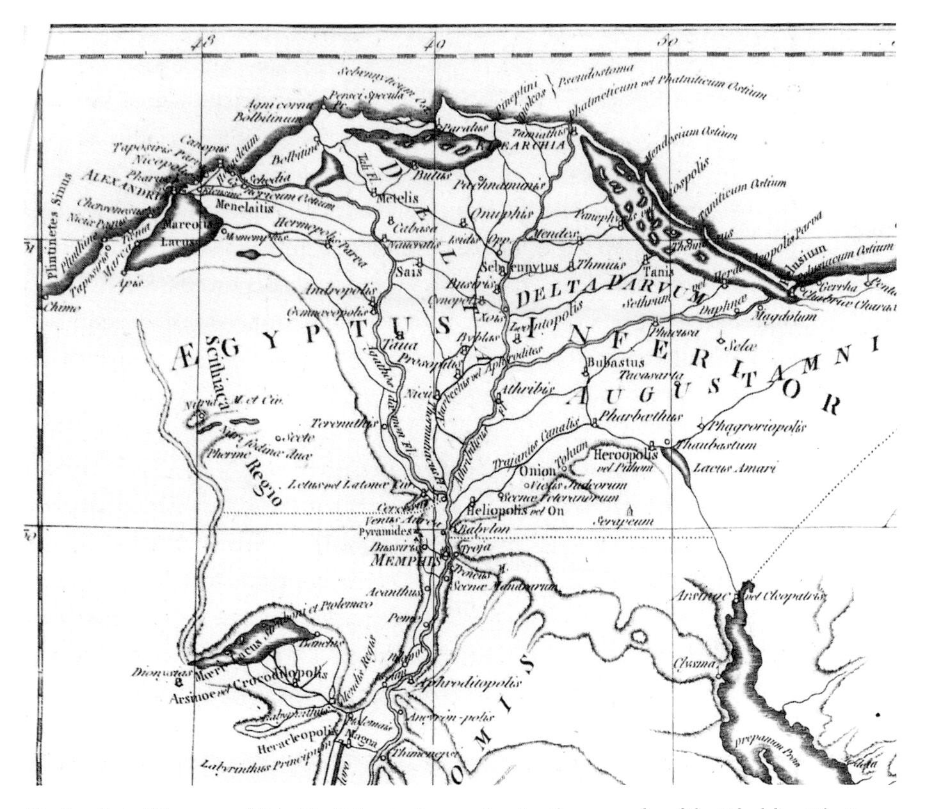

Fig. 6.4. Part of the map published by Dolomieu in 1793, showing the geography of the Nile delta at the time of Alexander the Great, as reconstructed by scholars and antiquarians. He used it to support his inference that the delta had begun to form at a finite time in the past, and hence his claim that the continents (here, north Africa) had not been above sea level for more than a few millennia. (By permission of the Syndics of Cambridge University Library)

composed of silt—but still about 1,000 square *lieues* (roughly 25,000 sq km)—was truly the product of the famous annual flooding of the Nile itself; he estimated that its rate of growth had gradually diminished as the fluvial system settled into a state of relative repose. In any case, for Dolomieu as for de Luc, the finite size of the delta pointed to a finite date in the past at which it had begun to form; the most recent

37. Dolomieu, "Constitution de L'Égypte" (1793), 41. His main antiquarian source was Fréret, "Accroissement de l'Égypte" (1751), though this prominent earlier scholar had *denied* that the land area was growing by deposition from the annual floods: see Grell and Volpilhac-Auger, *Nicolas Fréret* (1994). Like Fréret and other textual scholars, Dolomieu had to use Greek sources such as Herodotus, because the hieroglyphic inscriptions of the ancient Egyptians themselves could not be deciphered.

38. Dolomieu, "Constitution de L'Égypte" (1793), pt. 3. The textual problems concerned the position of the island of Pharos: according to the *Odyssey*, it was a whole day's sailing from Egypt, and yet—only a few centuries later—it was at the mouth of the harbor serving the city founded by Alexander the Great (and the site of the lighthouse that was one of the seven wonders of the ancient world).

39. Fig. 6.4 is reproduced from Dolomieu, "Constitution de L'Égypte" (1793) [Cambridge-UL: T340:2.b.6.45], part of map opp. 60; the southern part (not reproduced here) shows the Nile valley upstream as far as the First Cataract at Philae (now Aswan). The island of "Pharus" is near the city of Alexandria (top left), at the entrance to a vast lagoon (*Mareotis Lacus*); by Dolomieu's time further silting had turned the latter into a much smaller inland lake (now Bahra Maryut). The map was probably based on the work of the scholars in the Académie des Inscriptions in Paris. As already noted, de Luc's similar argument about the Rhine delta was much weakened by the lack of a map of this or any kind.

major revolution, whatever its exact character, could not have been indefinitely distant in deep time. He quoted with approval those textual scholars who had dismissed as mere fables the vastly ancient dates claimed by many early cultures; in sober fact, all could be reduced to the same modest age of a few millennia before the Christian era. The "physical constitution of Egypt" therefore provided powerful empirical evidence against those who claimed that the continents had been undergoing extremely slow erosion through unimaginable spans of time, without any major disruption. It was also a powerful argument against those literary scholars who claimed a similarly vast antiquity for human civilizations.[40]

The sciences under the Terror

By the time Dolomieu's paper on Egypt was fully in print, his premonition of political terror was proving all too accurate. The new Committee of Public Safety quickly became a body with vast and arbitrary powers (its Orwellian title marks it as the prototype of similar bodies elsewhere in later centuries). Once the Jacobin party gained supremacy under Robespierre, France became Europe's first truly totalitarian state. The Law of Suspects authorized the imprisonment of anyone denounced as having less than total loyalty to the Republic. Constitutional legality was abandoned as the new Revolutionary Tribunal prosecuted alleged crimes of opinion among suspected traitors and counterrevolutionaries, eventually without even nominal rights of defense. During the Terror of 1793–94, over 16,000 people were executed in France and about half a million imprisoned, after show trials or no trial at all. Lavoisier was among those guillotined, having been arrested along with other members of the powerful and greatly hated tax-collecting body [*Ferme Général*] of the Old Regime; it was his substantial income from this profitable position that had funded the scientific research he did in his free time. Savants as such were not targeted, but their social status as nobles, priests, or bourgeois made them vulnerable to persecution by a populist regime that was also increasingly lawless. Many savants fled abroad; Montlosier, for example, spent several years in London, where he edited a newspaper for emigrés. Others, such as Dolomieu, disappeared into the depths of the French countryside; or they went underground in other ways or just kept their heads down and hoped for the best.[41]

Nor were the effects of the Revolution confined to savants in France: Saussure in Geneva was ruined financially, for much of his investment income vanished in the collapse of the French economy. Not all savants were on the wrong side in the Terror: Soulavie, newly laicized and married, served as the Jacobins' representative in Geneva, and there were others who profited from the revolutionary situation. But in general its effect on the sciences was catastrophic. The Académie des Sciences itself was abolished, along with the other royal and learned academies; its former function as, in effect, the patents office of the Old Regime was particularly resented by the artisan class that was now in power. On the other hand, the Jardin du Roi escaped a similar fate by proposing a politically correct plan for its own reform. It was renamed the National Museum of Natural History (for which *Muséum* will be used here, the initial capital, accent, and Latin form distinguishing it from any other museum or *musée*). It was to be under the governance of twelve professors of equal status, taking annual turns to be director [*intendant*], the position that Buffon had occupied autocratically for almost half a century. For the sciences of the earth it was particularly important that one of the specified courses, and one of the professorships, were to be in "*géologie*", a science thereby formally recognized for the first time anywhere. The lectures were to instruct travelers on "the general theory of the globe, and more particularly of mountains", in order to help them discover new mineral resources; and the professor himself was required to make at least one field trip every year and to report on it to his colleagues. Faujas, who was already well-known for his studies of extinct volcanoes (§2.4), was appointed to this position; previously he had merely been an assistant to Daubenton, the curator of the royal museum (who now became professor of mineralogy and the first annual director).[42]

Equally important for the sciences of the earth was the survival of the small pre-Revolutionary Mining School [École des Mines] in Paris. In 1790 it had been threatened with closure on the grounds that a primarily agricultural nation could not afford such a luxury in a time of crisis. But it survived, and in 1794 the new Mines Agency [Agence des Mines] was given the practical task—now recognized as acutely important in the rapidly worsening economic situation—of coordinating the exploitation of mineral resources. Having a scientific function that was politically acceptable to the Jacobins, the agency provided effective cover for several savants who might otherwise have attracted attention on account of suspected disloyalty: among them was the aristocratic royalist Dolomieu; Faujas, also from a noble family and the former protégé of the deceased but now deeply suspect Buffon; and the distinguished crystallographer (and priest) René-Just Haüy (1743–1822).[43]

Conclusion

For clergy, as for nobles, Paris was now a dangerous place to be. As the Terror reached its peak, Catholicism was replaced by an official Cult of Reason; the ancient cathedral of Notre-Dame was assigned to this new religion, and Louis XVI's great new church of Sainte Geneviève became the "Panthéon" for the veneration of the Republic's secular saints. A new calendar was introduced, replacing the Christian starting point of the old with a new "Year One" dated retroactively from the foundation of the Republic; its months were named after the natural seasons, and the seven-day week—traditionally regarded as being modeled on the "days" of divine

40. Dolomieu, "Constitution de L'Égypte" (1793), pt. 2, 55–60; pt. 3, 212n.

41. Outram, "Ordeal of vocation" (1983), summarizes the impact of the Terror on those savants who had been members of the Académie des Sciences; she rightly criticizes those modern historians—many with Marxist commitments—who have played down the destructive effects of the Revolution on the sciences. Its atrocities were small in scale compared to those of later totalitarian regimes; but directly or indirectly the likes of Lenin, Stalin, and Mao found the French model inspiring.

42. Hamy, "Fondation du Muséum" (1893), prints its constitution (146–60), with its provision for *géologie* (150, 157–58).

43. Aguillon, "École des Mines" (1889), chap. 3.

creation—was replaced by a ten-day cycle [*décade*]. All these changes were signs of the Jacobins' cultural revolution, a systematic attempt to eradicate the Christian past.

For all savants, religious or not, and indeed for everyone without friends in high places or access to black markets, Paris was now not only a dangerous but also a hungry place, as a result of wartime shortages and hyperinflation. In the atmosphere of the Terror, terrifying not least in being so arbitrary and unpredictable in its impact, scientific activity in the world center of all the sciences came almost to a halt, to the shock and dismay of savants everywhere. At best it continued privately and even furtively, or else was confined to politically innocuous topics and those of immediate practical relevance. As a telling sign of the times, *Observations sur la physique* suspended publication, and La Métherie himself disappeared from view.

What would outlast the atrocities of the Jacobin regime and its Terror was the work done by savants such as Dolomieu, the disillusioned constitutional monarchist, who kept the scholarly ideals of the true republic, the Republic of Letters, alive in some of the darkest days yet seen in the history of the sciences in Europe. Specifically, Dolomieu's conception of a long geohistory punctuated by occasional revolutions in the form of mega-tsunamis, and his integration of its final phase with the few millennia of recorded human history, using the case of the Nile delta as a concrete example, would span the hiatus of the Terror and be available for discussion and evaluation when the community of savants in Paris, and in France as a whole, eventually revived.[44]

6.4 GEOTHEORY POLITICIZED (1793–96)

De Luc and Blumenbach

Whatever the shocking news from France, scientific activity elsewhere in Europe, and indeed beyond it, had not of course come to a halt. Towards the end of 1791, Blumenbach, as one of the leading naturalists in the Hanoverian part of George III's realm, had visited England as the guest of Banks, the president of the Royal Society. While in London, he took the opportunity to go to Windsor to see de Luc; they had first met many years earlier, when the Genevan was visiting Göttingen. This time their lengthy discussions about geotheory and its problems prompted Blumenbach to suggest that de Luc should publish an account of his system more concise than the one that was then unfolding in *Observations sur la Physique*. After his guest returned to Göttingen, de Luc wrote him seven long letters to add to those he had earlier addressed to La Métherie and to Hutton (§6.2). Blumenbach passed the texts to his brother-in-law Johann Heinrich Voigt, the professor of mathematics at Jena and editor of the *Magazine for the Latest in Physics and Natural History*, which had already published his own thoughts on "the natural history of the former world" (§6.1). So de Luc's "Letters to Blumenbach" appeared in this leading periodical over the next three years, in German translation, as he had doubtless hoped and expected. They made his geotheory even more widely known in central Europe (Fig. 6.5).[45]

Fig. 6.5. Jean-André de Luc at the age of about seventy: a print after a portrait by Wilhelmine de Stetten, painted around 1798. (By permission of the Bibliothèque Publique et Universitaire, Geneva)

De Luc started his new series as if indeed resolved to be more concise than before. He began with a definition: "*Geology* is principally distinguished from *Natural History*, which confines itself to the description and classification of the phenomena presented by our globe in the three kingdoms of Nature, insomuch as its office is to connect those phenomena with their causes." By itself, this would just have made geology a synonym for earth physics, as he had implied in his letters to Paris. But he now modified his definition once more by adopting an antiquarian analogy and thereby adding the dimension of geohistory. To ask "Why are there mountains on earth?" was, he wrote, like asking "Why are there pyramids in Egypt?": the geologist, like the antiquarian, had to try to explain what was observable in the present by giving a *historical* explanation that referred to events, processes, or "causes" in the past. Where great piles of strata could be seen, as in some of the mountains that were "nature's pyramids", it was "as easy to read the history of the sea as it is to read the history of man in the archives of any nation". The fossils of different kinds in successive formations, and the total absence of fossils in the lowest and oldest,

44. It is difficult, but not pointless, to imagine a modern parallel to the impact on the sciences of the Jacobin regime: perhaps a rigged election in the United States in which fundamentalists seize power in Washington, abolish the National Science Foundation and other sources of support for the sciences, and cause a large-scale emigration of scientists to Canada and Europe and the self-exile of others to the most remote areas. What in fact happened during the Third Reich, shocking though it was, was not comparable, because Germany in the 1930s was no longer the world center of the sciences, as it arguably had been in the later nineteenth century.

45. Fig. 6.5 is reproduced from a print engraved by Friedrich Schröder (Genève-BPU). De Luc, "Briefe an Blumenbach" (1793–96); on the relations between the two savants, see Dougherty, "Begriff der Naturgeschichte" (1986). The *Magazin für den neueste aus der Physik und Naturgeschichte* had first gained its well-deserved reputation under its earlier editor Georg Christoph Lichtenberg (1744–99), the professor of physics at Göttingen: see Heilbron, "Göttingen around 1800" (2002).

indicated a geohistorical sequence of diverse periods, during which the materials of the future continents had accumulated gradually on the seafloor. Then they had all been disrupted and turned into land areas. The resultant "chaos" of tilted and faulted strata was like the famous ruins at Palmyra in Syria: in both cases monuments dating from an earlier period of history had been broken up but not totally destroyed. As before, de Luc argued that the event that had produced the present continents had been quite recent, and he claimed that this inference "at one blow destroys all the systems of geology in which *slow* causes, acting through an innumerable sequence of ages, were used to explain their formation".[46]

De Luc's letters to Blumenbach can be described here quite briefly, because to a large extent they restated what he had expounded earlier, as was his intention. In fact, after a good start he soon relapsed into his customary verbosity, which he justified as matching the grandeur of his theme: "I do not believe I shall be accused of *longueur*, by those who recognize that I am here tracing—from its *monuments*—the fundamental basis of the ancient history of Men, since it concerns their *habitation*". He rehearsed the many reasons for inferring that the continents were of quite recent origin, citing in his support Dolomieu's newly published paper on Egypt as well as Saussure and other authors. He mentioned the erratic blocks that he himself had seen perched high on the hills of the Jura and strewn across the north German plain—the latter would have been familiar to many of his readers—as features that could not be explained in terms of "slow causes", no matter how much time was allowed. Conversely, he also cited his natural "chronometers", such as the lakes not yet silted up by incoming sediment, as proving that those same processes could not have been going on for an unlimited time. Once again, time itself was not at issue: de Luc readily conceded that before his decisive and geologically recent event there had been time enough and to spare, though it could not be quantified. Throughout, his quarrel was with speculative "philosophers" such as Hutton, who denied the reality of that event altogether and claimed that the continents *as land areas* were of an immense and unbroken antiquity.[47]

In the German periodical de Luc felt free to explain, more explicitly than in the French, how his own geotheory helped "to establish the *certainty* of the *revelation to Moses*" in Genesis. He criticized his opponents for being less than honest about their own determination to dismiss that revelation as valueless by treating it as a "fiction". He accused them of misusing their authority as established savants to mislead less informed readers, who might not perceive the ulterior motives behind what they were being told. For de Luc maintained that these claims to a vast age for the continents were not based on observation, but were a necessity imposed by an underlying theory: the combination of slow causes and unlimited time was invoked only in order to reach a conclusion desired for other reasons. Like Dolomieu, de Luc argued that it was the skeptics, not the believers, whose opinions were distorted by prejudice.[48]

All this, however, was preparatory to de Luc's exposition of the six periods into which, as before, he divided the history of the former world, starting with the addition of light to an inert globe, and ending with the sudden alteration of geography that had finally produced the present world. That he chose six periods was now related more explicitly to the six "days" of the Genesis story, though he insisted again that to equate the latter with ordinary days was as unwarranted in terms of textual exegesis as it was incompatible with the natural evidence. However, although he stressed that his letters were primarily concerned with "*natural history* and *physics*", with description and causation, his sixfold periodization was bound to suggest—in a general and nonliteral way—"the astonishing conformity between our geological *monuments* and all parts of this sublime *narrative*, in the very order in which they are found there". In any case the parallel established the feasibility of constructing a similar *narrative* of geohistory from the natural evidence.[49]

Only in the fifth of his long letters did de Luc's narrative reach the "birth of our continents" and his "proofs of the low antiquity of that *epoch*" or last great event in geohistory. This he now identified explicitly as the Flood recorded by Moses. His battery of natural chronometers was set out once more, but deployed now as evidence about the *history* of the world since that great revolution. Like Dolomieu, de Luc recruited antiquarian evidence, not only as potent metaphor but also once again as substantive material for estimating rates of erosion, deposition, and so on. His aim was to establish that the chronology based on nature was compatible with that derived traditionally from the biblical texts, though only for the period since the Flood, and only to the rough approximation that the natural evidence allowed. His tactic throughout was to show that observation in the field supported his conclusion, against what mere "imagination" concocted in the philosopher's study: "the attention of observers being fixed on this *physical chronology*, it will in the end eliminate all fabulous traditions of an immemorial [human] antiquity, and all the systems associated with it". As a concrete example he recalled the Roman ruins he had been shown many years before at Koblenz, found beneath eight feet of sand deposited by the Rhine during the subsequent centuries:

> Now the place that these *monuments* of the *Romans* occupy in the mass of *transported* materials—a mass that can have *begun* only at the *birth* of our *continents*—transforms these *historical documents* into *geological monuments*; they are examples of the *chronometric scales* that can be found along all rivers, consistent with each other and with all those provided by processes of other kinds, which do not allow the origin of our *continents* to be pushed back to an *epoch* more remote than that of the *Deluge* in the sacred *chronology*.[50]

46. De Luc, "Briefen an Blumenbach" (1793–96), letter 1 (1793), §§1–13; letter 2 (1794), §19. De Luc's numbered sections are given here, rather than pages, to facilitate reference to his original text published later as *Lettres à Blumenbach* (1798) or to the translation (see below) published as "Letters to Blumenbach" (1793–95); quotations are translated from the original. It was a commonplace to compare tilted strata with a ruined building (see Fig. 2.24), but the reference to Palmyra was significant because it was to ruins from a specific (Hellenistic) period.

47. De Luc, "Briefen an Blumenbach" (1793–96), letter 1 (1793), §§19–32; quotation from letter 3 (1794), §51.

48. De Luc, "Briefen an Blumenbach" (1793–96), letter 2 (1794), §§6–13.

49. De Luc, "Briefen an Blumenbach" (1793–96), letter 3 (1794), §1.

50. De Luc, "Briefen an Blumenbach" (1793–96), letter 5 (1796), esp. §55, §56; see also his *Lettres physiques et morales* (1779), 5: 498–99.

De Luc's sixth letter to Blumenbach had no parallel in his earlier series. It offered a detailed textual analysis of the Flood story in Genesis, to complement his previous account of the physical evidence for an extraordinary event. He was exploiting a rich and ancient tradition of biblical scholarship, but it was no old-fashioned exercise. Although he stated that his analysis was based on the "literal" meaning, in fact he contextualized the story in much the same way as Blumenbach's philological colleagues at Göttingen were doing with the Old Testament as a whole. He ingeniously integrated his geological conclusions into the exegesis, using them to clarify details that had traditionally been puzzling or obscure. For example, the olive branch brought back to the Ark, and Noah's ability to resume agriculture (and wine making) without delay, indicated to de Luc that a mild low-altitude environment had survived temporarily on mountains such as Ararat, where the Ark was said to have grounded. For on de Luc's geotheory the present mountain peaks had *not* been barren tracts of rock and ice before the revolution, but low-lying islands, where a temperate flora could have survived that otherwise destructive event. Most of the previous terrestrial fauna could also have survived there, for de Luc interpreted the text as meaning that the Ark itself had only carried a cargo of the species already domesticated. The stated universality of the Flood was likewise relativized into meaning the inhabited world as then known: beyond the antediluvians' horizon—both literally and metaphorically—had been the oceanic islands that were to become the mountain peaks of the new continents. If those remote lands were then uninhabited, the failure to find human fossils on the present continents was unsurprising and certainly no argument against the reliability of the textual record.[51]

Many of de Luc's interpretative moves were far from novel; what was original was the way they were integrated into his conception of the great revolution as a process of interchange between old land areas and new seabeds, old ocean floors and new continents. For de Luc, Moses was "the sacred *historian*", who had left a soberly factual record of an oral tradition that in his time stretched back to Noah himself. The biblical text was indeed "sacred" and of supreme human importance; but what also mattered was that it was ancient *history*, just as much as, say, Herodotus's account of the Greek world and Tacitus's of the Roman. De Luc's interpretation was almost as naturalistic as any Enlightenment philosophe could have wished: the crustal collapse, the subterranean caverns, the temporary violence of the waters, and so on were all accounted for in ways that were impeccably natural; if some of the explanations were also ingeniously ad hoc, they were no more so than in the systems of other geotheorists. Only at two points did de Luc suggest any supranatural element: in Noah's premonition of catastrophe to come, which led him to build his Ark in good time; and in a providence that brought the Ark safely to rest on Ararat rather than sweeping it down some black hole into the earth's interior as the Flood subsided.

De Luc integrated his sophisticated reading of the Flood narrative in Genesis with the origin stories of other ancient cultures, by adopting what was commonly accepted, at least in Britain, for the interpretation of pagan myths. Drawing on

Jacob Bryant's scholarly *Analysis of Ancient Mythology* and on more recent work by members of the Asiatick Society in Bengal, de Luc explained the differences between the biblical records and the more exotic pagan traditions, in anthropological terms of cultural divergence and corruption. In effect, he argued that ancient Jewish society had been conducive to the accurate transmission of historical records, whereas in pagan societies the same initial memories had been corrupted and exaggerated. Specifically, the sober truth of a relatively recent rebirth of all human societies had been corrupted by pagan cultures into myths of an immense antiquity, peopled with demigods and superhuman heroes: it was therefore "not surprising that their chronologies should have become in the end pure *fictions*". So de Luc inverted the claim of the deists and other skeptics, that Moses had merely borrowed from the pagans. On the contrary, he argued, geology now proved that it was the biblical record that had preserved the reliable history, from which the other traditions had diverged: "we now see that he [Moses] spoke only the truth, for our continents—this unalterable store of *chronometers*—confirm his chronology".[52]

De Luc concluded his letters to Blumenbach by recalling that he had devoted almost half a century to his geotheory. Unlike his opponents, he had made no secret of his underlying motives. He hoped his research would make religious belief more acceptable in an irreligious age: "God, by inviting us in his *revelation* to *study* Nature, has prepared in advance for the reestablishment of the faith, [just] when the distance of time [i.e., from the founding events of Christianity], and lapses in human imagination and emotions, might have given rise to unbelief". His own research on the great "revolution" at the dawn of human history would, he hoped, "gradually dissipate the shadows cast on Nature by the *fictions* spread by men who claim to enlighten the human race", and bring humans to recognize that their ultimate happiness lay in listening only to God. For de Luc as for Dolomieu, it was the prophets of a false Enlightenment, not the religious believers trusting in Moses' veracity, who were responsible for unfounded and speculative "fictions". In the end, human self-understanding was at stake, and that required the study not only of human history but also of geohistory:

> What can we say with certainty about the *origin* and the *nature* of Man, without knowing his history? How can we know the *history* of Man without being aware of that of the planet he inhabits? How can we learn the history of this planet, without doing research on the monuments of these *revolutions* and on all that physics can tell us about their causes?[53]

51. De Luc, "Briefen an Blumenbach" (1793–96), letter 6 (1796). His exegesis was not wholly original: on the much earlier tradition, see, for example, the classic work by Allen, *Legend of Noah* (1949), and Browne, "Noah's Flood, the Ark" (2003).

52. De Luc, "Briefen an Blumenbach" (1793–96), letter 6 (1796), esp. §44, §45. Bryant, *Ancient mythology* (1774–76), was intended "to give a new turn to ancient history; and to place it upon a surer foundation", by showing that pagan traditions were "all related to the history of the first ages, & to the same events, which are recorded by Moses" (1: xvii, xiv). The society in Calcutta had begun publishing its *Asiatick researches* in 1788.

53. De Luc, "Briefen an Blumenbach" (1793–96), letter 6 (1796), §51, §52.

Cultured despisers of religion

This was a clear statement of the theistic commitments that underlay de Luc's geotheory. It also explains what might otherwise seem puzzling: his almost obsessive focus on the most recent of the earth's many revolutions, and on its correct dating, rather than on the richly diverse events and indefinitely longer periods of geohistory that had preceded it. For "the place of Man in Nature"—to anticipate a much later phrase—was always at the top of de Luc's agenda. His geotheory was always a means to an end; the goal of geology was to understand how human beings were related to the earth, within a cosmology that related everything to God. However, far from that being the dominant view in his society, as later historical mythmaking has portrayed it, de Luc knew himself to be in this respect in a small minority, at least among the international community of savants. Nor was he alone in this assessment. The skeptical intellectuals whom he was criticizing in his seemingly interminable published letters were those for whom, a little later, the young German theologian Friedrich Schleiermacher was to write his famous *Speeches [on religion] to Its Cultured Despisers*, hoping to persuade them that religion was not a sop for the uncultured or a consolation for the aged, but the noblest ingredient in the intellectual life of mankind.[54]

Where "Science" and "Religion" were in potential conflict at this period, it was not always or everywhere that science was the underdog. Yet although de Luc's strong theistic beliefs put him in a minority among savants, it is important to emphasize once more that he was not a marginal figure in the Republic of Letters, but at its center. His work was published in two of the leading scientific periodicals of the time, La Métherie's *Observations* and Voigt's *Magazin,* and abstracted or criticized in many others in all the main languages of Europe. Nonetheless, there are unmistakable signs that the all-embracing character of his work, like that of other geotheorists, was passing out of fashion. The format of the Parisian periodical had already forced him to confine his letters to strictly "physical" topics (§6.2). And even Blumenbach—who was sympathetic to his underlying goals—suggested privately that the publication of his letters in the German periodical should be curtailed at a point that would have completely eliminated de Luc's attempt to integrate his geology with his biblical exegesis: "I dare to suggest this", wrote Blumenbach, "because I think it would suffice to give German naturalists a succinct outline of your strictly geological system, and that by contrast the rest—although extremely interesting—would not be desirable [*desideré*] in a journal of physics and natural history".[55]

In the event, however, Blumenbach must have changed his mind, or been overruled by Voigt, for the letters were published in their entirety. Nonetheless, Blumenbach's proposal indicated unease about de Luc's strategy, and he made his own position clear when he sent de Luc the latest edition of his book on the physical aspects of the racial diversity of the human species. He explained that publicly he had treated the topic "simply as an anatomist and naturalist", although privately he considered that his work confirmed the biblical story of the Adamic origin and consequent unity of all the races. In effect, a discreetly Nicodemite strategy was adopted:

not by a putative heretic in an illiberally orthodox society (as in Newton's case a century earlier), but by an orthodox believer within the illiberally skeptical society of the cultured despisers of religion. As Blumenbach put it to de Luc:

> I have refrained from mentioning this conformity explicitly and have not once cited Moses, because I know too well the unfortunate prejudices of that part of our public to which I hope to be of most use with these researches. These people would have believed me committed [*préoccupé*, i.e., to religion] if they had seen passages of revelation cited, and they would probably have dismissed my book without reading it and being instructed by it.[56]

The politics of Genesis

In Britain, by contrast, de Luc's letters to Blumenbach were given an extra religious gloss when they were translated for a second time—from German into English—and published in the *British Critic*, a new Tory periodical that was designed to counter the growing political influence of radical Revolutionary thinking in Britain. De Luc's "geology", the editor claimed, provided "demonstrative evidence against those who delight to calculate a false antiquity to the world, inconsistent with the sacred records". Readers who failed to wade through de Luc's own lengthy texts might well have assumed that for him the crucial word "world" referred to the whole universe; they might have concluded that de Luc followed Ussher, the tame savant of a much earlier British sovereign, in believing that the cosmos had been created in seven days flat in 4004 B.C. (§2.5). In fact, as emphasized repeatedly here, what was in question for de Luc was specifically the antiquity of the "present world" of human societies, which had followed the last major physical event in an unimaginably lengthy "former world" of geohistory. So in effect the *British Critic* recruited de Luc for the editor's own political agenda, giving his ideas a much more traditional gloss than the savant could have approved. However, this version of de Luc's letters, added to his earlier ones addressed to Hutton, did make his work still more familiar to anglophone readers.[57]

54. Schleiermacher, *Über die Religion* (1799); the final clause is quoted from the summary of Schleiermacher's work in Vidler's classic *Church in an age of revolution* (1961), 22.

55. Blumenbach to de Luc, 2 November 1795, printed in Dougherty, "Begriff der Naturgeschichte" (1986), 98; the suggestion was to stop at de Luc's own major break at letter 5 (1796), §10. There was in fact a hiatus of a whole year in the publication, after letter 4, probably while the matter was being negotiated. The published series also included letter 7 (1796), dealing with the problem of the generation—spontaneous or otherwise—of organisms; but de Luc explicitly treated this topic as strictly extraneous to his geology, although of course it was of great importance for what would later be called biology.

56. Blumenbach to de Luc, 3 June 1795, quoted in Dougherty, "Begriff der Naturgeschichte" (1986), 100; the reference was to Blumenbach, *De generis humani varietate nativa* (3rd ed., 1795), which was dedicated to Banks in London. Snobelen, "Isaac Newton, heretic" (1999), interprets the earlier savant as a "Nicodemite", after the Pharisee who sought a private interview with Jesus under cover of darkness. Blumenbach's scientific support for the "monogenist" position on race was of great political significance at a time when there were active moves, particularly in Britain, to outlaw the slave trade.

57. De Luc, "Letters to Blumenbach" (1793–95), editorial preface to letter 1 (1793), 231; letters 1–6 (but not letter 7) appeared after each was published in German. They were probably published in English without his permission—which was not difficult in an age without copyright conventions—for de Luc evidently did not release his original French text to act as the basis for the second translation.

That the authority of Moses as a historian was a political issue had by this time become starkly apparent from the notorious example of Thomas Paine (1737–1809). The English radical had long been known for his active support for the young American republic. His overtly republican *Rights of Man* (1790–92) had then made him a hero to those who sympathized with the Revolution in France and a dangerous subversive for those who opposed all it stood for. To escape prosecution for sedition he had fled to France, where he was made an honorary citizen. He later fell afoul of the regime, but he used his time in prison to write *The Age of Reason* (1794–95), which was published when he was released after the fall of the Jacobins. This work rejected the whole concept of revelation and thereby challenged not only the authority of the Bible but the intellectual foundations of the established political order in Britain. The interpretation of Genesis could no longer be regarded as just a matter of scholarly argument, in the light of comments such as Paine's:

> Take away from Genesis the belief that Moses was the author, on which only the strange belief that it is the word of God has stood, and there remains nothing of Genesis, but an anonymous book of stories, fables and traditionary or invented absurdities, or of downright lies. The story of Eve and the serpent, and of Noah and his ark, drops to a level with the Arabian Tales, without [even] the merit of being entertaining.[58]

In effect, Paine and other radicals politicized not only Genesis but also geotheory. In his letters to Blumenbach, de Luc had not identified those who were claiming a vast antiquity for the present continents and thereby impugning the authority of Moses as "the sacred historian"; but one of his chief targets, as he had made clear in his earlier letters in English, was a system being actively discussed in his adopted country, namely Hutton's. As already emphasized, his objection to Hutton's treatment of time was not to its magnitude but to its use as a principle of causal explanation (§6.2). But for de Luc it was even more important that behind Hutton's implicit dismissal of the historicity of the Flood there lay a radically ahistorical vision of the earth's revolutions; and that behind that in turn lay an eternalism that denied that the cosmos had an ultimately divine foundation (§3.4).

However, much of the criticism that Hutton's "Theory of the earth" received from other savants in the first years after its publication in 1788 had been directed not at its underlying metaphysics or theology but at its more implausible constituent physical ideas: of an inexhaustible store of intense heat in the earth's interior, of that heat as the immediate cause of the "stoniness" of rocks and the upheaval of mountains, and so on (§3.4). For example, early in 1793 the chemist and mineralogist Richard Kirwan (1733–1812) read a paper to the Royal Irish Academy in Dublin, on "the supposed igneous origin of stony substances", in which he criticized Hutton's views on that specific problem, forcefully but in a courteous tone. Only briefly did he refer at all to the broader implications of Hutton's geotheory. He suggested that it was the admittedly difficult problem of the chemistry of consolidation that had led Hutton into his implausible "igneous" explanation, and hence into his even more dubious system of "succession without a beginning". But was it really necessary, Kirwan asked, to "admit a process *ad infinitum*; an abyss from which human reason recoils?" He noted that "into this gulph our author however plunges"; and quoting Hutton's famous final sentence (Fig. 3.11), Kirwan concluded that "then this system of successive worlds must have been eternal". Like Hutton's other critics, Kirwan was in no doubt about the eternalism that underlay the Scotsman's system. It was Kirwan's paper that finally spurred Hutton into publishing his *Theory of the Earth* (1795), the long-delayed book containing the extensive "proofs and illustrations" of his system (§3.4); among its miscellaneous collection of supporting essays was his counterattack on the Irishman's chemistry; but Hutton's intemperate tone suggested that more than chemistry was now at stake.[59]

Kirwan's paper on stony substances had been read in Dublin only a few days after Louis XVI was executed in Paris and before the implications of that shocking news had been fully absorbed. Three years later, when Kirwan returned to geotheory with a new paper "On the primitive state of the globe and its subsequent catastrophe", the political dimension was, unsurprisingly, much more apparent: the Terror was still a recent memory—and a fearful one even outside France—and the Revolutionary wars were continuing unabated. Echoing de Luc, Kirwan asserted that "geology naturally ripens, or (to use a mineralogical expression) *graduates* into religion, as this does into morality". He argued that the earlier obscurity of geology had favored "various systems of atheism and immorality". This was a variation on a theme that had become common among critics of the Revolution: it was said that the political and social chaos had been caused ultimately by the philosophes' repudiation of traditional values and beliefs. So Kirwan ventured out of his usual fields of chemistry and mineralogy—and arguably also out of his depth—to consider the authority of Moses and the historicity of Genesis.

Whereas Kirwan's first paper had been primarily about earth physics, his second necessarily had a dimension of geohistory. Right at the start he set out the proper method for a *historical* science of nature, which would need to invoke appropriate causes to explain *past* events. These would be causes the reality of which was established by "actual observation", and which were known to be adequate to account for what was observed; in other words, they would be *present* processes, or what de Luc had named "actual causes". However, in addition to the physical traces of such processes, acting also in the past, Kirwan argued that it was absurd to reject human "testimony" to unique historical events: as absurd, in fact, as to reconstruct Roman history from the evidence of coins alone while ignoring textual sources such as Livy. Here the increasingly pervasive antiquarian analogy was neatly inverted: geohistory should take into account not only physical evidence ("nature's coins") but

58. Paine, *Age of reason*, pt. 2 (1795): 14. His deistic theology was far from original; it was its appearance in a brief and readable form, and in English, that made it alarming for British defenders of "revelation" (an early example of the *Lady Chatterley* argument, applied here to religion rather than sex).

59. Kirwan, "Stony substances" (1794), 63–64 (read on 3 February 1793); Hutton, *Theory of the earth* (1795), 1: 201–68; Laudan, "Problem of consolidation" (1977–78); Dean, *James Hutton* (1992), 61–62. It is beyond the scope of this book to offer even a summary of the debate between Hutton and critics such as Kirwan; it mainly concerned their respective ideas on matters of earth physics, not geohistory. Since Kirwan had the temerity to criticize the idol of many modern geologists, he has often been dismissed (though not by Dean or Laudan) as an ignorant backwoodsman; in fact he was a leading member of the small but lively community of savants in Ireland, and had quite a high reputation as a chemist.

also textual evidence—not least the early part of Genesis—provided that the two sources proved to be compatible.[60]

Most of Kirwan's paper was therefore a comparison between geology and Genesis, the natural and the textual evidence for geohistory. This was derivative and unoriginal and need not be reviewed in detail here. He claimed that the two sources were closely compatible; he even estimated—drawing on recent work on probability theory—that the chance of their agreement by coincidence was only one in ten million. In his conclusion he stressed once more the importance of his topic, because he had noticed "how fatal the suspicion of the high antiquity of the globe has been to the credit of the Mosaic history, and consequently to religion and morality"; but he claimed that he had now exposed the weakness of that skeptical position.[61]

Kirwan's conclusion shows that he, like the editor of the *British Critic*, was taking the correlation with Genesis much further than de Luc or any of the other savants whose work has been reviewed here. The antiquity that he saw as the crucial issue was not just that of the "present world" or the "present order of things", but that of "the globe". Lacking experience of the field evidence to the contrary—he was famously averse to outdoor life in any form—Kirwan assumed that a short timescale for the whole of geohistory was still tenable, as it had been a century earlier (§2.5). Stated in this case with all the authority of a reputable savant, this view was to confuse the debate on geohistory for many years to come. In Britain and Ireland it fostered among the educated public a "*scriptural geology*" that later anglophone savants—even if they were religious—would find it hard to combat. Yet, as the reviewer of one English work of this kind noted at the time of Kirwan's paper, the many otherwise incompatible geotheories had one thing in common: "they all suppose the world much older than the books of Moses represent". Unlike Kirwan, savants who were experienced *in the field*—now acknowledged as the primary site for the sciences of the earth—were all taking the vast "antiquity of the globe" for granted.[62]

Conclusion

The genre of geotheory continued to flourish in these years, as is shown by de Luc's letters to Blumenbach and by Hutton's arguments with both de Luc and Kirwan. Leaving Kirwan aside, Hutton and de Luc were well matched: both were regarded as intellectual heavyweights, and both offered comprehensive systems that set the earth within a context of human meaning, grounded in their respective theologies. Their disagreement was profound and beyond resolution, because it spanned every imaginable level from the empirical to the metaphysical: the nature and role of "actual causes"; the explanatory role of time itself (but not its vast magnitude, on which they were agreed); the nature of the earth's revolutions and the relation of such major changes to recorded human history; and ultimately the character of divine agency in relation to the cosmos, the earth, living things, and—above all—human lives. In the present context, however, the most striking contrast between the rival systems was in their relation to history. By equating the most recent revolution with the biblical Flood, de Luc tied geohistory into human history, guaranteeing, as it were, the *historical* character of the earth. In contrast, Hutton's profoundly ahistorical system was correctly seen as embodying a tacit eternalism (§3.4), and it left no room for any specific and unrepeated historical events such as the Flood, or for any unique or distinctive periods in the still deeper past.

Yet both de Luc and Hutton were beginning to look antiquated on two counts: not only because they tried to relate their systems to the most fundamental issues about human existence, but also because they claimed to account in principle for *everything* of importance about the physical system of the earth. Furthermore, their dispute revealed how such grandiose ambitions led geotheorists necessarily—though often not unwillingly—into involvement in the religious, social, and political conflicts of the age. In these circumstances an alternative began to look more attractive to many other savants, among whom Blumenbach has been mentioned here. Splitting the global issues into more closely defined problems, and tackling them piecemeal, seemed to offer a better chance of resolving them; and the same strategy suggested how the sciences of the earth might be detached from their contentious political and religious context and be pursued in peace. The genre of geotheory remained popular and secure in Hutton's and Kirwan's native lands, and in de Luc's adopted one; but in France—as the sciences began to revive after the trauma of the Terror—and elsewhere on the Continent, the value of geotheory was increasingly questioned and its future put in doubt.

6.5 "GEOLOGY" REDEFINED (1794–97)

The sciences after Thermidor

Over in France, the Terror had not lasted. In July 1794 Robespierre overreached himself politically and was overthrown in the coup d'état that came to be known by the Revolutionary month in which it took place: *Thermidor*, the hottest time of the year. After Thermidor, the apparatus of the Terror was dismantled or allowed to lapse, though the new regime, formalized under a five-man Directorate [*Directoire*], proved highly unstable and there were further less dramatic coups d'état. And the conditions of life hardly improved: food shortages and communal violence continued inside the country, and outside it the Revolutionary wars. For the sciences, however, Thermidor marked a decisive change for the better. The new regime returned to many of the Enlightenment ideals of the early, politically moderate phase of the Revolution, and promoted or at least permitted the revival of scientific life in what had been the very center of the scientific world.

60. Kirwan, "Primitive state" (1797), 233–36; read to the Royal Irish Academy on 19 November 1796.

61. Kirwan, "Primitive state" (1797), 269, 307–8. He reprinted this and his earlier paper, together with other material, in his *Geological essays* (1799), which, with an edition in German, made his ideas widely known.

62. Anonymous review of Howard, *Scriptural history* (1797), in *Analytical review*, 1797, 2: 238–47.

For example, the relaxation of the Jacobins' press censorship led to the proliferation of new publications of all kinds. Among them were scholarly and scientific periodicals. Taking inspiration from the great *Encyclopédie*, a substantial new monthly *Magasin Encyclopédique* covered the whole range of high literate culture, from mathematics to poetry, with a profusion of reviews (or at least notices) of books from all over Europe and short original articles; this offered savants an attractively swift way to make their ideas and discoveries widely known. Even more important for the natural sciences was that La Métherie's *Observations* reappeared, newly entitled *Journal de Physique* but unchanged in format. Many other scientific periodicals sprang up in Paris in the years that followed. Despite wartime conditions they reached readers well beyond the frontiers of France; and their contents were often reported, summarized, or even translated in full in periodicals in other languages.

For the sciences of the earth another important sign of revival came from the Mines Agency set up under the Jacobins. Soon after Thermidor it began to publish a monthly *Journal des Mines*. The very first article reported on a rich deposit of iron ore that Faujas had recently discovered in the new *département* of Ardèche; the committee responsible for mineral resources noted with prudent political correctness that "Nature seems, by a new kindness, to smile on the French revolution". The new periodical focused of course on practical mining matters such as this, and there were translated excerpts from relevant German and English works, among them those of Werner. But there were also articles on less directly utilitarian topics. For example, one early issue printed a letter that Dolomieu had sent back to Paris from his fieldwork for the agency, describing the layer of stalagmitic material that he had seen covering the floors of some caves. He was amazed that he had previously failed to realize what this implied: it indicated once again "the low antiquity [of] the present state of our continents"; for given what was known about the steady accumulation of this material, the present world could not be "of an unlimited or extremely remote antiquity, as some *géologues* have claimed".[63]

In sharp contrast to the populist ideology of the Jacobins, the leaders of the new regime promoted the reorganization of French higher education on technical and meritocratic lines. A powerful new École Polytechnique, with many leading savants among its teachers, was to provide a common foundation of scientific training for an array of more advanced and specialized technical schools. Among the latter was the revived School of Mines, which had already offered its first courses and enrolled its first students before the end of 1794. Still more important for scientific life in France was the foundation, late in 1795, of a new Institut National, which was designed to replace and revive all the learned academies that had been abolished by the Jacobins. All branches of knowledge were to be included under the umbrella of a single body; the inspiration came again from the polymathic ideals of the *Encyclopédie*. The Institut was divided into three "Classes". Significantly, the "First Class" gave pride of place to the "mathematical and physical [i.e., natural] sciences"; the Second and Third Classes covered the social sciences and the humanities respectively. In effect, the First Class of the Institut revived the Académie Royale des Sciences under another name, shorn of its royal title; many of those assigned places in it were former members of the Académie, some of them now returned from exile or emerged from their rural retreats.

The First Class was divided in turn into ten sections for different kinds of natural science; among those appointed to the section for natural history and mineralogy, for example, were Desmarest, Dolomieu, and Haüy. Lamarck was among those in the section for botany and plant physiology, Daubenton in the section for anatomy and zoology. The titles of those sections indicate the changing character of the tacit map of knowledge that the plan for the Institut embodied. For example, mineralogy was now distinguished from the rest of natural history, though still placed in the same section; and the causal science of plant physiology was still distinguished from the classificatory science of botany, though now directly associated with it (compare with Fig. 1.16). Of the savants just mentioned by name, Lamarck and Daubenton were also professors at the Muséum, which had survived the fall of the Jacobins without major crisis. Together, the Muséum and the First Class of the Institut soon became the main locus for the revival of the natural-history sciences in Paris, and therefore in the scientific world as a whole.

Desmarest's survey of geotheories

Among the scholarly and scientific projects that had been suspended during the Jacobin era was the *Encyclopédie Méthodique*. This was designed to expand and update the famous midcentury *Encyclopédie*, but with volumes classified on topical lines rather than alphabetically. Back in 1781, Desmarest had been commissioned to write on physical geography. In view of his well-known antipathy to speculation, it was a surprise that his first massive volume, the first part of which was published soon after Thermidor, consisted entirely of a review of other savants' geotheories. He gave detailed accounts, with his own critical comments, of no fewer than forty earlier systems, ranging in date from Classical Antiquity through the time of Woodward and Leibniz into his own century. In such a notoriously contentious field, he prudently concentrated on deceased authors. He recorded in his preface that he had not originally intended to deal at all with "theory of the earth", on the grounds that the genre was related to true physical geography in much the same way that mere fables were related to real history. Yet he had changed his mind, though he intended to confine himself to what could be salvaged from earlier systems for truly scientific use: "I have placed in these articles no notice that is not capable of being made instructive, either to demonstrate a truth, or to point out an error, or to open up a heuristic line of enquiry [*une route féconde en découvertes*], or to deflect false views and illusory trends."[64]

63. *Journal des mines* 1 (1794–95), 17; Ardèche was the part of the old province of Languedoc in which both Faujas and Soulavie had earlier studied extinct volcanoes (§2.4, §4.4). Dolomieu, "Passage d'une lettre" (1795); he had evidently not noticed de Luc's similar use of stalagmite as one of nature's "chronometers" (§6.2).

64. Desmarest, *Géographie physique* 1 (1794–95): 1–2; the supplement containing articles on Lavoisier and Hutton was probably not published until 1798, after both had died (I am indebted to Ken Taylor for this information). On the *Méthodique*, see Darnton, *Business of Enlightenment* (1979), chap. 8, and the list of contributors drawn up by the editor Panckoucke in 1789 (603); on Desmarest's original commission, see Taylor, *Nicolas Desmarest* (1968), 287.

Desmarest was no naive Baconian, as a later generation—in a travesty of the ideas of Francis Bacon—would often refer to those adopting a crudely inductive strategy for scientific work. He did not think that the true "theory of the earth" would emerge automatically once enough observations had been made and enough "facts" reliably recorded. On the contrary, he had a profound sense—derived directly or indirectly from the real Bacon—of how facts themselves needed to be established through rigorous observation and their meaning found in their relation to others, so that successively higher-level "constant principles" could be extracted, leading to successively broader syntheses. Only by that laborious route, Desmarest argued, could the goal of a more complete explanation of the earth be reached. The way of geotheory, by contrast, was falsely alluring because it promised a shortcut by restricting the savant's view to those facts that supported the particular system being propounded while ignoring the rest. In his final summary, he noted that "one ought to talk at this point about *géologie* as a new science", alluding to the current vogue for that term, as shown for example by Faujas's new professorship at the Muséum and La Métherie's annual reviews of the sciences. But Desmarest claimed not to know how the principles of this putative science differed from those of physical geography. He implied that "geology" was either a pretentious and redundant synonym for his own favorite science or else a label for an inherently inconclusive kind of overambitious theorizing.[65]

Desmarest's volume makes it clear that he felt that although many of the geotheories he reviewed contained valuable ideas and information, none could be accepted as an adequate explanation of how the earth works. After wading through, or at least dipping into, over 800 pages of dense print, his readers might well have agreed that the genre of geotheory was chronically inconclusive. In Desmarest's view the science of physical geography was the only—and laborious—route by which an adequate "theory of the earth" would ever be reached. The articles in his subsequent volumes would summarize what was known on specific topics and what needed to be investigated more closely; but evidently he regarded a satisfactory geotheory as a very distant goal.[66]

La Métherie's geotheory

However, Desmarest's reservations about geotheory did not inhibit other savants from continuing to add to the array of rival systems on offer. In the same year that Hutton published his two volumes of *Theory of the Earth* in Edinburgh, La Métherie published three in Paris with the same standard title in French; unlike Hutton's work the Frenchman's sold out so quickly that a second edition enlarged to five volumes appeared only two years later (Fig. 6.6).[67]

La Métherie's massive work offered a systematic mineralogy, "being the basis for a theory of the earth"; it was followed by a review of "the general phenomena of physics and cosmogony"; and finally, on those double foundations, "the explanation of geological phenomena". The strategy was "to develop the mechanism of the particular [i.e., local] and general formation of the different mineral substances, in order to rise then to that of the globe itself". Nothing could show more clearly how

Fig. 6.6. An engraved portrait of Jean-Claude de La Métherie, dated Year III (1794–95), which decorated his *Theory of the Earth.* He was shown with nothing but books in the background; but this was appropriate, because his massive geotheory was based mainly on his wide reading rather than any fieldwork. The larger books were probably volumes of the *Journal de Physique,* successor to the famous *Observations sur la Physique,* of which he had long been the editor. (By permission of the Syndics of Cambridge University Library)

the goal of geotheory, for La Métherie as for most other theorists, remained that of earth physics writ large: to explain the earth's development in terms of physical laws, rather than to reconstruct its history.[68]

Like all serious geotheorists, La Métherie took the vast timescale of the earth for granted; "countless ages [*siècles*] would be required", for example, for shellfish to produce the thick beds of Secondary limestone. Noting that human observation of terrestrial processes covered only two or three millennia, he asked rhetorically, "what is that duration, relative to such great phenomena?" But he attempted no geohistorical reconstruction of any kind, beyond the standard idea of an initially global proto-ocean, in which crystallization—for him the fundamental physical process—had taken place. Noting Lamanon's and Burtin's reports of human artifacts in Secondary rocks (§5.4), he was even skeptical about the general assumption that humans were newcomers in the history of life. He was highly critical of those such as de Luc who argued for a major catastrophe in the relatively recent past; and he rejected claims for a mass extinction at that time, doubting whether any fossil species were truly "lost". When he reviewed the ancient human records of major inundations, he pointedly omitted any mention of Noah's Flood; and when he

65. Desmarest, *Géographie physique* 1 (1794–95): "Considérations générales et particulières", 792–808, and summary, 842: all the sciences of the earth were tacitly subsumed under "physical geography".

66. In the event, even the first of the subsequent volumes did not appear until 1803, and dealt only with topics beginning with the letter "A" (§8.3); with ever briefer coverage, a third such volume (*Géographie physique* 4) reached "N" in 1811; but the work was still unfinished when Desmarest died in 1815 at the ripe age of ninety (it was completed by others in 1828, rather cursorily, but by then the whole project was seriously out of date).

67. Fig. 6.6 is reproduced from La Métherie, *Théorie de la terre,* 2nd ed. (1797), 1 [Cambridge-UL: L.46.24], frontispiece; the artist was Claude Jacques Notté.

68. La Métherie, *Théorie de la terre,* 2nd ed. (1797), 1: xii, xvii–xviii.

summarized the systems proposed in Antiquity he dismissed the biblical one with the briefest item in his whole survey. It read in full, "The system of the Hebrews, as reported in Genesis, is the same as that of the Egyptians and Chaldeans; Moses had derived these ideas from the priests of Egypt."[69]

The historical significance of La Métherie's *Theory of the Earth* lies in its very typicality. Apart from some indoor work on minerals and their chemistry, it was based entirely on textual sources; there was no sign of any fieldwork of his own. He reviewed more than twenty systems from his own century (in addition to others going back to Antiquity), adding his own critical comments to each; and his survey was right up to date, including living savants such as de Luc, Hutton, and Saussure. But the overall impression, as from Desmarest's survey, was bound to be that geotheory was a hopelessly inconclusive project. In fact, La Métherie conceded as much, concluding rather lamely that many more facts needed to be collected, errors corrected, mineral analyses improved, and so on. Yet his goal remained supremely ambitious: to "embrace the entire system of the universe".[70]

Saussure's geotheory and Agenda

If La Métherie represented the continuation of the long tradition of speculative geotheory and Desmarest the growing current of skepticism about the genre, Saussure's distinctive trajectory lay somewhere between the two. At just this time he was struggling to formulate the geotheory that he had been working towards throughout his career, on the basis of his almost unrivaled range of fieldwork.

Saussure had started out with as much ambition as any other prominent savant, to discover the one true "theory of the earth". He had focused his attention on the Alps because he believed that mountain regions would yield decisive evidence: Pallas's analogous system based on the Urals (§3.5) later became the one that he studied more thoroughly than any other. While he was touring Italy back in the 1770s he had speculated freely about the origin of the earth and even of the cosmos. But around the same time he had also begun compiling a series of notes entitled "Agenda", which listed the specific points on which he needed to make further observations. This was a strongly purposive search for the kinds of evidence that might help discriminate between the various geotheories of which he was already aware. He had even drafted an article "On mountains", marking the manuscript "Outline of results on the Theory of the Earth"; he began by noting that "for many years I have made the Theory of the Earth my principal study" and stated that he was planning a large illustrated work—the future *Alpine Travels*—containing "observations" that would form the basis for his system: as with Hutton, the theoretical model would have been expounded separately from the detailed evidence (§3.1, §3.4).[71]

In the first volume of his *Alpine Travels* (1779) Saussure had noted that physical geography was the only reliable basis for geotheory (§3.1); he had also described, for example, how he always made notes on the spot on the specimens he collected, choosing "above all those that had offered me some fact important for the Theory". Comments scattered through his *Travels* show that the kind of system that he found most plausible was, in broad outline, the familiar standard model (§3.5). But the complex structures that he observed in the Alps—above all the huge S-shaped fold (Fig. 2.25) that he saw every time he traveled between Geneva and Chamonix—had led him to adopt a far more dynamic concept of the earth than in most other variants of the standard model (which is why Hutton later found the *Travels* such a rich quarry of proxy evidence for his own system). Yet over the years Saussure had become more and more puzzled by what he saw in the course of his extensive fieldwork, and the prospect of finding a satisfactory geotheory had constantly receded. In 1780, however, having read de Luc's first set of letters (§3.3), he had sketched an outline of his own rival system; in 1786 he had drafted a brief table of contents; and in the same year the second volume of his *Travels* had promised that the final one would contain his "Theory".[72]

However, in 1794 a serious stroke brought Saussure's fieldwork career to an end and must have made him aware that he might have little time left. In 1796 he wrote to Pierre-Simon Laplace (1749–1827), the most prominent mathematical astronomer in Paris, whose celebrated *System of the World* had just outlined for the literate public his hypothesis for the natural origin of the solar system and its planets. Saussure explained how he himself now wanted to complement that work by describing the subsequent history of the earth. In August 1796 he revised his earlier table of contents, listed thirty-six projected chapters, and began a draft text entitled "Theory of the earth". But he wrote less than four chapters: the manuscript ends

Fig. 6.7. Saussure at the age of fifty-six, after a stroke had ended his fieldwork career. This was sketched by Jean-Pierre Saint-Ours in 1796 in preparation for his painting, which portrayed the naturalist more flatteringly in his prime (Fig. 1.7). The same year a second stroke cut short Saussure's belated attempt to formulate his own "Theory of the earth"; only his "agenda" for further research in that direction was ever published.

69. La Métherie, *Théorie de la terre*, 2nd ed. (1797), 5: 142–43, 305–31, 370, 386, 434.

70. La Métherie, *Théorie de la terre*, 2nd ed. (1797), 5: 404–523, 533–35.

71. Saussure, "Idées cosmogéniques et géogéniques" (Genève-BPU, Saussure MSS, dossier 28), undated but at the end of notes on Naples and Pompeii, and after his ascent of Etna in 1773; "Agenda" (dossier 81, cahier 4), with title page dated 1774 but evidently added to over subsequent years; "Projet d'un opuscule sur les montagnes" (dossier 59), dated 1776 in BPU catalogue. See also Carozzi, "Géologie" (1988).

72. For his brief published inferences about the temporal development of the earth in the Alpine region, see for example Saussure, *Voyages* 2 (1786): 339–40.

abruptly and poignantly in the middle of a page dated 21 December, the day on which he suffered a second and paralytic stroke, from which he never recovered; he died in 1799. Saussure had clearly intended from the start to offer his own geotheory, but he had left it too late (Fig. 6.7).[73]

Saussure's draft of his geotheory shows that he—like de Luc, La Métherie, and many others—planned to begin with the "general principles" of natural philosophy, basic issues of the physics and chemistry of matter. Next he would summarize the main phenomena of physical geography, in a broad sense that embraced the world of living organisms. Then would come "the Theory" itself; but even here the treatment was to be primarily topical, with the theory of the formation of mountains in the most prominent position. Only after such essays in earth physics would there have been a grand overall synthesis. Only in the final chapters did Saussure—alluding to Buffon's epochs—plan to deal with the temporal reconstruction of the earth's past, and with the "changes to be expected in the future". This focus on the causal explanation of all the major features of the earth, changing and developing through times past, present, *and future*, indicates that Saussure's system would have belonged unmistakably in the mainstream of geotheory. Had it been published in time, Desmarest might well have added it to the forty other systems he reviewed: Saussure's towering reputation would surely have qualified him to be, like Pallas, an exception to Desmarest's self-imposed rule to restrict himself to the safely departed.

What Saussure did publish in time was a lengthy "Agenda" for the sciences of the earth. He sent this to Paris for the *Journal des Mines*; in the same year it was also published where he had earlier promised that his full "Theory" would appear, at the end of the final volume of his *Alpine Travels*. These two printings ensured that his agenda gained a wide readership throughout Europe and beyond it, as he evidently intended. Even if his own system remained unpublished, at least he would leave a scientific legacy from which others could benefit.[74]

In the present context three parts of Saussure's agenda are of particular significance. He must have been thinking increasingly of geotheory as a *historical* project (as indeed he described it to Laplace), for at a late stage of drafting the text he inserted a major section on "Monuments historiques", immediately after his introductory review of physical and chemical principles. He claimed that the earth's major revolutions had preceded "all histories and all monuments of art", all human textual records and artifacts. This indicated his belief that most of geohistory was prehuman (or at least prehistoric) and implied that he rejected de Luc's equation of the most recent revolution with the Flood recorded in Genesis. Yet Saussure also emphasized the importance of searching human "traditions" for evidence about less dramatic events and processes: about the "progressive retreat of the waters" that had made the continents habitable, about "deluges or great floods, their epochs and extent", and about climatic changes within historic times. Clearly he realized the potential value of historical records as evidence for what de Luc called "actual causes", and hence as a vital key to the deeper past.[75]

A later section of Saussure's agenda gave special attention to the great spreads of "rolled pebbles" [*cailloux roulés*] and the larger erratic blocks as potential evidence for the most recent major physical event in the earth's history. He suggested specifically that a close study of the heights of these materials above the present valleys could "give indications of the direction, volume, and force of the currents produced by the earth's major revolutions" and show whether the blocks could have been transported by the mega-tsunamis that Dolomieu had suggested (§6.3), or by some other means. Both this section and the one on "historical monuments" highlighted the strategic importance, as it were, of research on the *last* major revolution and its aftermath as a key to understanding geohistory as a whole.[76]

Three later sections posed shrewd questions about the evidence for still earlier geohistory, in the "primitive", "secondary", and "tertiary" (i.e., Superficial) rock masses or formations [*montagnes*]. In his draft, Saussure dealt with them in that true geohistorical order, but in the printed text he inverted them—like Desmarest with his volcanic epochs (§4.3)—into the methodological and heuristic order that probed from the clearer to the more obscure, from the relatively recent into the deeper past. Even more significant, however, was Saussure's section on fossils. He suggested questions for research that would help discriminate among alternative explanations of the emplacement of fossils and clarify their use as evidence for the *history* of life. He highlighted the importance of studying precisely which fossils were found in which formations in order to discover "the relative ages and epochs of appearance of different species". Above all, like Blumenbach (§6.1), he urged that naturalists should "compare fossil bones, shells, and plants exactly with their living analogues" in order to determine whether they were precisely the same, or just varieties, or truly distinct species; and, if the same, whether they still lived in the same regions or in different climates. Such questions summarized effectively the inconclusive research of the previous years, which had raised these problems but signally failed to resolve them (§5.2, §5.3).[77]

Saussure's agenda ended with a review of the errors that might be made in the field by inexperienced observers and detailed recommendations for the equipment and clothing that were needed for the kind of fieldwork that he had pioneered in the Alps. He concluded with a rousing call to the new generation: "From this review it can be seen that *géologie* is made neither for sloths nor for the sensual; for the life of the *géologue* is divided between tiring and perilous journeys, on which one is deprived of almost all the conveniences of life, and varied and profound studies in the museum [*cabinet*]." The comment was notable not only for Saussure's acceptance

73. Fig. 6.7 is reproduced from Freshfield, *Life of Saussure* (1920), frontispiece. Saussure to Laplace, 1796, quoted (without exact date) ibid., 425; it is not clear from the published excerpt whether the letter was written before or after the publication of Laplace, *Systême du monde* (1796). The fragmentary draft text is Saussure, "Théorie de la Terre" (Genève-BPU, Saussure MSS, dossier 59, cahiers 8, 15). Carozzi, "Saussure's unpublished theory" (1989), prints translations of the 1786 and 1796 tables of contents and reproduces the first page of the latter; see also "Géologie" (1988).

74. Saussure, "Agenda" (1796); *Voyages* 4 (1796): 467–538; an English translation was published in 1799.

75. Saussure, "Agenda" (1796), §3; that this was a late insertion is shown by his MS draft (Genève-BPU, Saussure MSS, dossier 43, cahier 1, 628) in which all subsequent sections are renumbered.

76. Saussure, "Agenda" (1796), §8. An editorial footnote mentioned the possibility of flotation on icebergs: a tiny hint towards the glacial theory proposed several decades later, which became the foundation for the modern interpretation of these puzzling features.

77. Saussure, "Agenda" (1796), §§13–15, §17.

of "geology"—in the still novel sense of the word—as a genuine empirical science, but also for his acknowledgment that indoor museum work was the necessary complement to outdoor fieldwork, though clearly his heart remained with the latter. Above all, his agenda reflected his continuing conviction that progress in the sciences of the earth—and the eventual formulation of a satisfactory geotheory—would come only by the arduous route of detailed research.[78]

Publishing his agenda in the *Journal des Mines* was the best way in which Saussure could ensure that it would have maximal impact on those who might carry on his research in future years. But he had a further motive for sending it to Paris, for with his wealth destroyed by the Revolution he hoped to find scientific employment under the Directorate. He had already sent tentative enquiries to Göttingen and St Petersburg, and possibly also to Jefferson, who was looking for savants for his college in Virginia. Later in the year Saussure applied to teach natural history at the new high schools [*écoles centrales*] in Paris; but Dolomieu, who was already doing so, discouraged "l'illustre Saussure" from taking such a humble position. Dolomieu told him there were plans to invite him to join Laplace and other eminent savants at the new École Polytechnique, but nothing came of this. He was put forward for a place at the Institut, but as a nonresident in France he was ineligible. So Saussure remained, impoverished, in his native Geneva, and his second stroke finally put an end to any further hopes of active scientific life.

Dolomieu on "geology"

Dolomieu's hope that that the great savant would join the lively discussions on *géologie* that he reported from Paris remained unfulfilled; but he flattered Saussure—with evident sincerity—as the one who had led the way in turning that new science of the earth in a *historical* direction:

> I could never forget that it is you, Sir, who have taught us to interrogate nature and to ask it to account for events far greater and more important than any of those that the history of men has been able to transmit to us, and of a much earlier date. It is in your works that we have found models of good observation; it is they that have shown the futility of the old geological fictions [*romans*] and that have made us realize the difficulties of a problem for which the data are extremely complex.[79]

Dolomieu himself was not only teaching physical geography to teenagers in the *écoles centrales*, but also *géologie* to older students at the newly revived School of Mines. The notes taken by one of his first auditors at the latter show how he for one was now using that malleable word to denote the whole range of the sciences of the earth, in all their complex diversity, rather than any overambitious geotheory. It also gave him an opportunity to expound, in his penultimate lecture, his own ideas about the emplacement of erratic blocks and other Superficial deposits by "a violent and universal movement" of water, rather than by "the ordinary and peaceful work" of the sea, as Lavoisier had supposed (§2.4).[80]

The following year Dolomieu introduced his course on the "occurrence [*gissement*] of minerals" with an eloquent opening lecture "On the study of geology", which La Métherie published in the *Journal de Physique*. Dolomieu said he would prefer to teach geology not in the classroom but in the field, because it was only there that the relevant phenomena could be seen. Echoing—and indeed citing—Saussure's agenda, Dolomieu depicted geology as an outdoor field-based science for the adventurous in mind and body, not an indoor text-based project for the studious. The youthful science was sharply distinguished from the narrow dogmatisms that had characterized the pursuit of each savant's favorite system. As he put it, in what could have been a veiled criticism of La Métherie, "those who, in nourishing their meditations on this interesting subject, confine themselves to collecting citations and authorities in travelers' accounts, arranging and combining them in some way to form what they then call 'system of the world', cannot be regarded as geologists, although engaged in geology." True geology needed the patience, perseverance, and courage shown by miners and mountain peoples, but also "a spirit devoid of prejudices, passionate for the truth alone, and—above all—stranger to the desire to defend or to overthrow systems". The empirical foundations of geology were to be discovered in mountain regions such as Saussure's Alps; but Dolomieu reserved his best purple prose for emphasizing the vast panorama of *geohistory* that the science disclosed, dwarfing the whole of human history:

> Only the study of nature itself, lifting the imagination to the level of geology's high conceptions, can discover in the combination of circumstances the history of times long before the existence of the peoples who have figured on the world's great stage, long before even the existence of the human race and of all organisms . . . Bursting the limits [*durée*] of all historical times, and scorning as it were the brevity of epochs relative to the human species, he [the geologist] walks in the immense space that preceded the organization of matter [i.e., as life] in order to find there the epochs of those great events of which he observes the monuments.[81]

Conclusion

The contentious word "geology" has now been traced all the way from de Luc's original and hesitant use of it in the 1770s as a synonym for geotheory (§3.1) to its increasingly widespread adoption at the end of the century for what was claimed to be a *new* science: a science that in some way would incorporate and combine all the

78. Saussure, "Agenda" (1796), §§22–23.

79. Dolomieu to Saussure, 5 October 1796 (Genève-BPU, Saussure MSS, dossier 8, 334–37); see also 1 December 1796, printed in Lacroix, *Déodat de Dolomieu* (1921), 2: 124–27; Freshfield, *Life of Saussure* (1920), 385.

80. Brochin, "Leçons de géologie", MS notes for 3–28 ventôse IV [22 February to 18 March 1796] (Paris-EM, MS 50), leçon 7.

81. Dolomieu, "L'Étude de la géologie" (1797), 256, 262; this "discours d'ouverture" was delivered in ventôse V (February–March 1797), but printed in a volume of *Journal de physique* that was falsely dated 1794, perhaps to gloss over the break in publication at the time of the Terror.

earlier sciences of the earth, or at least parts of them, and also give them all a newly *geohistorical* dimension. However, the word remained controversial, and anyway prescriptive sketches such as Dolomieu's had yet to be filled with empirical content. Furthermore, his emphasis on the primacy of fieldwork would need to be qualified—as Saussure acknowledged at least in passing—by conceding that indoor studies of minerals and fossils might have a vital and complementary role in the new science.

Above all, however, the shifting meaning of "geology" was serving to distance it from the genre of geotheory. Indeed geotheory itself—as embodied in works entitled *Theory of the Earth* such as Hutton's and La Métherie's—was beginning to look decidedly antiquated in its ambition to explain every major feature of the earth; Saussure's lengthy struggle, and final failure, to formulate his own geotheory might be taken as a symptom of a wider malaise. The multiplicity of mutually incompatible "systems" now suggested that the fault lay not with any specific theory but with the entire genre, and that the inconclusiveness of geotheory was endemic and the genre itself a dead end. In its place, a redefined "geology" seemed, at least to some savants, to offer an overarching framework for the piecemeal investigation of specific focal problems, each of which on its own might be truly soluble and around which a consensus might coalesce among competent savants. The next chapter traces the development of research on one specific focal problem, which from an unexpected direction provided a template for the process of treating the earth more consistently as a product of nature's own *history*.

CHAPTER 7

Denizens of the former world

7.1 A MUSHROOM IN THE FIELD OF SAVANTS (1794–96)

Fossil bones as a focal problem

By the last decade of the century, "geology" was no longer being used simply to mean the genre of geotheory, but increasingly to denote a cluster of focal problems, either causal or geohistorical in character, that could best be tackled separately. But many of the causal problems, although continuing to attract great attention from savants, seemed unlikely candidates for real progress, at least in the near future. Those surrounding the formation and consolidation of rocks, for example, depended on matters of basic or general physics and chemistry that were unsettled and highly controversial. The problems of accounting in general terms for earthquakes and volcanoes also seemed intractable, or inescapably conjectural, because so little was known about the physical processes in the earth's deep interior.

Two other areas of research appeared, in contrast, to be much more promising. Both were *geohistorical* rather than causal in character. Fossils, as the record of life on earth, offered persuasive evidence for the sheer "otherness" of the deep past, allowing the possibility of reconstructing a history of the earth, unless one adopted the increasingly implausible assumption that *all* the animals and plants found as fossils were still alive somewhere (§5.2). As many naturalists pointed out, however, fossils needed much closer attention and more accurate identification than they had yet received, before they could fulfill their explanatory potential in geohistory. Second, the main argument in geohistorical interpretation was between those who saw traces of a major "revolution" or "catastrophe" in the relatively recent past and those who denied that any such event had interrupted the slow pace of everyday physical processes. In effect, this focused attention on the most *recent* history of the earth—the period covered by human records and immediately preceding them—rather than on the much earlier periods in which the Secondary formations had been deposited, let alone the extremely remote times when the enigmatic Primary rocks had been formed. The Primaries had the romantic attraction of extreme antiquity, and the philosophical interest of approaching the origin of the earth itself; but the materials dating from far more recent times seemed more promising because—being closer to the known present—they might be easier to decipher.

Where these two areas overlapped, a specific focal problem had already been recognized as one of exceptional interest and significance. This was the problem of fossil bones (§5.3). Most of the bones that could be identified seemed to belong to *terrestrial* animals such as elephants and rhinos, so they might throw light on the continents of the deep past, and perhaps on their transition to, or replacement by, those of the present world. And most fossil bones were found not in the "regular" beds of the Secondary formations, but in the Superficial deposits that were widely strewn over them, which clearly dated from a relatively recent period: perhaps from the most recent revolution of all, whatever that had been. By contrast, the fossil remains of marine animals such as mollusks, found in such abundance in many Secondary rocks, evidently dated from the deeper past; but they were more difficult to interpret in geohistorical terms, because it was not clear how many of them might have been lost and how many might survive and even flourish in remote parts of the world or in the ocean depths (§5.2).

For the terrestrial animals to which most of the fossil bones belonged, the crucial issue was again that of their relation to species still alive. Did the bones belong to known species, or were they distinct? If they had exact "analogues" in the living fauna, had they flourished in the same regions and climates or in quite different parts of the world? If they were distinct, had they become extinct? Or had they "degenerated" or been transformed (in modern terms, evolved) into the living forms? Or had they migrated to more remote regions where they were still alive, as yet undiscovered? These questions had been widely debated in earlier decades, but without conclusive results (§5.3). Yet it was obvious that there was much about the history of the earth that would become clearer, if only they could be resolved. More recently, naturalists such as Blumenbach (§6.1) and Saussure (§6.5) had recognized explicitly that what was needed was a much more accurate comparison between the fossil bones and those of living species. That in turn would require a deep understanding of animal anatomy in general and a wide experience of the *comparative* anatomy of different forms. But the latter could be gained only by naturalists with access to extensive museum collections of the skeletons of living species.

With this kind of newly precise research, it might of course turn out that what was true for one kind of fossil bones was not true for others. By the 1790s the "Ohio animal", with its elephant-like tusks and hippo-like teeth, was regarded as the most obviously distinct from any known living species, though it remained uncertain whether it was truly extinct or still alive in the barely explored interior of North America. The fossil elephants and rhinos from Siberia, in contrast, looked very similar to the living species in the tropics, and the main puzzle was that of accounting for their location in an arctic climate; the puzzle was only slightly lessened in the case of the similar bones found in the temperate climate of Europe. The bones found in caves far inland in Bavaria looked like those of some kind of bear; if it was the polar bear, which was known to live close to arctic coastlines, it raised questions about both climatic and geographical change. Then there was the huge deer or "elk" found in Irish peat bogs: certainly extinct in Ireland, and indeed elsewhere in Europe, but still possibly extant in the wilds of North America. And much further back in time were the rarer fossil bones from Secondary formations: the doglike "Montmartre animal" from the gypsum quarries outside Paris; the "Maastricht animal", perhaps either a crocodile or a toothed whale, from the chalk mines outside that Dutch town; crocodile-like forms from Honfleur in Normandy and Whitby in northern England; and a few others. As already emphasized, however, research on all these fossil bones was marked above all by dispute, uncertainty, and inconclusiveness (§5.3).

By the time the Revolution in France plunged much of Europe into war, research on fossil bones had already come almost to a halt for other reasons. William Hunter had died in London several years earlier, Buffon in Paris little more than a year before the storming of the Bastille, Camper in The Hague just a year after Buffon. These three had been among the most significant contributors to the debate, not only because they were all fine anatomists but also because each had access to a major collection: Hunter to that of his brother John, Buffon at the Cabinet du Roi, Camper to his own. Among other naturalists who had studied fossil bones in earlier decades, only Pallas in St Petersburg was in the same league and still alive and active, but he was much involved in other work. Several lesser players such as Esper had also died, and some such as Daubenton had long ago shifted their attention to other areas of natural history. John Hunter did emulate his elder brother by studying the German cave bones that had been sent to the Royal Society by the margrave of Anspach (in whose territory the caves lay); he noted that the bones were larger than those of even the polar bear. But he died later in 1793 before his paper was published.[1]

Few naturalists of a younger generation had even begun to fill the gaps left by these losses, and thereby to revive the problem of fossil bones as a crucial focus for understanding geohistory. Blumenbach was the most prominent, but his teaching position at Göttingen made his massive textbook of natural history his highest scholarly priority; and what time he had for truly original research was devoted increasingly to his major project on physical anthropology. As already mentioned (§6.1), he treated the Ohio animal as an "*incognitum*" that was probably extinct, and certainly not identical to either the Indian or the African elephants, which he regarded as species distinct from each other. But this, as he must have recognized, only touched the fringes of the problem of fossil bones (Fig. 7.1).[2]

Only two other potential recruits to the problem of fossil bones appeared around this time. One was the young physician Johann Christian Rosenmüller

1. Anspach, "Caves in the principality of Bayreuth" (1794); Hunter (J.), "Fossil bones presented to the Royal Society" (1794).

2. Fig. 7.1 is reproduced from Blumenbach, *Abbildungen* (1796–1810) [Cambridge-UL: MB.46.75], pt. 2 (1797), no. 19.

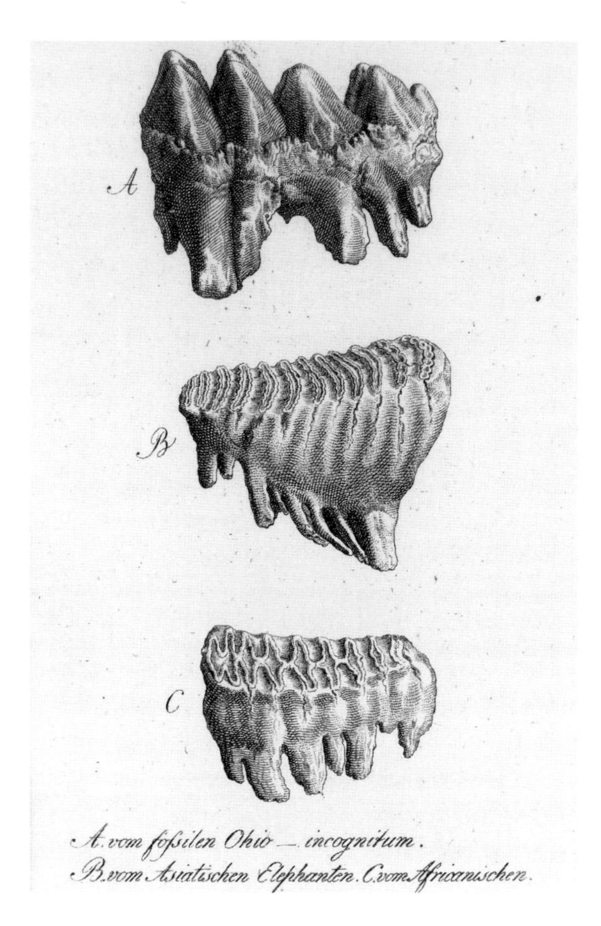

Fig. 7.1. Blumenbach's comparison between the molar tooth of the fossil "Ohio-incognitum" (A) and those of the Indian and African elephants (B, C), published in 1797 among his illustrations of notable objects of natural history, but already described verbally in his textbook of natural history. Like his left-spiraled fossil whelk shells (Fig. 6.1), which were pictured on the very next plate, these fossil teeth suggested the reality of extinction; but the evidence was not conclusive, since even the Ohio animal might be living undiscovered somewhere in North America. (By permission of the Syndics of Cambridge University Library)

(1771–1820), who in 1794 made his debut with a brief dissertation at Leipzig, in the customary academic Latin, followed by a longer version in German. He dealt with the fossil bones found in the Bavarian caves; like Esper he combined outdoor fieldwork in the caves themselves with an indoor study of the anatomical details of the bones. But he rejected Esper's idea that they had been washed in by some kind of flood, and he interpreted the caves instead as the dens in which the animals had lived. Furthermore, he displayed the fruits of his medical training by making a careful study of the comparative anatomy of bears, building on earlier work by Pallas and Camper. Like John Hunter he concluded that the fossil bones had belonged not to the polar bear but to an even larger and distinct species that was unknown alive. While mentioning the possibility that it had become extinct, Rosenmüller favored the alternative explanation that it had since "degenerated" into a smaller species, either the polar bear or more probably the brown bear still living in European forests (Fig. 7.2).[3]

This was a promising beginning, adding new and precise information about one of the important cases of fossil bones, but it went no further. The work was not lost from sight, because Blumenbach duly noted it in the next edition of his textbook, but Rosenmüller never followed it up with studies of other fossil bones. Having successfully defended his dissertation, he was appointed to an academic position in the medical faculty at Leipzig and set up in practice in the city. Although he later

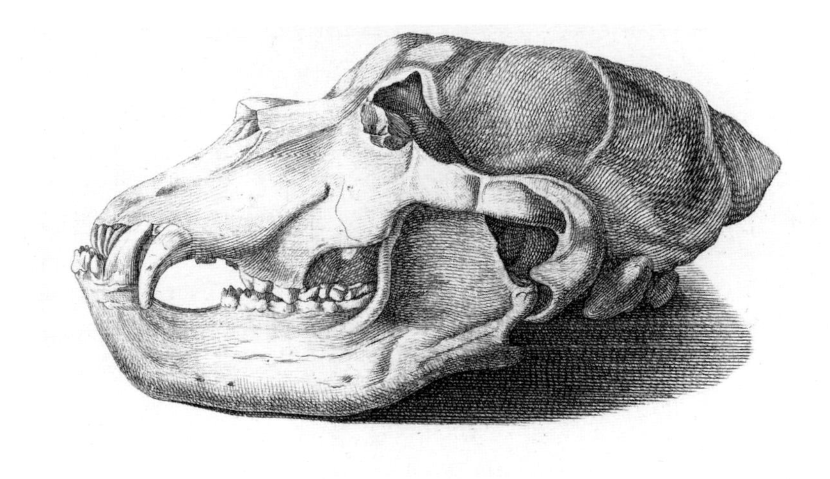

Fig. 7.2. The skull of a fossil bear from one of the caves in Bavaria, "drawn from nature" by Johann Christian Rosenmüller and published in 1794 to illustrate his medical dissertation at Leipzig. He claimed that the bears had been larger than any species known alive and that they had used the caves as their dens, rather than their carcasses being washed in during some kind of flood. He concluded that it either was an extinct species or had transmuted into one of those still extant. (By permission of the British Library)

published fuller accounts of the Bavarian caves and their fossils, he evidently had no time for more extensive research; anyway, Leipzig had no major museum collection of the skeletons of living species to act as a basis for wider comparisons with fossil bones.[4]

The young Cuvier

Another young naturalist landed in a more propitious situation at almost the same time and soon came to dominate the focal problem of fossil bones throughout Europe and beyond it. Georges Cuvier (1769–1832) had been born into a modest bourgeois family in Montbéliard, a small Lutheran but francophone enclave within eastern France, belonging to the duchy of Württemburg. In Stuttgart, the duchy's capital city, he had received a rigorous education that included a fine grounding in the natural sciences; he had also become fluent in German, which later gave him an important advantage among francophone savants. At the time the Revolution broke out, he was employed as a tutor by a Protestant noble family in Normandy and was indulging his passion for natural history in his free time. Like many others of his generation he was at first enthusiastic about the Revolution, but turned against it after witnessing scenes of mob violence in Caen. His employers prudently

3. Fig. 7.2 is reproduced from Rosenmüller, *De ossibus fossilibus* (1794) [London-BL: B.356(7)], pl. opp. 34 (engraved by the university's engraver Johann Friedrich Schröter); *Kenntnis fossiler Knochen* (1795); *Abbildungen merkwürdiger Hölen* (1796) described and illustrated some of the caves: see Rupke, "Caves, fossils" (1990), 245–46.

4. Blumenbach, *Naturgeschichte*, 5th ed. (1797), 702; Rosenmüller, *Beschreibungen der fossilen Knochen* (1804) and *Gegend um Muggendorf* (1804).

retreated to their rural chateau, and Cuvier with them, but they were not persecuted; Cuvier even held a minor local position under the new regime. When in 1793 faraway Montbéliard was annexed by France, he found himself a French citizen.

Cuvier was nothing if not ambitious for a career as a naturalist, and he sent zoological papers—on woodlice, limpets, and flies—to be published in Paris, where they might demonstrate his competence. Even before the old Cabinet du Roi was transformed into the new Muséum, he wrote to one of its curators, Bernard, count de Lacépède (1756–1825), who was becoming deeply involved in Revolutionary politics, and boldly offered to replace him. The older naturalist declined the suggestion, but as a result Cuvier was lucky to avoid being in Paris during the Terror. When the Muséum was being formed Lacépède was nominated as its professor of zoology but was forced into political exile before he could take up the position. Around the same time another exiled savant met Cuvier in Normandy and wrote recommending him as a young man of outstanding promise. Early in 1795, only a few months after Thermidor, Cuvier took the risk of moving to Paris, without any certainty of finding a position there. However, with the support of Étienne Geoffroy Saint-Hilaire (1772–1844), the even younger naturalist who had replaced Lacépède, Cuvier was appointed understudy (*suppléant*) to the elderly professor of animal anatomy, Antoine-Louis-François Mertrud. The Muséum then became not only Cuvier's place of work but also his home, since he lodged first with Geoffroy and then in Mertrud's professorial house in the grounds of the Muséum, the Jardin des Plantes. He was to remain there for the rest of his life. In effect, he was a beneficiary of the meritocratic policies of the new regime, which promised "careers open to talent" in place of the complex webs of privilege and patronage that had characterized the old.[5]

At first, after he settled into his new position, Cuvier's scientific work had nothing to do with fossil bones: when, a little later, he had his portrait painted, it was with jars of preserved animal parts in the background, not fossil specimens (Fig. 1.11). As the topics of his earliest papers suggest, his main interests were in the anatomy of what would soon be called invertebrates, including the marine fauna that he had studied intensively while living near the coast of Normandy. He focused his attention particularly on the mollusks, the anatomy of which was poorly understood; they gave full scope for his outstanding manual and visual skills, in both dissection and zoological drawing. But during his first year in Paris he extended his anatomical studies to "quadrupeds" and specifically to the mammals. He collaborated with Geoffroy on papers dealing with rhinoceroses and elephants, the tarsier and the orangutan; the two young naturalists were evidently exploiting the riches of the collections in the Muséum. They even proposed a new classification of all the mammals; for example, the "order" of "pachyderms" or thick-skinned mammals—of great significance in Cuvier's later work—was defined as including not only the elephants, rhinoceroses, and hippopotamuses, but also tapirs and pigs.[6]

Cuvier's broad outlook on zoology—covering the comparative anatomy of the whole animal kingdom—was also fostered by the demands of his teaching duties. Soon after his arrival in Paris he was appointed to join Dolomieu and many others in teaching in the new high schools, while at the Muséum he could teach at a more advanced level. Towards the end of 1795 he stood in for Mertrud by giving a course of public lectures on comparative anatomy in the Muséum's auditorium. His opening "discourse" stated plainly how he regarded animals as complex but functionally integrated "machines": this was a conception that—while not original to him—was to guide his subsequent work in zoology and to be a key to his geohistorical inferences. Only a week later, and perhaps in consequence of his performance, he was appointed a member of the section for anatomy and zoology within the First Class of the Institut, which had been set up less than two months earlier; he became its youngest member.[7]

This was just the first step in Cuvier's meteoric rise to prominence in the scientific world, in Paris and therefore internationally. Years later, he recalled how Daubenton told him he had sprung up "like a mushroom", but that he was a good one. Unlike a mushroom, however, Cuvier's rapid upward trajectory was not merely "natural", the result of being recognized as a savant of exceptional talents. It was also achieved through his own hard micropolitical work, forging alliances with patrons and clients and waging discreet campaigns against critics and rivals. Given his wide interests and broad interpretation of animal anatomy, the division of labor and of knowledge embodied in the Muséum brought him into potential conflict with several of his new colleagues: with his collaborator Geoffroy, if he seemed to be competing in the realm of mammals or birds; with Lacépède—who had returned from exile after Thermidor, being appointed to a special new position—if he strayed into the realm of reptiles; with Lamarck, if he dealt with "insects and worms" (or what Lamarck was soon to define as "invertebrates") in a way that the older naturalist disapproved; or with Faujas, if he started considering the implications of fossil animals for *géologie*, the title of that older colleague's chair. Cuvier's situation in Paris was far more favorable for research than, say, Rosenmüller's in Leipzig, but also much more risky. The potential advantages of working in Paris, and specifically of having access to the great collections at the Muséum, had to be set against the potential hazards of an institutional minefield.[8]

5. On his early life, see Negrin, *Georges Cuvier* (1977), pt. 1, and Outram, *Georges Cuvier* (1984), chaps. 1, 2; Desjardins-Menegali, "Georges Cuvier à Fécamp" (1982); Taquet, "Premiers pas d'un naturaliste" (1998); and, for his early work with fossils and on geology, Rudwick, *Georges Cuvier* (1997), chaps. 1–4. His earliest papers (published 1792) are listed in an indispensable bibliography, Smith, *Georges Cuvier* (1993), nos. 1–3. Lacépède to Cuvier, 25 September 1791 (Paris-IF, MS 3215/5, listed as 215/5 in Dehérain, *Manuscrits du fonds Cuvier* [1908]), records Cuvier's first attempt to get a position at the Muséum; see also Hahn, "Du Jardin du Roi au Muséum" (1997).

6. Geoffroy and Cuvier, "Orang-outangs", "Nouvelle division des mammifères", and other papers listed for 1795 in Smith, *Georges Cuvier* (1994), nos. 4–18. On their collaboration, see Appel, *Cuvier-Geoffroy debate* (1987), chap. 2; on Cuvier's work on classification, Daudin's classic study, *Cuvier et Lamarck* (1926). In subsequent years Cuvier published a stream of important papers on mollusks, which will not be cited in the present narrative because they have only a tenuous link with his work on fossils.

7. Cuvier, "Discours d'anatomie comparée" (1795); *Tableau élémentaire* (1798) was the textbook he wrote later. The auditorium of the Muséum survives unaltered in the Jardin des Plantes; its modest size gives a good sense of the relative intimacy of the courses.

8. Cuvier, MS autobiographical fragment (Paris-IF: MS 2598(3)), printed (with omissions) in Flourens, *Éloges historiques* (1856), 1: 105–96. His active construction of a career as a naturalist and his outstanding skills as a politician are graphically analyzed in Outram, *Georges Cuvier* (1984). Faujas's conception of *géologie* included work on fossil bones, making him Cuvier's most immediate potential rival: see his "Sur les dents d'éléphans", probably published in 1797 *after* Cuvier's early papers (§7.2).

Cuvier's early research on mammals, collaborating with Geoffroy, confronted him at first hand with what he would already have known from his wide reading: that the problem of fossil bones was both important and unresolved. He had certainly read *Observations sur la Physique* regularly during his years in Normandy, for he had summarized de Luc's letters to La Métherie (§6.2) for a friend in Germany at that time; and it is almost inconceivable that he would not have read Blumenbach's textbook on natural history as its successive editions appeared, since he—unlike his Parisian colleagues—could read German as easily as his native French. His and Geoffroy's joint paper on the two-horned rhinoceros, which was prompted by a brief report in the Royal Society's *Philosophical Transactions*, did not refer to the *fossil* rhinos that Pallas had found in Siberia. But their paper on "the species of elephants" not only followed Blumenbach and other naturalists in treating the Indian and African ones as separate species and the Ohio animal as much more distantly related; it also claimed that the Siberian fossil bones, although similar to those of the Indian elephant, "differ from it enough to be considered as a distinct species". This apparently modest comment was in fact pregnant with *geological* implications, and it marked the start of Cuvier's involvement with geohistory.[9]

Once he had been appointed a member of the Institut, Cuvier may well have felt that he no longer needed his tactful tactical alliance with Geoffroy, his junior in years but nominally his senior at the Muséum; or they may have fallen out, scientifically or personally or both. Anyway, from this point onwards all his zoological work was presented and published under his own name alone. Early in 1796, about a year after he arrived in Paris, two of his papers on fossil bones created a sensation in savant circles, first in Paris and then throughout the Republic of Letters. In both cases, Cuvier's work was made possible by serendipitous events, but his own talents—and not least his skill in scientific rhetoric—then enabled him to make the most of his good fortune.[10]

The megatherium

Late in 1795 a French diplomat and amateur naturalist, visiting Madrid in connection with the ceding of Santa Domingo to France, saw a remarkable fossil newly exhibited in the royal natural history museum [Real Gabinete]. He wrote to the Institut in Paris, describing it as the nearly complete skeleton of a large animal, over twelve feet long and six in height, which had been reconstructed from an assemblage of bones found in Spanish South America. The bones had in fact reached Madrid in 1789, and more recently had been pieced together and mounted by Juan-Bautista Bru de Ramón (1740–99), the museum's "painter and dissector", who had already mounted the skeleton of an Indian elephant in similar style. A set of Bru's unpublished plates was also sent to Paris; among them was a picture of the reconstructed skeleton (Fig. 7.3).[11]

The youngest member of the First Class of the Institut, and its newly recognized expert on comparative anatomy, was asked to report on the engravings from Madrid. Cuvier tactfully acknowledged that his elderly colleague Daubenton, and the deceased Hunter and Camper, had made a good start in the new science of

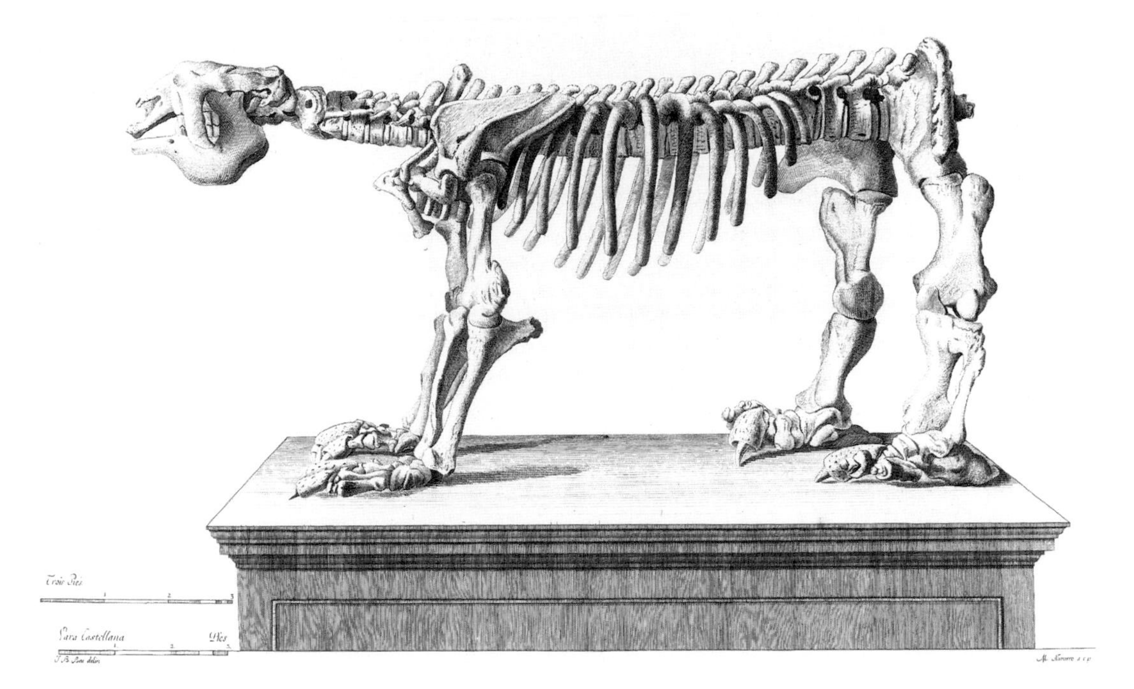

Fig. 7.3. An engraving of the fossil skeleton from South America in the royal museum in Madrid, after a drawing by Juan-Bautista Bru de Ramón, the curator who had reconstructed and mounted it. It was the first time that a reconstructed *fossil* skeleton had been depicted, in a style that had long been customary for living species. Together with engravings of many of the individual bones, this copy of the print was sent to the Institut National in Paris, where in 1796 Cuvier referred to it (mistakenly) as "the Paraguay animal" and formally named it the *Megatherium*. The scales are in French and Castilian feet, an example of the multiplicity of measures that the Revolutionary "metric system" was designed to eliminate. (By permission of the Bibliothèque Centrale du Muséum National d'Histoire Naturelle, Paris)

"comparative osteology", but he had ambitions to take it much further. He rejected the idea that the skeleton was the same as that of the "Ohio animal" or any other known species, living or fossil; it was totally new. Mistaking where it had been found, Cuvier referred to it as "the Paraguay animal", but this was soon eclipsed by the Latin names that he also proposed: *Megatherium fossile*, the "huge fossil beast" (renamed in the published paper as *americanum*, the "huge American beast"). Giving a Linnean binomial to a *fossil* animal—this was probably the first fossil mammal to be so named—was neither casual nor neutral; it deliberately embodied the potentially controversial claim that the fossil was distinct from any living species.

9. Geoffroy and Cuvier, "*Rhinocéros bicorne*" (1795) and "Espèces d'éléphans" (1795); that they knew of Bell, "Rhinoceros of Sumatra" (1793), indicates how news of scientific work in Britain continued to reach France (and vice versa) in spite of the war that had broken out between the two countries. On his reading while in Normandy, see the letters in Cuvier, *Briefen an Pfaff* (1845); his summary of de Luc's geotheory is translated in Rudwick, *Georges Cuvier* (1997), 9–12.

10. Cuvier and Geoffroy gradually diverged in their conceptions of living organisms, culminating in their famously vituperative dispute in 1830, which has often been mistakenly treated as an argument about evolution: see Appel, *Cuvier-Geoffroy debate* (1987).

11. Fig. 7.3 is reproduced from the copy [in Paris-MHN, MS 634(2)] that was used by Cuvier in his report to the Institut; the complete skeleton had probably been put on display only recently, because the Danish

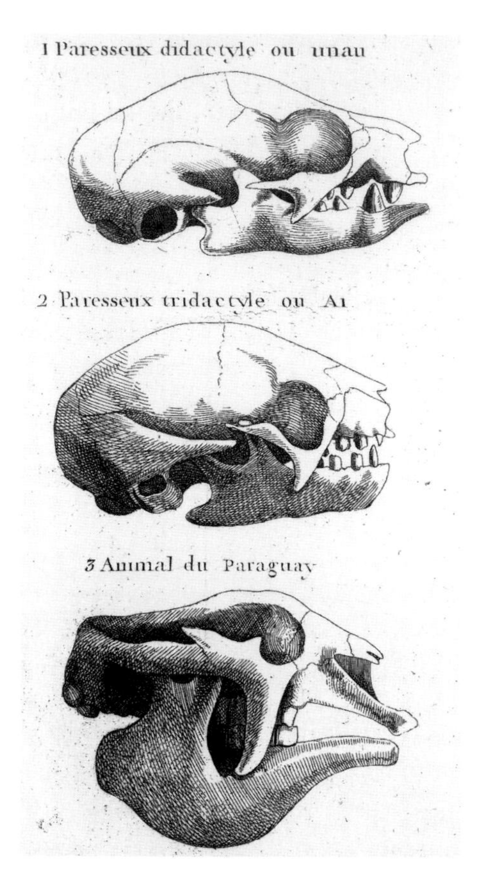

Fig. 7.4. Cuvier's drawing of the skull of the fossil "Paraguay animal" (3)—the part most likely to reveal its affinities—reduced to the same size on paper as those of the living *unau* or two-toed sloth (1) and the *ai* or three-toed sloth (2), in order to highlight their anatomical similarities. This engraving, published in 1796 to illustrate his paper, helped to make persuasive Cuvier's startling claim that the fossil *megatherium* was a huge edentate, and by inference an extinct denizen of a vanished former world. (By permission of the Syndics of Cambridge University Library)

Cuvier therefore focused on the osteology of the fossil, apologizing for the unavoidable "dryness" of technical details. Although his analysis was based entirely on proxy pictures, not on any real bones that he had seen, he insisted that the engravings were trustworthy and that fraud could be ruled out. Since the bones had all been found together, there was also little danger that what had been reconstructed had been cobbled together from bits of different animals. He concluded that the unique combination of characters in the fossil showed that its affinities [*rapports*] were with the edentates such as the sloths and anteaters. This was a startling claim, for the fossil was far larger than any of those exotic but rather humble creatures, and at first glance looked more like a rhinoceros or an elephant. It was of course a comparison that depended on the availability, in the Muséum, of skeletons of most of the known species of living edentates. Cuvier made his claim persuasive with a piece of consummate visual rhetoric (Fig. 7.4).[12]

Cuvier's anatomical comparisons were not only structural but also functional. From the massive jaws, for example, he inferred that like an elephant this strange animal had torn down the branches of trees rather than just eating their leaves; the nasal bones made him suspect that it had had a trunk, though a short one. In effect, Cuvier aspired not only to fix the affinities of the animal but also to *reconstruct* it as a living animal, bringing it back to life, as it were, in the mind's eye, just as antiquarians aspired to reconstruct the Classical world (§4.1). In doing so he would have been aided by the menagerie that had been established during the Revolution. The royal collection of live animals at Versailles had been transferred to the grounds of the new Muséum, where it was supplemented by the miscellaneous animals that had been expropriated, supposedly in the public interest, from their Parisian owners. The menagerie—almost literally on Cuvier's doorstep—complemented the collections of bones and skeletons in the Muséum itself by giving him a collection of living mammals (though not yet any edentates) in which he could observe comparative anatomy in action.[13]

Cuvier's conclusion was twofold. First, the megatherium confirmed the natural "laws" that he believed underlay the diversity of animal anatomy, the laws of the "combination" and "subordination of characters" that enabled natural affinities to be perceived reliably and the animals themselves to be classified correctly. Those principles, though not altogether original to him, were to guide all his zoological work. Second, however, the fossil had geological implications. It was taken by Cuvier as further evidence that the quadrupeds of "the former world" were *all* distinct from the species now alive:

> One of the most striking results of this study [of fossil bones of all kinds]—which was least expected and was even contrary to the original goal of [earlier] observers—is the general fact that there is no perfect analogy whatever between the bones found in the earth's interior and the same parts of the animals that are known to us [alive]. The proofs of this important fact accumulate daily and become more palpable: there is no longer—as there was thirty years ago—a single animal [i.e., that of Ohio] that could be confused with the elephant or be supposed to exist in countries that Europeans have not yet covered. They [i.e., the fossils] are so numerous, and so remarkable in form and size, that one can scarcely suppose that men who have collected and described the smallest insects from the least accessible climates would not yet have seen such substantial animals; and on the other hand they are so different from all known forms that one can still less suppose that they are [just] varieties or degenerations of them.[14]

---

astronomer Peter Christian Abildgaard, who had visited the museum in 1793, illustrated only the skull and some limb bones in his *Kongelige Naturalcabinet* (1796). On its place in the history of reconstructions of fossil animals, see Rudwick, *Scenes from deep time* (1992), 27–32; its huge impact at the time is difficult to recapture, now that more lifelike images from the deep past have become commonplace. The bones had been found in the banks of the Lujàn River west of Buenos Aires: see López Piñero, "Juan Bautista Bru" (1988), and *Bru de Ramón* (1996), 13–26, 87–104; the latter reproduces all Bru's plates, first published in Garriga, *Descripción de un quadrúpedo* (1796). The great building erected as the natural history museum in Madrid was later converted into an art museum, now well-known as the Prado. Bru's fossil skeleton is now on display in Madrid-MCN, Zona de Geologia; its modern curators—showing admirable respect for its historical importance—have refrained from remounting it according to modern conceptions of its posture as an almost bipedal animal. There is a fine historical display of other natural history exhibits from Bru's time, including his mounted skeleton of an Indian elephant and its stuffed exterior, in Madrid-MCN, Zona de Zoologia.

12. Fig. 7.4 is reproduced from Cuvier, "Squelette trouvé au Paraguay" (1796) [Cambridge-UL: Q900:2.d.20.7], pl. [2]. Geoffroy and Cuvier, "Orang-outangs" (1795), had already used the same visual rhetoric, probably on Cuvier's initiative, to compare primate skulls. By ignoring real sizes, it entailed the unusual procedure of deliberately *reducing* the factuality of proxies: see Rudwick, "Cuvier's paper museum" (2000).

13. See Burkhardt, "La Ménagerie" (1997).

14. Cuvier, "Squelette d'un très grand quadrupède" [1796], in Paris-MHN, MS 634(2). This paragraph near the start of Cuvier's MS report to the Institut expresses ideas that are found more briefly near the end of the published text, "Squelette trouvé au Paraguay" (1796). Conversely, de Luc's phrase "l'ancien monde" is only in the published text. On Cuvier's zoological principles, see the classic works by Russell, *Form and function* (1916), chap. 3, and Coleman, *Georges Cuvier* (1964); also Outram, "Uncertain legislator" (1986).

These were bold claims: too bold, perhaps, to be expressed fully in print. Cuvier had yet to show clearly for all the other known fossil mammals what he reckoned to have demonstrated for his new megatherium. There might have been something like Blumenbach's *Totalrevolution* separating de Luc's ancient or former world from the present one (§6.1), but Cuvier had yet to prove it. While not explicitly declaring himself in favor of extinction as the true reason for the contrast, Cuvier did firmly reject not only Bruguière's alternative claim that all fossil species might survive undiscovered in the modern world, but also that of invoking the "degeneration" (or evolution) of ancient species into modern ones. Anyway, a rewritten version of his paper, published promptly in the *Magasin Encyclopédique*, gained immediate attention from other savants. Above all, the little engraving attached to it, redrawn from Bru's larger picture of the mounted skeleton in Madrid, made the reality of this strange creature vivid and immediate, beyond what any amount of text could convey.[15]

Cuvier's paper made a great impression, and not only in France. The London *Monthly Magazine* chose it as the one and only paper to be translated (shorn of its "dry" anatomical detail) immediately following its summary of all the papers that had been read at the first two public meetings of the new Institut; and Cuvier's version of Bru's picture of the whole skeleton was engraved yet again to form the one and only illustration in that issue. A Spanish translation was included in the booklet that made public both Bru's engravings and his text about the bones, and a German translation followed later. All these ensured that Cuvier's claims about the megatherium were soon known throughout the learned world: in faraway Virginia, Jefferson—who had left Paris at the end of 1789, thinking the Revolution was already over—was among those who read the English translation and saw its picture of the "huge [South] American beast" (§7.2).[16]

The mammoth

Cuvier's second serendipitous present had arrived at the Muséum shortly before Bru's engravings from Madrid. Soon after Thermidor, in the wake of victories by the Revolutionary armies, an official team of French savants followed them into the Austrian Netherlands (now Belgium) to confiscate by "right of conquest" the province's valuable or useful "objects of sciences and arts". In addition to an architect, a bibliophile, and a horticulturalist, the team included the naturalist Faujas. They were primarily required to exploit the economic resources of the region, by collecting seeds from the best strains of plants, samples of agricultural and industrial tools and machines, and so on; but cultural and natural riches were also in their sights. From Brussels they sent back to Paris five wagons loaded with valuable books, manuscripts, plants, and fossils, and further convoys followed. The fossils included Burtin's finest specimens (§4.2); he himself had fled, having been prominent in the Austrian administration, so his collection was regarded as fair game. The team then moved on into the northern Netherlands, where the Patriot party had made peace with France after the Stadhouder William of Nassau had fled to England. Private collections in Holland were therefore spared, though van Marum had to plead with Faujas, whom he had met in Paris before the Revolution, not to pillage Teyler's Museum. But by the terms of the Treaty of The Hague the choicest items from the Stadhouder's great collection were taken to Paris. On arrival, all this loot was divided among the public museums: for example the new industrial museum [*Musée des Arts et Métiers*], the museum for the fine arts newly installed in the former royal palace of the Louvre, and of course the Muséum d'Histoire Naturelle.[17]

At the Muséum, 150 crates of specimens arrived—at about the same time as Cuvier—and were unpacked in the auditorium. The skeleton of a rhino, for example, was a rare prize that was noted at once; but the new acquisitions also included two elephant skulls, the incidental products of the far-flung Dutch commercial empire. One skull came from the Cape of Good Hope at the southern tip of Africa, the other from Ceylon (now Sri Lanka) off the southern tip of India. They must have been finer specimens than any already at the Muséum, for they became crucial evidence in Cuvier's studies—continuing what he had started with Geoffroy—of the species of elephants. They enabled him to claim again, but now more convincingly and in detail, that the Indian and African elephants were quite distinct species. But they also enabled him to take the comparisons a decisive further step. Using the *fossil* specimens that were recorded in earlier publications, the Dutch skulls gave him what he needed to argue that the celebrated elephant remains from Siberia belonged neither to the Indian nor to the African species, but represented a *third* species, as unknown in the present world as the megatherium from South America. The conclusion was not wholly new: for example, the Parisian physician Philippe Pinel, already famous for his humane treatment of the insane, had suggested the previous year that a fossil tusk in the Muséum came from a "race" of elephants that was "perhaps extinct thousands of centuries [*siècles*] before the foundation of Rome"; but a passing remark was not the same as a detailed study of comparative osteology.[18]

Cuvier made his case in a paper on "the species of elephants, both living and fossil". This was read at an ordinary meeting of the Institut early in 1796, and then

15. Cuvier, "Squelette trouvé au Paraguay" (1796), includes a small version of Bru's drawing of the skeleton, reversed by the reengraving process; it is reproduced in Rudwick, *Georges Cuvier* (1997), 28. Bruguière, who might have pressed the relevance of his parallel claim that all fossil *mollusks* would be found to have exact "analogues" in living faunas (§6.1), could not speak for himself, having left France in 1792 on an expedition to the Ottoman lands.

16. See Smith, *Georges Cuvier* (1993), nos. 27, 80. Since Bru had failed to publish his work, the Spanish savant José Garriga purchased his plates and text and published them, with his own translation of Cuvier's article, in his *Descripción de un quadrúpedo* (1796); Cuvier later included a French translation of Bru's anatomical description in the full version of his own paper, "Sur le megathérium" (1804). López Piñero, "Juan Bautista Bru" (1988), interprets the whole episode in terms of plagiarism by Cuvier; a more balanced study is Hoffstetter, "Bru, Cuvier et Garriga" (1959). Rather than feeling upstaged, Bru may have valued Cuvier's opinion on the affinities of the fossil, which he himself was hardly qualified to question; conversely, Cuvier's paper shows that he knew almost nothing of its geological context.

17. See Boyer, "Conquêtes scientifiques" (1971) and "Muséum sous la Convention" (1973).

18. Pinel, "Tête de l'éléphant" (1793). By the time Cuvier's paper was published (see below), a pair of *live* elephants—a great rarity in Europe—had arrived from The Hague to be added to the menagerie at the Muséum, giving him an opportunity to study the anatomy in action; see Burkhardt, "La Ménagerie" (1997), which describes inter alia the concert of popular music given for the entertainment of the elephants on their arrival. When, later, one of them died, Cuvier dissected it—in all senses a major operation—and gained valuable new insights into its functional anatomy.

again at the very first of its quarterly *public* meetings, which were to be occasions on which research of particular interest could be presented to a wider audience. It was then promptly published in the *Magasin Encyclopédique*, which ensured that it received as much attention as the earlier paper on the megatherium. Cuvier briskly rejected the conclusions of his many distinguished predecessors on the grounds that their anatomical observations had been insufficiently precise. He also rejected any idea that the fossil "mammoth" might have "degenerated" into either of the living species: to abandon the concept of the stability of natural kinds would, in his opinion, subvert the very foundations of natural history as a classificatory science. Using the skulls from The Hague as his decisive evidence, he argued that the two living species of elephants were as different in their anatomy as the horse from the ass or the goat from the sheep: far more than could be attributed to natural (in modern terms, intraspecific) variation. A similar comparison established that the mammoth was different again, subtly but unequivocally (Fig. 7.5).[19]

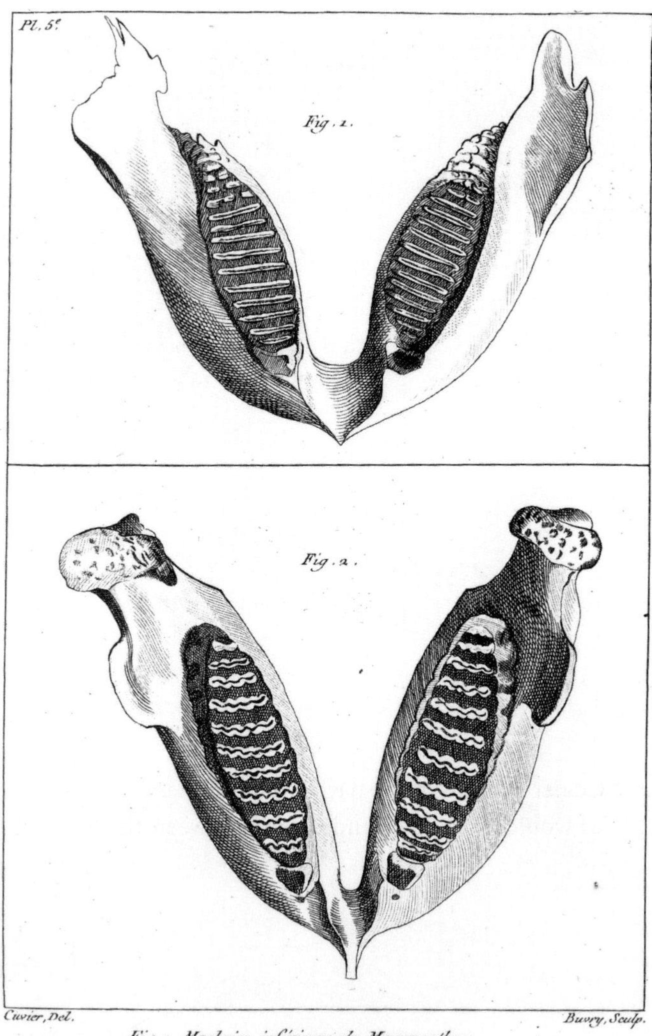

Fig. 7.5. Cuvier's drawings of the lower jaws of a mammoth (above) and an Indian elephant (below), showing differences in their molar teeth and in the form of the jaw itself. They supported his conclusion that the fossil species was distinct from any living elephant, and a member of a wholy vanished fauna. He displayed the specimens themselves when he read his paper on elephants to the Institut National in Paris in 1796; this engraving was one of those published with the full text in 1799. (By permission of the Syndics of Cambridge University Library)

However, the purpose of Cuvier's paper was not only to sort out the species of elephants but also to use them as the basis for making a bold assertion about "geology". That contentious science, he claimed, now needed to look to *anatomy* "to establish in a sure manner several of the facts that serve as its foundations". Decoding the allusion, Faujas, the professor of geology at the Muséum—who may well have been in the audience and would certainly have heard about what had been said—would need to defer to this young upstart to set his own work on the right lines. Only by the skilful comparison of subtly different bones, Cuvier implied, could one tell whether the species were distinct or the same, and hence what kind of history the earth had had. For Cuvier defined Faujas's science as concerned not just with the earth's static structure, but also with *geohistory*: as he put it—with the customary antiquarian metaphor—geology "collects the monuments of the physical history of the globe, and tries with a bold hand to sketch a picture of the revolutions it has undergone".

Cuvier's paper embodied a tacit challenge not only to Faujas but also to Faujas's former patron, the deceased director of the museum under the Old Regime: his claim that the mammoth was a distinct species knocked the bottom out of Buffon's model of geohistory. Buffon had assumed that the Siberian bones were those of the living species of elephant and rhino; only if they were identical, with the same tropical habitat, could they be used as evidence for the slow cooling of the globe (§3.2). If on the contrary the mammoth was a distinct species, it might have been adapted to the very same arctic climate in which its bones were now found, and Buffon's model would be deprived of its most persuasive evidence. Cuvier conceded that his rejection of that model left geology with new problems in its place. But he boldly claimed that *all* the well-known kinds of fossil bones, including now the mammoth and the megatherium, were distinct from any living species. Noting the absence of any authentic *human* fossils (§5.4), he concluded:

> All these facts, consistent among themselves, and not opposed by any report, seem to me to prove the existence of a world previous to ours, destroyed by some kind of catastrophe. But what was this primitive earth? What was this nature that was not subject to man's dominion? And what revolution was able to wipe it out to the point of leaving no trace of it except some half-decomposed bones?[20]

Conclusion

With a show of modesty, Cuvier left it to "more daring philosophers"—self-styled "geologists" such as Faujas would have been understood between the lines—to

19. Fig. 7.5 is reproduced from Cuvier, "Espèces d'éléphans" (1799) [Cambridge-UL: CP340:2.b.48.110], pl. 6; it was engraved by Buvry. The MS text read at the Institut (Paris-MHN, in MS 630(1)) refers explicitly to his displaying the specimens while reading his paper.

20. Cuvier, "Espèces d'éléphans" (1796), 444, read at an open meeting on 15 germinal IV [4 April 1796]; the full version, "Espèces d'éléphans" (1799), records the earlier reading at an ordinary (closed) meeting on 1 pluviôse IV [21 January 1796], which was *before* the paper on the megatherium (though the latter was the first to be published).

pursue his rhetorical questions. But there could be no doubt that he was adopting a binary model of geohistory close to that of de Luc and Dolomieu, with a prehuman former world sharply distinct from the present human world. Initially, in fact, he made the distinction even more boldly, echoing Blumenbach's idea of a total revolution, for he added that "several learned conchologists claim that none of the shellfish now existing in the sea are found among the abundant petrifactions with which the continents are filled". But after he had read the paper at the Institut, those experts—in fact probably his senior colleague Lamarck, who had begun just such research on mollusks (§7.4)—must have told him that the claim was untenable; for it was excised from the published text, and the contrast was confined to the *terrestrial* faunas of the present and former worlds. Even with that restriction, however, Cuvier's study of living and fossil elephants, coming hard on the heels of the megatherium paper, suggested that the binary model of geohistory had found a new and powerful advocate.[21]

More importantly, Cuvier had now shown that he could bring to the debate on fossil bones a potentially decisive technique, applying comparative anatomy with unprecedented precision to settle issues that had long been contentious and unresolved. If his early papers were reliable straws in the wind, the conclusion might be that extinction was a real phenomenon in nature, that a mass extinction had marked the end of the former world, and hence that the deep past was truly "other" than the present. It would follow that it might indeed be possible to reconstruct a detailed and reliable *history* of the earth before the first appearance of the human species.

7.2 CUVIER OPENS HIS CAMPAIGN (1797)

Cave bears and fossil rhinos

Cuvier's paper on elephants was published in the *Magasin Encyclopédique* at almost the same time as Saussure's influential "Agenda" (§6.5) in the *Journal des Mines*. To any savant who read both, it would have been obvious that the young newcomer had already adopted one of the Genevan's most significant recommendations. What was needed, if fossil bones were to yield their full potential as decisive evidence for geohistory, was a close comparison with the bones of living animals. Cuvier had shown himself able to do this with great precision, thanks not only to his skills in anatomy but also to the resources available to him in the Muséum. When he gave his course on comparative anatomy for the second time, he inserted a new lecture on fossil animals, immediately after his opening discourse and before launching into his advertised topic. This was a clear sign that he intended to make the problem of fossil bones a high priority. He then lost no time in extending his research to other examples.[22]

All competent naturalists now agreed that the Ohio animal, for a start, was distinct from any known living species; its bones were listed as the Muséum's highest priority, when an American naturalist proposed an exchange of specimens with Paris. Cuvier had claimed to add the mammoth and the megatherium. Even before he reported to the Institut on the latter, he applied the same methods to the well-known bones from German caves. In the less formal setting of the Société d'Histoire Naturelle, he rejected what Esper had suggested (§5.3) and claimed that the bones belonged to an animal as distinct from the polar bear as the mammoth was from the living elephants. The cave bear brought the number of distinct fossil species to four.[23]

A year later Cuvier gave the Institut a paper on a fifth case, that of the rhinos, again extending work he had begun with Geoffroy. He distinguished four living species, but emphasized that the fossil bones—well-known from European localities as well as Siberia—were distinct from those of any of them; the case paralleled the elephants almost exactly. Like them again, it was work that he could hardly have done without the resources of the Muséum, including the loot from the Netherlands. He noted cryptically that "various geological considerations" followed. In fact a German correspondent—one of the first in what soon became an extensive network of informants—had recently told him more about the fossil bones in Siberia and about how they had been interpreted by Pallas and others. There can be little doubt that Cuvier was by now speculating, at least in private, on the kind of natural revolution that might have caused all these fossil species to be "destroyed".[24]

Dolomieu and de Luc as Cuvier's allies

Cuvier might have found allies in either of the older savants who had already proposed binary models of geohistory and from whom he is likely to have derived his own ideas of a former world separated from the present by a major revolution. One of these was Dolomieu (§6.3), who had emerged after Thermidor as a prominent figure among Parisian savants. At the School of Mines, Dolomieu proposed a reformed geology that would reject the "fictions" of geotheory in favor of a sober search for the causes of the earth's revolutions, based on intensive fieldwork (§6.5). He then practiced what he preached by spending the summer of 1797 traveling extensively through central France and across the Alps into Italy, almost always "on foot and with hammer in hand". Explicitly emulating Pallas and de Luc, and above all Saussure, he focused his attention on the Alps, "those ancient monuments of the globe's catastrophes" that recorded events of "an epoch well before the times of [human] history".

However, Dolomieu first traversed the volcanic regions in Auvergne and Vivarais that Desmarest (§4.3) and Soulavie (§4.4) had made almost as famous as the

21. Cuvier, "Espèces d'éléphans" (1796) (Paris-MHN, MS 630(1)); see Burkhardt, *Spirit of system* (1977), 129. I am much indebted to Richard Burkhardt for discussion of the important differences between the MS and printed texts.

22. Brongniart, MS notes on Cuvier's course starting 20 pluviôse V [20 January 1797] (Paris-MHN, MS 2323/3), leçon 2; Alexandre Brongniart later became Cuvier's collaborator in geognosy (§9.1).

23. Lamarck and Geoffroy (in their official capacity) to Charles Willson Peale, 30 June 1796 [printed in Dean (B.), "Origin of Species" (1904)]. Cuvier, "Têtes d'ours fossiles" (Paris-MHN, in MS 634(1)), read 7 germinal IV [27 March 1796]; he had probably not yet heard of Rosenmüller's recent work nor seen John Hunter's posthumous paper, which both pointed to a similar conclusion (§7.1).

24. Cuvier, "Espèces de rhinocéros" (1797), read 15 floréal V [4 May 1797]; Reimarus to Cuvier, 6 April 1797 (Paris-IF, MS 3219/3).

Alps. Those two naturalists, and more recently Montlosier (§6.1), had attributed the topography to a continuous slow erosion of the valleys by the streams that still flow in them. Dolomieu disagreed, claiming that no present processes [*moyens actuels*] could have been responsible for such deep and broad excavations through solid rocks. Like de Luc, he regarded them instead as the products of a drastic catastrophic event, or series of events, separating the remote periods at which the older basalts had been erupted from the far more recent times in which the younger lavas had flowed down the newly eroded valleys. "These modern volcanoes", he argued, "are quite evidently later than the last crisis that has left our continents more or less as we see them". In other words, Dolomieu, visiting the region that had been the prime exemplar of uninterrupted fluvial erosion, reinterpreted it as further evidence for his conception of a violent revolution in recent geohistory.[25]

This was a conception based on evidence quite different from Cuvier's, yet clearly compatible with it. But before they could explore their common ground, Dolomieu removed himself from the scene. The young French general Napoléon Bonaparte—Cuvier's almost exact contemporary—was planning a military expedition to occupy Egypt in order to cut the British line of communication to India and the East (the conflict was rapidly becoming the first truly *world* war). A large team of savants of all kinds was to accompany his army; Dolomieu was one of those invited, though at the time he was unaware of their secret destination. So when the savants left Paris in the spring of 1798, the School of Mines was deprived of its geologist and Cuvier of a potential ally. (Another of those invited was Geoffroy, so while the Muséum was deprived of its official expert on mammalian zoology Cuvier was relieved of one of his potential rivals.) On their way to Egypt, Bonaparte ordered Dolomieu to demand from the Knights of Malta—his own former community—the surrender of their strategically placed island: this acutely distasteful task turned the savant against the general and made him want to curtail his participation in the expedition even before it reached what was now disclosed as its destination. Dolomieu did not vanish from the savant scene completely; for example, he reported back to Paris that his interpretation of the history of the Nile delta (§6.3) was confirmed by seeing the harbor of Alexandria at first hand. But for the time being he was out of the mainstream of scientific debate.[26]

Cuvier too had been invited to go to Egypt, but had declined. "My calculation was soon made", he recalled years later; "I was at the center of the sciences and in the midst of the finest collection, and I was sure to do better work there, more sustained and systematic, and to make more important discoveries, than on [even] the most fruitful journey." With a modest but secure position in Paris, and with the riches of the Muséum at his disposal, Cuvier decided to develop his career as a conventional museum naturalist—a *naturaliste sédentaire*, as he put it later—at the cost of declining the more exciting but risky potential of being a *naturaliste voyageur* in an exotic land.[27]

Cuvier may well have been more ambivalent about his other potential ally. De Luc's binary geotheory had been well-known for years among savants, first from his early *Physical and Moral Letters* (§3.3) and more recently from his lengthy letters in La Métherie's periodical (§6.2). His distinctive theistic agenda, openly emphasized in the former, had been soft-pedaled in the latter; but it was highlighted again when his letters to Blumenbach—already published in German and English translations (§6.4)—appeared in 1798 in Paris in their original French. The unnamed editor of the book explicitly commended de Luc's work as a powerful defense of revealed religion against "the attacks of the unbelievers"; he attributed the neglect of de Luc's geology in France to the political and cultural power of those skeptics; and de Luc himself was quoted as describing his letters as giving "a complete synopsis of geology" that would show that the science was compatible with Genesis. The point was further strengthened by the inclusion of de Luc's Haarlem prize essay on the theistic foundations of human morality as a "preliminary discourse"; and the volume also included his essay on the interpretation of the Flood narrative in Genesis, which Blumenbach had tactfully suggested omitting from the German edition (§6.4). To cap it all, the subtitle of the book described it unambiguously as "containing new geological and historical evidence [*preuves*] of the divine mission of Moses".[28]

This new edition of de Luc's work greatly strengthened the perceived link between *any* kind of binary geohistory and the highly contentious issue of the authority of the Bible. Moreover, the political dimension of that issue was accentuated by de Luc's own position. His fieldwork tour through the German lands in 1797 was partly a cover for an intelligence mission on behalf of George III's government, to help coordinate British policies with those of other regimes opposing the Revolution. Of course this was not generally known, but in 1798 his appointment as the first professor of "geology" at Göttingen—George III's Hanoverian university—highlighted his political status. He himself explained to Lichtenberg, the professor of physics, that he wanted to use his honorary position to propagate his notion of *géologie* as a science that would support the concept of divine revelation and counteract the skeptical influence of the Revolution and its catastrophic social and political effects. So Cuvier, by adopting a model of geohistory similar to de Luc's, could hardly avoid being suspected in Paris of having counter-Revolutionary sympathies.[29]

25. Dolomieu, "Voyages minéralogiques et géologiques" (1797) and "Voyages de l'an V et de l'an VI" (1798); he commended Montlosier's book as the best factual account of the Auvergne volcanoes, although he disagreed with the local naturalist on the question of erosion. What attracted even greater attention—though it is not directly relevant here—was Dolomieu's inference, based on his fieldwork, that the volcanic magma had risen *through* the granite of the region, implying that granite could not be the foundation of the earth's crust and that there were vast stores of deep-seated heat in the earth. This added to the growing sense among *géologues*—quite independently of Hutton's geotheory—that the interior of the earth was more dynamic than had generally been allowed.

26. Dolomieu, "Lettre au citoyen Lamétherie" (1798); Lacroix, *Déodat de Dolomieu* (1921). See the essays in Bret, *Expédition d'Égypte* (1999); also, on Dolomieu's part in it, Cooper, "From the Alps to Egypt" (1998).

27. Cuvier, MS autobiographical fragment (Paris-IF: MS 2598[3]), 39 (printed in Flourens, *Éloges historiques* [1856], 1: 185). Outram, *Georges Cuvier* (1984), 60–63, perhaps underestimates the extent to which a career as a museum naturalist was the better established and generally more prestigious option (§1.3).

28. De Luc, *Histoire physique de la terre* (1798): see "advertissement de l'editeur" (iii–xxii), "discours préliminaire" (xxvii–cxxviii) and "lettre VI" (287–337); he had failed to win the prize offered by the Hollandsche Maatschappij.

29. De Luc to Lichtenberg, 13 December 1797 (Göttingen- NSU, Cod. MS. Lichtenberg III.49.14 [printed in Joost and Schöne, *Lichtenberg Briefwechsel* (1983–92), 4: 794–95]). The planned work to which he alluded was probably *Cosmologie et géologie* (published in Brunswick, 1803) or perhaps *Traité élémentaire de géologie* (1809). On his intelligence work, see Tunbridge, "Jean André de Luc" (1971), 24–26.

Cuvier's research program

However, in the summer of 1798 Cuvier announced at the Société d'Histoire Naturelle what had now become his major research project, taking precedence over his beloved mollusks. It was to identify all the fossil bones he could, to reconstruct the animals to which they had belonged, and to compare them with species known to be alive in the present world. His score of those that were distinct from any known living forms had now risen to twelve. He listed the Ohio animal, the mammoth, the megatherium, the fossil rhino, and the cave bear; but he now added others, such as the Irish "elk" and a fossil hippopotamus, and also the Montmartre animal, which he identified as being similar to a dog. A little later, after further work, he amended that last item, assigning the bones from the gypsum quarries to a genus that was totally unknown alive, present in three distinct species of contrasting sizes. This accentuated the diversity of what was becoming an impressive fauna, and Cuvier now claimed explicitly that all the species were truly *extinct*. Rejecting Buffon's theory of slow climatic change (§3.2), Cuvier neatly shifted the explanatory burden on to the likes of Faujas: "after this it is up to the geologists to make such changes and additions to their systems as they find necessary to explain the facts that he [Cuvier] has thus set out". This striking report, and the research project it outlined, attracted immediate attention. A summary of Cuvier's paper was published promptly in the *Bulletin* of the Société Philomathique, and then in the *Magasin Encyclopédique* and the *Journal de Physique*; and it was translated in the *Monthly Magazine* and another English periodical, and later in a German one too. Naturalists throughout the Republic of Letters were given a clear summary of what this ambitious newcomer had already done and left in no doubt about what he intended to do in the future.[30]

At the next open meeting of the Institut, Cuvier explained this work in a way that made it interesting and accessible to an even wider public. He used as his example the bones from the gypsum quarries, because they came from Montmartre and other nearby villages that were familiar to his Parisian audience. These particular bones had further advantages for his purpose: they were clearly his own intellectual property, since earlier naturalists had given them little attention; and being found in solid gypsum rather than loose in river gravels, their extraction and assembly displayed his technical skills to the full. Above all, Cuvier showed that they came from creatures that were much *less* like any living species than the more familiar fossil bones. He concluded that they belonged to animals that were strangely intermediate in character between the pachyderms and the ruminants, or more precisely between the tapir and the camel; they helped to fill a major gap in the traditional "scale of beings" [*échelle des êtres*] that ideally linked all animals in a linear series. Their very strangeness underlined the distinctive character of the fossil fauna that his research was disclosing (Fig. 7.6).[31]

Cuvier described how he was piecing together the skeletons of these unknown animals, using the articulating facets on the scattered bones as a key to their correct assembly. He claimed that he could go even further, for his anatomical principles of the "conditions necessary for existence" ensured that what he was reconstructing

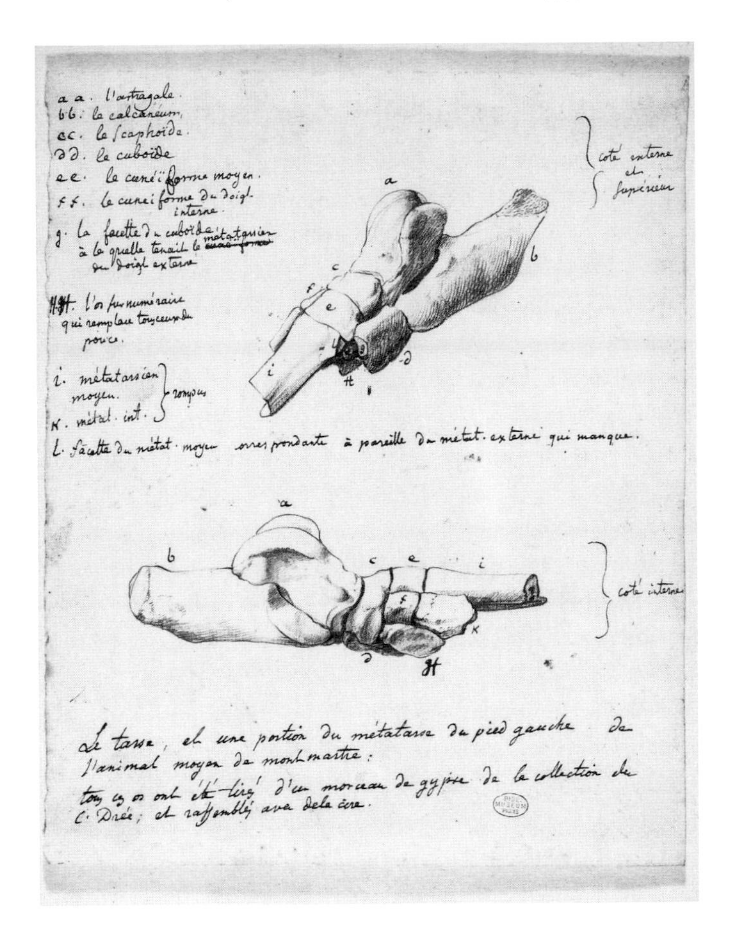

Fig. 7.6. Cuvier's drawings of some fossil bones from the quarries outside Paris, excavated from their gypsum matrix, assembled with the use of wax and drawn from two angles. The specimen was borrowed from "Citizen Drée", a democratized former marquis (and Dolomieu's brother-in-law) whose private collection was one of the finest in Paris. Cuvier's notes identified the bones as those of the left ankle region of "the middle [sized] Montmartre animal". This he first distinguished in 1798 as one of three species of a totally new genus of mammal (which he later named the *Palaeotherium* or ancient beast); it accentuated the strangeness of the fossil fauna that he was studying, which was quite different from the living mammals of the present world. The high quality of the drawings, which is typical of all his work, shows his outstanding talents as a scientific artist; the drawings were working visual notes but also drafts of what he would later publish (see Fig. 7.11). (By permission of the Bibliothèque Centrale du Muséum National d'Histoire Naturelle, Paris)

would indeed have been functionally integrated "animal machines", well able to follow a specific mode of life:

> The bones being well-known, it would not be impossible to determine the forms of the muscles that were attached to them; for these forms necessarily depend on those of the bones and their ridges. The flesh being once reconstructed, it would be straightforward to draw them covered by skin, and one would thus have an image not only of the skeleton that still exists [i.e., as fossil bones] but of the entire animal as it existed in the past. One could even, with a little more boldness, guess [*deviner*] some of its habits; for the habits of any kind of animal depend on its organization, and if one knows the former one can deduce [*conclure*] the latter.[32]

30. Cuvier, "Ossemens fossiles de quadrupèdes" (1798) and "Ossemens de Montmartre" (1798); for the translations of the former, see Smith, *Georges Cuvier* (1993), nos. 44, 50, 59, and 84.

31. Fig. 7.6 is reproduced from an undated loose sheet (Paris-MHN, MS 628, folder "Palaeotheriums et Anoplotheriums"). It must have been drawn at an early stage in his research, after he recognized three species but before he named a new genus for them. Rudwick, "Cuvier's paper museum" (2000), fig. 3, reproduces his sketch of another specimen annotated successively—as he returned to it in the course of his research—as the jaw of "un animal voisin du chien", "un animal semblable au tapir", and a "palaeotherium medium", making it ever more strange.

32. Cuvier, "Ossements dans la pierre à plâtre" (Paris-MHN, MS 628), read at Institut National on 15 vendémiaire VII [6 October 1798], translated and transcribed in Rudwick, *Georges Cuvier* (1997), 35–41, 285–90. For one of his later pictorial reconstructions of the Montmartre fossils, see Fig. 7.23.

Forestalling the objections that he expected from critics such as Faujas, Cuvier claimed that all this was no *more* speculative than what "geologists" tried to infer in their "systems". In the case of the Montmartre animal, he inferred that it had been a thick-skinned herbivore, with a short trunk like a tapir. Clearly his aim was to render his reconstructions authoritative and to "revive" [*ressusciter*] these strange animals in the mind's eye—just as the antiquarians tried to bring Pompeii back to life (see Figs. 4.4, 4.5)—making them as vividly real to his audience as the live mammals that they could see for themselves in the menagerie at the Muséum. In fact, the antiquarian analogy was the key to his argument, and he urged that naturalists should imitate the antiquarians:

> The former will have to go and search among the ruins of the globe for the remains of organisms that lived at its surface, just as the latter dig in the ruins of cities in order to unearth the monuments of the taste, the genius, and the customs of the men who lived there. These antiquities of nature, if they may be so termed, will provide the physical history of the globe with monuments as useful and reliable as ordinary antiquities provide for the political and moral history of nations.[33]

Cuvier also consolidated his research project at this stage by publishing the most notable of his specialized papers in the most prestigious form, namely in the Institut's new *Mémoires*, which had in effect replaced those of the old Académie des Sciences. His enlarged paper on living and fossil elephants repeated the conclusions of the original (§7.1), but with more osteological detail and—most importantly—a set of engraved illustrations to make them persuasive (Fig. 7.5). In a postscript added as the paper went to press, Cuvier noted how his own work on elephants went well beyond Blumenbach's. He also reported that he had found the Ohio animal among bones from both France and Peru, and two fossil tapirs—one of them the size of an elephant—in the south of France; both claims strikingly confounded previous assumptions about the distribution of animals between the Old World and the New. And he described the Montmartre animal as occupying "the middle between the tapir, the rhinoceros, and the ruminants", which further underlined its oddity.[34]

Hostile critics

However, Cuvier did not enjoy an unopposed triumphal march through the ranks of the Parisian savants. La Métherie, for example, had been deeply skeptical in his multivolume geotheory about the idea of a major catastrophe or mass extinction (§6.5); and he repeated that opinion in the first of his annual reviews of the sciences to be published in the *Journal de Physique* since the height of the Revolution. He reported that Lamarck had found that at least some fossil mollusks did have exact "analogues" in the living fauna—and had therefore *not* become extinct—though he thought others might have perished through changes in purely local circumstances. As for fossil mammals, La Métherie claimed that Cuvier's alleged specific differences were well within a range of variation comparable to that of living species such as the dog, and he implied that much of Cuvier's work had already been done (though not published) by the deceased Camper. He insisted that the real problem was that of explaining the change of *climate* that was indicated by the tropical-looking fossils—both mammals and mollusks—found in regions that are now temperate or cold.[35]

There were in fact at least three alternative explanations for the differences that Cuvier was describing between fossil bones and those of living mammals, any of which would avoid his geohistorical inference that many fossil species had been wiped out in some rather recent revolution. The first was La Métherie's, namely that the anatomical differences were no greater than those observed *within* certain living species. But by citing the dog as his example, La Métherie had broached another currently contentious problem, that of assessing the effects of environment, domestication, and breeding on animal form. It was open to Cuvier, and to those who agreed with him, to respond that under *natural* conditions the differences between, say, the mammoth and the living elephants were too great to be merely the result of a difference of habitat or climate.

The second alternative explanation was put forward at this time by Faujas: it was to concede that the fossil species were indeed distinct from any known living ones, but to claim that they were quite likely to be flourishing undiscovered in some other part of the world. This "living fossil" argument had long been a cogent one among naturalists, particularly when it was applied to marine animals, since the more remote seas and the ocean depths were still so poorly known (§5.2). Bruguière had used it for fossil mollusks a few years earlier with all the authority of an expert conchologist (§6.1), but he could not take it any further: tragically, he died in 1798 on his way home from his expedition to the Levant. However, Faujas argued the same case in a massive work, which began to appear the following year, on all the fossils from the famous chalk mines at Maastricht. Although he was concerned primarily with marine animals, Faujas extended his argument to cover the terrestrial ones on which Cuvier was basing his contrary claims; and he he snubbed his junior colleague, mentioning him only once and in passing (see below).

Faujas's book was a handsome production (and for twice the normal price one could buy it in deluxe folio format). Its fine large plates of engravings—many based on drawings by the Muséum's own scientific artist—made the fossil fauna accessible everywhere in proxy form. The specimens themselves had reached the Muséum as part of the spoils of war from the conquered Netherlands; Faujas himself had supervised the removal of fine collections from Maastricht. Most of the fossils he described and illustrated were of mollusks, sea urchins, corals, and so on. There were

33. Cuvier, "Ossements dans la pierre à plâtre" (Paris-MHN, in MS 628); the revealing verb *ressusciter* is used in a fragmentary draft of the lecture (Paris-MHN, in MS 627). His wording suggests that the idea of "nature's antiquities" was not yet generally familiar, or at least that it could not be taken for granted.

34. Cuvier, "Espèces d'éléphans" (1799), 21–22; postscript dated 6 vendémiaire VII [15 October 1798]. The paper also gave formal Linnean names and diagnoses to all four species of the genus *Elephas*, clearly distinguishing the true mammoth of Siberia (*E. mammonteus*) from the misnamed "mammoth" of Ohio (*E. americanus*).

35. La Métherie, *Théorie de la terre* (1797), 5: 196–219, and "Discours préliminaire" (1798), 75; see also Corsi, *Age of Lamarck* (1988), 68.

also some of the well-known but puzzling forms that were quite unknown alive, such as ammonites and belemnites (§5.2). Sharks' teeth and the rare remains of turtles completed this unquestionably *marine* fauna. Faujas considered himself qualified to assert that at least some of the fossils had true and exact "analogues" alive in the present world, and he listed forty-one examples from Maastricht and other localities. He noted that his colleague Lamarck, who since Bruguière's untimely death had begun to study fossil and living mollusks in detail, agreed with him on this point. The clear implication was that in time, with further exploration, *all* the fossil forms might turn out to be alive in the present world. Like La Métherie, Faujas concluded that the main task was to explain why so many of the analogues of the fossil species were now living in quite different climates; a major shift in climatic distribution was the big problem, not extinction.[36]

Faujas's supreme example was the famous "Maastricht animal", which Camper had identified as a toothed whale (§5.2; see Fig. 2.6). But Faujas claimed that before his death Camper had changed his mind and decided that it was a crocodile. This was Faujas's own view, and one that he supported with an anatomical comparison with museum specimens of crocodiles of living species. Above all, however, Faujas made his case with the finest known example of the enigmatic fossil, which had been the most highly prized of all the specimens that he had acquired

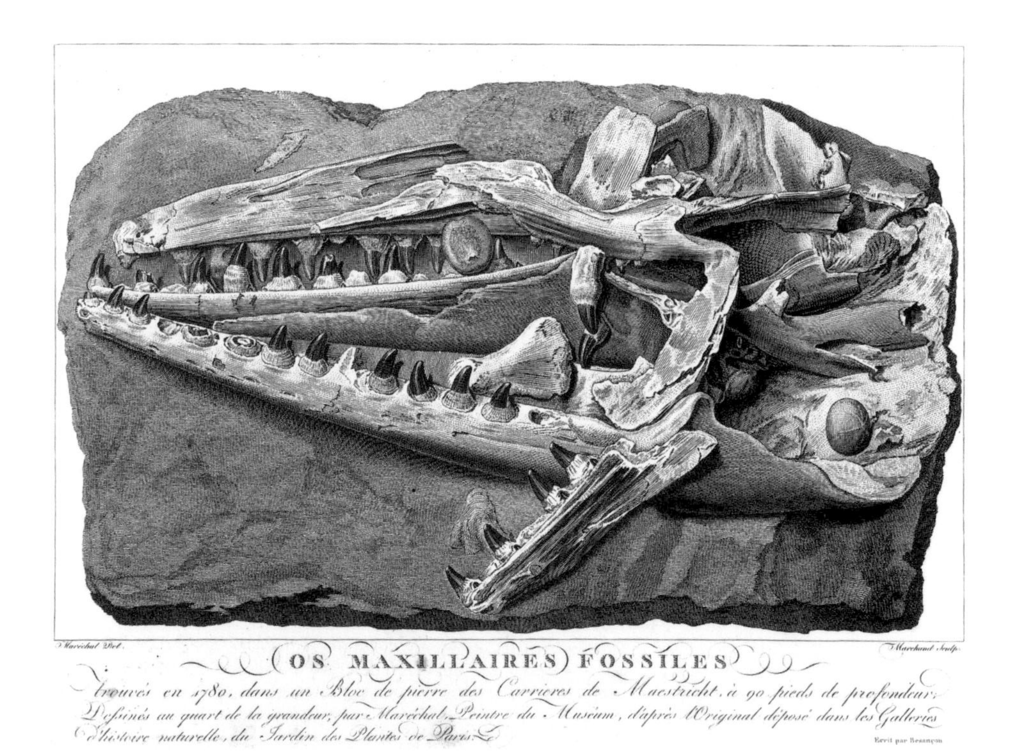

Maréchal Del.
Marchand Sculp.
(OS MAXILLAIRES) FOSSILES
trouvés en 1780, dans un Bloc de pierre des Carrieres de Maestricht, à 90 pieds de profondeur;
Dessinés au quart de la grandeur, par Maréchal, Peintre du Muséum, d'après l'Original déposé dans les Galleries
d'histoire naturelle, du Jardin des Plantes de Paris.
Ecrit par Besançon

Fig. 7.7. The finest specimen (over a meter in length) of the jaws of the famous fossil animal from Maastricht; it was acquired by Faujas in 1794 in the newly captured Dutch town, taken to Paris and placed in the Muséum, and treated as the centerpiece of his lavish descriptive account of all the Maastricht fossils (1799). Camper had identified the huge animal as a toothed whale, but Faujas claimed it was a gigantic crocodile. The two oval objects are sea urchins, preserved by chance among the bones and witnessing to the marine origin of the deposit. (By permission of the British Library)

in the conquered Netherlands. He decorated the title page of the deluxe edition of his book with a romantic reconstruction of its discovery many years earlier (Fig. 2.5), and he inserted a second engraving of the specimen itself, drawn at a larger scale and with more detail than the first, when he realized its potential value as crucial evidence for his argument (Fig. 7.7).[37]

Identifying the Maastricht fossil as a crocodile rather than a whale turned it from a marine into a terrestrial (or rather, continental) animal. Although he assumed it had been swept out to sea after death, Faujas imagined it laying its eggs on the banks of the rivers or lakes in which it swam, like the crocodiles that his artist portrayed living by the present Ganges, Nile, or Amazon. The new identification also turned the famous specimens into material more closely comparable to Cuvier's fossil mammals. Faujas, like Cuvier, attributed the unsatisfactory state of geology to the failure of comparative anatomy to resolve the problem of fossil bones, and he praised Camper for having got as far as he had with this problem. He mentioned several other savants who had at least made a start, and he noted finally that "Pinel and Cuvier in Paris are following the same course". Pinel had indeed published one minor paper on elephants in addition to his important medical research (§7.1), but to couple his name with Cuvier's in this offhand manner was a calculated insult to the latter, the more so since this was Faujas's *only* reference to his junior colleague. He was in fact concerned with the same fundamental question that Cuvier had been tackling with such a high profile for the past three years, namely the reality of extinction. But he relegated to a mere footnote his opinion that it was premature to count any of these fossil animals as extinct, since so little was yet known about the interior of continents such as Africa. Nevertheless, this was potentially a lethal objection to Cuvier's conception of the recent mass destruction of a whole terrestrial fauna.[38]

Jefferson's megalonix

That Faujas's living fossil argument might be supported with new American material became apparent to European naturalists when the latest volume of papers from the American Philosophical Society, published the same year in Philadelphia,

36. Faujas, *Montagne de Saint-Pierre* (1799), "Des coquilles fossiles en général", 84–92, and pls. 18–42; "Catalogue" of analogues, 92–98 (page citations are to the folio edition). According to Boyer, "Le Muséum sous la Convention" (1973), the collections were purchased rather than confiscated, but under wartime conditions the transaction must have been highly constrained.

37. Fig. 7.7 is reproduced from Faujas, *Montagne de Saint-Pierre* (1799) [London-BL: 39.i.15], pl. 51; he noted that the specimen was so important that it deserved this second illustration—a drawing by Nicolas Maréchal, the Muséum's official scientific artist, engraved by Jacques Marchand at one-quarter natural size—in addition to one already made on a smaller scale (pl. 4). This famous specimen is still on display at Paris-MHN. Faujas's engaging tale about its acquisition, that the French commander had offered his soldiers a reward of six hundred bottles of wine for its safe recovery when the town was captured (*Montagne de Saint-Pierre*, 59–62), is, sadly, of contested authenticity. Faujas's book also includes engravings of Camper's and van Marum's specimens (pls. 5–6); the sea urchins (echinoids) are shown in their own right in pl. 29.

38. Faujas, *Montagne de Saint-Pierre* (1799), 11–13, 25n, and section on fossil crocodiles (153–62). Faujas may have drafted this passage before Cuvier had published much on fossils; but he could easily have revised it, since the book did not start to appear until late 1798 at the earliest: its first fascicle was announced in the *Journal de physique* for nivôse VII [January 1799]).

crossed the Atlantic. It contained a report by Jefferson on bones that he had been sent from localities in rural Virginia, which he had identified as those of a new "incognitum". Although there were very few bones to go on, he inferred from its massive claw that it was a huge carnivore, distinct from the Ohio animal and perhaps its predator, and more than three times the size of a lion; he named it informally "the Great-Claw or Megalonix". When he read the paper in Philadelphia two years earlier he had added a postscript at the last moment, having just seen the translation of Cuvier's description of the megatherium in the latest *Monthly Magazine* (§7.1); he accepted that this was not a carnivore, although it had similar claws, so he regarded his own new North American animal as distinct. However, any well-informed naturalist would have seen that Jefferson had written his paper without reference to most of the recent research on fossil bones: apart from travelers' reports he cited little but Buffon's now quite antiquated work.[39]

The limitations of Jefferson's paper were compensated to some extent by an anatomical study of the bones themselves, published in the same volume. It was by Caspar Wistar (1761–1818), the vice president of the society in Philadelphia and a physician who also taught anatomy in the university there. He noted their similarity to the bones of sloths, but he too thought the megalonix was probably distinct

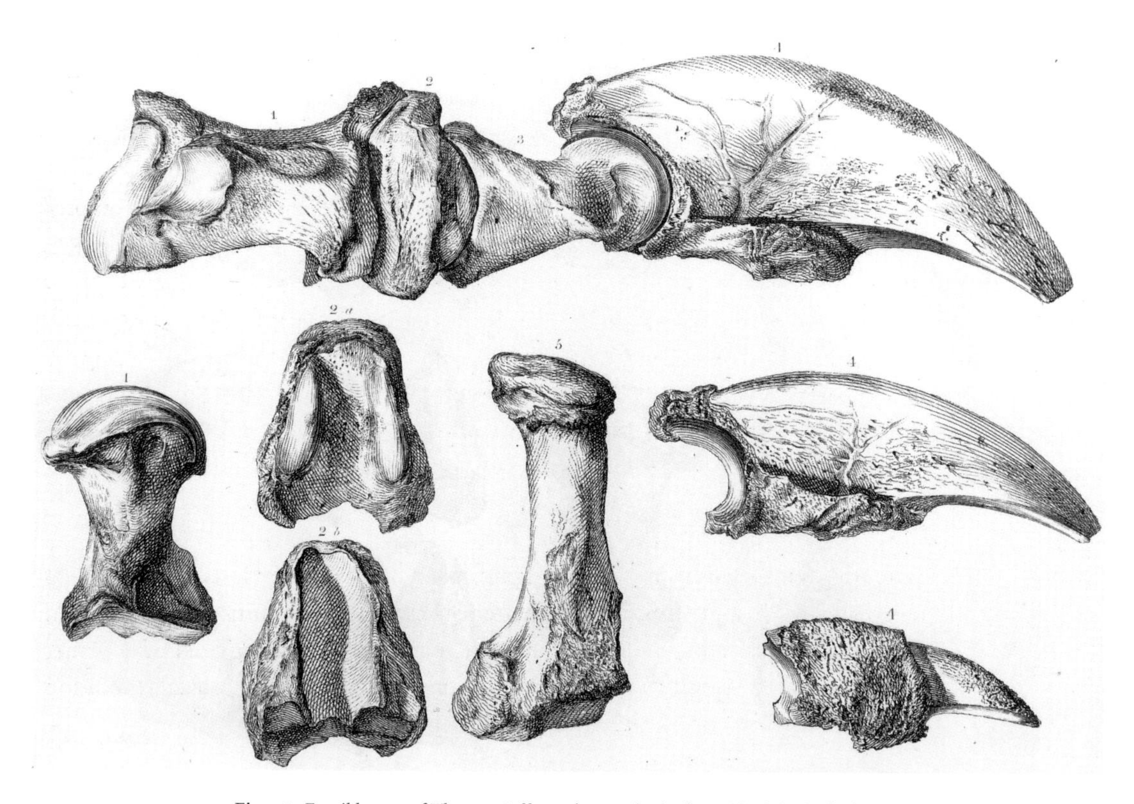

Fig. 7.8. Fossil bones of Thomas Jefferson's megalonix from Virginia, including its eponymous claw, as assembled by Caspar Wistar, identified by him as some of the bones of the foot, and published in Philadelphia in 1799 by the American Philosophical Society. Jefferson claimed that it was a carnivore, distinct both from the Ohio animal and from Cuvier's newly described megatherium from South America; he argued, partly on the basis of Native American oral traditions and artifacts, that it might still be alive in the unexplored interior of North America. (By permission of the Syndics of Cambridge University Library)

from Cuvier's megatherium. However, although Wistar's anatomical analysis was certainly competent, he himself conceded that he was severely limited by a lack of access to any major collection: in comparing the new bones with those of living edentates, for example, he had had to rely on old descriptions by Daubenton; and for the megatherium, of course, on the rather poor third-hand picture in the *Monthly Magazine.* So the most important impact of Wistar's paper—and hence of Jefferson's too—was that of its fine engravings: they made the Virginian bones available in proxy form on the other side of the Atlantic, giving Cuvier and others the evidence for adding the new megalonix to the growing list of quadrupeds known only as fossil bones (Fig. 7.8).[40]

However, the megalonix could not be taken as further evidence for a mass extinction without confronting Jefferson's claim that it might still be roaming the unexplored interior of North America, perhaps preying on herds of the Ohio animal (which he, like other anglophone naturalists, misnamed the "mammoth"). For Jefferson interpreted his new megalonix, as he had the Ohio animal some years earlier (§5.3), as a terrestrial case of the living fossil effect (§5.1). As before, he rejected even the possibility that it might be extinct, primarily on the grounds of nature's plenitude: "For if one link in nature's chain might be lost, another and another might be lost, till this whole system of things should evanish by piece-meal." He concluded that "if this animal then has once existed, it is probable on this general view of the movements of nature that he still exists, and rendered still more probable by the relations of honest men applicable to him and to him alone." That final clause alluded to a more specific reason for believing that the megalonix might still be alive. Jefferson believed that certain reports by early English explorers, and certain drawings carved on rock by Native Americans as well as their oral traditions, witnessed to the recent live existence of a huge and fearsome beast, even in the former colonies on the east coast. So it might simply have retreated westwards as the settlers advanced inland and might still be living in the poorly explored interior. Although Jefferson himself had not seen that remote hinterland, he was certainly in a better position than most European naturalists to appreciate its vast extent, and hence the plausibility of the living fossil argument in this case.[41]

Conclusion

The wider implications of Cuvier's work would have been obvious by this time to all savants with an interest in such matters. He was aligning himself with the kind of geohistory championed by de Luc and Dolomieu: a binary model of a former world, sharply separated from the present world by some kind of radical revolution

39. Jefferson, "Quadruped of the clawed kind" (1799), read on 10 March 1797, with postscript of the same date.

40. Fig. 7.8 is reproduced from Wistar, "Description of the bones" (1799) [Cambridge-UL: T340:1.b.90.4], pl. 2, engraved by James Akin after drawings by W. S. Jacobs; the other plate (1), showing bones identified as the ulna and radius, was drawn by Titian Peale, one of the three sons—all saddled with similar painterly names—of the artist Charles Willson Peale of Philadelphia.

41. Jefferson, "Quadruped of the clawed kind" (1799), 255–56.

or natural catastrophe. But Cuvier was claiming that the vanished world of the deep past could be described much more concretely than those older savants had supposed. His application of rigorous comparative anatomy was enabling him to reconstruct some of its most striking animals and to claim that the whole fauna had been distinct from that of the present world. He was explicitly imitating the antiquarians in their reconstruction of the vanished human world of Herculaneum and Pompeii; he claimed that his own reconstructions could similarly bring a vanished *natural* world back to life. Cuvier eschewed the ambitious goals of geotheory, prudently leaving it to "geologists" to speculate on the cause of the event that had "destroyed" his strange animals and brought the former world to an end. Instead, he was focusing on the *geohistorical* task of describing what that world—and specifically its animals—would have looked like, had anyone been there to see it.

However, Cuvier's confident inferences were not uncontested. His critics deferred to his remarkable skills in comparative anatomy, at least to the extent of conceding that he was right to claim that the fossil mammals had been distinct from any known alive. But this might be due to differences of environment, acting on species that were highly variable even under natural conditions; or the fossil species might be real enough, but still be living undiscovered in remote parts of the world; or—a third alternative not yet clearly articulated—they might have been transformed into known living species without any extinction at all. The combined plausibility of these alternatives was such that Cuvier's geohistory was far from being immediately accepted by all who heard him or read about his work. The ambitious research project that he had outlined would need much more hard work, both empirically and rhetorically, before he could convince naturalists, throughout Europe and even beyond it, that it was feasible to reconstruct the mammalian fauna of a vanished prehuman world and thereby take geohistory to a new plane.

7.3 THE NAPOLEON OF FOSSIL BONES (1798–1800)

Savants in wartime

The debate about the significance of fossil bones developed during the last years of the century with little outward sign of the political turbulence and wartime conditions in which many savants were now working. In France the sciences had revived with astonishing rapidity, as if making up for lost time, after Thermidor brought the Jacobin regime to an end (§6.5). The political situation remained highly unstable, and further coups d'état had followed, but they scarcely affected the flourishing scientific life of the capital. Yet the French Republic was at war, at first defending itself against counter-Revolutionary threats from the rest of Europe and virtual civil war at home, but later expanding aggressively by conquest and forced annexation, north into the Netherlands, east into the Rhineland, and south into Savoy. In 1798 the annexation of the independent city-state of Geneva must have saddened the last days of Saussure, who died the following year, and it may well have heightened de Luc's counter-Revolutionary zeal. A bewildering sequence of shifting coalitions and countercoalitions among the major powers—France, Prussia, Austria, Britain, and Spain, with Russia, Sweden, and the United States on the periphery—kept most of Europe continuously at war.

Yet even between France and Britain—France's most persistent opponent, and in naval warfare the most powerful—communication among savants was never cut completely. They were certainly restricted by being unable to visit one another freely; but scientific publications and even correspondence did manage to circumvent the hostilities, often by being routed through neutral territory. A substantial periodical was founded in Geneva in 1796, specifically to make anglophone publications of all kinds known throughout the Continent in spite of the wars. The *Bibliothèque Britannique* reported, for example, on the full version of Hutton's geotheory (§3.4), and it continued its work unaffected by the city's annexation by the French. Among savants, then, the negative effects of war were relative rather than absolute. Scientific life throughout Europe remained proudly and self-consciously aloof from political and military conflict in a way that would become unthinkable by the early twentieth century.[42]

In this respect one personal outcome was a tragic exception. Dolomieu was returning early from Egypt, before the rest of Bonaparte's cultural team, when his ship was forced off course by a storm, and he was captured and imprisoned in Messina. Even a letter from Cuvier, signed by thirty-eight other members of the Institut, to Banks, the president of their sister body the Royal Society in London, failed to persuade the British authorities—or rather, their allies in Sicily—to release the savant on scientific and humanitarian grounds; Dolomieu's role in the surrender of Malta to the French had not been forgotten or forgiven. Remaining nearly two years in solitary confinement, he was totally excluded from all the scientific action in Paris and the rest of Europe. He spent his time stoically, writing on mineralogy; had he been released, it is likely that his further work would have been in that direction, rather than pursuing his earlier geohistorical research (§6.3).[43]

More substantive effects of the war on the course and content of scientific work are harder to find, at least among the sciences of the earth. One modest example would be the way that wartime conditions induced van Marum to shift the focus of his research and teaching in Haarlem from electricity to geology. After the French invasion of Holland, Faujas might have taken the finest objects in Teyler's Museum into "safe custody" in Paris (§7.2). Van Marum averted that threat, but with his museum's finances in disarray he turned from the expensive business of electrical experiments to making full use of its fine fossil collection, including its prize exhibit of the second-best specimen of the Maastricht animal (Fig. 2.6). In the winter of 1796–97 he gave a lecture course on "geology" for the first time, followed in the next two winters by a longer one on "fossil animals". Although the lectures contained little that was original, his notes show that he was well-informed about current research, interpreting it for his bourgeois audience in terms of the earth's lengthy

42. Hutton's *Theory of the earth* (1795) was reviewed in *Bibliothèque britannique* 5: 53–73, 262–74 (1797).

43. Cuvier to Banks, 8 vendémiaire VIII [30 September 1799], and related documents (Paris-MHN, MS 226); De Beer, *Sciences were never at war* (1960), 81–107, prints some of the international correspondence about Dolomieu's plight; see also Lacroix, *Déodat de Dolomieu* (1921), 1: 1–62.

history and recent "revolution" [*omwenteling*]. And he became still better informed when he made an extensive tour through the German lands in 1798, in the course of which he met Blumenbach in Göttingen and Goethe in Weimar, among many other savants, and studied the fossils in all the main museums. Although he did not repeat his pre-Revolutionary visit to Paris, van Marum's life as a savant was by no means brought to a halt by the wartime conditions.[44]

A much more important example of the impact of the wars on the sciences was the work of the French savants in Egypt. The expedition's most celebrated find was the Rosetta stone, on which the trilingual inscription was immediately recognized as offering a potentially decisive key to the decipherment of the ancient hieroglyphs, and hence to the history of the oldest literate civilization then known. But the stone was no sooner found than it was captured by the British, who under Horatio Nelson had defeated the French in the great naval battle of Aboukir Bay and were also fighting them on land. It was taken in triumph to the British Museum in London (where it remains), though casts of the inscription were made available to savants elsewhere. For the sciences of natural history, however, the mummified animals found in the ancient tombs were equally important; here too the scientific value of the finds was recognized immediately, for a detailed comparison with living animals could provide evidence for any change in organic forms in the last few millennia (§7.4). Like the deciphering of hieroglyphs, however, this was work that would have to be done back in Paris or other European centers, not in Egypt.[45]

However, the most important and immediate sequel to Bonaparte's expedition was that in the summer of 1799—as he (and coincidentally Cuvier too) turned thirty—he handed its command to a subordinate, slipped past the British naval blockade, and secretly landed back in France. After he reached Paris, the coup d'état of Brumaire (November 1799) ousted the five-man Directorate and replaced it with a three-man Consulate. But Bonaparte had no intention of sharing power, and soon established himself as First Consul, taking command of all the French forces. He proclaimed that after ten turbulent years the Revolution was over. But it was replaced not by the monarchy that other European powers had hoped to reinstate but by a new kind of authoritarian regime. As First Consul, Bonaparte became in effect the dictator of France; as a brilliant military commander, his seizing of power raised the stakes in the continuing wars.

Cuvier and the First Consul

Brumaire made little difference in the scientific world. Cuvier was not the only savant to adapt to the new regime as he had to earlier political changes and as he would to later ones. His attitude, which has been condemned censoriously by some modern observers as that of a gallic Vicar of Bray, was not in fact unusual. In his case it was reinforced by the theological principle that was traditional among Lutherans—the group with which he remained culturally identified, whatever his personal religious beliefs—that citizens ought to serve the established civil power, even if it was personally uncongenial, for the greater good of maintaining a peaceable social order. Anyway, when the elderly Daubenton died on the last day of the

1700s—added to the death of Saussure earlier in the year it marked the end of an era for natural history—the relevant savants lost no time in canvassing to succeed him in his various positions. In the event, Dolomieu was appointed in absentia to the chair of mineralogy at the Muséum, and Cuvier canvassed successfully to gain the prestigious chair of natural history at the Collège de France. Cuvier then offered La Métherie one-third of the lucrative salary of that position in return for giving the required lectures, while reserving the right to give them himself whenever he wanted to reach the Parisian social and cultural elites. La Métherie may have been grateful for the income, but the arrangement was humiliating, since he was senior to Cuvier and had hoped to get the position himself.[46]

Another new appointment soon increased Cuvier's prominence among savants, and his power in the world of the sciences, still further. He became one of the two executive "secretaries" to the First Class at the Institut (the other, the astronomer Jean-Baptiste Delambre, was responsible for the "exact" sciences). This brought him into direct contact with Bonaparte, who fancied himself as a patron of all the arts and sciences. The First Consul got himself elected to the First Class—its members could hardly refuse him, though his scientific achievements were not obvious—and took his turn in acting as its president. In that capacity Cuvier came to know him personally, a contact that certainly helped his career in the years that followed. One of Cuvier's new duties was to compose and deliver elaborate obituaries [*éloges*] on deceased members, which in practice enabled him to promote his own ideas about the kinds of scientific work that most deserved to be honored and emulated. His very first *éloge*, read in Bonaparte's presence, was on Daubenton, whom he praised not least for the politically dégagé attitude that he himself so notably exemplified.[47]

Meanwhile, in an increasingly busy life, Cuvier was also pursuing his own multifarious research projects, notably that of strengthening the empirical support for his ideas on fossil bones. He used the *Bulletin* of the Société Philomathique and the *Journal de Physique* to get preliminary accounts of his work into print without delay and thereby to establish his priority. One case illustrates how the goals of his research were as much geological as zoological. The Italian naturalist Alberto Fortis (1741–1803) wrote to La Métherie (who published the letter in the *Journal de Physique*), criticizing Lamanon for having been far too credulous when—almost twenty years earlier—he had claimed that a human artifact had been found in the gypsum quarries at Montmartre (§5.4). Fortis rejected the vast human antiquity

44. Van Marum, "Geologische leszen" (1796–97) and "Fossiele dieren" (1797–98, 1798–99), were followed in 1799–1800 by a course on mineralogy (all are in Haarlem-HM, MS 8). The last lecture on fossils (MS 8–29) is dated 22 March 1799. Forbes et al., *Martinus van Marum* (1969–76), 2: 90–131, transcribes his travel diary (trans. 273–311); see also Palm, "Van Marums contacten" (1987).

45. The bicentenary of the expedition produced some fine accounts: see, for example, the essays in Laissus, *Savants en Égypte* (1998), and Bret, *Expédition d'Égypte* (1999), and the well-illustrated introduction in Murat and Weill, *Expédition d'Égypte* (1998).

46. See Corsi, *Age of Lamarck* (1988), 19, 87, 167; Cuvier later increased La Métherie's share to two-thirds. On Cuvier's Lutheran background, see Taquet, "Georges Cuvier" (1994).

47. See Negrin, *Georges Cuvier* (1977), 139, 499; on Cuvier's *éloges*, see Outram, "Language of natural power" (1978).

famously implied by that claim and argued effectively that Lamanon's putative iron key could not have been genuine. He also rejected Lamanon's fossil bird (Fig. 5.16), which did indeed look suspiciously well-preserved, and denied that there were any genuine "*ornitholithes*" and therefore any evidence that birds existed at that time. The former director of the School of Mines then recalled how the quarrymen had often brought him similarly spurious "fossils", hoping for a reward: one such specimen had adorned his desk until the stink of its rotting corpse forced him to throw it away.[48]

Significantly, however, Cuvier wrote a short paper for the Institut, arguing in effect that both sides were mistaken: not for the last time, he adopted the middle ground, and with good reason. He agreed that many alleged fossil birds were spurious, but not all. He gave a detailed osteological description of one newly discovered specimen, which he interpreted as a bird's leg. Even a single specimen, identified with his well-earned authority, was enough to prove that birds had indeed been around at the same time as the mammals of the gypsum. This in turn showed again how the former world had been both strange and familiar: strange in the specific forms of the mammals (and perhaps of the birds too), but familiar in the broader kinds of animal organization that had already been present and in the physical conditions that would have made their life possible. Cuvier's paper illustrates how his aim was not merely to assign fossil bones to their correct taxonomic category, or even just to reconstruct them in the mind's eye as living animals, but also to integrate those vanished beings into a unified reconstruction of a specific period in the past *history* of the earth (Fig. 7.9).[49]

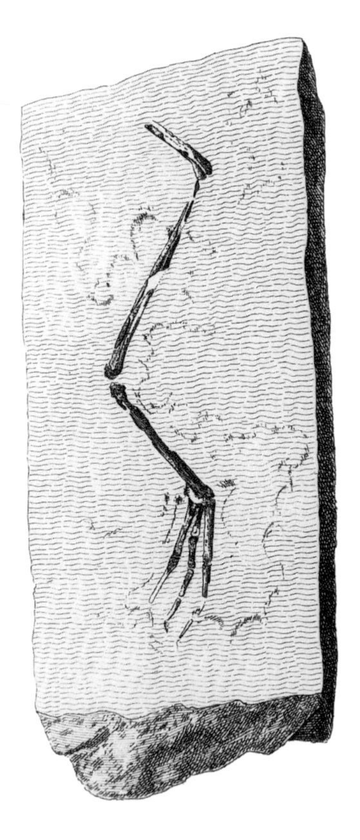

Fig. 7.9. Small fossil bones found in 1800 in the Gypsum formation outside Paris, described by Cuvier and identified by him as the leg of a bird. This specimen—almost unique at the time—was of great importance because it proved that there had been bird life as well as mammals in the former world represented by the Gypsum fossils. (By permission of the Syndics of Cambridge University Library)

Cuvier's network of informants

Cuvier's main strategy for his research, however, was simply to enlarge his database (to use an appropriate modern term) by assembling as many fossil bones as possible, from all available sources, in order to strengthen his theoretical argument with a cumulative weight of evidence. As soon as his early papers became widely known—either in their original French or in translation—other naturalists began to write to him, reporting on specimens that might interest him or offering suggestions about their significance. For example, the Hamburg physician Johann Reimarus (1729–1814), who also taught physics and natural history at the high school [*Gymnasium*] in that city, had written to him as early as 1797, telling him about Pallas's ideas on the geohistory of Siberia. Two years later Reimarus described German field evidence for the action of a "major revolution", as shown by fossil plants that were relics of the former world; and he reported that German savants agreed with Cuvier that the fossil bones were of species genuinely extinct. This was just the start of what soon became a flood of correspondence spanning most of Europe. For example, Giovanni Fabbroni (1752–1822) sent Cuvier an unknown fossil tooth from the grand-ducal museum in Florence, of which he was the deputy director: not the precious specimen itself, nor even a cast—which he explained was too large to send—but a drawing of it. This was an early example of what became a rapidly expanding traffic in paper proxies as well as verbal information. In at least one instance Cuvier himself took the initiative in this traffic, asking a scientific society in Berlin to act as intermediary in getting information from London about some important fossils found in Gibraltar, since the war made it difficult for him to contact the British directly.[50]

One specific contact, however, must have impressed on Cuvier how much his project could be advanced by recruiting the help of other savants and making full use of other collections throughout Europe. Like the specimens that had triggered the start of his work on fossil bones (§7.1), this contact began serendipitously. In the summer of 1799 the young Genevan naturalist Augustin-Pyramus de Candolle [or Decandolle] (1778–1841) visited the Campers' home in the northern Netherlands on his way to Paris: probably not so much to meet Adriaan Gilles Camper

48. Fortis, "Des morceaux de fer" (1800), referring to Lamanon, "Fossiles de Montmartre" (1782); see also Ciancio, *Autopsie della terra* (1995), 279. Sage, "Prétendu ornitholithe" (1800), recalled receiving a lizard ingeniously enclosed in a crystal of selenite.

49. Fig. 7.9 is reproduced from Cuvier, "Pied d'oiseau fossile" (1800) [Cambridge-UL: T340:2.b.16.54], pl. 1; the specimen came from quarries at Clignancourt, in the same formation as those at Montmartre. Cuvier's paper was printed immediately following Sage's note; his others published at this time are listed in Smith, *Georges Cuvier* (1993), nos. 70, 75, 76. There was a second specimen in the Camper collection, but fossil birds remained extremely rare. Leg bones alone were inadequate to judge whether this fossil bird was distinct from living species, as the Gypsum mammals clearly were.

50. Reimarus to Cuvier, 6 April 1797, 31 May 1799 (Paris-IF, MS 3219/3, 3221/10); Fabbroni to Cuvier, 25 May 1800 (Paris-IF, MS 3222/13), printed in Outram, "Storia naturale e politica" (1982), 196–97; Cuvier to Gesellschaft Naturforschender Freunde zu Berlin, 15 thermidor VIII [3 August 1800], printed in Théodoridès, "Lettre inédite de Cuvier" (1969). Dehérain, *Manuscrits du fonds Cuvier* 1 (1908), lists chronologically Cuvier's vast incoming correspondence in Paris-IF, giving an invaluable overview of his network of informants; Outram, *Letters of Georges Cuvier* (1980), lists his less abundantly preserved outgoing letters. On the Berlin society, see Heesen, "Natural historical investment" (2004).

(1759–1820) but rather to see the famous private collection that Adriaan had inherited when his father Petrus died ten years earlier. Camper (as he may now be called without confusing him with his deceased father) had evidently heard what Cuvier was doing with fossil bones, for he asked Candolle to prepare the ground for starting a correspondence with him. After Candolle reached Paris he duly did so, probably assuring Cuvier of Camper's competence; so when the latter wrote to suggest a collaboration on fossil bones, Cuvier responded promptly, positively, and with unusual cordiality. This was not in fact surprising, because he would have seen at once, from Camper's initial description of the specimens in his father's collection, that the Dutchman had much to offer him. In effect he anticipated a fair exchange. He outlined his own research, emphasizing how much progress he had already made; and while requesting drawings of particular items in Camper's collection he offered in return some choice duplicate specimens to fill some of its gaps. This was just the start of an intensive and fruitful correspondence (conducted of course in French).[51]

In the following months the relationship blossomed, and on a tacit basis of equality. Camper gratefully accepted Cuvier's offers of duplicate specimens and sent pictures of fossil bones as requested. He urged the Parisian—unrealistically, in view of the war—to visit Russia and America, in order to complete his great work on "these antiquities, the only ones that can throw light on the astonishing revolutions that the earth has undergone". In return, Cuvier confided in the Dutchman, giving him a detailed progress report on his research. For example he described his work on the puzzling mammals from the Parisian Gypsum, and his provisional interpretation of them, and asked Camper to check it against his own specimens. He mentioned that he agreed with Rosenmüller (whose work he now knew) that the fossil bears in the Bavarian caves belonged to a species distinct from any known alive. He noted that Faujas claimed to have found (in the museum at Darmstadt) the bones of seals among specimens from the same caves, but he commented tartly that "he is not knowledgeable enough for me to be able to believe his word about it". In return, Camper discussed Cuvier's ideas about the mammoth, the Ohio animal, and many others, with a clear sense of being equally competent to make the necessary comparisons. At many points Petrus Camper made his posthumous presence felt, for Adriaan possessed not only his father's specimens but also his many unpublished papers. He explained that he had hoped to publish them, had it not been for the French invasion: "my country is ruined!", he exclaimed to the one he addressed ironically as "Citizen Friend of the Sciences".[52]

Most striking, however, was Cuvier's summary of the trend of his research, and no less so for being phrased impersonally: "the more one examines these fossil bones and the more one finds them extraordinary, the more one is persuaded that they belonged to a *création* wholly different from today's". His use of the word "creation", like Blumenbach's (§6.1), does not necessarily imply that he thought any supernatural event was involved; it just expressed his growing sense that the two faunas—fossil and living, former and present—were sharply distinct. Moreover, Camper showed himself as well aware as Cuvier that what they were doing was closely analogous to the work of antiquarians such as those who had been in Egypt, and therefore *geohistorical* in character: "our researches, Citizen, have as their goal the most interesting of antiquities; the remains of extinct animals are our coins, our bas reliefs, our indecipherable annals" (to call them "indecipherable" was of course an exaggeration, for any identification was to some extent an elucidation of the meaning of the bones). Their exchange of letters recorded that process of decipherment, as provisional identifications were corrected in the light of further specimens or better comparative material. The savants' relationship might have seemed to be tilting out of balance when Camper abandoned both his own identification of horse bones among the Montmartre fossils and his father's interpretation of the Ohio fossil as a tuskless animal (§5.3), in consequence of the drawings and other information that Cuvier sent him: "you are right, my savant friend!", he exclaimed. But the balance was restored—though not at once—after Camper read a report of Faujas's new book on the Maastricht fossils and learned that the Frenchman identified the most famous animal as a crocodile, in direct contradiction to his father's conclusion that it was a toothed whale (§7.2).[53]

Cuvier told him that Faujas was probably wrong, and his criticism sharpened as he sent Camper a copy of the work: "you will see that there are lots of stupidities [*sottises*] in it, above all about the turtles, but that's usual for the author when he wants to talk anatomy". Camper quickly agreed, telling Cuvier that what Faujas described as the antlers of deer were just bits of the undersides of the turtles' carapaces. This was important—though Camper did not make the point explicitly—because it deprived the Maastricht fauna of one of its two putative mammals. The other then came up for similar scrutiny, when Cuvier reported that new specimens of "the celebrated whale or crocodile" had just been found near Maastricht and were about to be inspected by Lacépède, who was traveling in the area. In return, Camper told Cuvier that on reexamining his own specimens—in the hope of refuting Faujas's claim—he had to admit that his father had been mistaken, and that it was definitely reptilian; he promised to send a paper with illustrations to establish this, and asked Cuvier to get it published for him.[54]

However, while deserting his father's opinion, Camper did not embrace Faujas's. The paper that he sent to Paris—which was read at the Société Philomathique (to which he was then elected) and published in the *Journal de Physique*—argued that the strange animal was "*an unknown species of saurian reptile*", closer to the lizards

51. Camper (A.) to Cuvier, 12 November 1799 (Paris-IF, MS 3221/1); Cuvier to Camper, 30 brumaire VIII [21 November 1799] (Amsterdam-UB, MS X48a); these and most of their later letters (except some of Camper's in Paris-MHN) are listed and summarized in Theunissen, "Briefwisseling tussen Camper en Cuvier" (1980), which also gives a fine evaluation of the relationship. The mere nine days that separate this first pair of letters indicate not only Cuvier's keenness to cultivate the contact, but also the remarkable speed of the mails, even in wartime; and Camper's little town of Franeker was not a major center like Amsterdam or Haarlem.

52. Camper to Cuvier, 6 December 1799, 5–6 January 1800 (Paris-IF, MS 3221/2, 3222/12; Paris-MHN, MS 630, letter 23); Cuvier to Camper, 25 frimaire VIII [16 December 1799] (Amsterdam-UB, MS X48b,c; Cuvier's copy is in Paris-MHN, MS 630/2). The letters cited, here and in following footnotes, are just the more important ones in this intensive exchange.

53. Camper to Cuvier, 1 May, 9, 14 June 1800 (Paris-IF, MS 3222/5; Paris-MHN, MS 627, letter 135, MS 630/2); Cuvier to Camper, 16 germinal, 27 floréal VIII [6 April, 17 May 1800] (Amsterdam-UB, MS X48d,e).

54. Camper to Cuvier, 19 July, 3 August 1800 (Paris-IF, MS 3222/7,8); Cuvier to Camper, 8 messidor VIII [27 June 1800], and undated [July 1800?] (Amsterdam-UB, MS X48f,g).

than to any crocodile; the size of the skull suggested that it might have been some twenty-four feet long. This made it much more strange than Faujas had implied: a huge *marine lizard* without parallel among living reptiles. Cuvier reported that Lacépède, having seen the new specimens, thought they were vertebrae of a whale, but on learning more about them Camper insisted that they too fitted his lizard identification. In this case even Cuvier himself was confused for a time, but in the end he accepted correction and adopted Camper's conclusion: a rare occasion on which he deferred to another savant on a matter of comparative anatomy. If they were right, both Petrus Camper and Faujas had been mistaken: the Maastricht fauna contained no mammals at all and no crocodiles; it belonged more clearly than ever to a distinct former world.[55]

The strange Maastricht animal was just one item among the many that Camper and Cuvier discussed in the course of this remarkably intense correspondence on fossil bones. They had never met—Cuvier had not yet arrived in Paris when Camper visited the city before the Revolution—and he was flattered when Camper asked him to send a portrait of himself; yet the relationship became almost intimate, and on Cuvier's side unusually candid about the progress of his research. It did not last: their correspondence tailed off, and Camper later reproached his pen friend for failing to answer his letters. Cuvier blamed his ever increasing burdens of teaching and administration, and doubtless with some truth; but it is also likely that he came to feel that Camper had helped him as much as he could, and that he needed a similar collaboration with many others in order to exploit an even wider range of material.

Cuvier's international appeal

After Cuvier and Camper had been corresponding for barely a year, the Parisian gave the Institut a revised progress report on his research on fossil bones; and he widened his catchment area dramatically by appealing to savants and collectors everywhere to send him material to make it as complete as possible. Cuvier set his research in an unambiguously geohistorical context, taking it for granted that "everyone now knows that the globe we live on displays almost everywhere the indisputable traces of vast revolutions", most of them long before "the empire of mankind". As on previous occasions (§7.2), he used the analogy with the work of antiquarians to explain what naturalists could do with this evidence for prehuman geohistory. Cuvier rejected the fanciful constructions of the geotheorists and praised instead "the Saussures, Pallases, and Dolomieus", because "they recognized that the first step in divining the past was to establish the present securely" by careful fieldwork.[56]

Cuvier then introduced fossil bones as a class of evidence that had been relatively neglected yet was potentially of decisive importance. For the crucial issue—which he couched in terms clearly borrowed from Dolomieu and de Luc—was "to know the extent of the catastrophe that preceded the formation of our continents". More precisely it was to know "whether the species that then existed have been entirely destroyed, or solely modified in form, or simply transported from one climate to another": a clear statement of the three alternative explanations of extinction, transmutation, and transport (the last either by slow migration or by a mega-tsunami). Cuvier argued that fossil mammals were exceptionally promising material for this purpose, because the class was of limited size and already well-known, with skeletons of most species available for comparison with fossil bones; unlike marine animals such as mollusks and fish, perhaps living in distant seas or deep oceans, it was relatively unlikely that any *large* mammals remained alive but still undiscovered.

This acknowledged tacitly that any argument about the relation between living and fossil species was bound to be *probabilistic* in character, and therefore that the only realistic strategy for strengthening the argument would be a *cumulative* one. The more fossil mammals that could be shown to be distinct from any known living species, the more *likely* it would be that the true explanation for the difference was that the older species had become extinct (or, less welcome to Cuvier, that they had transmuted into living species), rather than having merely migrated elsewhere. And if extinction was the true cause, it would imply that there had indeed been a "revolution"; not merely a major change but specifically a sudden and even violent "catastrophe" or "upheaval" [*bouleversement*]. For only such a major event could have overwhelmed and wiped out the well-adapted "animal machines" that Cuvier believed all these mammals to have been.

As in his earlier lecture to the Institut (§7.2), Cuvier used the fossils in the Parisian Gypsum to explain his methods for identifying and reconstructing putatively extinct species from their scattered bones. He insisted that in *all* the cases he had been able to study at first hand the fossils were distinct; all those in which other naturalists had claimed that they were identical to living species had evaporated in the light of a critical comparative anatomy. The only apparent exceptions were some fossil ruminants—but it was notoriously difficult to distinguish even the living species of that group from their bones alone—and some bones from peat bogs that were not truly fossils at all. Cuvier's tally of extinct species had now risen to twenty-three, of which he himself took credit for no fewer than eleven. But this was not just to boast of his own achievement: he pointed out that if he, a single naturalist, had found so many in so short a time, with the help of only a few informants, far more might be found, given more time and the cooperation of a wider circle of naturalists and fossil collectors (Fig. 7.10).[57]

For it was to the "savants and amateurs" of the whole civilized world—far beyond the membership of the Institut—that Cuvier's paper was explicitly directed.

55. Camper to Cuvier, 9 August 1800 (Paris-MHN, in MS 629.I), published as Camper (A. G.), "Ossemens fossiles à Maëstricht" (1800); and 12, 27 August, 9 September, 2 November, 31 December 1800 (Paris-IF, MS 3222/1,2,9–11); Cuvier to Camper, 24 thermidor, 6 fructidor, 26 vendémiaire VIII [12, 24 August, 18 October 1800] (Amsterdam-UB, MS X48h–j). Cuvier's vacillation over the vertebrae from Sichem is described, and the "*mosasaur*" (as the Maastricht animal was later named, though not by Cuvier) is analyzed as a test case for the application of his anatomical principles, in Theunissen, "Mosasaurusvraagstuk" (1984) and "Cuvier's *lois zoologiques*" (1986).

56. Cuvier, *Espèces de quadrupèdes* (1800).

57. Fig. 7.10 is reproduced from Banks's copy [London-BL, B.352.(7)] of Cuvier, *Espèces de quadrupèdes* (1800); Cuvier to Banks, 15 frimaire IX [6 December 1800] (London-BL. Add. MS 8099, f.154), was the covering letter.

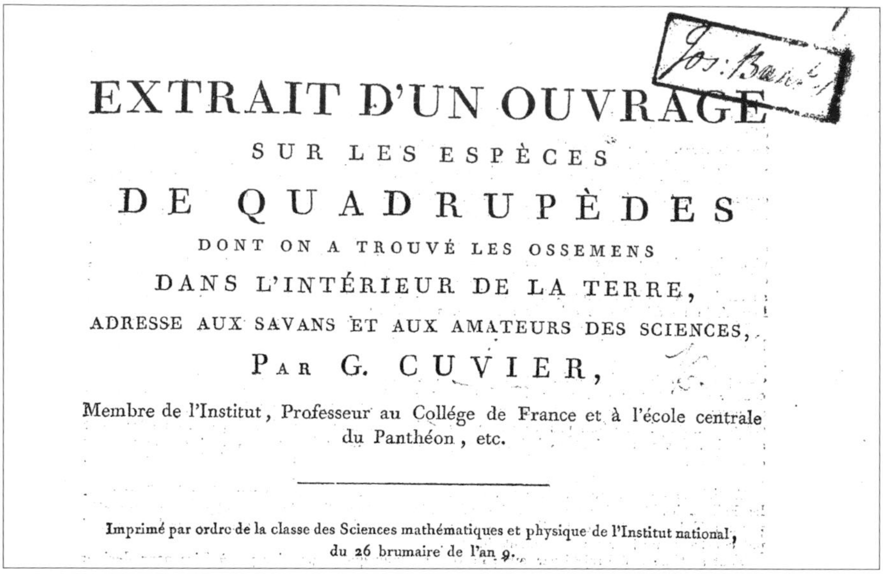

EXTRAIT D'UN OUVRAGE
SUR LES ESPÈCES
DE QUADRUPÈDES
DONT ON A TROUVÉ LES OSSEMENS
DANS L'INTÉRIEUR DE LA TERRE,
ADRESSE AUX SAVANS ET AUX AMATEURS DES SCIENCES,
PAR G. CUVIER,
Membre de l'Institut, Professeur au Collége de France et à l'école centrale du Panthéon, etc.

Imprimé par ordre de la classe des Sciences mathématiques et physique de l'Institut national, du 26 brumaire de l'an 9.

Fig. 7.10. Part of the first page of Cuvier's appeal to the whole Republic of Letters for help in his research project "on the species of quadrupeds, the bones of which are found in the earth's interior". This pamphlet was published by the Institut National almost immediately after he read the paper in Paris in November 1800. It was distributed internationally, addressed both to established naturalists [*savans*] and to the owners of fossil collections [*amateurs*]. Cuvier appealed to them to send him drawings of their specimens, in return for which he offered to identify the bones and to acknowledge their assistance in the major work that he was planning, of which this was just a preliminary "extract". This particular copy was sent to Sir Joseph Banks, the president of the Royal Society in London (and marked with his library stamp), despite the major war between the two countries. (By permission of the British Library)

He asked them to send him information about the fossil bones that they possessed or knew about, and particularly to send drawings of them, to enlarge his database with proxy bones. (Tacitly he took it for granted that they would not want to entrust the specimens themselves to the nation that had so rapaciously looted its conquered territories in the preceding years and was continuing to do so under Bonaparte.) In return he offered to identify their specimens with all the authority that he was assuming for himself, and to give them in print the "glory" that would be their due for having assisted in the great work that he planned to publish. As he put it, "this reciprocal exchange of information [*lumières*] is perhaps the most noble and interesting commerce that men can have". To establish his own credentials and to encourage wider collaboration, Cuvier listed all the colleagues, collectors, and foreign savants who had already helped him. To discourage any rivals, he emphasized that his project was already far advanced; in the crucially important and expensive matter of engraving, he had no fewer than fifty plates ready for publication (Fig. 7.11).[58]

The Institut ordered that Cuvier's paper be printed at once at its expense, and it was duly distributed as a pamphlet to naturalists, collectors and institutions throughout the learned world. Cuvier's position as the relevant secretary of the First Class doubtless helped in getting it this unusually favorable treatment. It was also published promptly in the *Journal de Physique* and the *Magasin Encyclopédique*, and extracts were translated in German, Italian, and English periodicals. At the

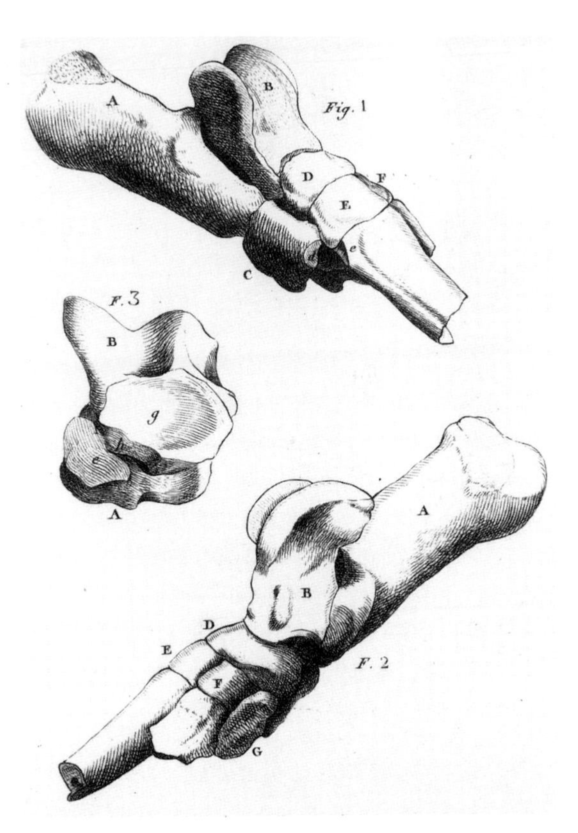

Fig. 7.11. "Paris fossils": engravings that were ready for publication when Cuvier appealed for international cooperation with his work on fossil bones; his stockpile of fifty such plates established his claim to priority as the authoritative center for such research. These particular bones from the Paris gypsum were engraved by Cuvier himself, from his drawings reproduced here as Fig. 7.6 (reversed by the engraving process); he had been taught the craft by the Parisian engraver Simon Miger, who himself engraved many of Cuvier's later plates. (By permission of the Syndics of Cambridge University Library)

start of the new century, as Bonaparte's armies expanded the boundaries of the French Republic and its conquered territories, puppet states, and allied powers, his almost exact contemporary in the Republic of Letters made a bid to dominate the focal problem of fossil bones with a similar but intellectual appropriation of material from the rest of Europe and even beyond it.[59]

Conclusion

The Revolutionary wars that engulfed most of Europe in the last years of the eighteenth century had relatively little impact on the work of scientific savants. Their travels were greatly restricted, but they continued to exchange ideas, information, and specimens—the last often in the more convenient form of paper proxies—almost as readily as in times of peace. Dolomieu's incarceration as a prisoner of war

58. Fig. 7.11 is reproduced from Cuvier, *Ossemens fossiles* (1812), 3 [Cambridge-UL: MD.8.67], "Pieds de derrière", pl., first published as "Restitution des pieds" (1804), pl. 4, annotated "Cuvier del. et Sc.". Its early date is indicated by the description of the bones as those of one of the as yet unnamed "Fossiles de Paris". On Cuvier as an engraver, see Rudwick, "Cuvier's paper museum" (2000). A notorious recent case of Bonaparte's acquisitiveness in the field of natural history was his expropriation in 1797 of the largest collection of Monte Bolca fossil fish (see Fig. 5.7), for deposit in the Muséum in Paris: see Frigo and Sorbini, *600 fossili per Napoleone* (1997).

59. Cuvier, *Espèces de quadrupèdes* (1800); translations, etc., are listed in Smith, *Georges Cuvier* (1993), nos. 60, 98, 105, 106, 125. See also Rudwick, "Cuvier et le collecte des alliés" (1997). Viénot, *Napoléon de l'intelligence* (1932), used the parallel in the title of his biography of Cuvier.

was quite exceptional; more typical was van Marum's switch from electricity to fossils as a convenient (and less expensive) topic for his scientific lectures. Bonaparte's military expedition to Egypt, although strategically a failure, was a scientific and cultural success; and his subsequent seizure of power as First Consul and virtual dictator of France also had little immediate impact on the savant world. In particular, the carefully apolitical Cuvier continued his meteoric rise in influence and power in the world of the natural-history sciences. He cultivated a growing network of informants to help in his research on fossil bones. His intensive dialogue with Camper's son Adriaan was particularly fruitful, although conducted wholly by letter. This may have prompted him to issue an appeal to "savants and amateurs" everywhere, to send him paper proxies of relevant specimens in their collections. Their contributions were intended to make his projected study of fossil bones as comprehensive as possible, and thereby to enable him to arrive at a definitive solution to the problem that they posed. In place of the vague speculations of geotheory, Cuvier hoped to prove conclusively that a major physical "revolution" at the dawn of human history had wiped out a large and varied fauna of quadrupeds. Extinction would then be recognized as part of the course of nature, and the *history* of nature would be established as accessible to human knowledge.

7.4 LAMARCK'S ALTERNATIVE (1800–1802)

The threat of transformism

Just as Cuvier was consolidating his power in the savant world and appealing internationally for help with his research, a major threat to his conclusions about fossil bones appeared unexpectedly. His senior colleague Lamarck announced publicly that he had now adopted the idea that all organisms were being continuously transformed in the course of time, so that "species" were in the long run no more than arbitrary points on a continuum. The differences between fossil and living species might simply reflect this process of endless flux, rather than extinction (or origin) or any other contingent event in geohistory. This was the greatest challenge to Cuvier's argument about fossil bones. La Métherie's claim, that the differences between living and fossil forms fell within the range of intraspecific variation, might be refuted by careful study of variability under natural conditions. The suggestion by Faujas and others, about the possible survival of apparently extinct species as "living fossils", might become less and less plausible in the course of further exploration around the world. But claims about transmutation would be more difficult to refute, if it was a process that unfolded too slowly to be observed on a human timescale. And such ideas could no longer be dismissed as fanciful and speculative, as they had been in the past, once they were being put forward by a savant with a solid reputation as a naturalist.

Before the Revolution Lamarck had made his name as a botanist at the Jardin du Roi. However, when it was transformed into the Muséum he did not get the new botanical position and was instead appointed professor for "insects and worms" (or, to use the modern term that he soon invented, all the invertebrates). Far from being demeaning, however, or a mere consolation prize, this position gave him a welcome opportunity to study some of the simplest organisms, which he believed might hold the key to understanding life itself. For in addition to his fine taxonomic work, Lamarck regarded himself as a wide-ranging natural philosopher, and he aspired to understand the causal "physics" of *everything* in the natural world. In this respect, however, the papers that he read at the Académie des Sciences had been judged too speculative to be published by that august body (and his opposition to Lavoisier's "new" chemistry had not helped his cause). So the first of his major works of this kind—covering in modern terms much of physics, chemistry, and biology—did not appear until 1794, after the Académie had been abolished. His later papers were perhaps also spurned by Parisian publishers, for in 1797 Lamarck published them privately. Anyway, in these volumes he rejected the kind of developmental theorizing that Buffon had used in *Nature's Epochs,* and treated the natural order of things as moving from the complex to the simple, not the other way round. At this point Lamarck was not, in modern terms, an evolutionist of any kind.[60]

However, in 1800, in the opening lecture of his annual course on zoology at the Muséum, Lamarck reversed what he had expounded in his earlier works—and even on the same occasion the previous year—and claimed that the "marche de la nature" was from simple to complex, and that time and favorable circumstances were the principal means by which complexity developed. The implication was that organisms were in a state of continual flux, and that any one species would inevitably be transformed into another, given the lapse of enough time. This view of life was associated with a corresponding view of continually shifting continents and oceanic basins, which would provide organisms with ever-changing environments. Lamarck had sketched this theory to the Institut the previous year, treating the features of the earth's surface as the products of ordinary physical processes acting through immense spans of time. It was a familiar kind of geotheory: like many others, such as La Métherie's (§6.5), in its reliance on ordinary processes and a vast timescale; specifically like Hutton's in its endless steady state, though without the internal heat that powered Hutton's dynamic earth machine (§3.4).[61]

Lamarck's all-embracing concept of global flux, organic and inorganic, belonged to a well-established tradition of natural philosophy. But as in the case of geotheory, what was now esteemed by leading savants was theorizing that was less speculative, perhaps less globally ambitious, and certainly more rigorously buttressed by concrete observational evidence. Lamarck was well aware of this, and he proposed to provide such evidence through a close study of fossil and living mollusks. He promised a work on the "elements of conchology", which might settle the long-running dispute on the relation between living and fossil species. He was

60. Lamarck, *Recherches sur les causes* (1794) and *Mémoires de physique* (1797). See Burkhardt, *Spirit of system* (1977); Corsi, *Age of Lamarck* (1988), esp. chap. 2; and, more generally, Jordanova, *Lamarck* (1984); more recent research is covered by the essays in Laurent, *Jean-Baptiste Lamarck* (1997). It is beyond the scope of the present book to deal with Lamarck's work on what he was almost the first to define as "*biologie*" and its underlying philosophy.

61. Both texts were revised and published later, in Lamarck, *Animaux sans vertèbres* (1801) and *Hydrogéologie* (1802). See Corsi, *Age of Lamarck* (1988), 103–4.

well-placed to do so. He had a fine shell collection of his own, and access not only to those at the Muséum but also to those of Parisian collectors, among which that of Jacques Defrance was exceptionally rich. When Bruguière had left Paris on his expedition to the East (in the event, never to return), Lamarck had taken his place as the leading conchologist, continuing the younger naturalist's work on mollusks for the *Encyclopédie Méthodique* (§6.1). It was probably this new taxonomic research that precipitated Lamarck's rather sudden conversion to the idea of transmutation. Like many others at this time, he rejected the possibility of extinction as contrary to the fundamental character of nature's economy and could not conceive how it could ever happen naturally, without human intervention. Yet his work on mollusks convinced him, against his expectations, that not all fossil shells had exact "analogues" among living species. So transmutation, supplementing the appeal to the living fossil argument that Bruguière had championed (§5.2), became the only explanation of the *non*-identity of past and present that Lamarck could accept.[62]

At the time that Cuvier issued his general appeal to "savants and amateurs", Lamarck's new views were still barely sketched in outline. But his ideas about the transmutation of organisms, combined with his steady-state geotheory, clearly challenged Cuvier's research project. Not only did it offer an alternative to any putative mass extinction; more fundamentally, it denied that the earth and its life could have had any true *history*. For Lamarck combined an endless cycle of environmental change with a process through which extremely simple forms of life [*monades*] were continually being generated "spontaneously" from nonliving matter and thereafter slowly transformed into ever more complex forms. Such a model necessarily implied that at no point would the system be distinctive or characteristic *of that specific time*. The potential disagreement between Lamarck and Cuvier therefore ranged all the way from differing opinions on the relation between particular fossil species and their living counterparts to incompatible views about the nature of nature itself. The relationship between the two naturalists appears to have been cordial in the early years after Cuvier's arrival in Paris: although the younger man was doing substantial research on the comparative anatomy of mollusks, Lamarck could well have regarded it as usefully complementary to his own purely taxonomic work on the same animals. But in the longer run a clash between them was almost inevitable.

The publication of Lamarck's *Animals without Backbones* (1801) made his ideas known far beyond the circles of his Parisian auditors. It printed his opening lecture of the previous year as an introduction to the systematic review of invertebrate animals that occupied most of the book. But Lamarck also included, as a kind of appendix, a brief but important essay "On fossils". He described fossils as "extremely precious monuments of the state of the revolutions that different points on the surface of the globe have undergone". This much was uncontroversial. But he added immediately that fossils were also valuable traces "of the changes that living beings have themselves successively experienced [*éprouvés*] there". This was far more problematic, since he was clearly referring to a process of gradual transmutation in the organisms themselves, rather than to changes in the composition of successive faunas.[63]

Most important, however, was Lamarck's insistence that finding even a small number of extant species among fossils—as he claimed to have done with the mollusks—was enough to refute any explanation of the contrast in terms of "a universal upheaval [*bouleversement*], a general catastrophe". For if every organism was continually changing in form, albeit insensibly slowly, the apparent differences between fossil and living species could be due to the passage of time and changes of environment; the supposedly "lost" species could simply have changed in appearance. Indeed, Lamarck affected surprise that *any* fossils were identical to living species, and he suggested that they were those that had not yet had time enough to change. This turned Cuvier's argument on its head: the greater the contrast between fossils and living forms, the more—according to Lamarck—it proved the ubiquity of transmutation in the organic world and the vast scale of time, rather than any catastrophic extinction. With Cuvier clearly if covertly in his sights, Lamarck argued that to suggest a catastrophe was to abandon the search for the orderly regularities of nature that ought to direct the practice of every part of causal "physics". The mistake, he charged, was both profound and culpable: "a universal upheaval, which necessarily regularizes nothing, and confuses and disperses everything, is a highly convenient means for those naturalists who want to explain everything, and who take no trouble at all to observe and study the course that nature follows in regard to its productions and all that constitutes its domain."[64]

The scope of Lamarck's ambitions was clearly expressed in his plans for a major work on "physique terrestre", in three parts dealing with meteorology, geology, and what he termed *biologie* (or, in modern terms, respectively with the atmosphere, hydro- and lithosphere, and biosphere). The second part, his *Hydrogéologie* (1802), was the first to be published; as the title implied, it was a geotheory centered on the action of water, and particularly the oceans. In that respect it contrasted with Hutton's, which Lamarck would have known about from Desmarest's substantial account (§6.5) or other reports in French; but like Hutton's it was a theory that assumed a *balance* between antagonistic processes, maintaining the earth in a steady state. Lamarck argued that in general there was steady erosion on the eastern coasts of continents, balanced by sedimentation on the western ones, so that in the long run the ocean basins were being slowly displaced westwards around the globe, in an endless rotation; the model was not unlike the one originally advocated by Lamarck's former patron, before Buffon switched to the developmental model of *Nature's Epochs* (§3.2). For Lamarck, as for Hutton, the balanced processes operated imperceptibly slowly; in the Enlightenment style that was common to them both, Lamarck insisted that "for nature, time is nothing". The scale of time was strictly incalculable; but he suggested that the formation of a new continent would take

62. This follows broadly the interpretation offered in Burkhardt, *Spirit of system* (1977), chap. 5, which remains in my opinion the most convincing explanation of the genesis of Lamarck's evolutionism.

63. Lamarck, *Animaux sans vertèbres* (1801), 406; the "Discours d'ouverture" (1–48) and the note "Sur les fossiles" (403–11) are translated in Newth, "Lamarck in 1800" (1952). The quoted phrase illustrates yet again how, in the contemporary use of the word, *any* major changes were "revolutions", even though for Lamarck they were insensibly gradual and certainly not sudden or violent.

64. Lamarck, *Animaux sans vertèbres* (1801), 407.

"an enormous multitude of centuries", while a complete cycle might last perhaps three million (i.e., in modern terms, 300Ma).[65]

Such vast guesstimates give Lamarck's geotheory an apparent modernity that is as spurious as in Hutton's case. By the time it was published, *Hydrogéologie*, like other geotheories, already looked somewhat outmoded; it is not surprising that—as Lamarck himself complained—it received little critical attention. Nor was his rejection of catastrophes and his insistence on the total adequacy of present processes (de Luc's "actual causes") a matter for any comment, since such explanations had been an acceptable kind of theorizing ever since Desmarest and Soulavie used them to interpret the topography of central France (§4.3, §4.4). However, unlike those savants (but like Hutton), Lamarck in effect denied geohistory: in his geotheory the continents and oceans at any one time would have been the same in general character as those at any other time (though of course the details of geography might never have been repeated exactly). Cuvier, in sharp contrast, was extending the ideas of de Luc and Dolomieu, and claiming that the former world of his extinct mammals was distinctly different from the present world of living species, proving that life on earth had had a *history* of its own, just as much as the human history recovered by the antiquarians. This, rather than any catastrophe, was the most important point that Lamarck was rejecting.[66]

The response to Cuvier's appeal

Cuvier was far too astute not to recognize the challenge posed by his prominent colleague's conversion to transformism and adoption of an ahistorical geotheory. But rather than confronting it directly, he first continued to consolidate his own interpretation of fossil bones.

There were two complementary ways in which his sharp distinction between the faunas of the former and present worlds could be confirmed. The first was simply to extend his detailed studies of specific kinds of fossil bones and show by careful osteological comparisons that all of them really were distinct from the bones of any living species. His appeal to other naturalists had of course been designed to enlarge his database for just this purpose. In addition to the copies sent out by the Institut, Cuvier himself sent the pamphlet to his existing informants around Europe, asking them to give it further publicity in their own circles of "savants and amateurs". For example he asked Camper and Fabbroni to get it translated for Dutch and Italian periodicals respectively; and Gotthelf Fischer von Waldheim (1771–1853), who was teaching natural history in Mainz, sent Cuvier a list of seventeen German savants, among them Goethe, to whom he suggested that further copies should be sent.[67]

Cuvier's appeal began at once to yield a rich harvest: reports and—far more usefully—proxy specimens in the form of accurate drawings of bones arrived at the Muséum from all over Europe and even from beyond it. A few early examples are enough to illustrate the point. The authority that was attributed to Cuvier by most of his informants was well expressed when Fabbroni sent him a drawing of one fossil tooth and added that "it's for you to instruct us about the animal that bore it". A later letter enclosed a picture of "the remains of an unknown creature" found by a friend; Fabbroni was sure that its accompanying explanation in Italian would give Cuvier no problem, since "you're familiar with all languages" (an exaggeration, of course, but not in terms of those of scientific importance). Diedrich Karsten (1768–1810), who taught mineralogy at the mining school in Berlin, forwarded a note about the mammoth, by a traveler in Siberia, which had been read at the local scientific society to which Cuvier had earlier appealed for help with getting information from Britain. The same society later sent him a report on fossil bones of an animal that "seems without doubt to have had a pre-Adamic existence"—the adjective echoed Blumenbach's earlier work (§6.1)—and Karsten followed this up with details of the cave in Westphalia in which they had been found, "hitherto unknown to the republic of letters". Knowing that these remains of "a creation that has perished" might be important, Karsten got an artist to make no fewer than ten plates of drawings of the bones and had their accuracy checked by de Luc (who happened to be in Berlin) before they were sent to Paris. As a final example of all this savant traffic, and a typical example of what Cuvier was receiving in addition to letters, Blumenbach himself sent a drawing of a tooth of the Ohio animal from his collection in Göttingen (Fig. 7.12).[68]

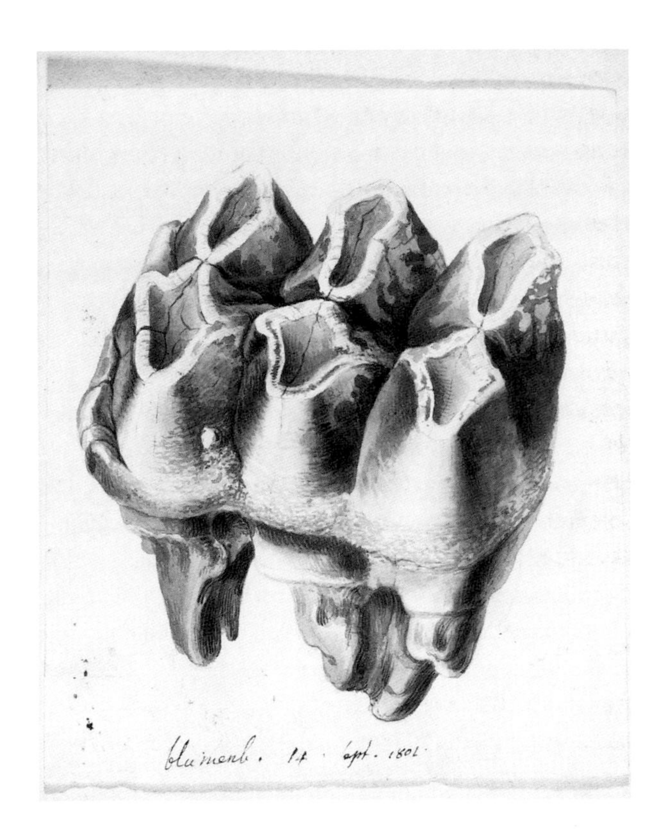

Fig. 7.12. A drawing of a molar tooth of the "Ohio animal", the original of which was in Göttingen, sent to Paris by Blumenbach in response to Cuvier's international appeal for proxy specimens to enlarge his database for his research on fossil bones. Cuvier's annotation, "blumenb[ach] 14 sept[ember] 1801", recorded the date of the covering letter; he later published the drawing as an engraving, to illustrate his paper (1806) on what he then named the *mastodon* and interpreted as an extinct distant relative of the elephants. (By permission of the Bibliothèque Centrale du Muséum National d'Histoire Naturelle, Paris)

65. Lamarck, *Hydrogéologie* (1802), 86–90, 178–80; and *Physique terrestre*, dated Year X [1801–2] (Paris-MHN, MS 756); see also Gohau, "L'Hydrogéologie" (1997). It is almost inconceivable that Lamarck did not read Desmarest's review of geotheories while he was contributing to the same encyclopedia; the lack of explicit reference to either him or Hutton means little, in view of contemporary practices on citation (or rather, noncitation) of sources.

66. The third part of Lamarck's grand design appeared later the same year, as his *Recherches sur les corps vivans* (1802); it amplified and clarified his earlier statements of his transformist view of life, using his geotheory as the inorganic foundation for his "biology".

67. Cuvier to Camper, 20 nivôse IX [10 January 1801] (Amsterdam-UB, MS X48k); Cuvier to Fabbroni, 21 pluviôse IX [10 February 1801] and Fabbroni to Cuvier, 30 September 1801 (Paris-IF, MS 3223/35), both printed in Outram, "Storia naturale e politica" (1982), 198, 200; Fischer to Cuvier, 26 nivôse IX [16 January 1801] (Paris-IF, MS 3223/16).

68. Fig. 7.12 is reproduced from a drawing in Paris-MHN, MS 630(2), published as an engraving in Cuvier, "Grand mastodonte" (1806), pl. 49, fig. 5. Of the vast number of drawings preserved in Cuvier's research files

The accumulation of such materials was enlarging Cuvier's database of fossil bones, in real or proxy form, and thereby strengthening his evidence that all the species represented differed from any living animals. The second way in which he tried to sharpen the contrast between the present world and that former world of putatively extinct species was to show that even the earliest *textual* accounts of animals, however oddly they might read, referred exclusively to species that were known alive (unless they were purely mythical beasts). In effect, this was to trace the record of animals backwards from the present, as far as possible towards the revolution that marked the boundary between the two worlds, just as his work with fossils could hope to trace it forwards out of the deeper past towards that same event.

In this respect as in so many others, Cuvier was the right naturalist in the right place at the right time. He and his colleague Lacépède were assigned the task of reporting to the Institut on papers submitted by a prominent Classical scholar, dealing with the identity of certain animals described in Antiquity. And he and Lamarck were appointed consultants on natural history for the French edition of the first volumes of *Asiatick Researches*, the great periodical published by the scholarly Asiatick Society in British India (§1.2), in which many of the articles dealt with Sanskrit texts from a quite different Antiquity. Both tasks led Cuvier to deploy his skills as a naturalist in the field of textual criticism: the naturalist from the First Class was contributing to the humanistic scholarly work of the Third. In effect, Cuvier himself became an antiquarian: not just analogically, as he was when trying to reconstruct geohistory, but quite literally, in evaluating in scientific terms the textual records of early *human* history.[69]

Mummified animals from Egypt

At the same time, Cuvier had the opportunity to check the fauna of Antiquity more substantially, by making use of the mummified animals that had been brought back from Egypt. One animal in particular, which he studied even before the expedition returned, exemplified his method of combining a careful study of specimens with a critical analysis of ancient textual and pictorial sources. Its subject was the sacred ibis of the ancient Egyptians, one of the animals most commonly mummified and most frequently represented in their art. Initially, Cuvier used a single mummified specimen sent back to Paris by one of the French generals in Egypt. He claimed that the sacred bird had been misidentified as Linnaeus's *Tantalus ibis*, a stork, when in fact it belonged to a previously undescribed species of curlew, which Cuvier named *Numenius ibis*. His argument was supported not only by a detailed study of the osteology and plumage of the two species, but also by an equally rigorous analysis of the descriptions of the sacred bird in Herodotus, Plutarch, and other ancient authors, and its depiction on ancient monuments in Egypt, at Herculaneum, and elsewhere. Cuvier made it clear that, without such careful comparative research, many errors would be left uncorrected and any wider conclusions would be unreliable (Figs. 7.13, 7.14).[70]

Such wider conclusions soon became apparent. When the expedition returned from Egypt, after the French forces were defeated by the British, the naturalists at

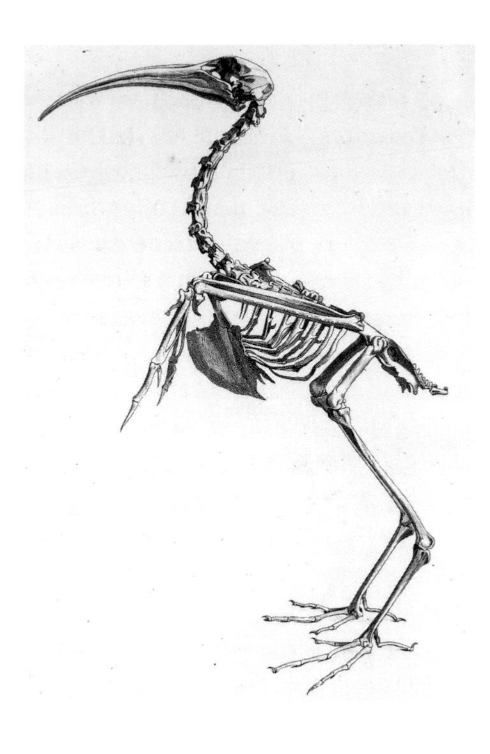

Fig. 7.13. "Skeleton of the ibis, taken from a mummy from Thebes in Egypt": an engraving used by Cuvier to illustrate the full account (1804) of his claim that the species treated as sacred by the ancient Egyptians had been misidentified and was in fact a bird still living in the region. Mummified animals such as the ibis represented a three-thousand-year-old fauna that extended natural history back towards the putative catastrophe separating the present world from the far older former world of Cuvier's fossil bones. (By permission of the Syndics of Cambridge University Library)

Fig. 7.14. "*Numenius ibis*, the bird that I consider to be the true ibis of the Egyptians": an engraving of the living species of curlew that Cuvier in 1800 identified as the sacred bird preserved as mummies in the tombs and depicted on the walls of the temples of ancient Egypt. Cuvier argued that the identity between the mummified and the modern bird refuted Lamarck's claims about a continuous slow transmutation of animal form. (By permission of the Syndics of Cambridge University Library)

(in Paris-MHN), this is one of the few that is dated and can be matched confidently with those mentioned in his incoming letters, in this case Blumenbach to Cuvier, 14 September 1801 (Paris-IF, MS 3223/26). Other letters cited above are Fabbroni to Cuvier, 27 July, 26 December 1801 (Paris-IF, MS 3223/33, 36), printed in Outram, "Storia naturale e politica" (1982), 199, 204–5; Karsten to Cuvier, 4 November 1801 (Paris-IF, MS 3223/16) and 24 May 1802 (Paris-MHN, MS 634, letter 35/40); Gesellschaft Forschender Freunde to Cuvier, 16 February 1802 (Paris-MHN, MS 634, letter 41). See also Rudwick, "Cuvier et le collecte des alliés" (1997), and particularly the distribution map (fig. 2) of his informants before and after his appeal.

69. Cuvier's reports, on papers by Gail on the animals known in Antiquity as *Panther* and *Pardalis* and on two species of hare described in Xenophon, were read on 7 September 1799 and 9 August 1800: see Smith, *Georges Cuvier* (1993), nos. 52, 63. Jean-Baptiste Gail (1755–1829) published editions of Classical Greek texts and lectured at the Collège de France (and was later a member of the Institut). Translating and editorial work for Labaume, *Recherches asiatiques* (1805), must have begun around this time or soon afterwards.

70. Figs. 7.13, 7.14 are reproduced from Cuvier, "Ibis des anciens Égyptiens" (1804) [Cambridge-UL: Q382.b.11.4], pls. 52, 53, drawn by Thérèse Baudry de Balzac (1774–1831), an artist whose work (primarily botanical) was much used in the Muséum's publications. A third plate (54) included an ancient drawing of the bird "from one of the temples in Upper Egypt". Cuvier's preliminary account, "Ibis des anciens Égyptiens" (1800), was illustrated only with an earlier version of Fig. 7.14; it was based on a specimen thought to come from Senegal, in the collection confiscated from The Hague, but the species was known to be still living in Egypt.

the Muséum were instructed to report on the collections that Geoffroy and his colleagues brought back with them. Lacépède, who wrote the report, gave the mummified animals the prominence they clearly deserved, for the collection was in effect a three-thousand-year-old "cabinet of zoology", an ancient analogue of the Muséum itself. It owed its preservation to the "ignorant adoration" of certain animals by "these bizarre men"—Lacépède had no regard for modern political correctness—but it was outstandingly valuable because it offered a sample of the Egyptian fauna as it had been some three millennia in the past. "For a long time it has been desirable to know whether species change their form in the course of time," he noted with masterly understatement; "this question, apparently futile, is in fact essential to the history of the globe", and it could now be resolved as a result of Geoffroy's thorough collecting, for "these animals are perfectly similar to those of today". It was a tacit snub to Lamarck's claims about the continuous transmutation of animal forms in the course of time. Readers of the report were left in no doubt that the mummified fauna marked a kind of milestone, a short way back in time towards the far older and truly *fossil* fauna: "So one day it will be interesting to see, arranged in three series, today's animals, these others [from Egypt] already so ancient, and lastly those of an origin incomparably more remote, hidden in the better sealed tombs of the mountains over which our globe's terrible catastrophes extended."[71]

Although this report was written by Lacépède, it was also signed by his colleagues Lamarck and Cuvier. Cuvier would have been pleased at this confirmation of his own recent work on the ibis: the fauna of the present could be traced back, at least as far as human records went, without any sign of there having been any transmutation of species. Lamarck, on the other hand, must have cringed at Lacépède's suggestion of "terrible catastrophes" in the deep past, and he certainly did not accept the report's implied dismissal of his own claims about transformism. As he argued soon afterwards, the Egyptian case proved nothing, because three millennia were far too short a span of time over which detect any change in organic form.

Lamarck's Parisian fossils

The report on Egypt was published in the inaugural volume (1802) of the Muséum's own *Annales*, a new outlet that offered all the professors ready publication in a handsome format, with a generous allowance of the engraved plates that were so valuable in any branch of natural history. Later in the same volume Lamarck opened a discreet counteroffensive against Cuvier, by publishing the first installment of a major work on fossil mollusks. It was not the comprehensive "elements of conchology" that he had promised earlier, but instead a monograph dealing specifically with the fossil mollusks found in the region around Paris. He listed several localities, but by far the largest number of specimens came from a single site, the small but prolific quarry at Grignon, which had long been a favorite spot for Parisian fossil collectors such as Defrance. Anyway this work gave full scope to Lamarck's great talents as a taxonomist, though its impact was blunted by the fact that he published no illustrations of the fossils until 1805, when he was halfway through the eight-year project (Fig. 7.15).[72]

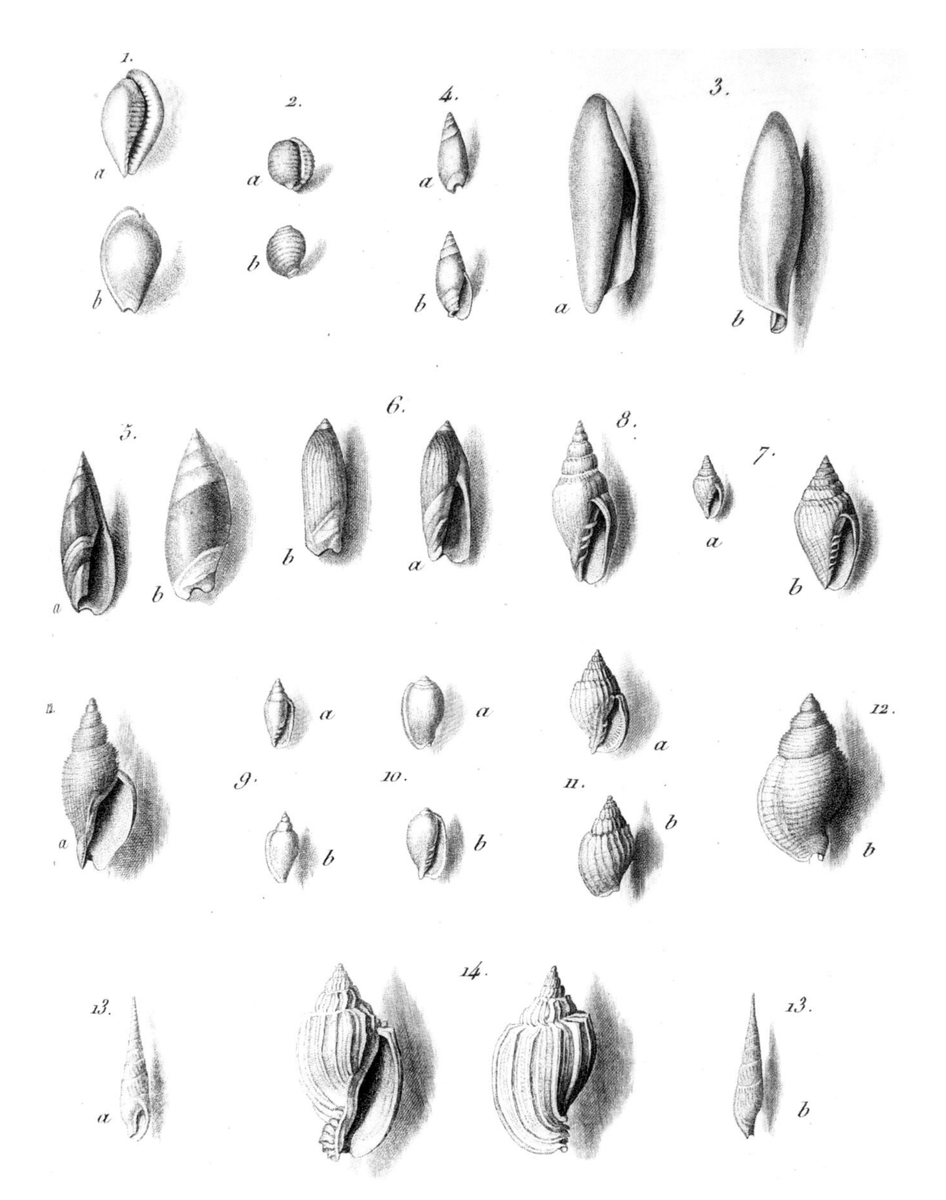

Fig. 7.15. Engravings published by Lamarck in 1805 to illustrate his great work on "Fossil shells from around Paris" (1802–09). This plate, which is characteristic in style, shows species of *Cypraea* (figs. 1, 2), *Terebellum* (3), *Oliva* (4), *Ancilla* (5, 6), *Mitra* (7, 8), *Marginella* (9, 10), *Cancellaria* (11), *Buccinum* (12), *Terebra* (13), and *Harpa* (14). All were reproduced exactly at natural size (except in fig. 7, where a very small shell was also shown enlarged), in order to facilitate direct comparison with further specimens. All but one of these species were known only from the famous locality at Grignon; many of the specimens illustrated came from Lamarck's own collection. Although lacking their original colors, all these shells were so well preserved that Lamarck could check confidently whether they were identical to species known alive. Elsewhere he attributed the contrast between the fossil and living assemblages to the slow transmutation wrought by changes of circumstances and the passage of time, and not to any extinction at all. (By permission of the Syndics of Cambridge University Library)

71. Lacépède et al., "Rapport des professeurs" (1802), dated 9 fructidor X [27 August 1802], 234–36. The figure of three thousand years was only a round guess, because with the hieroglyphic inscriptions undeciphered the exact chronology of ancient Egypt remained obscure and controversial. But it is worth noting that Lacépède, like any other competent naturalist of his generation, assumed that the timescale of geohistory was "incomparably" vast in relation even to this most remote human history.

72. Fig. 7.15 is reproduced from Lamarck, "Fossiles des environs de Paris" (1802–9), pl. 44 in *Annales* 6 (1805) [Cambridge-UL: Q382.b.11.6], in the first installment to include any illustrations. Up to this point Lamarck had

However, Lamarck's monograph was more than an important contribution to molluscan taxonomy. In its opening essay he cited his neglected *Hydrogéologie* and stressed the importance of fossils as "*monuments* of the slow revolutions of the earth's surface" and hence for "the true theory of our globe". He insisted that these mollusks, with their shells superbly preserved, must have lived where they are found and that they were all marine (apart from a few freshwater forms that could have been swept out to sea). That much was uncontroversial. But he also claimed that the many species belonging to genera that are now tropical in habitat witnessed to a significant change of climate, reflecting an ever-changing configuration of continents and oceans. He cited his recently published *Researches on Living Beings* for his theory that organisms are continually undergoing modifications in their form in response to such environmental changes. His clinching argument, however, was that all these changes happen too slowly to be appreciated by human observers. Even the few thousand years of human records were "an infinitely small duration, relative to those that see the effecting of the great changes that the surface of the globe undergoes". The organic stability that the Egyptian evidence clearly suggested to Cuvier, and indeed to Lacépède and others, was thus implicitly circumvented and indeed rejected. Cuvier's broader claim, to be describing a former world that was sharply distinct from the present and separated from it by a sudden and even violent catastrophe, was even more clearly challenged and indeed repudiated altogether.[73]

Cuvier would have been unconvinced by Lamarck's argument, not least because he was more deeply impressed by the example of his colleagues practicing the "exact" sciences at the Institut. Laplace, perhaps the most powerful and prestigious savant in France at this time, was in the middle of publishing his impressive volumes of *Celestial Mechanics* (1799–1805), with his rigorous demonstration of the adequacy of Newton's laws of gravitation for explaining all the movements of the solar system; he was even improving on Newton by resolving what had been some puzzling anomalies. Since before the invention of telescopes, this kind of mathematical astronomy had been based on precise observations of planetary movements: even the lengthy orbits of the outer planets could be calculated by extrapolation from observations made on the limited timescale of human lives. If Cuvier applied the conventional parallel between the vastnesses of astronomical space and geological time—as he certainly did later—he would have recognized that the Egyptian mummies were the analogue of this kind of accurate telescopic observation. If there had been no perceptible change in animal forms in the three millennia since the ancient Egyptians, there would be no good grounds for inferring that a far longer lapse of time would show any greater change: zero multiplied even by infinity would still be zero. Anyway, whether or not Cuvier consciously applied the astronomical analogy at this point, he certainly remained convinced that Lamarck's explanation of the difference between fossil and living species was deeply flawed and that his own concept of a mass extinction was far more plausible.[74]

Conclusion

Lamarck's conversion to the idea of transformism posed the greatest challenge to Cuvier's interpretation of recent geohistory. For if all organic forms were in a state of continuous flux, the apparent differences between living animals and their fossil counterparts might be simply a product of the lapse of time and in no way evidence for the reality of extinction as a natural process. Furthermore, Lamarck's steady-state geotheory denied in effect that the earth had had any true history, since there would have been nothing distinctive about the configuration of continents and oceans, or the environments that they provided for organisms, at any point in the deep past. Cuvier did not respond at once to this challenge. Instead he consolidated his own position, both by accumulating further fossil specimens (including paper proxies) sent in response to his international appeal, and by his analysis of the textual evidence about the animals known in Antiquity. In addition, the mummified specimens brought back from Egypt provided him with what he regarded as decisive evidence that there had been no transmutation at all in the past few millennia. For Lamarck, however, this was too short a span of time in which to expect to see any change. His detailed study of the abundant fossil shells found around Paris established that many of them had no living "analogues"; but he interpreted the difference not as evidence for any extinction but as further proof that they had transmuted into the living species of the present world.

The argument between Cuvier and Lamarck was therefore unresolved. It was of outstanding importance because it encompassed divergent views about the stability or mutability of species, about the reality or otherwise of extinction—particularly in a putatively recent and catastrophic revolution—and above all about the *historical* character of the earth and of the life that it sustained.

7.5 ENLARGING A FOSSIL MENAGERIE (1802–4)

A peaceful interlude

In the spring of 1802, after some ten years of war, France and Britain agreed to an uneasy truce, formalized in the Treaty of Amiens. This gave savants a welcome opportunity to travel freely in both directions across the Channel and to learn at first

---

referred only to the set of manuscript drawings that the Muséum's artists had made for its great (and formerly royal) collection of *vélins*, but of course this was not accessible except at the Muséum itself. The facsimile reprint (1978) of Lamarck's published text, and the drawings reproduced in their entirety in Palmer, "Vélins of Lamarck" (1977), reflect the enduring importance of his work for modern paleontologists. Grignon is about 14km west-northwest of the Palais de Versailles.

73. Lamarck, "Fossiles des environs de Paris" (1802–9), first installment, *Annales* 1 (1802): 299–312.

74. Laplace, *Méchanique céleste* (1799–1805). There seems to be no documentary evidence from the earliest 1800s for this reconstruction of Cuvier's reasoning; but a few years later (in his *Ossemens fossiles* of 1812, dedicated to Laplace) he certainly did refute Lamarck's argument by appealing to this analogy between space and time (§9.3).

hand what the other side had been doing in the sciences. Cuvier was among those who planned to take advantage of the peace and thereby to extend his catchment area for fossil bones. He knew that the collections in London, particularly those at the British Museum and the Royal College of Surgeons, contained many specimens of great potential importance for his project: for example the jaw from Big Bone Lick (Fig. 5.12) that William Hunter had used some thirty years earlier to argue that the Ohio animal was not a known living species (§5.3). But English specimens had not been matched by English research in this field, at least since the death of John Hunter several years earlier (§7.1).

Everard Home (1756–1832), Hunter's brother-in-law and his successor as surgeon at St George's Hospital in London, was almost the only English savant who was doing any serious research on comparative anatomy, and he became Cuvier's chief contact in Britain. Home had recently read papers to the Royal Society on the teeth of various mammals, mainly using specimens from Hunter's great collection. In one paper he did deal with fossil material, but when Cuvier read it in the *Philosophical Transactions* he would not have been much impressed. For Home referred to "the animal incognitum" as if the Ohio case was the only one; he mentioned the South American find but none of Cuvier's other work; and he admitted that he had studied the fossils only "as far as can be done from the specimens preserved in this country". Partly as a result of the wartime isolation of Britain, Home's comparative anatomy was no match for what was being done on the Continent. Still, the outbreak of peace did improve the flow of information into Britain: for example, the Royal Society received a copy of *Hydrogéologie* from Lamarck, and from the French Mines Agency a complete set of its *Journal des Mines*. The flow in the other direction had been less constricted, thanks, for example, to the circumvention of Bonaparte's blockade by the editors of the Genevan *Bibliothèque Britannique* (§7.3).[75]

In the event, Cuvier was unable to join the other French savants who were visiting London. As he told Home, he had been appointed to supervise the reorganization of secondary education in the south of France, which required him to travel there for several months. He therefore missed seeing with his own eyes one of the first fairly complete fossil skeletons to be found since Bru's from South America (see Fig. 7.3). The American portrait painter Charles Willson Peale had established a private natural history museum in Philadelphia for the entertainment of a paying public, having been inspired in part by the task of drawing some of the Ohio bones for Michaelis (§5.3). In 1799 he heard that huge fossil bones had been unearthed on a farm in upstate New York. After tortuous negotiations he purchased them, and in 1801 he organized an excavation to recover more of the bones. With Wistar as his anatomical consultant and his son Rembrandt as his technician, Peale then assembled and mounted two fairly complete skeletons of what Americans still called the "mammoth"; that is, of the "Ohio animal" that Europeans such as Blumenbach and Cuvier distinguished sharply from the true mammoth of Siberia. Peale kept the best skeleton in his museum; he also used it for a brilliant publicity stunt when he entertained a dozen guests to dinner *inside* the skeleton, at which patriotic songs and toasts indicated its adoption as an icon of cultural nationalism.

In 1802, taking advantage of the peace, Peale sent his second-best skeleton on a tour of Europe in the care of Rembrandt. It went first to London, where it joined the plethora of public shows for which the city was famous; it was an early example of the showbiz use of fossils (later to be transferred to the as yet undiscovered dinosaurs). The younger Peale published a booklet, dedicated to Banks, to accompany the exhibit, in which he summarized his own and his father's interpretation of the monster. They inferred that it had been a huge carnivore adapted to a cold climate (a scrap of wooly hide had been found with the bones). They attributed its "extirpation" to a "violent and sudden irruption of water", after which the bones had been buried under deposits that must have been "the production of a long succession of ages". In other words, the Peales argued for a standard kind of diluvial catastrophe. As already mentioned, Cuvier was unable to join the crowds that paid to stare at the New York "mammoth" in London. By the time he returned from the

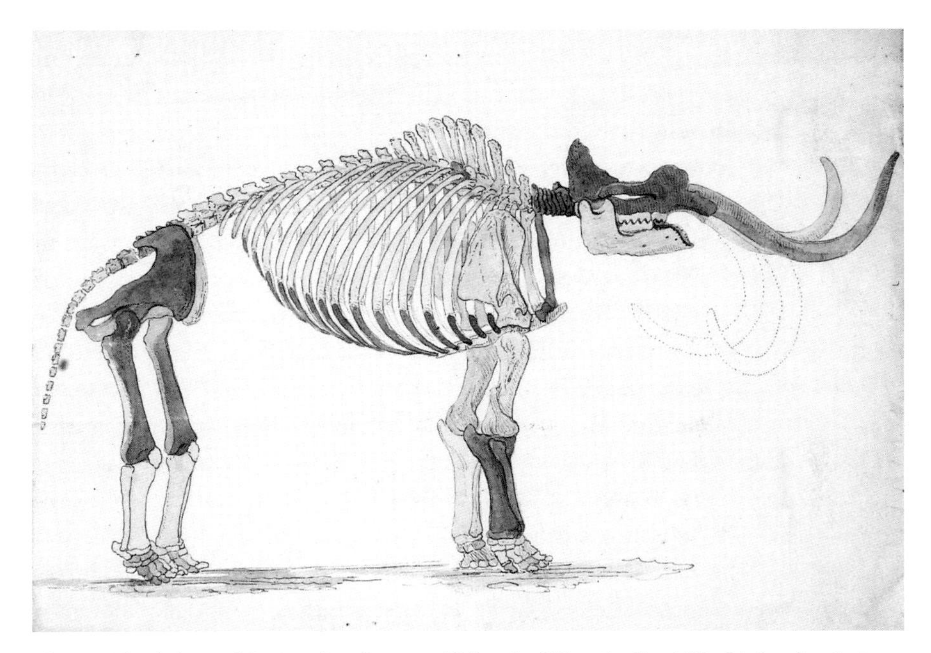

Fig. 7.16. The skeleton of the American "mammoth" (i.e., the Ohio animal), exhibited in London during the Peace of Amiens by Rembrandt Peale, whose father, the painter Charles Willson Peale, had masterminded its excavation in upstate New York and its reconstruction in Philadelphia. This drawing was sent by the London anatomist Everard Home to Cuvier in Paris; it highlighted the missing bones for which Rembrandt had carved replacements in wood. It was, as Home put it, "only for your own eye, as making it public would expose the imperfections of the skeleton, which I have no wish to do". Cuvier later had the drawing engraved—honoring Home's scruples—for his article (1806) on what he then named the *Mastodonte.* The most important part missing altogether was the roof of the skull. The drawing showed the upward curve of the tusks that Home thought more likely to be correct than the downward curve (in dotted outline) of the Peales' reconstruction. (By permission of the Bibliothèque Centrale du Muséum National d'Histoire Naturelle, Paris)

75. Home, "Teeth of graminivorous quadrupeds" (1799) and "Observations on the grinding teeth" (1801). He was widely suspected of plagiarizing Hunter's unpublished papers, of which he had custody and which he later destroyed. The Royal Society's "presents" for 1802 are listed in *Philosophical transactions* 92 (1802): 529–35. Lamarck sent his book even before the peace was formalized; the *Journal* had already been reaching Britain, but not by this official route.

south of France the fragile peace had collapsed and the war had resumed; his planned visit to London and the monster's to Paris were both cancelled, and Rembrandt and his bones returned to Philadelphia. However, Home later sent Cuvier a drawing of the Peale skeleton, which became important evidence in the Frenchman's work on the Ohio animal (Fig. 7.16).[76]

As with Bru's South American monster, Cuvier had to be content with a proxy of the Peales' North American one. However, his bureaucratic trip to the south of France had compensations: the salary was generous and, as he told Karsten in Berlin, he could make good use of his free time to study fossil collections in provincial museums. While he was away from Paris, two new appointments finally ensured his long-term financial security. On Mertrud's death he became his successor at the Muséum and had the chair redefined as "*comparative* anatomy"; and a few months later his position at the Institut was upgraded into one of two "*permanent* secretaries" of the First Class, at an enhanced salary that raised eyebrows among envious colleagues. Cuvier may have made enemies in Paris, but he also had powerful friends.[77]

However, Cuvier's supremacy even in his own field was again not uncontested. Before he left Paris to take up his duties in the south of France, his nemesis Faujas reasserted his rights to much of the same intellectual territory. He introduced his course on *géologie* at the Muséum with a lecture on the present state of the science throughout Europe, and defined its object as nothing less than "the theory of the earth". As in his book on the Maastricht fossils (§7.2), however, some of his ideas on geohistory were not unlike Cuvier's. For example he interpreted the great spreads of gravel he had seen in the Alpine regions as "irrefutable witnesses [to] sudden displacements of the sea that can only have been produced by terrible catastrophes", which in turn had punctuated long "periods of calm". He had read the work of Hutton's friend James Hall on the basalt problem with sufficient care to notice Hall's footnote outlining a theory of mega-tsunamis that might account for these catastrophes (§10.2); but he pointed out that it was much like the one that Saussure, Dolomieu, and he himself had been developing for more than twenty years.

All this might have been music to Cuvier's ear. But Faujas also argued that "the great question relative to the existence of analogues, one of the most remarkable and important for geology", was unresolved, and would remain so at least until Lamarck's work on the Parisian fossil mollusks (§7.4) was fully published; and he claimed that it was premature to infer any extinction at all, until the distribution of living species was better known. In his later lectures, his target became clear: Faujas dealt with all the main cases of fossil mammals, including the elephants, directly challenging his junior colleague and explicitly rejecting his conclusions. The rivalry must have been all the more galling for Cuvier, since he had such a low opinion of Faujas's competence as an anatomist. It was probably around this time that he took to referring to him—though not, presumably, to his face—as "Faujas sans fond" [Faujas without foundation], in spoken French a clever if unkind pun on his real family name of Faujas de Saint-Fond (the latter now restored from its enforced democratic abbreviation during the Revolution).[78]

Faujas continued his attack in the Muséum's own *Annales*. One of his articles (though not one on fossils) was given pride of place as the very first in the inaugural volume, published soon after he gave his lectures. In the second volume, which appeared around the time that his lectures were published (and that the Peace collapsed), there were no fewer than four of his articles, all of them on fossils and three on fossil bones: it was another clear sign that he was claiming intellectual rights to Cuvier's field. One paper, for example, dealt with the bones of fossil cattle found widely in Europe and also in Siberia and North America. Faujas followed Pallas in attributing those in Siberia to a "diluvian revolution" or mega-tsunami sweeping north out of India, thereby tacitly rejecting all that Cuvier had written about the (true) mammoths with which those bones were found (§7.1). He appealed as usual to the living fossil argument to combat what he treated as an undesirable tendency to invoke extinction: "I believe it will not be necessary to admit lost species [*espèces perdues*] definitively until one has exhausted all possible means to check that they do not exist in remote and unfrequented parts of the globe, such as the interior of Africa or New Holland [Australia]."[79]

A cumulative case for extinction

Cuvier could not ignore such direct challenges both to his detailed research and to his general conclusions. The following year he launched in the *Annales* a series of articles that, like a massive broadside in one of Nelson's naval battles, might blow his rival out of the water. In his appeal for international cooperation (§7.3) and in his correspondence, he had made it clear that he planned to produce a major work on fossil bones. Without waiting to complete this project he began publishing the constituent parts separately, in effect as preprints. As he finished his analysis of the bones of each kind of animal, he put it into the public realm without delay. He also had a large number of extra copies printed off. Some of these he used at once in his "noble commerce" with other savants, in effect as currency to help pay for their further cooperation. But most copies he held in reserve, ready to be incorporated in the fuller work (§9.3). For each article, he first assembled all his proxy pictures of the bones that he attributed to a specific kind of animal, both his own drawings and those he had received from his informants. He then selected those that most deserved publication and wrote the osteological descriptions of them and his analysis of their significance. The selected drawings were mounted and sent

76. Fig. 7.16 is reproduced from a drawing enclosed with Home to Cuvier, 17 July 1804 (Paris-MHN, MS 630(2), letter 24); engraved in Cuvier, "Grand mastodonte" (1806), pl. 53. Peale (R.), *Skeleton of the mammoth* (1802), 35, 38; Peale (C. W.), "Lettre de M. Peales [*sic*] au citoyen Geoffroy" (1802). Peale also sent *casts* of some of the New York bones, and of Jefferson's from Virginia: Peale to Cuvier, 16 July 1802 (Paris-MHN, MS 630(2), letter 30). On Peale's work, see also Semonin, *American monster* (2000), chap. 13.

77. Cuvier to Home, 24 floréal, 14 messidor X [14 May, 2 July 1802] (London-RCS, Home papers), the latter printed in Eyles (J. M.), "Banks, Smith" (1985), 44–45; Cuvier to Karsten, 9 thermidor X [28 July 1802], Berlin-SB, Lc 1801(3), 21–22; Cuvier, autobiographical fragment (Paris-IF, MS 2598(3)); Negrin, *Georges Cuvier* (1977), 142, 322, 423. He resigned from the educational position after only a year, probably because with his new sources of income its financial rewards no longer compensated for its drain on his time.

78. Faujas, "Discourse" of 1 May 1802, printed in his *Essai de géologie* 1 (1803), 3–7, 22–23, referring inter alia to Hall, "Whinstone and lava" (1799), 67–68n. See also his list of fossil shells with living analogues, and his chapter on living and fossil elephants, in *Essai* 1: 58–75, 237–314.

79. Faujas, "Deux espèces de boeufs" (1803), 195.

to an engraver and later returned for the engravings to be approved before the plate was printed. In effect, Cuvier set up an assembly line for discrete papers on fossil bones, which incorporated the riches of his vast collection of paper proxies and indeed *appropriated* those he had been sent from all over Europe and even from

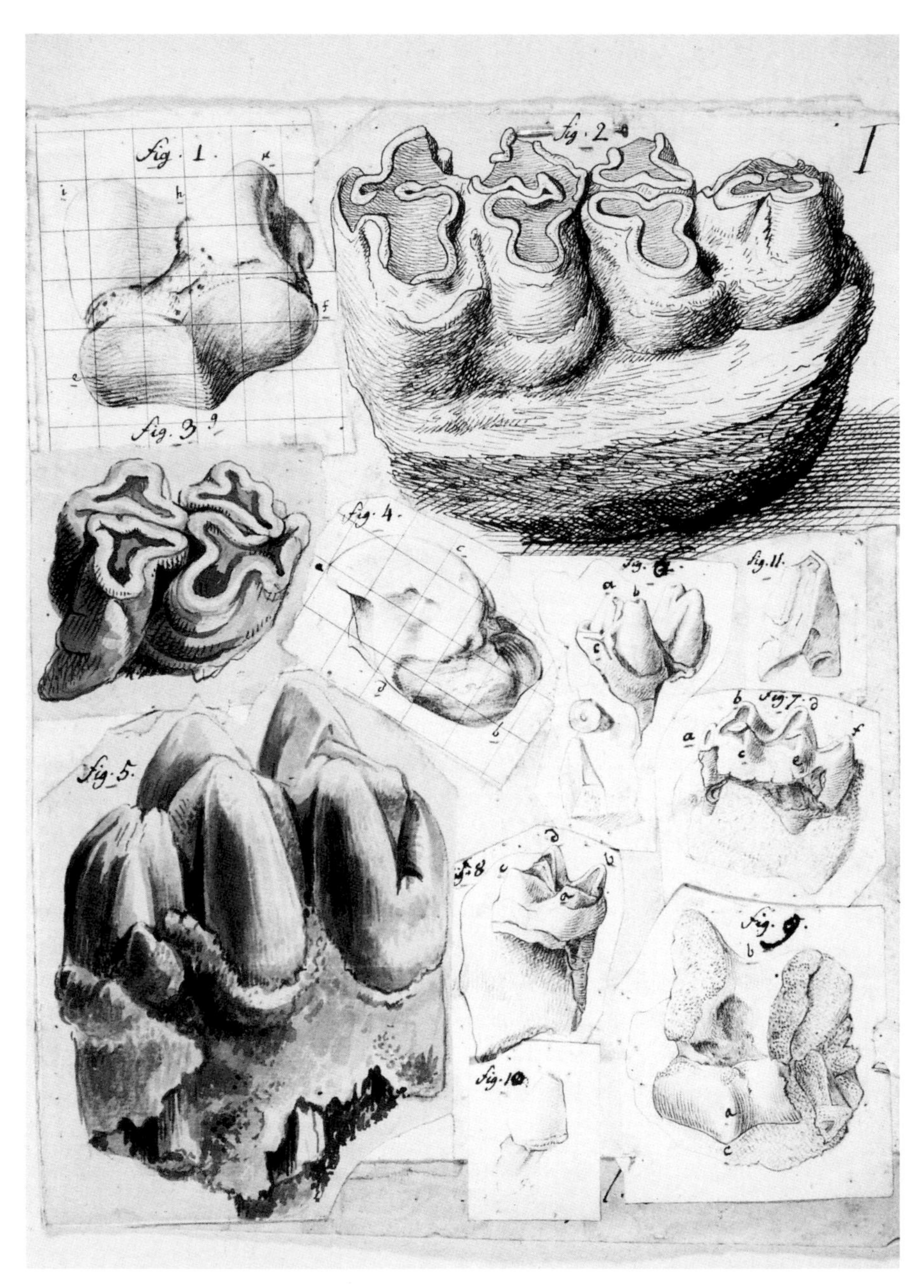

Fig. 7.17. A typical plate of illustrations destined for Cuvier's work on fossil quadrupeds: it shows drawings of bones and teeth that he attributed to two fossil species of hippopotamus, cut out and mounted ready to be sent to an engraver. The drawings by others are easily distinguished by their style from those by Cuvier himself: for example, two teeth of the larger species (figs. 3, 5) had been sent from Florence by Fabbroni. Two of Cuvier's own drawings (figs. 1, 4, two views of an astragalus of the larger species) are marked with a grid of ruled lines, a standard technique used by artists when redrawing an original to a different size. (By permission of the Bibliothèque Centrale du Muséum National d'Histoire Naturelle, Paris)

beyond it. His working space at the Muséum became a kind of factory, the like of which had rarely been seen before in natural history, and certainly not in work on fossil bones (Figs. 7.17, 7.18).[80]

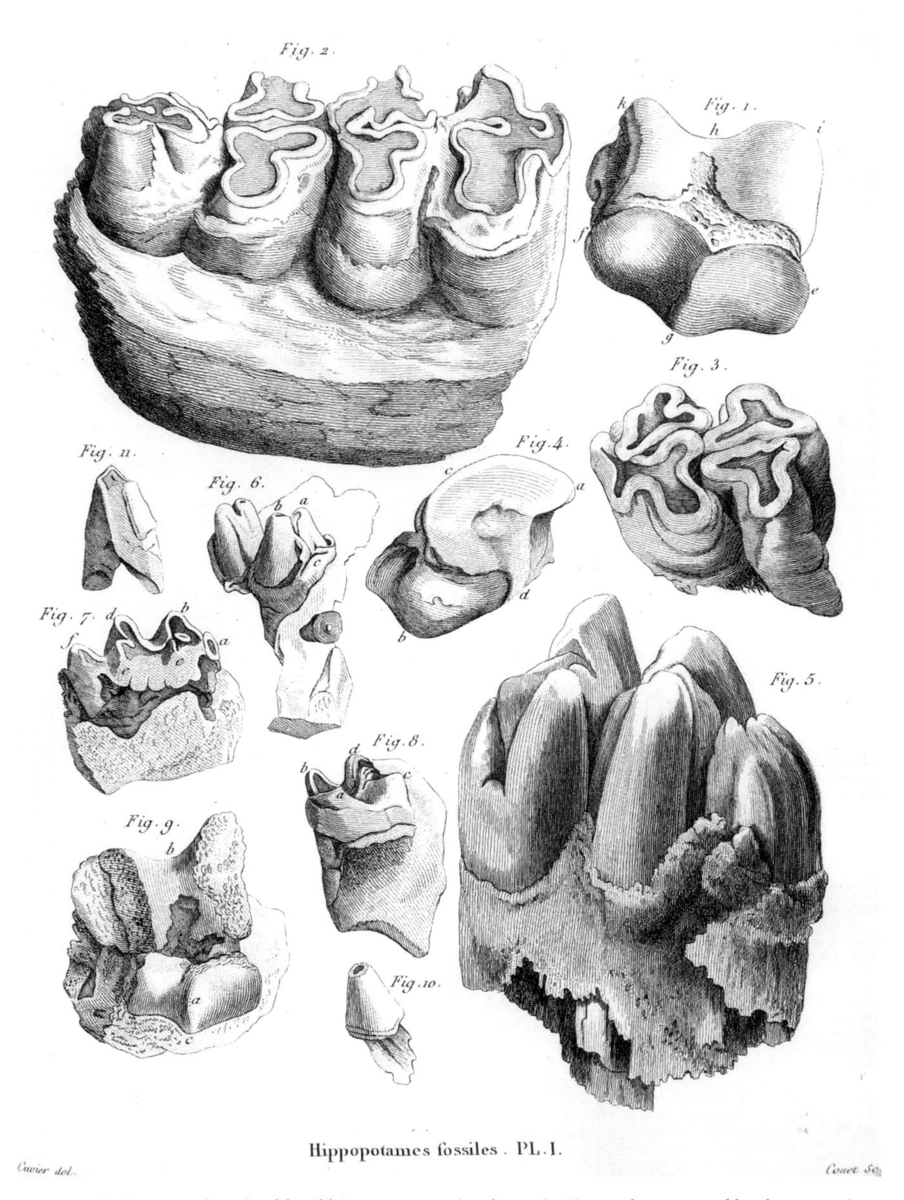

Fig. 7.18. The bones and teeth of fossil hippopotamus (as shown in Fig. 7.17 but reversed by the engraving process), published by Cuvier in 1804. In this particular case, he even marked the plate as if he himself had made all the drawings ("Cuvier del."), although in fact he had not; he thereby literally appropriated the work of his informants. The specimens came from his "large" (figs. 1–5) and "small" or pygmy fossil species (figs. 6–11), both of which he claimed were distinct from any hippopotamus known alive. (By permission of the Syndics of Cambridge University Library)

80. Fig. 7.17 is reproduced from a set of mounted drawings in Paris-MHN, MS 628. Fig. 7.18 is reproduced from Cuvier, *Ossemens fossiles* (1812), 2 [Cambridge-UL: MD.8.66], 6e mém., pl. I, first published in his "Ossemens fossiles d'hippopotames" (1804), pl. 9; it was engraved by Couet. On Cuvier's working methods, see Rudwick, "Cuvier's paper museum" (2000).

Cuvier fired many further broadsides during the following years. In 1804 alone, truly an annus mirabilis for him, three successive volumes of the *Annales* contained a total of sixteen articles on the bones of living and fossil vertebrates (not to mention half a dozen on molluscan anatomy, and other papers in other periodicals). All showed the fruits of his international appeal, and he was punctilious about acknowledging the assistance of his informants. He dealt in critical detail with earlier studies of the fossils, and where appropriate he evaluated animal descriptions from the literature of Antiquity, to show that they all referred to known living species, not to the fossils. Most of these articles were translated or at least summarized in other periodicals, not only in France but also in Germany and, less frequently, in England and other countries. So the progress of his research project could be followed closely by interested savants and amateurs throughout the Republic of Letters.[81]

At first glance the order in which Cuvier's articles on fossils were printed looks almost random, but there was an underlying strategy. Most importantly, he made his actualistic method explicit by describing the osteology of *living* mammals—in cases where it was poorly known—before setting out his material on their fossilized predecessors: he showed clearly that he regarded the present as the key to the past. For example, almost his first paper was on the living tapir, an exotic and rather obscure animal known only from South America. This served as a baseline to substantiate his striking claim, in the paper that followed immediately (his first on fossil material), that other tapir species—one of them of great size—had formerly roamed France itself. With characteristic panache, Cuvier argued that "it is clear that this destroys all hypotheses founded on the Asiatic origin of our [European] fossils": it eliminated the kind of diluvial explanation that Faujas, following Pallas, had just revived. Indeed Cuvier insisted again that Faujas's science of *géologie* needed the "touchstone" that his own science of anatomy could offer, in order to avoid the illusory certainties of "systems" that ignored awkward facts while claiming to explain almost everything (Fig. 7.19).[82]

Other living mammals that received a similar treatment included the one-horned rhinoceros (Cuvier's very first paper in the *Annales*), the hippopotamus, the sloths and the anomalous *hyrax* (the "coney" of the then standard English translation of the Bible, but no rodent). Like the tapir, all these served in effect as keys to fossil bones. The hippo's osteology, for example, was followed by an article on two fossil species, both distinct from the living one (see Figs. 7.17, 7.18). In the case of the smaller of these, he and his rarely mentioned "*aides*" excavated the specimens with difficulty from a block of hard rock that he had found in a storeroom at the Muséum with no record of its original location. During his trip to the south of France he found another block of the same rock in a private collection in Bordeaux, but to his frustration this too was unlocated. Still, the bones did witness to the former existence of a remarkable pygmy hippopotamus somewhere, and probably in France. It was important because it showed that not all the putatively extinct species had been larger than those now alive, though most of them were. The article on the living sloths was followed likewise by two on their fossil relatives. Cuvier accepted the distinctiveness of the megalonix (Peale had sent him casts of the specimens from Virginia, to supplement Wistar's published pictures), but he rejected

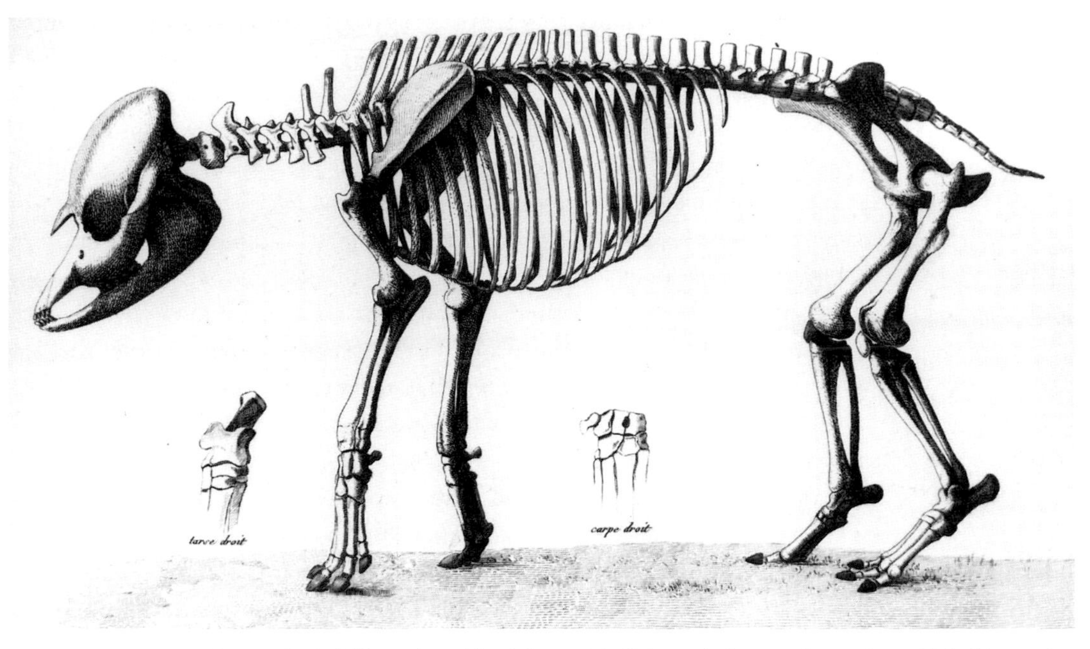

Fig. 7.19. Cuvier's illustration of the skeleton of the living tapir from South America, published in 1804 in an article immediately preceding one on the fossil teeth and bones that he attributed to tapirs that had once lived in France itself. The bones of the living species formed part of the vast reference collection available in the Muséum in Paris, which enabled him to claim that all the fossils belonged to related but distinct species, which he inferred were now extinct. (By permission of the Syndics of Cambridge University Library)

Jefferson's interpretation of it: he classified it as a giant herbivorous sloth and doubted if it was still alive in the American interior. In the case of the megatherium, he greatly enlarged his early paper (§7.1) and added a translation of Bru's original description; he argued that its anatomy showed it had been well adapted to its mode of life, so that "the causes of its destruction" could hardly have lain in any flaw in its animal organization.[83]

Further articles must already have been on Cuvier's assembly line in 1804, though they did not appear in the *Annales* until a year or two later. Among them was one on fossil hyenas; others enlarged and updated his early papers on rhinoceroses, cave bears, elephants, and the still controversial Ohio animal. The hyena was a

81. For a full list of his papers for 1804 see Smith, *Georges Cuvier* (1993), nos. 137–65; see also Rudwick, "Cuvier et le collecte des alliés" (1997).

82. Fig. 7.19 is reproduced from Cuvier, *Ossemens fossiles* (1812), 2 [Cambridge-UL: MD.8.66], 7e mém., pl. I, first published as "Description ostéologique du tapir" (1804), pl. 10; it was drawn and etched by Cuvier himself, before being engraved by Couet. The tapir fossils were published in "Quelques dents et os" (1804). Cuvier's case would have been somewhat less impressive had he known of the living tapirs of southeast Asia, but they were not discovered until later in the century. The range of mammalian specimens on display in the Muséum at this time is described in Fischer, *Nationalmuseum der Naturgeschichte* (1802–3), 2: 92–147.

83. Cuvier, "Ossemens fossiles d'hippopotames" (1804), 112; "Sur le megalonix" (1804), 361; "Sur le megathérium" (1804), 399; and other papers listed in Smith, *Georges Cuvier* (1993), nos. 137, 140, 145, 147. Spanish was one language that Cuvier seems not to have known, or not fluently: Bru's text was translated for him by the botanist Aimé Bonpland, who had just returned with Alexander von Humboldt from his great expedition to Latin America.

case in which he was hampered by lack of comparative material. Even the Muséum had no complete skeleton of the living species from south Africa, which he thought the fossil bones resembled more closely than the hyenas of the Levant. Since the fossils indicated an animal the size of a bear, he evidently suspected that as usual it had been a species distinct from any living one. "Whatever it may be," he concluded, "a skeleton of the Cape hyena needs urgently to be obtained [by the Muséum], in order to complete the comparative [natural] history of the fossil hyena". Cuvier was well aware that in such cases his conclusions were only provisional, and some of his articles were explicitly "supplements" to earlier ones, as and when important new specimens came to hand. There was none of the overweening dogmatism for which he has often been censured by modern historians and scientists.[84]

The fossil rhinoceros was "a monument of such an extraordinary kind and date" because it threw particular light on the event that had preserved most of these fossils. The famous specimens found in frozen ground in Siberia proved, in Cuvier's view, that they could not have been swept there from India in a mega-tsunami, as Faujas (following Pallas) claimed, but must have been living more or less on the spot. They must have been wiped out "not by slow and insensible changes but by a sudden revolution", perhaps by a sudden deep freeze; the wording hinted at his rejection of Lamarck's ideas of extremely gradual change. The skin preserved on these exceptional specimens showed that the Siberian rhinoceros had been a long-haired animal, which confirmed that it was a species quite distinct from the living ones. The implication was that it had been adapted at least to temperate conditions if not to the present subarctic climate of the region.[85]

The character of the enigmatic "revolution" received further consideration in Cuvier's major enlargement of his earlier accounts of living and fossil elephants (§7.1). Like so many other fossil bones, those of fossil elephants were always found in the Superficial deposits, "*pêle-mêle*" with other bones. But some of Cuvier's specimens were encrusted with oyster shells, indicating that for some time they had been submerged in *sea* water. Yet they were well-preserved, and could not have been rolled vast distances in any mega-tsunami; like the Siberian rhinoceros, the animals must have lived near where their bones were now found. Cuvier inferred that they had been buried by one of the most recent "causes" to affect the earth's surface, "yet a physical and general cause": a natural and global event, not due to human action, and certainly a *sudden* event. Since he believed the bones were never found on high ground, the event had apparently been confined to low-lying areas. As he put it in his summary of all the fossil pachyderms he had so far described, "the catastrophe that has buried them was thus a major marine inundation, but a transient [*passagère*] one"; it was "the last, or one of the last, of the globe's catastrophes". Any suggestion that the fossil species, rather than being wiped out, had just been transformed in the course of time into living ones was dismissed by referring to the Egyptian finds: "we shall see, from a study of the oldest mummies, that no certain fact justifies a belief in changes as large as those that must be invoked for such a transformation, above all in wild animals".[86]

This conclusion also covered the contentious "Ohio animal", which was now at last given a thorough analysis on the basis of a larger array of specimens than any earlier naturalist had been able to study. Cuvier reviewed in detail the long and complex history of its investigation; his authoritative conclusion was that it had been quite like an elephant in size and general anatomy, and not least in its tusks, yet so different in its teeth that it deserved to be put in a distinct new genus, the "*Mastodonte*" (named in allusion to the breastlike protuberances on the grinding surfaces of its molar teeth, as in Figs. 3.4, 7.12). The North American fossils belonged to the largest of no fewer than five fossil species found in both the New World and the Old. And he claimed that although it was truly extinct, its habits and habitat could be inferred with some confidence:

> All this description implies . . . that it fed more or less like a hippopotamus or a boar, choosing by preference the roots and other fleshy parts of plants; that this kind of food attracted it to soft and marshy ground; that it was not made to swim or live much in water like a hippopotamus, but was truly a terrestrial animal; that its bones are much more common in North America than anywhere else; that they are better preserved and fresher than any other known fossil bones; and yet that there is not the least evidence [*preuve*] or authentic witness that might properly suggest that in America or anywhere else any individual is [still] alive.[87]

Earlier and stranger mammals

To the general public that took an interest in fossils, size mattered (as it still does to the dinosaur-loving public today): huge beasts such as the misnamed "mammoth" that the younger Peale had exhibited in London caught the public imagination as relics of a strange world of vanished monsters. For more discerning observers, however, size was not everything. In parallel with all these articles on the bones found in river gravels and other Superficial deposits, Cuvier also started a long series dealing with those from the gypsum strata around Paris. These were of special interest, not only because they were evidently older, but also because they became progressively *more* strange the more closely Cuvier studied them.

One new specimen from the gypsum quarries was particularly striking and certainly showed that size did not always matter. This fossil was tiny, yet of outstanding interest. From its teeth Cuvier suspected that it was a marsupial, a family known mainly from Australia though also from America, but not at all from the Old World. To confirm this, Cuvier made a risky prediction and staged a dramatic test. In the presence of competent witnesses, he used a fine steel needle delicately to excavate the gypsum, even sacrificing part of the backbone of the fossil, and duly

84. Cuvier, "Ossemens fossiles d'hyènes" (1805), 143.

85. Cuvier, "Rhinocéros fossiles" (1806), 50, 52.

86. Cuvier, "Éléphans vivans et fossiles" (1806), 266–69; "Dents du genre des mastodontes" (1806), 420–24; the concluding sections of both papers are translated in Rudwick, *Georges Cuvier* (1997), 91–97.

87. Cuvier, "Grand mastodonte" (1806), 311–12. His pl. 53, an engraving of an almost complete skeleton, was based on the drawing of Peale's specimen that Home had sent him (Fig. 7.16). Important new specimens from Big Bone Lick, sent by Jefferson to the Muséum in 1808, were too late to be included: see Rice, "Jefferson's gift" (1951).

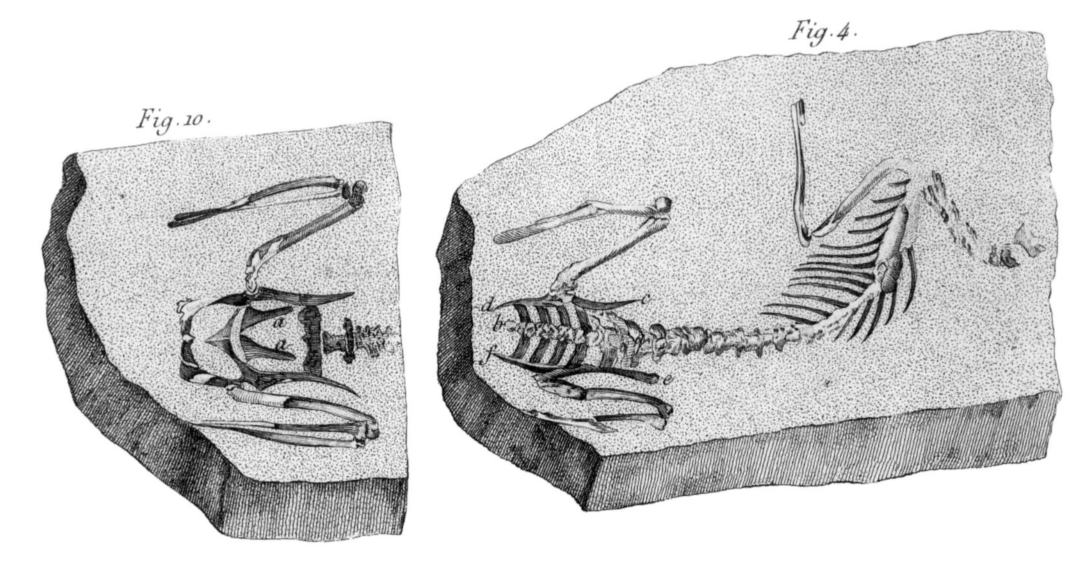

Fig. 7.20. Engravings of part of a unique specimen of a fossil opossum [*sarigue*] from the Paris gypsum, before (fig. 4, right) and after (fig. 10, left) Cuvier had excavated below (i.e., ventral to) the backbone (b) to expose the marsupial bones (a, a) that he had predicted in advance. It was a persuasive demonstration of the heuristic power of his anatomical principles, but it also accentuated the strangeness of the Parisian fossil fauna. (By permission of the Syndics of Cambridge University Library)

revealed the characteristic marsupial bones that he had predicted in advance. It was a spectacular vindication of his principles, which he thought would help raise anatomy to the prestigious predictive heights of more exact sciences. He concluded that the fossil was related to the living opossums of America, which Geoffroy had been studying, yet it was distinct from any of them. Cuvier stressed the marvel of its preservation through "thousands of ages [*siècles*]": like his contemporaries, he clearly took for granted a vast if barely quantifiable timescale for geohistory. He also claimed that it showed the futility of geotheoretical "systems" that could be "destroyed" by a single "fact" such as this, for the implications of finding a fossil marsupial in Europe could not be ignored (Fig. 7.20).[88]

This surprising Parisian marsupial remained unique. Far more common were the bones of the Montmartre animal, which Cuvier had initially regarded as some kind of dog. But he had quickly concluded that it was much more peculiar: a new kind of pachyderm, apparently intermediate between tapirs and camels (§7.2). These fossils now demanded much fuller treatment; unlike, say, the tapir or hippopotamus, they required not one article but many. Cuvier gave priority to separate articles on skulls, jaws, and feet, the parts most likely to reveal the affinities of the animals, before dealing with their other bones. He defined three species of *Palaeotherium* ("ancient beast") and four of *Anoplotherium* ("unarmed beast"): two new genera and seven species were added to his fossil menagerie. He recorded how, after describing so many strange unknown animals, he had been quite relieved to find that his bones also included (though rarely) those of a more familiar doglike carnivore, although as usual it too turned out to be distinct from any living species. With that exception, and the opossum, *all* the mammals of the ancient world of the Parisian gypsum were pachyderms. It was rather like the almost exclusively marsupial fauna of the present world of Australia: the parallel, he added cryptically, helped "establish some conjectures" about the earth at the remote "epoch" of the Paris fossils.[89]

Cuvier's "reconstruction" [*restitution*] of these strange mammals was both aided and confirmed by the discovery of further material, even while his research was in progress: there was an intrinsically cumulative element to the project, which justified its publication in installments. For example, "chance" came to the rescue when, as he put it, "a specimen was discovered, precisely appropriate to enlighten me on most of the points I had hitherto lacked". Workers in the gypsum quarries at Pantin, not far from Montmartre, found the first relatively complete skeleton, which was described in the newspapers as that of a ram. One of the Parisian amateur fossil collectors alerted Cuvier, and the precious specimen was duly presented to the Muséum. It was far more complete than any previously discovered, and it enabled Cuvier to confirm the tentative reconstruction that he had based on more fragmentary specimens (Fig. 7.21).[90]

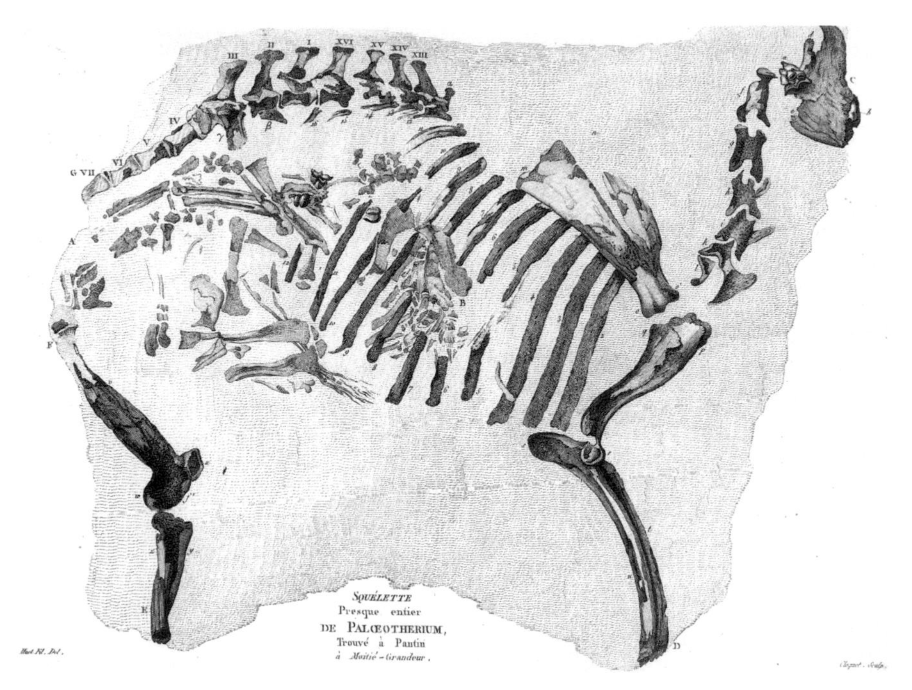

Fig. 7.21. "Almost complete skeleton of Palaeotherium", found in 1804 in the gypsum quarries at Pantin (then just outside Paris) and published by Cuvier the same year. It confirmed the validity of the reconstructions he was making at just that time on the basis of disarticulated bones, and helped establish that the palaeotherium was a strange form unlike any living mammal. (By permission of the Syndics of Cambridge University Library)

88. Fig. 7.20 is reproduced from Cuvier, *Ossemens fossiles* (1812), 3 [Cambridge-UL: MD.8.67], 10e mém., pl. [1], engraved by Couet, first published as pl. 19 in "Genre de sarigue" (1804). This famous little specimen is still on display in Paris-MHN.

89. Cuvier, "Os fossiles de Paris" (1804–8), sections on "Restitution de la tête", "Examen des dents", and "Restitution des pieds" (all 1804).

90. Fig. 7.21 is reproduced from Cuvier, *Ossemens fossiles* (1812), 3 [Cambridge-UL: MD.8.67], 5e mém., 1re partie, pl. [1], first published in "Os fossiles de Paris" (1804–8), section on "Sur les os du tronc" (1804), pl. 46.

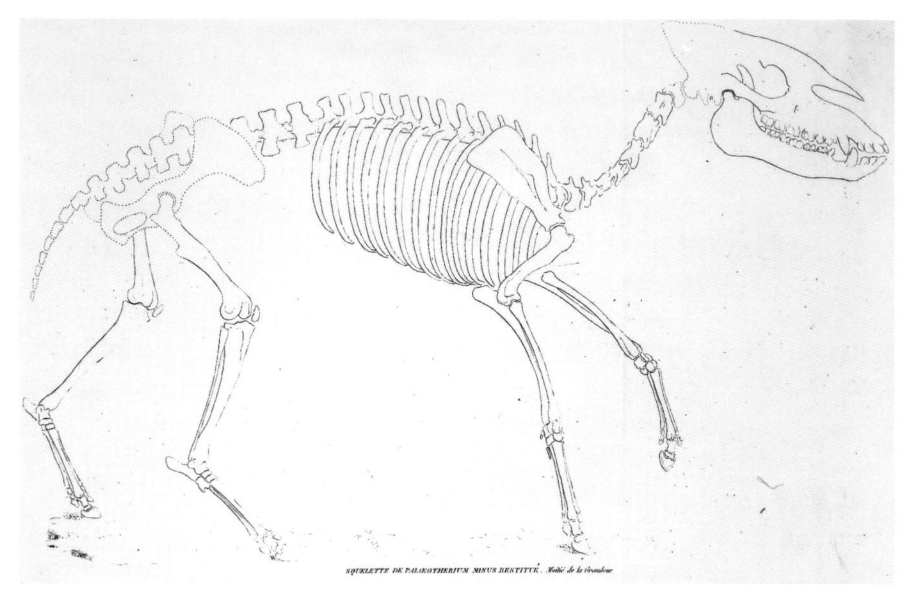

Fig. 7.22. "Skeleton of *Palaeotherium minus* restored": Cuvier's drawing of one of the mammalian species that he distinguished among the fossil bones from the Gypsum formation around Paris; this species would have been somewhat over two feet from head to tail. Cuvier was careful to show exactly which bones he had found and which he had inferred from the homologous parts of related species; the latter (e.g., pelvis and tail) were shown with dotted outlines—a convention borrowed explicitly from cartography—thus circumscribing the conjectural element. This engraving was published in Cuvier's full *Researches on Fossil Bones* (1812) but was probably drawn several years earlier. "If we could have brought this animal back to life as easily as we have reassembled its bones," he wrote, "we would have thought that what we were seeing running was a tapir as small as a sheep, with light and spindly limbs." The lively pose helped to make the extinct species as credible as any living mammal; the former world of which it had been a part became likewise a credible period in the earth's history. (By permission of the Syndics of Cambridge University Library)

With his ever larger database of relevant specimens, Cuvier reached the point where he felt confident about reconstructing the skeletons of some of the species in pictorial form, in the style that had long been customary in comparative anatomy. Unlike the somewhat wooden pose of Bru's megatherium (Fig. 7.3), Cuvier's fossil skeletons were lively and lifelike. They show how profoundly he understood the dynamics of mammalian bodies, an insight doubtless enhanced by lengthy contemplation of the animals in the Muséum's menagerie (of which his younger brother Frédéric had just been put in charge). Although he delayed publishing these reconstructions until his project was completed, they were probably drawn around the time of his great series of articles about the Parisian bones and were presented explicitly as their culmination (Fig. 7.22).[91]

Privately, however, Cuvier's reconstructions went one stage further. As in his first public lecture on fossil bones (§7.2), he claimed that on the basis of the skeleton he could go on to "reconstitute" more or less confidently the musculature and the whole body form of a fossil animal, and even infer its habits and habitat. In fact he drew some astonishingly lifelike pictures of the animals from the Paris gypsum. Depicted with the internal anatomy visible, as if caught by an X-ray camera, they showed the skeleton with musculature attached, the external body form, and even

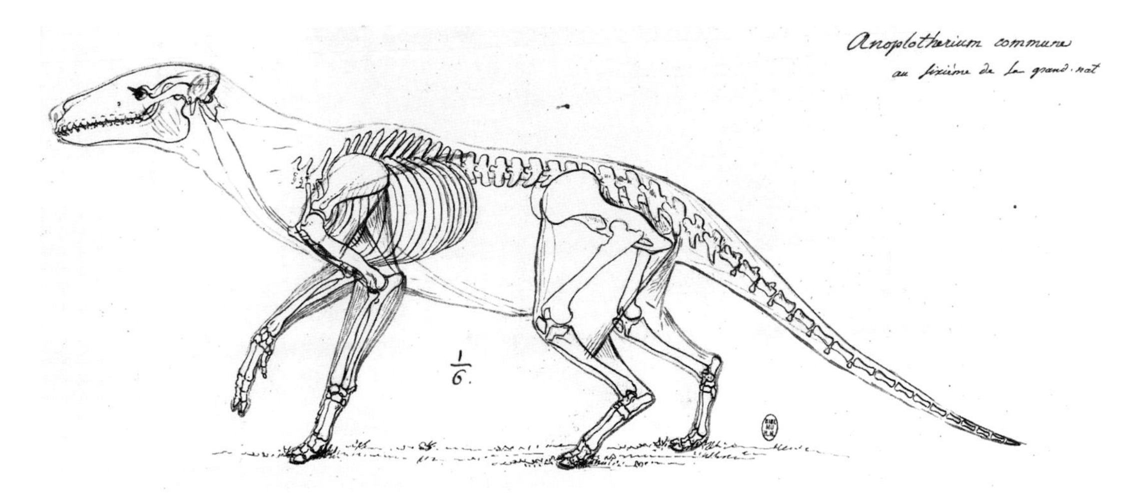

Fig. 7.23. Cuvier's reconstruction of the *Anoplotherium commune*, one of his fossil mammals from the Gypsum around Paris, shown in lifelike pose with its skeleton, musculature, and body form, and even its eyes, ears, and snout; the ground beneath its feet hints at its habitat. This and similar drawings of other species were never published, but they do show how Cuvier aspired to bring his fossil animals back to life, at least in the mind's eye: they were creatures as real as any living mammals, yet also strange denizens of a vanished former world. (By permission of the Bibliothéque Centrale du Muséum National d'Histoire Naturelle, Paris)

such inferred details as the eyes and ears. He may have shown these drawings to trusted colleagues and possibly even displayed them when reading his papers in Paris; but he never published them, probably because they might have laid him open to the charge of unwarranted speculation that he was so zealous in pinning on others. Nonetheless, they illustrate the goal of all his research on fossil bones: not merely to identify or classify fossil mammals, nor even to prove that they were all distinct from living species, but above all to bring them, as it were, back to life—as vividly as the living mammals in the menagerie round the corner from his house in the Jardin des Plantes—and thereby to reconstruct the whole former world of which they had been a part (Fig. 7.23).[92]

Publicly, Cuvier claimed that what he had achieved with his hundreds of disarticulated bones (and, rarely, more complete assemblages) was almost beyond human capacities: "it was almost a resurrection in miniature, and I did not have the almighty trumpet at my disposal". Instead, however, he had unchanging organic laws, and so "at the voice of comparative anatomy each bone—each fragment of bone—took its place again". The valley of dry bones became a scene from deep

91. Fig. 7.22 is reproduced from Cuvier, *Ossemens fossiles* (1812), 3 [Cambridge-UL: MD.8.67], 7e mém. ("Rétablissment des squelettes"), pl. [1], quotation on 72–73; trans. Rudwick, *Georges Cuvier* (1997), 63–67. Lacépède and Cuvier, *Ménagerie du Muséum* (1801–5), described and illustrated its living mammals; Frédéric Cuvier, "Du rut" (1807), 119, referred to menageries as the analogues of chemists' laboratories.

92. Fig. 7.23 is reproduced from an undated drawing in Paris-MHN, MS 635; Cuvier's verbal reconstruction of the species is in *Ossemens fossiles* (1812), 3, 7e mém., 66–67, trans. Rudwick, *Georges Cuvier* (1997), 64–65. From the same set, the drawing of *A. medium* is reproduced in Rudwick, *Georges Cuvier* (1997), 66; that of *Palaeotherium minus*, in Coleman, *Georges Cuvier* (1964), 122; see also Rudwick, *Scenes from deep time* (1992), 32–37. The drawings may not have been made until a few years later, but the aspiration was certainly there from an early stage in Cuvier's research.

time, with the varied animals of the Montmartre gypsum foraging around the shores of a vanished lake. This evocative scene of secular resurrection, activated by a scientific last trump, was embedded in the introduction to Cuvier's volume on the mammals from the Paris gypsum. But it was duly noticed by his readers and could serve as a precedent for even more ambitious reconstructions. Cuvier showed, by such examples, how particular fossils could be made the empirical basis for an imaginative time machine that could transport the naturalist back into the depths of geohistory. Nature's "antique monuments" could be restored in all their original lively glory; in Cuvier's famous later phrase, the naturalist could "burst the limits of time" (§9.3).[93]

Conclusion

Even at this point, quite early in what was to prove a long and laborious research project, Cuvier was putting into the public realm an astonishing series of detailed studies of a wide range of fossil mammals and their living relatives. His stream of articles in the Muséum's *Annales* went far beyond what earlier naturalists had been able to achieve, in both quality and quantity. His anatomical skills, and especially what must have been an outstanding visual memory for the forms of animal bones, combined with the unparalleled collection of real bones in the Muséum and the proxy bones that he received from his informants, enabled him to assert with great confidence that *all* the fossils he had studied belonged to species—or even genera—that were distinct from any known alive in the present world. Collectively, they represented what appeared to be a whole terrestrial fauna of the former world, wholly vanished. That they were indeed extinct was a conclusion that would be reinforced when the famous expedition under Meriwether Lewis and William Clark, sent by President Jefferson to explore a land and river route across North America to the Pacific, returned to civilization. On the eve of the expedition, Jefferson had still hoped for a positive result: "it is not improbable that this voyage of discovery will procure us further information of the Mammoth [i.e., mastodon], & of the Megatherium [now equated with his megalonix]", he told Lacépède in Paris; "there are symptoms of its [the latter's] late and present existence". But Lewis and Clark failed to find any herds of large mammals hitherto unknown alive, or any rumors of them among the Native Americans whose territories they had traversed.[94]

Cuvier's massive broadsides duly sank his rival Faujas, who continued to publish geological articles in the *Annales* but never again on fossil bones. Faujas's alternative explanation for the faunal difference between past and present, using the living fossil argument, could not be disproved, at least until all the continents had been much more thoroughly explored. But Cuvier's cumulative tally of unknown species made it less and less plausible: the more fossil species that he showed were distinct in their anatomy, the more likely it became that they had indeed been "lost" altogether from the earth. Lamarck's third alternative, that they had simply been transformed into living species without extinction, remained a possibility; but Cuvier's case of the Egyptian ibis, also published in full in 1804, made that option more questionable than ever.[95]

All in all, Cuvier was seen to be establishing the reality of the former world as never before. De Luc, Dolomieu, and others had focused their attention on defining and dating the boundary event that separated it from the present, but the former world itself had remained relatively indistinct and only vaguely characterized. Cuvier's detailed studies of fossil bones, in contrast, peopled it with animals as distinct and lifelike as those described by travelers to distant lands. Cuvier was assembling a fossil menagerie as credible as the one in the grounds of the Muséum; the denizens of the former world were almost coming back to life. With the former world defined so concretely by its mammalian fauna, it was now clear that the earth had indeed had a *history*; and with Cuvier's extinct animals playing the role of the ancient Greeks and Romans, that history became *knowable* in detail.

However, neither Cuvier nor anyone else had yet attempted to integrate this astonishing fossil fauna into any broader picture of geohistory. That would require combining the study of fossil bones with the hitherto separate study of the rock formations and other deposits in which they were found. Fossil anatomy—still, in its methods, part of the indoor museum science of "mineralogy" (§2.1)—would have to be integrated with the outdoor field science of geognosy (§2.3). The next chapter traces how that combination was forged in scientific practice.

93. Cuvier, *Ossemens fossiles* (1812), 3: Introduction, 3. The allusions to the prophetic vision in Ezekiel 37: 1–10 and to the eschatological one in 1 Corinthians 15: 52, would have been obvious to Cuvier's biblically literate generation.

94. Jefferson to Lacépède, 24 February 1803, printed in Jackson, *Lewis and Clark expedition* (1978), 15–16. The king of Spain's similar but earlier instruction, that his subjects should search for traces of Bru's great animal surviving in *South* America, had been equally fruitless.

95. Cuvier, "Ibis des anciens Égyptiens" (1804), expanded his earlier paper, with much better illustrations (Figs. 7.13, 7.14). The significance of the case was still not fully explicit, but would have been obvious to anyone who had read the Muséum's more general report on the Egyptian collections, written by Lacépède but also signed by both Cuvier and Lamarck (§7.4).

CHAPTER 8

Geognosy enriched into geohistory

8.1 THE "ARCHAEOLOGY" OF THE EARTH (1801–4)

Geognosy and fossils

Cuvier had been aware of Werner's work on geognosy (§2.3) while still in Normandy, but at that time he had not yet read it for himself and had probably not appreciated its structural or three-dimensional character. For example, he had told a German friend that one part of Normandy was underlain by the distinctive Chalk formation [*Kreideberge*] and another by limestones that provided the famous building stone of Caen; but his account gave no hint that he knew that the former lay *above* the latter, or that it might be important to determine the structural relation between the two.[1]

Likewise, a few years later, Cuvier had listed the kinds of fossil bones that he was planning to study, without any hint that they might not all be of the same age: specifically, that those from the solid gypsum strata in the hills around Paris might be far older than those found in the silts and gravels in the valleys between those hills (§7.1). He may well have seen the familiar Parisian topography through the eyes of Lamanon (§4.3), whose earlier paper he had certainly read. Lamanon had interpreted the gypsum as a precipitate from a former lake that had drained away in prehistoric times, leaving the plateaus around the city as vestiges of its former bed, now high and dry above the valleys that had been eroded through it subsequently (see Fig. 4.11). This reconstruction did not imply any great difference in age between the gypsum and the valley gravels or, therefore, between their respective

1. Cuvier to Pfaff, October 1791, printed in Cuvier, *Briefen an Pfaff* (1845), 245–48, trans. Rudwick, *Georges Cuvier* (1997), 7–9.

fossils. Indeed, Lamanon used the freshwater mollusks living in the present rivers, of the same kinds as the fossil shells in the gypsum strata, as evidence for the lacustrine origin of that older deposit. Nor was Cuvier unusual in having paid little attention, at the start of his career, to the structural relations of the rocks in which his fossils were found. Faujas, his senior colleague at the Muséum, likewise listed the fossil shells that he claimed were identical to modern species, with no hint that those he was describing from Maastricht might not be of the same age as those from Grignon or other famous localities scattered across northwest Europe (§7.2).

Perhaps more telling than Faujas is the example of Cuvier's almost exact contemporary the mineralogist Alexandre Brongniart (1770–1847), who was soon to become one of his most important collaborators (§9.1). Brongniart had known Cuvier since the latter's arrival in Paris and was perhaps as much his personal friend as any of his colleagues. His family was prominent in the Parisian *haute bourgeoisie*; his father had been a successful architect in the city since before the Revolution, and his uncle was the Muséum's professor of applied chemistry, having taught pharmacy at its forerunner. Inspired by the latter's example, the teenaged Alexandre had become interested in natural history and had joined savants such as Lavoisier on weekend field excursions. He had been to England before the start of the Revolutionary wars, traveling, mainly on foot, as far as the great mining area of Derbyshire. There he had met the mineralogist White Watson, who was famous for making and marketing solid model sections of the local formations, constructed out of the rocks themselves. Watson's three-dimensional geognostic outlook was evidently unfamiliar to Brongniart at that time, for after being taken by him into the field the young Frenchman noted: "what seemed to me quite astonishing is that he would show me on the earth's surface those different formations [*couches*] that he had told me were one above the other".[2]

Back in Paris during the Revolution, Brongniart had been appointed to the Mining Corps and was sent to report on mineral resources in Normandy, the Alps, and Provence. When Dolomieu joined the expedition to Egypt, Brongniart had taken his place at the Mining School, and in his first lecture had echoed the older savant's insistence on the centrality of fieldwork; he also censured "the geologist" for turning limited observations into "a novel [*roman*], which he terms *theory*". This repudiation of geotheory, and a preference for sober fieldwork, were to be characteristic of all Brongniart's scientific work.[3]

At the turn of the century Brongniart was given an exceptional career opportunity, happily coinciding with his need for a higher income on getting married. The state porcelain factory at Sèvres, just outside Paris on the road to Versailles, had been a symbol of French technical and artistic supremacy under the Old Regime, but also a symbol of conspicuous consumption by royalty and aristocracy alike. During the Revolution it had become a politically incorrect anachronism; but it had not been closed, though it was reduced to turning out ceramic expressions of populist Revolutionary sentiments. After Brumaire, however, its reform and revival became part of Bonaparte's expansive cultural policy. Brongniart, at the age of only thirty, was appointed its director; like Cuvier's meteoric rise, it was an example of the policy of making "careers open to talent". Brongniart was known to be a competent mineralogist and chemist and was certainly well-connected, but he had no experience of administration. However, he proved to be highly effective, swiftly restoring the factory's financial viability, not least by selling off its now unfashionable stock from the Old Regime at reduced prices and employing artists and designers who soon made Sèvres a leader in a new decorative style.[4]

Brongniart's work at Sèvres gave his mineralogical interests a practical focus, namely that of finding new sources of ceramic materials. Kaolin (china clay), essential for the production of porcelain, was an acute need, since the wars had cut off supplies from Cornwall; but many other minerals were potentially valuable for expanding the range of colors that could be used on Sèvres wares. So Brongniart began to explore the Paris region, the part of France closest to the factory: as early as 1801, for example, he made notes on outcrops of clays and on local potteries encountered on a trip with his father-in-law, Charles Coquebert de Montbret, the editor of the *Journal des Mines* and a former member of the Mines Agency. However, there is nothing in his notes to suggest that Brongniart was as yet seeing the country in geognostic terms, as the surface expression of a three-dimensional structure.[5]

In fact, most French naturalists, young and old, with interests in rocks and fossils, continued to take for granted a distributional approach that went back to Rouelle's influential lectures half a century earlier, which had set the field study of rocks within the conceptual framework of physical geography (§2.2). Their tacit assumption was that the various "masses" [*amas*] of rock were characteristic of particular localities or regions, not that they were piled on top of one another in a determinate structural order. In this respect Lavoisier and Soulavie had been lonely exceptions, each recognizing in his own way that he had an uphill task to convey to others his three-dimensional understanding of rock masses, and even more the temporal inferences that could be drawn from it (§2.4, §4.4).[6]

Meanwhile, however, the field science of geognosy, in which three-dimensionality was constitutive (§2.3), had not stood still, particularly in the German lands. Werner's colleagues and his former students from various countries had continued

2. Brongniart, notebook on trip to England in 1790 (Paris-MHN, MS 2351/1), also for 1788–90 (MS 2350/1); see Lejoix, "Alexandre Brongniart scholar" (1997); Launay, *Les Brongniart* (1940), gives valuable biographical information on his father's, his own, and his son's generations. Ford, "White Watson" (1960), describes the Englishman's solid sections.

3. Brongniart, "Discours" (1798), 179–80.

4. See Préaud, "Brongniart as administrator" (1997). The new style—not only in ceramics—came to be known as "Empire" after Bonaparte turned himself into Napoleon (§8.3). Chemists had been employed at Sèvres ever since its foundation in 1756. Without his appointment there, Brongniart's research might have gone in another direction: his "Classification des reptiles" (1800) distinguished Sauria, Batrachia, Chelonia, and Ophidia (i.e., roughly, lizards, frogs, turtles, and snakes) by their anatomy, and this classification was soon adopted by others; but with both Cuvier and Lacépède already working on reptiles Brongniart would have known that it was an overcrowded field.

5. Brongniart, notebook for 1801 (Paris-MHN, MS 2350/4), recording a journey to Gisors. His booklet *Couleurs obtenues des oxydes* [1802] displayed his technical competence as an applied chemist at the start of his career; his massive *Traité des arts céramiques* (1844) summarized his experience four decades later, and reflects his lifelong outlook in having a strong historical element. See Préaud, "Brongniart as technician" (1997).

6. It is significant that the French word *terrain*, which was often used (and still is, by modern francophone geologists) to denote rock masses or formations, such as the Chalk or the Paris gypsum, is etymologically a term denoting two-dimensional extension, not three-dimensional solidity.

to produce important work on the rock masses [*Gebirge*] of specific regions, particularly those in which the Primary rocks were of economic importance on account of their mineral veins. And Werner's conception of his science was being applied globally, for example by his former student the Prussian naturalist Alexander von Humboldt (1769–1859), who reported from South America on the pile of rock formations that he was finding there. Werner himself had published a *New Theory of the Origin of Veins* (1791), in which their structural relations—noting which veins cut which others—were interpreted in temporal terms and used to support his version of the standard model of geotheory (§3.5). Characteristically, this theoretical work was, in the words of his subtitle, "applied to mining, particularly to that of Freiberg". It was just this close and fruitful link between geognostic theory and mining practice that the new Mines Agency in Paris had appreciated, when it was set up during the Revolution (§6.3). The French had recognized that they needed to absorb as rapidly as possible what geognosts in the German lands had been doing in recent years. The Mining School, as reconstituted after Thermidor, even employed a teacher of German so that the students could learn to read publications on geognosy for themselves.[7]

Tragically, the French naturalist who might have absorbed and even extended the German work most effectively died just before he had a chance to do so. Early in 1801, after France made peace with the kingdom of the Two Sicilies, Dolomieu was at last released from prison, and returned to Paris to take up his position as Daubenton's successor at the Muséum. In the summer he made a field trip to the high mountains that he believed held the key to a true understanding of the earth, and in Geneva he was feted as Saussure's obvious and worthy successor. He had

Fig. 8.1. Dolomieu as Saussure's rightful successor: a symbolic scene drawn by Wolfgang-Adam Töpffer, showing the prefect of the département of Léman (i.e., the city-state of Geneva, recently annexed by France) handing Dolomieu a quill pen with which to write his sequels to Saussure's volumes. The latter are ranged at the foot of a memorial to "the illustrious Saussure, first [natural] historian of the Alps" (in Revolutionary iconography a pyramid denoted the immortality of the works of the deceased). Mont Blanc rises in the background. Sadly, Dolomieu died soon after his visit to Geneva, and his potential for the sciences of the earth was never fulfilled.

planned to travel on to Freiberg to confer with Werner, but his health was failing and he died before he could get there (Fig. 8.1).[8]

Dolomieu was mourned as widely as he had been respected: at the Institut in Paris Lacépède delivered a memorable *éloge*, praising the deceased naturalist for having worked so much *in the field*. However, on the evidence of what Dolomieu published before the fateful Egyptian expedition (§6.3), it is likely that had he lived his further scientific research would have been in the classification of minerals and rocks and in the causal explanation of their origin: in other words, in the synthesis of mineralogy and earth physics for which the term "geology" was increasingly being used throughout Europe (§6.5). In retrospect, his most significant foray into true *geohistory*—his analysis of the Nile delta, exemplifying a "new method" of combining natural with *human* historical evidence (§6.3)—had been no more than a promising and suggestive beginning.

Werner and geohistory

A year after Dolomieu's death, the Treaty of Amiens made it possible for savants throughout Europe to travel freely for the first time for a decade (§7.5). Werner took the opportunity to visit the world center of almost all the sciences other than his own. The circumstances of his trip to Paris are unclear, but it is likely that he was invited there by the Mines Agency or by some influential savant. It is probably no coincidence that a French translation of his book on mineral veins was published in Paris in the same year. The French were certainly keen to hear more about geognosy, best of all from the horse's mouth (Fig. 8.2).[9]

Fig. 8.2. A portrait of Werner at the age of fifty-two, published shortly before he visited Paris in 1802 and explained his conception of geognosy to French naturalists.

While he was in Paris, Werner was interviewed by La Métherie for the *Journal de Physique*. He defined geognosy unambiguously as a branch of mineral natural history, concerned above all with the structural situation and spatial relations of rock masses. He firmly excluded "hypotheses", for example about the origin of the earth, arguing that such speculations were quite different from

7. Humboldt, "Tableau géologique" (1801), esp. ideal section, 60; Werner, *Entstehung der Gänge* (1791). Reuss, *Lehrbuch der Mineralogie* (1801–6), was a representative textbook: five volumes on mineralogy were followed by two on geognosy.

8. Fig. 8.1 is reproduced from Freshfield, *Life of Saussure* (1920), 439; see also Buyssens, "Saussure mémorable" (2001), 10–11. On the last phase of his life see Lacroix, *Déodat de Dolomieu* (1921), xv–xliv.

9. Fig. 8.2 is reproduced from a print dated 1801, engraved by Christian-Friedrich Stoelzel after a painting by [Christian-Leberecht?] Vogel. Werner, *Formation des filons* (1802).

the sober inferences that followed directly from what could be observed. But he claimed that it was legitimate to reconstruct the earth's temporal development, and he did so once again along the lines of the standard model. As the global sea level fell progressively, the "chaos" that had characterized the Primary epoch, marked by the chemical precipitation of granite and then of other Primary rocks, had changed gradually into the more orderly Neptunian epoch of the successive Secondary formations, comprising both precipitates and detrital sediments. (Werner had inserted a new category of "Transition" rock masses [*Übergangsgebirge*] a few years earlier, to cover intermediate rock types such as greywacke and to make it clear that the passage from Primaries to Secondaries was *gradual.*) Increasingly abundant marine fossils showed that by this time living organisms "had begun to appear"; the cautious phrasing disavowed fruitless speculations about their causal origin. Finally the alluvial (or Superficial) deposits had accumulated on the continents that had eventually emerged as the sea level fell still further. Mentioned last, and not clearly located within this temporal sequence, were the basalts and true volcanic rocks. Werner continued to interpret basalts as Neptunian or aqueous precipitates, holding out against the Vulcanist opinions of many of his colleagues; and he still attributed the true volcanic rocks, in sharp contrast, to the subterranean combustion of older coal deposits, not to any deep-seated cause, so that by implication they were all of relatively recent origin.[10]

Much of Werner's conception of geognosy was thus almost unchanged from earlier years (§3.5). But a shift of emphasis on at least two counts can be detected within La Métherie's brief report. First, his summary suggested Werner's growing interest in the Secondary formations, not at the expense of the Primaries but relative to them. This almost necessarily entailed giving more attention to fossils, and Werner had in fact already begun to include fossils in his lectures at Freiberg. Second, his classification of rock masses was more temporal or developmental in character than it had been before: the major categories now explicitly denoted inferred temporal epochs as well as observed types of rock.[11]

This shift of emphasis would have been no surprise to readers of the *Journal de Physique.* Three years earlier, in his annual review of the sciences for 1798, La Métherie had summarized how four of Werner's former students had jointly distinguished a sequence of major Secondary formations [*couches*] that could be traced all the way from Moscow to Cadiz, that is, right across Europe. Most notable in this sequence of seven formations was the *recurrence* of three major rock types: two formations were composed predominantly of sandstone, two included gypsum deposits, and three were massive limestones. However, the sequence was treated not just as a structural pile of rocks, but as a record of successive temporal events. Not only were the formations numbered in chronological order from first to last, from bottom to top, but more significantly the repeated rock types were defined as "old" and "new" rather than the customary "lower" and "upper". But fossils were mentioned only in passing and in general terms.[12]

This pan-European sequence was evidently understood as embodying, in modern terms, a *correlation* between the sequences observed in several regions (§2.3). Although ambitious in its geographical range, the sequence was clearly preliminary and provisional. Moreover, it was not dependent on the standard model of geotheory; indeed, the repetitions of the three main rock types could not be accommodated within the standard model without ad hoc adjustment. For they implied an irregular *recurrence* of specific conditions of precipitation or sedimentation, which did not fit easily within a model of regular and directional physicochemical change (§3.5). In other words, in tracing this sequence across much of Europe Werner's former students were not only shifting his kind of geognostic practice still further from the classificatory towards the temporal, but also from the developmental towards the truly *geohistorical* (§4.5). Although Werner himself may not have fully appreciated it, his followers were beginning to construct a complex geohistory bottom-up from detailed field study of specific rocks in specific regions, not deducing a simple sequence a priori from an overarching theory about the earth's development.

Cuvier and the history of life

However, even this modified geognosy gave no great attention to fossils: if the work reported by La Métherie had used fossils systematically to help define the seven successive formations, that fact would surely have been noted even in a brief summary. Fossils had indeed been used in that way in the preceding decades, to describe some *local* sequences of Secondary formations. Arduino had noted that in the foothills of the Italian Alps some formations were marked by specific sets of fossils (§2.5); similarly, Soulavie had described how three successive limestones in Vivarais were each characterized by a different set of fossil shells showing a progressive approach to living forms (§4.4); and more recently de Luc's fieldwork had made him aware of the distinctive assemblages of fossils in the formations on the south coast of England (§6.2). But each of these was a brief report embedded within much more extensive descriptions and was easily overlooked; it is not surprising that they had not been widely adopted by other naturalists as models for their own practice. In a broader sense, of course, the match between formations and their fossils had long been recognized: many of the lower and therefore older Secondaries contained ammonites and belemnites, for example, which were totally absent in the younger formations lying above them. Provided the living fossil argument was discounted, this suggested a real history of life, showing a gradual approach to the present fauna (§5.2).

10. La Métherie, "Idées de Werner" (1802), published in the *Journal* for frimaire XI [November–December 1802]. The exact timing of his visit to Paris—which would have affected whom he was able to meet—is unclear. It was certainly in the same year that his star student the Prussian geognost Leopold von Buch (see §10.2) made a field study of the classic region of Auvergne and became convinced that his former teacher was wrong to reject Desmarest's (and other French naturalists') view that basalts were simply ancient volcanic lavas. This proved a turning point in favor of the Vulcanists in their long-running argument with the Neptunists; but it affected the specific problem of basalt rather than the broader questions of geognosy. See Buch, *Deutschland und Italien* 2 (1809): letters to Karsten, 225–311.

11. On Werner's use of fossils, see Guntau, "Biostratigraphic thinking" (1995).

12. La Métherie, "Discours préliminaire" (1799), section on *géologie* (63–67), reporting on work by von Buch, Humboldt, Grüner, and Freiesleben; no original source was cited, and unlike most of the items reviewed by La Métherie their work had not been published in the previous year's *Journal*.

At the turn of the century, Cuvier gave this idea further support on the basis of his new evidence from fossil bones. As already mentioned, his first public summary of his research had treated them as if they were all of the same age. He had adopted a simple binary model from de Luc or Dolomieu, taking for granted a singular "former world": as he put it in his first report on elephants, fossil bones "seem to me to prove the existence of a world previous to ours, destroyed by some kind of catastrophe" (§7.1). However, by the time he issued his international appeal a few years later (§7.3), he had begun to be aware that his fossils were traces of a more differentiated history of life: "I even believe I have noticed a fact still more important, which has its analogies in relation to other fossils: namely that the older the beds in which these bones are found, the more they differ from those of animals that we know today."[13]

This was no more than a passing remark, and Cuvier still listed the different kinds of fossil bones simply in the order in which they had been discovered, and by his degree of confidence in his identifications. Nonetheless, he had evidently recognized by this time that they fell into at least three groups, by relative age. He was in correspondence with Adriaan Camper about the putative crocodile from the Chalk at Maastricht (§7.3), and had just begun working on what he identified as crocodile bones from Honfleur in Normandy. Both came from strata that he now knew were lower in position, and therefore older, than those from the Paris gypsum; and the latter in turn were clearly older than the river gravels in the valleys eroded through that formation. In effect, the older Secondary formations appeared to contain only reptiles; the younger Secondaries had mammals, but of genera quite different from any known alive; and the Superficial deposits had mammals of mostly familiar genera but of species that all differed from living ones. Cuvier did not put it that way in public, but he did hint that in this respect his bones were analogous to the fossil shells studied by others: in quadrupeds as in mollusks, there seemed to have been some kind of progressive approximation to the present fauna.

However, Cuvier's empirical basis for this inference about quadrupeds was as yet very frail. To establish it on firmer foundations would obviously demand a new kind of research, paying much closer attention to the different rock formations in which his various fossils had been found. He was unexceptional in having initially treated the science of mineralogy (broadly defined) as quite distinct from the science of geognosy: fossils, like specimens of rocks and minerals, were studied indoors in a museum (§2.1); rock masses in their three-dimensional relations were studied outdoors in the field (§2.3). In the very first years of the new century, however, two new developments—widely separated conceptually, geographically, and even socially—began to bring the detailed indoor study of fossil specimens into a newly close conjunction with the detailed outdoor study of rock formations. The rest of this section describes the first of these new developments; §8.2 deals with the second.

Blumenbach's geohistory

Blumenbach in Göttingen was prominent among the savants to whom Cuvier had sent an appeal for further specimens, and he obliged by promptly sending a proxy tooth of the Ohio animal to Paris (see Fig. 7.12). With it he sent a printed summary of a lecture he had just given to mark the fiftieth anniversary of the Royal Society of Sciences in Göttingen, which was in effect the research wing of his distinguished university. His lecture offered an important framework for understanding the significance of fossils of all kinds in a strongly *geohistorical* perspective. The full text was published in Göttingen in 1803, and a French version appeared in the *Journal des Mines* soon afterwards, making it well-known and accessible throughout Europe.[14]

It is hardly surprising that fossils were the focus of Blumenbach's new picture of geohistory. Over the past two decades, the successive editions of his highly esteemed textbook on natural history had given progressively greater attention to fossils, as he himself had become aware of their potential value for understanding life on earth. He had first divided them into those of species "known" and "unknown" alive, inferring with growing confidence that the latter, the *Incognita*, were probably extinct. And he had turned this analysis into geohistory, when he had claimed that a "*Totalrevolution*" in the relatively recent past had been marked by a complete faunal replacement: the "unknown" species of the "pre-Adamitic" or prehuman world had all become extinct and had been replaced by the "known" species of the present world (§6.1).

Major support for this kind of interpretation had come from de Luc's claim that the earth's own history was broadly in accord with the Creation story in Genesis, and its most recent "revolution" with the story of the Flood. Blumenbach himself had remained prudently Nicodemite about that scriptural correlation, fearing that any overt adoption of it might cause the "cultured despisers of religion" among his readers to reject his scientific views altogether (§6.4). But he would hardly have urged de Luc to send him a new account of his theory for translation and publication in Germany had he not been in substantial sympathy with the Genevan's biblical and therefore profoundly historical outlook. And he must surely have lent tacit support, at the very least, to de Luc's appointment as honorary professor of "geology" at Göttingen, which made him formally Blumenbach's colleague and gave the honorary Englishman an academic platform from which to promote his ideas (§7.2). However, a further source for Blumenbach's ideas had been implicit in his adoption of historical metaphors (§4.2). To treat fossils as "the most infallible documents in nature's archive" had come particularly easily to a naturalist working at Göttingen, for several of Blumenbach's colleagues were famously bringing new rigor to the writing of human history—and further prestige to their university—by their critical analysis of human documents both sacred and secular (§4.1). Perhaps no other naturalist anywhere in Europe was in such a favorable environment for turning a traditionally static "natural history" into a dynamic *history* of nature, based on a comparably rigorous analysis of "nature's documents".

13. Cuvier, "Espèces de quadrupèdes" (1801), 260; trans. Rudwick, *Georges Cuvier* (1997), 52.

14. Blumenbach, "Specimen archaeologiae telluris" (1801) and *Specimen archaeologiae telluris* (1803); his *Beyträge zur Naturgeschichte*, 2nd ed. (1806–11), reused much material from this and earlier publications. Heron de Villefosse, "Considérations sur les fossiles" (1804) enthusiastically paraphrased the *Specimen*.

Around the turn of the century Blumenbach began to modify his strongly geohistorical but still conventionally binary conception into a more differentiated history of life on earth. In the last edition of his textbook published in the old century, he had inserted a third category of "doubtful" [*zweifelhafte*] fossil species between the "known" and what he now called the "completely unknown" [*völlig unbekannte*]; among fossil mammals, for example, the "doubtful" cave bear was distinguished from "the colossal terrestrial monster" from the Ohio, the latter almost certainly extinct. This revised classification made more explicit the *degrees* of confidence entailed in treating fossils as extant or extinct.[15]

In his celebratory lecture late in 1801, however, Blumenbach also turned it into a criterion of geohistory. His affinity with his historical colleagues at Göttingen was evident in the title he chose for his formal address. With the Latinity appropriate to a German academic occasion, he called it *Specimen Archaeologiae Telluris*, a sample of the earth's archaeology; the subtitle explained that the sample was the territory of Hannover, in which Göttingen lay. At this time the word *archaeologia* denoted the study of Antiquity as a whole, including—but not confined to—the description and analysis of artifacts retrieved by excavation (or archaeology in the modern sense). So for example the Society of Antiquaries in London, the capital of the Hanoverian George III's British kingdom, called its own prestigious periodical *Archaeologia*, in order to denote that broad range of studies (Hamilton had used it to report on the excavations at Pompeii: see Fig. 4.2). So the title of Blumenbach's lecture indicated that he would try to reconstruct the earth's history in just the same way that antiquarians used all the relics of Antiquity to reconstruct ancient human history.

According to the brief account of the lecture published in Göttingen a month later, Blumenbach revealed "the oldest history of our planet" and "the quite distinct catastrophes" it had undergone in the course of time, by means of "a new view and effective application of the science of fossils [*Petrefactenkunde*]". Fossils were to be the key to a newly differentiated geohistory that went beyond any simple binary division between past and present; and this was explicitly related to the rock formations in which the fossils were found. In effect, Blumenbach programmatically linked the museum science of fossils with the field science of geognosy in order to reconstruct the history of life in retrospective order from the present back into the deepest past; at the same time, he clearly distinguished this kind of reconstruction from any attempt to quantify the timescale or to explain the history in causal terms:

> He [Blumenbach] infers from the general classification [of fossils] a chronological arrangement, which is founded first on their critical comparison with the organisms of the present creation and second on their stratal position [*Lagerstätte*] and the respective relation and the age determinable therefrom. Starting from these [fossils] of newer date he goes back to those of older origin, and ends with the very oldest monuments [*Denkmahlen*] of an organized creation on our earth . . . But of course it would as yet be scarcely possible to establish with any confidence a firm chronological subdivision of the successive periods in which it happened, let alone to state the causes.[16]

In effect, Blumenbach transformed his three categories of fossils—the known, the doubtful, and the unknown—into nature's documents or monuments dating from three successively older and stranger *periods* of geohistory. These in turn were described explicitly as the natural analogues of the traditional major periods of *human* history. Blumenbach's first and most recent period of geohistory was characterized by organisms of "known" living species: this was analogous to the "historic" period of human history, with quite straightforward and reliable records, such as those of Classical Antiquity. The second and preceding period of geohistory was marked by fossils of "doubtful" species, similar to known living ones but not exactly the same, and in many cases found in quite different climatic regions: this was analogous to the "heroic" period of human history, of which the scanty surviving records had to be interpreted with great care and more than a pinch of salt. The third and oldest period was that of the completely unknown species, which Blumenbach inferred were truly extinct, having been wiped out by some kind of global catastrophe: this was analogous to the oldest period of human history, the "mythical", of which the putative records were so obscure that their historical value was highly controversial and their meaning difficult to discern.[17]

This extended analogy enabled Blumenbach to explain to his fellow savants how a new and more rigorous "natural history" of fossils might yield a reconstruction of the "*history* of nature", even back into the deepest past before the human species existed, just as some of his colleagues were constructing a new and more rigorous picture of ancient human history. It also suggested how the most effective method for investigating prehuman history was indeed to start with the known present world and work backwards by degrees into the more obscure past; in effect, Blumenbach was prescribing, in relation to fossils, the "analytical route" that Desmarest had long been practicing on the volcanoes of Auvergne (§4.3).

If taken literally, however, Blumenbach's analogy was too defeatist or at least too modest. For it implied that what he was doing with the fossil remains of "completely unknown" species could yield a history of life analogous only to the mythical stage of early human history, with its extreme uncertainty and obscurity. It suggested that fossils might be incapable of ever making the former world of nature as knowable as the former human world of Antiquity. Yet Blumenbach was aware that Cuvier was already using a newly rigorous comparative anatomy to lift the study of fossil bones out of any such morass of obscurity, making the putatively extinct mammals as concretely knowable as the ancient Greeks and Romans themselves (§7.5). Furthermore, although Blumenbach's threefold periodization of geohistory

15. Blumenbach, *Naturgeschichte*, 6th ed. (1799), 688–708.

16. Blumenbach, "Specimen archaeologiae telluris" (1801), 1978, 1983. The lecture was delivered on 14 November; the summary (written in the third person probably by Blumenbach himself) was published in the *Göttingische Anzeigen* for 12 December. In modern geology the word "Lagerstätte" is used in a more specific sense, to denote the exceptional kind of formation (e.g., the Solnhofen lithographic limestone in Bavaria and the Burgess Shale in Canada) in which even the "soft parts" of organisms are preserved in the fossil state.

17. Blumenbach, *Specimen archaeologiae telluris* (1803). On the use of the three periods in earlier human historiography up to the time of Vico, see, for example, Rossi, *I segni del tempo* (1979).

modified the conventional binary division between present and past, it hardly allowed for much differentiation of the former world itself. It did little to suggest that further research on fossils in relation to their respective rock formations might lead to a detailed history of the many revolutions that he claimed the earth had undergone even before the most recent such event.

Blumenbach's lecture on the "archaeology" of the earth was therefore most important as an inspiring programmatic sketch of how fossils might be used in the future to construct a geohistory analogous to human history. Given his prominent position among European naturalists, the geohistorical perspective that had been marginal in previous generations (see Chap. 4) was now set to take a more prominent place in the practice of natural history. However, Blumenbach himself could hardly hope to fulfill the promise implicit in his lecture. As a busy teacher and leading member of his university, he did not have the time for extensive original research enjoyed by, say, Cuvier in Paris; and anyway his main efforts in research were increasingly focused on one specific question within geohistory, namely the origin and diversity of the human species itself. He tacitly left others to use the correlation between fossils and the rock formations in which they were found as evidence for tracing the longer history of life in greater detail. In fact, one of his former students was already well-advanced with just such a study.

A former world of plants

Ernst Friedrich von Schlotheim (1764–1832) had been trained in law at Göttingen—and had also been inspired by Blumenbach's lectures on natural history—before going on to study at the mining school in Freiberg. He had then embarked on a successful career as a civil servant in the duchy of Gotha, his native state, while maintaining his interests in the sciences of the earth. In the latter he soon focused his attention on the neglected topic of fossil plants. In 1804, the same year that an account of Blumenbach's lecture in French reached a wide international audience, Schlotheim published in Gotha the first part of a finely illustrated monograph entitled "Noteworthy leaf impressions and plant fossils"; its subtitle described it more memorably as "a contribution to the flora of the former world [*Vorwelt*]". It was a sign of the times, and of his own intellectual loyalties, that Schlotheim dedicated his work not to some princely personage but to three of his seniors in the science of geognosy: to Werner and his colleague Charpentier at Freiberg, and to their early student von Trebra. Above all, he had recruited an excellent engraver, who made the work a valuable paper museum of proxy specimens (Fig. 8.3).[18]

In his introduction Schlotheim dated serious interest in fossils from the great compilation by Knorr and Walch half a century earlier (see Figs. 5.2, 5.3, 5.9). He mentioned how fossils had been attributed first to the biblical Flood, then to an earlier and more general inundation of some kind, and more recently to an immensely extended series of events. He cited Blumenbach's lecture, Cuvier's appeal, and other recent publications, showing himself to be well aware of current research on fossils; echoing Blumenbach's phrase, he called them "the remains of an earlier so-called pre-Adamitic creation", an assemblage that was almost certainly extinct.

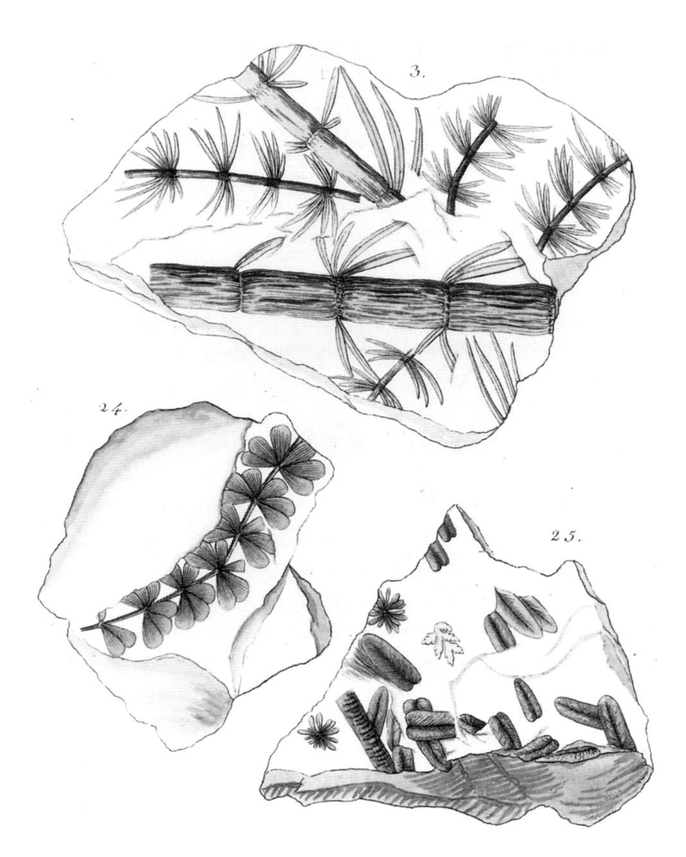

Fig. 8.3. Some of the strange fossil plants from the Coal formation of Thuringia: a typical plate from Schlotheim's *Plant fossils* (1804). It was engraved in 1801 by Johann Gapieux of Leipzig, a leading botanical artist who, unconventionally, was explicitly thanked by the author for his fine work. Schlotheim tried to compare his fossils with living plants classified by Linnaeus, but remained puzzled by their differences. Since the Coal strata were known to be much lower and therefore far older than some of the other well-known Secondary formations, Schlotheim's work helped to focus the attention of naturalists on the geognostic positions of fossils, and hence on their relative ages; his reconstruction of this strange flora provided an enriched sense of the *sequence* of geohistorical periods within the former world. (By permission of the British Library)

More surprisingly, perhaps, he also gave a prominent place to Werner, expressing the hope that "this astute naturalist [*Naturforscher*] and creator of the newer mineralogy [i.e., geognosy]" would soon publish his ideas about fossils:

> As is well-known, a close scrutiny of the local relations in which fossils occur led him, even early on, to characterize more of the Secondary strata [*Flötzlager*] further by their specific fossils; and judging by remarks made to me more recently he is not only more and more in a position to confirm this by continued observations and to determine the relative age of the different Secondary and Transition formations [*Flötz- und Übergangsformationen*] with greater confidence, but also, from the level at which fossils in general occur on the earth's surface, he is led to other information of the highest interest.[19]

Schlotheim may have been flattering his former teacher, or at least wanting to recruit a powerful ally to his cause; but his report on Werner's activities is likely to have been well-informed. Since Werner had certainly started lecturing on fossils, he may indeed have been giving ever more attention to the fossil contents of successive

18. Fig. 8.3 is reproduced from Schlotheim, *Pflanzen-Versteinerungen* (1804) [London-BL: 441.g.24], pl. 2, described on 30–32, 57–58. The plant in no. 3 (top) was noted as abundant in coal mines near Ilmenau, but specimen no. 24 (bottom left) was unlocated, the plant being known to Schlotheim only from collections; he noted that the plant in no. 25 (bottom right) was perhaps a fern.

19. Schlotheim, *Pflanzen-Versteinerungen* (1804), 9–10.

formations as a valuable criterion to add to those of lithic character and geognostic position (§3.5). And he may well have become convinced, at least in recent years, that what Arduino, Soulavie, and others had claimed long before on a local level was of more general validity: namely that certain fossils were *characteristic* of specific formations, and that formations could therefore be recognized by their fossils even from one region to another. Finally, it is compatible with what Werner told La Métherie while in Paris, that he should now be interpreting the sequence of fossils geohistorically, not only in terms of the relative ages of the formations in which they were found, but also as the record of a general history of life on earth.

However, all this remained unsubstantiated, unless and until Werner published an account of his research—as Schlotheim hoped he would—or at least until he communicated his ideas informally but in greater detail to other naturalists. Meanwhile, Schlotheim himself was contributing to what would necessarily have to be a collaborative research effort, by publishing an account of some of the fossil plants found in one specific Secondary rock mass, the Coal formation [*Steinkohlengebirge*] of his native region. He gave a fine description of the geognostic position of his fossils and of their botanical affinities. On the latter he tried to compare his plants with those in Linnaeus's classification; as with Cuvier's megatherium a few years earlier (§7.1), this highlighted their puzzling relation to known living species.

Schlotheim concluded provisionally that among his fossils the ferns were closest to those now living in the East Indies or in the Americas, yet they were not identical. When the floras of all parts of the world were fully known, he thought it might be possible to decide whether the fossils represented "mere lost plant species of our present creation", the victims of the earth's past revolutions, or alternatively "the enigmatic documents of a distinctive earlier creation". Schlotheim was hopeful that intensive international research might soon resolve this question, but meanwhile he strongly suspected that his plants were truly and completely extinct. The argument was of course closely parallel to Cuvier's with fossil mammals; for Schlotheim too the case for extinction was necessarily probabilistic but also cumulative. Above all, his inferences were as profoundly *geohistorical* as Cuvier's, not only in their telltale use of the powerful metaphor of nature's "documents", but also more substantively in that he too tried to use his fossils to imagine what the world might have been like at this even more remote period in its history. His own part of Europe, he concluded, had been the site of vast swampy forests of palms, tree ferns, and giant horsetails, quite different from anything known in the present world, even in the tropics.[20]

This first part of Schlotheim's monograph described leaves and small branches; a promised second part, which was to deal with larger stems and also to review more generally the theories that tried to account for "the remarkable appearance of southern [i.e., tropical] animals and plants in our northern regions", never appeared. Nonetheless, like Cuvier, Schlotheim showed by detailed description that his fossils were distinctly different from any species known alive, and he did so with plant fossils, a category that had previously been neglected. And in one respect, compensating for the incompleteness of his work, Schlotheim took the interpretation of fossils beyond what Cuvier had so far done. His fossil plants from the Coal formation were not merely traces of "the flora of the former world", as he put it modestly in his subtitle. They could be located more precisely as the flora of a specific period of geohistory, the period of the Coal formation, which was known to be well down in the pile of older Secondary rocks.

Conclusion

From different starting points, and using quite different empirical materials, Werner, Blumenbach, Cuvier, and Schlotheim all illustrate two distinctive trends during the earliest years of the new century. The first was towards paying more attention to the fossils that many formations contained and using the science of geognosy more consistently as raw material for geohistory. The second trend was towards replacing a simple binary geohistory with one that was much more differentiated, in which "the former world" was recognized as a sequence of many different "worlds", each characterized by a distinctive set of animals and plants. Schlotheim's study of the Coal plants found in his own region was at least a straw in the wind: it was a suggestive model of how geohistory might be enriched if—*in detail*—the indoor study of fossil specimens were to be combined with the outdoor study of the geognostic positions of the formations in which they were found. The next section describes how that combination of indoors and outdoors, fossils and formations, was first taken on to a new level of detail, in work that—while not itself geohistorical at all—would soon make possible a vastly improved practice of geohistory.

8.2 THE ORDER OF THE STRATA (1801–6)

The isolation of Britain

Blumenbach at Göttingen and his former student Schlotheim at Gotha both exemplify the way that German naturalists were following what was going on in Paris—and indeed in other lesser centers of scientific activity throughout Continental Europe—despite the constraints of the seemingly interminable wars. In contrast, Britain remained relatively isolated, not in all the sciences but certainly in those concerned with the earth. In London, for example, Everard Home had known little of what Cuvier in Paris had been doing, and in the event Cuvier was unable to visit London during the Peace of Amiens to enlighten him (§7.5). The wartime conditions certainly aggravated the intellectual isolation of Britain but do not wholly account for it. In Britain the sciences of the earth had been affected by the problems of relating new knowledge to traditional ways of thinking—specifically of course about the interpretation of the Creation and Flood stories in Genesis—far more directly than anywhere on the Continent. While the Revolution in France was at its height, for example, debates in Britain and Ireland about Hutton's steady-state

20. Schlotheim, *Pflanzen-Versteinerungen* (1804), 15, 63–68. His "palms" were not, in modern terms, angiosperms at all. Only in a later work, Schlotheim, *Petrefaktenkunde* (1820), did he actually give fossil plants Linnaean binomials.

geotheory had become hitched to political issues that linked his eternalism with alleged impiety and hence even with suspected sedition (§6.4). A decade later, when Bonaparte's seizure of power had made the Revolution a thing of the past, the sciences of the earth still remained, in some quarters in Britain, under that same cloud of political suspicion.

The backwardness—as it was perceived at the time—of British research on fossils, for example, was starkly revealed by a volume on fossil plants, which by coincidence was published in the same year as Schlotheim's. Its author, James Parkinson (1755–1824), was an apothecary in Hoxton in east London, and in his free time a keen amateur fossil collector. In 1804 he published the first of three volumes of *Organic Remains of a Former World*; it was the first substantial work on fossils in English. He was aware of some earlier publications on the subject; he also knew personally some of the leading English naturalists, such as Banks, the president of the Royal Society. But his intended readers (and purchasers of his volumes) were not primarily the savants at the Royal Society but rather his fellow amateur fossil collectors, and particularly beginners. He offered them a series of leisurely didactic "letters" to an unnamed and probably fictitious correspondent, illustrated with proxy specimens in the form of colored engravings. This format alone would have struck Continental naturalists as being very old-fashioned. The verbose text was in the traditional epistolary form that they had abandoned long ago—even de Luc, the elderly emigré Continental, had by now dropped it—and the pictures resembled in style those in the great paper museum by Knorr and Walch half a century earlier.

Parkinson's text was also profoundly old-fashioned in a more significant sense. He continued to interpret fossils of all kinds as the relics of the biblical Flood, a view that most serious naturalists on the Continent had rejected, or transcended, at least half a century earlier (§2.5). That he referred in his title to a singular "former world" was unremarkable (in fact his wording matched Schlotheim's singular *Vorwelt*). But Parkinson's subtitle described it more specifically as "the antediluvian world", and he intended that phrase to be understood literally. His strongly scriptural tone may have been a protective cover to distance himself from his own earlier radical and pro-Revolutionary activities, which might have laid him open to charges of sedition. But this is not incompatible with inferring that the opinions he expressed were also sincerely held. After some preliminary letters explaining the nature of fossils in general, the rest of his first volume was devoted entirely to fossil plants, most of them from the Coal strata (and therefore similar to those described by Schlotheim). But the frontispiece of the volume—which in any book was expected to be a visual epitome of the contents—depicted unambiguously his claim that *all* classes of fossils were relics of the biblical Flood that had been emplaced in the rocks during that brief historical event (Fig. 8.4).[21]

"No one would think this writer had ever wandered further than the sound of Bow Bells" was the fair comment that one early reader wrote in the margin of Parkinson's volume. By that traditional criterion of the true Cockney, the demands of Parkinson's medical work did indeed keep him in London, giving him few opportunities for travel or fieldwork, and he seems to have built up his collection mainly by purchase and exchange. His book was therefore based on the traditional

Fig. 8.4. The frontispiece of the first volume (1804) of James Parkinson's *Organic Remains of a Former World*, showing symbolically the "diluvial" origin that he attributed to fossils of all kinds. Noah's Ark has been beached by the subsiding Flood on an islet in the distance; stranded in the foreground are shells, including ammonites, which will become some of the fossils that his work was to describe and illustrate. The design closely resembles an image used almost a century earlier in Scheuchzer's well-known *Herbarium of the Deluge* (1709); by Parkinson's time it would have seemed utterly outdated to naturalists—including serious amateur fossil collectors—in Continental Europe. (By permission of the British Library)

practice of studying specimens indoors, in his own cabinet and those of his friends, and in public collections such as the British Museum. So it is not surprising that there is little if anything in his book to suggest that he had any clear conception of rock formations or geognostic structures. In effect, he treated all his fossils as being of essentially the same age, as indeed his "diluvial" interpretation would have led

21. Fig. 8.4 is reproduced from Parkinson, *Organic remains* (1804–11), 1 [London-BL: 458.b.16], frontispiece, drawn by Richard Corbould and engraved by Samuel Springsguth; ammonites or "snake-stones" were the first fossils mentioned in the text (1: 2–3); on the effects of the Flood, see especially letter 24 (1: 246–56). Rudwick, *Scenes from deep time* (1992), chap. 1, sets this picture in the context of the earlier image in Scheuchzer, *Herbarium diluvianum* (1709); Morris, *James Parkinson* (1989), describes his early radical politics (chap. 3) and his later work with fossils (chap. 11) as well as giving a full account of his medical career. Parkinson first described (in 1817) the disease that now bears his name.

him to expect. However, although his work was very old-fashioned by Continental standards, its success—there was a market for two subsequent volumes—shows how well it appealed to the many amateur fossil collectors among the leisured classes in Britain, who in turn depended on the skillful eyes of laborers and quarrymen to supply them with many of their finest specimens (§1.2).[22]

Smith the surveyor

Around the time that Parkinson's first volume appeared, fossils were brought to the attention of several other social groups in Britain—particularly that of landowners keen on improving the productivity of their rural estates—through the work of one of their social inferiors. William Smith (1769–1839) was Cuvier's (and Bonaparte's) almost exact contemporary. But he came not from their bourgeois background—let alone from that of Schlotheim the upper-class *Freiherr*—but from the humble level of rural artisans. In his native Oxfordshire, Smith had learned the skilled trade of land surveying through an informal apprenticeship. In the early 1790s, work on an estate in the Somerset coalfield had given him experience of surveying the structures of solid rocks. Unlike the ore-bearing veins in Primary rocks—the classic site of Continental geognosy (§2.3)—this was a world of layered Secondary rocks, among them the valuable seams of coal itself. But it too was a world of three-dimensional structures, the dimension of depth being of course directly observable in the shafts and adits of the mines themselves (see Fig. 2.15).[23]

Smith's next employment, in the later 1790s, had been on one of the great civil engineering projects of the time: the construction of inland "navigation", the expanding network of canals and navigable rivers by which coal and other heavy goods could be transported cheaply from producer to consumer in an increasingly urban and industrial Britain. Smith's specific task as resident engineer with the Somerset Coal Canal Company was to survey the line for two new canals and supervise their construction. In a lowland region such as most of southern England, the construction of any canal would have offered favorable opportunities for observing the rocks normally concealed by soil and vegetation. Temporarily, until a new canal was filled with water, its excavation created a long and almost continuous exposure—comparable to a line of low coastal cliffs—in contrast to the scattered exposures provided by the small pits and quarries that at this time dotted the English countryside. But Smith's particular canals had been even more favorable, because they happened to be cut through an area of diverse and gently dipping strata, some of them containing abundant fossils. That the strata were indeed gently dipping could be clearly seen because of course a canal had to be dug along a precisely horizontal line (except where a lock was planned); the continuous exposure made it possible to observe unambiguously the *order* in which the formations lay on top of one another; and the unmechanized excavation—by the manual laborers known as "navigators" or "navvies"—gave optimal chances of finding good fossil specimens (see Fig. 10.14). And, of course, Smith had been there on the spot when they were found.

Smith's work on the Somerset Coal Canal had enabled him to extend to these soft clays, sandstones, and limestones what the miners in the coalfields believed about the harder rocks of the Coal formation: that individual beds could often be recognized across wide areas by their individual character, and that they lay in the same sequence or *order* wherever they were found. The miners applied those ideas within the limited space of a few adjacent mines and to coal seams and other distinctive strata often only a foot or two thick. Smith's careful scrutiny of the canal excavations convinced him that they also applied on a much larger scale, to beds that were tens of feet thick or more, over spaces measured in miles. Above all, he had noticed that many of these beds could be identified individually by their distinctive fossils, most of them shells. Early in 1796—almost the same time that Cuvier first arrived in Paris—Smith had written a brief private note of the conclusions he had reached during his work on the canal. It referred to "that wonderful order and regularity with which Nature has disposed of these singular productions [i.e., fossils] and assigned to each Class its peculiar [i.e., specific] Stratum". That emphasis on the regularity of nature and on the *order*—in the sense of both orderliness and sequence—of its strata was to remain as characteristic of Smith as his fossils were of their respective formations.[24]

Smith was no illiterate peasant. Although he knew little of the work of metropolitan savants, on fossils or anything else, his skill and success as a surveyor ensured that he had become a respected member of the flourishing provincial world of agricultural improvement. At Bath—the nearest large town, England's most fashionable spa, and a major center of cultural life—Smith had been elected to the quite prestigious Bath and West of England Society; at its meetings and agricultural shows he came into contact with locally and even nationally important figures. He had given two amateur naturalists (and parsons) a list of the twenty-three formations he had distinguished, ranging from the Chalk at the top to the Coal at the base. In return, one of them had supplied Smith with scientific names for the fossils that he had found to be characteristic of many (but by no means all) of these formations. Smith had also drawn a small map showing the outcrops of the formations around Bath. None of these items had been published. But yet another local parson described and acknowledged Smith's discoveries in print—Smith had arranged his fossil collection for him—when he wrote a brief chapter on "mineralogy and

22. Thackray, "Parkinson's *Organic Remains*" (1976), records the quoted marginal comment, gives a full publication history, and notes that in later years Parkinson may have collected some of his fossils in the field, during occasional travels in England (§9.4). On fossil collecting at a slightly later period, see Knell, *Culture of English geology* (2000).

23. The *DSB* article on Smith (1975) by Joan Eyles is a valuable summary of the biography that this leading Smith scholar did not, sadly, live to complete; it can be supplemented by, for example, her "Smith: a chronology" (1969), "Smith: life and work" (1969), and "Banks, Smith" (1985). Other important articles to which this section is indebted include Fuller, "Industrial basis of stratigraphy" (1969); and Torrens, "Commemoration of Smith" (1989), "Banks and the earth sciences" (1994), "Timeless order" (2001), and "Life and times" (2003). Cox, "New light on Smith" (1942), prints important MSS by and about Smith. I am particularly grateful to Hugh Torrens for sharing with me his vast and still largely unpublished knowledge of Smith's work.

24. Part of Smith's "Swan" memorandum (so named from the inn where he wrote it) is reproduced in Cox, "New light on Smith" (1942), 12.

fossilogy" for a book on the history and topography of Bath and its environs. On this local level, Smith's work was becoming well-known.[25]

Smith as a mapmaker

By the turn of the century, Smith's ambitions for his work were expanding rapidly. He had earlier been on a brief trip to the north of England to study methods of canal construction and had become convinced that his inferences about formations might be valid over a far wider area than just the county of Somerset; he had sketched more than one small map showing provisionally some of their outcrops across the whole country. When his canal work came to an abrupt end in a dispute with his employers, he launched himself as a self-employed agricultural surveyor and a consultant on land drainage and improvement. His income became variable and all too insecure, but on the other hand his freelance work gave him almost unlimited opportunities for travel. In 1801—just as Cuvier was issuing his international appeal about fossil bones—Smith issued a prospectus appealing for subscribers to what he described as "Accurate delineations and descriptions of the natural order of the various Strata that are found in different parts of England and Wales, with practical observations thereon." He planned to publish a large map showing the outcrops of the Secondary formations (which he always called "Strata"), with an accompanying text explaining that many of them could be recognized by their characteristic fossils. The final phrase in the title indicated that, like Werner's work on veins, it was intended to be of great utility.[26]

Smith's plans for publishing a treatise on the rock formations of England and Wales suffered the first of many setbacks when the intended publisher went bankrupt only a few months after the prospectus was issued. But Smith persevered with the steady improvement of his map, recording the outcrops of the rocks on a topographical map as he widened his coverage with further extensive travels. It was an astonishingly ambitious project. Wherever he went he harnessed the highly localized knowledge—of rock exposures and their fossils, for example—of people of widely different social classes; but for coordinating and interpreting all this information he was in every sense on his own.

The difficulty of the task that Smith had set himself should not be underestimated. He was mapping in regions where rocks of any kind were rarely visible except in small isolated exposures: in brick pits, stone quarries, road cuttings, stream banks, and so on. Lowland Britain was not like, say, the deserts of Arabia or the southwestern United States, where exposures may be almost continuous because there is little or no vegetation. Smith was mapping areas where even his expert eye—let alone the unpracticed eye of a potential patron—might at first glance see nothing but fields and woods, gentle hills, and broad valleys. He was tracing the course of what in general could *not* be seen; his map was a remarkable achievement because it disclosed what was otherwise hidden from view. A few naturalists on the Continent had used similar techniques many years earlier to produce maps of roughly the same kind (see Figs. 2.13, 2.16). But Smith's was remarkable for distinguishing so many discrete formations over such a wide area, ranging from the clays and sands around London all the way down to the Coal formation that was powering Britain's industrial growth; and for being the product of his own almost single-handed fieldwork.[27]

When tracing the outcrops of the formations across country, Smith was certainly guided by the landforms and characteristic vegetation that they produced at the surface. The steep scarps and bare grasslands of the Chalk, for example, could be recognized easily, even from many miles away; and Smith's increasingly experienced topographical eye—which must have been as outstanding as Cuvier's eye for the forms of fossil bones—could detect many other formations by more subtle features. But some formations showed no surface features at all, and even in those that did the scarps were often discontinuous. Vegetation could be deceptive, as where the Chalk hills were covered with a surface layer of clay and acid soil. Even the rock types could vary along the outcrop. So Smith's "characteristic fossils" were decisive. Not all his formations had abundant fossils and some had none at all. But fossils served to distinguish many otherwise confusingly similar formations from each other (for example, dark clays and shales) and enabled him to recognize in an unfamiliar area where he was, as it were, in the unvarying "order of the strata".[28]

Smith's map used an ingenious cartographical convention to depict the formations that rose successively to the surface, roughly from southeast to northwest over much of England. In place of the spot symbols used on the great "mineralogical map" of France (Fig. 2.12), or the uniform color washes on Charpentier's "geognostic map" of Saxony (Fig. 2.13), Smith's "delineation of the Strata" of England and Wales showed the outcrops with bands of color that were graded in such a way as to suggest the three-dimensional character of the pile of formations (Fig. 8.5).[29]

Smith himself evidently had such a clear sense of this three-dimensionality that—like any competent modern geologist—he could "see" the solid structure of the formations in his mind's eye just from contemplating his two-dimensional map on paper. But others, and particularly his potential patrons, could not be assumed

25. Warner, *History of Bath* (1801), 394–99, a typical chorographical work (§4.1).

26. Smith, "Prospectus" (1801); its title page is reproduced in Eyles (J. M.), "Smith: a bibliography" (1969), 89. Smith's earliest small MS map of England and Wales (1801) is reproduced in Cox, "New light on Smith" (1942), pl. 2.

27. On the international context of Smith's mapping, which has often been neglected by anglophone historians, leading sometimes to inflated and chauvinistic claims, see especially Ellenberger, "Cartographie géologique" (1985).

28. Laudan, "Smith: stratigraphy without palaeontology" (1976), proposed the revisionist thesis embodied in her title; it was forcefully criticized by, for example, Eyles (J. M.), "Smith: mere fossil collector?" (1979). Smith has indeed been the object of as much historical mythmaking as, say, Hutton, but in his case the extant documentary evidence suggests that Laudan's radical debunking was not justified.

29. Fig. 8.5 is reproduced from Smith, *Delineation of the strata* (1815) [London-BL: Maps 1180(20)], part of sheet 11; the brilliant colors—sadly lost in a reproduction such as this—were chosen to mimic, or rather to intensify, the real colors of the rocks; the same area is superbly reproduced in color as a frontispiece in the reprint (2003) of Phillips (J.), *Memoirs of Smith* (1844). The coalfields on this part of the map were "inliers", i.e., surrounded by overlying formations, so in their case the intense color around the edge represented not the base but the (local) top of the Coal formation. Smith's system of coloring made his map exceptionally expensive to produce, since the carefully graded watercolors demanded much more time and skill from the colorist than the simple washes used on most other colored maps. There are several important variants to the map, because Smith made corrections and improvements while it was on sale: Eyles (V. A. and J. M.), "Different issues" (1938).

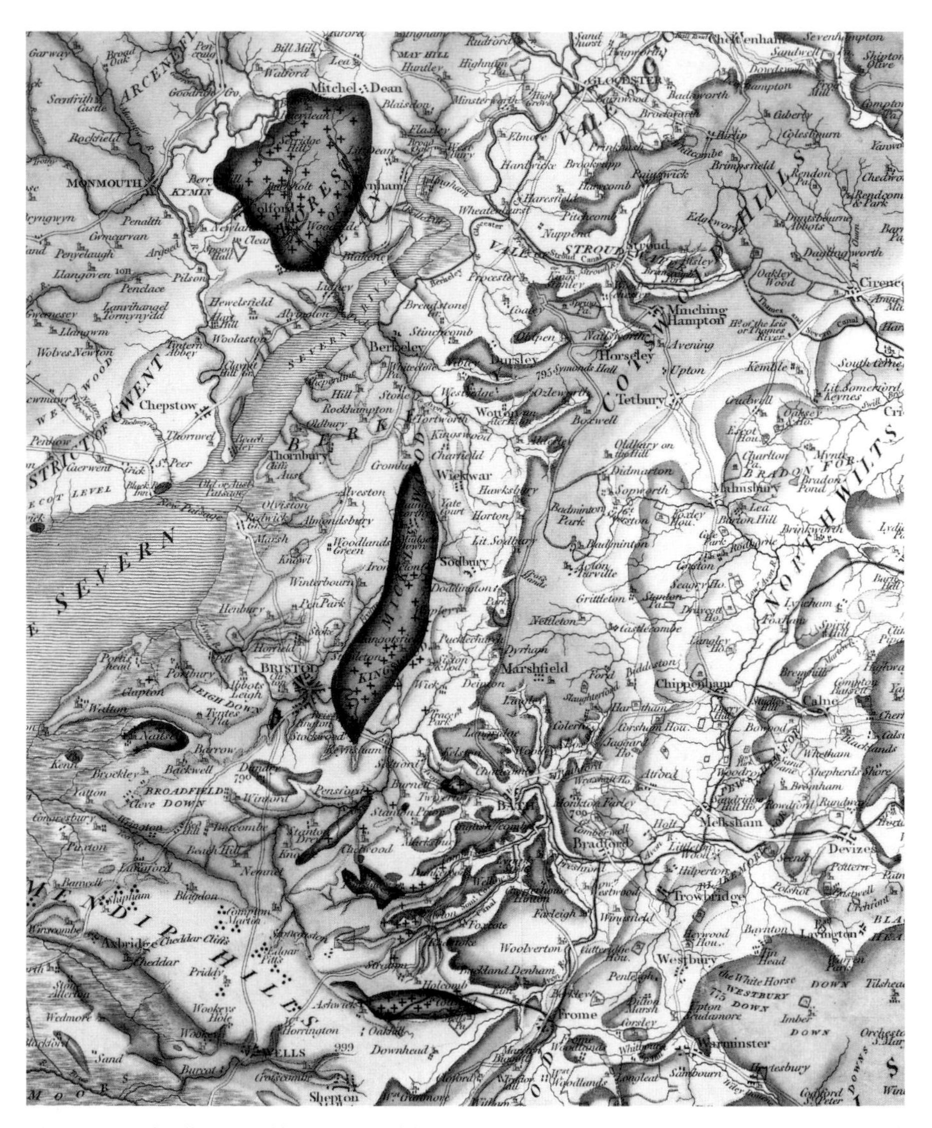

Fig. 8.5. Part of William Smith's great map of the outcrops of the "Strata" or formations found in England and Wales. Each color is most intense at the base of the formation it represents, fading away as the outcrop extends towards the next overlying formation: this conveyed a sense of the three-dimensional structure of the pile of formations, perhaps as vividly as was possible on a two-dimensional map. The small portion shown here is of the region (around Bristol and Bath) where Smith had begun his survey and was perhaps most confident about its detail. The uppermost formation (in the southeastern corner) is the Chalk of Salisbury Plain; the lowest is the Coal formation found in several small coalfields (with crosses to denote individual mines). The map was not published until 1815, after many delays beyond Smith's control; but earlier versions of it, which were probably the same in general appearance if not in detail, had been publicly exhibited in England on many occasions since 1802. (By permission of the British Library)

to share that distinctive visual skill or to appreciate so easily the structural significance of the pattern of outcrops on the map. So from an early date Smith also drew sections along selected lines across his map, to give a more direct visual representation of the pile of formations. Years later he published several such sections, which underline the three-dimensional or structural character of his project (Fig. 8.6).[30]

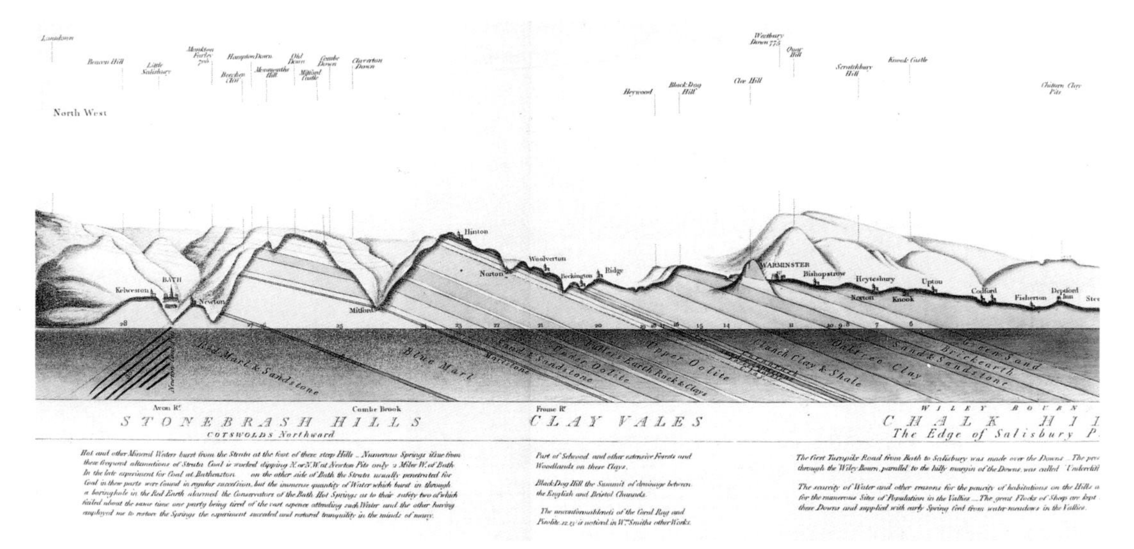

Fig. 8.6. Part of Smith's *Section of the Strata* (1819), showing the pile of formations cut along a line running from northwest to southeast from Bath through Warminster. As usual in such sections, the vertical scale—and therefore also the dip of the strata—is greatly exaggerated in order to clarify the structure and succession (see Figs. 2.16, 2.19). The "Stonebrash [Cotswold] Hills" and "Chalk Hills" sketched lightly *behind* the section heighten the pictorial similarity to an offshore view of coastal cliffs: the section is a "virtual" cliff depicting strata that in reality can be seen only in small exposures at the surface. Their inferred or extrapolated courses below that surface are drawn—as in some earlier English sections (see Fig. 2.24)—with straight *ruled* lines, as if they were precisely uniform courses of masonry (the darker part of the section is below sea level). The lowest formation is the Coal (30, with coal seams, left) underlying and abutting against "Red Marl" (28), the lowest in a regular sequence extending up to the Chalk of Salisbury Plain (right). The Cornbrash (see Fig. 8.7) is the thin layer (16) underlying the "Clunch Clay and Shale" (center). (By permission of the Syndics of Cambridge University Library)

Smith's prospects for publishing his map depended on recruiting the support and patronage of those landowners who were actively engaged in the improvement of their estates. Banks, the president of the Royal Society, who had estates in lowland Lincolnshire, was prominent among them. John Farey (1766–1826), a man of Smith's social class and like him a surveyor, acted as the crucial link between them. At the time, Farey was employed in drainage and irrigation works on the duke of Bedford's estates around Woburn. He had not previously given any particular attention to subsurface rocks; but in 1801 he met Smith at Woburn, and the two surveyors then made a tour of the surrounding region, at ducal expense, "to investigate the strata". Farey later reported on the trip to Banks, whose interest in mineral matters had been heightened after he inherited an estate in Derbyshire that included an important lead mine. Farey told Banks how greatly impressed he had been with Smith's large map—it measured some seven feet by five—and with the practical

30. Fig. 8.6 is reproduced from Smith, *Section of the strata* (1819) [Cambridge-UL: Atlas.5.81.8]; Fuller, "Strata Smith" (1995), has facsimile reproductions of this and other sections. Smith's colleague John Farey drew sections in a closely similar style from 1806 onwards, showing formations in other parts of England, which were widely known although unpublished: see Ford, "First detailed sections" (1967), and Torrens, "Banks and the earth sciences" (1994).

value of his methods when tested in the field. He explained "the pains Mr Smith has been at, by the peculiar fossils, or combinations of them, to identify each strata [*sic*] in different parts of its course on the surface"; and he urged Banks to support Smith's work.[31]

Banks was among the landowners present when Smith "exhibited his map in a very considerable state of forwardness" (as one newspaper reported it) at the annual sheepshearing meeting at Woburn later in 1802. The following year, while visiting London, Smith called on Banks and explained to him, presumably in greater detail, "my plan of Strata and arrangement of fossils". And at the Woburn meeting in 1804, Banks acted on Farey's earlier advice; having realized that Smith was in dire financial straits, he opened a subscription list for the map, contributing the substantial sum of £50, though his friends and acquaintances were slow to follow his lead.

The same year, however, Smith took the financially risky step of moving to London and renting a house off the Strand, not far from the Royal Society's premises. The house was large, because he planned not only to store all his fossils there but also to make it in effect a private museum in which he could demonstrate the results of his work, especially to the landowners who might commission him to survey their mineral resources or improve their agricultural land. What his potential patrons would have seen at his house represented a highly original synthesis of the indoor museum science of fossils and the outdoor field science of rock formations. Most fossil collections were arranged according to the kinds of organisms of which they were the remains—plants, mollusks, corals, sea urchins, and so on—with no relation to the fieldwork by which they had been obtained. Paper museums followed the same convention; Parkinson's volume on fossil plants, published the same year, was typical. In contrast, Smith arranged his fossils primarily according to the formations from which they came. Furthermore, he exhibited his fossils in an ingenious way, on a series of *sloping* shelves that mimicked the successive outcrops of the formations themselves. His museum therefore displayed, in condensed or diagrammatic form, the three-dimensional structure of the successive formations that he had surveyed in the field, by means of the fossils that he had used to identify them. Years later he published an album of engravings of his fossils, which conveyed in proxy form something of the impact that these shelves full of characteristic fossils must have had on those who visited his house to view them (Fig. 8.7).[32]

Smith's map was apparently exhibited quite widely in the early years of the new century, as he tried—with limited success—to collect further subscriptions to finance its publication; it was probably shown for a time in Banks's house, where it would have been seen by the intellectual and social elite that gathered there for his weekly salon. Together with the fossils that substantiated its practical value, the existence of the map must therefore have become widely known, even in its unpublished form: not only to the country landowners whose support Smith most needed, as they spent the winter social "season" in London or attended the summer agricultural meetings in the country, but also to metropolitan savants such as those at the Royal Society; Banks himself exemplified the substantial overlap between those two groups within the social elite.

In addition to landowners and savants, the visitors to Smith's house might have included some of the amateur collectors who had purchased Parkinson's new book on fossil plants. If so, they would have found the comparison instructive. They

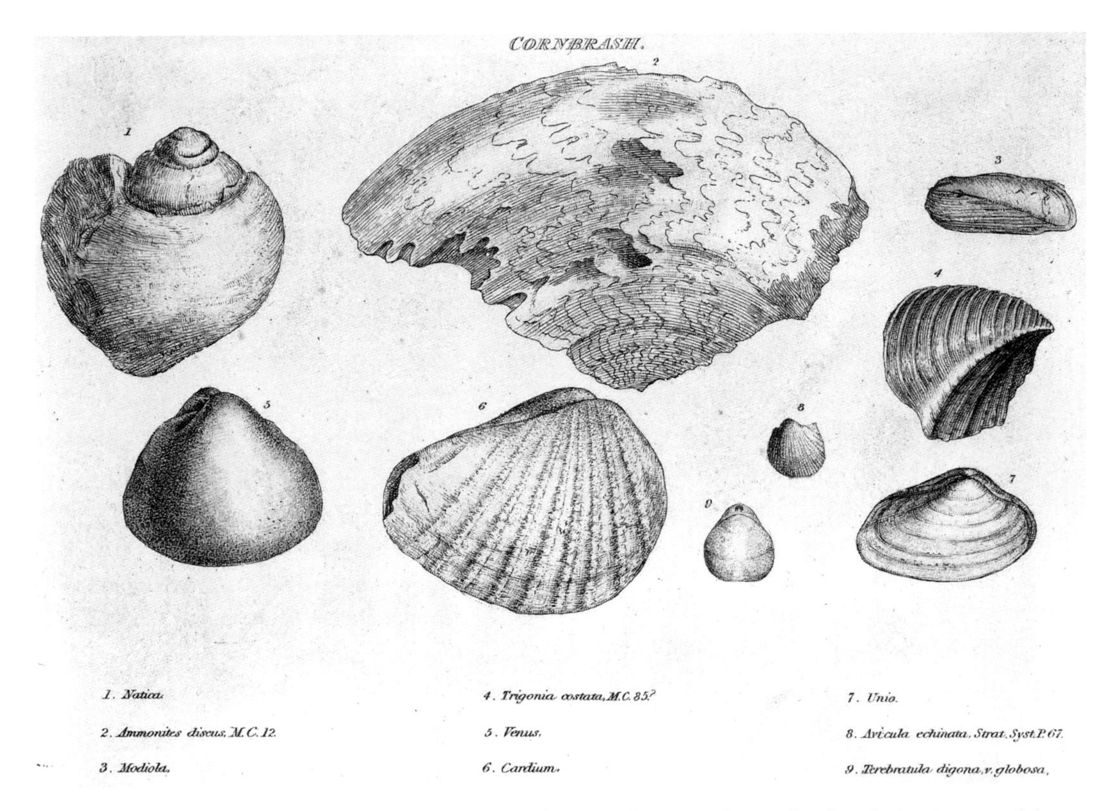

Fig. 8.7. Fossils characteristic of a specific English formation, the Cornbrash, as depicted on one of the plates in Smith's album of proxy specimens, published in 1816–19. Thc fossils for each formation were printed on paper colored to match the outcrop depicted on Smith's map, thus visually reinforcing the association between fossils and rocks. These specimens must be similar—perhaps even identical—to some of those he had displayed in his house in London, from 1804 onwards, on the sloping shelf representing the outcrop of the Cornbrash. By the time these engravings were made, he had been given Linnaean names for his specimens (in several cases just the genus); but his practical use of them had depended not on knowing their names but on his visual skill in recognizing them in the field. The marine origin of the formation would have been obvious to him: the *Cardium* (6) was of the same genus as the living cockle, and the *Modiola* (3) was close to the living mussel, both harvested on English seashores (and sold as fast food on London streets). The largest (and rarest) specimen is a fragmentary ammonite (2) with characteristic suture lines (see Fig. 5.2). The Cornbrash was a striking example of Smith's method because this very thin limestone (see Fig. 8.6) maintained its identity, its characteristic fossils, and its place in the "order of the Strata", all the way from Dorset in southwest England to Yorkshire in the northeast. (By permission of the Syndics of Cambridge University Library)

31. Farey to Banks, 11 February 1802 (Cambridge-FM, Percival MSS, H237); Torrens, "Banks and the earth sciences" (1994). Ford and Torrens, "John Farey" (1989), reviews the whole range of his work. Smith's most appreciative colleague had no doubt about the important role of fossils in his practical methods; see Eyles (J. M.), "Smith, Banks" (1985), 41–42.

32. Fig. 8.7 is reproduced from Smith, *Strata identified* (1816–19) [Cambridge-UL: Lib.3.81.31], pt. 4 (1819), plate [1]. He had long intended to publish illustrations such as this, and had even had a couple of plates engraved by 1804: see Cox, "New light on Smith" (1942), 16 and pl. 4. That the specimens were "*arranged on sloping shelves, to represent the strata*" is recorded (in italics, to emphasize the point) by his nephew and disciple John Phillips, *Memoirs of Smith* (1844), 79.

would have recognized most of Parkinson's fossils (similar to Schlotheim's: Fig. 8.3) among those that Smith displayed on his shelf representing the Coal formation, one of the lowest of all. On the other hand, the ammonites that Parkinson's artist had depicted explicitly as the relics of Noah's Flood (Fig. 8.4) would have been found on shelves representing formations far higher in the thick pile depicted on Smith's map, such as that for the Cornbrash. In effect, Smith's display would have introduced fossil collectors to a three-dimensional or structural setting for their specimens, which they could hardly have begun to glimpse from either the text or the plates in Parkinson's book.

Around the same time, the practical value of Smith's work began to be apparent in the most effective way, in terms of hard cash. All over Britain, hopeful or avaricious landowners were persuading their gullible friends and neighbors to invest substantial capital sums in the sinking of wildcat shafts or boreholes in the hope of finding coal, often on the basis of the most tenuous surface clues. In one such case, in Somerset in 1805, Smith told them bluntly that they were wasting their money. They imagined that their shaft was being dug in the Coal formation, and that it would soon strike actual coal seams. But among the strata they had already encountered was what Smith recognized, by its characteristic variety of fossil oyster, as his Kelloways Stone. This was a formation as distinctive as the Cornbrash, and it too lay right in the middle of his pile, many hundreds of feet above the real Coal strata. The credulous investors ignored him, to their cost. It was perhaps the first time that Smith used his "characteristic fossils" to such practical effect, but it was not the last. Farey proved an apt disciple, and tried in vain in the following years to discourage an even more expensive project in Sussex.[33]

In fact, Farey became Smith's spokesman and even his bulldog. In 1806 he sent the first of a stream of letters to the editor of the *Philosophical Magazine*—a scientific monthly that had started publication in 1798 as a rather pale imitation of the *Journal de Physique*—stressing the value and importance of Smith's work. He also referred to it around the same time in several articles for Rees's *New Cyclopedia*, a major reference work. Smith's map remained unpublished. But Farey's championing of his cause left no excuse for the scientifically literate, at least in Britain, to remain unaware of what Smith was doing. Farey emphasized unambiguously how Smith was concerned with the structural *order* of the formations. He explained that this could be determined by three criteria, fossils being added to the established criteria of superposition and rock type:

> The most complete and certain rules have been, or may in every instance be, deduced for ascertaining the relative position (which probably never varies) of each distinct stratum, however thin, with regard to those above and below it in the series (or *natural order of the strata*, as Mr Smith called it in his first printed prospectus); rules equally general have, or will on sufficient inquiry, be found, for identifying each particular stratum, either by the knowledge of its relative position with other known strata in its vicinity, [or] by the peculiar organized remains imbedded in it, and not to be found in the adjoining strata, or by the peculiar nature and properties of the matter composing the stratum itself.[34]

Smith as a geognost

Smith's work needs to be evaluated in relation to what was being done around the same time elsewhere in Europe. But this is difficult, because Smith, like Hutton (§3.5), has become an iconic father figure to later generations of anglophone geologists. Any attempt to set him in his context is too readily misinterpreted as an iconoclastic attempt to topple him from his pedestal. However, a preliminary step is to locate his work on the conceptual map of the sciences (§1.4), and particularly the sciences of the earth. It should be clear that Smith was, in all but name, a geognost; and that what he was doing with such conspicuous success was, again in all but name, geognosy (§2.3). As a robustly insular Englishman he would probably have found these foreign words repugnant, but what he was practicing was in fact the well-established science of rock structures, albeit an importantly novel variety of it.

Smith's work, like that of other geognosts, had begun in the context of practical mining; and when he extended it to the soft rocks of lowland Britain it remained strongly practical in orientation, first in the context of canal construction and later in that of agricultural improvement. Like Farey and some other land surveyors, Smith proudly identified himself as a member of a distinctive social group of "practical men". Within the class structure of British society, they were far inferior in social status to the aristocratic or gentlemanly landowners who employed or commissioned them to survey or improve their estates. But that was a contingent feature of the British social scene. What they were doing was in most respects no different from what Continental geognosts—some of them decidedly upper-class in status—were doing with an equally practical orientation. The English "practical men" had received their training in surveying by the traditional route of apprenticeship (in Smith's case informally), whereas the Continental geognosts had received theirs by the new route of technocratic education in the mining schools at Freiberg and elsewhere (§1.2). But both groups were equally well-equipped to make accurate surveys, not only on the surface but also—either literally or in the mind's eye—underground.

Smith's early experience was with the varied strata of a Secondary formation, namely the Coal rocks encountered in coal mines, whereas the geognosts typically worked with the ore-bearing veins in mining regions of Primary rocks (Figs. 1.6, 1.9). But as already mentioned (§8.1), Werner and his colleagues and former students were currently extending their attention and their fieldwork to the Secondary formations, so that Continental geognosy was enlarging its scope to deal more fully with just the kinds of rocks that Smith was surveying. What united them much more significantly was that they were all concerned with the three-dimensional

33. Torrens, "Smith et le projet" (1997), "Coal hunting at Bexhill" (1999), and "Timeless order" (2001).

34. Farey, "Stratification of England" (1806), 43; on his articles for the *New Cyclopedia*, see Cox, "New light on Smith" (1942), 39; also, more generally, Ford and Torrens, "John Farey" (1989). The *Philosophical magazine* covered the whole range of "natural philosophy", i.e., all the natural sciences. Its editor, Alexander Tilloch, chose La Métherie, his counterpart in Paris, as the scientific worthy whose portrait (re-engraved from the one reproduced in Fig. 6.6) decorated the volume for 1804.

structure of the rocks, rather than just with the surface distributions described by physical geographers (§2.2). So it is hardly surprising that Smith recognized independently what the geognosts had already adopted as their standard practice: namely that the results of any survey were best recorded and communicated by means of a combination of maps and sections (§2.3). Smith's map, like, for example, Charpentier's before it (Fig. 2.13), depicted the surface outcrops of the different formations by means of different colors. In format, Smith's was original only in its ingenious mode of coloring. But this was important, because to a trained eye it gave even a two-dimensional map a vivid sense of three-dimensional structure (though Smith did also appreciate the valuable role that sections could play in explicating the structure).

Conclusion

In one other respect, however, Smith's work was so original that it *enriched* the practice of geognosy in a decisive way. This was of course his use of fossils. He was not the first to recognize the correlation between formations and their fossils, despite later mythmaking and often blatantly chauvinistic claims that he was. As already pointed out more than once, several earlier naturalists, such as Arduino and Soulavie and more recently de Luc, had noted this—if only briefly or in passing—in their own fieldwork areas. On the more ambitious scale of the whole of Europe, Werner himself was (according to Schlotheim) actively investigating, at just this time, how far fossils could be said to be characteristic of specific formations (§8.1). But Werner had not yet made his conclusions public, even in the sense of disseminating them informally in conversation with his colleagues.

What distinguished Smith's work was that by about 1806 he had shown *in detail* that the concept of characteristic fossils was valid over a far wider area—most of England and parts of Wales—than ever before; and that it could be used with great reliability to distinguish a large number of discrete formations, some of them so thin and apparently insignificant that they would previously have been treated merely as subordinate beds within a single Secondary formation. And although Smith's results were unpublished in the sense that they were not yet in print, they were fully in the public realm, at least in England, and were widely known among those who mattered. Within a few years of the turn of the century, his manuscript map had been exhibited in several public places; the specimens that substantiated it were available for inspection at his house in London, arranged to display his methods and his conclusions in the most striking manner possible; and Farey's assiduous championing of his work, published for example in a scientific periodical and a major reference work, made it impossible to ignore.

Finally, however, it is important to note what Smith was *not* doing, simply because it was not part of the task he had set himself. Like any other geognost, Smith was not primarily concerned to explain in causal terms how his formations had come into being. He did privately have ideas about that—as early as 1802 he was trying to write on "the formation of strata and the effects of the deluge &c."—but they played no obvious part in his plotting of the outcrops of the formations or in his empirical use of their characteristic fossils. Geognosy was not in itself a causal science for Smith, any more than it had been for earlier geognosts.[35]

Still less was Smith's enriched geognosy a historical science. With a good general knowledge of natural history—not to mention his familiarity with the traditional London street cry, "Cockles and mussels, alive, alive-O!"—Smith was well aware that most or all of his fossil shells belonged to *marine* mollusks and that his rocks had therefore originated in seawater (see Fig. 8.7). But he showed little interest in reconstructing those vanished conditions, let alone with making a geohistorical narrative out of them. Unlike Lavoisier, for example, he never attributed the distinctive assemblages of fossils in his formations to *ecological* differences such as that between shallow and deep water habitats (§2.4). His three-dimensional pile of rocks, and the distinctive fossils they contained, were scarcely treated even as the record of a *temporal* sequence of events, let alone as nature's documents recording the *history* of life on earth. Even with its enrichment by fossils, Smith's science remained on the level of static structure, as the earth's "architecture". It was a science of structural *order*, to use his own favorite term, not temporal narrative, let alone geohistory.

8.3 TIMESCALES OF GEOHISTORY (1803–5)

Cuvier's Parisian lectures

While Smith was developing a geognosy enriched by the use of fossil shells (§8.2), Cuvier was developing a comparative anatomy enriched by being extended to fossil bones (Chap. 7). Apart from their interest in fossils, however, their projects had little in common. Smith was preoccupied with the empirical value of characteristic fossils for practical purposes, with no regard for geohistory. Cuvier was concerned above all to demonstrate the distinctive nature of his fossil mammals, and the reality of their extinction in a catastrophe that was the most recent major event in an eventful geohistory. But both excluded geotheory: Smith implicitly, and perhaps almost in ignorance of the genre; Cuvier explicitly and often scornfully, not least in his frequent criticism of the "geology" of his colleague Faujas. Smith's practice was limited, as were his fossils, to the Secondary formations; he was aware of the Primary rocks below and the alluvial materials above, but neither came within the scope of the task he had set himself, at least initially. Cuvier's practice did extend to the Superficial deposits, which indeed contained the best evidence for his putative catastrophe, but he had been scornful of all attempts to reach back to the origin of the earth, let alone to construct an explanatory model that would account for its entire history.

35. See the verbose MS preface (dating from about 1802–3) for his planned book on strata, printed in Cox, "New light on Smith" (1942), 81–90. Smith believed that the formations had been emplaced successively in a series of violent floods—in modern terms, something like a series of turbidity currents—each sweeping across England from the southeast but extending little further than their present scarps. He probably derived this causal interpretation from the geotheory of John Strachey half a century earlier: see Fuller, "Cross-sections by John Strachey" (1992), esp. fig. 5. But it had no relation to his work on characteristic fossils, except perhaps to provide post hoc justification for his treatment of each formation as a discrete unit, characterized by the same fossils throughout its horizontal extent and vertical thickness.

In the spring of 1805, however, Cuvier surprised his colleagues by giving a series of lectures that signaled a major change in his attitude on this point. The surprise was twofold. First, Cuvier chose "*géologie*" as the title of his course, thereby tacitly upstaging Faujas, the Muséum's professor in that science, but also indicating that he now regarded the subject as worthy of serious attention. Second, the lectures were given at the Athénée des Arts, which had been founded before the Revolution as a kind of adult education center for the Parisian bourgeois public. Cuvier had not previously tried to reach this wider audience: the Athenée, on the Right Bank (i.e., north of the Seine), attracted a public distinctly different from those at the Institut on the Left Bank (near the university or Latin quarter) and at the Muséum on the eastern outskirts of the city. For the first time, Cuvier offered an outline of his research in a broader theoretical context than had so far seemed appropriate at either the Muséum or the Institut (Fig. 8.8).[36]

Cuvier's own preliminary notes for his course show, unsurprisingly, that he intended to focus on the rocks containing fossils, because they provided "the strongest proofs that the globe has not always been as it is at present", or in other words that the earth had indeed had a *history*. Among the specific points that he planned to emphasize were that the oldest rocks, lacking all trace of fossils, showed that "organization [i.e., life itself] has not always existed"; that in the earth's subsequent history "there have been different ages, producing different kinds of fossils"; that fossils had usually been preserved "in tranquil water" where the organisms were living, rather than being swept in violently from elsewhere; that some of the earth's "revolutions" had nonetheless been sudden; and that the physical processes now active on earth were inadequate to account for the larger changes in the deep past. None of this was novel; what was significant was that it was Cuvier who planned to propound it, and to an audience of the general public.[37]

Fig. 8.8. Cuvier at the age of about thirty-six, around the time of his first public lectures on geology (1805): an engraving by Simon Charles Miger, after a painting by François André Vincent.

According to the notes taken at the lectures by one of his auditors—the young Italian naturalist and nobleman Giuseppe Marzari Pencati (1779–1836) from Vicenza—Cuvier duly reviewed in nontechnical terms the various fossil mammals that he was currently describing in the Muséum's *Annales* (including some not yet published), but also mentioned a wide range of other fossils such as fish, mollusks, and plants. This review comprised the bulk of his lectures, but he prefaced it with a brief summary of earlier theorizing about the earth, ranging from Genesis through Descartes to Buffon, and he concluded with an equally brief sketch of six successive "epochs" in its history. Although the language of epochs recalled Buffon's geotheory (§3.2), Cuvier's outline was closer to the standard model (§3.5): he began with the earth covered by a primal ocean and finished with the relatively recent "epoch of our [present] continents".

Within this generally unproblematic framework, Cuvier staked out his own position on three contentious issues. The first would have been no surprise to anyone familiar with what he was currently publishing: he firmly rejected the claims of Lamarck and his followers about the transmutation of species (§7.4). In particular, Cuvier cited the varied mummified animals found in Egypt, all of them identical to their living counterparts, and the absence of forms intermediate between, for example, the fossil cave bear and any living bears. He concluded that "species do not change by degrees" (*per gradi*, as his Italian auditor recorded it). According to Marzari Pencati, Cuvier even joked about the Lamarckian ideas and openly ridiculed them.

The second issue, in contrast, was one that Cuvier had refrained from mentioning in the papers he had read at the Institut and published in the Muséum's *Annales*, having tacitly treated it as inappropriate in those savant settings. This was the question of the relation between his new research and the traditional interpretation of Genesis. However, when at the start of his lectures he cited Genesis as an early example of theorizing about the earth, he added almost casually (as Marzari Pencati paraphrased it) that it "is always—for whatever reason—in accord with geological monuments"; he claimed that, "geologically", fossil fish had preceded fossil mammals, while humans, "the last and newest creatures", were missing altogether from the fossil record. In other words, the fossil record conformed in broad outline to the Creation story in Genesis, provided that there was the usual assumption that the biblical "days" were long stretches of time. This interpretation was similar to what de Luc had been arguing since before Cuvier arrived in Paris (§6.2, §6.4, §7.2).

The third contentious issue was also related to the biblical record. Cuvier claimed that the last of the earth's major "revolutions", the one that had wiped out many of his fossil mammals, was quite recent: "the epoch of our [present] continents does not date from more than 10,000 years ago". Taking his cue from de Luc or Dolomieu (§6.3), this was the first time that Cuvier had publicly put any figure

36. Fig. 8.8 is reproduced from Bultingaire, "Iconographie de Cuvier" (1932), pl. 2. Miger also taught his craft to Cuvier (see Fig. 7.11) and engraved many of Cuvier's drawings of fossil bones.

37. Cuvier, "Cours du Lycée de l'an XIII. Géologie" (Paris-IF, MS 3111), translated and transcribed in Rudwick, *Georges Cuvier* (1997), 84–86, 290–91 (during the Revolution the Athénée had temporarily been renamed the Lycée Republicain); these brief notes are the only extant record of the lectures in Cuvier's own hand.

on the *date* of the decisive event that marked the start of the "modern world"; it was of course a date compatible with the one computed by the textual chronologers for the catastrophic event recorded in ancient human records, notably (though not only) as Noah's Flood. In effect, therefore, Cuvier claimed that this story too was broadly confirmed by geology.[38]

Cuvier's lectures were evidently a great success with his new public; as Marzari Pencati reported to a friend in Geneva, the "geologists" among the Parisian savants—Faujas would surely have been among them—had been scornful in advance about such a popularization, and then regretted that they had not subscribed to the course and been able to hear at first hand what Cuvier had to say. His auditor commented, not unfairly, that the title of "geology" was a misnomer, because the lectures were mostly on Cuvier's comparative anatomy, albeit "*applied to geology*". Marzari Pencati also judged that Cuvier had been on shaky ground "when he wanted to go farther down than the *thin crust with fossils*" (i.e., the Secondary formations), because he knew little of the Primary rocks in the field and had merely "made a world in his study [*cabinet*]" when dealing with the earth's earliest history. But he conceded that this did not affect the importance of what Cuvier had said about the earth's most recent history. Marzari Pencati had been surprised to hear Cuvier following Dolomieu and de Luc on the origin of the present continents: "the Holy man [*le Saint homme*] doesn't assign them even ten thousand years". The epithet was ironic, because Cuvier's reputation in Paris was as one who was "hardly very devout"; so his auditor reckoned that this newly disclosed opinion represented "an appalling loss for the atheists", who had been boasting in advance about getting the support of "the great Cuvier". Indeed Marzari Pencati had suspected at first that Cuvier, by starting his lectures with "his altogether Mosaic opinion" about the historicity of the Creation story and the Flood, was (as he put it jokingly) "on the lookout for a cardinal's hat" or merely trimming his sails to the prevailing political wind; but he had later concluded that no bad faith was involved and that Cuvier's opinion was sincere.[39]

The politics of the timescale

Marzari Pencati's somewhat gnomic comments—natural enough in a private letter to a friend who would have been familiar with Parisian gossip—require some unpacking. In France "geology" was less burdened with suspicion than in Britain, but Cuvier was not exploring this science or any other in a political vacuum. Bonaparte's regime was generally benign towards the natural sciences represented by the First Class at the Institut, because he had faith in their potential value for technocratic purposes. But like some later authoritarian figures he was deeply suspicious of the social sciences. He may not have believed that there is no such thing as society, but he had regarded the Second Class as a potential hotbed of subversive political thinking; and in 1802, in the reform of the Institut after he gained power, it had been abolished, leaving only the politically innocuous humanities cultivated in the Third Class to complement the technocratic virtues of the First. In this climate, although "geology" was among the natural sciences, what it was doing was open to question, because it claimed to have authority to pronounce on the relation of the human—and therefore social—world to the world of nature. Specifically, of course, it appeared to challenge more traditional origin stories, particularly those embodied in the Bible, and thereby raised questions about the authority of the church in civil society.

These were not abstract matters at the time of Cuvier's lectures. In 1801 Bonaparte had agreed to a concordat with the Vatican, marking a decisive break with the anti-Catholic campaigns during the Revolution; and the following year a new law had given French Protestants (such as Cuvier) greatly improved civil rights. Bonaparte regarded religion of any stripe as a usefully cohesive element in the firmly regulated society he was trying to establish; the Terror was remembered as an awful warning of the social costs of repudiating religion altogether. Skepticism about religion, of the kind that had been propagated by the more radical philosophes before and during the Revolution, was therefore ripe for repression. Conversely, any concordat between the sciences and religion was likely to be welcomed by the regime. Bonaparte had consolidated his own position by promoting himself to First Consul *for life*; and in 1804 he severed the last connection with his own Revolutionary past by pronouncing himself emperor, aping old-style monarchs by using his first name, Napoléon. Later that year, in a consummate piece of political theater, Napoleon (as he must now be called) crowned himself in Notre-Dame; Pius VII was present to witness the ceremony but was pointedly not asked or allowed to perform it.

It was in this counter-Revolutionary atmosphere that Cuvier gave his lectures on "geology". With the pope in Paris, Marzari Pencati had suspected at first that Cuvier's loosely Mosaic interpretation of the science might have been intended to curry favor with the regime—hence the joke about Cuvier the Protestant hoping to be made a cardinal—or at least to avoid political criticism. But his auditor was probably right, on reflection, to concede Cuvier's sincerity. There is in fact much to suggest that Cuvier was trying to stake out a place for the science of geology, in the face of two equal and opposite threats to its legitimacy, and that the new political situation gave him the opportunity to expound his own distinctive position more clearly than before.

The first threat came from the religious traditionalism that had resurfaced in French cultural life in the wake of Napoleon's rapprochement with the papacy. This was exemplified by the five-volume *Genius of Christianity*, which the prominent Parisian writer François Auguste Réné de Chateaubriand (1768–1848) published just after the concordat. This best-selling work propounded a highly traditional

38. Marzari Pencati, "Corso di geologia all'Ateneo nel 1805" (Vicenza-BB, MS S.C.28 (7)); I am grateful to Pietro Corsi for showing me a copy of this manuscript. See Corsi, *Age of Lamarck* (1988), 182–85.

39. Marzari Pencati to Gosse, 10 floréal XIII [10 May 1805], excerpt printed in Plan, *Henri-Albert Gosse* (1909), lxxxii–lxxxiv, and translated in Rudwick, *Georges Cuvier* (1997), 86–88. Corsi, *Age of Lamarck* (1988), 180–82, focuses on Marzari Pencati's report that Cuvier also poured scorn and ridicule on the transformist theory propounded by "materialists" such as Lamarck; but the text suggests that the cause of his auditor's astonishment was not only this attack on his senior colleague but also Cuvier's dating of the last revolution. The wording also suggests that the overt target of Cuvier's jokes was La Métherie's recent *Êtres organisés* (1804), though Lamarck's earlier *Corps vivans* (1802) was doubtless covertly in his sights: see Rudwick, *Georges Cuvier* (1997), 87.

version of Catholicism, and it expressed in religious terms the nostalgia felt in some quarters for the vanished Old Regime. It included a major section on the "Truths of scripture", which dealt explicitly with "objections to the Mosaic system"—that is, with objections to the traditional literalism based on Moses as the inerrant author of Genesis. Among those objections were the scientific arguments casting doubt on the historicity and universality of Noah's Flood. Chateaubriand dismissed any such skepticism by referring, for example, to the carcasses of Indian elephants found in Siberia: he was unaware of, or chose to ignore, the debates surrounding Cuvier's well-publicized claims that the bones were *not* those of Indian elephants and therefore no evidence for any Flood sweeping out of the tropics (§7.1, §7.5). Chateaubriand also dealt briefly with the contentious issue of the age of the earth; he dismissed what "geologists" were claiming, arguing that a mere appearance of antiquity had been imposed by the Creator at a far more recent moment of Creation. His was an unambiguously "young" earth.[40]

Such criticisms, if endorsed by Napoleon's regime, might well have threatened the legitimacy of geology and hampered its practice in France. It was probably in response to this potential threat that earlier scientific arguments in favor of an inconceivably vast timescale for the earth were being put forward with renewed vigor at just this time. One straw in the wind, and probably no coincidence, was that Montlosier's influential little book on the extinct volcanoes of Auvergne, invoking "an infinity of ages [*siècles*]" for their history (§6.1), was republished in the same year as Chateaubriand's work appeared. On a more technical level, Montlosier's book also reinforced the volcanic interpretation of basalt, in opposition to the Neptunist claims that Werner had repeated while visiting Paris the same year (§8.1). The basalt question was also at issue when, in 1804, the now elderly Desmarest gave the Institut an account of his own even earlier research in Auvergne (§4.3), more or less in full, at last, after a quarter-century's delay. But the great unpublished map he displayed, and his analysis of the three widely spaced "epochs" of volcanic activity he had detected, would have left his audience in little doubt that an even more important issue concerned the vast timescale for the whole earth that was indicated by his intensive fieldwork in this one region (see Figs. 4.10, 6.2).[41]

In fact Desmarest had put his views on the timescale unambiguously into the public realm the previous year, when the first of his substantive volumes for the *Encyclopédie Méthodique* appeared at last, long after his initial volume had reviewed the genre of geotheory (§6.5). His assessment of "the duration of time" was part of a lengthy essay "Nature's Anecdotes". This provocatively borrowed its title—and a lot of its material—from the work that Boulanger had long ago left in manuscript (§4.1), and that both Buffon and Desmarest himself had earlier found so inspiring. In effect, Desmarest generalized from his work in Auvergne by arguing that the "monuments of nature" recording the earth's many revolutions needed to be analyzed in "chronological order" to reconstruct the earth's history: as in his earlier work (and in Boulanger's), metaphors drawn from human history were pervasive (§4.3). But an equally pervasive theme was the vast timescale implied by all these observable "anecdotes" of physical geography: "The depth of the abyss of time [*l'abîme des tems*] into which our mind is thereby obliged to plunge seems so immense, so little in accord with our way of thinking, that it is not surprising that most people are little inclined to believe in these revolutions, although they are confirmed by many monuments."[42]

Desmarest's essay analyzed a wide range of physical features in the light of his conviction that geohistory had been played out over inconceivably lengthy periods of time. Unsurprisingly, in view of his own work in Auvergne and of Dolomieu's contrary interpretation (§7.2), the time implied by the erosion of valleys was particularly prominent (fossils, on the other hand, were notably absent). Again and again Desmarest emphasized how such features remained unintelligible unless a vast timescale was invoked. He contrasted the conclusions that naturalists had reached with those of textual scholars [*érudits*], who continued to attribute everything to the effects of "the unique revolution of the Flood" within a total timescale of no more than about sixty-four centuries. Much of Desmarest's text, in this essay as in the rest of his volume, seems to have been written long before and taken from a proverbial bottom drawer, but there can be little doubt that he selected it in deliberate response to the challenge represented by Chateaubriand's fashionable revival of biblical literalism. Unlike the parallel challenge exemplified in England by Parkinson's diluvial interpretation of fossils (§8.2), Chateaubriand's could not easily be dismissed by French savants; for he, although no naturalist, was a literary figure at the very center of their cultural life, and his counter-Revolutionary religious position was looked on with increasing favor by Napoleon's regime.

The vast timescale invoked by naturalists such as Montlosier and Desmarest was suspect, however, not only or even principally because it contradicted a traditionally literal exegesis of Genesis. Much more significantly, as in the case of Hutton in Britain some years earlier (§6.4), it was widely suspected of being a scientific cover for an *eternalism* that would subversively deny the divine origin and grounding of the world altogether. And more specifically, those who claimed a vast timescale for the earth as a whole were suspected of believing that the history of the human race had been equally extended, perhaps even infinite, in its duration. Certainly Montlosier had written—maybe just for rhetorical effect—of the *infinity* of the ages disclosed by his fieldwork in Auvergne (§6.1); and Desmarest argued that the "annals of civil history" would be found to need a similar expansion of time, implying that human history might not have been limited to the most recent portion of the vast timescale that his epochs represented. That kind of speculation had been revived by the recent discovery, by savants attached to Napoleon's expedition to

40. Chateaubriand, *Génie du Christianisme* (1802), pt. 1, bk. 4, esp. chaps. 6, 7 (1: 155–62); these passages should, however, be kept in proportion, as just a few pages out of five volumes. On the anti-philosophe party at this period, and Napoleon's relation to it, see McMahon, *Enemies of the Enlightenment* (2001), chap. 4.

41. Montlosier, *Volcans d'Auvergne* (1802), reset but otherwise unaltered from the original (1789); it is not clear whether the author, or an opportunistic publisher, was responsible for the new edition. Desmarest, "Détermination des trois époques" (1806)—the full version of his "Détermination de quelques époques" (1779)—was read at the Institut on 1 prairial XII [21 May 1804].

42. Desmarest, "Anecdotes de la nature", in *Géographie physique* 2 (1803): 532–86; esp. "La durée des tems" (560–62); quotation on 536–37.

Egypt, of zodiacal inscriptions that were being claimed as proof of the advanced level of astronomical knowledge at an extremely early date (see §10.1). Whether or not Montlosier and Desmarest were among the "atheists" to whom Marzari Pencati alluded in his report on Cuvier's lectures, their scientific work might well have been seen as supporting some kind of eternalistic view of the world. It was certainly seen as dispensing with the need to invoke any kind of geologically recent catastrophe of a kind that could be equated with the biblical Flood.[43]

The challenge of Lamarck

More seriously, Cuvier's senior colleague Lamarck was continuing to argue for a steady-state model of the earth, which was virtually eternalistic in its implications. His earlier book on *Hydrogéologie* (§7.4) could have been dismissed as just another example of geotheory or fanciful science fiction; but in 1805 he expounded the same ideas on "the theory of the globe" in—of all places—the Muséum's respectable *Annales*, in the same volume as the latest installments of his soberly descriptive study of the fossil mollusks from around Paris. His article arose from Baudin's recent voyage to the East Indies and Australia: the expedition's naturalist, François Péron (1775–1810), had found fossil coral reefs perched 1,500 feet up on the island of Timor, but doubted if this showed either the elevation of the earth's crust in that region or a worldwide subsidence of the sea. Lamarck agreed and claimed that Péron's observation was further evidence for his own model of the imperceptibly slow and unceasing displacement of continents and oceans around the globe, and that it supported his rejection of all alleged sudden catastrophes in the deep past.[44]

Furthermore, Lamarck had supported this steady-state model from another direction, when in the *Annales* for the previous year he described a new mollusk that Péron had found on a shore off the south coast of Australia. It was a find of outstanding importance, because Lamarck—the leading authority on such matters—identified the shells as those of a new and living species of *Trigonia*. This was a distinctive genus of which many fossil species were already well-known from the older Secondary formations, being sometimes found with ammonites and belemnites. Lamarck suggested that the living species might be one adapted to shallow coastal waters, while all the others might still be alive and well and flourishing at greater depths, where they had so far escaped detection. In other words, Péron's modest little shells powerfully revived the plausibility of Bruguière's earlier use of the living fossil argument (§6.1), reinforced Lamarck's own rejection of extinction, and thereby threw doubt on the whole interpretation of the fossil record in terms of a *history* of life on earth (Figs. 8.9, 8.10).[45]

Lamarck's continued advocacy of his geotheory, which flatly denied the reality of extinction and major catastrophes, suggests why Cuvier might have regarded an indefinitely lengthy timescale as a threat to his conception of geology, no less than the traditional short timescale revived by Chateaubriand. Cuvier evidently had no objection to a long timescale as such: this was implicit in his lectures, notably in the way he treated the Secondary formations and their fossils, and in his sketch of occasional revolutions punctuating long periods of tranquility. But his conviction

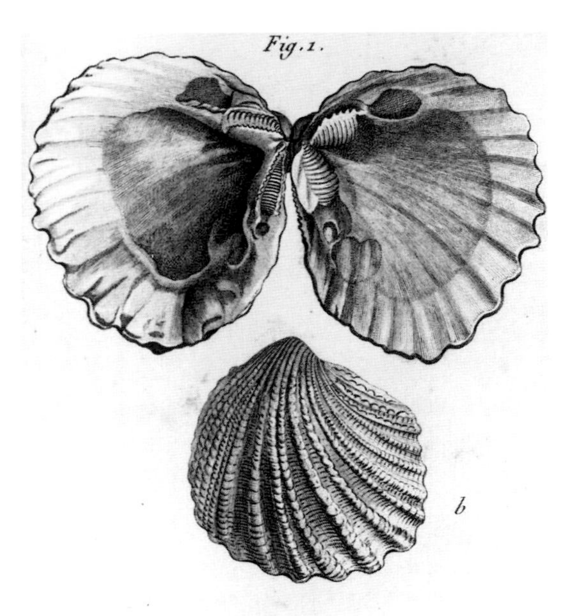

Fig. 8.9. The shell of a living species of the mollusk *Trigonia*, found by Péron on the shore of King Island (off Tasmania) during Baudin's voyage, and described and illustrated by Lamarck in 1804 in the Muséum's *Annales*. The find was of great importance, because Lamarck regarded the new species as a "living fossil", which threw doubt on the reality of *any* extinction (except by human agency) and supported his steady-state model of a quasi-eternal earth.

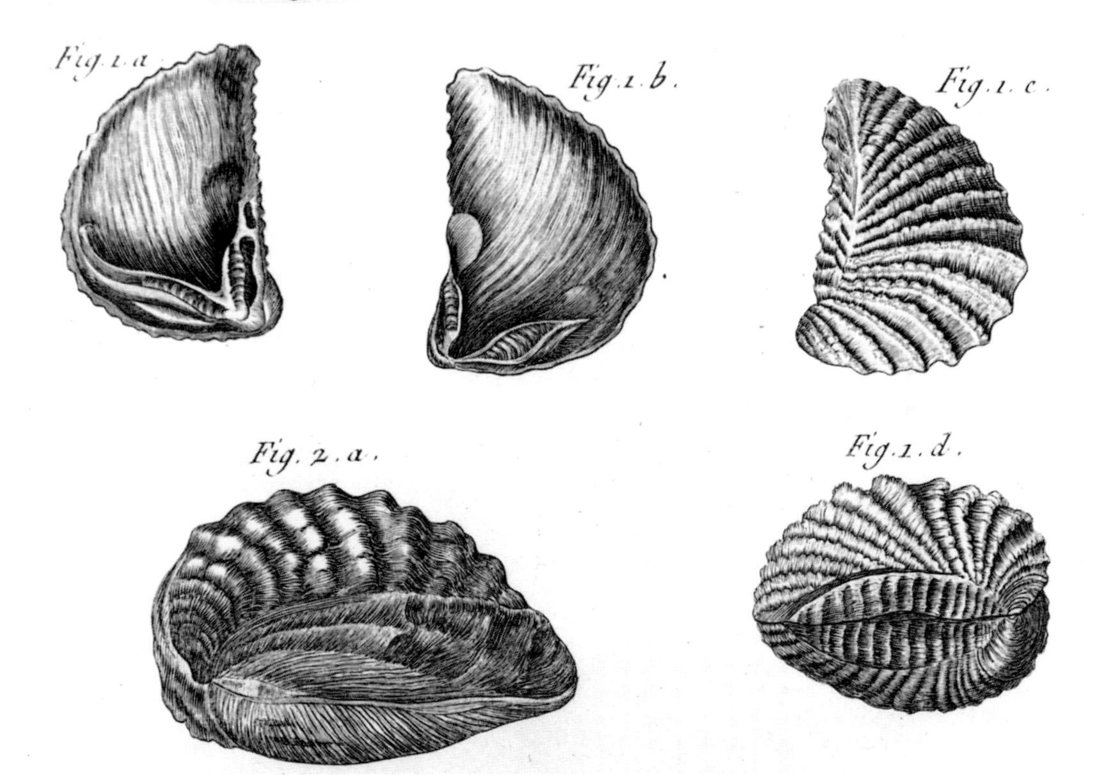

Fig. 8.10. Two of the many fossil species of *Trigonia* found commonly in some of the older Secondary formations, as illustrated by Bruguière for the *Encyclopédie Méthodique* (1797). The triangular shape of the shells (hence the generic name) and the large grooved "teeth" that held the two halves of the shell together were generic features that can be seen in both the living and fossil species shown here. (By permission of the Syndics of Cambridge University Library)

43. Denon, *Voyage dans l'Égypte* (1802), xlii–xliii and pl. 130, on the inscription at "Tentyris" (Dendera); see Buchwald, "Egyptian stars under Paris skies" (2003).

44. Péron, "Faits zoologiques" (1804), and Lamarck, "Faits applicable au théorie du globe" (1805). Baudin's expedition was competing with Matthew Flinders's British one for control of the south and west coasts of Australia; on the handling of its specimens back in France, see Burckhardt, "Unpacking Baudin" (1997).

45. Fig. 8.9 is reproduced from Lamarck, "Nouvelle espèce de Trigonie" (1804), pl. 67, fig. 1. Fig. 8.10 is reproduced from Bruguière, *Vers testacées*, pl. 4 (1797) [Cambridge-UL: XXVI.1.123], part of pl. 237; the living species was much smaller than these fossils (here shown reduced in size). Another fossil *Trigonia* is among the specimens on Smith's plate of Cornbrash fossils (Fig. 8.7, no. 4, center right).

that a real *history* of the earth could be reconstructed was threatened by Lamarck's profoundly ahistorical model, with its denial of any true otherness about the deep past. More specifically, a long timescale that supported a steady-state and even eternalistic view of the earth undermined Cuvier's conviction that the world he was reconstructing in his research on fossil bones was a *prehuman* world, and that extinction was an integral part of nature and not caused only by human action (§7.5).

Cuvier's middle way

In this climate of conflicting claims about the earth's timescale—Chateaubriand's extremely short one versus Lamarck's indefinitely or even infinitely long one—Cuvier's lectures staked out a position *between* the two: he said, in effect, a plague on both your houses. On the one hand, he rejected Chateaubriand's biblical literalism and the very short timescale entailed in adopting the chronologers' traditional interpretation of the Creation story in Genesis. The field evidence to the contrary was now overwhelming (though Cuvier knew it largely at second hand), and geology would become impossible under the extremely restricted timescale advocated in Chateaubriand's popular volumes and others like it. Hence it was important to explain *to the general public* that a short timescale was not the only option, indeed that it was no longer a scientifically acceptable option at all. On the other hand, Cuvier also wanted to distance himself from the eternalism associated—rightly or wrongly—with the virtually unlimited timescales advocated by some of his savant colleagues. For these threatened to undermine his argument that the earth had had a true *history*, and that the deep past could only be known by patient investigation of nature's own monuments.

As for Genesis, Cuvier rightly treated its contentious relation to the new geology as conflating two distinct issues. On the one hand he conceded that the record of rocks and fossils seemed to run roughly parallel to a nonliteral reading of the Creation story: more recent research had broadly confirmed what de Luc, for example, had tirelessly argued a few years earlier (§6.4). But apparently Cuvier mentioned this almost casually and in passing, as if it was a matter of little importance to him. His audience could take it or leave it as they wished; but by alluding to it as an option he skillfully deflected political criticism of the science and tacitly indicated its legitimacy. On the other hand the story of the Flood (later in the Genesis text) was far more significant for him, because it was a human record of a natural event that he claimed had been crucial in the recent history of the earth. The Flood story, interpreted in a far from literal fashion, was important because it helped confirm the sheer historicity of the earth's most recent revolution, which he claimed had wiped out his fauna of large fossil mammals. For Cuvier, as for de Luc and Dolomieu before him, the Flood defined the interface between the present and former worlds; it was the crucial link between human history and the geohistory that had preceded it. So in his lectures Cuvier adopted the claims that those older naturalists had made for the dating of the event: unlike the vast spans of undateable geohistory, the "last revolution" could be dated as having taken place less than ten thousand years ago.

Cuvier's lectures at the Athénée introduced the general bourgeois public to the results of his latest research and its broader implications for a scientific understanding of the world. He thereby staked out a claim for the legitimacy and indeed the respectability of the new science of geology, as he understood it, among influential elements of the social elite in Paris. Immediately after this course was completed, he began another on the other side of the city, at the Collège de France, setting out much the same view of geology for a different (though probably overlapping) public. The notes taken by two of his auditors show that Cuvier covered much of the same ground, though apparently in greater technical detail. In effect, he swallowed his own earlier criticisms of the genre of geotheory and indulged in it himself, outlining his own variant of the standard model (§3.5) and relating it to the contrasting model put forward more than thirty years earlier by Buffon, his famous predecessor at the Muséum (§3.2).

Cuvier explained at the start that he aimed to use "facts" or observable features "to trace the history of the globe's revolutions in their natural sequence": geohistory was once again at the heart of his conception of geology. As usual, "revolutions" simply denoted major changes, though Cuvier certainly claimed that some of them had been sudden. He began with an earth covered with a global "unknown liquid", which had changed in composition as the materials of the various strata were precipitated out of it. Beds of conglomerate (consolidated gravel) recorded occasional violent episodes, but these had in general become less intense and more local in the course of time, as the global sea level subsided. Cuvier noted, but left aside, the causal problem of accounting for this subsidence; but anyway it had not been uniform, for "the surface [of the sea] has been lowered and raised several times". At a certain point in this complex history, life had begun; again, any causal questions about its origin were set aside. Thereafter, it had had a distinctively directional history, with the time of the ammonites and belemnites preceding that of any terrestrial fauna; and among the latter, strange animals such as those from the Parisian gypsum had preceded more familiar kinds. "Finally, mankind was born", for no human bones were known as fossils.[46]

Since these lectures covered much the same ground as the earlier course, only three points need to be emphasized here. First, Cuvier envisaged a *complex* geohistory: his concept of a sea level that had oscillated irregularly while subsiding generally was far from the simplest and purest form of the standard model (§3.5). His was a version that would allow, indeed anticipate, an equally complex record of varied formations in the Secondary rocks. A second and related point is that Cuvier envisaged a geohistory in which the later revolutions (except apparently the last of all) had been increasingly local in their action and confined to separate

46. Anonymous, "Leçons de géologie au Collège de France" (Paris-IF, MS 2378/6); notes by the young naturalist Jean-Baptiste d'Omalius d'Halloy (1783–1875) from the southern Netherlands (now Belgium), are described by Grandchamp, "Deux exposés de Cuvier" (1994). The two sets of notes do not seem to be closely parallel, and it is possible that Cuvier gave *two* separate courses at the Collège de France. Omalius's notes are on lectures given—rather surprisingly—as an introduction to Cuvier's course on *physiology*, from 20 floréal to 2 prairial XIII [10–22 May 1805] (i.e., the first was given coincidentally on the day that Marzari Pencati wrote to report on the completed course at the Athénée). The anonymous notes are dated only with the year XIII [1804–5].

regional "basins". This too implied a complex particularity in the record of geohistory. Third and last, Cuvier claimed that, in the absence of fossil evidence of the human species, "the epoch of its origin is not very remote", and less than ten thousand years in the past; but here he supported this date by referring to the quite brief historical records of *all* ancient cultures (discounting those that were merely fabulous). Cuvier's review covered the ancient Greeks, Egyptians, Chinese, and Indians, reflecting his awareness of the textual scholarship of the Third Class at the Institut; it was at just this time that the French edition of the English *Asiatick Researches*, for which he had acted as a consultant, was finally published (§7.4). It is not clear whether the ancient Jews were also included, or pointedly omitted; but even if Cuvier did mention the biblical records, it would certainly have been in a multicultural context that implicitly relativized them into just one ancient literature among many. Cuvier was distancing himself decisively from the biblical literalism of Chateaubriand and his ilk, no less than from the grandiose eternalistic theorizing of Lamarck and his.[47]

Conclusion

Cuvier's lectures in 1805 widened his audience beyond the savants at the Institut to the general educated public in Paris. He used these lectures not only to expound his own conception of the eventful history of the earth, and particularly its records in the form of fossils, but also to defend the legitimacy of the science of geology in the new political atmosphere marked by Napoleon's concordat with the papacy and self-elevation to the rank of emperor. In Cuvier's view his own conception of geology was threatened on two fronts. The biblical literalism represented by Chateaubriand's resurgent Catholicism—like that of modern Protestant fundamentalism—would make the practice of geology impossible, by denying it the lengthy timescale that the observable features clearly demanded. But the eternalism represented by Lamarck's steady-state geotheory, and the universal organic transformism associated with it, would deprive geology of its claim to reconstruct the *history* of the earth by denying that any periods in the deep past had been distinctive in their organisms or anything else. Even the vast timescale advocated by Desmarest and Montlosier might betray a covert eternalism. Cuvier therefore tried to define a middle way between the young earth of the literalists and the unimaginably ancient and possibly eternal earth of some of his colleagues. His lectures explained to the general public how, in his opinion, geology disclosed a more plausible third option. The earth had had an extremely lengthy and eventful history, but it was not eternal; and it could only be known through patient research on its physical "monuments", not by imposing on it an all-explanatory geotheory.

8.4 A NEW AGENDA FOR GEOLOGY (1806–8)

André's geotheory

Cuvier's lecture courses in Paris in 1805 marked a striking change in his attitude towards geology, in that he presented the science in a much more positive light. His prominent position and ever growing prestige throughout the savant world was such that what he thought about geology was likely to have a major impact. At just this time, in the spring of 1806, in the middle of continuing intense warfare between Napoleon's self-styled empire and Britain and its Continental allies, Cuvier was elected a Foreign Member of the Royal Society in London (an honor he shared with his colleague Lacépède).

A few months later, Cuvier reported to the Institut in Paris on a forthcoming book by a French naturalist and used the occasion to set out his own ideas about how geology should be practiced. The book was by "Mr. André", but the author was better known as Fr. Chrysologue de Gy (1728–1808), for until he was laicized during the Revolution he had been a Capuchin monk. He had earlier earned a respectable reputation as a naturalist. For example, he had published a competent account of the physical geography of the Franche-Comté region, in which he had discussed, among other problems, the puzzling one of accounting for the erratic blocks perched on the hills of the Jura, in relation to what other savants such as Saussure and de Luc had written. Now an elderly layman—though he was not as old as Desmarest—André had synthesized a lifetime's extensive field observations and offered a global interpretation of them. The title of his book, *Theory of the Present Surface of the Earth*, identified it clearly as an essay in geotheory, though he inverted the usual structure (§3.1) by starting with an account of his own observations, followed by a review of relevant work by others such as Buffon, de Luc, and Dolomieu. He reserved to the end his own conclusion that to produce the present topography "a general, uniform, violent, and sudden cause" must have been required. André's geotheory was thus one that gave a central role to a geologically recent catastrophe, not unlike the one that Cuvier had been claiming on quite different grounds.[48]

André submitted his book to the Institut in advance of publication, as was the common practice, in the hope of winning its formal endorsement; as was also the custom, the Institut set up a committee to report on it. Cuvier, as the relevant permanent secretary of the First Class, had himself appointed, and drafted the report; the other two members seem to have played only a minor role (one of them was Haüy, since Dolomieu's death the professor of mineralogy at the Muséum, and also, since the concordat, an honorary canon at Notre-Dame). Cuvier used the occasion quite openly as an excuse for presenting his colleagues with his own ideas on "geology": it was, he said, the first time he had had a chance to do so. In fact he dealt with the substance of André's work quite briefly, and only at the end of his report. He praised the ex-Capuchin for having followed his order's rules of poverty by doing all his extensive fieldwork on foot, rather than taking the less arduous but costly option of hiring a horse or a carriage; for this made his "Enlightened" [*éclairé*] field observations even more reliable, regardless of the "system" they were used to support. As for that theorizing, Cuvier noted that André agreed with de Luc

47. Labaume, *Recherches asiatiques* (1805). The last page of the anonymous set of notes on Cuvier's lectures is missing, and may possibly have included a final biblical flourish.

48. Chrysologue, "Franche-Comté" (1787); [Chrysologue] André, *Théorie de la surface* (1806).

and Dolomieu in inferring that the earth's present topography was of recent origin and owed its form to an enigmatic catastrophe. He added that he and his colleagues on the committee were "personally" in sympathy with André on this point, but that they would refrain from commenting on it in the august arena of the Institut.[49]

Cuvier's reason for drawing this boundary would have been apparent from what he had said in the earlier part of the report, for which his evaluation of André's work was merely a convenient peg. He noted that one major science of the earth—mineralogy, in its newly restricted (and modern) sense—had now achieved the same status and precision as the other physical sciences: the allusion was particularly to Haüy's distinguished recent work on the solid geometry of crystal forms, which had dramatically improved the classification and identification of minerals. The other main science of the earth, according to Cuvier, was the one that studied "the superposition of mineral [masses]"; that is, what others called geognosy. Cuvier claimed that this science too was capable of reaching exactness and "logical rigor". But as yet it had signally failed to achieve that potential.[50]

Cuvier pinned the blame on geotheory. He reckoned that there were already more than eighty distinct "geological systems" on offer: "we see new systems hatched every day, and the scientific journals are full of the attacks and defences that their authors make against each other". Echoing Desmarest's earlier argument (§6.5), Cuvier claimed that the reason for all this fruitless controversy was that too many theories were chasing too few facts: the systems were more or less incompatible with one another, because they were the product of hypothesizing that was unconstrained by any adequate knowledge of the effects to be explained. Ironically, fossils—Cuvier's own favorite part of the evidence—were particularly to blame: they were so striking that they had encouraged irresponsible and premature theorizing: "In a word, they have changed it [geology] from a science of facts and observations into a fruitless web of hypotheses and conjectures, so much at odds with one another that it has become almost impossible to mention its [geology's] name without provoking laughter."[51]

This was an indictment calculated to put influential noses out of joint—not least that of Faujas, the Muséum's professor of *géologie*—particularly because it was uttered in the prestigious setting of the Institut's meeting room and in the presence of France's leading savants (and perhaps of Faujas himself). But Cuvier's assessment of geology was not all negative. He noted that the Primary rocks were now well-known—he praised naturalists such as Pallas, Saussure, de Luc, Dolomieu, and Werner for this—in comparison with the Secondaries. Yet only the latter could provide the key to one of the crucial unsolved problems for geotheory. This was of course the relation between the organisms found as fossils in the Secondary rocks and their relatives alive in the present world. Were they identical or not? And if not, was that due to extinction, migration, or transmutation? Taking as an example just the rock formations nearest at hand, those around Paris, Cuvier noted that there were about a dozen alternative hypotheses on offer—"explanations thought up calmly in the study [*cabinet*]", as he put it scornfully—to account for them. Yet none took any account of Lamarck's painstaking current work on the fossil mollusks of Grignon (§7.4), most of which were of species unknown alive, or of his own work on the even stranger fossil quadrupeds from the Gypsum (Cuvier carefully praised Lamarck for his descriptive work on fossils, not for his geotheory).

The moral was clear, and Cuvier presented his colleagues with an unambiguous prescription. The Institut should put its institutional weight behind encouraging factual descriptions and withhold the stamp of its authority from mere systems: "these castles in the air" would soon dissolve in the face of "the more solid edifice of facts and induction". Geology should be made "a science of facts" before tackling "the great problem of the causes" that had brought the earth to its present state. Cuvier's prescription was not naïve. He did not simply advocate the "Baconian" piling up of observations, but rather the pursuit of a carefully planned observational *agenda*. He offered a list of nine problems towards which research ought to be directed and claimed that "there is not one of them about which anything is absolutely certain"; the castles of geotheory had been built on air, because it had been falsely assumed that the answers to them were already known.

Cuvier's first and last topics, on the structure of mountain ranges and valleys, recalled the wide scope of the famous "agenda" that Saussure had offered a decade earlier (§6.5). But all the other topics focused on the Secondary formations, and particularly on the relation between these rocks and their fossils. There was the question of the constancy of the sequence of formations in different regions, and of the fossil assemblages that they contained; of the points at which specific kinds of fossils appeared and disappeared, and whether they ever reappeared in the sequence. Above all, it was important "to compare fossil with living species, with more rigor than hitherto; and to determine whether there is a correlation between the age [*ancienneté*] of beds and the resemblance or nonresemblance between fossil and living organisms." Nothing could show more clearly that Cuvier recognized that the Secondary formations and their fossils now constituted the focal problem of greatest strategic importance for the future of the science. It was an ambitious program, but not an impossible one: "The series of problems has been proposed; nothing more than an enlightened perseverance is needed to fill out the framework, the ensemble of which will constitute the science [of geology]." And at least implicitly, it would be the achievement of many naturalists in collaboration, not of a single geotheorist in isolation.

The significance of Cuvier's agenda can hardly be overestimated, because it suggested new lines of fruitful research that focused on a specific cluster of solvable problems. Although occasioned by an evaluation of a rather minor work by a rather minor savant, the agenda attracted immediate attention: the review was recognized as being far more important than the work reviewed. The report by Cuvier (and, nominally, his colleagues) was published the following year in the *Journal des Mines* and in the Institut's own *Mémoires*, ensuring it wide circulation. Both the relevant volumes would have found their way in due course, despite the wars, to the Royal

49. Cuvier et al., "Rapport sur André" (1807), trans. Rudwick, *Georges Cuvier* (1997), 101–11. The manuscript of the report (Paris-IF, MS 3160) is in Cuvier's hand throughout, but its extensive alterations suggest that his initial draft was modified in the light of comments by his colleagues.

50. See Haüy, *Traité de minéralogie* (1801).

51. Cuvier et al., "Rapport sur André" (1807), 417.

Society in London and to other learned academies around Europe; a translation in the *Philosophical Magazine*, for example, gave the essay still wider circulation in the anglophone world.

Cuvier's apparently ambivalent treatment of André's book becomes intelligible in the light of the research program embodied in his agenda. He wanted the Institut to praise the elderly naturalist's detailed fieldwork, while suspending judgment on his geotheoretical conclusions. André's model, centered on a fairly recent major catastrophe, was in itself highly attractive to Cuvier and compatible with what he was inferring from his studies of fossil bones. Nonetheless, he recognized that its epistemic status was as yet rather shaky, and he wanted the premier scientific body to draw a sharp line between grand speculative geotheory and more limited conclusions founded on well-tested empirical research. What was permissible in other settings, for example in popular lectures, was inappropriate in the exalted arena of the Institut, because the intellectual authority of that body defined in effect what should *count* as good scientific work. So Cuvier considered himself at liberty to expound his own geotheoretical ideas at the Athénée, or even at the Collège de France, but not at the Institut. The papers he read at the Institut and published in the Muséum's *Annales* were not, in his judgment, the right place to mention his broader ideas on geology, except in passing; or, to put it the other way round, his popular lectures gave him an opportunity to expound ideas that he felt constrained from expressing so fully in his role as a leading savant.[52]

The progress of the sciences

Central to Cuvier's careful drawing of intellectual boundaries was his changing conception of the science of geology, which he had criticized so relentlessly in earlier years. Although de Luc had in effect defined *géologie* as geotheory (§3.1), the meaning of the word in ordinary usage had been shifting towards some kind of synthesis between the descriptive practices of mineralogy, physical geography, and geognosy, and the causal interpretations of earth physics (§6.5). So Cuvier was conceding—somewhat belatedly—that "geology" had now become a convenient label for the whole cluster of sciences dealing with the earth, without necessarily implying any ambition to provide a comprehensive explanation of the terrestrial system. "Geology" had therefore shifted across Cuvier's tacit boundary, to become a respectable branch of empirical enquiry and no longer just a laughing matter.

Just at this time, Cuvier was assigned a task that gave him the opportunity to define this new sense of the place of geology on the intellectual map of the sciences. Back in 1802, Bonaparte (as he then was) had ordered the Institut to produce a review of the progress of all the sciences since the start of the Revolution, in part to demonstrate to the world the further progress that they were now making under his own benevolent rule. The troublesome social sciences having been eliminated, the task was divided between the two remaining Classes, and, for the First, assigned to its two newly appointed permanent secretaries. Delambre was assigned the "mathematical" sciences; Cuvier, the "natural" sciences of chemistry and natural history, together with the applied sciences. Even that division of labor left each with

RAPPORT HISTORIQUE
SUR LES PROGRÈS
DES SCIENCES NATURELLES
DEPUIS 1789,
ET SUR LEUR ÉTAT ACTUEL,

Présenté à SA-MAJESTÉ L'EMPEREUR ET ROI, en son Conseil d'état, le 6 Février 1808, par la Classe des Sciences physiques et mathématiques de l'Institut, conformément à l'arrêté du Gouvernement du 13 Ventôse an X;

RÉDIGÉ par M. CUVIER, *Secrétaire perpétuel de la Classe pour les Sciences physiques.*

IMPRIMÉ PAR ORDRE DE SA MAJESTÉ.

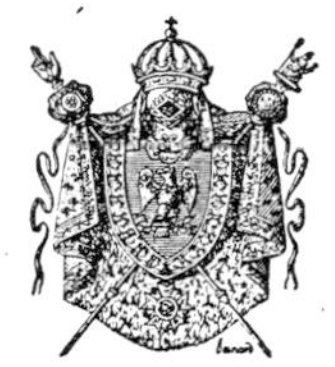

A PARIS,
DE L'IMPRIMERIE IMPÉRIALE.

M. DCCC. X.

Fig. 8.11. The title page of Cuvier's report to Napoleon on the progress of the "natural sciences" since the start of the Revolution in 1789. Published in 1810 by Napoleon's "Imperial Press", it matched another volume by the mathematician Delambre on the "mathematical" sciences and a third by the historian Dacier on the humanities. As the style of the title page and its imperial logo suggest, the reports were intended to serve the cultural politics of Napoleon's self-proclaimed Empire, but in the event all three savants reviewed their sciences from a fully international perspective. Cuvier's was notable, among other things, for his newly positive evaluation of "geology". (By permission of the British Library)

a vast field to cover, though they used their colleagues as informants on particular subjects. The preparation of their reports must have made these savants, at this moment in history, two of the most widely knowledgeable in the world. Early in 1808 their reports were presented to Napoleon (as he had by then become) at a meeting of the Institut; the report on the humanities, by the historian Bon-Josephe Dacier, was relegated to a later meeting. The contents of all three soon became widely known in savant circles, although their publication was delayed another two years (Fig. 8.11).[53]

The arrangement of the reports from the First Class reflected the map of natural knowledge (§1.4), or the classification of the sciences, as it had developed by this time. Chemistry, for example, was still grouped among the "natural" sciences and therefore covered by Cuvier, whereas the whole of physics (in its modern sense) was now "exact" enough to be covered by Delambre along with pure and applied mathematics. The bulk of Cuvier's volume, however, was on "natural history", and

52. See Outram, "Language of natural power" (1978).

53. Fig. 8.11 is reproduced from Cuvier, *Rapport historique* (1810) [London-BL: 446.e.11]; the other two reports were identical in design.

it surveyed in turn the sciences dealing (in modern terms) with the atmosphere, hydrosphere, lithosphere, and biosphere. As in his report on André's work, Cuvier divided the sciences of the solid earth into mineralogy proper and *geology*: the latter, which he had so persistently scorned and even ridiculed, was at last formally brought in from the cold.

Ignoring the nationalistic goals that had motivated Napoleon's original commission, Cuvier reviewed "the progress of the sciences" from a truly international perspective. In compiling his section on geology, as for much of the rest of the volume, Cuvier relied heavily on his informants. Such sources enabled him to cast his net as widely as the sciences themselves. For example, he was sent a long list of relevant books and periodicals in the library of the Mines Agency: most of them in German, some translated into German from Swedish, some in Italian. (There were few works in English on this list, or in the text that Cuvier wrote with its help: not out of nationalistic prejudice or plain ignorance, but because there was relatively little to record.) He probably used this list to guide his own reading, which must have made him far better informed about the sciences of the earth than he had been before.[54]

Cuvier combined his usual criticisms of geotheory with praise for the fieldwork of a highly international cast of heroes past and present. The scope of their science was clearly identified as that of physical geography enlarged by the structural perspective of geognosy; he added a reference to fossils that deftly recruited Lamarck in support of his own unmentioned research:

> This knowledge of the different positions of mineral masses [as described by Werner and his followers] has become the object of a true science, which now directs the research of the best people [*les bons esprits*] and replaces those illusory conjectures that not long ago bore the pompous name of Geology. The Pallases, Saussures, Desmarests, Dolomieus, Werners, de Lucs, Ramonds, and Humboldts have given it this new aspect; their arduous travels and scrupulous studies have made known to us the true structure of that part of the earth's crust that we can penetrate, while at the same time leading us to despair of ever discerning its origin. And this crust is full of the remains of organisms, irrefutable evidence [*preuves*] of major revolutions and objects worthy of the attention of naturalists; the Pallases, Campers, and Lamarcks have studied them and found them largely different not only from those that now live in the same climates but also from all those that have been collected [anywhere] on the earth's surface.[55]

Cuvier's virtual definition of geology as geognosy became even clearer when the full text was published. In his section on geology, he distinguished between descriptive [*positive*] and explanatory [*explicative*] geology. He treated the former as "a wholly modern science", dating only from the pioneering fieldwork of naturalists such as those he had listed. He reviewed a wide range of "particular" studies of specific regions, emphasizing the practical value of those already sponsored by the Mines Agency in France itself. He argued that this kind of local research was the basis on which a more global or "general" geology was gradually being constructed. But all this was sharply distinguished from explanatory geology or earth physics, which was virtually equated with the speculative excesses of geotheory. Although Cuvier conceded in passing that "systems have had the merit of giving an incentive for research into facts", he claimed in the end that they had more often been used to foreclose further investigation and therefore had no proper place in his survey of scientific progress. In effect, therefore, Cuvier defined the scope of geology as being that of the structural science of geognosy (§2.3)—into which the distributional practice of physical geography (§2.2) was now more or less assimilated—while the causal goals of earth physics (§2.4) were acceptable only at a low level, and not when they were enlarged into the global aspirations of geotheory (§3.1).[56]

The Geological Society

Cuvier's definition of the proper scope of "geology", enunciated in the authoritative setting of the Institut and in Napoleon's presence, came just too late to influence directly a parallel decision to adopt that formerly contentious word; but anyway Cuvier was merely endorsing—somewhat belatedly—a usage of "geology" that had begun to be quite widely adopted among savants since the turn of the century (§6.5).

In Britain, in sharp contrast to France, the pursuit of the sciences—and indeed of practical, social, and cultural goals in general—had long been regarded as the proper concern not of the state but of individuals and voluntary associations. So for example the state-supported and highly bureaucratic Institut in Paris was matched unequally by the independent and self-governing Royal Society in London: the former was small, select, and highly professional in outlook; the latter was large and diffuse, and the competence and commitment of its members were far more uneven. The Royal Society did in practice act as an advisory body to the British government on matters of science and technology, particularly through Banks, in effect its president for life; but unlike the Institut it played this role quite informally (§1.2). Also unlike the Institut, the Royal Society confined itself to mathematics, the natural sciences, and the more fundamental aspects of technology and medicine: in effect, to the same range of sciences as the First Class alone (the Society of Antiquaries complemented it by covering at least some of the same range as the Third Class). Although Scotland and Ireland had their own Royal Societies, in England the only other major body of savants concerned with the natural world was the Linnean Society; but this confined itself to the study of systematic botany, a specialized practice that London's Royal Society did not attempt to cover.

Among the sciences that the Royal Society—and Banks in particular—did reckon to cover, the sciences of the earth were in principle included, though in

54. Anonymous, "Catalogue des livres" (Paris-IF, MS 3139, no. 1), among Cuvier's manuscripts for the preparation of his report; the works in English were by Kirwan, Jameson, Hutton, and Williams.

55. Cuvier, MS summary of *Rapport historique* (Paris-IF, MS 3140), probably read out at the Institut when the volume was presented to Napoleon. The proper names were all in the plural, to suggest that these individuals were representative of many more. Ramond had published an important account of the Pyrenees; the Prussian naturalist Humboldt had recently returned from Latin America and was living in Paris.

56. Cuvier, *Rapport historique* (1810), 131–51, trans. Rudwick, *Georges Cuvier* (1997), 115–26.

practice its meetings and its prestigious *Philosophical Transactions* did not often pay them any attention. However, they and their potential for useful application were much discussed among the social and political elites in London around the turn of the century. A British Mineralogical Society had been formed to promote the practical application of that science, but its activities had petered out within a few years. Late in 1807 an attempt was made to fill the gap with a new body of the same kind. What was surprising and significant was that when its thirteen founders met in a London tavern to discuss their plans, they decided to call it the "Geological Society".

The founders were a mixed bunch (Parkinson was the only one who has been mentioned already in this narrative). Half of them were Fellows of the Royal Society, and Banks himself was elected a member a few weeks later; there was clearly no intention that the new society would compete with the long-established one. Some of the early members, among them the young chemist Humphry Davy (1778–1829), simply wanted a small informal social club in which to exchange ideas about the mineral sciences. Some were subscribers to a large and handsome monograph (on calcite and related minerals) by the count de Bournon, an emigré French mineralogist, and may have seen a role for the new body in promoting the publication of similar original research. Some had interests—not least financially—in mining and agriculture and looked to the infant society to coordinate research that would benefit those practical activities. These goals were not entirely compatible, and soon led to tensions: within a couple of years, as the Geological Society began to talk of acquiring its own premises and publishing its own periodical, Banks and Davy saw it as a threat to the Royal Society, in a way that the Linnean Society was not. They resigned from the new body, but it survived that crisis and grew steadily in size and prominence in the scientific life of London, and indeed on the national level. Its meetings, far less formal than those of the Royal Society, soon became famous for their liveliness and conviviality.

In the present context the most important aspect of the new society was the decision—apparently almost at the last moment—to call it "geological" rather than "mineralogical". In spite of the war, some of the founders were well aware of what was going on in Paris and would have known that "geology" was being used increasingly to cover most or all of the sciences of the earth. That Continental connection was in fact embodied in Bournon, who was certainly in touch with his compatriots; his subscribers must have been aware of his opinion that the science of mineral specimens [*oryctognosie*]—exemplified by his own work—needed to be expanded into the study of mineral masses in their wider setting, or *géologie*.[57]

However, the decisive impulse to call the society *geological*, and thereby to commit it to an emphasis on outdoor fieldwork, probably came from the man who became its first president and was soon its driving force. The young George Bellas Greenough (1778–1855) had followed Blumenbach's lectures in Göttingen and had been as far as Berlin and Vienna (but not to Werner's Freiberg). He had later taken advantage of the Peace of Amiens to visit Paris, where he had attended meetings at the Institut and met Bonaparte; continuing a geological Grand Tour he had traveled to Geneva to see erratic blocks, to Chamonix to see Mont Blanc and the high Alps, to Naples to see Vesuvius, and to Sicily to see Etna. When the peace collapsed, he had been imprisoned in Palermo but eventually reached Gibraltar in a naval convoy under Nelson, and so got safely home. With the Continent closed off, he had then turned to his own country: in 1805 he traveled widely through the north of England and Scotland, and the following year toured Ireland in the company of Davy. Although, like Davy, he was still under thirty when the Geological Society was formed, he was probably the most widely traveled of its founders and the one with the greatest firsthand experience of the major objects of "geological" attention.[58]

In 1804 Greenough, an orphan, had inherited a huge fortune from a London apothecary grown rich on the manufacture of patent medicines, and he had settled in London as a wealthy young gentleman, choosing to devote his time and money to the sciences. Earlier in 1807 he was elected both to the Royal Society and to Parliament (the latter under the notoriously unreformed franchise of the time). It was probably under his forceful leadership that the Geological Society was soon publicly committed to a policy of "Baconian" fact collecting and to the repudiation of the pretensions of geotheory. When he toured Scotland he had been deliberately comparing the Neptunist and Vulcanist explanations of rocks such as granite and basalt, testing them against his own observations. Finding the evidence almost equally balanced, he had decided that such high-level theorizing was best suspended until more extensive fieldwork had been completed. This was a conclusion closely parallel to Cuvier's, although reached quite independently.[59]

It was reinforced by Greenough's strongly negative reaction to what he saw of the savant scene in Edinburgh: the supporters of the deceased Hutton were engaged in an acrimonious running battle with those who claimed to represent Werner's approach to the sciences of the earth. Prominent among the former was John Playfair (1748–1819), the university's professor of mathematics and natural philosophy. Playfair had published a volume of *Illustrations of the Huttonian Theory of the Earth* (1802), in which he claimed to present Hutton's ideas in a more accessible style; as in Hutton's own subtitle, the word "illustrations" referred only to textual examples (the book contained not a single visual image). Playfair's text was certainly much shorter than Hutton's, but it described much the same range of phenomena and marshaled them in support of the same steady-state geotheory. However, the alleged increase in readability was achieved at the price of suppressing what Hutton had regarded as most fundamental about his theory, namely its proof of the wise design of the whole terrestrial "system" (§3.4). Playfair thoroughly expunged all the deistic metaphysics from his hero's geotheory, offering the new century a bowdlerized version that reduced it to a purely mechanical system of physical processes interacting in dynamic equilibrium. This was closely analogous to

57. Geological Society, MS Minutes 1 (London-GS, OM/1); Bournon, *Chaux carbonatée* (1808), xi–xiii, xxxvi; Rudwick, "Foundation of the Geological Society" (1963); Weindling, "Prehistory of the Geological Society" (1979) and "British Mineralogical Society" (1983).

58. Greenough, notebooks for 1798–1806 (London-UC, Greenough MSS); Kölbl-Ebert, "George Bellas Greenough" (2003), gives valuable biographical information; see also Wyatt, "George Bellas Greenough" (1995), for his early links with Coleridge and his circle.

59. Rudwick, "Hutton and Werner compared" (1962), prints relevant excerpts from Greenough's notebook for 1805 (London-UCL, Greenough MSS).

the dynamic steady state attributed to the Newtonian cosmos, which, as the product of the prestigious science of mathematical astronomy, was central to Playfair's conception of his university chair. But his reinterpretation of Hutton did not go unchallenged, and the arguments to which it gave rise had split the Edinburgh savants into rival camps. To Greenough it seemed an example to be avoided at all costs: in the new Geological Society, geotheory should be ruled firmly out of bounds.[60]

Like Cuvier, Greenough did not advocate an indiscriminate piling up of facts, but the pursuit of a planned agenda. Only a few weeks after the new society was founded, its members—prompted probably by Greenough himself—decided to draw up "a series of questions relating to the most essential points in Geology"; after discussion and revision, they were to be distributed not only to the paying "Ordinary Members" resident in London, but also to the rapidly growing number of "Honorary Members" in the provinces, whose collaboration was regarded as essential to the success of the society's project. The little booklet entitled *Geological Inquiries* (1808) was its very first publication. Claiming that geology was "a sublime and difficult science", the booklet made an eloquent plea for information, based on a division of labor between those who could report relevant local observations and those who alone could generalize from them and find their true explanation:

> To reduce Geology to a system demands a total devotion of time, and an acquaintance with almost every branch of experimental and general Science, and can only be performed by Philosophers; but the facts necessary to this great end, may be collected without much labour, and by persons attached to various pursuits and occupations; the principal requisites being minute observation and faithful record. The Miner, the Quarrier, the Surveyor, the Engineer, the Collier, the Iron Master, and even the Traveller in search of general information, have all opportunities of making Geological observations.[61]

It would be easy to condemn this as politically incorrect elitism or as a patronizing exploitation of provincial "practical men" by metropolitan savants. In fact, however, the society's plea for collaboration recognized that the spatially extended character of many of the features of greatest scientific interest demanded some such sharing of the observational task, particularly in an age of relatively slow and expensive travel. In the event the booklet yielded only a meager harvest of new information, and much of it was sent in personal letters to Greenough himself rather than as more formal communications to the society. But the network of provincial informants, encouraged by being invited to be Honorary Members, was certainly of great importance to the active core of metropolitan Ordinary Members, when in subsequent years they themselves traveled around the country doing geological fieldwork.[62]

Most of *Geological Inquiries*, however, was devoted not to general exhortation but to detailed questions about the whole range of features of geological interest. The format and much of the contents were clearly modeled on the agenda that Saussure had published a decade earlier (§6.5). In the present context, what is striking about the Geological Society's questions is the modest space—at the very end of the agenda—devoted to the fossils in the Secondary formations. In contrast to the more specific agenda that Cuvier had just proposed in his report on André's book, fossils were treated as quite peripheral to geology, and most of the eleven questions about them concerned their mode of emplacement and preservation. Only one question raised the issue that Smith had been pursuing so assiduously: "Do particular shells, &c. affect particular strata?" Only one dealt with the problem over which Cuvier had been wrestling with his colleagues in Paris: "Are any analogous living species now found, or known to have been formerly found in their vicinity or elsewhere?" The relation between fossils and their formations, which on the Continent was regarded increasingly as crucially important for the future of geology, was given no such priority by the leading members of the new Geological Society in London.[63]

It is therefore not surprising that those leading members were less than impressed by what they saw and heard when, in the same year, they were invited by Farey to visit Smith's house and inspect his collections (§8.2). For a start, their agenda shows that their interests were focused far more on the Primary rocks than on the Secondaries, and more on the geology of highland regions than on the lowlands that Smith had surveyed: they were, in the informal jargon of modern geologists, "hard-rock", not "soft-rock", men. Furthermore, they may well have disliked Smith's dogmatic claims about the unerring certainty of his principle of "characteristic fossils", which would have struck Greenough, for one, as presumptuous and premature. Nor would they have liked his evident suspicion that the world was out to steal the credit for his work and the financial rewards that he hoped it would yield, or his consequent reluctance to put his results fully into the public realm. In fact, Smith's suspicions were not unfounded, for Greenough soon had his own plans for a rather similar geological map of the whole country, which would be untainted by any "theoretical" bias such as Smith's insistence on the validity of his fossil criterion, and which would represent not the labor of one man but the result of collaborative work by all the members of his society.[64]

However, it would be wrong to interpret the strained and even hostile relations between Smith and the society in simplistic terms of class struggle, as evidence of a split between lower-class "practical men" and more gentlemanly geologists, or between practical interests and more theoretical ones. Some of the surveyors were indeed excluded from the early Geological Society by its rules: Smith and Farey, being resident in London, were not eligible to be Honorary Members and were not gentlemanly enough to be Ordinary ones (and could not have afforded the fees).

60. Playfair, *Illustrations* (1802), was answered in print by, for example, [Murray], *Comparative view* (1802); Playfair's book is well summarized in Dean, *James Hutton* (1992), 102–25. The arguments between Huttonians and Wernerians in Edinburgh cannot be followed any further here, since they were tangential to the theme of the present book.

61. Geological Society, *Geological Inquiries* (1808), 2; Ordinary Minutes, 1 January to 4 March 1808 (London-GS, MS OM/1).

62. See Rudwick, "Foundation of the Geological Society" (1963), 330, for a distribution map of the early Honorary Members, based on data in Woodward, *Geological Society* (1907), 268–73.

63. Geological Society, *Geological inquiries* (1808), 19–20.

64. The records of this encounter between Smith and members of the society are frustratingly sparse. Greenough may also have doubted—with good reason—whether Smith's map would ever be published.

But the membership as a whole was certainly not uninterested in their practical activities, still less hostile towards them, and the society included many whose chief attachment to geology was certainly utilitarian. The relations between the two groups were indeed bedeviled by the English class system, but in practice there was plenty of communication between some of the surveyors and some of the society's members. Banks, for example, displayed Farey's map and section of Derbyshire in 1808, at one of the early meetings of the society, and did much to champion the practical value of his survey, though that particular role model was lost when Banks resigned the following year.[65]

Conclusion

These early years of the new century saw a final transformation in the meaning of "geology", from denoting the avowedly speculative project of geotheory—the attempt to find a comprehensive explanatory "system" for the earth—to its consolidation as a useful label covering all the sciences of the earth *except* that increasingly questionable genre. The change was marked with particular authority in Cuvier's survey of all the natural sciences, which finally brought "geology" in from the cold and defined it as a soberly empirical science devoted to detailed description and careful inference. But the change was also signaled with striking clarity in the decision by the founders of a new society in London to call it the *Geological* Society, because that choice was coupled with a research policy that explicitly excluded the pretensions of geotheory and honored an adherence to strictly "Baconian" observation. The shift in meaning proved to be stable and permanent—at least for the rest of the century—so that from this point onwards in the present narrative the words "geology" and "geologist" can at last be used without anachronism.[66]

More directly relevant to the main theme of this book—the historicization of nature in what can now be called geology—was Cuvier's new and characteristically perceptive understanding of the importance of the Secondary formations and their fossils. His opportunities to expound his ideas about the broad sweep of geohistory, in lectures to the general public, had already shown him that a more rigorous knowledge of these rocks and fossils was essential if geohistory and its varied revolutions were ever to be known with any confidence or precision. That little was yet known about them was a conclusion derived from his crash course in the whole geological literature, which was demanded by his obligation to review the progress and current state of the science. These contingent opportunities then combined with his belief that scientific research should be a collaborative activity, which he was successfully applying to his own research on fossil bones (albeit with himself in a firmly dominant role). From all this he derived the notion of an explicit *agenda* for research on the Secondary formations and their fossils, and the conviction that it should be proposed in public for his colleagues and contemporaries to pursue. Saussure had earlier left his agenda as a last testament at the end of his life, covering the entire range of what was now being called geology (§6.5). Cuvier, in contrast, proposed a much more sharply focused agenda, and did so at the height of his career, when he could still hope to see it substantially fulfilled.

In fact, Cuvier himself was already deeply involved in research that was designed to act as a model or sample of what needed to be done; it was also an example of collaboration with a colleague, under more equal conditions than in his work on fossil bones. This collaboration, focused on the rocks around Paris that were yielding some of his most important fossil mammals, is the subject of the first section of the next chapter of this narrative. In the event, it became the spearhead of much wider research on the *younger* Secondary formations (those soon to be redefined as *Tertiary*). As Cuvier recognized, these formations could play a crucial role in the task of reconstructing the whole of geohistory, because they clearly represented the period closest to the present world and contained fossils least unlike the animals and plants known alive. In short, they could act as a key to the still deeper past. The next chapter will trace how this line of research was pursued, by Cuvier and a widening international cast of geologists, in the years that followed the formulation of his agenda.

In those same years, however, another aspect of Cuvier's research was equally fruitful. This was his focus on the alleged boundary event of the "last revolution" in the earth's history, perhaps a dramatic "catastrophe" and possibly one of unparalleled magnitude. The problem was to understand how the "present world" of living organisms—and, above all, of human beings—was related historically to even the most recent rock formations and fossils, dating from a "former world" that had apparently been prehuman. The question of the earth's most recent "revolution" formed in effect another line of research, only loosely linked to the first. It will be the subject of the last chapter of the present narrative. In other words, the narrative part of this book splits at this point into two parallel narratives, each tracing in a separate chapter one of these two distinct lines of research. The book will conclude with a brief look forwards in time, to suggest how the two focal problems were later brought together again (a story that I hope to trace in a sequel to the present volume).

65. Geological Society, Minutes, 2 December 1808 (London-GS, OM/1); Torrens, "Banks and the earth sciences" (1994).

66. It was only in the twentieth century, as "physical geology" was redefined as "geophysics" and its practitioners contrasted their own high-tech methods with the allegedly antiquated practice of fieldwork, that the nineteenth-century unity of "geology" was broken up. When eventually the need to recapture that unity became inescapable, new umbrella terms such as "earth sciences" or "geoscience" had to be invented, both of them neologisms that duplicate the nineteenth-century meaning of "geology".

CHAPTER 9

The gateway to the deep past

9.1 THE GEOHISTORY OF PARIS (1802–8)

Brongniart as a geognost

By the time Cuvier presented Napoleon with a report on the progress of the natural-history sciences, with its rehabilitation of geology, he was in fact already engaged in research that was designed to exemplify the agenda he had set out in his earlier report on André's geotheory (§8.4). The preliminary results of his new research were offered to the Institut only two months later, but it was not his alone. The work embodied a collaboration much more equal than that between himself and his many informants on fossil bones. His collaborator was the mineralogist Brongniart, and their research was a geognostic study of the Paris region.

When Brongniart was appointed director of Sèvres at the turn of the century, he—like other French naturalists—had shown no clear grasp of the structural analysis of rock masses, as used by geognosts in the German lands (§8.1). However, in 1802 either or both of two events gave him a new insight into this three-dimensional way of seeing, which was essential for his collaboration with Cuvier. First, he took advantage of the Peace of Amiens to revisit England, now primarily in his role as director of Sèvres. The French recognized that the English had become world leaders in the mass production of ceramics for the expanding middle classes, but still hoped to regain supremacy at the luxury end of the market. Brongniart duly purchased samples of Wedgwood and other wares for his museum (see below), but he never got beyond London and did not visit Wedgwood's famous factory, one of the major sites of the burgeoning industrial revolution. However, his trip to England was also made in his role as a savant, and a well-connected one at that: his young wife accompanied him, and while in London they probably stayed with her father Coquebert, who at the outbreak of peace had been appointed the French consul-general. Brongniart was invited to meet the inner circle of the Royal Society at their dining club; he also visited Banks and other leading London savants at their homes (they would of course have had no difficulty conversing with him in French). In one way or another, Brongniart may well have heard about William Smith's work, and may even have seen his unpublished map: although Smith had not yet moved to London, he had already explained his ideas to Banks, and his map was exhibited in several places around this time (§8.2). In any case the gossip among the London savants may at least have alerted Brongniart to the value of looking at the surface topography of a lowland region in terms of its underlying formations and to the possibility that their fossils—if studied closely—could be used to distinguish one formation from another.[1]

The second event that would have heightened Brongniart's sense of three-dimensional structure in the practice of geology was Werner's visit to Paris towards the end of the year (§8.1). It is inconceivable that Brongniart, as a leading Parisian mineralogist, would not have been among the savants who spent time talking with the famous geognost from Freiberg, or who at least heard him speak. Later, Brongniart was almost certainly one of Cuvier's informants about German geognosy, and indeed about geology in general, while the latter was compiling his report for Napoleon (§8.4). Among other points, he urged Cuvier to mention Werner's book on mineral veins: "it's in this work that he determines *the age of ores* [*métaux*] in a precise manner, by the way the veins are cut". If, even after his trip to London, Brongniart still did not fully appreciate the geognostic treatment of formations as three-dimensional structures, and their interpretation in terms of relative age, then Werner's visit and his own related reading—both for his own teaching and to help Cuvier with his report—would surely have impressed it upon him.[2]

At the same time, Brongniart's work at Sèvres was giving him unanticipated firsthand experience of the practice of *history*, analogous to Cuvier's experience as a consultant for the French edition of *Asiatick Researches* (§7.4) and to Blumenbach's as a colleague of the distinguished historians at Göttingen (§8.1). Among the properties confiscated from the aristocracy and clergy during the Revolution were valuable collections of ceramics, including many pieces of great historical importance. Brongniart soon decided to use these to create a national museum of ceramics at Sèvres, to act as a reference collection for his designers and technicians as well as an artistic and educational resource for the general public. Although the museum was intended primarily to display and explain the different kinds of ceramics (earthenware, porcelain, etc.) and the techniques used to decorate them, it also had a strong historical dimension. It showed the distinctive artistic styles and techniques that were *characteristic* of different "epochs" from Antiquity to modern times. This was a historical perspective that, like Cuvier's and Blumenbach's, could readily be imported into the world of fossils and formations; certainly it would have predisposed Brongniart to be sympathetic to transposing culture into nature in this way.[3]

Anyway, only a couple of years after Brongniart's visit to London and Werner's to Paris, the Frenchman began work on a systematic survey of the Paris region, with a fully three-dimensional or geognostic approach, and with Cuvier's collaboration perhaps from the start. In addition to his practical reasons for the survey, related to the needs of Sèvres, Brongniart wanted to put Paris on the map—both literally and metaphorically—in the science of geology, just as he hoped to put Sèvres back on the international map in the world of luxury ceramics. He evidently suspected that the formations around Paris were *not* those that Werner and his students had been mapping elsewhere in Europe, so that he had a chance of making a significant scientific impact by enlarging the geognostic sequence as a whole. By 1807, when he published a textbook on mineralogy commissioned by Napoleon's regime for use in French high schools, he was sure on this point: he claimed that the Gypsum formation around Paris could not, by its geognostic position, belong to either of the two that Werner and his followers had distinguished (§8.1). So a survey of the Paris area was likely to yield important results for the science of geognosy well beyond its regional or even national significance.[4]

Cuvier's reasons for collaborating in Brongniart's survey have already been mentioned. The fossil mammals from the Gypsum formation were growing in importance for him at just this time; he was transforming them from a usefully local example of his methods (§7.2) into the prime instance of his sensational results (§7.5). But he could not place them confidently in any broader picture of geohistory until the geognostic position—and therefore the relative age—of the Gypsum was much clearer. Specifically, those strange mammals needed to be located geognostically in relation both to the fossil elephants and other species from the Superficial deposits and to the fossil reptiles from the older Secondary formations (§8.1). So Cuvier the anatomist had strong reasons for wanting to recruit the expertise of Brongniart the mineralogist and geognost; in return he could offer his collaborator the weighty authority of his powerful position at the heart of the scientific establishment.

Brongniart and Cuvier both had onerous official duties, and apparently most of their fieldwork had to be done at weekends. They made a series of traverses radiating from Paris in different directions. This was a good strategy, because the city turned out to be more or less at the natural center of what they were mapping. Paris was (and is) almost surrounded by broad tracts of the distinctive Chalk, which was overlain nearer the city by the various other rocks in which they were interested. They were not working from scratch. They must surely have read

1. Brongniart, manuscript notebooks on trip to London, 23 September to 26 October 1802 (Paris-MHN, MSS 2351/5, 2340), support—or at least are compatible with—the link between Smith and Brongniart suggested by Eyles (J. M.), "Banks, Smith" (1985). On his return to Paris Brongniart's technicians successfully imitated Wedgwood's famous and fashionable "black basalt" ware.

2. Anonymous, manuscript notes on *minéralogie* and *géologie* (Paris-IF, MS 3139, no. 12), among Cuvier's notes for his *Rapport historique*; a marginal note on this manuscript suggests, plausibly, that it is by Brongniart. La Métherie's interview with Werner (§8.1) was published in the *Journal de physique* for frimaire XI [November–December 1802], so Werner's visit to Paris was almost certainly *after* Brongniart's trip to London.

3. Brongniart and Riocreux, *Musée céramique de Sèvres* (1845), described—with superb chromolithographic illustrations—what the museum had become by the end of Brongniart's career, but its basic arrangement was probably similar in its early years; it remains one of the most important public ceramics collections in the world. See also Millasseau, "Brongniart as museologist" (1997).

4. Brongniart, *Traité de minéralogie* (1807), 1: 177n. The circumstances of the start of his and Cuvier's collaboration are obscure; the preliminary report that they presented jointly in 1808 (see below) stated that it was the product of four years' work, which would date it from 1804.

Lavoisier's earlier paper (§2.4), but they did not list him among their forerunners, probably because they would not have approved his highly conjectural kind of causal explanation. One naturalist whom they did acknowledge was Jacques Michel Coupé (1737–1809), a former parish priest who had represented the clergy in the early phase of the Revolution but who had later been laicized (like Soulavie and André). Coupé must already have been making his own survey of the Paris region, for in 1805 he published in the *Journal de Physique* a long paper on "the subsoil [*sol*] of the environs of Paris", describing a sequence of five formations. Brongniart and Cuvier adopted this, or at least agreed with it, as the basis for their own more elaborate inventory. But some of Coupé's interpretations, like Lavoisier's, were too speculative to appeal to them: for example, Coupé attributed the tropical appearance of the fossil shells that Lamarck was describing (§7.4) to "a simple [*sic*] permutation of the equator", a change in the earth's axis of rotation, which was just the kind of explanation that they were determined to avoid.[5]

Brongniart and Cuvier would usually have discussed their fieldwork face to face, leaving no documentary traces of their collaboration, but two known letters are probably typical. At one point Cuvier referred to the Paris region as "an enormous bowl [*chaudière*] sunk in the Chalk", which shows that he understood that the large-scale structure was what geologists ever since have called the *Paris Basin*. He also looked forward to their joint production of "one of the finest geological works that has yet been composed on the Secondary formations", which shows that he regarded the formations filling the basin (among them, the Gypsum) as belonging, like the Chalk, to the "regular" Secondaries, rather than to the Superficial deposits as Lamanon's earlier conjecture might have implied (§4.3). On a later occasion, Brongniart told Cuvier that the man he employed to collect fossil bones and purchase them from the workers had found a bed of clay full of "fluviatile shells" above the Gypsum at Montmartre. This find was important, because it supported Lamanon's claim that the area had once been a freshwater lake (see Fig. 4.11) and that its history had therefore not been one of continuously marine deposition, as both Lavoisier and Coupé had assumed and as might have been expected on the standard model of geotheory.[6]

Geognosy of the Paris Basin

By 1808, when Brongniart and Cuvier read a preliminary report of their work at the Institut, they were able to display a draft of a colored map of the region around Paris, which summarized their research in the most vivid manner possible. Like Smith in England at the same time, they had surveyed a gentle landscape of fields and forests, in which exposures of any rocks were usually small and scattered; like Smith, they had mapped what in general could *not* be seen (§8.2). Their map covered a wide area, though much less extensive than Smith's (theirs was similar in area to southeast England, south of London and east of Portsmouth). It also depicted the outcrops of only nine formations, many fewer than Smith's, and more crudely. But where it was most detailed—in the area close to Paris—it was an impressive achievement (Fig. 9.1).[7]

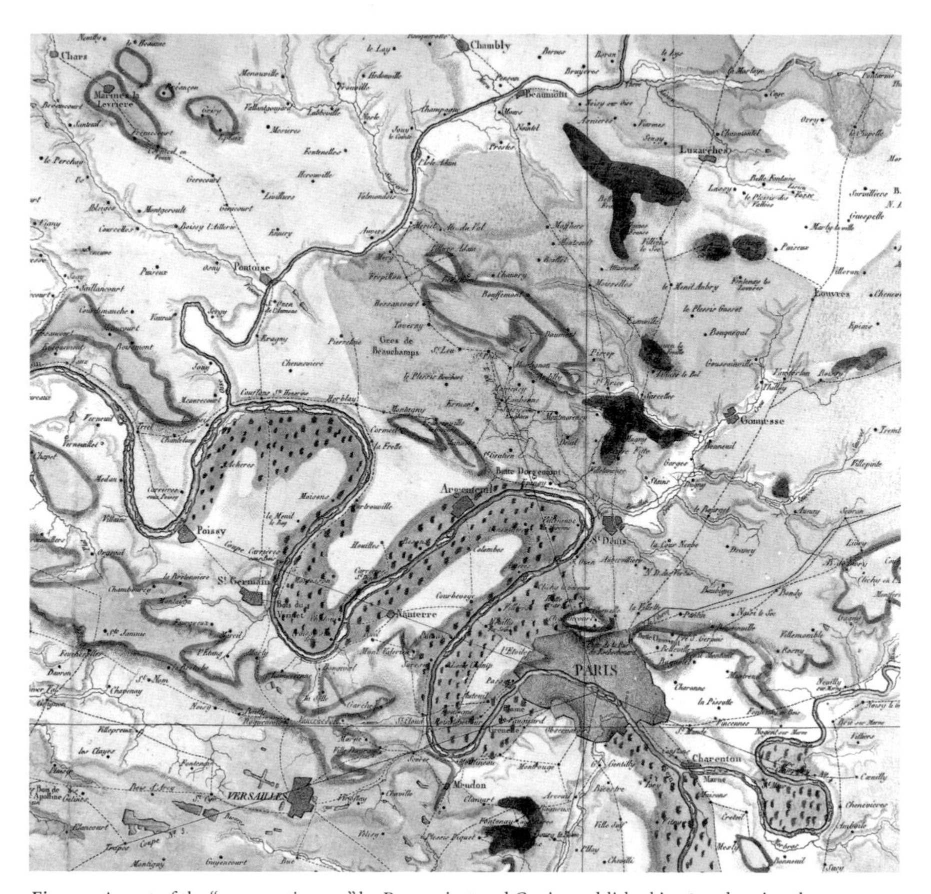

Fig. 9.1. A part of the "geognostic map" by Brongniart and Cuvier published in 1811, showing the area (about 45km square) immediately around Paris; a manuscript version was displayed at the Institut National when their preliminary paper on the "mineral geography" of the region was read in 1808. Their map used the ordinary convention of flat color washes, as in Charpentier's earlier map of Saxony (Fig. 2.13), not the graded coloring used later by Smith (Fig. 8.5). In this reproduction the darkest tone represents the Gypsum formation, outcropping on the tops of hills (the irregular patches) or on their flanks (the irregular loops); Montmartre is the small loop immediately north of the city. The stippled areas represent the Detrital Silt along the meandering valley of the Seine. The straight lines radiating from the center of Paris are the lines of their measured traverse sections (see Fig. 9.6). Sèvres is on the outer side of the first bend of the river southwest of the city; Grignon, the most prolific locality for fossil shells, is on the left edge, due west of Paris. (By permission of the Syndics of Cambridge University Library)

5. Coupé, "Sol des environs de Paris" (1805), 381; the formations were described in upward order as chalk, blue clay, building stone, gypsum, and fine sand, which can easily be equated (see below) with some of those in Cuvier and Brongniart, "Géographie minéralogique" (1808); on Coupé, see also Gaudant, "L'exploration du Bassin Parisien" (1993), 23–25. Brongniart's name is put in first place throughout the present account, to emphasize that—as Cuvier acknowledged with unusual generosity—most of the fieldwork was due to him; but in their publications Cuvier's name had pride of place, perhaps because Brongniart conceded that this would give their joint production a higher profile and much greater authority.

6. Cuvier to Brongniart, 9 September 1806, and Brongniart to Cuvier, 20 October 1807, quoted in Launay, *Les Brongniart* (1940), 110–12.

7. Fig. 9.1 is reproduced from Cuvier, *Ossemens fossiles* (1812), 1 [Cambridge-UL: MD.8.65], part of "Carte géognostique" dated 1810, first published in Cuvier and Brongniart, "Géographie minéralogique" (1811): the whole map covers an area of 142km by 124km at the scale of 1:200,000; eleven formations are distinguished. It is superbly reproduced in color in Anonymous, "L'Essor de la géologie française" (2003), cover. The unpublished version is referred to in the text read at the Institut on 11 April 1808 (Paris-MHN, MS 631) and published as "Géographie minéralogique" (1808), 299–300, trans. Rudwick, *Georges Cuvier* (1997), 137. See also Ellenberger, "Cartographie géologique" (1985), 41–44.

When Brongniart and Cuvier presented their paper at the Institut, they made it clear that it was primarily a piece of *geognostic* research. They entitled it a "mineralogical geography" of the Paris region, using the usual French phrase for what in Germany had long been called geognosy; and when their map was engraved two years later they entitled it "geognostic". Most of their paper duly described nine formations in order from the lowest to the highest (with one exception, to be mentioned below). For each formation, geographical details of its outcrop were combined with details of its constituent rocks. Much of this text, and particularly the notes on the practical uses of the rocks—as stone for building, clay for pottery and so on—was almost certainly written by Brongniart alone. So far the format was entirely conventional. However, what was striking about the paper, even at this descriptive level, was that the distinctive Chalk was the first and *lowest* formation to be described, whereas it was commonly treated in geognostic works as one of the *highest*, if not the highest of all, in the pile of "regular" formations. This alone implied that, as Brongniart had already concluded, most of the other formations described in the paper were *additions* to the top end of the general geognostic sequence. Converted from structural into temporal terms, this demonstrated that there were many new formations even *younger* than the Chalk (though older than the Superficial deposits), implying in turn that the geohistory represented by the pile of "regular" formations must also be extended.[8]

A second feature of the paper that was clearly innovative was the detailed attention given to the fossil contents of the formations. As already emphasized, this was not wholly novel, but earlier accounts of Secondary formations had generally noted their fossils only briefly and in passing, if at all; Coupé's recent account of these same Parisian rocks was a case in point. Brongniart and Cuvier, in contrast, described the fossils found in each formation in some detail even in this preliminary paper and promised more in the fuller account to follow; and if a formation contained no fossils at all they made a point of noting that fact, and had evidently tried hard to find some. In this sense their paper was clearly intended to exemplify the agenda that Cuvier had set out two years before, in his report on André's book (§8.4). It followed that agenda more specifically, however, in using the fossils to help *define* the formations and to distinguish them. The fossils in the Chalk, for example, were contrasted with those in the Coarse Limestone [*calcaire grossier*]—the next formation to contain any—as a distinguishing character that could be added to those of geognostic position and rock type, jointly making the two formations unmistakably distinct and impossible to confuse. This was a striking new instance of what Arduino, Soulavie, and others had noted long before, and what Smith was busy doing—possibly with the French savants' knowledge—at just this time (§8.2). It made their account of the Paris region, unlike Coupé's, but like Smith's still unpublished account of England and Wales, a clear example of a newly *enriched* kind of geognostic practice.[9]

In the case of one formation, Brongniart and Cuvier maintained that the principle of characteristic fossils was valid at a greater degree of detail than even Smith might have claimed. Smith's formations included some (e.g., Cornbrash and Kelloways Rock) that were only a few feet thick, yet had characteristic fossils that were found throughout their wide geographical extent (§8.2). The French naturalists claimed, however, that even *within* one of their formations (the Coarse Limestone again) the thin constituent beds maintained the same sequence over wide areas, and could be recognized by fossils that were characteristic of each:

> This constancy in the order of superposition of the thinnest beds, over an extent of at least twelve myriameters [120km] is, in our opinion, one of the most remarkable facts that we have noted in the pursuit of our researches. The consequences that should flow from it, for the [practical] arts and for geology, should be as much more interesting as they are more certain. The means we have used to recognize a bed already observed, amid such a large number of limestone strata, is drawn from the nature of the fossils enclosed in each bed. These fossils are always generally the same in the corresponding beds and show quite notable differences of species from one set [*système*] of beds to another set. This is a sign of recognition that so far has not misled us.[10]

Brongniart and Cuvier were of course in a much more advantageous position than Smith, in being able to compare their fossils with those in some of the finest collections available anywhere, and to draw on the expert knowledge of one of the finest conchologists—their colleague Lamarck—to identify what they found. Although they deferred such details for their full report, the most important implications of their close attention to fossils could be set out even in their preliminary account. The geognostic pile of formations was simultaneously interpreted as a record of the *geohistory* of the Paris region; the formations, described in turn from lowest to highest, became the evidence of a temporal sequence of events from earliest to most recent. And the decisive evidence for constructing this geohistorical narrative came from the nature of the fossils. The paper was therefore a strikingly novel example of a *doubly* enriched geognostic practice: enriched by the use of fossils not merely to characterize and identify different formations, but also to reconstruct the conditions in which they had been formed and the sequence of events that they represented. As might be expected, this second enrichment seems to have been Cuvier's work more than Brongniart's: it was Cuvier who had already transposed a strong sense of human history into the world of nature, in his reconstructions of fossil mammals and in his inferences about the event that had made them extinct (§7.5).[11]

8. Cuvier and Brongniart, "Géographie minéralogique" (1808), trans. Rudwick, *Georges Cuvier* (1997), 133–56. Smith, although not knowingly in the geognostic tradition, likewise treated the Chalk as one of the highest of his formations (see Fig. 8.6). The Secondaries above and younger than the Chalk were those that were later renamed *Tertiary* formations (§9.4), the informal term that modern geologists continue to use.

9. Rudwick, "Cuvier and Brongniart" (1996), analyzes further the comparison with Smith.

10. Cuvier and Brongniart, "Géographie minéralogique" (1808), 307–8, trans. Rudwick, *Georges Cuvier* (1997), 143. The Coarse Limestone was quarried around Paris for buildings in the city: many of the houses in the older streets on the Left Bank still display this distinctive and beautiful stone.

11. Cuvier and Brongniart, "Géographie minéralogique" (Paris-MHN, MS 631). The whole manuscript is in Cuvier's hand, but the geohistorical comments read as if he had inserted them into a purely descriptive draft by Brongniart.

The uncontroversial baseline for any geohistorical reconstruction was provided by the fossils found in formations such as the Chalk and the Coarse Limestone. For the former, Brongniart and Cuvier could refer to Faujas's earlier account of the fossils from the Chalk at Maastricht (§7.2). For the latter they made extensive use of Lamarck's detailed work (which he was completing at just this time in the Muséum's *Annales*), nominally on all the fossil shells from the Paris region, but in fact mainly those from the prolific quarry at Grignon (Fig. 7.15). In contrast, Brongniart and Cuvier claimed that this fauna was characteristic of an extensive formation, not just of that single locality. Most of the shells in the Coarse Limestone were easily identifiable as genera—and a few of them, according to Lamarck, even as species—that were known to be alive in present seas (§7.4). With them were fossil corals and sea urchins, organisms that were never found alive in the present world except in seawater (see Fig. 9.2 below). So there could be little doubt that the formations containing all these fossils had been deposited while the Paris region was submerged by the sea. This was unsurprising and uncontroversial because most naturalists took it for granted—as the standard model of geotheory led them to expect—that *all* the "regular" formations everywhere were marine in origin. In the specific case of the Parisian rocks, Coupé had emphasized that assumption by giving his paper the Homeric motto "The ocean is the origin of everything" [*Oceanus genesis omnium*].[12]

Freshwater formations

However, this simple assumption was challenged by persistent claims that some of the Parisian formations contained shells closely similar to those that in the modern world live only in the fresh waters of lakes and rivers, or even on dry land. Lamanon's reconstruction of a vast former lake on the present site of Paris (Fig. 4.11) had been based on an early claim of this kind, though he seems to have assumed that the shells came from a Superficial deposit not far removed in time from the mollusks still living in the rivers nearby (§4.3). Brongniart and Cuvier revived this idea and claimed that these fossils could be used as reliable indicators of former freshwater conditions. Yet they regarded the Gypsum formation as clearly Secondary. This made their claim highly controversial because it flew in the face of the assumption built into the standard model of geotheory, that the whole pile of Secondary formations had been deposited on the seafloor as global sea level declined and the present continents emerged as dry land (§3.5).

Lamarck, with his eternalistic model of ever-shifting ocean basins (§7.4, §8.3), was equally opposed to the new interpretation. He was well aware of the problematic genera and had just been describing their smooth and delicate shells along with those of the much more diverse genera of unquestionably marine origin: both kinds were represented in his museum collections from Parisian localities. But he minimized the significance of the allegedly freshwater and terrestrial shells: he suggested that the fossil ones belonged to marine species of genera that happened now to be exclusively freshwater; or that they had been swept out to sea from their freshwater habitats; or that they were too well-preserved to be genuine fossils at all. He therefore concluded that these distinctive fossils did *not* indicate that freshwater conditions had prevailed at any time during the deposition of the Parisian formations.[13]

Faujas had just offered another explanation, coming to the same conclusion. In 1806 he claimed in the Muséum's *Annales* that "these shells, which at first glance have a fluviatile appearance, are [in fact] marine". He described how those found around Mainz (which had been annexed by France during the Revolution) were abundant in thick beds that must have been "the result of a series of slow successive deposits" on the seabed. He suggested that they were similar to species now living in fresh water, simply because in those early times the seawater might have been low in salinity or even fresh in composition; a later rise in salinity might then have been the cause of their extinction. This speculative idea of changing salinity—a typical piece of geotheorizing—was tacitly but firmly rejected by Brongniart and Cuvier, when they reported on their joint research two years later; they inferred that these shells were true indicators of truly freshwater periods in the long geohistory of the region.[14]

In reaching that conclusion Brongniart and Cuvier drew not only on Lamanon's much earlier work. They were probably swayed more by a recent report of significant fieldwork by a local naturalist outside Paris: since his report was read at the Institut and then published in the *Journal de Physique* they would certainly have known about it. Jean Louis Marie Poiret (1755–1834), who taught natural history at the high school in Soissons, had discovered near that town a bed of lignite [*tourbe*] lying above beds containing the freshwater shells, and below beds with the shells of marine mollusks. Poiret turned this modest piece of geognosy into *geohistory*: "nature discloses to the view of the observer one of the oldest pages of its archives". His reconstruction of the sequence of beds was explicitly based on an actualistic comparison with "what happens every day under our eyes". During "a long series of ages [*siècles*]" the area had been one of lakes, rivers, marshes, and forests—rather like the virgin American landscape, he suggested—which had later been submerged beneath a returning sea. This was a striking conclusion: Poiret claimed that a period of truly freshwater or terrestrial conditions had been *followed* by a marine period, so that the local geohistory could not be subsumed under any simple model of a progressive retreat of the sea and emergence of the continent.[15]

12. Lamarck, "Fossiles des environs de Paris" (1802–9), was fully published by 1808, except for the final few plates; the comment on Grignon is in Cuvier and Brongniart, "Géographie minéralogique" (1808), 309–10, trans. Rudwick, *Georges Cuvier* (1997), 144; see also Coupé, "Sol des environs de Paris" (1805), 364n.

13. Lamarck, "Fossiles des environs de Paris" (1802–9): see for example his comments under *Cyclostoma*, *Lymnaea* and *Planorbis* (all 1804).

14. Faujas, "Coquilles fossiles de Mayence" (1806).

15. Poiret, "Tourbe pyriteuse" (1800–1803), esp. pt. 1 (1800); his "Formation des tourbes" (1804) dealt with the origin of the lignite itself. But his "Principes de la géologie" (1805)—a sketch of a planned *Treatise on geology*—might have struck Brongniart and Cuvier as far too ambitious; anyway they did not include Poiret in their list of predecessors. See also Ellenberger, "Lignites du Soissonais" (1983), 11–13, and Gaudant, "L'exploration du Bassin Parisien" (1993), 22–23.

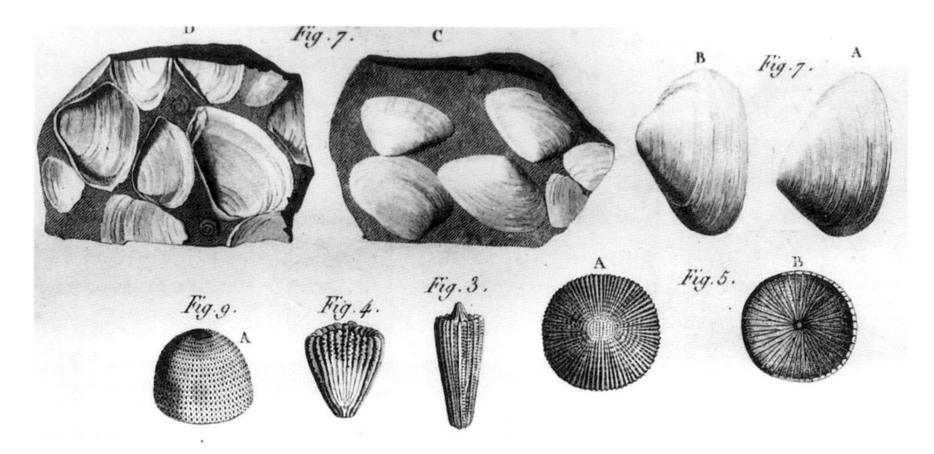

Fig. 9.2. Parisian fossils of special geohistorical significance, illustrating the full report by Brongniart and Cuvier published in 1811. In the upper row are the small shells found immediately above the Gypsum formation, which they took as evidence for a reversion to marine conditions after a long lacustrine episode. In the lower row are small corals from the Coarse Limestone, which—since living corals were known to be strictly confined to seawater—supported the authors' inference that that formation had been deposited during an earlier marine phase. (By permission of the Syndics of Cambridge University Library)

Poiret's local example was an inspiration for, or at least supported, the more general interpretation that Brongniart and Cuvier seem to have fashioned gradually in the light of further fossil finds. Rare freshwater shells in the Gypsum formation, and the complete absence of marine shells, suggested to them that it represented a freshwater period, following the marine period of the Coarse Limestone. Then they found, almost immediately above the Gypsum, a very thin but remarkably extensive bed full of the small shells of a marine genus, which they took to mark the abrupt return of seawater; it was followed by beds containing shells of the same marine genera as those in the Coarse Limestone much lower down. In effect, these beds were interpreted as the record of a change closely parallel to the one that Poiret had inferred; but it was expanded into a record of a freshwater *interlude* between two separate marine periods. Among the few fossils that were illustrated in the full version of the Parisians' account were those providing evidence for this striking piece of geohistory (Fig. 9.2).[16]

The geohistory inferred by Brongniart and Cuvier did not end there. At the very top of the whole pile was a formation—for the discovery of which they claimed particular credit—full of the freshwater shells, and they were so sure about its origin that they named it the Freshwater Formation [*terrain d'eau douce*]. Indeed, so confident were they of this overall reconstruction that they inferred that the Plastic Clay [*argile plastique*], lying between the Chalk and the Coarse Limestone, represented yet another and much earlier freshwater episode, although they could find in it no fossils at all. In short, Brongniart and Cuvier became convinced that there had been an *alternation* between marine and freshwater conditions: they claimed that the pile of formations they had mapped around Paris represented no fewer than three separate freshwater episodes alternating with three periods of marine conditions. Their reconstruction can be followed most easily on the "ideal

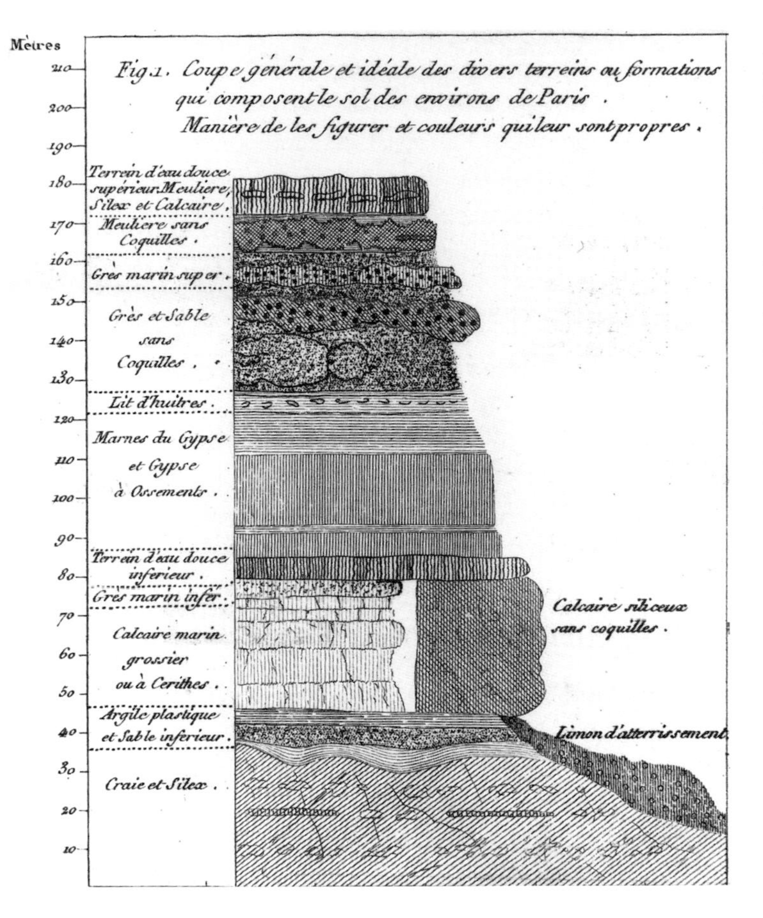

Fig. 9.3. The "general and ideal section" of the formations [*terrains*] around Paris, published by Brongniart and Cuvier in the full version (1811) of the paper they first presented at the Institut in 1808. The Detrital Silt [*Limon d'atterrissement*] is on the floor of an imaginary valley (right) eroded through all the other ("regular") formations. The latter lie horizontally in an invariable order; but two, the Coarse Limestone [*Calcaire grossier*, here labeled *marin grossier*] and the Siliceous Limestone [*Calcaire silicieux*] are shown rather awkwardly side by side (in modern terms, as lateral equivalents of contrasting facies). Two formations are marked as containing shells of marine genera (*cerithes*, and *huîtres* or oysters), and two others are also assigned a marine origin. Conversely two formations are not merely attributed to fresh water but named as a lower and upper Freshwater Formation [*terrain d'eau douce*]. The Gypsum formation [*Gypse*] yielded the fossil bones [*Ossements*] that had already become the centerpiece of Cuvier's research. All these annotations reflect and describe the *geohistorical* inferences that were drawn from the evidence of the rocks and fossils. The vertical scale shows a total of about 150m of strata above the Chalk. (By permission of the Syndics of Cambridge University Library)

section" that they used to illustrate the full text of their paper. It did not represent a slice though the rocks along any specific line, even in the stylized fashion of Smith's and Farey's English sections (see Fig. 8.6). Instead it showed all the formations piled in the correct order, but as if they were all exposed at a single locality on the side of a single imaginary valley. The section was purely geognostic: it simply depicted the three-dimensional relations of the rock masses, with their typical thicknesses. But the annotations alongside the formations summarized the fossil contents, which in turn provided the key to the conditions of their deposition and hence to the geohistory of the Paris region (Fig. 9.3).[17]

16. Fig. 9.2 is reproduced from Cuvier, *Ossemens fossiles* (1812), 1 [Cambridge-UL: MD.8.65], "Géographie minéralogique", part of pl. 2, first published in Cuvier and Brongniart, "Géographie minéralogique" (1811). In their preliminary report (1808) the shells had been referred to the genus *Tellina* [*tellines*], which was also listed among those found in the Coarse Limestone; but in this full report they assigned the shells to another marine genus, *Cytherea*, and identified them as *cytherée bombée*.

17. Fig. 9.3 is reproduced from Cuvier, *Ossemens fossiles* (1812), 1 [Cambridge-UL: MD.8.65], "Géographie minéralogique", pl. 2, fig. 1, first published in Cuvier and Brongniart, "Géographie minéralogique" (1811). This style of columnar section (in modern terms a "vertical" section), although not quite original to them, was an important addition to the visual repertoire of the science: see Rudwick, "Visual language" (1976). The caption shows that the section was intended to be colored to match the map, though this copy is uncolored.

Environmental geohistory

Although Brongniart and Cuvier used fossils to help characterize the various formations, they were doing so in a quite different manner from Smith. For the French naturalists, fossils were not just pragmatically useful to distinguish one formation from another; far more importantly they also indicated past environmental conditions and hence were evidence for reconstructing a complex and eventful *geohistory*. In practice this took precedence over any Smithian use of the fossils. The beds above the Gypsum formation, for example, had many of the same fossils as the Coarse Limestone below it, but the two formations were not for that reason taken to be the same. The same set of fossils recurred in the higher and later formation simply because the same environmental conditions had returned: formations were *not* to be identified just by their characteristic fossils.

This emphasis on reconstructing past environments and geohistory—unlike anything Smith had done or was planning to do—was expressed vividly in one of the interpretative passages that Cuvier probably inserted into Brongniart's descriptive draft text. Halfway through their geognostic inventory of formations, the published text paused, as it were, to recapitulate the story so far. It used the present tense to reconstruct a narrative of the changing scene, in a style that recalls the famous "vision" of Alpine geohistory that Saussure had experienced long ago on the summit of Crammont (§4.5):

> One pictures [*se représente*] first a sea that deposits on its floor an immense mass of chalk and of mollusks of distinctive species. This precipitation of chalk and of the accompanying shells suddenly ceases. Beds of a quite different nature succeed it, and only [Plastic] clay and sand are deposited, without any organisms. Another sea returns: this one sustains a prodigious quantity of shelled mollusks, wholly different from those of the Chalk. Massive beds [of the Coarse Limestone] are formed on its floor, consisting in large part of the shelly coverings of these mollusks; but little by little this production of shells diminishes and ceases completely. Then the ground is covered with fresh water; and alternating beds of gypsum and marl are formed, which envelop the debris of the animals that these lakes sustained, and the bones of those that lived on its shores. The sea returns a third time [etc.].[18]

This was a narrative of the geohistory of the region set out on an implicitly lengthy timescale of tranquil deposition. The transitions between the marine and freshwater periods were not all interpreted as sudden changes. As this quotation shows, the transition from the sea of the Coarse Limestone to the lakes of the Gypsum formation was described as having been as gradual as the changes in the strata and the fossils themselves. The boundary between the Chalk and the Plastic Clay was indeed abrupt, yet it too was interpreted as not having been as sudden as it might seem: at one locality the base of the clay was marked by a "breccia" composed of chunks of chalk, which proved that the chalk had already been solid, so there was "perhaps a long span of time, between the deposition of the chalk and that of the clay". On the other hand, the equally sharp boundary between the gypsum beds and the distinctive overlying bed with marine shells (Fig. 9.2) was interpreted as "the sudden start of a new marine formation". But even here the crucial word "sudden" [*subit*] was added while the article was in press and seems to reflect a conviction—probably Cuvier's rather than Brongniart's—that at least some of these ancient changes of environment and geography had indeed been revolutions in the narrower sense of *sudden* changes.[19]

However, one striking feature of the pile of formations depicted on the "ideal section" that summarized all these local details shows the limitations of the geohistorical inferences that Brongniart and Cuvier were prepared to draw from their description of the Parisian formations. The Siliceous Limestone [*Calcaire silicieux*] was said to have "a geological situation parallel, as it were, to that of the marine [i.e., Coarse] limestone: it is situated neither below it, nor above it, but beside it" (Fig. 9.3). Both formations were apparently sandwiched between the Plastic Clay below and the lower of the two Freshwater Formations above, which implied that they had been deposited more or less *simultaneously*. Such a situation—both geognostic and geohistorical—was utterly unexpected, and Brongniart and Cuvier evidently had great difficulty accepting what their fieldwork showed. The implication was that two contrasting environments had existed at the same period: the unfossiliferous Siliceous Limestone must have been deposited across a large area to the southeast of Paris towards Fontainebleau, at the same time that the richly fossiliferous Coarse Limestone was accumulating on the floor of a sea covering the rest of the basin. For the time being, however, such a reconstruction (in modern terms, of two contrasting facies of roughly the same age) was a step too far.

Even without it, however, what Brongniart and Cuvier offered was a highly complex geohistorical narrative for the Paris region. Their imagined picture of successive seas, alternating with periods in which the area was covered with lakes or rivers, did not portray the inexorable—and therefore in principle predictable—working out of fixed physical principles through the vast spans of deep time. Rather it was a geohistory just as unpredictable, complex, and contingent as the turbulent political history that both authors had lived through during the previous two decades. It was a complex story of successive "revolutions"—some of them perhaps as sudden in their own way as the coups d'état of Thermidor and Brumaire—pieced together from "nature's documents", the concrete evidence of often small and subtle features that they had observed in the field. It was a contingent geohistory, constructed unmistakably bottom-up, not deduced top-down from some overarching geotheory (see §4.5).

Throughout their account of the rocks of the Paris Basin, Brongniart and Cuvier abstained from causal explanations as consistently as they suggested geohistorical ones. They offered no causal reasons for the alternation of marine and freshwater conditions, or for the changes in physical geography that it implied.

18. Cuvier and Brongniart, "Géographie minéralogique" (1808), 320, trans. Rudwick, *Georges Cuvier* (1997), 152.

19. Cuvier and Brongniart, "Géographie minéralogique" (1808), 306, 317, trans. Rudwick, *Georges Cuvier* (1997), 142, 149.

Their paper was an essay in geognosy, doubly enriched by its use of fossils to help define formations and to reconstruct geohistory; it was not an essay in earth physics, let alone in geotheory.[20]

Conclusion

The previous chapter traced the way in which, during the first years of the new century, the well-established outdoor science of geognosy (§2.3) began to be enriched by a far closer indoor attention to the fossils contained in the "regular" Secondary formations. At a pragmatic level, Smith the English civil engineer or "practical man" discovered that many of these formations contained characteristic fossils that could be used to trace their outcrops and establish their exact order of succession, even in lowland regions where rock exposures were rare. Although the principle was not new, Smith applied it with unprecedented precision and detail. But his map and sections remained unpublished, and his expert knowledge—which he correctly regarded as valuable intellectual property—was not available outside the circles of his patrons and associates.

Brongniart may have heard of Smith's work while visiting England during the Peace of Amiens, or he may have developed the technique independently, but in any case he and Cuvier found it valuable for surveying the formations around Paris. However, they took it on to a quite new plane when they used the fossils to indicate environmental conditions. This enabled them to begin to flesh out Blumenbach's sketch of a global geohistory that would be directly analogous to the "archaeology" of the human race (§8.1). Their fieldwork around Paris, and the alternation between freshwater and marine conditions that they inferred from the fossils, led them to reconstruct a detailed local geohistory for the Paris region and for the relatively recent portion of geohistory that the Parisian formations evidently represented. The rest of this chapter traces the further development of this work and its huge impact and influence on other savants in the years that followed the first announcement of their results.

9.2 CONSOLIDATING GEOHISTORY (1808–12)

Beyond the Paris Basin

Cuvier treated his and Brongniart's famous joint paper as the prime exemplar of the agenda he had proposed for geology two years earlier, in his report on André's book (§8.4), which must in fact have been formulated in the light of what their collaboration was already yielding. A few weeks after their paper was read, he set it in a much broader context in two sets of new lectures at the Collège de France. The longer course, entitled significantly "Principles of geology", was divided into two parts: first a review of all the different classes of fossils, beginning with the quadrupeds; and then a review of their geognostic setting, from the oldest formations to the most recent, treated as the basis for reconstructing the relative "chronology" of the earth's history. This clearly indicated Cuvier's juxtaposition of two scientific practices that had usually been pursued in isolation: the study of fossil specimens indoors in museums (§2.1) and the study of rock formations outdoors in the field (§2.3). These practices, he urged, ought to be combined; his and Brongniart's new work demonstrated, implicitly, the value of doing so:

> In general, this study [of fossils] has not been carried out at all in the manner that would have been most advantageous. Zoologists have only been concerned to recognize [fossil] species, without studying their respective positions in the bosom of the earth; while geologists, above all those of Werner's school, have not taken note of the shells that the beds contain. However, their [natural] history, if studied under this double relation, would be of the greatest importance.[21]

This was just what Cuvier knew that his collaborator was already doing. Brongniart too treated their joint paper as a model for a new kind of practice, which he lost no time in extending beyond the Paris Basin. A few days after their paper was read at the Institut, he left Paris on a two-month field trip to the south of France. His former student Constant Prévost (1787–1856) went with him, as companion and assistant, and also of course to get some practical training. The main purpose of Brongniart's fieldwork, which took him and Prévost as far south as the Pyrenees before they turned east to the Massif Central, was to explore sources of kaolin for Sèvres. But this could be combined—legitimately, in terms of his expense account—with more general studies of the rocks and fossils that they encountered on the way or were shown by local naturalists.

Brongniart utilized his recent field experience to identify similar formations far from the Paris Basin. At Bordeaux, a local naturalist showed him some fossils collected nearby, and he noted that "these shells are absolutely similar to those at Grignon [i.e., in the Coarse Limestone], though some peculiar species are [also] found there; the sea urchins and other chalk fossils are found only in a bed some feet below". Two familiar Parisian formations were thus recognized *by their fossils*, in their familiar positions although hundreds of miles from Paris. Later, near Biarritz, he saw a hard gray limestone that "appears to belong to the chalk formation": an identification based again on its distinctive fossils, despite its *contrast* with the soft white limestone found around Paris. After he left the Pyrenees he recorded seeing formations that ranged from Superficial deposits down through "Coarse Limestone" and older Secondaries with belemnites all the way to the "primitive terrain" of the Massif Central. This kind of geognostic fieldwork was novel only in

20. The brief final section of the paper, dealing with the even more recent event that had excavated the valleys and created the Superficial deposits, will be described at the start of the next chapter, on the "last revolution" (§10.1).

21. From "leçon 12" of Cuvier's nineteen lectures given from 2 June to 11 August 1808 as part of his course on "histoire naturelle", as paraphrased by Omalius, who had also audited Cuvier's earlier lectures on geology (§8.3): see Grandchamp, "Doctrines de Cuvier" (1994). Another set of manuscript notes (Paris-IF, MS 3103), by an anonymous auditor, summarizes four lectures on "géologie" given, from 7 June 1808, as part of a course on "philosophie de l'histoire naturelle". These were apparently two separate sets of lectures, delivered in parallel, though the contents overlap. That he chose to give so much attention to geology in his lectures for 1808 is significant in itself.

the degree of attention that Brongniart gave to fossils. But it would have put his and Cuvier's recent studies into perspective, by highlighting how the Parisian formations had extended the total sequence at the top end; in temporal terms, they prolonged the record of geohistory before whatever final event had produced the Superficial deposits.[22]

In Cantal (south of Auvergne) Brongniart and Prévost were joined by Desmarest's son Anselme-Gaëtan (1784–1838), who showed them what Brongniart identified as the Freshwater Limestone, complete with its Parisian fossils. Later, in Auvergne, they toured the extinct volcanoes that the elder Desmarest had made famous thirty years before (§4.3) and which were back in the spotlight as a result of his recent full publication of that work (§8.3). Here they found further broad tracts of the Freshwater Limestone, capped in places by some of the ancient lava flows. So the distinctive formation that topped the sequence in the Paris Basin (see Fig. 9.3) now appeared to be widely distributed in France.[23]

Controversial freshwater fossils

Brongniart's joint paper with Cuvier was published in the Muséum's *Annales* and in the *Journal des Mines*, just around the time that he returned to Paris, well satisfied with his fieldwork and with his young companions. His first priority was then to complete the mass of local details—and particularly the lists of identified fossils—that would substantiate their main conclusions. Before the whole work could be published, however, he had to attend to the Achilles' heel of his and Cuvier's argument. Their geohistory of the Paris Basin depended crucially on the validity of their inferences about the fossils that they claimed were indicators of freshwater conditions. If those fossils could not be relied on in that way, their concept of alternating marine and freshwater periods would be in jeopardy.

That interpretation was just what Faujas hoped to undermine, in order to salvage his own alternative explanation in terms of the slowly changing salinity of the world's oceans (§9.1). He had hitched that notion to the fossils found near Mainz, but a comparable study of those nearer home would challenge his rivals' ideas more effectively. Having just been assigned a junior assistant [*aide-naturaliste*] at the Muséum, Faujas set him to work on that very topic. The young Parisian naturalist Cyprien Prosper Brard (1786–1838) duly produced the first installment of his first publication, in the Muséum's *Annales* for 1809, supplementing Lamarck's work by giving a detailed description of the contentious fossils. Brard claimed that these mollusks "never lived in the great freshwater lakes that—as has been supposed—existed in these same places at very remote times". He thought it more likely that the deposits were "the results of a great diluvial quake [*secousse*]" that had mixed up shells of all kinds.[24]

Prévost and the younger Desmarest sprang to the defense of their mentor—in an equally oblique manner—in a joint paper read to the Société Philomathique, which was continuing to function as a lively forum for younger savants; printed in the *Journal des Mines*, it formed their own first publication. Studying the Gypsum formation in the Montmartre quarries in detail, bed by bed, they found marine shells in its lowest part and freshwater ones in the highest. These finds confirmed and refined the geohistorical interpretation that Brongniart and Cuvier had given to the gypsum, by locating more precisely the freshwater episode that came between the end of the previous marine period (represented by the younger naturalists' new discovery) and the start of the next (the thin bed with marine shells, Fig. 9.2). Less detailed fieldwork or less careful collecting might have led to the false conclusion that the formation contained a *mixture* of marine and freshwater shells; Prévost and Desmarest claimed that in fact they were found in quite distinct parts of the sequence. The clear implication was that Brard's other alleged cases of mixing might likewise be resolved by more accurate fieldwork.[25]

Tacitly rejecting Brard's claims, Brongniart therefore focused on what he and Cuvier had boldly called "freshwater formations" [*terrains d'eau douce*]. In the light of his recent fieldwork these formations seemed more significant than ever and might even prove decisive. In 1810 he published in the Muséum's *Annales* a substantial paper on these rocks and especially their fossils. He described in detail how there were *two* distinct freshwater formations in the Paris Basin, separated by marine strata (see Fig. 9.3), which implied at least two distinct periods of freshwater conditions. He emphasized that "the singularity of this succession" could easily be checked, for "this phenomenon is exposed to the eyes of all the distinguished savants who live in or visit one of the most enlightened cities in Europe, at the very gates of Paris [i.e., at Montmartre]". Brard was dismissed in a footnote.[26]

Brongniart forestalled any suggestion that the freshwater formations might just be the relics of "marshes dried up in early historical times" by emphasizing that they were overlain by marine strata (around Paris) or by ancient volcanic rocks (in Auvergne) and therefore clearly belonged to the prehistoric "former world". He concluded that the constancy of the assemblage of fossil mollusks, and their close similarity to those now living exclusively in fresh water, was more than enough to justify the name that he and Cuvier had given to these formations. The apparent mixture inferred from Lamarck's museum work was resolved when tested in the field: Brongniart claimed that at Grignon the freshwater and marine shells were not in fact found in the same beds. And he argued that the vast extent of these formations—as revealed by his fieldwork in the Massif Central and by published reports from other regions—was no reason not to attribute them to fresh water, since in the present world the Great Lakes of North America were of comparable size. As usual, the present was the best key to the deep past.[27]

22. Brongniart, notebooks for 17 April to 18 June 1808 (Paris-MHN, MS 2341), notes for 17 April, 6, 21–25 May; Brongniart to Cuvier, 27 April 1808, cited in Launay, *Les Brongniart* (1940), 112.

23. Brongniart, notebooks for 17 April to 18 June 1808 (Paris-MHN, MS 2341), notes for 26 May to 18 June. The elder Desmarest had seen the same strata but had assumed that they were marine in origin and had not given their fossils any special attention (§4.3).

24. Brard, "Coquilles fossiles du genre Lymnée" (1809), 440.

25. Prévost and Desmarest (A.-G.), "Corps marins à Montmartre" (1809).

26. Brongniart, "Terrains d'eau douce" (1810), 362–64.

27. Brongniart, "Terrains d'eau douce" (1810), 397–405.

When Brongniart turned to detailed descriptions of the freshwater shells that characterized these formations, he used Lamarck's work as a starting point, but he ignored what Faujas's assistant had begun to do and in effect duplicated Brard's systematic description of the controversial fossils. However, he defined and named a new genus *Potamides* ("river dweller"), which he had found in the Freshwater Formation not only around Paris but also far away in Cantal and Auvergne. But he gave that name to shells that, as he admitted, were indistinguishable from Bruguière's marine genus *Cerithium* [*cérites*], except that these ones were found in beds with putatively freshwater shells (*Lymnaea*, *Planorbis* etc.) and not in those with the usual marine ones (Fig. 9.4).[28]

The circularity of Brongniart's argument was all too obvious, making the validity of the genus dubious and the ecological interpretation questionable. Brard lost no time in questioning it: the old hostility between Cuvier and Faujas seemed set to continue between their proxies. Brard pointed out the circularity in Brongniart's

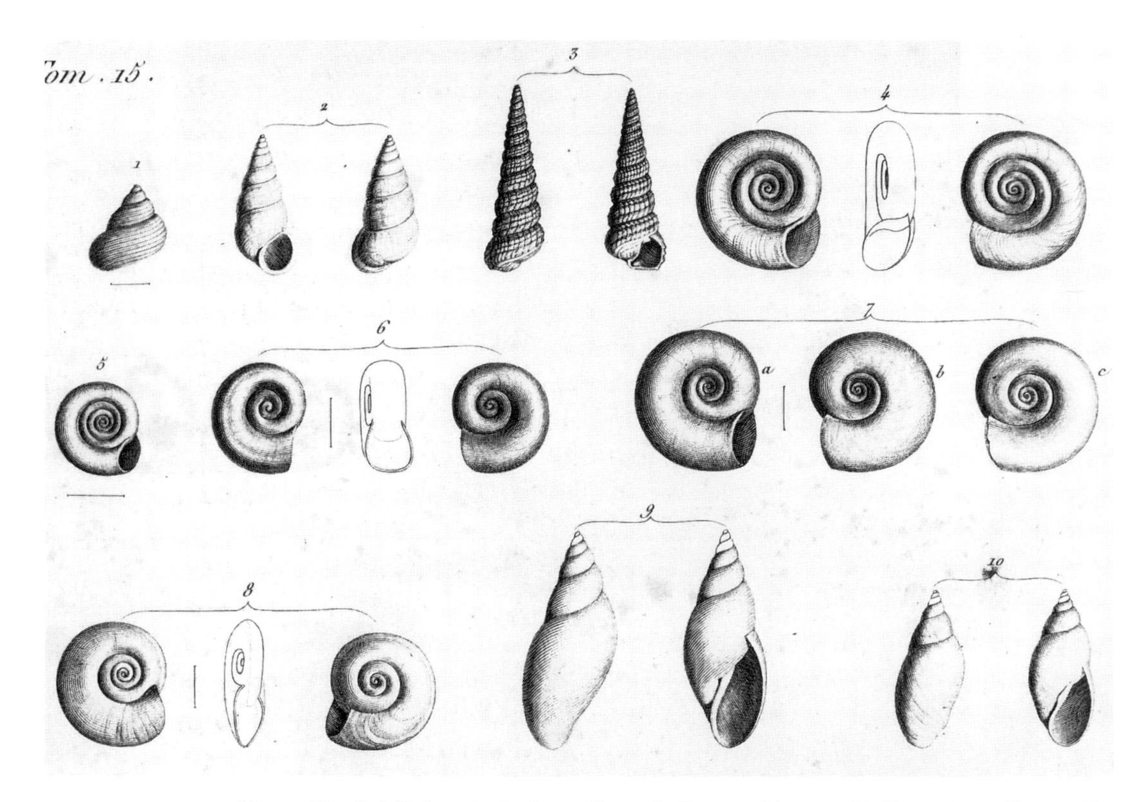

Fig. 9.4. "Fossil shells from the Freshwater Formation": some of those used by Brongniart to illustrate his paper (1810) on the formations that he and Cuvier attributed to freshwater periods in the geohistory of the Paris region. The smooth and delicate shells that he identified as *Cyclostoma* (figs. 1–2), *Planorbis* (figs. 4–8), and *Lymnaea* (figs. 9–10) were closely similar to those found living in fresh water at the present day. But Brongniart's new genus *Potamides* (top center, fig. 3) was quite different in appearance, and much more problematic, because he could not distinguish it from the marine genus *Cerithium* by any characteristic *except* its association with these other shells; to give it another name therefore begged the question of the putatively freshwater environment in which the formation had been deposited. He named the controversial species *Potamides lamarcki*: a dubious compliment, since Lamarck would hardly have approved what Brongniart was doing in his own domain of systematic conchology. (By permission of the Syndics of Cambridge University Library)

definition of *Potamides* and proposed a rival interpretation, both of the fossils themselves and of the environmental geohistory that they recorded. His was a "hypothesis of one and the same liquid", in which *all* the Parisian fossil mollusks had lived and all the formations had been deposited; he contrasted this with Brongniart's claim that there had been "a double concourse of two different liquids [that] came and returned several times". In other words, young Brard had the temerity to reject the environmental criterion on which Brongniart's (and Cuvier's) entire reconstruction of a highly eventful geohistory for the Paris region was based. La Métherie joined in the argument on Brard's side; his admonition that "the discussion ought to be *calm*, made in *good faith*, and *without jest*" suggests that this was just what it was not.[29]

Brard argued that the supposedly freshwater formations lacked the bivalve mollusks (for example, freshwater mussels) characteristic of freshwater faunas in the modern world; and he denied that the fossils found in these formations were true and exact "analogues" of their modern counterparts. His main argument, however, was that, once the problematic *Potamides* was recognized as a synonym for the marine *Cerithium*, the allegedly freshwater shells were always found mixed with marine ones, so that all must have lived in "a unique global [*général*] fluid, of a flavor unknown to us [i.e., in the present world]". As his work unfolded it became clear that Brard had adopted Faujas's notion that the composition of the earth's oceans might have changed gradually in the course of geohistory. He argued that the truly freshwater mollusks of the present world might have diverged equally gradually in the course of time from related but significantly different ancestors; the latter would have lived in seas of unknown chemical composition, alongside those mollusks that were ancestral to the species found in modern seas.

In other words, Brard suggested a gradual differentiation of two distinct faunas from a common earlier one. As evidence for this kind of environmental lability he cited recent experiments in which living freshwater mollusks had been artificially adapted to increasing salinities. So Brard solved the riddle by invoking a process of gradual transmutation; unsurprisingly, his argument had the support of Lamarck as well as that of Faujas and La Métherie. The publication of Lamarck's *Zoological Philosophy* at just this time, with its full-length exposition of his theory of transformism (see §10.1), ensured that the wider implications of Brard's study did not go unnoticed. Whereas Lamarck questioned the stability of species within a wide-ranging natural philosophy, Brard's study was a more focused one that might have undermined Brongniart's and Cuvier's reconstruction of relatively recent geohistory. But in the event it did not: Brard was young and without great influence, and not long afterwards he was appointed to a position in the Mining Corps and left Paris for a career in the south of France.[30]

28. Fig. 9.4 is reproduced from Brongniart, "Terrains d'eau douce" (1810) [Cambridge-UL: Q382.b.11.15], part of pl. 22. He treated his "*potamide de Lamarck*" as superseding Brard's name "*cérite tuberculée*" for the same shells (368): modern rules for priority in zoological nomenclature were not yet in force.

29. Brard, "Lymnées fossiles" (1810), 421; La Métherie, "Observations sur les terrains" (1811), 470.

30. Brard, "Troisième mémoire" (1811) and "Quatrième mémoire" (1812); after he left Paris in 1813, his *Coquilles terrestres et fluviatiles* (1815), on the living species of the Paris region, repeated his earlier interpretation

A more weighty reason for Brard's objections to be brushed aside by his elders (and, in their own estimation, betters) was that a small but decisive piece of new evidence soon made their freshwater interpretation—and with it their reconstruction of an eventful geohistory—more plausible than ever. Lamarck had given the name *gyrogonite* to a tiny fossil (about the size of a pinhead) that was found in huge numbers in the alleged Freshwater Formation around Paris. He had tentatively identified it as some kind of miniature mollusk. When Brongniart and his companions found it in the apparently equivalent formation in the Massif Central it became still more clearly "characteristic". In 1810 the younger Desmarest described it in detail to the Société Philomathique, as a modest appendix to Brongniart's paper, but its affinities were too obscure for it to be used on either side of the argument about the environmental origin of these rocks.

It was only in 1812—just too late to feature in the monograph by Brongniart and Cuvier—that Desmarest's brother-in-law, like him an amateur naturalist, noticed that the *gyrogonite* was closely similar to the calcified fruits of the living *Chara*. It was of course conceivable in principle that this freshwater plant might have been transformed in habitat in the course of time, as Brard had argued in the case of the controversial mollusks; but it was far simpler to infer that like them it was truly an indicator of truly freshwater conditions, in the former world as much as in the present one. Desmarest incorporated this discovery in a revision of his earlier paper; when this was published in the *Journal des Mines* with illustrations of the fossil and its close "analogue" from the modern world, the identification became almost irrefutable. With the *gyrogonite* as nature's diminutive witness, the interpretation of the Freshwater Formation as indeed freshwater in origin became almost impregnable (Fig. 9.5).[31]

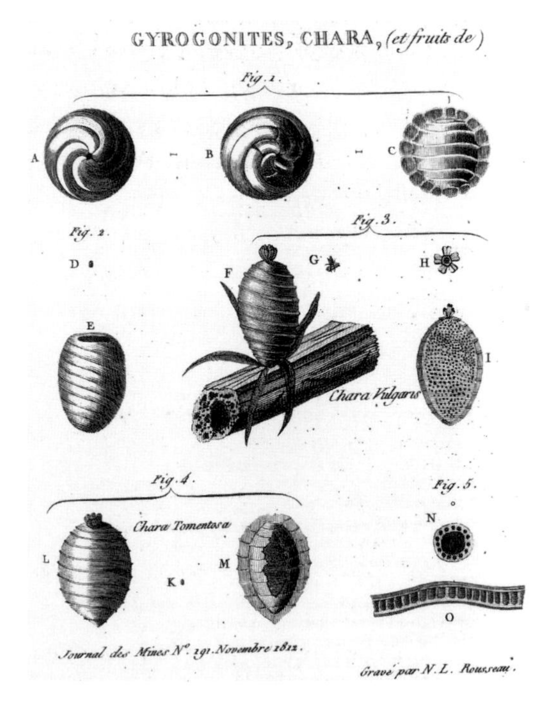

Fig. 9.5. The previously enigmatic fossil *gyrogonite*, which was abundant in the Freshwater Formation of the Paris Basin and elsewhere, compared with the calcified fruit of the living freshwater plant *Chara*, as illustrated by Anselme-Gaëtan Desmarest (the son of the veteran naturalist Nicolas Desmarest) in his report establishing their generic identity (1812). The engravings show two views of the hollow spherical fossil and a section through it (fig. 1 A,B,C); a similar but elliptical fossil from Dalmatia, redrawn from an earlier picture by Fortis, at natural size and enlarged (fig. 2 D,E); the fruit of the living *Chara vulgaris* on its stem, with its five sepals, enlarged and at natural size, and a section through it (fig. 3 F,G,H,I); a living *Chara tomentosa* shaped more nearly like the fossil forms, at natural size, enlarged and in section (fig. 4 K,L,M); and Brongniart's picture of a section through the Parisian fossil, redrawn at a lesser enlargement, with a piece of the tubular fossil stems found in the same rocks but not previously identified as parts of the same plant (fig. 5 N,O). Apart from those that showed the real size of the tiny objects (D,G,H,K), all the drawings must have been made with a low-power microscope. (By permission of the Syndics of Cambridge University Library)

The Parisian case in full

Cuvier had left Brongniart on his own to defend their joint interpretation, and indeed to revise their work for publication, for he himself was far away in Italy. As Napoleon continued to enlarge his empire by conquest and annexation, he had designed an Imperial University that would bring all its higher education under centralized French control. Cuvier, who had proved himself an effective administrator, was appointed to oversee the integration of the existing universities outside France into this new structure; it was a task analogous to his earlier one, which had prevented him from visiting England during the short-lived Peace (§7.5), but applied to a higher level of education and on a broader stage. Whatever his reasons for serving Napoleon's illiberal regime and adding to his own overcrowded professional duties—they seem to have included the hope that he could protect and strengthen the Enlightened character of the foreign universities—Cuvier's new position certainly gave him welcome opportunities for extensive travel at state expense. Like Brongniart in his search for sources of kaolin, Cuvier's educational appointment was employment on the back of which he could hope to further his own scientific research.[32]

Cuvier's first mission, which took him away from Paris the year after his and Brongniart's joint work was first made public, was to the puppet kingdom of Italy (nominally ruled by Napoleon's stepson). He duly used the opportunity to enlarge his paper museum of proxy fossil bones by visiting Italian collections and making accurate drawings of their more important specimens. Writing to Brongniart with news of what he was doing, Cuvier thanked him for taking care of "a work that I no longer dare to call ours, for you will soon have made it wholly your own". It is not too cynical to detect behind this unusually generous comment a discreet hint that he would in future withdraw from any further detailed involvement with this kind of research. Having put the stamp of his authority on an exemplary piece of doubly enriched geognosy, he might well have felt that he could leave to Brongniart and others the further development of the research agenda that he himself had suggested. In the same letter from Florence, he promised to give the necessary time to "the study of outcrops [*gisements*]"; but it is not clear whether in fact he did any fieldwork while in Italy, whereas he certainly did a lot of work where his heart lay, namely in museums full of fossil bones.[33]

---

of the fossil shells (13–18), but by that time it carried little weight. The anomaly posed by Brongniart's *Potamides* was later resolved by the discovery that similar mollusks in the modern world were unusually tolerant of a wide range of salinities, and it became a reliable indicator of brackish conditions.

31. Fig. 9.5 is reproduced from Desmarest (A.-G.), "Sur la Gyrogonite" (1812) [Cambridge-UL: CP432.c.16.32], pl. 8, engraved by Nicolas Louis Rousseau; see also Léman, "Sur la Gyrogonite" (1812). Desmarest inferred that the engraver of Fortis's picture, in *Viaggio in Dalmazia* (1774), pl. 7, figs. 8, 9, had failed to reverse the image, which had therefore been printed with the spiral in the wrong direction, obscuring its similarity to the Parisian fossils.

32. See Negrin, *Georges Cuvier* (1977), 334; Outram, *Georges Cuvier* (1984), 79–80.

33. Cuvier to Brongniart, 14 January 1810 (Paris-MHN, MS 1997, 1: 87–88). Cuvier's research files on fossil bones (in Paris-MHN) contain drawings and notes on specimens from his time in Italy, but no field notes.

Brongniart greatly amplified their joint paper with lists of fossils and local details; he completed its map (Fig. 9.1), and designed the "ideal" section that summarized their conclusions in visual form (Fig. 9.3). He also added a remarkable series of sections, which were based on accurate barometric measurements of the topography, of just the kind that Lavoisier had planned to make in the service of his own interpretation of these rocks before he became a victim of the Terror. Since the strata were usually almost horizontal, the sections were drawn with a bold and unprecedented degree of vertical exaggeration. The loss of verisimilitude was more than compensated by the depiction of thin and variable beds with an accuracy that was likewise almost unprecedented. The sections displayed the sheer detail of the authors' evidence in the most impressive way imaginable (Fig. 9.6).[34]

Brongniart and Cuvier presented the whole work at the Institut in 1810, and it was published in its prestigious *Mémoires* the following year (and also in book form). With some adjustment of boundaries and definitions, their nine formations had now become thirteen. The relation of two of the formations as lateral equivalents, which they had earlier found so puzzling, was confidently confirmed and shown as such on their ideal section (Fig. 9.3), though they still gave it no clear explanation in terms of geohistory. On the other hand, Brard's objections (not yet fully in print) to their inference about the beds with freshwater molluscan shells were in effect ignored, though not without reason. Detailed collecting of the kind that Prévost and the younger Desmarest had done at Montmartre, and Brongniart

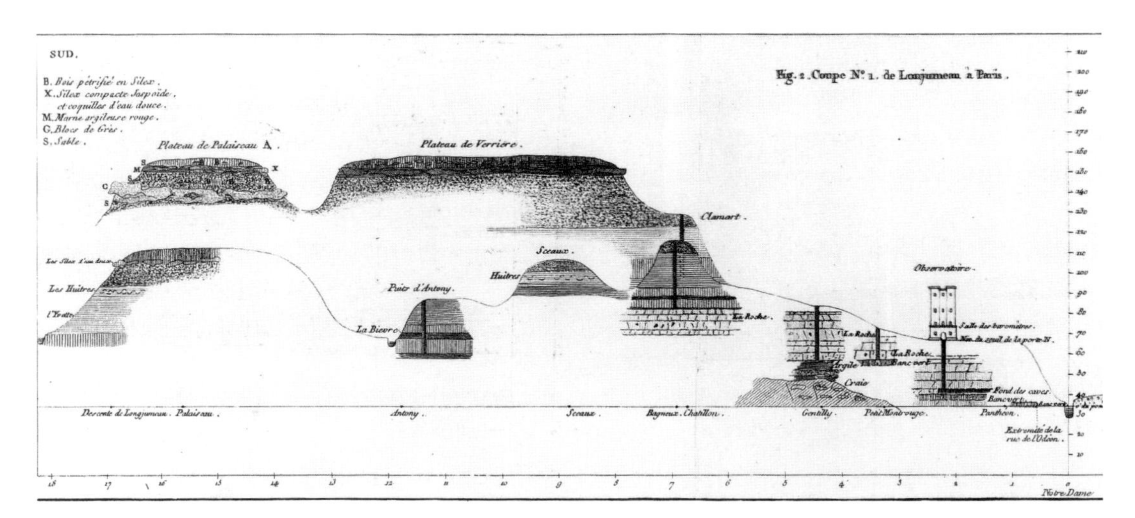

Fig. 9.6. Part of one of the accurately measured traverse (in modern terms "horizontal") sections that illustrated the full version of the monograph on the Paris Basin published in 1811 by Brongniart and Cuvier. The dramatic vertical exaggeration—by a factor of 35—allowed much detail to be shown. This part of the section runs from the bed of the Seine in the center of Paris (far right), through the famous observatory—the reference point for all the barometric leveling—and southwards to the plateau above Longjumeau (left). The section shows the complex sequence of beds, particularly in the shafts of wells; where the rocks were not known the section was left blank. The lowest formation is the subterranean Chalk [*craie*] at Gentilly; oysters [*huîtres*] are marked about halfway up; and at the top are "petrified wood" (B) and "freshwater shells" (X). Annotations on other parts of this section showed much further evidence for the authors' geohistorical interpretation of the strata as the record of an *alternation* of marine and freshwater conditions. The vertical scale is in meters, the horizontal in kilometers, both measured from the Seine at Notre-Dame. (By permission of the Syndics of Cambridge University Library)

himself at Grignon, had eliminated most of the alleged cases of mixtures of marine and freshwater shells, so that the evidence in favor of several periods of truly freshwater conditions seemed increasingly persuasive. But the earlier inferences about the ancient environments were repeated without further elaboration. And at the very end of the paper, after suggesting briefly one final reconstruction, the authors—or more probably Brongniart—seemed to draw back from any such exercise: "These scenes [*tableaux*] of what our ancient landscape [*sol*] must have been like pander too much to the imagination; they would lead us in spite of ourselves to violate our self-imposed rule, to describe only the facts." Such caution suggests that geohistorical reconstruction was still tainted by a perceived association with speculative geotheory.[35]

As soon as their work was fully launched, Brongniart used his next field trip to consolidate its wider significance. His official purpose was once again to explore sources of kaolin for Sèvres, this time in Normandy; but once again he expected to see much more at the same time. Just before leaving Paris he sent Cuvier an advance copy of their monograph, hot from the press, and told him how his forthcoming trip "should be rich in facts and observations instructive for me and useful for the continuation of our work". Whatever Cuvier's intentions may have been, Brongniart evidently expected, or at least wanted, the collaboration to go further. He told Cuvier that he hoped to be able to see at different localities the "passage" from the Parisian formations to the Chalk, and thence down through the hard Secondary limestones with ammonites, all the way into the shales and sandstones of the Coal formation, and "indeed finally on to the granite" and other Primary rocks. This shows again how he intended to use their work on the Paris Basin as an exemplar for much broader research, applying its enriched geognosy—much as Smith was doing at this time in England and Wales (§8.2)—to the whole pile of formations from the most recent back to the most ancient. For example, even this brief summary indicated how, with this fieldwork, the strange fossil plants of the Coal formation, as described by Schlotheim and Parkinson among others (§8.1, §8.2), might be set more firmly in their proper place in the geognostic sequence, in relation to the formations with ammonites and those of the Paris Basin.[36]

Reactions outside France

Meanwhile Britain remained out of bounds for French naturalists, and vice versa. Yet, as emphasized repeatedly in this narrative, communication continued in spite

34. Fig. 9.6 is reproduced from Cuvier, *Ossemens fossiles* (1812), 1 [Cambridge-UL: MD.8.65], "Géographie minéralogique", pl. 1, fig. 2, part of "Coupe No. 1", first published in Cuvier and Brongniart, "Géographie minéralogique" (1811). The extension of this section to the north (right), through Montmartre, is reproduced in Rudwick, *Georges Cuvier* (1997), 148. Leaving unknown parts of the section blank, just as they had done on their map, reflected their scrupulously careful methods of inference; Cuvier's use of dotted lines on his reconstructions of the fossil mammals from the gypsum (Fig. 7.22), probably drawn at about the same time, showed the same kind of caution. The barometric survey was done after the paper was first read in 1808, so it is unlikely that they displayed even a draft version of these sections on that earlier occasion.

35. Cuvier and Brongniart, "Géographie minéralogique" (1811), 278.

36. Brongniart to Cuvier, 12 September 1811 (Paris-IF, MS 3318/29); Brongniart, field notebooks, 14 September to 8 October 1811, and official instructions (Paris-MHN, MS 2342, 2351/6).

of the wars, at least in publication and even in correspondence. The striking new study of the Paris Basin immediately attracted international attention. Brongniart and Cuvier probably sent offprints of their preliminary article to favored correspondents, and anyway both the *Journal des Mines* and the *Annales du Muséum* were continuing to reach other countries. For example, the article was picked up (from the *Annales*) by the *Philosophical Magazine* in England, where it was published in translation and then reviewed by Smith's self-appointed bulldog Farey, vigilant as ever to ensure that his colleague was given due recognition (§8.2). Farey had good reason to want to be seen as championing Smith, because the latter was indignant that Farey had recently gained a valuable commission that Smith had assumed would be his. The government's Board of Agriculture had in effect despaired of Smith's ability to deliver anything on time and had therefore asked Farey to report on the mineral and agricultural resources of Derbyshire, which might be just the first in a series of such county surveys.[37]

The French naturalists had mentioned in their paper that "certain circumstances" had impelled them to present it at the Institut in preliminary form, in advance of its definitive version. Farey jumped to the conclusion that they had heard of Smith's progress with his map and wanted to stake their claim to priority before it appeared. In the chauvinistic atmosphere generated by the war, the charge might have seemed plausible. But there was a much more likely reason for their haste, which academic protocol prevented them from stating openly. Napoleon's Imperial University was to be represented in Paris by a Faculté des Sciences, for which the professors were about to be chosen. Brongniart was a candidate for the chair of mineralogy, but he urgently needed to have some important scientific work in the public realm before the decision was made (in the event he was appointed).[38]

On the other hand Brongniart and Cuvier had indeed claimed too much when they asserted that their map was "a first [*sic*] attempt at mineralogical maps in which each kind of formation is highlighted by a particular color". Among the important German works that Cuvier had been told about while he was preparing his report for Napoleon (§8.4) was Charpentier's classic book on the geognosy of Saxony, and he—or at least Brongniart—must surely have seen its closely similar map (Fig. 2.13); a superb new map of the Alps in the same style, by the Swiss geognost Johann Gottfried Ebel, may have reached Paris just too late. On the other hand, although Brongniart may well have heard about Smith's map while he was in London, or even seen it in draft form (§9.1), an unpublished work counted for as little in priority stakes at this time as it does in the modern scientific world. Again, on the more important issue of the use of "characteristic fossils" in relation to formations, Brongniart may have heard about Smith's practice while he was in London, and certainly he and Cuvier followed it, knowingly or not. But as emphasized already, they could well have developed the idea independently, simply by refining what earlier naturalists such as Arduino and Soulavie, and more recently de Luc, had already done more locally or in more general terms. And anyway, as has also been emphasized, they took the idea on to a quite different plane, far beyond Smith, by using fossils as indicators of ancient environments, not just of particular formations.[39]

Claims to priority apart, Farey reviewed the substance of the French work quite fairly, after introducing it with faint praise as "the hasty but valuable Memoir by our able and industrious neighbours on the continent". He noted that Brongniart and Cuvier had defined the Parisian formations above the Chalk by their "peculiar fossils", just as Smith had used the same method to distinguish those below that distinctive marker. He put the French formations into proportion by noting that the lowest of those that he himself was currently mapping in Derbyshire lay some *three miles* (about 5,000m) vertically below the Chalk. Such was the magnitude of the pile of formations in England that both he and Smith were mapping, relative to which the Parisian ones were just a thin upper crust (as Brongniart, after his recent fieldwork, would surely have agreed).

More importantly, Farey suggested that the French rocks might be represented in England, if the London Clay and other English formations above the Chalk turned out to be "the Paris strata in a modified state". Brongniart had identified one of the Parisian formations in the Massif Central; Farey was now suggesting that they might also be recognized across the Channel and thus have even wider significance. In fact the fruitfulness of the French work as an exemplar of a new research agenda now became clear: Farey suggested that the members of the recently founded Geological Society (§8.4) should investigate these younger formations in England and try to compare them with the French. He pointed out that English conchologists would need to compare their fossils with those described by Lamarck, and emulate the accuracy and detail of his work. On the other hand, Farey was less convinced by the French naturalists' geohistorical interpretation. He doubted that the Paris gypsum could have been precipitated in fresh water; like Brard he questioned whether the associated fossils could be trusted to indicate such conditions: "has it or can it be proved, that [shell]fish nearly or exactly resembling . . . those that are *now* peculiar to either *fresh-* or *salt-water*, may not have had other powers and habits in the *old world*?"[40]

Communication between savants in the warring countries continued not only by exchange of publications but also more directly by letter. De Luc, for example, used his brother in Geneva to bypass the Continental blockade, and he wrote to Brongniart to comment on the Frenchman's paper on the freshwater formations. He claimed that it was compatible with his own conception of them, and even that it reinforced his "system". Unlike Farey he accepted the genuinely freshwater origin of the deposits, but he assumed that they had been formed on islands in the ancient oceans rather than in any "basin" within an ancient continent. It was evidently inconceivable to him that they could be truly intercalated with other deposits of

37. Farey, "Cuvier and Brongniart's memoir" (1810), and *Derbyshire* (1811); see also Ford and Torrens, "John Farey" (1989), and Torrens, "Banks and the earth sciences" (1994).

38. On Cuvier's role in the university and faculté, see Outram, *Georges Cuvier* (1984), 80. Another possible reason for their haste, namely a wish to forestall Coupé, is suggested by Gaudant, "L'exploration géologique du Bassin Parisien" (1993), 24.

39. Ebel, *Bau der Erde in dem Alpengebirge* (1808), map 2, of the western Alps, in ten colors.

40. Farey, "Cuvier and Brongniart's memoir" (1810), 134. The far older formations that he alluded to were parts of the Mountain (later, Carboniferous) Limestone, low down in the Secondaries: Farey, *Derbyshire* (1811).

marine origin, as Brongniart and Cuvier claimed, or that they could represent periods of freshwater conditions *alternating* with marine periods. Even allowing for the octogenarian's increasingly inflexible outlook—he summarized his system in terms almost unchanged from twenty years earlier (§6.2)—de Luc's reaction does underline the innovative character of what Brongniart and Cuvier were doing. Anyway, Brongniart replied courteously, assuring de Luc that the belated publication of the latter's *Travels*—which, de Luc insisted, contained the empirical foundations for his system—would indeed be valuable for geology; and he promised to send a copy of the full version of the Paris monograph. In practice, however, de Luc's views were no longer being taken seriously by younger naturalists, or even by the now middle-aged Brongniart and Cuvier. Having been central to savant discourse twenty or thirty years earlier, de Luc was now being left on the sidelines; he died a few years later, in 1817, at the ripe age of ninety.[41]

Like their preliminary report, the full version of the Paris Basin monograph by Brongniart and Cuvier gained international attention; published in the Institut's *Mémoires*, it reached all the other learned academies throughout Europe. It was made more readily accessible in the German lands when Ludwig Wilhelm Gilbert (1769–1824), the professor of physics at Leipzig, published a substantial summary in his prestigious *Annalen der Physik*. In his editorial introduction, Gilbert noted perceptively that the French research had now made the "youngest Secondary formation" [*jüngste Flötzformation*] even more interesting and important than the better known older ones: "it informs us of a series of Secondary strata [*Flötzgeschichten*] about which hitherto we knew next to nothing". Simply as geognosy—leaving aside its innovative use of fossils—it extended the known sequence of formations upwards, and hence went some way towards closing the gap between them and the present world.[42]

Parkinson's new look

When Farey urged that the Parisian research be extended to England, Parkinson, one of the Geological Society's founders, was probably among those he had in mind. The first volume of Parkinson's *Organic Remains* (1804) had been extremely old-fashioned by Continental standards (§8.2), but it had sold well enough among British fossil-collecting amateurs for him to publish a second volume in 1808, illustrating and describing a variety of fossil sponges, corals, and other lowly animals. The general character of the work was almost unchanged, and he admitted that it had no overall plan. His third and final volume (1811) completed his survey of familiar fossils by illustrating a selection of sea urchins, mollusks of all kinds (including ammonites and belemnites), trilobites, and a few common vertebrate fossils such as sharks' teeth and mammoths' molars. However, tucked in among all these British specimens were redrawn versions of Bru's megatherium and Faujas's Maastricht jaws (Figs. 7.3, 7.7). These signaled a new input from Continental sources, despite the war. Right at the start of his third volume Parkinson explicitly credited Cuvier and Lamarck for his own dramatically revised views on geology; the reference was to their papers in the Muséum's *Annales*, on vertebrate and invertebrate fossils respectively, which Parkinson had now read with care and digested thoroughly.[43]

Parkinson praised Lamarck for his systematic fossil conchology, which made it possible to give reliable names to his own fossil shells (he would probably have found Lamarck's theorizing about transformism abhorrent had he known of it). And he praised Cuvier for revealing a former world of fossil quadrupeds different from those of the present, not least because this provided a bulwark against the eternalistic theorizing of the unmentionable Hutton. After describing the work of both French naturalists at length, he concluded that "the formation of the exterior part of this globe, and the creation of its several inhabitants, must have been the work of a vast length of time, and must have been effected at several distant periods". The comment was utterly unremarkable by Continental standards, but it expressed a striking change from Parkinson's highly traditional interpretation only a few years earlier, and it is one that was clearly due to his belated discovery of French research. However, his outlook remained traditional in other ways: the new perspective of a lengthy geohistory was equated—in a way that de Luc would have approved (§6.4)—with the successive "days" of the Creation story in Genesis: "it becomes only necessary to consider these periods as occurring at considerable indefinite lengths of time, to provide an exact agreement between that particular history [in Genesis] and those phenomena which appear on examining the stratification of the earth".[44]

The accommodation of Genesis to a long timescale was itself highly traditional (§2.5), thus allowing Parkinson—and others in Britain who shared his outlook—to pursue their studies of fossils without fear of being suspected of atheism or political sedition. Parkinson's final volume at last alerted anglophone readers to almost two decades of distinguished French work on fossils, and might in the future help to narrow the conspicuous gap in standards between British naturalists dealing with fossils and their counterparts on the Continent. However, Parkinson based his summary of the *sequence* of fossils not on any foreign work but on Smith's, though his knowledge of Smith's list of formations came at second hand from what Farey had just described in his report on the geology of Derbyshire.[45]

By the time the last volume of Parkinson's *Organic Remains* appeared, the first volume of the Geological Society's new *Transactions* (1811) had just been published. Its preface set out the policy on which the society had been founded, of encouraging collaboration in the collection of empirical "facts" that could then be

41. De Luc to Brongniart, 10 July 1811, and Brongniart to de Luc, 6 December 1811 (Paris-MHN, MS 1965/182 and 182a [copy]); de Luc, *Geological travels* (1810–11).

42. Cuvier and Brongniart, "Versuch einer mineralogischen Geographie" (1813), with Gilbert's introduction (229–32) and a redrawn version of their ideal section. The terminology of geognosy was still fluid and imprecise: Gilbert treated *all* the "regular" Parisian rocks above the Chalk as a single *Formation*, but recognized the varied *Geschichten* within it.

43. Parkinson, *Organic remains* 2 (1808), preface; 3 (1811), preface, pl. 19, fig. 1, pl. 22, fig. 1; on the work as a whole, see Thackray, "Parkinson's *Organic remains*" (1976).

44. Parkinson, *Organic remains* 3 (1811), 449–53. Lamarck's classification of mollusks is followed in letters 6–15; Cuvier's fossil quadrupeds are summarized in letters 18–31.

45. Parkinson, *Organic remains* 3 (1811), 441–49; Farey, *Derbyshire* 1 (1811), 108–17.

used to test theoretical "opinions" in geology (§8.4). In order to avoid the disputes that were considered to be polluting the savant atmosphere in Edinburgh, between supporters and opponents of Hutton's geotheory, the Londoners declared their society to be neutral: "In the present imperfect state of this science, it cannot be supposed that the Society should attempt to decide upon the merits of the different theories of the earth that have been proposed". Like Cuvier and others on the Continent, those with power in the Geological Society, headed by its president Greenough, regarded "systems" as generally obstructive, and certainly premature until far more empirical work had been done in the field. Anyway, their inaugural volume was a handsome production modeled on the *Philosophical Transactions*, and it was sold to members at a price to match, probably beyond the reach of "practical men" such as Smith and Farey. As in the Royal Society's volumes, there were many untranslated quotations in French, for it too was aimed at savants and well-educated amateurs.

However, in terms of its contents the Geological Society's new periodical was undistinguished. Most of the papers were conventionally mineralogical or traditionally geognostic; there were few illustrations and few maps, and none of the latter was truly geological. Most of the geognostic papers were also conventional in that they dealt with Primary rocks. Only two—neither with any map, section, or other illustration—described any Secondary rocks or mentioned any fossils. Nonetheless one of them, by Parkinson, was a straw in the wind, in that it dealt with English formations *above* the Chalk. It represented a first modest attempt to describe the English equivalents of those that Brongniart and Cuvier had described around Paris; and they in turn might then be used to extend across the Channel the French naturalists' conception of that most recent era in geohistory (see §9.4).[46]

Conclusion

Brongniart was beginning to extend his and Cuvier's doubly enriched geognosy to the rest of France and to relate the Parisian rocks and fossils to the much older formations that had long been studied by more traditional methods. He defended their interpretation of the Freshwater Formations—as fully deserving that name—against criticisms that the relevant mollusks might have changed their habitat in the course of geohistory; he insisted that on the contrary they were reliable indicators of freshwater conditions, and therefore valid evidence of a highly eventful geohistory of alternating periods of marine and freshwater conditions, within the quite recent era represented by the younger Secondary formations. This research received wide attention internationally, despite the disruptions caused by Napoleon's continuing wars. Across the Channel, for example, Farey noticed the French naturalists' work; and although he suspected that they were trying to upstage his colleague Smith, he did suggest that it might be possible to extend their research to England. Farey may also have alerted Parkinson to Continental work on fossils; certainly the latter—one of the founding members of the new Geological Society in London—began to take note of it. A later section (§9.4) will describe how Farey's hint was taken up, by Parkinson and others. First, however, this narrative returns (in the next section) to France, in order to trace the origin of Cuvier's most famous work on fossils and to illustrate how and why he assigned a crucial role to research on the younger Secondary formations, in the broader strategy of reconstructing the whole of geohistory.

9.3 CUVIER'S *FOSSIL BONES* (1809–14)

Research on fossil reptiles

Just as Brongniart extended the Parisian fieldwork into the rest of France, and Farey hinted that it might also apply across the Channel, so Cuvier used it to extend his studies of further kinds of fossil bones. By 1808, when his and Brongniart's joint paper was first read at the Institut, the Muséum had already published in its *Annales* most of Cuvier's many papers on the mammals found in the Superficial deposits, and almost all his detailed studies of the bones from the Paris gypsum (§7.5). His output in the *Annales* continued unabated, but its direction changed significantly in the very next volume. As a result of his and Brongniart's fieldwork, the fossil reptiles found in the Secondary formations could now be placed confidently at a much earlier period of geohistory than the strange mammals in the gypsum, let alone the extinct but more familiar ones from the Superficial gravels. These ancient reptiles became the subject of a new series of articles.

Continuing his earlier policy of describing living forms before their fossil relatives, Cuvier published two papers on living species of crocodiles, the second followed immediately by one on the bones of fossil crocodiles; as before, the sequence highlighted his method of using the present as the key to the deep past. He described bones from Honfleur in Normandy, a few similar ones from Whitby in England, and some strange "saurians" from Thuringia; at this stage he had to rely on proxy pictures of the foreign specimens. However, he noted that they all "belong to very ancient beds among the Secondaries", dating from long before the mammals from the Parisian gypsum beds. Cuvier shed appropriate crocodile tears at the necessity of engaging in "polemical discussions": he scornfully rejected what Faujas had written about crocodiles in his book on Maastricht and failed to withdraw in his later work. In the next paper in the same volume Cuvier again dismissed Faujas's work as worthless, giving his own more authoritative analysis of the "Maastricht animal" (see Fig. 7.7). Cuvier defined it as a giant marine lizard, acknowledging the younger Camper's earlier role in that fruitful interpretation (§7.3).[47]

The following year, the *Annales* contained a paper on two further notable fossil reptiles. Cuvier was well aware that they, and those from other famous localities in the Secondary formations such as Maastricht, could "hardly be [all] from the same

46. Geological Society, *Transactions* 1 (1811), preface dated 28 June, priced at £1 12s; Parkinson presented his *Organic remains* 3 (1811) on 20 December (London-GS, OM/1, 128).

47. Cuvier, "Espèces de crocodiles vivans" (1807); "Ostéologie des crocodiles vivans", "Ossemens fossiles de crocodiles", and "Grand animal fossile de Maestricht" (all 1808). The latter was later named (though not by Cuvier) the *mosasaur*, or lizard from the Meuse or Maas, which flows past the town.

system of beds, or have been buried at the same epoch". But he realized that their relative ages would remain uncertain until there had been far more field research, of just the kind that both Smith and Brongniart were doing: "how many comparative observations ought to be made", he exclaimed, "in order to recognize the relations of superposition between such distant beds?". From those at Oeningen—famous for their beautifully preserved plant fossils (Fig. 5.9)—came the celebrated specimen that Scheuchzer a century earlier had ascribed to "a Man a witness of the Deluge", and that van Marum had purchased for Teyler's Museum in Haarlem (§5.4). Cuvier knew it only from illustrations, but these were enough for him to confirm his suspicion that it was in fact a giant salamander. This was marvel enough, and in line with all the other giant fossil species that Cuvier had already described. But it debunked a marvel of another kind, which he dismissed with Enlightened disdain: these beds were nothing to do with any biblical Flood, and the fossil had nothing to do with any human being (Fig. 9.7).[48]

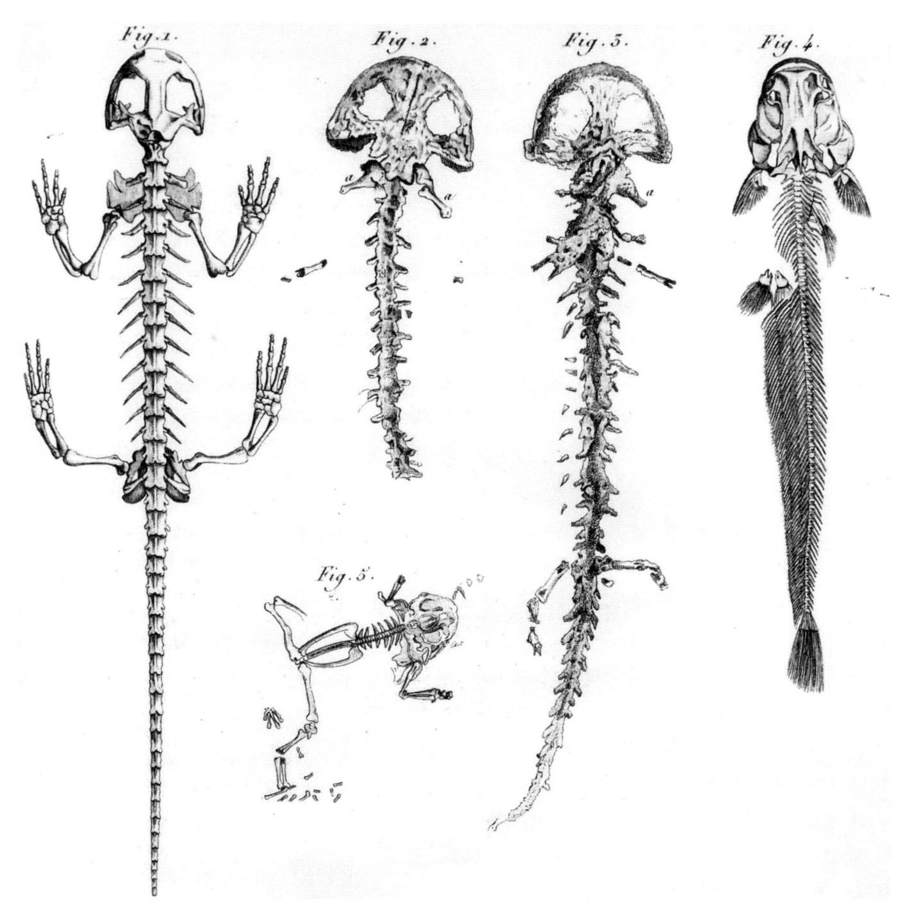

Fig. 9.7. The "Proteus of Oeningen", which the Swiss naturalist Johann Jacob Scheuchzer almost a century earlier had claimed as "a man a witness of the Deluge", but which Cuvier identified more prosaically—but still strikingly—as a giant salamander. These engravings from Cuvier's article (1809) show Scheuchzer's specimen (fig. 2) and a more complete one found more recently (fig. 3), both greatly reduced in size in order to highlight their similarities to the lifesize skeletons of a living salamander (fig. 1) and frog (fig. 5); the skeleton of a fish (fig. 4) is included to show its dissimilarity. (By permission of the Syndics of Cambridge University Library)

Cuvier soon saw this famous specimen for himself, although just too late for inclusion in his paper. His second educational mission, a year after the one to Italy, was to the Netherlands—now another puppet state, with Napoleon's brother as its king—and to the allied or annexed German territories along the Rhine. He was certainly impressed by the quality of their existing educational systems (among the institutions he inspected was his own alma mater in Stuttgart, which he revisited for the first time since he was a student). As in Italy, he took the opportunity in his free time to study fossil bones in public museums and private collections, where all his tact was needed to assure their owners that he was not about to appropriate their best specimens for Paris, as Faujas had done during the Revolution (§7.1). Most notable was his visit to Teyler's Museum in Haarlem, where he staged a performance similar to his famous risky prediction with the fossil opossum from the Parisian gypsum (Fig. 7.20). This time, with van Marum's approval and in his presence, he excavated the precious specimen from Oeningen, with the skeleton of a salamander alongside for guidance and comparison. He revealed the well-preserved forelimbs of the fossil for the first time and thereby confirmed his earlier conclusion that it was indeed a giant salamander.[49]

Another specimen, almost equally celebrated, was known to Cuvier only by proxy. It had been described a quarter-century earlier by the Florentine naturalist Cosimo Alessandro Collini, the then curator of the princely museum at Mannheim (and previously Voltaire's private secretary). Since it came from the limestone near Solnhofen in Bavaria, which was famous for its superbly preserved marine fossils (Fig. 2.20), Collini had assumed that this too was an unknown marine creature. Unfortunately, the specimen had been lost or mislaid when, during the wars, Mannheim's museum was moved to Munich; and Collini himself had died in 1806. Still, his superb engraving was sufficiently accurate to enable Cuvier to analyze the bones of the enigmatic animal in detail. He rejected Blumenbach's conclusion that it was a bird and other suggestions that it was a bat. He claimed instead that, like Scheuchzer's supposed victim of the Flood, it was in fact a reptile, but—sensationally—a flying form, which he named the "*ptero-dactyle*" (wing-fingered animal): he claimed in triumph that it confirmed the validity of his anatomical laws, even in "this inhabitant of a world so different from ours" (Fig. 9.8).[50]

48. Fig. 9.7 is reproduced from Cuvier, *Ossemens fossiles* (1812), 4 [Cambridge-UL: MD.8.68], part 5, 5e mém., part of unnumbered pl., first published as "Quadrupèdes ovipares" (1809), pl. 30. The older specimen was copied from the large engraving in Scheuchzer, *Homo diluvii testis* (1726); the newer, from Karg, "Steinbruch zu Oeningen" (1805), pl. 2, fig. 3 (Karg, a local naturalist, thought it a fish). By modern standards, the Secondary localities mentioned by Cuvier were, as he suspected, highly diverse in age, ranging from the Miocene of Öhningen down to the Permian of the Thuringian *Kupferschiefer*. What modern zoologists distinguish as the Amphibia (frogs, newts, salamanders, etc.) were treated by Cuvier and his contemporaries—as for example in Brongniart's classification (§8.1)—as one of the four major groups within what they defined as "reptiles".

49. The staged event was just in time to be mentioned in the "discours préliminaire" (see below) prefixed to the first edition of Cuvier's *Ossemens fossiles* (1812), but the feat itself was not described in print until the second edition, 5(2) (1824), 436–37: the specimen prior to Cuvier's work on it is shown on pl. 25, fig. 2 (reprinted from the first edition), and its improved appearance on pl. 26, fig. 2. It remains prominently on display at Haarlem-TM.

50. Fig. 9.8 is reproduced from Cuvier, *Ossemens fossiles* (1812), 4 [Cambridge-UL: MD.8.68], part 5, 5e mém., unnumbered pl., first published as "Quadrupèdes ovipares" (1809), pl. 31; redrawn from Collini, "Zoolithes

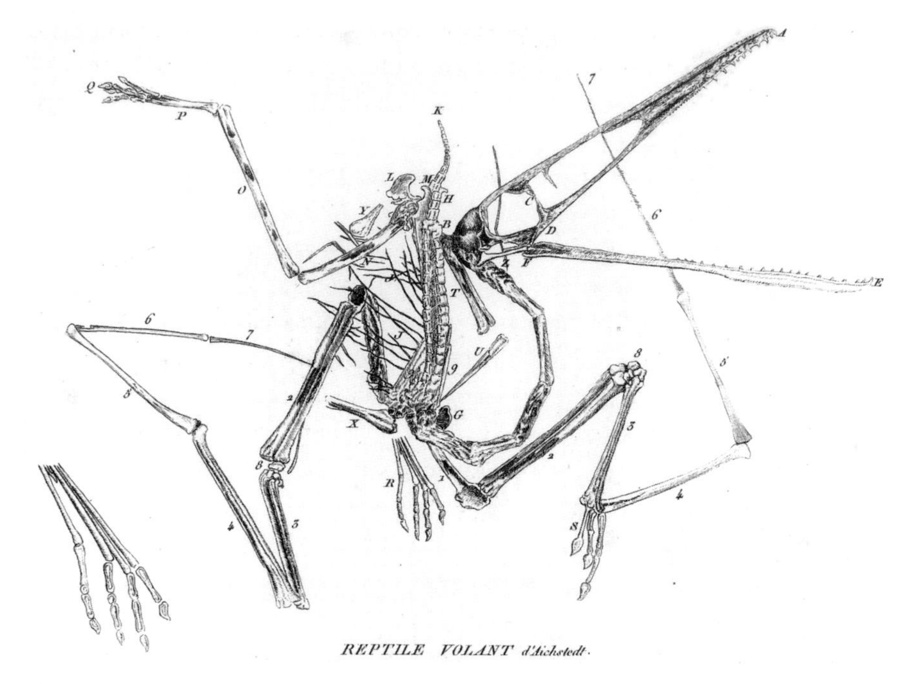

Fig. 9.8. Cuvier's illustration (1809) of a famous and (at the time) unique fossil from near Solnhofen in Bavaria, which he named the "ptero-dactyle" (wing-fingered animal) and interpreted as a flying reptile (in modern terms, a pterosaur). It was known to him only as a proxy picture, the specimen having been mislaid, but it was accurate enough for him to recognize its reptilian anatomy. Along with the giant lizard from Maastricht and other fossil reptiles, it greatly accentuated the strange character of the fauna of the older Secondary formations. (By permission of the Syndics of Cambridge University Library)

There was still more to come, though nothing quite so striking. In a study of turtles, Cuvier exposed in public yet another of Faujas's mistakes: as the younger Camper had pointed out in private, the supposed antlers of deer [*élan*] that Cuvier's senior colleague had reported from the Chalk at Maastricht were nothing but fragments of the underside of the carapace of marine turtles (§7.3). Though Cuvier did not make the point explicitly, the correction eliminated the last alleged case of fossil mammals in the older Secondaries (i.e., from the Chalk downwards). As he had suspected since early in his research (§8.1), the vast span of geohistory that these formations represented seemed to have been an age of reptiles: truly a former world, and one without mammals.[51]

Collected papers on fossil bones

Cuvier published these, the last of his articles in the Muséum's *Annales*, despite potential or actual criticisms of his work from various quarters (see §10.1). He was convinced that his studies on fossil bones of all ages would collectively substantiate his inferences about geohistory. He now prepared to assemble all his articles in a form that would make them not only more accessible but also a more impressive monument to his achievement. As each was published, he had held back a stock of extra copies, ready to be bound up in volumes that could be purchased by naturalists who did not have access to the periodical or who wanted a complete collection of their own. Since he could now arrange the articles how he chose, regardless of the order in which they had been published, his plans for these volumes give important insight into his conception of his work.

Back in 1807 Cuvier had sent a batch of his latest offprints to Karsten in Berlin, as he was doing to other favored correspondents. In a covering letter, he had outlined his plan for republishing his articles on fossil bones in three volumes. The first would contain those on elephants, mastodons, and other pachyderms, the most striking members of the geologically recent fauna from the Superficial deposits. The second would include the many papers on the bones from the Paris gypsum, and principally his strange palaeotherium and anoplotherium; an accompanying "geological description"—the geognostic paper that he and Brongniart were then preparing (§9.1)—would establish the geohistorical position of this older fauna. The third and last volume would be miscellaneous, including, for example, the bears, hyenas, and megalonix. Allowing for the fact that not all the fossils from the Superficial deposits could be crammed into the first volume, and that the Secondary reptiles (which he had not yet written up) would go in the third, it is clear that Cuvier was planning the arrangement of the work primarily on *geohistorical* rather than zoological lines. The three volumes would represent, as nearly as was practicable, a sequence backwards in time from the most recent set of fossil quadrupeds to the most ancient.[52]

When, five years later, Cuvier published his great *Researches on Fossil Bones*, it was in four substantial quarto volumes. Apart from a new introductory volume, the other three were much as he had outlined them to Karsten. The primarily geohistorical arrangement was accentuated by the inclusion of all the Secondary reptiles in the final volume. The papers on the bones from the Parisian gypsum—intermediate in age between the most recent fossils and the most ancient—were now substantial enough on their own to fill what had become the third volume. Since their "geological description" had been hugely enlarged into the work already published by the Institut (§9.2), this was moved (as a massive reprint) into the introductory volume. The whole work opened in the customary manner: with astute political sense, Cuvier dedicated his magnum opus to Laplace, the most powerful scientific figure in Napoleonic France. He presented his work explicitly as a project that aspired to the same scientific rigor as Laplace's great *Celestial Mechanics*, which was taken to have perfected Newton's mathematical analysis of the solar system. But the flow of prestige was not all one way, as Laplace would have appreciated when he accepted the dedication. If Cuvier's work was acclaimed in its own sphere, Laplace would gain the credit of being associated with research that

---

du Cabinet à Mannheim" (1784), pl. 1. Cuvier's interpretation remained controversial, and was not accepted generally until later: see Wellnhofer, "Cuvier and the first known pterosaur" (1982), and Padian, "Bat-winged pterosaur" (1987). Taquet and Padian, "Restoration of a pterosaur" (2004), reproduces the manuscript drawings that Johann Hermann of Strasbourg had sent Cuvier in 1800, which had first drawn his attention to Collini's specimen.

51. Cuvier, "Ossemens fossiles de tortues" (1809).

52. Cuvier to Karsten, 15 March 1807 (Berlin-SB, Lc1801(3), 23–24).

achieved a rigorous understanding of nature in the dimension of deep time, just as his own work had in that of deep space. Cuvier's *Fossil Bones* was, and was intended to be, an impressive work (Fig. 9.9).[53]

In the first volume, preceding his and Brongniart's detailed monograph on the Paris Basin, Cuvier printed a long essay pitched at the general educated public. He called it a "preliminary discourse", echoing that of the great *Encyclopédie* (§1.4) and other major works from the century of the Enlightenment. As in his first public lectures on geology (§8.3), on which the new essay was based, Cuvier appealed

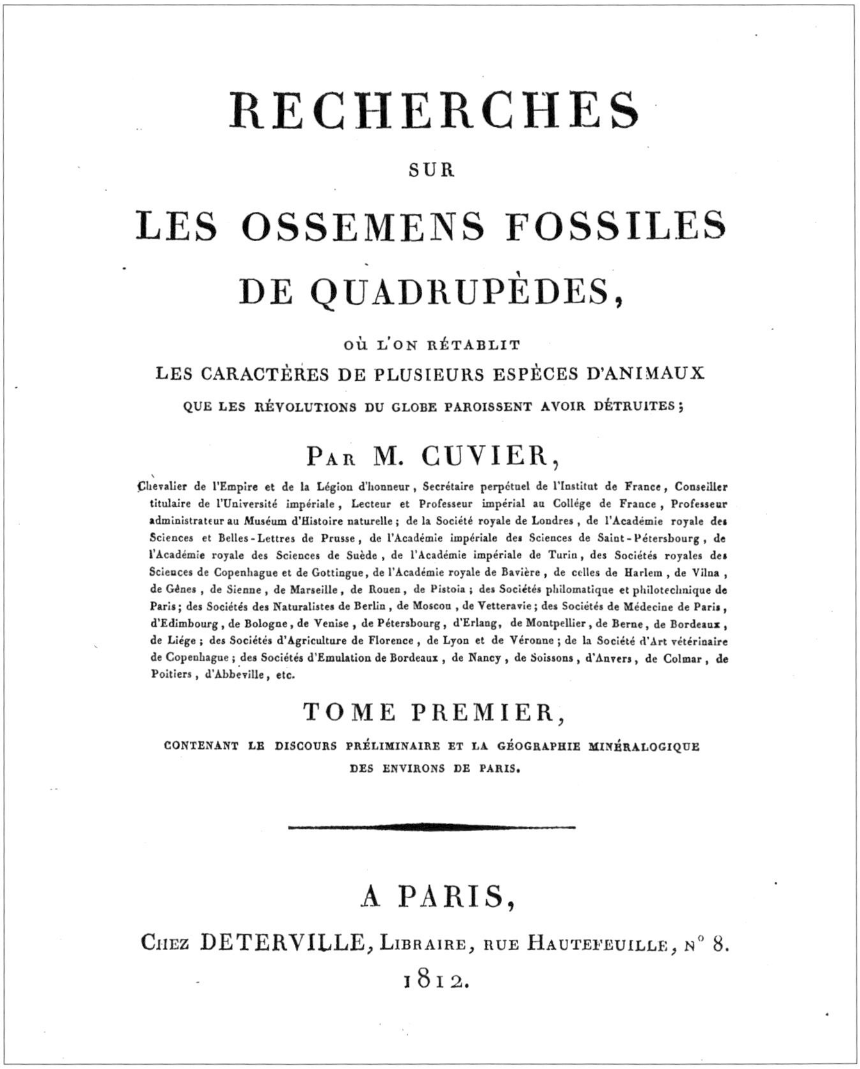

RECHERCHES

SUR

LES OSSEMENS FOSSILES

DE QUADRUPÈDES,

OÙ L'ON RÉTABLIT

LES CARACTÈRES DE PLUSIEURS ESPÈCES D'ANIMAUX

QUE LES RÉVOLUTIONS DU GLOBE PAROISSENT AVOIR DÉTRUITES;

PAR M. CUVIER,

Chevalier de l'Empire et de la Légion d'honneur, Secrétaire perpétuel de l'Institut de France, Conseiller titulaire de l'Université impériale, Lecteur et Professeur impérial au Collége de France, Professeur administrateur au Muséum d'Histoire naturelle; de la Société royale de Londres, de l'Académie royale des Sciences et Belles-Lettres de Prusse, de l'Académie impériale des Sciences de Saint-Pétersbourg, de l'Académie royale des Sciences de Suède, de l'Académie impériale de Turin, des Sociétés royales des Sciences de Copenhague et de Gottingue, de l'Académie royale de Bavière, de celles de Harlem, de Vilna, de Gênes, de Sienne, de Marseille, de Rouen, de Pistoia; des Sociétés philomatique et philotechnique de Paris; des Sociétés des Naturalistes de Berlin, de Moscou, de Vetteravie; des Sociétés de Médecine de Paris, d'Edimbourg, de Bologne, de Venise, de Pétersbourg, d'Erlang, de Montpellier, de Berne, de Bordeaux, de Liége; des Sociétés d'Agriculture de Florence, de Lyon et de Véronne; de la Société d'Art vétérinaire de Copenhague; des Sociétés d'Emulation de Bordeaux, de Nancy, de Soissons, d'Anvers, de Colmar, de Poitiers, d'Abbeville, etc.

TOME PREMIER,

CONTENANT LE DISCOURS PRÉLIMINAIRE ET LA GÉOGRAPHIE MINÉRALOGIQUE DES ENVIRONS DE PARIS.

A PARIS,

CHEZ DETERVILLE, LIBRAIRE, RUE HAUTEFEUILLE, N° 8.

1812.

Fig. 9.9. The imposing title page of Cuvier's *Researches on the Fossil Bones of Quadrupeds* (1812), "in which the characters of many species of animals that the revolutions of the globe appear to have destroyed are restored": Cuvier's claims about extinction, and his goal of reconstructing the vanished past, were thus emphasized from the start. After mentioning his positions in Paris, the small print listed the many foreign academies and societies to which he had been elected—from Edinburgh to Florence, Liège to St Petersburg, but with the Royal Society in London (the capital city of France's most tenacious wartime enemy) in first place—as well as others in provincial France: they reflected his reputation throughout Europe and served as a tacit claim to scientific authority.

over the heads of his fellow savants to readers with a general interest in the natural world and its significance for human life. Three and a half quarto volumes of dense scientific detail were thus prefaced by an outstanding example of what the French still (untranslatably) call *haute vulgarisation.* It is not clear whether this increased the sales of Cuvier's massive work, though the hope that it would do so may have been among his reasons for including it. With an eloquently readable introductory essay, cultured people with the necessary purchasing power may well have decided to add some impressive volumes to their libraries, even if they never intended to penetrate the technical arcana of the osteology of mammoths and mastodons.[54]

Fossil bones and geohistory

Cuvier's "preliminary discourse" opened with a bold and vivid claim. The focus was not on the wonders of nature, nor—as would have been expected in an earlier age—on the wisdom of its Creator, but on the savant himself. Using the *historical* metaphors he had adopted almost from the start of his research (§7.2), Cuvier presented himself as "a new species of [natural] antiquarian": new, because he had focused his work not on valleys or volcanoes but on the relatively neglected "monuments" of fossil bones. His task had been to learn how to "decipher" nature's hieroglyphs by understanding the principles of comparative anatomy. This alone had enabled him to identify fossil species from their scattered remains, and hence to "restore" nature's monuments and "reconstruct" the animals themselves; for his ultimate goal was to trace "the ancient history of the earth". He claimed that this, just as much as ancient human history, was worthy of the attention of Enlightened people: the "extinct nations" of the remote human past were matched by his extinct animals, "the traces of revolutions previous to the existence of every nation". The language could not have indicated more clearly that Cuvier wanted his work to be regarded above all as *prehuman geohistory.*[55]

Cuvier's main inspiration for geohistory came from the practice of human history, but his confidence in its possibility was boosted by the precedent and example of a quite different science. As he had hinted in his dedication, Cuvier compared his research to the rigorous mathematical astronomy that Laplace had recently constructed on Newtonian foundations. The analogy was so important, indeed decisive, that it has been used to provide the present book with its title and its epigraph:

53. Fig. 9.9 is reproduced from Cuvier, *Ossemens fossiles* (1812), 1, title page; the dedication is translated in Rudwick, *Georges Cuvier* (1997), 168–69; see also Outram, *Georges Cuvier* (1984), 149–50. The offprints from the *Annales* were unchanged except that the pages and plates were renumbered.

54. The discourse continued to be reprinted frequently through the rest of the century, for its *literary* merits, long after it was scientifically outdated. *Haute vulgarisation* is not only untranslatable—"high-level popularization" is a poor approximation—but also apparently difficult to export: good modern French examples of the genre (of which there are plenty) resemble most works of anglophone "popular science" about as much as *haute cuisine* resembles fast food.

55. Cuvier's "discours" is translated in full in Rudwick, *Georges Cuvier* (1997), 183–252. On his style, see Outram, *Georges Cuvier* (1984), chap. 7, and Cohen, "Stratégies et rhétorique" (1997).

> We admire the power by which the human mind has measured the movements of the globes, which nature seemed to have concealed forever from our view; genius and science have burst the limits of space, and observations interpreted by reason have unveiled the mechanism of the world. Would there not also be some glory for man to know how to burst the limits of time, and, by observations, to recover the history of this world, and the succession of events that preceded the birth of the human species?[56]

The achievement of the astronomers had lain not so much in discovering the literally inconceivable magnitude of deep space—which had long been a commonplace—but rather in showing that the workings of the universe (or at least of the solar system) were knowable with great precision by human beings confined to one small planet. Likewise the "limits of time" that Cuvier's work might help to burst were not those of the traditional short timescale for the earth—which for savants, if not for the wider public, had long since ceased to be any constriction at all (§2.5)—but rather the limits of human *knowledge* about the deep past. The intense excitement generated by Cuvier's work arose from the prospect that human beings confined to one brief moment in cosmic history might be able to reconstruct events that took place long before there was anyone there to witness them; it was to burst through the limits set by the oldest human records and gain reliable knowledge of *prehuman* (or, more precisely, preliterate) history. Here again Cuvier defined his work unmistakably as geohistory.

Just in passing, Cuvier also described his discourse as an "essay on a small part of the theory of the earth", thereby conceding that geotheory was a legitimate though distant goal for the science of geology. He claimed that his own research would contribute in a small but decisive way, because it would establish the reality of the earth's past "revolutions", and so provide the geohistorical basis on which any *causal* explanations would have to be founded. He retreated somewhat from his own earlier remark, which had so incensed Faujas and his allies, to the effect that such theorizing had made "geology" into a laughing matter (§8.4): he protested rather unconvincingly that he had merely reported a common opinion, not given it as his own.

However, Cuvier's review of earlier "systems", ranging in date from Descartes to his own contemporaries, showed that his attitude to geotheory had mellowed little. He cited, with barely concealed scorn, not only de Maillet's notoriously speculative *Telliamed* from the previous century (§2.5) but also the recent works of his own colleague Lamarck. His most entertaining joke at the expense of the transformists was deleted, however, before the discourse was published, perhaps because he was persuaded that it would be imprudent and counterproductive. But another comment in his draft text was not in fact unfair to Lamarck and others like him: "we have here not only geology; there is also an entirely new chemistry, mineralogy, botany, and physiology; and in fact these creators [*sic*] of the earth usually have yet another concern, that of re-creating all the sciences". Among many reasons for repudiating transformism, Cuvier regarded it, not unjustifiably, as an integral part of a much broader threat to the fruitful pursuit of the sciences of nature.[57]

However, even those geotheorists who—unlike Lamarck—had kept their speculations within the ordinary principles of physical science had still produced a bewildering variety of incompatible explanations. Not for the first time, Cuvier argued that this was because the relevant empirical conditions had not yet been established, so that the theorizing was (in modern terms) grossly underdetermined. He therefore set out once more an agenda that would help introduce some empirical constraints. As in his report on André's book (§8.4), it focused on the need to understand the Secondary formations and their fossils; as in his most recent lectures (§9.2), he pointed out that it would require a new kind of collaboration, integrating the museum study of fossils with the field study of the rocks in which they were found (at this point he did not mention the shining example of his own work with Brongniart). Above all, fossils needed to be treated as "historical documents", far more consistently than they had been hitherto.

Much of Cuvier's discourse was devoted to showing how fossil quadrupeds could provide crucial evidence for geohistory, in a way that Lamarck's mollusks and other fossils could not. Much of his argument focused on the methods he had used in comparing living species with their counterparts in the most recent fossil fauna (in the Superficial deposits) and on his interpretation of the differences in terms of a geologically recent episode of mass extinction. His reasoning can be followed most readily in the wider context of debates about that "last revolution" in geohistory, and its analysis will therefore be deferred to the following chapter (see §10.3). Taking for granted, for the moment, Cuvier's conclusion that his fossil bones represented genuinely extinct animals, his argument can be picked up again at the point where he drew further inferences from his fossil bones.

Cuvier reported that during his many years of assiduous research he had identified forty-nine species of quadrupeds "definitely unknown to naturalists until now", compared with only eleven or twelve that were indistinguishable from species known alive, and sixteen to eighteen that were still doubtful. It was an impressive tally. He also summarized the numbers assigned to mammals and reptiles, ruminants and carnivores, and so on. Far more important, however, was "to know in which beds each species is found" and whether there were any "general laws" relating their zoological character to their relative age as judged by their position in the pile of formations. In other words, Cuvier hoped to use geognostic information to trace the *history* of quadruped life through the vast spans of deep time. This was the "definitive object" of his work, for it "establishes its true relation to the theory of the earth". Confirming his earlier hunch (§8.1), he was now "certain

56. Cuvier, *Ossemens fossiles* (1812), 1: "Discours", 3; trans. Rudwick, *Georges Cuvier* (1997), 185. Modesty was not among Cuvier's virtues, but his suggestion (later in the same paragraph) that "natural history [might] also have its Newton one day" is unlikely to refer to himself. The context implies that he saw his own role as that of a Kepler, preparing the ground for a later genius: perhaps Charles Darwin would have fulfilled his vision, had Cuvier lived long enough to read—and to understand—Darwin's theory of "common descent", so radically different from Lamarck's unacceptable idea of transformism.

57. Cuvier, *Ossemens fossiles* (1812), 1: "Discours", 25–31, trans. Rudwick, *Georges Cuvier* (1997), 199–202; the joke from the MS text (Paris-MHN, MS 631) is quoted there (200n42) from Burkhardt, *Spirit of system* (1977), 199; it may be one of the jokes he made during his earlier lectures (§8.3), which he initially intended to put on record for the delectation of the wider world.

that oviparous quadrupeds appeared much sooner than the viviparous", or reptiles long before mammals. The Thuringian "monitor" came from what geognosts regarded as "among the oldest beds of the Secondary formations"; the crocodiles from Normandy and England and the strange flying reptile from Solnhofen were much younger, but still below the Chalk; and from the latter came the giant lizards and turtles of Maastricht. In the Coarse Limestone of Paris were the first mammals, but they were all marine ones such as seals; only in the subsequent gypsum deposits had he found the first abundant remains of terrestrial mammals.

The geognostic sequence of quadruped fossils was thus translated into a true history of quadruped life, in which reptiles had preceded mammals and marine mammals had preceded terrestrial ones. Cuvier also explained how even the latter had their own history, since the Paris gypsum contained genera unknown alive, whereas the still younger Superficial deposits recorded a fauna of mostly known genera, yet all of unknown species. Throughout this reconstructed history of quadrupeds, Cuvier was carefully noncommittal about the causes of these faunal changes: he simply inferred from their fossil record that the new forms had "appeared" or "begun to exist" at certain times. He also pointed out that any such geohistorical synthesis was necessarily provisional, because many of his fossils had been collected without adequate information on their geognostic position or even any record of their location. Here was ample scope for a new collaboration between practitioners of museum work and fieldwork, which he had prescribed as essential for future progress.

Cuvier emphasized that the fossil animals to be described in the body of his great work bore witness not just to the earth's "last revolution" but to other earlier ones too: "judging by the different kinds of animals of which the remains are found, they [the continents] had perhaps suffered up to two or three invasions by the sea". This alluded to the detailed geohistory that he and Brongniart had inferred from the alternation of marine and freshwater formations that they had detected in the Paris Basin (§9.1). Significantly, Cuvier opened the concluding section of his discourse—"ideas for research still to be carried out in geology"—by suggesting that "these alternations now appear to me to be the most important geological problem to be resolved". This was not a call for their causal explanation, which might be "an enterprise of quite another difficulty", but that their nature and extent as geohistorical events or periods should be "clearly defined and circumscribed". The priority was not earth physics, but geohistory. So for the third time, as in his report on André's book (§8.4) and earlier in the discourse, he singled out the detailed study of the Secondary formations and their fossils as the topic that "calls imperiously for the attention of [natural] philosophers".

Cuvier contrasted the "aridity" of the Primary rocks that formed the staple diet of traditional geognosts (§2.3) with the exciting prospects of research on the Secondary formations: the latter could "satisfy, as it were, the most ardent imagination", and, being relatively recent, were likely to be the key to the former. "These ideas", Cuvier confessed, "have pursued me—I could almost say, tormented me—while I carried out the research on fossil bones that I now present to the public in collected form". His craving to understand "the earth's penultimate age" had, he explained, driven his collaborative fieldwork with Brongniart; their joint monograph—for which Cuvier gave his colleague full credit—was now to follow in the same volume because he regarded it as "an integral part" of his discourse, and "certainly the best proof" of it. Nothing could show more clearly the importance, in Cuvier's strategy, of their joint exploration of the Paris region, as an exemplar of the doubly enriched geognosy that a closer study of the Secondary formations could yield, and hence of a more reliable and detailed geohistory.

So Cuvier referred briefly to the Secondary formations known in other regions: those of apparently the same age as the Parisian ones and those evidently older. But the former were the higher priority: in Cuvier's grand strategy "to burst the limits of time" by penetrating back from the known present into the unknown past, the *younger* Secondaries were in a crucial position. Cuvier pointed specifically to those on the flanks of the Apennines, which he had probably seen during his mission to Italy, and which were famous for their abundant and well-preserved fossil shells (see §9.4). For he suspected that they, perhaps even more than the Parisian formations, might serve to connect the "older and more solid formations" with the "recent alluvium". In other words, the younger Secondaries held the promise of filling the yawning gap between the familiar present world and the deeply unfamiliar world of the older Secondaries with their ammonites, giant lizards, and ptero-dactyles. Cuvier urged the importance of finding answers to the many questions posed by all these formations and their fossils; it was an agenda almost unlimited in its prospects for fruitful discovery. By contrast, the "contradictory conjectures" about the earth's origin and earliest history, as put forward by geotheorists, were sterile, because they could not be connected with "our present physical world".

Cuvier ended his discourse as he had begun it, by drawing an analogy with human history. For geologists to focus their attention on the Primary rocks, which contained no fossils, was as if scholars were to lose interest in early French history at just the point when the arrival of the literate Romans first provided it with ample documentation. Cuvier's research agenda proposed a more fruitful way forward: to penetrate back from the present into the geologically quite recent past before trying to tackle the more obscure earlier periods of geohistory. Above all, Cuvier fleshed out his historical metaphors: the geologist should use fossils as the historian uses documents: to piece together an authentic history of the earth and of life at its surface. "And man, to whom has been accorded only an instant on earth," he concluded, "would have the glory of reconstructing [*refaire*] the history of the thousands of ages [*siècles*] that preceded his existence, and of the thousands of beings that have not been his contemporaries!". To "burst the limits of time" was to reconstruct *prehuman geohistory*.[58]

58. Cuvier, *Ossemens fossiles* (1812), 1: "Discours", 110–16, trans. Rudwick, *Georges Cuvier* (1997), 249–52. Even if, improbably, Cuvier intended *siècles* to mean literal "centuries", the timescale he envisaged would still have been one that dwarfed the totality of human history and breached the short cosmic timescale of the old chronologers (§2.5).

Cuvier in English

Cuvier intended his discourse to be an integral part of his larger work, and he did not issue it as a separate small volume until many years later. But in the absence of international copyright—not to mention the major war that was in progress—he could not prevent the prompt publication of an English translation, even if he wanted to do so. It was edited by Robert Jameson (1774–1854), Edinburgh's professor of natural history, and translated by Robert Kerr, who many years earlier had made Lavoisier's revolutionary work on chemistry available to anglophone readers in the same way. In the event, Kerr's translation of Cuvier was rather poor; more seriously, Jameson's editing distorted Cuvier's objectives in ways that were to have an enduring effect on the anglophone reception of the French research.[59]

As a young man, Jameson had studied under Werner at Freiberg and had made his name with two mineralogical volumes based on extensive travel and fieldwork in the Scottish Highlands and Islands; a German edition, with an introduction by the translator presenting the work as a fine embodiment of Werner's methods, shows how his reputation had already extended where it mattered most in the science of geognosy. Soon after his appointment at Edinburgh in 1804, he had published two volumes of systematic mineralogy and one on geognosy, modeled on Werner's practice; unsurprisingly, he had taken Werner's Neptunist side in the specific argument over the origin of basalt. Like his Saxon hero he was strongly opposed to geotheory: "by Geology", he noted approvingly, "Werner understands idle and imaginary speculations respecting the formation of the earth". Jameson was particularly hostile to the speculative "system" promoted by Hutton and his Scottish followers such as Playfair (§8.4), and he had led like-minded naturalists in seceding from the Royal Society of Edinburgh and founding a rival new Wernerian Society.[60]

Jameson's edition of the "preliminary discourse" was not incompetent. In the longest of his editorial "notes", he summarized the rest of Cuvier's volumes effectively, including excerpts from specific articles, so his readers could get a fair impression of the whole range of the Frenchman's research. He also summarized the monograph on the Paris region, noting correctly that it was primarily concerned with "geognostic structure". He minimized its originality, however, and what has here been dubbed its enrichment, by calling it an example of "Wernerian geognosy" and claiming that Werner had paid more attention to fossils than the French naturalists seemed to realize.[61]

However, Jameson's editorial work gave Cuvier's essay a distinctive slant that was deeply misleading. First impressions matter, and the distortion was apparent even on the title page. Jameson ignored Cuvier's repeated scorn for the genre of geotheory—including his barely retracted assertion that it had turned "geology" itself into a laughing matter—and entitled his edition *Essay on the Theory of the Earth* (1813). As already mentioned, Cuvier had indeed referred to his work in passing as an "essay" towards the geotheory that might emerge in the very distant future. So in itself Jameson's epithet might have been justified. But to put "theory of the earth" into the title distorted Cuvier's whole project, for it replaced its

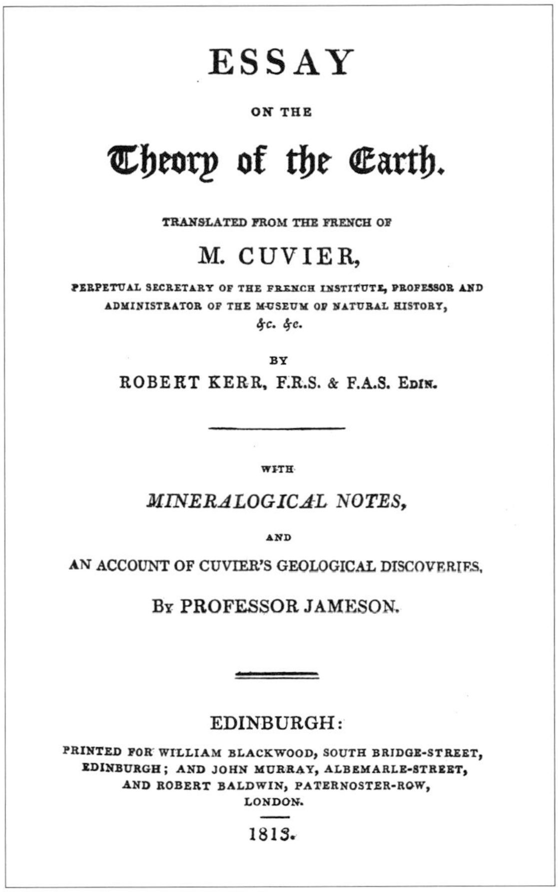

ESSAY

ON THE

Theory of the Earth.

TRANSLATED FROM THE FRENCH OF

M. CUVIER,

PERPETUAL SECRETARY OF THE FRENCH INSTITUTE, PROFESSOR AND ADMINISTRATOR OF THE MUSEUM OF NATURAL HISTORY, &c. &c.

BY

ROBERT KERR, F.R.S. & F.A.S. Edin.

WITH

MINERALOGICAL NOTES,

AND

AN ACCOUNT OF CUVIER'S GEOLOGICAL DISCOVERIES,

By PROFESSOR JAMESON.

EDINBURGH:

PRINTED FOR WILLIAM BLACKWOOD, SOUTH BRIDGE-STREET, EDINBURGH; AND JOHN MURRAY, ALBEMARLE-STREET, AND ROBERT BALDWIN, PATERNOSTER-ROW, LONDON.

1813.

Fig. 9.10. The title page of Robert Jameson's edition of the "preliminary discourse" to Cuvier's *Researches on Fossil Bones*, published in Edinburgh only a year after the French original. Not by accident, Jameson's title implied a parallel with Hutton's earlier book, tacitly transforming Cuvier's work from an innovative research program for geohistory into a geotheoretical "system" to rival Hutton's. The Gothic typeface chosen for the crucial phrase "Theory of the earth" had antiquarian and ecclesiastical associations, hinting at the biblical slant that Jameson's editing gave to Cuvier's work (see §10.4).

primarily *geohistorical* goals with geotheory. In Jameson's local context, however, his flaunting of the phrase is easily understood, as a tactical move in his argument with Playfair and others in Edinburgh who were championing Hutton's work of the same name (Fig. 9.10).[62]

Conclusion

Cuvier's excursion (with Brongniart) into fieldwork and geognosy gave him a firm temporal framework on which to plot the history of the quadrupeds, confirming

59. A copy of *Ossemens fossiles* was presented to the Institut in Paris on 7 December 1812, and probably went on sale at the same time. Jameson's edition is dated 1813, and must therefore have been completed—from the arrival of the original to the publication of the translation—within about twelve months.

60. Jameson, *Mineralogy of the Scottish Isles* (1800); *Mineralogische Reisen* (1802), with introduction (i–xlviii) by Heinrich Wilhelm Meuder; *System of mineralogy* (1804–8): 1 (1804), xx n. Jameson had still been using "geology" in its older sense, to mean geotheory, rather than in the sense embodied soon afterwards by the new Geological Society in London (§8.4).

61. Cuvier, *Theory of the earth* (1813): Jameson's "Note I, Werner's views of the natural history of petrifactions" (225–27), and "Note K, Cuvier's geological discoveries" (227–65). Jameson made no reference to Farey's claims on behalf of Smith for priority in the use of characteristic fossils, though he must surely have read Farey's article in the *Philosophical magazine* (§9.2).

62. Fig. 9.10 is reproduced from the first edition of Cuvier, *Theory of the earth* (1813).

his earlier hunch that an age of strange reptiles had preceded the age of mammals and that the latter had become progressively more similar to living forms. He embodied this geohistory in the eloquently readable "preliminary discourse" with which he introduced the weighty volumes of his collected papers on fossil bones. He presented himself as nature's antiquarian, explicitly drawing on parallels with the practice of human history. But he also linked his project to "burst the limits of time" with what his powerful patron Laplace had done in a more prestigious science: he aspired to make the depths of geological time as reliably knowable as the depths of astronomical space. In his strategy for reconstructing prehuman geohistory, Cuvier rejected the pretensions of geotheory in favor of a more cautious exploration, penetrating backwards from the known present into the increasingly obscure past, and renouncing any ambition to explain the earth's earliest origin. The most recent major set of fossil-bearing formations—the younger Secondaries—therefore assumed a crucial role, because they bridged the gap between the familiar present world and the strange and unfamiliar world of the fossils in the older Secondaries, and still more the world of the obscure Primary rocks with no fossils at all. Cuvier's discourse appeared promptly in English; but its editor, the Scottish geognost Jameson, gave it a highly misleading title, implying that the French research belonged to the genre of geotheory rather than being an innovative program for geohistory.

The next section describes how a few geologists outside France had begun to recognize the crucial importance of the younger Secondary formations and their fossils, even before Cuvier made it eloquently explicit in his discourse, and how their research made this most recent major period of geohistory truly international.

9.4 PARISIAN GEOHISTORY BEYOND PARIS (1811–14)

Fossils at the Geological Society

In the inaugural volume of the Geological Society's *Transactions* (§9.2), only two brief papers dealt with the Secondary formations that were the object of so much interest and attention on the Continent. One of these was by Parkinson. In the years since he began to publish his descriptions of common English fossils, based on indoor study of specimens in his own and other collections (§8.2), he had come to realize that outdoor fieldwork was also indispensable (and he had evidently found the time and money to do some). The study of fossils, he argued, was no longer just an appendix to botany and zoology: if "these memorials of the old world" were studied in relation to the "strata" they could become important for "geological inquiry". Clearly he appreciated the new kind of enriched geognosy that was being developed on both sides of the Channel, but patriotically he claimed English priority for it:

> This mode of conducting our inquiries was long since recommended by Mr. W. Smith, who first noticed that *certain fossils are peculiar to, and are only found lodged in, particular strata*; and who first ascertained the constancy in *the order of superposition*, and *the*

> *continuity of the strata in this island* . . . these observations have lately also occurred to Messrs. Cuvier and Brongniart whilst examining into the nature of the strata of the neighbourhood of Paris.[63]

Parkinson's paper was on the formations around London. It had no map, section, or any other illustration, though he did refer to his larger work as a source of pictorial proxies for the fossils he mentioned by name. London, like Paris, was (and is) flanked by hills—the Chilterns and North Downs—composed of the distinctive Chalk; but the formations between them, overlying the Chalk, were some that Smith's unpublished map indicated only sketchily, so there was scope for Parkinson to describe them in detail. In effect he was adopting Farey's suggestion that members of the Geological Society should search for the English equivalents of the formations in the Paris Basin (§9.2). Like Smith he described his formations from the top downwards—in the *reverse* of the geohistorical order that they represented—which suggests that he had absorbed Smith's enriched geognosy based on characteristic fossils, rather than the French naturalists' further enrichment into geohistory.

The highest of Parkinson's four formations was composed of sands and gravels with fossil bones and teeth, but also fossil mollusk shells: some of the latter he thought were identical to "living analogues" but others "probably lost". Below these Superficial sediments was the thick "Blue clay stratum" underlying London, well-known for its fine fossils, particularly those from the Isle of Sheppey on the Thames estuary; following Smith's fossil criterion, Parkinson equated it with the formation that yielded the fossils described by Brander long ago from the Hampshire coast (Fig. 2.3). Below the clay were highly variable sands and clays, some with shells that Parkinson considered similar to those described by Lamarck from Grignon, although embedded in quite different material. The lowest formation that Parkinson described was the Chalk itself, with fossils similar to those that Brongniart and Cuvier had described. In sum, Parkinson thought there was certainly a "continuity of the stratification" between the two countries. Some of the apparent differences might be due simply to poorer exposures in England; but others could perhaps be attributed to local circumstances "such as the existence of fresh or salt water lakes, at the period of the drying up of a former ocean". The comment suggests that Parkinson, rather like de Luc (§9.2), was trying to fit the French concept of freshwater episodes into the standard model of the emergence of continents from an ancient ocean (§3.5).[64]

In the Geological Society's volume, the only other paper that dealt with Secondary formations was in striking contrast to Parkinson's, in that its author paid

63. Parkinson, "Strata and fossils of London" (1811), 325n, where a quotation from the *Annales* version (1808) of the French paper shows that he had not just relied on Farey's translation; his paper was "acknowledged" at the meeting on 21 June 1811, the last before the summer recess, but apparently not read, perhaps for lack of time (London-GS, OM/1, 117–18).

64. Parkinson's four formations can be identified by the localities and fossils that he cited: in modern terms, they were (1) Pleistocene gravels in the London area and the marine "Crag" sediments of Plio-Pleistocene age in Essex and Suffolk; (2) the London Clay of Eocene age; (3) the Woolwich and Reading beds, etc., of earlier Eocene or Palaeocene age; and (4) the Chalk of late Cretaceous age.

little attention to fossils and was apparently quite unaware of what Smith was doing. It was by Jean-François Berger (1779–1833), a Genevan physician and amateur mineralogist who had migrated to London in 1809 in search of a medical career. He had soon joined the society, and his fieldwork was being supported financially—with due gentlemanly discretion—by its members. His sketch of the geology of Dorset and Hampshire on the south coast was centered on the conspicuous ridge of Chalk, acting as usual as an unmistakable benchmark, which runs through the Isles of Wight and Purbeck (the latter in fact part of the mainland). One of the formations that Berger described *below* the Chalk was a "Coarse Shelly Limestone", which—as he noted in one of his few comments on fossils—contained the distinctive "*Trigonia* of Lamarck" (Fig. 8.10). But on the grounds of its rock type he explicitly equated this limestone with the Coarse Limestone [*calcaire grossier*] of Paris, although the latter lay of course *above* the Chalk (§9.1). As for the English strata in that position, Berger simply dismissed them: "Over the chalk lie several beds of strata of a later formation, the relative age of which I shall not now presume to determine, as their alternation with each other appears to be several times repeated". So the complexity that for Brongniart and Cuvier had been the key to reconstructing an eventful geohistory was for Berger just a source of confusion. Altogether it was an undistinguished piece of work, even by the traditional standards of geognosy, which gave little attention to fossils (§2.3). It illustrated the limitations of that science without the new enrichment that fossils could supply.[65]

Webster on the Isle of Wight

The formations above the Chalk on the south coast of England, which Berger neglected, became for another member of the Geological Society an exciting opportunity to search for the Parisian formations on English soil; and the circumstances were more favorable there than for Parkinson around London, because the rocks were well-exposed in the sea cliffs. Years earlier, they had attracted the attention of a wealthy antiquarian, Sir Henry Charles Englefield (1752–1822), while he was on holiday on the Isle of Wight. He was particularly struck by the brilliant white Chalk, with its regular bands of contrasting black flint nodules, because in some places the beds were in an almost *vertical* position. He was planning to publish a full description of the geology and antiquities of the island—in effect, a local chorography (§4.1)—illustrated with engravings based on his own fine drawings of its scenery and ancient buildings. A decade later, however, his work was still unfinished because he himself had become unfit to do the necessary fieldwork, and he was also aware that he had not kept up with "that part of natural science lately called Geology". So he commissioned Thomas Webster (1773–1844), an architect and artist who had earlier been employed on building the Royal Institution in London, to go to the Isle of Wight to study the rocks on his behalf and make some further accurate drawings for him (Fig. 9.11).[66]

Although Webster had joined the Geological Society two years earlier, he had little experience of the science; but as a topographical artist he had a fine visual sense, and as an architect an acute sense of three-dimensional structure. He began

Fig. 9.11. Strata of white chalk with lines of black flints in the Needles, at the western tip of the Isle of Wight: an engraving published in Sir Henry Englefield's *Beauties of the Isle of Wight* (1816), after a drawing made in 1811 by the artist Thomas Webster (who was soaked to the skin while sketching for two hours on a rock close to sea level). The Chalk strata are almost vertical on the knife-edge ridge (left), but curve over to a lower angle in the bay on the right, forming part of what Webster later described as a huge fold structure (see Fig. 9.13 below). The style of the picture made it not only a "highly sublime" scene that would appeal to Englefield's general readers, but also at the same time a proxy of great scientific value, accurately depicting features of importance to geologists. (By permission of the Syndics of Cambridge University Library)

fieldwork on the Isle of Wight in the summer of 1811 and frequently reported back to Englefield. After checking in at the military headquarters on the island—to avoid being mistaken for a French spy—he hired a boat to take him round its coastline to get the best offshore views of the rocks in the cliffs. Like Englefield, he was astonished to see the almost vertical Chalk at both ends of the island; like Berger, he tried to relate it to other formations to north and south. The south of the island was relatively easy to interpret: the Chalk curved round to a lesser angle and other formations (of sandstone and clay) emerged from beneath it; near the south coast the hills were capped with an "outlier" of horizontal Chalk, confirming that the other formations were indeed below it (Fig. 9.12).[67]

65. Berger, "Hampshire and Dorsetshire" (1811), 251; MacArthur, "Berger of Geneva" (1990), gives invaluable biographical information and describes the "travelling fund" that supported him. Berger's formations, like Parkinson's, can be identified by their localities: his "Coarse limestone" was clearly the Portland (of late Jurassic age), famous for its fine building stone.

66. Fig. 9.11 is reproduced from Englefield, *Isle of Wight* (1816) [Cambridge-UL: Ll.11.60], pl. 25; see i–iii, 161–63. Pls. 1–9 are by Englefield himself, pls. 10–13 are unattributed; the remainder (pls. 14–50) are by Webster; all were engraved in 1815 by the brothers William Bernard Cooke and George Cooke of London. Edwards (N.), "Thomas Webster" (1971), uses inter alia his important manuscript autobiography (London-RI, MS 121).

67. Fig. 9.12 is reproduced from Englefield, *Isle of Wight* (1816) [Cambridge-UL: Ll.11.60], part of pl. 50. The map extends much further to the west, to include the mainland Isle of Purbeck (and the geological sections of both areas); on its topographical base, see iii–iv. The island is 36km from west (the Needles) to east.

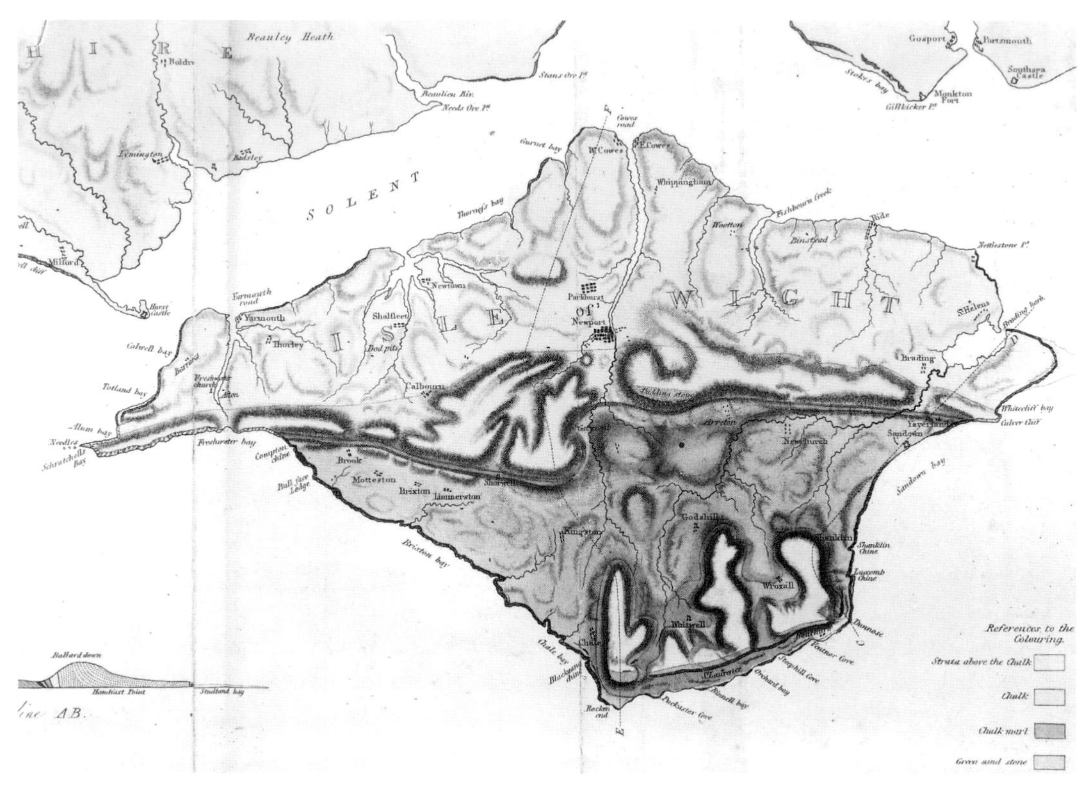

Fig. 9.12. Webster's geological map of the Isle of Wight, based on fieldwork in 1811–13, engraved in 1815 and published in Englefield's *Isle of Wight* (1816); it shows the narrow ridge of Chalk across the center of the island, with lower formations to the south and higher ones to the north. In its coloring convention it follows the map of the Paris Basin by Brongniart and Cuvier (1811) rather than Smith's of England and Wales (1815). Unlike either, Webster had his sections engraved on the same plate as the map (only a fragment is visible here, bottom left), thereby enabling a viewer with a practiced eye to construct a mental image of the complex three-dimensional geognostic structure. The base map was compiled from coastal charts provided by the Admiralty, because the "great survey" by the military Board of Ordnance had not yet published its sheet covering this area; Portsmouth, the home port of the Royal Navy, is on the mainland in the northeast corner of the map. (By permission of the Syndics of Cambridge University Library)

North of the ridge of Chalk were the real surprises and puzzles. Webster found that the Chalk in the Needles was followed by an astonishing sequence of vertical strata of sands and clays. Among them was a clay containing fossils that Webster recognized as those that Brander had described long ago from horizontal strata at Hordwell on the mainland coast (Fig. 2.3). So whatever process had tilted the Chalk into its extraordinary vertical position had also affected still higher formations, those that Parkinson had just suggested as equivalent to the clay underlying London: the process or event must therefore have been even more recent. Finally, still further north, on Headen Hill, and still higher in geognostic position, were other strata lying horizontally. It was now clear to Webster that there was a single sequence of formations throughout the island, but the central zone of high angles and even vertical strata was difficult to understand. He later cruised round the cliffs of the Isle of Purbeck (on the mainland), and found similar vertical and tightly curved Chalk strata; it was all so puzzling that he returned to that area the following summer to check his fieldwork.[68]

After that second field season, Webster wrote to give Englefield his considered interpretation of all he had seen; it showed that he had rapidly turned himself into a more than competent geologist. True to the Geological Society's policy, he put the "principal facts" into first place, but added "a few speculative remarks on the probable changes which have taken place" in the area he had explored. He inferred that the formations had "once formed a sort of arch, or vault", which had since been partly eroded away "by the same denudating causes which have swept away such large portions of the earth in other places". Significantly, he introduced this tentative piece of geohistory with metaphors that his antiquarian patron would have grasped at once:

> In connecting together, and, as it were, restoring (to use the language of the antiquary) this series of strata, which, though now in ruins, was probably much more entire in some former condition of the earth, a singular conclusion presents itself to the imagination, and which would seem to force itself upon our conviction, with nearly the same certainty which we feel in putting together the fragments of an ancient temple.[69]

Webster hedged his bets on the vast amount of erosion that seemed to be indicated, implying that it might have been due to a sudden and violent event—Farey's account of the "denudations" of the Derbyshire formations was probably his immediate source—or to the slow and long-continued action of rain and rivers; or "perhaps several causes may have combined in producing this deficiency in the strata". Distinct from any such speculation in earth physics, and in Webster's opinion much more reliable, was his *geohistorical* reconstruction of events in "several successive æras", which could be inferred from the observable features. First had been the deposition of all the strata, horizontally on the bed of a vanished ocean; then their distortion—whether by upheaval or subsidence—into a remarkable vaultlike fold structure; then the filling of the "basin" or hollow to the north of the "vault", with further sediments; then the "denudation" itself; and finally "the retiring of the denudating cause", leaving the topography as it now is. The evidence for all this was condensed into two remarkable sections (through the Isles of Wight and Purbeck respectively) drawn accurately to true scale, which displayed vividly the magnitude of the folding, of the inferred erosion, and of the hidden depths to which the rocks must extend (Fig. 9.13).[70]

The Ordnance Survey had been set up in 1791 under military command, for strategic purposes; it produced (from 1809) the first topographical maps of England to match the accuracy of the great Cassini survey of France half a century earlier (see Fig. 2.11). Webster's map, added to Cuvier and Brongniart's of the Paris Basin (§9.1), Ebel's of the Alps (1808), and Maclure's of the United States (1809), consolidated Charpentier's style as the *standard* for geological cartography.

68. Webster to Englefield, 21 May to 16 June 1811, 3 June to 1 July 1812 (letters 1–10), printed in Englefield, *Isle of Wight* (1816), 117–98; these letters are here taken to be a reliable record of Webster's fieldwork, although their texts were probably tidied up for publication. Englefield may have instructed him to limit his survey to the Chalk and the formations adjacent to it, for Webster turned back at Durlstone Bay in Purbeck, where "from the dip, I could perceive that I was arriving at inferior [i.e., lower] strata" (174).

69. Webster to Englefield, 2 August 1812 (letter 11), 201, printed in Englefield, *Isle of Wight* (1816), 198–225.

70. Fig. 9.13 is reproduced from Englefield, *Isle of Wight* (1816) [London-BL: L.50/141], pl. 47, fig. 1; the strata are shown extending about 1,000m above and 1,500m below the present land. The other section (fig. 2),

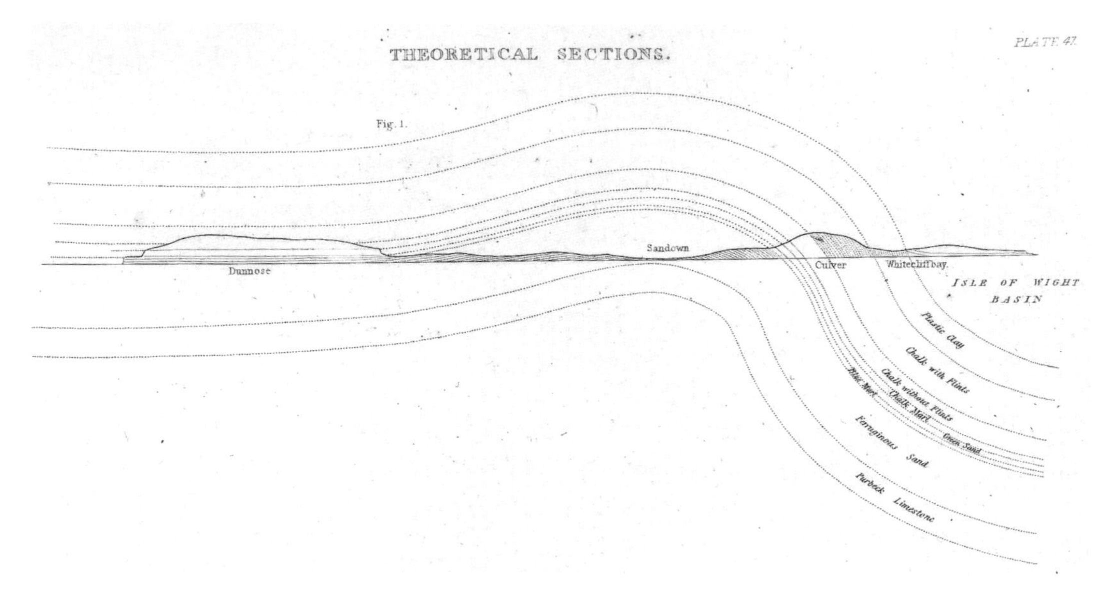

Fig. 9.13. One of Webster's "theoretical sections", drawn at true scale (without vertical exaggeration) through the eastern part of the Isle of Wight, and published in Englefield's *Isle of Wight* (1816). The strata visible in the coastal cliffs (and extrapolated inland to the line of the section) are confined to a thin strip between sea level and the land surface. Dotted lines indicate their inferred extension down into the depths of the earth and up into the air, the latter implying an amount of subsequent erosion that some of Webster's contemporaries found literally incredible. To the north (right) is the southern margin of what he called the "Isle of Wight Basin", filled with sediments younger than the Chalk, which he equated with those of the Paris Basin; but the folding had affected at least some of them (see the "Plastic Clay"), and therefore had to be even more recent. (By permission of the British Library)

However, the most striking implication of the fold structure that Webster had inferred was not its magnitude but its apparent date. Other folds had long been known elsewhere, about as large and even more dramatic in form: Saussure's huge fold at Nant d'Arpenaz was a famous case in point (Fig. 2.25). But that Alpine structure affected limestones that could well be very ancient on the scale of all the Secondary formations. The fold running through the Isles of Wight and Purbeck, in contrast, affected not only the Chalk—widely regarded, until the recent Parisian work, as the uppermost of the Secondaries—but also even younger formations. As already mentioned, Webster had recognized among the vertical strata some that contained fossils like those in the London Clay, which in turn were like some of those in the Parisian formations. The folding must of course have been more recent than their deposition. Just how recent depended on a detail that Webster had found difficult to resolve in the field. To the north of the vertical sands and clays (i.e., further from the vertical Chalk) he had found, as already mentioned, horizontal beds that clearly overlay the others and were therefore younger still. But the exposures near the junction between the two sets of beds were poor, and he was not sure whether a slight tilting of the horizontal ones was due to their deposition against the edge of the vertical ones (i.e., an unconformity) or to their being slightly affected at that point by the fold itself. But even if the horizontal strata had been deposited after the folding, the buckling of the earth's crust must have been astonishingly recent by geological standards.

In 1812 the Geological Society appointed Webster to be their salaried multifunctional official: their librarian, the curator of their museum, and a draftsman to turn their rough sketches into publishable pictures and diagrams (later he was made, in addition, the society's secretary). He was never happy in the position—as an employee he was probably made to feel socially inferior—but he needed the income. Soon after his appointment, some important specimens were donated by the count de Bournon, the French emigré whose mineralogical book had been one of the factors in the foundation of the society (§8.4). The specimens, which had somehow circumvented Napoleon's naval blockade, came from Bournon's friend Brongniart and were intended to illustrate the now famous paper on freshwater formations (§9.2). Seeing them prompted Webster to ask the society for a leave of absence to return to the Isle of Wight to search more closely for English equivalents of the French formations. And so, months after sending Englefield what he had obviously intended to be his final summary for inclusion in the book, Webster surprised him by writing again with news of his latest fieldwork, which dramatically extended the significance of his research.

Among the horizontal beds on Headen Hill, which he had not studied in detail previously, Webster found a sequence that sensationally paralleled the French: at the base were some sands with the freshwater shells *Planorbis* and *Lymnaea* (see Fig. 9.4); then came clays with shells "entirely marine", identified for him by Parkinson; then a limestone containing the same freshwater shells again, and finally a topping of gravelly "alluvium". Leaving aside that Superficial deposit, but counting in the beds with marine shells in the still lower sequence of vertical strata, he had two "Marine Formations" alternating with two "Freshwater Formations", just as the French naturalists had found them around Paris; he even found in the freshwater strata the telltale *gyrogonites* (Fig. 9.5). Webster thought the striking "correspondence" with the Paris Basin was "extremely curious" and its implications "astonishing": it showed that "the same place has been a lake, and a part of the sea in succession", not just once but twice. The match was not exact—there was no gypsum on the Isle of Wight, for example—but it was good enough (Fig. 9.14).[71]

With Englefield's big book still delayed by the slow and expensive engraving of its many plates, Webster reported these important results at the Geological Society without delay; his paper "On the freshwater formations in the Isle of Wight" was published in 1814 in the second volume of its *Transactions*. He argued that his work proved the alternation of marine and freshwater periods to have been widespread, "tending to throw some light on the later [i.e., relatively recent] revolutions which our planet has undergone" and showing that the circumstances "were subject at distant places to the same general laws". The French and English sequences established "a series of epochs" in geohistory, subsequent to the time of the Chalk. The

---

not reproduced here, shows the similar fold through the Isle of Purbeck further west, but the highest formation there was the Plastic Clay.

71. Fig. 9.14 is reproduced from Webster, "Freshwater formations" (1814) [Cambridge-UL: Q365.b.12.2], pl. 11; Webster to Englefield, 11 February 1813 (letter 12), printed in Englefield, *Isle of Wight* (1816), 226–31. Headen Hill is now spelled Headon.

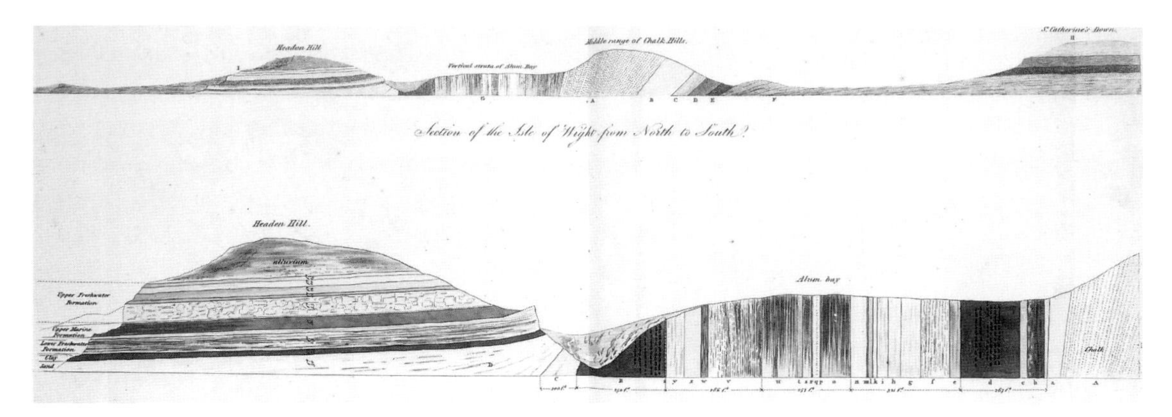

Fig. 9.14. Webster's sections through the Isle of Wight, published in 1814 and based on exposures in the cliffs on its west coast. Above, a general section from the horizontal formations on Headen Hill on the north (left), through the vertical strata in Alum Bay and the highly tilted "middle range of Chalk hills" (and the Needles, Fig. 9.11), to the horizontal formations on St Catherine's Down in the south (right). Below, an enlargement of the crucial portion between Headen Hill and Alum Bay, with the poorly exposed and therefore problematic transition from horizontal to vertical strata, on which depended the dating of the event that had produced the massive folding (Fig. 9.13). The horizontal beds on Headen Hill show, below the alluvium, two "Freshwater Formations" separated by an "Upper Marine Formation" (the lower one being represented among the vertical beds), thus indicating an alternation closely similar to that described by Brongniart and Cuvier in the Paris Basin. (By permission of the Syndics of Cambridge University Library)

English one underlined the "unfathomable antiquity" of Cuvier's mammals from the French gypsum formation, and indeed of the freshwater formations themselves; even the Upper Freshwater Formation—now the youngest known regular formation—was "anterior to the great event which gave the last shape and surface to our land", namely its erosion to form the present topography. Webster's work gave a heady perspective on the more recent part of geohistory, quite without precedent in the anglophone world. It could not be dismissed as empty speculation, because he carefully abstained from causal geotheory and rooted his geohistory in the kind of detailed fieldwork that the society approved. Finally, he made the newly international dimension of the research visually explicit, in a map that showed his Isle of Wight Basin in relation to the London Basin described by Parkinson and the Paris Basin that was the model and exemplar for both. The latter was no longer unique: "basins" now seemed to be the natural habitat of the youngest Secondary formations (Fig. 9.15).[72]

Webster sent Brongniart an offprint of his paper as soon as he could, praising the Parisian research, "which indeed has given rise to many of the observations I have described". He regretted that "circumstances"—the war, of course—had prevented him from visiting Paris in person, "to have viewed [i.e., in the field] those formations which have enabled us to extend the analogies of Geology". When Brongniart replied to thank him, he asked to be sent English specimens in return for his own. Research on the youngest Secondary formations and their fossils was now as international as "circumstances" allowed.[73]

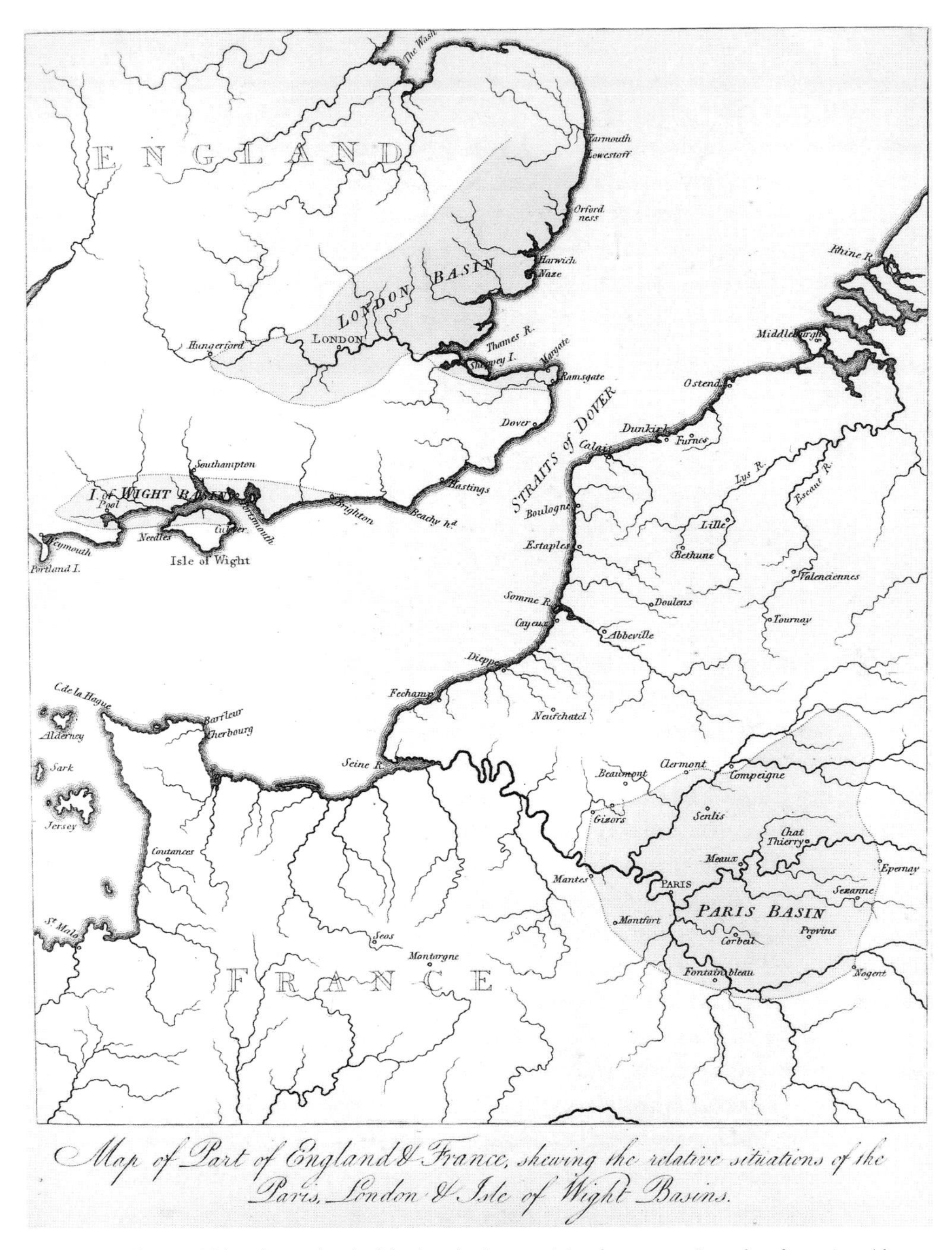

Map of Part of England & France, shewing the relative situations of the Paris, London & Isle of Wight Basins.

Fig. 9.15. Webster's map (1814) of the three basins containing the youngest Secondary formations (those above and younger than the Chalk), linking his own work on what he called the Isle of Wight Basin with Parkinson's on the London Basin and with what Brongniart and Cuvier had already described as the Paris Basin. (By permission of the Syndics of Cambridge University Library)

72. Fig. 9.15 is reproduced from Webster, "Freshwater formations" (1814) [Cambridge-UL: Q365.b.12.2], pl. 10* (the numbering implies that this map was inserted at a late stage, perhaps as an afterthought). Webster's basin was later renamed the Hampshire Basin in recognition of its much wider outcrop on the mainland (and anyway the island was and is a part of that county). Two other maps (pls. 9, 10) outlined the Hampshire and London Basins on a larger scale, but without showing the outcrops of the constituent formations.

73. Webster to Brongniart, 10 December 1814, with Brongniart's note on his reply on 8 March 1815 (Paris-MHN, MS 1968/795).

Brocchi on the Subapennine fossils

While Webster was applying the exemplar of the Paris Basin to the two English basins, and thereby extending its geohistorical significance to the north, it was also being applied and extended to the south, to Italy. Lamarck's work with the Parisian fossil shells (§7.4), which had given Brongniart and Cuvier the key to their geognostic use of fossils (§9.1), had been facilitated by the great fossil collection amassed by the Parisian amateur Defrance. In just the same way, the Italian research was grounded in an equally outstanding collection made by Giuseppe Cortesi, a lawyer and amateur naturalist in Piacenza. It consisted mainly of fossil shells from his native region, which were as abundant and varied as those from Grignon and even more superbly preserved. There were also some other important specimens, such as a complete skeleton supposedly of a dolphin, and the bones of large quadrupeds. Apart from the latter, which were found as usual in Superficial deposits, all Cortesi's fossils came from the strata of the "Subapennine" formation on the northern flanks of the Apennine mountains, along the edge of the vast alluvial plain of the Po; they were also known to form similar foothills along the southern flanks of the same range.

In 1808, Cortesi had offered his great collection for sale. Cuvier had tried to get the Muséum to acquire it for Paris, but the patriotic Cortesi wanted it to remain in Italy. In the end, after lengthy negotiations, the government of the puppet kingdom of Italy had provided the funds to buy the fossils. In Milan, its capital city, a new museum of natural history was founded in 1809 with Cortesi's fossil collection as its core. Other collections were soon added—for example, minerals and rocks arranged on Werner's principles were acquired in 1810 from Voigt of Weimar—and there were plans for a botanic garden and a menagerie, all on the model of the Muséum in Paris. This new institutional setting was highly favorable for the sciences of natural history in Italy.[74]

Immediately after Cortesi's collection was purchased and the museum founded, Giovanni Battista Brocchi (1772–1826) was appointed its curator. He was a good choice. A native of Bassano in the hinterland of Venice, Brocchi had been sent to Padua to study law, but he had found natural history and the Classics more appealing. As a young man he had spent much time in Rome studying art and antiquities, and his first publication was on ancient Egyptian sculpture. As with Cuvier, Blumenbach, and Brongniart among others, it was historical experience that would have made a geohistorical perspective seem natural and congenial, when later he was making his career in the sciences of the earth. Brocchi had entered the Venetian mining service, and a few years later he made his name with two volumes on the geology of an important iron mining district. His introductory essay adopted the same line as Cuvier (§8.4): recent progress in *geologia* had been due to careful fieldwork, whereas "the spirit of system", or geotheory, was inimical to its progress. His notes show that around the same time he was reading Saussure, de Luc, Dolomieu, and Faujas, among others, and works as recent as Brongniart's. Probably as a result of his growing scientific reputation, Brocchi had been appointed to a more senior position in Milan only a few months before he became in addition the first curator of the new natural history museum.[75]

Soon after his appointment, Brocchi had set out his research plans for the minister of the interior. Stressing the great scientific importance of Cortesi's collection, he had proposed to publish exact descriptions of all its fossils, using comparative anatomy to identify the vertebrate bones and noting where the molluskan species with modern analogues were now living: he thought all the fossils were tropical in appearance. But he had also wanted to study "the nature of the rock [*suolo*]" in which they were found and the "geognostic circumstances" of their emplacement; and he would set out "the best established conjectures" to explain how these tropical organisms could have come to be fossilized in northern Italy. Brocchi's ambitious plan was obviously drawn up with full knowledge of the published work of both Cuvier and Lamarck, as well as the already famous geognostic work on the Paris Basin (§9.1).[76]

By 1811 Brocchi's plans were more ambitious. He proposed to extend his Subapennine fieldwork over the full length of their outcrops on both flanks of the Apennines; to explore the older rocks of the Apennines themselves; and to study volcanic rocks as far south as the active volcanic area around Vesuvius. The Mines Council in Milan granted him three months' leave, but he stayed away much longer, despite their demands for his return. However, he made good use of his time, working in the field, meeting other naturalists in every town and city, and inspecting their museum collections (including, for example, that of the deceased Spallanzani: see Fig. 1.12). What is most important in the present context, however, is that his notes show that at least at one point he was thinking *geohistorically* about what he saw. In the ancient hill town of Volterra (on the far side of the Apennines from Cortesi's Piacenza), he climbed the cathedral tower in order to grasp "the extent of the shelly [Subapennine] formation". He interpreted the "singular spectacle" of the vast panorama below him as showing "the confines of this ancient bed of the sea": hills composed of a much older limestone would have been islands in the vanished sea; and the lower valley of the Arno down towards the coast at Pisa would have been "the opening by which the Mediterranean penetrated and flooded all this terrain". Although not as vivid as Saussure's vision on the summit of Crammont long ago (§4.5), Brocchi too was certainly reconstructing the former state of the area in his mind's eye and imagining its subsequent history.[77]

74. See Visconti, "Raccolte del Museo Reale" (1987), 143–47; Cortesi's fossil shells are still in Milano-MSN, but his vertebrate fossils (which had been moved to another museum) were destroyed during World War II.

75. Brocchi, *Trattato mineralogico* (1807–8), 1: lix–lx; see also Berti, "Formazione di Brocchi" (1987), 13–29, and *Giambattista Brocchi* (1987), chaps. 1, 2.

76. Brocchi's letter of 15 August 1809 is quoted in Visconti, "Raccolte del Museo Reale" (1987), 147–48.

77. Brocchi, field notes for 10 October 1811 (Bassano-MC, MS 502, 31-A-20–1, ff. 20–22); see also Pancaldi, *Darwin in Italia* (1983), 31; Visconti, "Raccolte del Museo Reale" (1987), 149–51; Berti, *Giambattista Brocchi* (1987), 91–97. Brocchi wanted to see active volcanoes and their products, not least because he was taking the Vulcanist side of the continuing basalt controversy: see his *Valle di Fassa* (1811), also published in German in 1817; also Vaccari, "Geologia in Italia settentrionale" (1992) and "Mineralogy and mining in Italy" (1998).

At some point Brocchi evidently decided to confine his work on Cortesi's Subapennine collection to its fossil shells; tacitly he left its vertebrates to Cuvier, whom he had almost certainly met during the latter's first educational mission to Italy (§9.2). Further fieldwork in 1813 completed what Brocchi needed to do away from his museum, and his work on *Subapennine Fossil Conchology* was finally published by the royal press in Milan early in 1814. Unfortunately for Brocchi, however, the tides of war turned again soon afterwards, when Napoleon was forced to abdicate and was exiled to Elba. Among the many consequences of these momentous

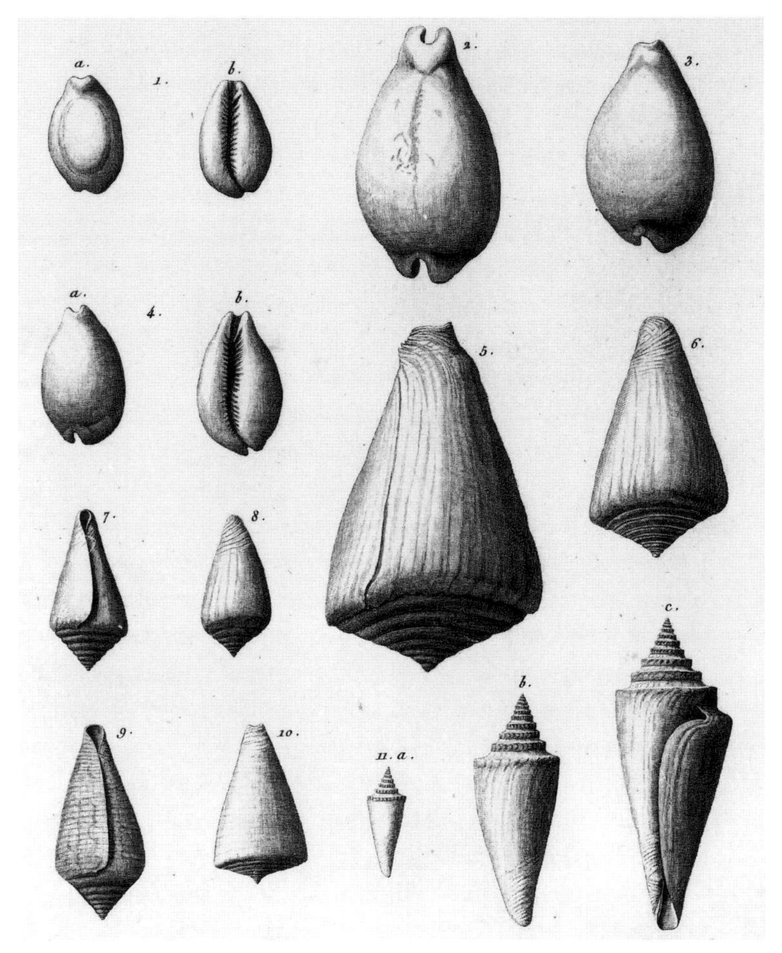

Fig. 9.16. One of the plates in Brocchi's *Subapennine Fossil Conchology* (1814), showing fossil shells of four species of cowries [*Cypraea*] (figs. 1–4) and seven of cone shells [*Conus*] (figs. 5–11), most of them collected from the Subappenine formation near Piacenza. Of the species depicted on this plate, Linnaeus's *Cypraea annulus* (fig. 1) was known to Brocchi to be still living, in warm waters off Egypt and in the East Indies. Bruguière's *Conus antediluvianus* was shown with its juvenile, mature, and elderly [*invecchiato*] shell forms (fig. 11a, b, c), indicating that Brocchi knew how to avoid giving spurious new names to what were merely the growth stages of a single species. Apart from these two species, all those illustrated here were newly defined and named by Brocchi himself. In the opinion of one reviewer, the Scottish naturalist Leonard Horner, the engravings on Brocchi's plates were the finest images of fossil shells yet produced. With their accompanying text, they portrayed a fauna that matched or even excelled Lamarck's Parisian one in its richness, variety, and superb preservation, and therefore provided a point of reference of outstanding importance in the reconstruction of geohistory. (By permission of the Syndics of Cambridge University Library)

events, the kingdom of Italy collapsed and Lombardy returned to Austrian rule. Brocchi, as a servant of the former regime, feared for the future of his position; he moved for a time to Rome, but in the event he was not dismissed. The dramatic political changes may have hampered the distribution of his work around the rest of Europe; but Cuvier had received a complimentary copy by the end of the year, praised it, and noted its connection with the English research.[78]

Leaving aside for a moment the first volume of Brocchi's handsome work, it was the second that contained its empirical core: the systematic description, analysis, and classification of the fossil shells from the Subapennine formations. This would have been valuable in itself, even if Brocchi had published nothing else, for it described a collection of fossil shells as rich and varied as Lamarck's Parisian one. Furthermore, it was illustrated with superb engraved plates that, combined with the text, made this outstandingly important fossil fauna accessible everywhere in proxy form (Fig. 9.16).[79]

Brocchi's first volume, which contained his lengthy interpretation of the fossil shells described in the second, included a massive compilation of all previous publications on the fossils of the region, arranged chronologically from the earliest reports in the sixteenth century right up to his own time. It was an appropriately historical introduction to a work that he defined at the start as *geohistory*: he would treat his fossils as "documents", and use them to elucidate "the ancient history [*l'antica storia*] of the globe". Like Cuvier and all other serious naturalists, he took a vast timescale for granted; just in passing he assured his readers in the conventional manner that there was no conflict with the story in Genesis provided its "days" were understood to denote periods "of indeterminate length". His method for reconstructing geohistory was to be actualistic, in that his first point of reference would be the mollusks now living in the nearest sea (the Adriatic, off the east coast of Italy), although he would of course also compare his fossils with Parkinson's from London and above all with Lamarck's from Paris. His Subapennine beds were so distinct from the older Secondary strata of the Apennines themselves that Brocchi called his Subapennine formation "*Tertiary*". This term will be used henceforth in the present narrative (as it still is by modern geologists) to denote the younger formations—those lying above the Chalk or its equivalents—that had previously been treated as the youngest of the Secondaries.[80]

78. Brocchi, *Conchiologia fossile subapennina* (1814); Cuvier to Brocchi, 24 December 1814 (Bassano-MC, MS 525). See also Visconti, "Raccolte del Museo Reale" (1987), 146; Vergani, "Brocchi funzionario" (1987), 74; Berti, "Formazione di Brocchi" (1987), 32–38, and *Giambattista Brocchi* (1987), 97–111.

79. Fig. 9.16 is reproduced from Brocchi, *Conchiologia fossile subapennina* (1814) [Cambridge-UL: MF.51.7], 2: pl. 2, explanation on 679; [Horner], "Brocchi, *Conchiologia*" (1816), 180. Brocchi's shell collection is still in Milano-MSN, and some of his specimens are in the museum's historical display.

80. Brocchi, *Conchiologia* (1814), 1: Introduzione, 1–11, 21; the historical *Discorso* follows in a separately paginated section (i–lxxx). Arduino had defined as "*monti terziari*" the youngest set of formations (apart from the alluvium) on the north side of the Po plain (§2.3), but the name had not been generally adopted. Yet by Brocchi's time the need for some such distinction had been tacitly recognized in the widespread observation that the older Secondaries contained ammonites and belemnites, for example, whereas the younger ones never did. Modern geologists still use the word Tertiary in Brocchi's sense, to denote rocks formed in any part of the Cenozoic era—the most recent major portion of geohistory—other than in the geologically very brief, recent (and human) Quaternary.

Reflecting Cuvier's prescription (§8.4), Brocchi argued that the study of geohistory should give priority to the Tertiaries, not to the more remote Secondaries (as now redefined) and still less to the obscure chaos of the Primaries. So he dealt quite briefly with "the structure of the high Apennines" and their hard and highly folded Secondary limestones with ammonites. This was just a prelude to his description of the soft marls and sands of the horizontal Subapennine formation (see Fig. 9.21). Likewise his reconstruction of the geohistory of the region started with the Apennines already in place, as it were, more or less as they are today, except that they were largely submerged below sea level. The subsequent geohistory was a story of a sea that subsided progressively, though in fits and starts, uncovering ever more land—as in the process that he had seen in his mind's eye while at Volterra—until finally the present topography of Italy emerged and the Mediterranean was confined to its present limits. In effect, this was an adaptation of the standard model (§3.5), restricted to the time represented by the Subapennine formation and the even more recent alluvial deposits to the north and south of the Apennine range (in the plains of the Po and, for example, the Arno).[81]

Brocchi expressed himself unsurprised by the differences between the Tertiary sediments of Paris and London and those of his own region, attributing them to differences in local circumstances. He based this interpretation on similar observations on a smaller scale. Twenty years earlier, his fellow Venetian the naturalist (and priest) Giuseppe Olivi (1769–95) had studied the marine zoology of the Adriatic—he was one of the first naturalists to make systematic underwater observations—and had reported great variation in the nature of the sea floor and its living fauna. Brocchi used Olivi's work as the key to understanding the deep past of the Subapennines. The argument became clear when he turned to the Subapennine fossil shells themselves, for he reported that they were usually found in specific assemblages or "families" of species:

> I believe that not much reasoning is needed to demonstrate that this union in families is just what applies today with the various species of shells in the sea: whether because by their particular character they live in a kind of society [*sorta di società*], like so many other gregarious animals, or because they have a preference for certain kinds of seafloor.[82]

This interpretation—in modern terms, as a difference of facies and paleoecology—was a geohistorical insight analogous to the French naturalists' distinction between freshwater and marine faunas. Yet it was more subtle, since most of Brocchi's fossils were unquestionably marine: he had not found any of the alternations reported in the Paris Basin, and he thought that in the Subapennines the rare freshwater shells had simply been drifted out to sea. When he turned to the specific composition of the Subapennine fauna, he reported that more than half of its species were known alive, a much higher proportion than among the Parisian fossils (he thought Parkinson's were probably similar to Lamarck's, though less well-described). And most of these still lived in the Adriatic: some had been assumed prematurely to be exotic and tropical simply because the Mediterranean fauna had not been studied closely enough. "The cause of all these differences is not, it seems to me, very obscure", he concluded; "considering the formations [*terreni*] of the last and modern origin [i.e., the Tertiaries], it is clear that the shells they contain correspond to those that live in the nearby seas". In effect, he treated all the Tertiaries as being of roughly the same age: in his opinion the Parisian fauna differed from that of the Subapennines for the same geographical or climatic reasons that the modern north Atlantic fauna differed from that of the modern Mediterranean.[83]

Brocchi's last (and subsequently most famous) chapter contained his "reflections on the loss [*perdimento*] of species". It dealt with the possible cause of the differences between fossil and living faunas in general, and between the Subapennine mollusks and their living counterparts in particular. He conceded that Bruguière's argument for "living fossils" (§5.2) was plausible in the light of discoveries in distant seas. But he was convinced that many fossil species were truly lost or extinct: Cuvier's quadrupeds certainly were, but probably many of his own and Lamarck's mollusks too. However, Cuvier claimed that his fossil bones were wholly those of extinct species, whereas Brocchi's molluscan fauna was a *mixture* of extant and extinct species (as was Lamarck's, in a different proportion). Lamarck had explained this in terms of continuous transmutation, the few identical species being those that had not yet had time to be transformed into something different (§7.4). Brocchi rejected transformism in favor of a conjecture of his own. It was that species, like individual organisms, have intrinsically limited life spans: "why not admit that species perish in the same way that individuals do, and that like them they have a fixed and determinate period for their existence?" Although speculative, the significance of this idea was obvious: it proposed a mechanism for extinction that was as natural in character as Cuvier's catastrophes, yet intrinsic and piecemeal rather than extrinsic and sudden; and it respected the reality of species as discrete entities or natural kinds, rather than dissolving them in an endless flux of transmutation. It also suggested, though less explicitly, that the *origin* of species might have an equally natural yet episodic mechanism, analogous to the birth of individuals.[84]

Finally, Brocchi applied his ideas on organic change to the longer perspective of geohistory. The ammonites found as fossils in the Secondary rocks of the Apennines were, he argued, "documents" that recorded "the world's former age"; he believed that neither they nor the other Secondary fossils were still alive in the ocean depths. His conjectural process of piecemeal births and deaths of species scattered through the vast expanses of deep time would naturally generate an ever-changing specific composition of the faunas found as fossils, gradually approaching that of the present day. Brocchi's historical analogy, and the conclusion he drew, echoed Cuvier's similarly antiquarian sentiments (§9.3):

81. Brocchi, *Conchiologia* (1814), 1: introduction, 19–21, 35–36; chaps. 1, 2.

82. Brocchi, *Conchiologia* (1814), 1: 143; referring to Olivi, *Zoologia Adriatico* (1792).

83. Brocchi, *Conchiologia* (1814), 1: 166; chap. 4.

84. Brocchi, *Conchiologia* (1814), 1: 227. Brocchi's theory (in his chap. 6) is discussed in its ultimately Darwinian context in Pancaldi, *Darwin in Italia* (1983), chap. 1, and "Teoria delle specie di Brocchi" (1987); it was first set out briefly in Brocchi, *Trattato mineralogico* (1807–8), 2: 256–57.

> Like, one could say, the obelisks of the Egyptians, on which the chronological history of the country is carved, they [i.e., fossils] present that of organic creation, as it were, provided one knows how the characters should be interpreted. . . . That there is a relation between the age of strata and the kinds of species, and that the more remote their origin the greater the number of shells they contain that are unlike those we know [alive], is an evident fact already attested by many naturalists.[85]

Conclusion

Brongniart and Cuvier had described the geognosy of the Paris region in a way that made it an exemplary reconstruction of the more recent portion of geohistory (from the time of the Chalk onwards). Parkinson, acting on a hint from Farey, applied Smith's methods to similar formations around London, noting their similarities to the French sequence. Webster then investigated a much clearer case on the south coast of England, defining what he termed the Isle of Wight Basin to match Parkinson's London Basin and the exemplary Paris Basin. Webster found spectacular evidence for a major movement of the earth's crust at this relatively recent time in geohistory, for some of these sediments (even younger than the Chalk) had been tilted into a vertical position. After Brongniart sent French specimens to London for comparison, Webster also found on the Isle of Wight an alternation of marine and freshwater formations that closely matched the Parisian sequence, thereby extending the same kind of eventful geohistory well beyond France.

At almost the same time, Brocchi's research showed that the Subapennine formation in Italy was comparable to the rocks in the "basins" further north; he defined them all as *Tertiary* formations, overlying (and therefore younger than) the Chalk and its equivalents. He attributed the relatively minor differences between the French (and English) fossils and those in Italy to contrasts of climate and geography analogous to those in the present world. On the other hand, the much larger differences between all the Tertiary faunas together, and the fossils found in still older formations, were of course attributed to a major contrast in age. However, Brocchi argued that such differences could have accumulated piecemeal in the course of time. His suggestion that species might have intrinsically limited life spans was highly speculative, but it did explain organic change in a way that might turn out to be more satisfactory than either Cuvier's notion of occasional catastrophes and mass extinctions or Lamarck's vision of the continual transmutation of organisms without any extinction at all.

Brocchi's work, added to that of Parkinson around London and Webster on the Isle of Wight, demonstrated within a few years the fertility of the doubly enriched geognosy that Brongniart and Cuvier had pioneered in the Paris Basin. This research showed on an international level that the Tertiary formations represented a distinctive major period of geohistory, more recent than the time of the Chalk. Furthermore, the differences in detail between the Tertiary sequences in different regions pointed to a highly contingent kind of geohistory: the "former world" might have been just as varied as the present, and its geohistory could certainly not be deduced from any overarching geotheory. Finally, as Cuvier had noted with his customary acuity (§9.3), the period represented by these Tertiary formations and faunas was of unique importance, because it could bridge the gap between the obviously "antique" or "former world" of the Secondary formations (as now redefined) and the "present world" of human history. The further development of this already fruitful line of research will be followed in the final section of the present chapter (§9.6); but first it must be set in a broader context, with a review of the changing practices of geognosy in the years immediately following the downfall of Napoleon.

9.5 GEOGNOSY INTO "STRATIGRAPHY" (1814–23)

Europe at peace

Back in 1812, while Cuvier's magnum opus was in press (§9.3), Napoleon had been leading his "Grande Armée" triumphantly into Russia. But the following year the emperor had retreated from Moscow in disarray, his army decimated not so much in battle as by the Russian winter. However, this military catastrophe had not destroyed the Napoleonic regime. Cuvier, for example, had been sent on a second educational mission to Italy and had again used his official travel for scientific purposes, collecting further fossil bones—or rather, making drawings as proxies for those he saw in museums—for the planned second edition of his *Fossil Bones*. But in 1814 his official trip to the Rhineland was aborted when he met the French army retreating, after further defeats, back into France itself. Later he witnessed the allies' victorious entry into Paris, which was swiftly followed by Napoleon's abdication and banishment to a comfortable exile on the Italian island of Elba, while the Bourbon monarchy was restored in the person of Louis XVIII.

Early in 1815, however, Napoleon escaped from Elba, landed back on the south coast of France, marched to Paris at the head of a growing force of supporters, and seized political control after the king had fled. The war resumed, and the xenophobic sentiments endemic in Britain plumbed new depths. As a small geological example, Jameson (or perhaps his publisher) apparently feared that Cuvier's nationality would detract from sales of the second edition of his own work, for he tried to dissociate the savant from the enemy, stating that "Cuvier is a native of Mumpelgardt in Germany". It was a tendentious claim, for Montbéliard—as its inhabitants always called it—had not been German territory, nor Cuvier a German subject, since it was annexed a quarter-century earlier (§7.1).[86]

Anyway, Napoleon remained in power only for what became known as the Hundred Days, before a combination of Austrian and British armies finally defeated him at Waterloo, not far from Brussels. He abdicated for the second time, and was dispatched to a far more remote (and final) exile on St Helena in the

85. Brocchi, *Conchiologia* (1814), 1: 234. In fact, it was still *not* known how to interpret the hieroglyphs, and hence Egyptian history.

86. Cuvier, *Theory of the earth*, 2nd ed. (1815), Jameson's preface, vii, n; with Napoleon finally defeated, the third edition (1817) silently deleted this footnote.

south Atlantic. After its false start the Restoration of the French monarchy then looked secure; so were the other European powers, as they negotiated among themselves to put the political clock back, as nearly as could be, to where it had been before the start of the Revolution.

For savants everywhere, all these political storm clouds had a silver lining in that the peace enabled them to move freely around Europe for the first time for more than a decade. Even during what turned out to be a false dawn, naturalists lost no time in traveling in both directions across the Channel. Cuvier was apparently too busy politicking to do so: his dutiful service under Napoleon, far from counting against him, was rewarded when Louis XVIII promoted him to be a full councilor of state. But some of his colleagues from the Muséum, including his brother and Lamarck, did visit London in 1814; among other naturalists' delights, they were able to see the great collection formerly owned by John Hunter, which had still been inaccessible at the time of the brief Peace of Amiens.[87]

Traveling in the opposite direction, Greenough, who until the previous year had been the first president of the Geological Society (§8.4), started out for France the very day after the first peace was signed: "to see Kings and Emperors, and Cuviers and Crocodiles", as a younger member of the society put it. In fact Greenough's primary goal seems to have been the mineral collections at the School of Mines; but he probably went also to meetings at the Institut and to Cuvier's celebrated *salon*, and anyway he returned with Parisian specimens for the Geological Society's nascent museum.[88]

Such shuttling to and fro was just a foretaste of the free exchange of ideas that savants throughout Europe hoped the peace would make possible; they could confidently rebuild the networks of personal acquaintance that had characterized the Republic of Letters long before. More specifically, the geologists among them could at last plan more ambitious travels, not only to meet others but also to extend their firsthand experience of the features in the field and the specimens in museums that were at the heart of geological debate. Anticipating such renewed fraternal contacts, the Geological Society created a new honorific category of "Foreign Members"; the first batch of seven unsurprisingly included Brongniart, Cuvier, and Werner.[89]

Brongniart on the fossil criterion

Within geology, the project likely to gain most from the new international contacts was the geognostic one of correlating the diverse rock masses that were now being classed as Primary, Transition, Secondary, and Tertiary, and of resolving them into a single sequence of general validity. Geognosts in many parts of Europe had detected clear sequences in the field, particularly in the Secondary formations, but correlations between one region and another remained uncertain and controversial (§2.3). Distinctive rock types, such as shelly limestone, red sandstone, and gypsum, seemed to recur at different levels. In such cases it was not clear which formation in one country or region should be correlated with which in another: Brongniart, for example, had claimed that the Paris gypsum did not correspond to any of the gypsum formations that Werner and his former students had described elsewhere (§9.1).

Independently or not, Smith and Brongniart had both found that fossils could help to distinguish between otherwise similar formations if they were given closer attention than had been customary among geognosts (§8.2, §9.1). Their new emphasis on fossils did greatly enrich their practice, but usually it remained geognosy, rather than being used as the basis for earth physics or geohistory. Their primary goal was to determine the true structural *order* of the formations, rather than to infer the perennial causal processes that had produced them or to construct a narrative of the distinctive events that they represented. Brongniart did enrich his geognosy into geohistory, but he seems to have had to be nudged in that direction by his collaborator Cuvier, who was no geognost (§9.1). Left to himself, Brongniart remained a fairly traditional geognost, as is shown both by his extensive travels in the German lands before the fall of Napoleon and by his subsequent lectures in Paris.[90]

The new and more settled peacetime situation now enabled Brongniart to broaden the scope of his geognosy, but it led him into difficulties that severely tested the new methods based on fossils. His position among Parisian savants had been strengthened when in 1815 he was appointed to fill the vacancy at the Institut left by the death of the elder Desmarest. The following year Louis XVIII reconstituted that Republican body into four new or revived academies under royal patronage; Brongniart, with Haüy and others, was assigned to a new section for mineralogy within a new Académie Royale des Sciences: it was in effect a royalist reincarnation of the First Class of the Institut and, still further back, of the old royal academy that had been abolished at the height of the Revolution (§6.3).

Brongniart continued to direct the porcelain factory at Sèvres and to lecture in Paris, and it was not until 1817 that he was able to embark on major new fieldwork. He then took his wife and their teenage son Adolphe on a long tour through Switzerland. He made good use of local geologists, both in the field and in their cabinets, and noted carefully the field evidence for the dramatic folding of the formations in the Jura and elsewhere. He visited famous fossil localities such as the quarries at Oeningen, where the strata—which he interpreted as Tertiary in age—had yielded so many fine fossils (Fig. 5.9), including the famous human "witness of the deluge" that Cuvier had turned into a giant salamander (Fig. 9.7). As he reported to Cuvier, the Freshwater Formation that they had jointly defined around Paris seemed to be not only very widespread but also as much involved in

87. See Outram, *Georges Cuvier* (1984), 84–89; Negrin, *Georges Cuvier* (1977), 141–46; Sloan, "Muséum de Paris vient à Londres" (1997).

88. Greenough, journal for 19 March to May 1814 (London-UCL, Greenough MSS); Buckland to Conybeare, April 1814, quoted in Gordon, *William Buckland* (1894), 14–15.

89. Geological Society, minutes for 3 February, 5, 19 May, 16 June 1815 (London-GS, OM/1). The very first Foreign Member was the Prussian geognost Leopold von Buch (see §10.2), who was already in England; a few other foreigners, among them Faujas, had previously been Honorary Members and were transferred to the new category.

90. Brongniart, journal for 11 August to 8 November 1812 (Paris-MHN, MS 2343); and Omalius's notes on his lectures in 1813, transcribed in Grandchamp, "Cours de géognosie par Brongniart" (1999).

subsequent folding as the far older Jura rocks, for he had found it in a near-vertical position that made it look much more *ancien* than it really was.[91]

However, Brongniart's growing sense that in such cases fossils were the most reliable criterion of relative age was challenged dramatically by the most important discovery of his whole trip. While he was in the Alps, he and a local naturalist (with a villager as their guide) climbed high into the massif of the Buet, on the far side of Chamonix from the still higher massif of Mont Blanc. On the narrow crest of the Rochers de Fis, they climbed as far as any rocks were visible beneath the remnants of winter snow. Brongniart noted that "all this terrain is certainly what is called Transition", an inference probably based on the types of rock he saw. Yet near the summit he found a hard black limestone containing fossils unmistakably characteristic of the Chalk; most notable was the "turrilite", an unusual and distinctive ammonite coiled in a helix (like a snail shell) instead of the usual plane spiral. This was doubly unexpected. If the fossils were to be believed, marine sediments dating from the time of the Chalk—only just before the start of the Tertiary period—had subsequently been upheaved over 2,500m above sea level; and the rock itself was about as different as it could be from the soft white limestone of the Chalk of northern France and southern England. The find posed a crucial problem for the new geognosy in the most striking way imaginable. Could fossils be given priority over the more traditional criteria of topographical position and rock type (§2.3), or could they not? Was the rock truly of Chalk age, or did it really come from a Transition formation with unexpected fossils?[92]

When, a few days later, Brongniart wrote to Cuvier to tell him what he had been doing, he failed to mention this striking discovery of Chalk fossils at high altitude and in an unfamiliar matrix. He was probably reluctant to disclose his own great uncertainty about it. Two years passed before he described it to the new Académie, and two more before he published in the *Annales des Mines* a paper that drew out its wider implications. The paper was on the fundamental issue of "the zoological characteristics of formations", taking the Chalk formation as a now crucial test case. The Rochers de Fis became the exemplar of a methodological rule that took the new style of enriched geognosy into situations far beyond anything that Smith had had to tackle. Brongniart admitted that his discovery was so unexpected that he had doubted how it should be interpreted. But he argued that it confirmed his previously tentative inference that, unless "*superposition évidente*" proved otherwise, the criterion of characteristic fossils should be given priority over all others. The hard black limestone that he had found high in the Alps might look like a Transition rock, in both its appearance and its situation, but Brongniart claimed that it was really the local representative of the Chalk, for it contained the same characteristic fossils.[93]

In itself, this was just a working rule for the more effective practice of geognosy, but it contained the seeds of dramatic implications for geohistory. To find Chalk rocks on top of the Alps implied that there had been major crustal movements even within the Tertiary period; and hence that the earth had remained, in geologically quite recent times, as dynamic as even Hutton could have desired. Brongniart was soon confronted with a still more striking case of an apparent exception to the rule of characteristic fossils, which could be resolved only by inferring even more startling kinds of upheaval in the earth's crust.

In 1820, three years after his Alpine fieldwork, he made a much more extensive tour of northern Italy. This was in preparation for the revision of his study of the Paris Basin, which he planned to enlarge into a general correlation of the Tertiary formations throughout Europe (see §9.6 below). On his way home, he crossed the Alps by the Simplon pass and then down the upper Rhône valley. On the way he visited Jean de Charpentier (1787–1855), the director of the salt mines at Bex and a son of Werner's colleague Johann von Charpentier of Freiberg (§2.3). The younger Charpentier showed him the fossils he had found in a hard dark rock, high up on the massif of the Diablerets that towers above Bex; but these fossils were as clearly of Tertiary age as those in the Buet massif had been of Chalk age. Brongniart climbed the mountain to see the spot for himself, but he seems to have regarded it, quite reasonably, as another case of the same kind. Although equally spectacular, and indeed involving even younger rocks, there was no reason for it to invalidate the policy of giving priority to the fossil criterion.[94]

Three years later, however, just as Brongniart was completing his book on the Tertiary formations of northern Italy, and using them to display the use of characteristic fossils in stratigraphical practice (see §9.6 below), the young Parisian geologist Léonce Élie de Beaumont (1798–1874) told him what he had seen during his own recent fieldwork on the Diablerets. Brongniart could hardly doubt what his highly competent former student reported, yet it seemed to imply a major failure in the fossil criterion that was proving so fruitful elsewhere. Élie de Beaumont had found that the strata with Tertiary fossil shells high on the mountain were *overlain* by a huge pile of ancient Secondary limestones and shales: "evident superposition" indicated that something was seriously amiss. Yet it turned out that the fossil criterion could be rescued; in classic terms, the phenomena could be saved. The superb exposures on the south face of the mountain convinced Élie de Beaumont that the older rocks had originally *underlain* those with Tertiary fossils, but had subsequently been thrust on top of them by an enormous sliding movement

91. Brongniart, "Voyage au Jura" notebooks, 7 July to 20 September 1817, and Brongniart to Cuvier, 14–19 August 1817 (Paris-MHN, MS 2344 and MS 627/134). He must have known that Webster had detected a similar case of geologically recent folding on the Isle of Wight, because although Webster's "theoretical sections" (Fig. 9.13) were only in Englefield's chorographical book, which Brongniart may not have seen, Webster had sent him an offprint of his paper, with its dramatic cliff sections (Fig. 9.14).

92. Brongniart, "Voyage au Jura" (see above), entry for 27 July 1817; the mountain and the fossil are depicted in *Terrains de sédiment supérieurs* (1823), pl. 2B, fig. 1, and pl. 7, fig. 3. Within the Buet massif, the narrow ridge of Rochers de Fis (2,733m) is about 10km northwest of Chamonix and 8km southwest of the peak of Le Buet itself (3,096m); visitors who are short of time (or energy) can get a superb view of it, and a distant appreciation of Brongniart's dilemma, by taking the cable car from Chamonix up to Saussure's favorite viewpoint of Le Brévent (2,524m). In modern terms, the rocks in question are part of the same huge Helvetic nappe that includes, further west, Saussure's dramatic fold structure at Nant d'Arpenaz (Fig. 2.25).

93. Brongniart to Cuvier, 14–19 August 1817 (Paris-MHN, MS 627, no. 134); Brongniart, "Caractères zoologiques" (1821). The *Journal des mines* had changed its name to *Annales* in 1816, after the Restoration, to mark the political break with the past, though its format remained unchanged. In principle, of course, the high altitude of the Chalk rocks could have been explained—on the traditional standard model (§3.5)—in terms of a falling sea level without any crustal movement at all, but in conjunction with the evidence for major folding that option had now become highly implausible.

94. Brongniart, journals for 1820 (Paris-MHN, MS 2345), cahier 4, notes for 5–11 September.

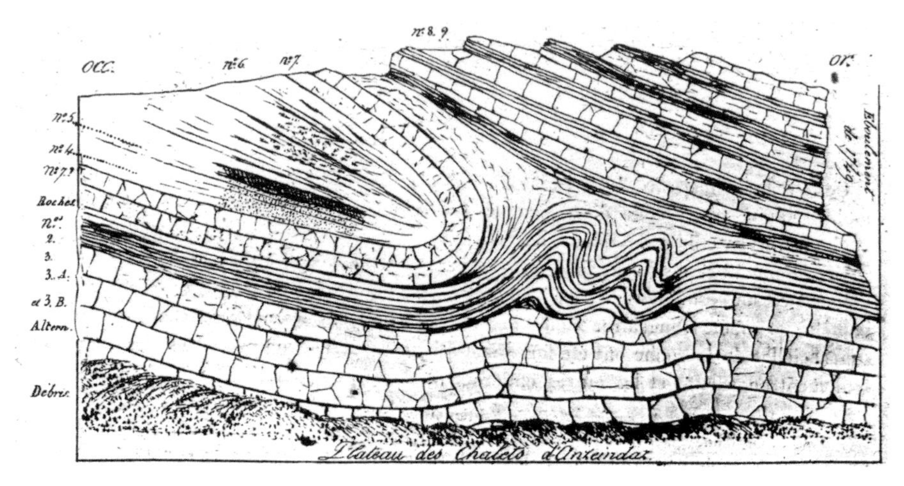

Fig. 9.17. The south face of the Diablerets in the Swiss Alps, as sketched by the young Léonce Élie de Beaumont and published in 1823 by his former teacher Brongniart. Traversing high up along the crest from west [*occ.*] to east [*or.*], the strata (no. 6) with Tertiary fossils would appear to underlie the Secondary formations (nos. 7, 8, 9). But the face of the mountain, seen from below, showed that "no. 7" was folded over, and "no. 8, 9" apparently thrust massively on top of the other rocks, so that the fossils were not after all deceptive or an invalid criterion of relative age. (By permission of the Syndics of Cambridge University Library)

[*glissement*] that had left some of them with "a remarkable overfold [*replis*]". So the case of the Diablerets, like that of the Buet, disclosed the magnitude of subsequent movements in the earth's crust, but the fossil criterion emerged unscathed and even strengthened (Fig. 9.17).[95]

Smith's "stratigraphy"

While Brongniart was testing the new style of enriched geognosy to its limits, in the evidently highly disturbed terrain of the Alps, Smith's earlier work in the gentler terrain of lowland Britain (§8.2) became—at long last—generally accessible. Since the British sequence of formations turned out to be exceptionally straightforward and undisturbed, it had the potential to serve as a standard of comparison for international and even global correlation.

In 1815, after almost endless trials and tribulations, Smith's great map (Fig. 8.5) was published with a brief textual explanation. It was followed the next year by the first installment of his set of paper proxies, displaying the fossils that he claimed were "characteristic" of each formation (Fig. 8.7); and in 1817, by his *Stratigraphical System*, explaining the basis of the work, and by the first of the sections that complemented his map (Fig. 8.6). Sadly, he had to sell all his specimens to the British Museum to raise much-needed funds (later, he spent time in prison for his debts). Banks, who until his death in 1820 remained Smith's loyal patron, helped with the negotiations. But the relevant curator, the German-born mineralogist Charles König (1774–1851), "showed an excessive contempt for fossils"—as one active woman collector told Greenough—and Smith's specimens were not even unpacked. Still less were they displayed to the public as they had been in Smith's

London house in earlier years (§8.2). So while Smith's map became potentially available to all, the primary evidence for the method on which it was based actually became less accessible than before.[96]

Still more sadly, sales of Smith's map and other publications were few and slow, perhaps in part because potential purchasers knew that Greenough's rival map was in an advanced stage of preparation. Some may have reckoned that the latter, as the fruit of collaborative effort by many of the members of the Geological Society, was likely to be more accurate and detailed. When Greenough's map was eventually published, five years after Smith's, that expectation was realized, at least in the opinion of some, though its compiler claimed tendentiously that his did not depend on the "theory"—about which he was highly skeptical—that formations were each characterized by their respective fossils.[97]

Smith's work has often been celebrated uncritically, so it is important to stress once more where its significance lay at the time of its publication. His *Memoir* made it clear that he himself regarded his map as important above all for showing the areal *distribution* of distinctive "Strata" (as he always called them) and their structural *order*. His tabular explanation of its coloring listed the formations "taken in succession from East to West, as the Strata occur", that is, in purely geographical order: not even from top to bottom, in their structural order, let alone from bottom to top or from oldest to youngest, in their geohistorical order. Although he emphasized the "regularity" of the formations, each with "properties peculiar to itself", the latter included not only fossils but also the physical character of the rocks and their chemical composition. The map, he said, depicted the formations "in the Order in which their edges successively terminate", a formulation that implied no temporality at all. More informally, he often likened the pile of formations to slices of bread and butter.[98]

It is no slight to Smith or his reputation to note the limitations of what he had done, indeed of what he had set out to do (§8.2). Like any other geognost—though he himself would have rejected that "foreign" label—he was concerned above all to

95. Fig. 9.17 is reproduced from Brongniart, *Terrains de sédiment supérieurs* (1823) [Cambridge-UL: 8340.b.23], 47; Élie de Beaumont and his colleagues always used his double-barreled family name in full. The spectacular face of the Diablerets (3,210m) that was the basis for his interpretation can be well seen from the path that crosses the Pas de Cheville (2,038m) to the south of the peak. Brongniart noted that other cases of large-scale folding were already known elsewhere in the Alps, but this one had been problematic on account of the fossil evidence.

96. Smith, *Delineation of the strata* (1815), *Memoir to the map and delineation* (1815), *Strata identified by organized fossils* (1816–19), and *Stratigraphical system* (1817); Etheldred Benett to Greenough, 22 June 1819 (London-UCL, Greenough MSS). See Eyles (J. M.), "Smith: sale of his collection" (1967), and Torrens, "Banks and the earth sciences" (1994), 69. Fortunately, geologists were not dependent solely on Smith's own proxy pictures, since they could also use his lists of named fossils in conjunction with the proxies in illustrated works such as Parkinson, *Organic remains* (1804–11), and above all the superb engravings in Sowerby, *Mineral conchology* (1812–46).

97. Greenough, *Geological map of England and Wales* (1820) and *Critical examination* (1819), 284–92. Modern geologists who feel inclined to revile Greenough and lavish posthumous pity on Smith should reflect on their own likely behavior in similar circumstances: for practical reasons they would surely purchase the later (and presumably more accurate) publication, even if its author had borrowed substantially—and with scant acknowledgment—from a more innovative predecessor.

98. Smith, *Memoir to the map and delineation* (1815), 2–3; on this and Smith's other publications, Eyles (J. M.), "Smith: a bibliography" (1969), remains invaluable. On the non-geohistorical character of his work, see especially Torrens, "Timeless order" (2001).

plot the three-dimensional *structure* of the land: the surface distribution of the rocks and their extension at depth, as depicted respectively on his map and on the sections that he intended to be "read" alongside it. Compared with this spatial and structural understanding, causal and even temporal questions were quite peripheral. The distinctive, and by now idiosyncratic, convention that he adopted for his map—with each formation colored most intensely at the base of its outcrop (Fig. 8.5)—displayed the three-dimensional structure very effectively. It also depicted incidentally what he believed was their original extent, since he thought they had been deposited from a sequence of discrete torrents, all flowing in the same direction, which had each stopped short where the corresponding deposit now outcropped, with little subsequent erosion. But any such causal speculations were in effect optional extras added to what was primarily a structural interpretation.

Smith certainly had used fossils, as he claimed, as an important criterion—and often the decisive one—for discriminating between otherwise similar formations and for tracing them across a fertile countryside in which exposures of any rocks were usually small and scattered. But fossils featured as just one criterion among several, even if they were the most reliable. Some of his rocks contained no fossils at all, or very few; but Smith regarded this as "fortunate", because the barren formations served to separate those with fossils from one other, and thus helped to discriminate between them. He did use historical analogies in passing, noting that "fossils are to the naturalist as coins to the antiquary", and describing fossils as "the medals of Creation, the antiquities of nature and records of time"; but these metaphors did no explanatory work and were no more than the clichés they had long since become.[99]

Smith remained unmistakably a geognost. But he was also an innovative one, in that he had tackled the Secondary formations rather than the ore-bearing Primaries; and he had transformed geognostic practice by demonstrating over an extensive area (the whole of England and Wales and a bit of Scotland) how a *detailed* study of fossils could lead to far more decisive results than ever before. At this point it is therefore appropriate to adopt Smith's own word "*stratigraphy*"—which is also the word used by modern geologists—to denote what has hitherto been called an enriched form of traditional geognosy. Smith chose his new word with care: stratigraphy was to be devoted to *describing* what he called the "strata", not to explaining them causally or interpreting them geohistorically.[100]

Geology at Oxford

Smith's relations with the early Geological Society had been strained for many reasons, not least his and Greenough's rivalry and mutual antipathy (§8.4). However, in the postwar years two younger members of the society adopted Smith's methods with great success, first in England and then internationally. Ironically, they owed their careers in geology, at least in part, to Greenough. Most of the Londoners who founded the Geological Society had been businessmen or professionals, not churchmen. However, in 1810 their president had begun to visit Oxford, the intellectual center of the established Church of England, to encourage a small group of dons who were showing interest in geology. Among the most active were William Buckland (1784–1856) and William Daniel Conybeare (1787–1857). Both were sons of Anglican parsons and had become fascinated by fossils at an early age; both were sent to Oxford to gain a degree, the necessary qualification for being ordained and following in their fathers' footsteps; and both chose a customary career path by staying on at the university after graduation, waiting for either a college fellowship or a "living" (i.e., parish appointment) to become vacant. Both attended the course on mineralogy by John Kidd (1775–1851), the university's "reader" in chemistry (and later in mineralogy), the only lectures being given at that time on any of the sciences of the earth. But unlike most of the society's early members they soon focused their attention on the Secondary formations. They may have sensed—from what they read and perhaps from talking with Parkinson and Webster—that this was a promising direction to take, and certainly it was the one that could be followed most easily from a base in Oxford.[101]

Buckland and Conybeare both began to do substantial fieldwork, around Oxford during term-time, and every summer all over England and as far afield as northern Ireland. Together or separately, they covered vast distances, largely on foot, plotting the outcrops of the various Secondary formations, establishing their sequence and collecting their fossils. At one point, for example, Conybeare told Greenough how he had solved the puzzling sequence on Shotover Hill just outside Oxford by finding exposures of a crucial formation that was "beautifully characterised by all its distinguishing fossils". His use of Smith's terms and Smith's method of enriched geognosy is unmistakable. In effect, they were replicating what Smith had done years before and on his own (§8.2). But Smith's map was still unpublished at this time, and they regarded their own work as contributing to Greenough's rival map, which probably seemed to them more likely to see the light of day.[102]

Buckland had been elected to a fellowship at his college, getting himself ordained in a hurry in order to qualify for the position; it gave him a home and a modest income—provided he remained unmarried—and enabled him to spend much of his time on geology. Conybeare took the alternative career path, leaving Oxford in 1814 to take up a country curacy and to get married; however, a generous

99. Smith, *Strata identified* (1816–19); *Stratigraphical system* (1817), ix–x.

100. Strictly speaking, Smith only introduced the adjective, as in the title of his *Stratigraphical system*; the noun (and its cognates in other languages) did not enter common usage until later in the century. In modern geology stratigraphy remains a primarily descriptive practice, though it is used as the basis for a variety of causal and geohistorical inferences. Much of what had earlier been included in "geognosy" was *not* transformed into stratigraphy, and was absorbed instead into other practices such as igneous, metamorphic, and tectonic (or structural) geology. It would therefore have been highly misleading to use the word "stratigraphy" earlier in the present narrative.

101. Buckland to Greenough, and Conybeare to Greenough, many letters for 1811–14 (London-UCL, Greenough MSS). Boylan, *William Buckland* (1984), has been used extensively as a source for the present narrative; it is based on massive manuscript sources and largely supersedes earlier studies, but see also Edmonds, "Patronage and privilege" (1978), and other sources cited in later footnotes. The meager secondary literature on Conybeare is quite incommensurate with his historical importance, but see North, "W. D. Conybeare" (1956), and the "fragment of autobiography" printed in Conybeare (F. C.), *John Conybeare* (1905), 114–45.

102. Conybeare to Greenough, undated but watermarked 1812 (London-UCL, Greenough MSS), referring to the distinctive Coral Rag formation.

legacy gave him a private income that enabled him to continue to pursue his geological interests in his free time. In 1812 Kidd had been appointed reader in anatomy—a position more compatible with his mainly medical interests—and had therefore looked for a suitable successor to take charge of his course on mineralogy. Conybeare, his first choice, probably intended to accept the position before deciding to get married instead. So in the event it was Kidd's second choice, Buckland, who was appointed to replace him as reader in mineralogy. The position had previously carried no income, apart from the fees paid by the few who attended the lectures; but Buckland successfully petitioned the Prince Regent (acting in place of the now pathetically insane George III) for a small salary, though he was expected to meet all his expenses out of it. It was a modest start to what was to become a career of outstanding importance for the natural sciences at Oxford.

Buckland lectured for the first time in 1814, on "the structure of the earth". The title signaled his intention to interpret the "mineralogy" of his readership in a broad sense that would encompass the new science of geology. The novelty of the lecturer and—in Oxford—of the topic generated a large audience of fifty-six subscribers. Buckland soon introduced two further novelties. He used visual aids such as maps, sections, and drawings of fossils, and solid specimens, in lecture rooms accustomed to nothing but the spoken word. And he took his class *into the field* outside Oxford, for example to Shotover, to demonstrate geology in the open air to academics accustomed to nothing but indoor studies (Fig. 9.18).[103]

These were not novelties elsewhere: in Paris, for example, Cuvier had routinely used visual aids in his lectures at the Institut since early in his career, and Brongniart was already supplementing his lecture course at the Faculté des Sciences with an all-day excursion outside the city. Even in England the use of visual aids was customary at the Royal Institution, and probably also at the Geological Society (it certainly was a few years later). But for Oxford the innovations were significant: they marked the point at which a science based on fieldwork, specimens, and

Fig. 9.18. A sketch of Buckland on horseback, lecturing to his geology class in the field, on Shotover Hill outside Oxford (the "dreaming spires" are in the far distance). Many of his audience—mainly dons rather than students—are properly attired for an academic occasion, in cap and gown. He appears to be using a bone to point out the geology, though this may just be a humorous touch by the artist.

images began to penetrate an intellectual heartland traditionally devoted almost exclusively to words and texts.[104]

As the title of his course implied, most of Buckland's lectures were devoted to a systematic description of the successive rock formations comprising "the structure of the earth", in the usual geognostic order from lowest to highest: he told Conybeare that "my lecture on the basin of Paris will be among the last in the set". To accompany his course he produced a large printed sheet entitled "Order of superposition of strata", listing the formations known in England; he displayed a map of Europe on which he incorporated material from maps of Germany, France, and the Netherlands that Conybeare had compiled from the published literature; and among his other visual aids was "a large section of Europe" that Conybeare had drawn, likewise relating the structure and formations of Britain to those on the Continent. There was nothing narrowly national about Buckland's (or Conybeare's) approach to the science; and a reference to a map of North America suggests that they were both thinking in global terms. With Conybeare's assistance, Buckland clearly intended to integrate the best Continental practice of geognosy—including the enriched variety that Brongniart and Cuvier had recently demonstrated—with what they and others had been finding in Britain by following, directly or indirectly, the similar example of Smith. But like most of those geognosts their primary goal was to determine the correct structural *order* of the formations—identified with or without the help of fossils—not to use them as nature's documents to trace nature's *history*. Both Buckland and Conybeare were practicing geognosy, not geohistory.[105]

A global standard for stratigraphy

The injection of Smithian methods into geognosy in England—in which both Buckland and Conybeare played leading parts—is neatly epitomized by the evolution of what began as a small popular book on geology. Its author, William Phillips (1773–1828), was a London publisher and founding member of the Geological Society. His *Outlines of Mineralogy and Geology* (1815), designed for the use of beginners and "especially of young persons", described the successive formations in the usual order from lowest to highest, granite to alluvium; in line with the society's policy, he attributed the recent progress of the science to its focus on "facts" and its abstention from geotheory. The following year a new edition added

103. Fig. 9.18 is reproduced from Gordon, *Life of Buckland* (1894), 29, reproducing in turn a sketch by the Viennese amateur geologist Count Breunner, probably while he was visiting Buckland in 1819; Buckland MSS, list of subscribers for 1814 (Oxford-MNH, Misc. mss. 13); Edmonds and Douglas, "Oxford geological lecture" (1976).

104. Brongniart's course for 1813–14 (Paris-MHN, MS 643) included a seventeen-hour excursion to St Ouen and Franconville, and was probably not the first occasion of its kind. The text-centered culture of Buckland's Oxford is rightly emphasized in Rupke, *Great chain of history* (1983).

105. Buckland, "Order of superposition", 1814 version, the first of eight versions (to 1821): see Torrens, "Banks and the earth sciences" (1994), 68. Buckland to Conybeare, April 1814, quoted in Gordon, *Life of Buckland* (1894), 14–15; Conybeare to Greenough, 24 January 1823 (London-UCL, Greenough MSS); see also his "Geological map of Europe" (1823). The American allusion was probably to the map in Maclure, "Geology of the United States" (1809).

(with permission) a simplified version of Smith's new map, and Phillips noted that it proved that England was a "regularly stratified country"; although greatly reduced in size, the map made Smith's achievement far more widely known than the large and expensive original. Like Webster (Fig. 9.12) and a rapidly growing number of other geologists, Phillips ignored Smith's style of graded coloring and produced a geological map on which the flat color washes emulated earlier maps such as that of the Paris region (Fig. 9.1).[106]

Two years later, having absorbed Smith's work more thoroughly, Phillips reorganized his material to form his *Selection of Facts*, which in effect equated geology with stratigraphy, and described the formations of England and Wales in Smith's order, from top to bottom. To illustrate their wider significance, Phillips also printed a version of Buckland's tabular "Order of superposition", showing a provisional correlation between Britain and the Continent, while almost apologizing that "this classification is necessarily theoretical in part". Conybeare, with characteristic generosity, then plied Phillips with so many suggestions for the further improvement and updating of his book that Phillips invited him to collaborate in a major revision. In the event, Conybeare's contributions were so substantial that their joint *Outlines of the Geology of England and Wales* (1822) was published with Conybeare, the younger man, rightly shown as its senior author. This completed the transformation of a little book for beginners into a substantial reference work for all serious British geologists, and a standard of comparison for their colleagues throughout the scientific world. "Conybeare and Phillips" soon became the epithet that was used by all—with a mixture of respect and affection—for this indispensable handbook.[107]

"Conybeare and Phillips" epitomized the way that Smithian or "stratigraphical" geology became, within a few postwar years, a major focus of research among the members of the Geological Society, to the point that "geology" itself became almost a synonym for stratigraphy. And since Smith's methods could not be applied to the Primaries (without fossils), and only with difficulty to the Transition rocks (generally with rare fossils), the effect was to divert attention away from all those older rocks and on to the Secondaries. Significantly, "Conybeare and Phillips" stopped short at the base of the Secondaries, and its planned "Part II" on the Transition formations and Primaries never appeared. More specifically, the formations most amenable to the use of characteristic fossils were those that Smith had mapped with greatest confidence and accuracy, namely those from the Chalk downwards as far as what Conybeare named—for the first time—the "*Carboniferous*" formations, those associated with the economically vital coal deposits. In fact these comprised *all* the British "Secondary" formations, as now redefined by the excision of the Tertiaries overlying the Chalk.

Even before Conybeare's collaboration with Phillips, he and Buckland were prominent among those who were developing stratigraphical geology in Britain. At the end of 1818, Buckland read a paper at the Geological Society on the Secondary formations in the area of the important Somerset coalfields near Bristol, in the heart of Smith's original territory (Fig. 8.5). He then persuaded Conybeare to collaborate in a more ambitious study, which they presented in London a year later. Their massive joint paper described *two* distinct sets of formations separated by a major unconformity, an inference that was supported by a detailed map and some superb sections. This structure implied that there had been an important episode of crustal disturbance and subsequent erosion *within* the time represented by the Secondaries (Fig. 9.19).[108]

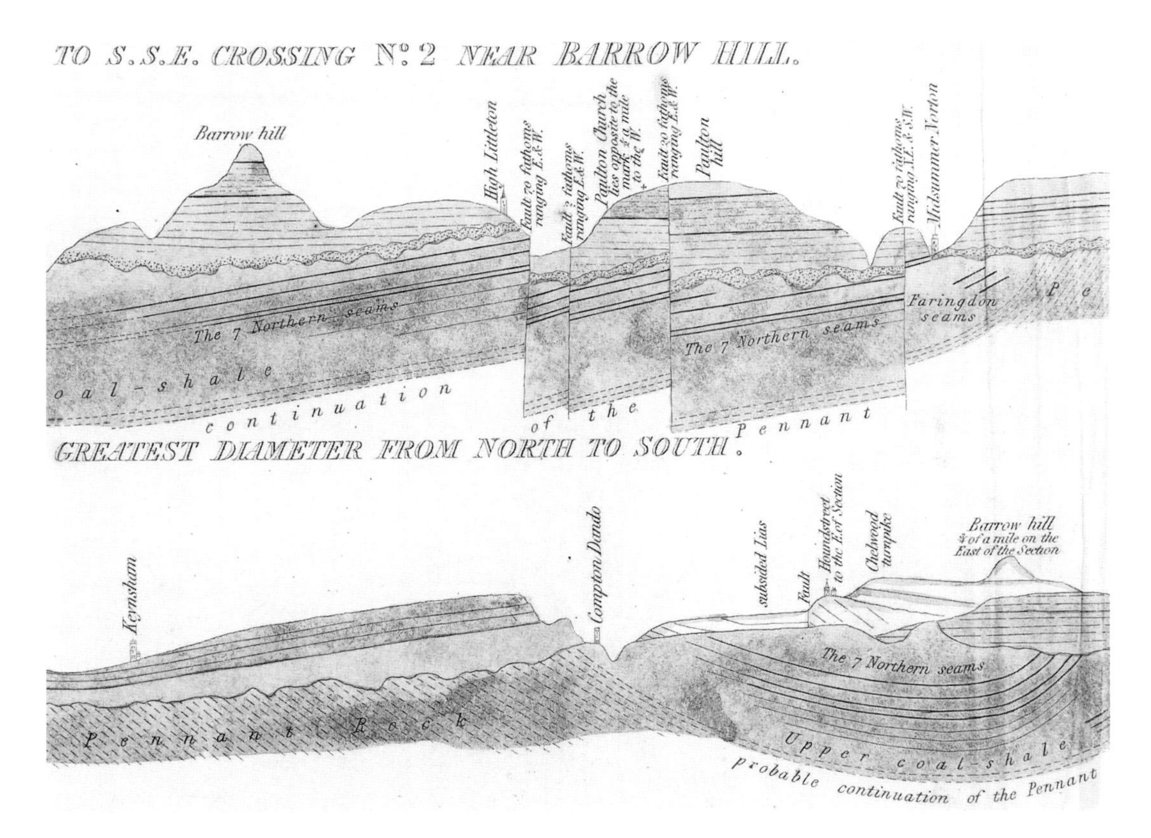

Fig. 9.19. Parts of two of the sections drawn (with conventional vertical exaggeration) through the Bristol region in the west of England, illustrating the paper by Buckland and Conybeare read at the Geological Society in London in the winter of 1819–20 (and published in 1822). It depicts the unconformity between the "Carboniferous" formation of the Bristol coalfield and the overlying "Oolitic" (in modern terms, Jurassic) formations. This showed that there had been a major episode of folding followed by erosion, *within* the time represented by all these Secondary formations, implying in turn a more complex and eventful geohistory than any simpler sequence would have suggested. Combined with their equally detailed map of the region, their sections exemplified the precision of the new "stratigraphical" geology based on the use of Smith's concept of characteristic fossils. (By permission of the Syndics of Cambridge University Press)

106. Phillips (W.), *Outlines* (1816), pl. 5. The latter convention became the standard for almost all geological maps (and remains so for modern ones): what they lost in the illusion of solidity was compensated by the geological sections that were becoming their equally routine accompaniment.

107. Phillips, *Selection of facts* (1818), 19; Conybeare and Phillips, *Geology of England and Wales* (1822).

108. Fig. 9.19 is reproduced from Buckland and Conybeare, "South-western coal district" (1822) [Cambridge-UL: Q365.b.12.6], part of pl. 32 (the corresponding map is pl. 38). Buckland's preliminary paper was read on 18 December 1818 and 1 January 1819; the joint paper, on 17 December 1819, 7, 21 January and 17 March 1820 (London-GS, OM/1). The style of these sections—drawn freehand but with much accurate local detail—is in contrast to the more diagrammatic style (with *ruled* lines) generally adopted by both Farey and Smith (Fig. 8.6).

The intensive survey of the Bristol region by Buckland and Conybeare was just one of a growing number of local studies that amplified Smith's map and Greenough's later one. Complemented by textual descriptions of all the formations, as summarized in "Conybeare and Phillips", this stratigraphical work made England the envy of Continental geologists and gave them an incentive to correlate their own formations with what was soon being treated in practice as an international standard of comparison. Buckland started this process, after meeting Werner during his first Continental tour (see §10.4), when he added Werner's list of formations to his printed tabular summary of English ones. On his next tour, in the summer of 1820, he made the correlation of British and Continental formations his main objective. He and Greenough went first to Paris to consult the leading savants and to gather suggestions of where to go and what to see. Leaving the capital they traveled south to see first the famous extinct volcanoes of Auvergne and then turned eastwards through the Alps. At some point Greenough went his own way, but Buckland continued as far as Prague before returning home to Oxford.[109]

The fruit of all this fieldwork was a paper that Buckland published the following year in the *Annals of Philosophy*—a recent addition to the burgeoning range of British scientific periodicals—which was immediately translated for the veteran *Journal de Physique* in Paris. It was "The structure of the Alps and adjoining parts of the Continent"—"structure" in the sense of geognostic sequence rather than tectonics—but above all on "their relation to the Secondary and Transition rocks of England". It set out a remarkably confident correlation of all the major formations, ranging from the Transition rocks upwards through the Secondaries to the Tertiaries and Superficial deposits; it traced them as widely as Buckland had seen them, not so much in the Alps themselves but rather on their flanks to north and south. For example, the Oolite formations of England (including many of the "oolitic" limestones widely used as building stones) were correlated with the "Jura" limestones of the Jura hills themselves and similar ones elsewhere (all, in modern terms, Jurassic in age). As was now becoming usual in geognosy, Buckland used a combination of rock types and fossils as the criteria for making his correlations, which on this level of generality were fairly straightforward. His paper had a huge impact on geologists throughout Europe and even beyond it, setting up a standard sequence that could be used internationally, but which at the same time was open to correction and improvement. It was complemented soon afterwards by Conybeare's sketch for a geological map of most of Europe, which synthesized all that could be known from the published record.[110]

Conclusion

The well-established science of geognosy continued to flourish in the early postwar years. At the same time, the primacy of characteristic fossils, treated as a new criterion of relative position in the geognostic pile, was put to the test by the puzzling cases revealed by Brongniart's extensive fieldwork in the Alps. But the criterion emerged unscathed and even strengthened, and seemed set to bring increasing order and precision to the regional and even global correlation of Secondary formations, and perhaps also to the Tertiary ones above and the Transition ones beneath. Brongniart's fieldwork highlighted the major movements of the earth's crust that had disturbed even the Chalk and Tertiary formations in certain regions, suggesting that the earth had remained dynamic throughout its history.

The use of fossils as a criterion for identifying and correlating formations was greatly strengthened when the belated publication of Smith's English survey made the example of his work generally available. Where the practice of geognosy could be enriched by the detailed use of fossils, it was swiftly transformed into what Smith called "stratigraphical" geology. Conybeare played a prominent role in codifying this new practice within Britain, and Buckland in correlating the English sequence with formations on the Continent. By the early 1820s the Secondary sequence for the whole of Europe was becoming well established, at least in outline. With this framework in place, research that was focused on the overlying Tertiaries could hope to fill the geohistorical gap between the Secondary formations and the modern world. This is the subject of the final section of this chapter.

9.6 THE TERTIARY ERA ESTABLISHED (1816–25)

Brongniart in Italy

In the first years of peace in Europe, the Secondary formations became the heartland of the new stratigraphical geology, defined by its systematic and detailed use of fossils; and the English sequence was rapidly accepted as the best standard for comparison with other parts of Europe, and even—more tentatively—with the rest of the world. For the most part, stratigraphy remained a descriptive science of structural order, like the geognosy from which it had evolved. But it could be, and was, used as a foundation for a further enrichment or transformation into geohistory. For example, Cuvier had shown that the Secondary formations contained the remains of strange reptiles such as the Maastricht lizard and the ptero-dactyle, but no mammals at all (§9.3); added to the ammonites, belemnites, and other distinctive fossils already well-known in the Secondaries, it became increasingly clear that they represented a "former world" that was strikingly different from the present.[111]

However, if stratigraphy were ever to be enriched more generally into geohistory, the focus would need to shift to the Tertiary formations, which alone could bridge the gap between the unfamiliar world of the Secondaries and the familiar

109. Boylan, *William Buckland* (1984), 99; the archival record of this tour is patchy and the itinerary not altogether clear. Breunner the Viennese amateur was a third member of the party, as far as Prague.

110. Buckland, "Structure of the Alps" (1821), and Conybeare, "Geological map of Europe" (1823), pl. 19; see Boylan, *William Buckland* (1984), 107, and Rupke, *Great chain of history* (1983), 125. The *Journal de physique* had continued under a new editor after La Métherie's death in 1817, though with increasing competition from new periodicals.

111. It is beyond the scope of this volume to describe the further reptilian fossils, notably the ichthyosaur and plesiosaur, and then the megalosaur and iguanodon, that strengthened still further this sense of the strangeness of the world of the Secondary era. The discussion of significant new specimens of these fossils was just beginning in the period covered by this section, but only came to fruition in later years. I hope to describe and analyze this research in a sequel to the present volume.

"present world" of human history. Cuvier had noted perceptively that the Tertiary formations of Italy, in particular, might play that crucial role (§9.3). Brocchi had promptly obliged by describing the rich Subapennine fauna in exemplary fashion; and further north, in England, Webster had used the Paris Basin as a model for defining an Isle of Wight Basin, which he had related in turn to the London Basin described by Parkinson (§9.4). In short, the pioneer study of the Paris Basin (§9.1) was proving immensely fruitful all over Europe, acting as a standard of reference for the Tertiaries in just the same way that Smith's English sequence was for the underlying Secondary rocks. So it is not surprising that Brongniart, the chief instigator of the Parisian work, was expanding the horizons of his research. Since Cuvier intended to publish a revised edition of his *Fossil Bones*, in which their originally joint work around Paris would again have an important supporting role, Brongniart had a strong incentive to describe and correlate *all* the Tertiary formations of Europe. He had already made a good start with this project when he embarked on fieldwork that traced the equivalents of the Parisian formations in other regions of France (§9.2) and then—when the peace made extensive travel easier—in Switzerland (§9.5).

In 1820 Brongniart expanded his fieldwork still further, with a major tour through Italy; he took his son with him again, but Adolphe was now a budding naturalist old enough to be a useful companion. As usual on such trips, Brongniart made full use of local naturalists, studying their collections of fossils and being shown the relevant rocks in the field. Near Le Puy in the Massif Central, one such naturalist had found a palaeotherium jaw in what Brongniart regarded as the equivalent of the Freshwater Formation around Paris. Later, down in Provence, Adolphe found the telltale gyrogonite (Fig. 9.5) in a similar formation near Aix, complete with gypsum beds that made it, as Brongniart told Cuvier, a close analogy to "our gypsum formation in Paris". So when they reached the Mediterranean at Marseille and Brongniart added that his fieldwork had confirmed "the geologi-

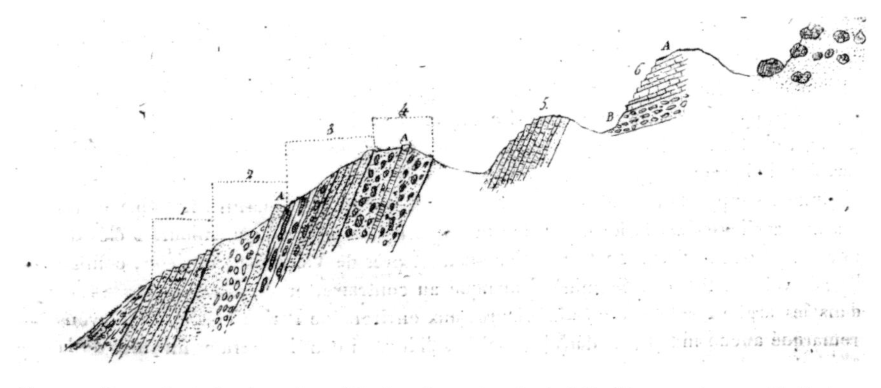

Fig. 9.20. Brongniart's sketch section of Tertiary formations in the hill of Superga, just outside Turin, which he was shown by a local naturalist during his Italian fieldwork in 1820; the abundant fossil shells were in bed 5, halfway up the hill. He published the section in 1823 in his monograph on the Tertiaries of northern Italy. It was further evidence that major crustal movements had taken place at a quite recent time in geohistory. The highly tilted strata also showed how commonsense assumptions could be misleading: climbing up the hill took one down the sequence. (By permission of the Syndics of Cambridge University Library)

cal principles that I have long professed", he was almost certainly alluding to the validity of characteristic fossils, now being tested on the Tertiaries over greater distances than ever. After crossing into Piedmont and reaching Turin, the city's professor of zoology took them to see a famous fossil locality nearby, with abundant Tertiary shells; the strata were almost as highly tilted as Webster's in the Isle of Wight (Fig. 9.14), which added another case of geologically recent upheaval (Fig. 9.20).[112]

Brongniart then turned south to trace the Subapennine formations down the spine of Italy. At Piacenza he studied the fine collection that Cortesi had been amassing in the years since his earlier one went to the Milan museum to be described by Brocchi (§9.4). On the other side of the Apennines, at Volterra, he noted a "freshwater formation as widespread as it is thick", with the usual associated gypsum beds, overlying the marine sediments with Subapennine fossils that Brocchi had described. In Rome he met Brocchi himself, who showed him further Ter-

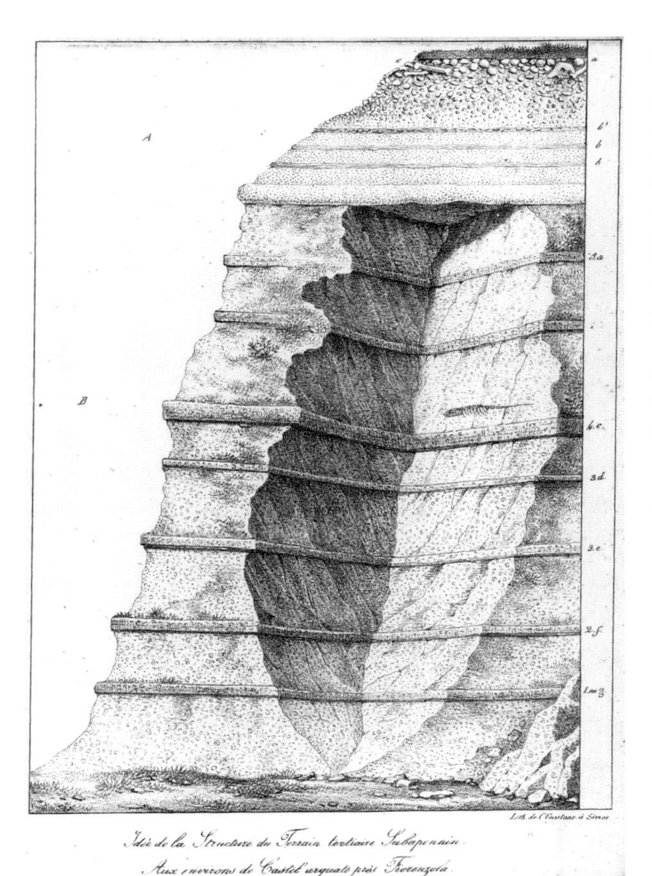

Fig. 9.21. "Idea of the structure of the Subapennine Tertiary formation": a lithographed drawing of a quarry that Brongniart visited in 1820 at Castell'Arquato (at the foot of the Apennines between Piacenza and Parma), showing the horizontal Subapennine beds (B) that yielded the superb fossil shells that Brocchi had described a few years earlier (Fig. 9.16); a fossil "whale" is also shown. These beds were overlain by Superficial sands and gravels (A) with the skull and bones of an elephant and other such fossils; Brongniart took it for granted that a vast span of time must have separated the two sets of deposits. Lithography—of which this is the first example to be reproduced here—had been invented at the turn of the century, but it was only in the early 1820s that it began to be exploited extensively for scientific purposes. It enabled an artist's drawing to be reproduced directly, without having to use an expensive copperplate or employ a skilled engraver, and was also better suited to the depiction of many geological subjects. (By permission of the British Library)

112. Fig. 9.20 is reproduced from Brongniart, *Terrains de sédiment supérieurs* (1823) [Cambridge-UL: 8340.b.23], 26, described on 27–33; bed 6 was the same as 5, but with a lower dip and evidently faulted from the rest. Brongniart, notebooks for 1820 (Paris-MHN, MS 2345), notes for 18–19, 25–26 April, 21 May; Brongniart to Cuvier, 26 April 1820 (Paris-IF, MS 3242/6). Their local informants in Le Puy and Turin were, respectively, the geognost Jacques Mathieu Bertrand-Roux (who then accompanied them for part of their tour), and the conchologist Franco Andrea Borelli.

tiaries; and he went as far as Naples before doubling back to Florence and Venice. He then worked his way westwards to Milan by way of the Tertiary formations in the foothills of the Alps, including the area that Arduino had made famous half a century earlier (Fig. 2.32) and the famous fossil locality of Monte Bolca (Fig. 5.7). Finally he crossed the Alps by the Simplon pass, saw the Tertiaries at high altitude on the Diablerets (§9.5), and traveled home by way of Geneva. This geological Grand Tour gave him ample material for a thorough review of all the Italian Tertiaries and their fossils. The following year he presented the Académie with an account of those in the north of Italy, which had been relatively poorly studied hitherto, and later expanded it into a substantial monograph. Above all, he saw with his own eyes, while still in the field, the Subapennine formations that Brocchi had made famous throughout Europe, and which—as Cuvier had suggested—were the most important Tertiaries to compare with those around Paris (Fig. 9.21).[113]

Prévost on the Vienna Basin

When Brongniart reached Geneva on his way back to Paris, he found a letter waiting for him from Prévost, his former student and companion on earlier field trips, announcing his own recent return to Paris. When the wars ended Prévost and a friend had migrated to Austria, hoping to make their fortunes in the burgeoning textile industry by setting up a mill for spinning linen yarn. In his free time Prévost had sustained his interest in geology by exploring the stratigraphy of what was already recognized as a small Tertiary "basin" with its outlet at Vienna, hoping to compare it with the Parisian one. He had had no competition from the Viennese geognosts, who were experts on the hard old rocks of the Austrian empire but had given little attention to the soft Tertiary formations in their own backyard. Prévost had made good progress with his survey, but in 1818 a disastrous fire had destroyed his house and most of his notes and specimens. He had abandoned his business and returned to Paris, where a generous father-in-law was supporting him while he tried to build himself a new career as a naturalist.

Prévost told Brongniart that his survey of the Vienna Basin was "a work that I undertook on your principles". Most of the species of fossil shells that he had collected were identical to Brocchi's and quite different from Lamarck's. This suggested to him that the formations of the Vienna Basin were equivalent to the Subapennines, not to the Coarse Limestone of the Paris Basin. Brocchi had inferred that the faunal contrast between his own shells and Lamarck's was due to geographical or climatic differences between the two regions, at the time when the mollusks were alive (§9.4); but Prévost, in contrast, attributed it to a difference of *relative age* between the deposits. Up to this point, all the geologists working on the Tertiary formations—including Brongniart—had been tacitly assuming that they contained just one major faunal assemblage, distinct from that of the Chalk and other Secondaries, and (to a lesser extent) from that of the present world. Prévost suggested on the contrary that there were *two* distinct and successive Tertiary faunas: Brongniart's principles could be applied even *within* the Tertiaries, which might therefore represent a more significant portion of geohistory than had previously been apparent.

However, Prévost was applying those principles in a subtly different way from Brongniart (or, had he known it, from Smith). Rather than looking for strikingly distinctive fossils that might be confined to the formation of which they were "characteristic", Prévost was treating the species of Tertiary fossils as if they were individuals in a population, so that a formation might be characterized by an assemblage with a certain overall composition, even if no single species was found exclusively in it. On this basis, Prévost suggested to Brongniart that the Viennese formations and the Subapennines might all be *younger* than the Parisian ones; or, if they were equivalent to any in the Paris Basin, it would be to the *upper* marine formation—above the distinctive gypsum beds—and not to the lower marine formation of the Coarse Limestone, from which most of Lamarck's fossils had come (see Fig. 9.3). Far from being a mere stratigraphical detail, this was a profoundly innovative suggestion, for it opened up the possibility of turning Tertiary stratigraphy into a truly *geohistorical* narrative, marked by a gradually changing marine fauna.[114]

Prévost told Brongniart that he had already expounded these ideas at the Société Philomathique in Paris. Soon after Brongniart's return to the city, and almost certainly under his patronage, a more substantial account was read in the more august setting of the Académie and then published promptly in the *Journal de Physique*. Like Brongniart and Cuvier, whose work he treated as the indispensable standard of comparison, Prévost used geognosy as a foundation for geohistory: he first described the Secondary formations that surrounded the Vienna Basin, and then the "Tertiary or modern formations" that filled its interior, the latter recording "the last revolutions" that the earth had undergone. As in the Paris Basin, the highest formation with fossils contained shells clearly of freshwater origin. Below it was one with abundant marine shells. Of all those that he had collected from the latter, he compared the sixty-three species that he had managed to rescue from the fire with those from Italy and from Paris; his fauna had many species in common with Brocchi's but only two in common with Lamarck's. Prévost drew from this a geohistorical inference: "*The modern formations observed in Austria and Italy should be considered as the simultaneous product of the last time that seawater covered the present continents, while the Coarse Limestone formation around Paris (of*

113. Fig. 9.21 is reproduced from Cuvier, *Ossemens fossiles*, 2nd ed. (1821–24), 2 (1822) [London-BL: 443.g.12], part 2, pl. 2a (illustrating Cuvier and Brongniart, "Couches des environs de Paris"), lithographed by Charles-Louis Constans, one of Brongniart's artists at Sèvres. The fossil shells were mostly in the numbered harder beds, but the "whale"—a prize exhibit in the Milan museum—had been found in one of the intervening softer beds of "clayey silt". Brongniart, notebooks for 1820 (Paris-MHN, MS 2345) and *Terrains de sédiment supérieurs* (1823). Brocchi, "Observations géologiques sur les Apennins" (Paris-MHN, MS 871), a manuscript translation of the geological part of his *Conchiologia subapennina* (1814), was certainly made at an early date, and probably around this time and for Brongniart's benefit.

114. Prévost to Brongniart, 8 August 1820 (Paris-MHN, MS 1967/599). He had been living and working at Baden, 25km south of Vienna and in the middle of his basin; Bork, "Constant Prévost" (1990), gives valuable biographical information.

which the shelly beds at Grignon are a part) must be regarded as belonging to a different epoch and a much more remote age of the earth".[115]

As he had already suggested to Brongniart, Prévost argued that his own and Brocchi's marine formations should be correlated with the marine strata *above* the Gypsum formation in the Paris Basin, whereas Lamarck's fossils came from those *below* the gypsum: "the Tertiary terrain comprises two quite distinct marine formations, between which there must necessarily have been a long sequence of ages [*siècles*]". The putative necessity followed from the model of organic change with which Prévost accounted for the faunal contrast. If the molluscan fauna had changed *gradually*, the sheer passage of time could have produced an almost complete turnover in species between the older formation and the younger ones. Prévost maintained that this simply confirmed the "law" or generalization that Cuvier and others had already formulated, namely "that fossil organisms . . . differ *the more* from those now existing, the more ancient the beds in which they are buried". He argued that, if this were true, the sea that returned to the Paris region after the freshwater episode (represented by the Gypsum formation) would have sustained a fauna that had inevitably changed during the interval since the much earlier sea that deposited the Coarse Limestone.

How exactly the fauna had changed remained obscure. Prévost had evidently read Brocchi's work with care and would surely have noticed his speculation that species might have inbuilt limited life spans, so that whole faunas might have changed in piecemeal fashion by the successive "births" and "deaths" of discrete species (§9.4). He must also have been aware of the equally speculative ideas circulating among Parisian naturalists, inspired by the now elderly Lamarck, about the possible slow transmutation of species themselves, without any extinction at all (§7.4). Prévost's language, for example his inference that "in the chain that forms the different degrees of organization there have been graduated and perceptible modifications", suggests that he may have favored the Lamarckian option. But in fact the mechanism of faunal change, however important to zoologists, was irrelevant to his geological claim. This was that the study of *an entire fauna*—not just a handful of "characteristic" fossils—could be used to determine the relative age of any formation: "at any given epoch since the [primal] creation, the sum of the similarities or differences [between any two assemblages] should always be in direct proportion to the time that has elapsed before or since that epoch". Although Prévost did not use de Luc's metaphor, and may not even have known of it, he was in effect suggesting that a *natural chronometer* could be derived from the ever-changing composition of fossil faunas. This might be valid, whatever the process by which one set of species was gradually replaced by another.[116]

Prévost applied these general ideas of gradual faunal change to the specific case of the Vienna Basin. He followed Brocchi in adapting the old standard model of geotheory (§3.5) to the more limited task of explaining the geohistory that had produced the Tertiary formations. Apparently unaware of the growing evidence for major crustal movements even within the Tertiary era (§9.5), he took for granted the stability of the earth's crust during that time, assuming simply that global sea level had fallen progressively, though not continuously and perhaps with minor fluctuations on the way. Taking note of barometric measurements of the altitudes at which the relevant strata were found, in France and Italy as well as in Austria, he sketched a narrative reconstruction of the gradual emergence of the present continent of Europe, as the sea retreated from the different basins and was replaced by bodies of fresh water and eventually by dry land. His conclusion was that the Tertiary formations were not to be dismissed or sidelined as merely local deposits of little interest: they were as widespread as any of the Secondaries and represented a major portion of geohistory, and one that could be reconstructed in some detail. Brongniart may have felt uneasy that his former student might be toying with transformist ideas—and his colleague Cuvier even more so—but he must have welcomed Prévost's careful account of yet another Tertiary basin to add to the roster of all those that had now been described around Europe.[117]

Brongniart on the Tertiary era

In fact Brongniart adopted most of Prévost's suggestions and conclusions, though with scant acknowledgment, when he published his review of Tertiary stratigraphy two years later. The occasion was the second edition of Cuvier's *Fossil Bones.* As already mentioned, a revised version of their joint account of the Paris Basin (§9.1) played the same role as before in this massive work, providing both a stratigraphical and a geohistorical framework for Cuvier's strange mammals from the Paris gypsum (§9.3). Like the larger work, the geological part was greatly enlarged, not just by the revision of Parisian stratigraphy but also by Brongniart's totally new descriptions of what he regarded as the equivalents of the Parisian rocks elsewhere. In effect, this turned a local monograph into a general account of the Tertiary formations throughout Europe (and even, briefly, beyond it). The new *Geological Description of the Beds around Paris* (1822), published as a separate volume as well as being incorporated in Cuvier's magnum opus, was recognized at once as a work of far greater importance for the science than its modest title suggested.[118]

Brongniart treated seven Parisian "*terrains*" as the local representatives of seven major "*formations*" that were valid at least across Europe and perhaps universally. He described each in turn, from lowest to highest, but treated that geognostic order as being at the same time a geohistorical order from oldest to youngest. The lowest and oldest formation was that of the Chalk, to which he gave close attention because it formed in effect the baseline for the Tertiary formations that followed.

115. Prévost, "Constitution physique et géognostique" (1820), 461 (the quoted sentence italicized for emphasis), read at the Académie des Sciences on 13 November 1820; the sole (unnumbered) plate comprises a rather crude lithographed map and section.

116. Prévost had subscribed to Lamarck's lectures in 1809, 1810, and 1812: Corsi, *Lamarck* (2001), 354. On his ideas of faunal change see ibid., 260–62 [or Corsi, *Age of Lamarck* (1988), 212–13].

117. Prévost, "Constitution physique et géognostique" (1820), 465–71.

118. Cuvier and Brongniart, *Environs de Paris* (1822), also published in Cuvier, *Ossemens fossiles*, 2nd ed. (1821–24), 2 (1822), part 2; the citations below are to the former, which differs from the latter in its title page and pagination. The revised monograph is treated here as being Brongniart's work alone (the descriptions of regions beyond the Paris Basin were explicitly so, and distinguished by a smaller typeface), but it was still Cuvier's name that came first on the title page.

He devoted several plates to illustrating its characteristic fossils, including those from high in the Alps. What he had found in the Buet massif (§9.5) was of immense significance, because in this case "the mineralogical characters disappear completely, the geognostic position is obscure, [and] nothing is left but the zoological characters". Since the fossil criterion was not contradicted by any evidence from superposition, he argued that it deserved to be trusted, and indeed treated as supreme. It was the ultimate test of the enriched geognosy that he had pioneered—whether or not he had borrowed the idea from Smith—in his survey of the Paris Basin some fifteen years earlier (§9.1).[119]

Brongniart's explanation of the fossil criterion was relegated to a long footnote, which belied its great importance. For it showed that, although he was still using Smith's method of treating certain distinctive fossils as characteristic of specific formations, for example in the case of the Chalk, he was supplementing it in the case of the Tertiaries with something much closer to Prévost's use of assemblages or populations of species. He explained that in the course of his research he had detected an empirical "rule", which he had applied "timidly at first" but then with increasing confidence: the fossils found in each set of deposits had a "distinctive collective resemblance" throughout its lateral extent, and a "general collective difference" from those of deposits above and below; and those differences were roughly proportional to the vertical separation of the deposits in the stratigraphical pile. This led to an inference about the steady pace of organic change throughout geohistory, which he made more plausible by implying that it was already consensual: "All geologists now agree that the generations of organisms that have successively inhabited the earth's surface are the more different from the present generation as their remains are found buried in the deeper beds of the earth, or—what comes to the same thing—as they lived in times more remote from the present."[120]

The Chalk was of course followed by the Tertiary formations, though Brongniart abandoned that already well-established term in favor of his own: for him they were the upper [*supérieur*] division of a tripartite sequence of sedimentary rocks [*terrains de sédiment*], the middle and lower divisions corresponding to what others called the Secondaries. Somewhat masked by this idiosyncratic terminology (which was never adopted by other geologists) was Brongniart's insistence that the Tertiaries comprised not only those of the Paris Basin but also "a great number of other *terrains* spread across the whole surface of the earth", which had been "almost entirely unknown to the geologists of the celebrated Freiberg school", that is, to Werner and his followers. The potential universality that Prévost claimed for the Tertiaries was at least suggested by Brongniart's inclusion not only of examples from other parts of Europe but also of reports even from America; but he wisely and explicitly focused his attention on those that he had seen for himself.[121]

It is neither feasible nor necessary to describe here in any detail how Brongniart dealt with all the Tertiary formations. In brief, he assumed that the alternation between marine and freshwater episodes shown by the deposits in the Paris Basin were the key to those elsewhere. Starting immediately above the Chalk, he noted that the

Plastic Clay—now confidently ascribed, by its fossils, to freshwater conditions—had been found by Webster in Hampshire, and that Buckland had refined Parkinson's description of it in the London Basin. The Coarse Limestone that followed had proved to be widespread across Europe, and its distinctive marine fauna had been recognized by the English (in the London Clay) despite a major contrast in the kind of sediment involved. On the basis of their fossils he also assigned the north Italian formations (of Turin, Monte Bolca, and elsewhere) to the same marine period, and described them in greater detail in the special monograph that he published the following year. In the Paris Basin, the Gypsum formation that overlay the Coarse Limestone, representing a second episode of freshwater conditions, had likewise been recognized widely, not least by his own work in the south of France and in many parts of Italy; it was particularly important for marking the first appearance of terrestrial mammals. The marine beds above the Paris gypsum, representing a second marine incursion, had been found by Webster on the Isle of Wight; and Brongniart also adopted Prévost's claim that this was where Brocchi's widespread Subapennines properly belonged. Finally, the Freshwater formation that topped the sequence around Paris also had its equivalents elsewhere, for example in the famous quarries at Oeningen and in the deposits that Brocchi had shown him at Rome. With the eventual draining of these extensive bodies of fresh water, the European landmass had taken more or less its present form, only to be ravaged in the geologically recent past by whatever event had been responsible for excavating the valleys and depositing the Superficial "*terrains de transport*".[122]

Overall, the thrust of Brongniart's systematic survey of the Tertiary formations was to emphasize their vertical thickness and geographical extent, and hence their potentially global significance. In effect, it assimilated the Tertiaries into the whole stratigraphical sequence in a way that accentuated their similarities to the Secondary formations and distanced them from the thin, patchy, and variable Superficial deposits, with which in earlier years they had often been confused. In Brongniart's analysis, the Tertiaries belonged unambiguously among the "regular" formations. It then became less surprising to find that they too had been

119. Cuvier and Brongniart, *Environs de Paris* (1822), 90–91; pls. 3–7 depict Chalk fossils, and pl. 7 includes those from the "Montagne des Fis" in the Buet massif.

120. Cuvier and Brongniart, *Environs de Paris* (1822), 91n; Brongniart, *Terrains de sédiment supérieurs* (1823), 26.

121. Cuvier and Brongniart, *Environs de Paris* (1822), 8n. Brongniart's novel nomenclature may have been in part a covert challenge to Buckland's recent proposals for one based on the English sequence (§9.5). The middle [*moyen*] division of his *terrains de sédiment* stretched from the Chalk down to the Jura limestones (in modern terms, Cretaceous and Jurassic); the lower [*inférieur*] extended from the "calcaire à gryphites" (also Jurassic) down to the base of the Secondaries; the underlying Transition formations (roughly, the Lower Paleozoic) were distinct from the *primordial* rocks ["terrains anciens ou *primordiaux*"] (very roughly, basement rocks of any age), which Brongniart treated in conventional manner as having been formed by crystallization rather than sedimentation.

122. Cuvier and Brongniart, *Environs de Paris* (1822); the Superficial deposits and the erosion of valleys were treated quite briefly, as they had been in the first edition, being tacitly recognized as a distinct and separate research problem (see Chap. 10). Brongniart, *Terrains de sédiment supérieurs* (1823). Buckland, "Plastic Clay" (1817), read at the Geological Society on 6 January 1816, had been almost his first published paper on geology.

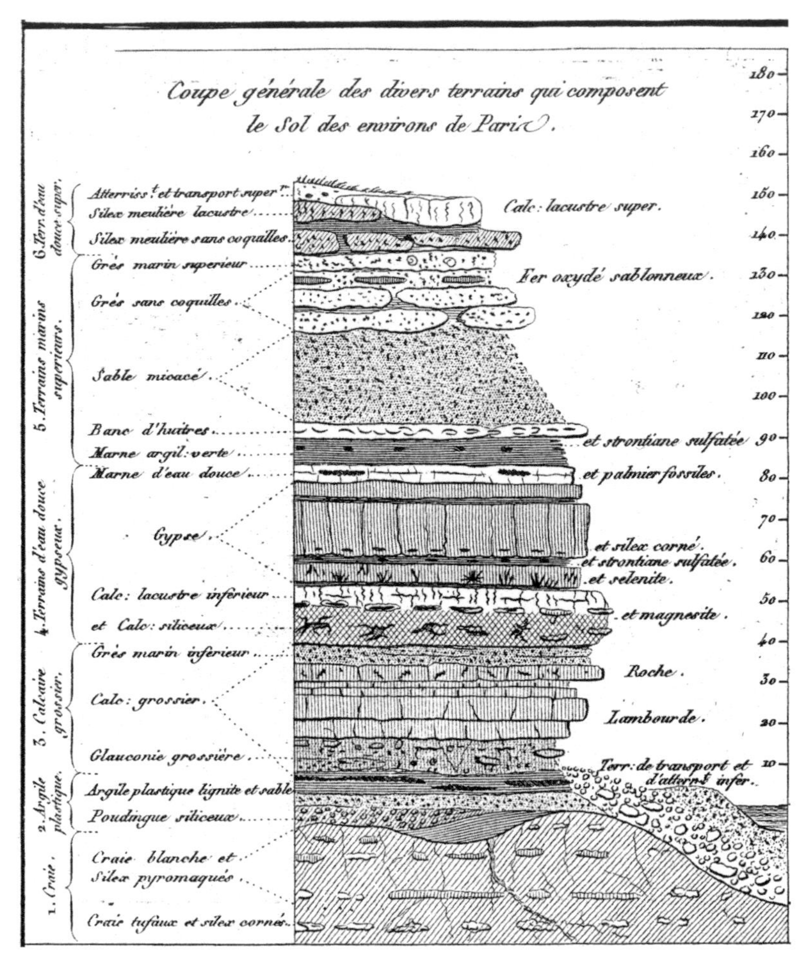

Fig. 9.22. Brongniart's new version (1822) of his "general section" of the Tertiary formations [*terrains*] (2–6) lying above the Chalk (1) in the Paris Basin. He resolved the earlier anomaly of two formations being lateral equivalents (Fig. 9.3): the Siliceous Limestone [*calcaire silicieux*] was now shown as the lowest subdivision of the Gypsum formation (4), lying *above* the Coarse Limestone [*calcaire grossier*] formation (3), not beside it. The change, which was said to be based on further fieldwork, had the effect of eliminating any suggestion that contrasting types of sediment—and therefore contrasting environments—might have existed in the Paris region at one and the same time. On the other hand the depiction of six numbered formations, each with several subdivisions, reflected the increasingly precise standards of stratigraphical practice. The scale (in meters) indicates the average or typical thicknesses of the various beds.

affected by the more recent movements of the earth's crust, so that they too had been recognized—by their fossils—as high up in the Alps as the Chalk and still earlier Secondary formations. Although he himself had been as surprised as anyone to find them there, Brongniart made it more credible by emphasizing the relatively small scale of the crustal movements that their altitude implied: in proportion, even the Alps would be no more than one millimeter high on a two-meter globe. But on the causal process that had raised even the Tertiary formations into Alpine altitudes Brongniart remained silent: he was engaged in the practice of geognosy or stratigraphy, extending it cautiously in the direction of geohistory; he was not tackling the quite separate questions of earth physics.[123]

The new and enhanced status that Brongniart gave to the Tertiary formations worldwide was indicated by the one and only significant amendment to his interpretation of the Paris Basin. He claimed that further fieldwork had resolved what had been a puzzling anomaly in the first edition. The Siliceous Limestone had appeared to replace the Coarse Limestone *laterally* in parts of the Paris Basin (§9.1); he now claimed that it lay *above* it, though he conceded that it almost replaced it to the southeast of Paris towards Fontainebleau. So the ideal section that had summarized the sequence in the first edition (Fig. 9.3) was redrawn with one major change: the two formations that had been depicted rather awkwardly side-by-side, as lateral equivalents, were rearranged and fitted into a simple and straightforward sequence (Fig. 9.22).[124]

Brongniart's further fieldwork may indeed have shown him that where both formations were represented in the same place the Siliceous Limestone was always found above the Coarse Limestone. But the change was accentuated in the summary diagram far more than such observations required. Two formations that had been treated as replacing each other completely in different parts of the basin were now treated as being—equally completely—in unambiguous sequence, one above the other. The change may seem minor or even trivial, but in fact its implications were profound. For it embodied a tacit claim that at each period of Tertiary geohistory the type of sediment being deposited, and hence the kind of environment present, had been uniform throughout the Paris Basin; and Brongniart's survey of regions further afield implied that this uniformity also applied quite generally. So Brongniart's efforts to have the Tertiaries recognized as a part of the total global sequence no less important than the Secondaries—and his own research as no less praiseworthy than, say, Buckland's—were bought at the price of sacrificing an important part of what had made the earlier version of his Parisian work so innovative, namely its profoundly *geohistorical* character. Of course his synthesis of Tertiary formations throughout Europe did enlarge and deepen the geohistory that could be derived from the stratigraphy, since it disclosed a complex sequence of events, particularly an alternation of marine and freshwater periods, against the background of a gradually changing marine fauna. What was lost, however, or at least relegated to the margins of the geohistory, was any sense of the diversity of environments that might have existed at one and the same time, more or less like those of the present world.

At the same time, however, Brongniart's revised account of the Paris Basin also reflected the increasing precision that now characterized stratigraphical practice. This generated a sense of the complexity of the pile of sediments, and an appreciation of how they differed *in detail* from one region to another. It was no longer considered sufficient to describe major formations as if they were undifferentiated in their physical character and fossil contents: Brongniart's new "general section" showed much more detail than his earlier "ideal" one, and his formations

123. Brongniart, *Terrains de sédiment supérieurs* (1823), i–iv.

124. Fig. 9.22 is reproduced from Cuvier, *Ossemens fossiles*, 2nd ed., 2 (1822), pl. 1A, unnumbered fig.

[*terrains*] each included some quite diverse subordinate beds. Potentially if not yet explicitly, this new precision allowed for much more detailed reconstructions of the *geohistory* that the beds represented. For example, within the "gypseous freshwater" formation (Fig. 9.22, no. 4), a thin "lacustrine limestone" underlying the gypsum itself might suggest that there had been an ordinary freshwater phase *before* the onset of the unusual conditions that had caused that mineral to be precipitated.

Conclusion

Within this context of rising standards, the importance of Brongniart's pioneering synthesis of Tertiary stratigraphy can hardly be overestimated. Above all, he used fossils as the primary evidence for correlation. And he did so not just empirically, as Smith had done, but as a criterion justified by an underlying conviction that the world's faunas had changed gradually—by whatever means—throughout geohistory, showing a steadily increasing approximation to those of the present day. The formations that he studied in the north of Italy, and which he described in detail in a separate monograph, offered a test case. For their fossil shells were much more similar to those of the Coarse Limestone of Paris, hundreds of miles away on the far side of the Alps, than they were to those of the Subapennines just on the other side of the plain of the Po. Brocchi's explanation of the difference between the Subapennine and Parisian shells, in terms of a regional contrast in Tertiary times between the faunas of northern Europe and the Mediterranean, had therefore lost its plausibility. In its place, Brongniart had adopted—albeit briefly—Prévost's suggestion that the contrast was due to a difference of *relative age.* The molluscan fauna of the Subapennines was similar to that of the present world because it was relatively recent; the fossils of the north Italian and Parisian formations were like each other, and less like modern shells, because both sets dated from the same more distant period.[125]

This was a profoundly important inference: combined with Brongniart's putative alternation of marine and freshwater periods, it expanded the Tertiary era—previously almost undifferentiated—into an immensely long and eventful part of geohistory. Brongniart's big book on all the Tertiaries (supplemented by his monograph on the north Italian formations) became as much an exemplar of the new fossil-based stratigraphy as its prototype had been some fifteen years earlier. But the new version was vastly enlarged not only in bulk but more importantly in scope. It showed *in detail* how fossils could be used to reconstruct a long and eventful geohistory that seemed to bridge most of the gap between the time of the Secondary formations and the world of the present day. It traced the history of life all the way from the time of the ammonites and belemnites and Cuvier's strange reptiles (in the Chalk and still earlier Secondary formations) to the time of mollusks that only an expert conchologist could identify as being, in some cases, different species from those alive in present seas (as in the Subapennines).

Yet all this research left unanswered the problems raised by the most recent period of all. What had been responsible for excavating valleys—which had been cut through even the most recent Tertiary formations—and for depositing the Superficial gravels on their floors and often on the hills above? Still more worryingly, what had caused huge erratic blocks to be moved tens of miles or more, and what had apparently wiped out Cuvier's striking fauna of mammoths and mastodons? To put it another way, all the research on Tertiary formations and their fossils that has been described in this chapter failed to close the last and most crucial gap: it did not show how the most recent phase of Tertiary geohistory—as represented, for example, by the Subapennine deposits—had been linked to the present world and the dawn of human history. The research that tried to tackle these problems is the subject of the last chapter. At this point, therefore, the narrative must double back to the time of Cuvier's rehabilitation of "geology", in his report to a Napoleon who was then still at the height of his power (§8.4).

125. Brongniart, *Terrains de sédiment supérieurs* (1823), 26. In the Italian case the fossil criterion was crucial, because the two sets of formations had not been found together in any single locality, so that "evident superposition" was not available to confirm or refute it. Surprisingly, Brongniart hardly commented on the striking contrast between the Coarse Limestone of Paris and the London Clay, although he accepted their equivalence on the basis of the close similarity of their fossil assemblages.

CHAPTER 10

Earth's last revolution

10.1 THE INTERFACE BETWEEN PAST AND PRESENT (1807–9)

The frozen mammoth

In their famous work on the Paris Basin, Brongniart and Cuvier had turned a pile of formations into geohistory, by interpreting the successive sets of fossils as traces of a complex alternation of marine and freshwater conditions (§9.1). This had swiftly become a fruitful model for further research across Europe on what soon became known as Tertiary formations (§9.4). Within the few years covered in the previous chapter it became clear to geologists that the Tertiaries represented a major portion of geohistory, characterized by faunas much more similar to those of the present world than anything preserved in the still older Secondaries (now redefined, by the excision of the Tertiaries, as those from the Chalk and its equivalents downwards). In the broad sweep of geohistory, the Tertiary era linked the unfamiliar Secondary world to the familiar present. But it did not fill the gap completely. Between the present world and what appeared to be the most recent of the Tertiary formations—notably Brocchi's Subapennines (§9.4)—were the traces of the earth's most recent major revolution. The character of this event, let alone its physical cause, remained obscure and highly controversial, but few geologists doubted its sheer historical reality. Something very peculiar seemed to have happened in the geologically recent past, at or near the dawn of human history. The final chapter of this book describes some of the research that tried to elucidate what this event or episode might have been. It was research that was only loosely linked to the work on Tertiary stratigraphy and geohistory, and carried out in parallel; so this chapter covers much the same span of years as the previous one.

The famous paper on the Paris Basin already showed signs of this bifurcation of research directions. Brongniart and Cuvier ended their account by tacking a last revolution onto the end of the long and eventful geohistory that they reconstructed from the pile of "regular" formations. This final phase was represented by the erosion of the present valleys through all those earlier formations, and by the deposition of the Detrital Silt on the valley floors (Figs. 9.1, 9.3), containing the bones of the large mammal species that had apparently perished at that time. Brongniart and Cuvier claimed that this Superficial deposit formed the interface, as it were, between the present world of human history and the former world of all the other formations: "although very modern in comparison with the others, it is still anterior to historical times". Almost in passing, they attributed the event to some kind of mega-tsunami, "a great irruption, coming from the southeast" (the direction being inferred from the form of the valleys); but in effect they set it aside as belonging to a different area of research. This first section of the present chapter describes how the enigmatic last revolution was being investigated at this time, in its own right, not least by Cuvier himself.[1]

By 1807, when he and Brongniart were well advanced with their fieldwork on the "regular" formations around Paris, Cuvier had already published most of his articles on the fossil mammals from the Superficial deposits (§7.5). Unlike the strange mammals from the Parisian gypsum, and a fortiori the reptiles from the still older Secondary rocks (§9.3), most of these more recent fossil animals could be assigned to familiar genera (the mastodon and the megatherium were exceptions). But Cuvier had claimed that they were all, without exception, of otherwise unknown *species*. At the level of species the anatomical differences were often slight and subtle, and it needed a skillful eye—supremely, in his opinion, his own—to detect them (Fig. 7.5). Nonetheless, once they were pointed out, particularly with the aid of persuasive pictures, Cuvier claimed that the differences were as unmistakable as those that separated living species of the same genus, such as the wolf and the fox.

Cuvier had attributed this contrast between fossil and living quadrupeds to a catastrophe that had wiped out the earlier forms. He had inferred that the event must have been a widespread but transient incursion of the sea—some kind of mega-tsunami—over the lower-lying parts of the continents (§7.5). But in his opinion it had not been on a scale large enough to reach high altitudes, as Dolomieu had supposed (§6.3). And it had been quite unlike the event described in Genesis, contrary to what de Luc had supposed (§6.4); for no human vessel—not even an Ark navigated by Noah—could have survived a surge that was powerful enough to carve out valleys in solid rocks. Nor had the catastrophe been the kind of permanent interchange between continents and oceans that de Luc had envisaged. On the other hand Cuvier's mega-tsunami had not swept animal carcasses from the tropics into temperate and even arctic regions, unlike the one suggested by Pallas (§5.3) and more recently by Faujas and others. Cuvier had insisted that these animals had been living where their bones were now found: they were not of the same species as those now living in the tropics, and their bones showed no signs of the wear and tear that transport by mega-tsunami would have inflicted on them. So Cuvier agreed with de Luc and Dolomieu about the historical reality of the catastrophe, but not about its physical character; and he agreed with Pallas and Faujas about its physical character, but not about its relation to the fossil bones found in the Superficial deposits that recorded the event.

One striking piece of new evidence reached Paris in 1807 and was immediately taken by Cuvier to strengthen his interpretation of the last revolution. The Russian botanist Mikhail Ivanovich Adams had recently returned from a long expedition through Siberia, partly following in Pallas's footsteps (§5.3). Back in St Petersburg, Adams reported in a Russian periodical (in French, of course) that a frozen mammoth had been found several years earlier by the indigenous Tungus people in eastern Siberia, near where the Lena flowed into the Arctic Ocean (see Fig. 3.3). Unfortunately the carcass had thawed out and most of its flesh had been eaten by local dogs or wild animals before what remained could be rescued for science. One early attempt at reconstructing it—by a Russian trader in Siberia—showed it as a bizarre cross between an elephant and a wild boar (Fig. 10.1).[2]

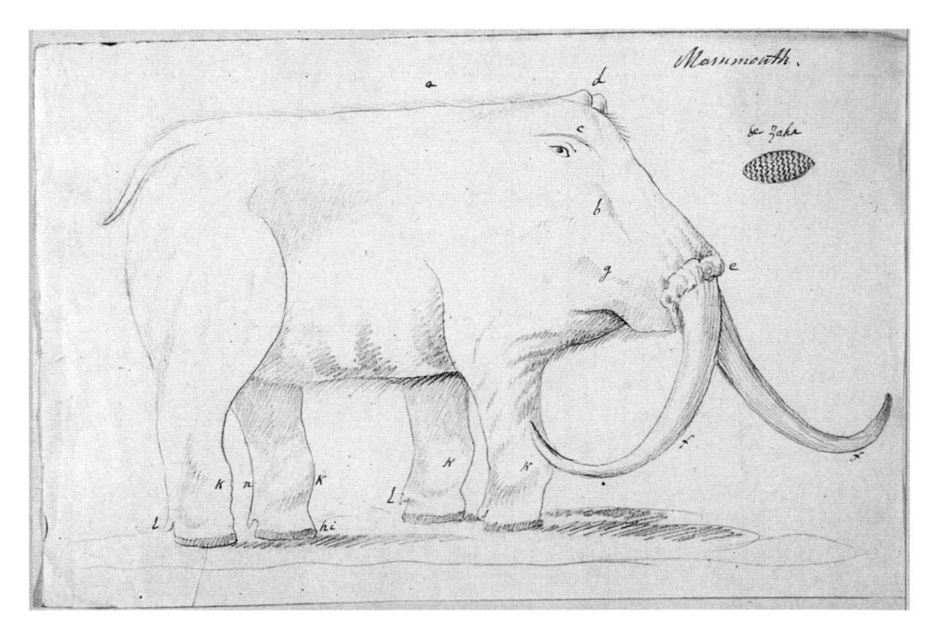

Fig. 10.1. A crude reconstruction of the frozen mammoth found in Siberia in 1799, drawn by the Russian merchant who had first heard of it (the small drawing shows the grinding surface of the molar tooth). Its bones, and some remaining scraps of skin, were brought back to St Petersburg in 1806 by Mikhail Adams; this drawing was sent to Paris and probably used by Cuvier in preparing his report to the Institut on this important find. The thick woolly hide supported Cuvier's claims that the mammoth had been a species adapted to a cold climate, not an elephant swept in from tropical India, and that it must have been annihilated in some "sudden revolution". (By permission of the Bibliothèque Central du Muséum National d'Histoire Naturelle, Paris)

1. Cuvier and Brongniart, "Géographie minéralogique" (1808), 326; trans. Rudwick, *Georges Cuvier* (1997), 156.

2. Fig. 10.1 is reproduced from an anonymous and undated manuscript drawing among Cuvier's notes (Paris-MHN, MS 630(1)), on which the German caption states that the mammoth was unearthed (in fact, first seen by Adams) in June 1806, thus identifying it as his specimen: see Cuvier, "Rapport sur le cadavre" (1807), 386. This drawing may have been copied from a more elegant one entitled "das russische *Mammut*", which is reproduced in Cohen, *Destin du mammouth* (1994), 28, dated 1804 and attributed to Roman Boltunov, a merchant in Yakutsk.

Adams knew of Cuvier's work on elephants (§7.1, §7.5) and accepted that the mammoth was distinct from the living species and a denizen of "a very ancient world". His formerly frozen specimen was important, particularly because the remaining fragments of hide showed that the animal had had a thick fleece, made up of two distinct kinds of hair, much like those of living mammals adapted to arctic climates; it recalled Pallas's earlier discovery of a frozen rhinoceros in the same region, which likewise had traces of a woolly hide (§5.3). News of Adams's report spread rapidly in spite of the wars. Karsten published a German version of his article in a Berlin periodical, and this reached Paris, where Cuvier and Lacépède were asked to report on it at the Institut. Cuvier argued that the new find not only confirmed his own earlier conclusions about the mammoth as a species distinct from the living elephants, but also proved that the animal had been well able to live in the arctic climate where its remains were found. He concluded that the Siberian mammoth witnessed to "a sudden revolution that caught these animals, destroyed their species, and froze the individuals that are now found so far in the north". He argued that the similar bones found near Paris and elsewhere in Europe were relics of the same event, and were of animals that had likewise lived where their remains are found. So the frozen individual was chiefly important because it provided further evidence that the event must have been sudden, since otherwise its carcass would

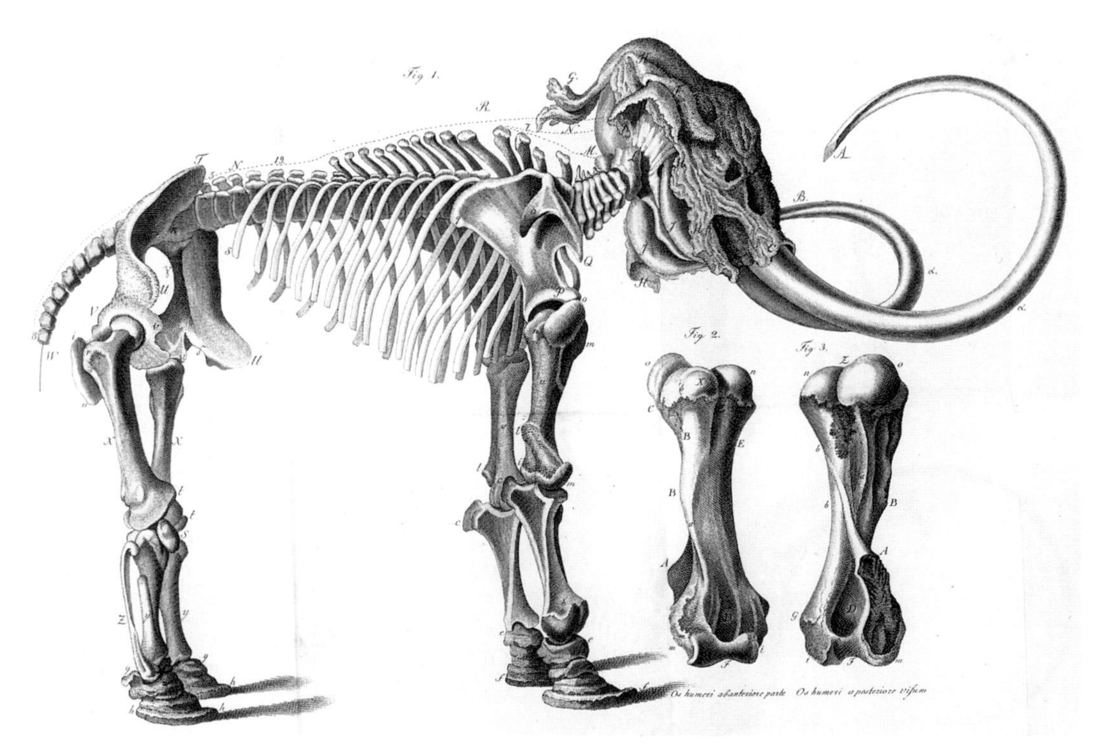

Fig. 10.2. The skeleton of the mammoth found frozen in arctic Siberia in 1799 and brought back by Adams to St Petersburg in 1806, with scanty remnants of its skin still attached to the skull and feet: a reconstruction by Tilesius, published in 1815 by the Academy of Sciences in St Petersburg. (The two larger views of the humerus helped to make full use of the expensive engraved plate.) (By permission of the Syndics of Cambridge University Library)

have decayed away long before burial, even in an arctic climate. When, a few years later, the German naturalist Tilesius (Wilhelm Gottlieb von Tilenau, 1769–1857) described and reconstructed the specimen along Cuvierian lines, Adams's mammoth made an impressive addition to the vanished fauna (Fig. 10.2).[3]

Borderline problems

Cuvier used Adams's frozen mammoth to support his argument about the sudden catastrophe that had wiped out all the geologically recent fossil mammals he had described. But other fossils threatened to spoil the elegant simplicity of his geohistorical inference. A relatively easy puzzle was that of the bones embedded in hard breccias, which were found within limestone fissures in several places around the Mediterranean, from Gibraltar in the west to the Greek island of Cerigo [Kithira] in the east. Piecing together evidence from his own specimens and from proxy pictures, Cuvier judged that most of these bones were those of living species, and even in some cases those of species still living in the same region. Although he considered that the breccias were ancient in relation to human history—he thought that nothing of the kind was currently being formed—he concluded that they belonged safely on the modern side of the great divide.[4]

A more difficult case was that of the bones from the famous caves in Bavaria and elsewhere (Figs. 5.14, 7.2). They were those of a huge bear, which was as distinct from any living species as the mammoth was from the living elephants, yet there was no trace in the caves of any aqueous incursion, let alone a marine inundation. Like Rosenmüller (§7.1), Cuvier inferred that the bears had used the caves as dens. But he left open the possibility that the species had become extinct at some later date, though still before historical times. If so, its demise could hardly be attributed to the same decisive event; its relation to the former and present worlds remained ambiguous (Fig. 10.3).[5]

The fossil bones of ruminant mammals such as deer and cattle were even more puzzling, though the most spectacular case of all was not. Cuvier argued that the huge "Irish elk" was as clearly extinct a species as the mammoth; and from published reports he inferred that its bones were always found in the same kinds of deposit, underlying the obviously recent Irish peats. So it belonged unambiguously

3. Fig. 10.2 is reproduced from Tilesius, "De skeleto mammonteo" (1815) [Cambridge-UL: Q340:9.a.1.67], pl. 10, explained on 490–99; Adams, "Mer Glaciale" (1807), 627–28. Tilesius had been in Russia at the time, on another expedition; as usual with the Russian Academy's more specialized work, his paper was published in Latin. His mounting of the skeleton has been subjected to several later "rectifications": see the photograph [1910?] in Cohen, *Destin du mammouth* (1994), 143. Almost all the rare later finds of Siberian mammoths preserved with "soft parts" are also in Russian museums, but some fragmentary specimens are on display in Paris-MHN. Of course neither Adams nor Cuvier (nor Tilesius) had the benefit of modern knowledge about the permafrost regime under which mammoth carcasses might have been emplaced.

4. Cuvier, "Brèches osseuses" (1809). His information on the best case—the British naval base of Gibraltar, which remained inaccessible to French naturalists during the wars—came from an earlier paper to the Royal Society of Edinburgh: Imrie, "Mountain of Gibraltar" (1798).

5. Fig. 10.3 is reproduced from Cuvier, *Ossemens fossiles* (1812), 4 [Cambridge-UL: MD.8.68], pt 4, 1e mém., pl. 2, first published in "Ossemens du genre d'ours" (1806), pl. 19. The size of the cave bear would have been somewhat less striking had he had a larger specimen of the polar bear for comparison.

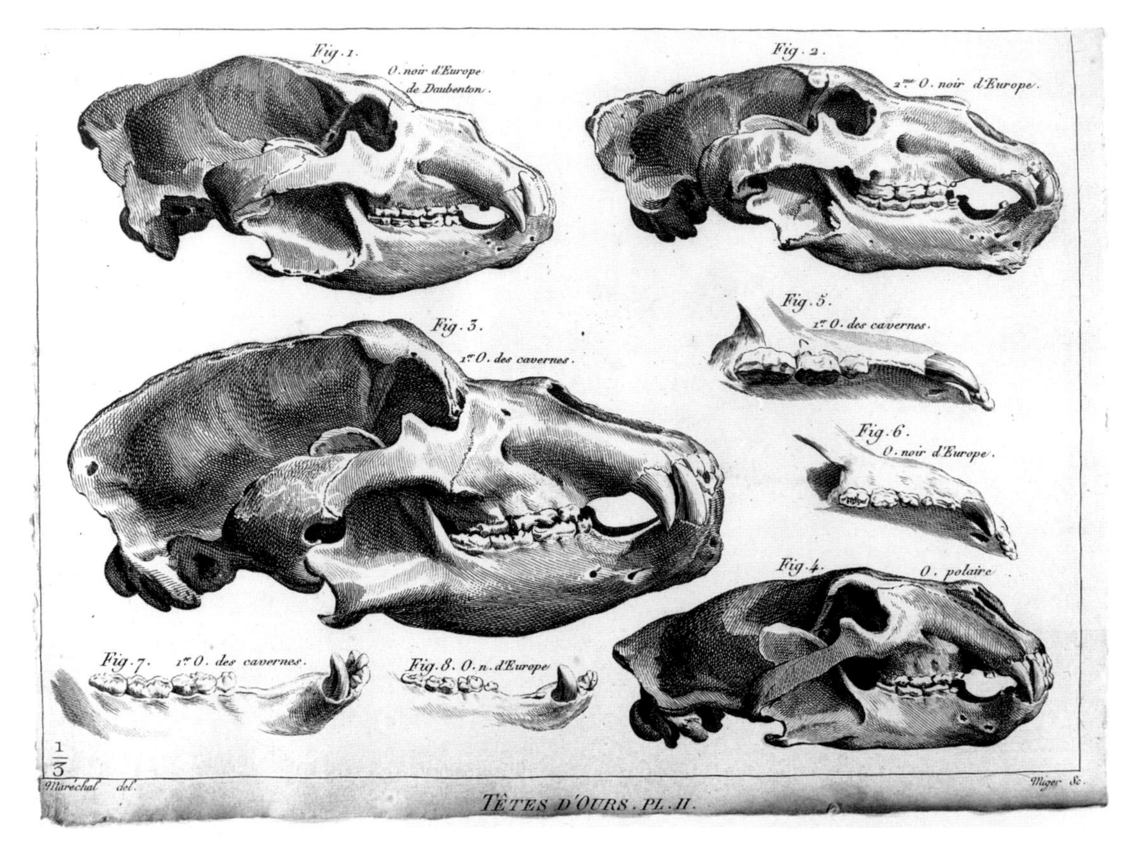

Fig. 10.3. Cuvier's comparison of the skulls and dentition of living and fossil bears (1806), all drawn at the same scale. The cave bear (figs. 3, 5, 7), which he inferred was extinct, was strikingly large in comparison to the polar bear (fig. 4) and the European bear (figs. 1, 2, 6, 8). (By permission of the Syndics of Cambridge University Library)

to the former world. Most fossil ruminants, however, were much more difficult to explain: their bones were often found in the same Superficial deposits as those of Cuvier's extinct species and had evidently been part of the same fauna, yet many of them were impossible to distinguish from the bones of living ruminants. At first sight these fossils seemed to count against Cuvier's claim that the last revolution had caused a total mass extinction among the mammals. But he had enough experience with comparative anatomy to see a plausible explanation for this otherwise worrying anomaly. It was well-known to naturalists that among living ruminants many unquestionably distinct species were distinct only in features such as the form of their horns and the coloration of their body, which were not preserved in the fossil state. So Cuvier argued that while the fossil ruminants associated with extant species belonged unambiguously to the present world, those found with extinct species were probably, like them, extinct inhabitants of the former world, even though it was impossible to prove the point from the bones alone. He expressed that conclusion not only in an appropriately provisional tone, but also with the historical metaphor of "documents" that for him was never "merely" metaphorical:

> The facts collected up to the present therefore seem to show—at least as far as such incomplete documents can do so—that the two sorts of fossil ruminants belong to two kinds of formation [*terrain*], and consequently to two different geological epochs: that the first kind were buried (and are still buried today) in the period in which we are living; while the other kind were victims of the same revolution that has destroyed the other fossils of the superficial deposits, such as the mammoths, mastodons, and all the pachyderms of which the genera live today only in the tropical zone.[6]

In dealing with all these problematic or borderline cases, Cuvier's objective was to demonstrate that the last revolution had caused a *total* disjunction between fossil and living species: the earlier fauna had been "destroyed"—his own favorite verb—in its entirety. Only that sweeping claim seemed adequate to establish the reality of the former world as genuinely distinct and show that the deep past really was, as it were, a foreign country. Otherwise, it might have been just a variant of the present world, merely with a changed biogeography: as Faujas and others continued to claim, the animals might just have migrated to new regions or been swept out of their usual ones, in either case without extinction (§7.2). The other threat to Cuvier's interpretation had come from Lamarck, who added the idea of transmutation to that of migration or transport: in cases where fossil species really were different from living ones, it might be simply because they had changed in form over the vast spans of deep time, so that once again no extinction was involved (§7.4). In the face of Lamarck's unquestioned expertise with fossil shells, Cuvier had in effect conceded that the last revolution had not impinged on marine mollusks as radically as on terrestrial mammals, and he had concentrated on making the best case for the drastic effects of the catastrophe *on land*.

Cuvier's tactical retreat, confining his argument to terrestrial animals, became explicit in one of the last of the papers that he published in the Muséum's *Annales*. It dealt with the osteology of the manatee or sea cow, and with fossil bones that he assigned to that animal and to seals. He thought them worth describing, to help complete his survey of fossil bones, but he also noted that strictly speaking such *marine* mammals fell outside the scope of his work. This he defined, significantly, as being concerned with those that could have been "destroyed" by a marine inundation (such as a mega-tsunami). In effect, therefore, he framed his research not by any purely zoological parameter but by its *geological* goal.[7]

Actual causes

However, Cuvier was not having everything his own way. He had to fight hard for his interpretation of the last revolution as a sudden and even violent event. He had

6. Cuvier, "Os fossiles de ruminans" (1808), 398; see Rudwick, *Georges Cuvier* (1997), 159–60.

7. Cuvier, "Ostéologie du lamantin" (1809). By the same criterion his studies of the fossil turtles and giant lizard from Maastricht were also tangential to his main objective, but they were clearly of a much earlier date than the last revolution.

to defend it against threats to its validity coming from three different directions. First, there were naturalists who claimed that the physical world of the deep past had graduated into the present world without any exceptional event at all. One of the topics at the center of this debate was the vexed question of the origin of valleys (§2.4). Cuvier himself had little relevant field experience of valleys, except in the gentle landscapes of Normandy and around Paris. When he attributed valley erosion to the rapid effects of some kind of mega-tsunami, he was simply adopting the opinions of other naturalists such as Dolomieu, who had seen a far wider range of valley forms (§6.3, §7.2). But the alternative explanation of very slow erosion by ordinary streams and rivers, which had long been championed by naturalists such as the elder Desmarest and Soulavie (§4.3, §4.4), had recently been brought back into prominence by the former. The octogenarian Desmarest was still a respected figure, and in 1806 the long-delayed publication of his account of the ancient volcanoes of Auvergne had again made a persuasive case for an interpretation that stressed the total continuity between present and deep past (§8.3).

Three years later, just as some of Cuvier's last papers on fossil bones were appearing in the *Annales*, the third volume of Desmarest's vast compilation on physical geography for the *Encyclopédie Méthodique* (§6.5, §8.3) brought him as far as the fourth letter of the alphabet, so that patient subscribers could at last read his thoughts on, among other topics, "Déluge". Here he criticized those who invoked extraordinary causes rather than ordinary ones, the latter having "always been such as we observe today, simple, regular, and obeying ordinary laws". It was of course a restatement of his long-standing claim that such everyday processes were adequate by themselves to explain all the relics of the deep past: given enough time, even small streams could excavate large valleys. Yet although Desmarest thus tacitly criticized Cuvier's kind of analysis, in one important respect the two savants were at one: both were consciously doing geohistory, and Desmarest continued to deploy the historical metaphors of nature's "monuments", "epochs", and "chronology" as consistently as Cuvier.[8]

The trouble with Desmarest's renewed claims for the slow erosion of valleys by the ordinary action of streams and rivers was that the very same features (in Auvergne) had been given a diametrically opposite interpretation, when viewed in the field by other competent observers such as Dolomieu (§7.2). Although Dolomieu himself was no longer alive to advocate his alternative—invoking rapid erosion by a sudden and powerful cause (§6.3)—the somewhat similar scenario envisaged by de Luc got a new airing in the same year. In the first restatement of his ideas since his published letters to Blumenbach more than a decade earlier (§6.4), de Luc—now another octogenarian—had at last abandoned the old-fashioned epistolary format; but his *Elementary Treatise on Geology* (1809) lacked any clear structure (or table of contents) and in style it was as rambling and verbose as ever. Once again, Cuvier may well have felt ambivalent about being seen to have de Luc as his ally. De Luc continued to argue for the reality of a geologically recent physical catastrophe, somewhat similar to what Cuvier inferred; but he also hitched that event to his relatively literal interpretation of the Flood story in Genesis and—more overtly than ever—to his own religious goals, thereby making the book an easy target for those who were hostile to any such project.[9]

Ironically, de Luc's book had the effect of reviving the argument about the adequacy of "actual causes" in geological explanation more effectively than Desmarest's rather obscure encyclopedia articles. For de Luc's chief target was no longer Hutton himself, who had died back in 1797, but his friend John Playfair, who had become Hutton's bulldog as much as Farey was Smith's. Playfair's *Illustrations of the Huttonian Theory* (1802) had restated Hutton's geotheory of ceaseless crustal upheaval and continental erosion, more briefly and in attractive style (§8.4). And he had made it still more acceptable by airbrushing the deistic metaphysics out of Hutton's argument, all in the name of clarifying the supposed obscurity of the original prose. But he had collected much empirical evidence in Hutton's favor, primarily to refute Kirwan's attacks (§6.4). In Britain, Playfair's book had provoked a lively debate about geotheory; on the Continent it had been treated as yet another contribution to an over-prolific and under-productive genre. However, de Luc's criticism of Playfair's work did serve to focus attention on the more specific issue dividing them, namely the adequacy or otherwise of actual causes, to explain the present topography of the earth's surface.[10]

It should be unnecessary by now to point out that all savants in this debate were agreed on the desirability—wherever possible—of using known and observable agencies to interpret the distant past. They agreed that the present was always the best key to the deep past and the first route to be explored; they were all, in modern terms, avowed actualists. Cuvier was part of this consensus, basing his studies of fossil bones explicitly on close comparisons with those of living species: he took it for granted that the principles of anatomy and physiology had been the same before the last revolution as they were in the modern world. The disagreement among savants was about the extent to which "actual causes" were *adequate* to explain *all* the observable features of the earth's surface and accessible interior. De Luc had coined the term in the course of arguing that their action must have begun at a finite time in the quite recent past; back beyond that time, in his opinion, a process of a different kind—though no less natural or physical in character—must have effected the great interchange between continents and oceans that had brought the former world to an end and ushered in the present world of human history (§6.2, §6.4). No savant of any significance proposed "causes" of kinds that

8. Desmarest, *Géographie physique* (1796–1811): "Déluge" (3 [1809]: 606–15); see also ibid., "Constitution extérieure du globe" (3: 454–68) and "Époques" (4: 39–49). Volume 4 (1811) brought him, with ever briefer coverage, as far as the letter *N*; he died in 1815 at the age of ninety, and the rest of the now superannuated work was published by others many years later. Soulavie also remained active in Paris, but as a historian, not a naturalist (§4.4); he died in 1813 at the age of sixty-one.

9. De Luc, *Traité élémentaire* (1809), published in Paris in spite of the war. A translation, published the same year in London, first brought de Luc's term "actual cause" into the English language, where it was later misunderstood—owing to the changed meaning of "actual" in English—to denote not a causal process observable in the present world, but one that was true or real rather than conjectural or imaginary: see Ellenberger, "Causes actuelles" (1987).

10. Playfair, *Illustrations* (1802). Basset, *Explication de Playfair* (1815), made available in French not only Playfair's book but also one of its more effective critiques, Murray's *Comparative view* (1802).

were unparalleled in the present (except for the very start of the terrestrial system, where everything was conjectural anyway). Dolomieu's putative mega-tsunami, for example, was scaled up from known tsunamis such as the one that had devastated Lisbon within living memory; it was certainly unparalleled in magnitude, but not in kind, and it was impeccably natural in character.[11]

Lamarck and transformism

Cuvier's interpretation of a geologically recent catastrophe was also under fire from a second direction, namely from Lamarck's continuing advocacy of transformism. This was not (as it became for Darwin half a century later) a theory about the origin of species; for Lamarck it was a part of an all-encompassing natural philosophy of flux and impermanence in all aspects of the natural world (§7.4). Cuvier was profoundly hostile to it, but not because it was contrary to the Creation story in Genesis. His primary reason was that it was incompatible with his conception of the stability of natural kinds in general and of the functional integration of animal bodies in particular, on both of which he believed that any reliable natural history depended. Derivatively, however, transformism was also incompatible with his *geological* interpretation of fossil bones. He attributed the disparity between fossil and living species to the *extinction* of the former during the last revolution to affect the earth's surface, whereas Lamarck denied the reality of any extinction whatever (except marginally by human agency) and attributed the disparity to the effects of changed environmental circumstances and the sheer lapse of time. Cuvier and Lamarck hardly differed at all as to the matter to be explained, but they differed profoundly as to the underlying cause.

Lamarck produced his long awaited magnum opus on his theory of transformism in the same year as Desmarest's encyclopedia articles and de Luc's book, and just as Cuvier was publishing the last of his articles on fossil bones. Its title, *Zoological Philosophy* (1809), made it clear that it was a work of natural philosophy or causal explanation (§1.4), with animal life as its subject; it focused on the properties of the organism that made it alive and distinct from nonliving entities in the natural world. Unlike what Cuvier was doing, Lamarck's work was not a contribution to the *history* of animal life. It is therefore not surprising that it contained almost no reference to any possible *fossil* evidence for the process of transmutation that it propounded. Lamarck repeated his earlier claims that it was premature to infer that any species had truly gone extinct by natural causes, while so little of the earth's surface had been fully explored: in other words, he invoked once again the classic "living fossil" argument (§5.1). He conceded that Cuvier's fossil mammals were indeed distinct from any living species, and probably extinct, but he attributed their disappearance to human agency alone. He conceded that the mummified Egyptian ibis was indistinguishable from the living bird—he had joined Cuvier (and Lacépède) in signing the report to that effect (§7.4)—but he argued that the lapse of time since ancient Egypt had been too brief, and any environmental change too slight, for any transmutation to have yet taken place. Finally, with Cuvier once again as his obvious though tacit target, Lamarck scornfully dismissed any catastrophe to account for extinction: "It is a pity that this agency [*moyen*]—so convenient for rescuing one from embarrassment when one wants to explain operations of nature the causes of which one has been unable to grasp—has no foundation other than the imagination that has created it, and cannot be grounded in any evidence [*preuve*]."[12]

In sum, Lamarck and Cuvier were talking past each other: not only because they were committed to profoundly different conceptions of the natural world (or to incommensurable paradigms), but also more prosaically in the sense that they were practicing different sciences, the one causal, the other historical. Nonetheless, the appearance of Lamarck's book was a challenge that Cuvier could not ignore, for Lamarck remained a powerful figure on the savant stage.[13]

Antiquarian researches

The third and last threat to Cuvier's conception of recent geohistory was perhaps more serious than either of the others; its impact has been largely overlooked by modern historians, probably because it came from right outside the sacred boundaries of "Science", namely from the *Wissenschaft* of human history. Cuvier had argued repeatedly that the animal extinctions had been caused by a geologically recent and widespread catastrophe; and in his lectures he had boldly claimed, like de Luc and Dolomieu, that the date of that "last revolution" could not be more than a few thousand years in the past (§8.3). Since the event appeared to form the boundary between the former world of extinct animals and the present world of human civilizations, the duration of the latter would necessarily be limited in the same way. That inference had brought Cuvier, like his predecessors, into the notoriously contentious field of historical chronology (§4.1). It was a field in which the stakes were high indeed, since any claim to a vastly extended human antiquity was liable to be treated as an attack not only on the Genesis story of human origins but also on the religious authority and trustworthiness of the Bible as a whole and hence on the moral foundations of society.

Whatever its supposed implications, however, the character of the argument was *historical*. It hinged on the interpretation of whatever texts and artifacts survived from the remote past history of human cultures. And that plural was decisive: it was a question of the records of many diverse human cultures or civilizations, no longer just of a privileged few (Jewish, Greek, and Roman). The intellectual elite of the earlier Enlightenment had relished the multiplicity of the human cultures, past

11. This interpretation of actualistic methodology, which is now rightly taken for granted among historians of geology, was first set out in the classic work by Hooykaas, *Natural law and divine miracle* (1959), and amplified in his *Continuité et discontinuité* (1970); see also Rudwick, "Uniformity and progression" (1971).

12. Lamarck, *Philosophie zoologique* (1809), 1: 75–76, 80. Taquet, "Lamarck en 1809" (1997), prints documents that indicate the author's successful efforts to ensure that his work was adequately publicized.

13. The stories of Lamarck's marginalization, and of the neglect of his work, have long been recognized by historians (but not always by scientists) as a myth propagated in the context of the evolutionary debates much later in the century, when Lamarck became *politically* useful—particularly in the conflicts in France over *laïcité*—as a supposedly progressive and secular genius who had been persecuted and downtrodden by a supposedly reactionary and religious Cuvier.

and present, that were coming increasingly into view with the progress of scholarship and the expansion of European trade and colonies. The science of chronology had likewise expanded its horizons, to include the scholarly analysis of cultures outside Europe and far from those recorded in the Bible (§4.1).

In the last years of the old century and the first of the new, there was a remarkable further efflorescence of scholarly research on all the known ancient civilizations. The erudite studies of Sanskrit texts being published by British scholars in India in their *Asiatick Researches*, and then the effort that went into the translation of those volumes for its French edition (on which Cuvier acted as a consultant: §7.4), were just one sign of this heightened interest among savants in the remote past of exotic cultures. Another was the burgeoning textual research on the Bible itself, studied in its original context of other ancient Near Eastern cultures, in which Blumenbach's philological colleagues at Göttingen were particularly prominent. A third example came from the New World. Humboldt and his French colleague Aimé Bonpland had become famous for their great scientific expedition to Latin America at the turn of the century. In Paris, where Humboldt had settled, they were beginning to publish the massive results of their research, not only on natural history but also (in modern terms) on the ethnography and archaeology of the regions they had visited. Their lavishly illustrated volumes were making the monuments of ancient Mexico, for example, familiar to Europeans as never before. The savants of Cuvier's time, more Enlightened than the conquistadors, were fascinated by this evidently complex culture planted in the New World, with no obvious connection with those of the Old.[14]

In exploiting some of these diverse cultural riches for historical purposes, one antiquarian, with whose work Cuvier was certainly familiar, can stand here as a representative of many more. As a young aristocrat, the marquis Agricole Fortia d'Urban (1756–1843) had spent years in Rome trying to recover his ancestral property in Avignon. This had at least given him firsthand experience of historical research in the service of genealogy. Having kept his head down in the provinces during the Revolution, Fortia had later settled in Paris, where his substantial private income enabled him to be a prolific scholar, perhaps too prolific for his own good. In 1805 he published the first of no fewer than ten volumes on early multicultural human history; the last appeared in 1809, the same year as Lamarck's latest book, and just as Cuvier was rounding off his papers on fossil bones. Two of Fortia's early volumes dealt fairly modestly with the pre-Roman history of his native region; but another expanded ambitiously into a review of the genre of geotheory, with discussions of the origin and antiquity of the world. Fortia then focused on the earliest history of China, before the supposed deluge associated with the shadowy figure of the emperor Yao; and two further volumes covered the origins of other ancient peoples and more on early Chinese chronology. The last three dealt more generally with the identity of the deluges of Yao in Chinese, Noah in Jewish, and Ogyges in Greek records, and with the legend of Atlantis. Fortia claimed that although Noah's Flood could not have been universal it was identical to Yao's in China, which he dated to 2297 B.C. and attributed to a comet deranging Venus. He also concluded

not only that the earth was far older than the traditional timescale allowed—which was no news to naturalists (§2.5)—but so was humankind.[15]

This brief summary does scant justice to Fortia's wide-ranging erudition, but it is sufficient to indicate its general character. Whatever the originality and validity of his conclusions may have been in the eyes of contemporary scholars, he was clearly practicing the textual science of chronology, expanded from its traditional base to include all relevant ancient cultures and particularly the Chinese. In the present context, what matters is that Fortia's volumes exemplified the claim by literary antiquarians to be able to make a distinctive contribution to the debate about the putative "deluge" in relation to human origins. A naturalist such as Cuvier could not ignore it, because he too was making claims about the very same supposed event, at or near the origin of human civilizations if not of the human species itself. If the last revolution had been a real event, and one as drastic and recent as Cuvier maintained, his evidence from fossil bones would have to be compatible with the antiquarians' evidence from ancient texts. The catastrophe was not only the temporal interface between the former and present worlds, but also the disciplinary interface between Cuvier and Fortia, between the natural sciences and the humanities, between the First Class of the Institut and the Third. But it was an interface not in the sense of a sharp demarcation or boundary line but rather as a zone of overlap: the supposed deluge was—almost uniquely—a matter on which the natural and physical evidence intersected the human and historical.

Almost contemporary with the travels of Humboldt and Bonpland in Latin America had been Bonaparte's military expedition to Egypt, with its train of savants in tow (§7.2). The latter had brought back to Paris not only the mummified animals so important for Cuvier's research (§7.4), but also a far larger mass of Egyptian antiquities. The publication of this material, illustrated as lavishly as the American travels, was one of the most conspicuous products of Napoleon's cultural policy. It offered, to those who were privileged to see the volumes, a vast and ongoing series of superb pictorial proxies for the experience of seeing the monuments in situ and at first hand. The Egyptian antiquities added immensely to the contemporary sense of the astonishing diversity of the early history of humankind, and Egyptian elements were prominent in artistic and decorative products in Napoleon's "Empire" style.[16]

The new Egyptian material was equally fascinating for scholarly antiquarians, but it was also tantalizing and frustrating. For the abundant hieroglyphic inscriptions remained undeciphered, so that the *history* of ancient Egypt remained almost as obscure as ever; the trilingual Rosetta stone was recognized as likely to provide

14. Humboldt and Bonpland, *Régions équinoxiales* (1805–34); among the thirty-four volumes, Humboldt's *Vues des Cordillères* (1810) described both the natural features and antiquities of Mexico.

15. Fortia, *Histoire ancienne* (1805–9); there is an invaluable summary of this bibliographically chaotic work (which twice changed even its series title) in the penultimate volume, *Histoire et theorie du déluge* (1809), 1–9. Guignes, *Voyages à Peking* (1808), was probably his source on Yao and Chinese chronology.

16. [Jomard], *Description de l'Égypte* (1809–28); the architectural plates are finely reproduced in Gillispie and Dewachter, *Monuments of Egypt* (1987).

the key, but the code had not yet been cracked. It therefore remained difficult to evaluate the long-standing claims (based at second hand on ancient Greek texts) that the records of Egyptian history extended without a break far further back than any traditional date for the Flood. For Cuvier this represented another potential threat to his concept of a geologically recent catastrophe; for if it had been drastic and widespread enough to have wiped out his "lost" mammals, it would also surely have destroyed any earlier cultures and demanded a new start for humanity.[17]

Furthermore, this potential threat did not wait for the cracking of the Egyptian code. Among the artifacts found in Egypt were several inscribed zodiacs, the most notable being from a temple ruin at Dendera in Upper Egypt. It was claimed that they portrayed the skies as they would have been some 15,000 years ago (taking into account the precession of the equinoxes), not for those of any more recent time. When this was first reported in Paris, Guillaume-Antoine de Luc—Jean-André's brother in Geneva and his zealous supporter—had immediately criticized its implied refutation of the traditional dating of the biblical Flood. In Paris, however, the issue had remained open and highly controversial among the scholars in the Third Class at the Institut. For example, the philologist and orientalist Joseph de Guignes published a review of the antiquity of the zodiacal system just as Cuvier was completing his papers on fossil bones, and Cuvier was certainly well aware of the argument. If the scholars were to decide that the Egyptian zodiacs were authentic and had been interpreted correctly, his dating for the last revolution (and that of the brothers de Luc) would again be in trouble.[18]

Conclusion

Cuvier's conception of the "last revolution" as a sudden and drastic event that had totally wiped out his most recent fauna of fossil mammals was dramatically reinforced by Adams's report of a woolly mammoth found frozen in Siberia. Cuvier disposed of several ambiguous cases, more or less convincingly, by assigning them to one side or the other of his great divide, as either extinct or living species from either before or after the decisive event. In the case of the ruminants, he argued that the ambiguity was intrinsic and unavoidable, because species could not be distinguished by bones alone; so fossils that appeared to be identical to living species might in fact belong to extinct ones.

However, Cuvier's argument was open to being undermined from any of three distinct directions. First, Desmarest continued to insist that valleys were being excavated slowly by the streams that still flow in them, not by any sudden and violent agency in the past; and Playfair's restatement of Hutton's similar claims was given prominence, ironically, by de Luc's latest restatement of his own argument for the limitations of such "actual causes" and for a putative catastrophe not unlike Cuvier's. Second, the full publication of Lamarck's theory of universal transformism brought into similar prominence his alternative explanation for the difference between fossils and living species, which excluded any extinction whatever by natural causes. And third, although the scholarly multicultural research of antiquarians could be taken to confirm the reality of some kind of global deluge at the dawn of human history, it could alternatively be interpreted as new evidence that took human antiquity much further back into the deep past without any such traumatic event.

All in all, the relation between human history and the unimaginably longer tracts of geohistory remained obscure, and the character of any boundary event at their interface was fraught with uncertainty. The next section describes a class of evidence that, although far from novel, was brought into prominence at this time, greatly boosting the plausibility of a massive catastrophe in the recent history of the earth.

10.2 THE PROBLEM OF ERRATIC BLOCKS (1810–14)

The problem posed

The character of the earth's last revolution had long been a matter of active debate among savants. Had the earth's last revolution been a sudden catastrophe—a "*Totalrevolution*", as Blumenbach had graphically described it (§6.1)—or had the former world of the deep past graduated imperceptibly into the familiar present world of recorded human history? Cuvier had brought fossil bones into this debate as powerful new evidence on the side of those who argued for a sharp disjunction between past and present; but he was of course well aware that this had been opposed by other naturalists. The erosion of valleys, for example, had long been a battleground between the two viewpoints, but the evidence was notoriously ambiguous. In the classic case of Auvergne, as recalled in the previous section, Desmarest's arguments for imperceptibly slow erosion had been well matched by Dolomieu's for a sudden scouring by some kind of mega-tsunami.

A quite different kind of physical feature was more decisive in this debate because it was much less ambiguous. It was generally considered to provide strong evidence for an exceptionally violent event; or at least, it was far more difficult to explain by any observable actual cause, no matter how vast the span of time that was invoked. The case was that of erratic blocks, which had somehow been moved tens or even hundreds of miles from their points of origin (§2.4). Erratics were extremely puzzling, because most of them were not close to river beds, where they might have been rolled along by exceptional winter floods. They were often perched high on hills and plateaus, and many were enormous, some even the size of a small house. It seemed inconceivable that they could have been moved by anything less than a huge and violent mass of water on a scale unparalleled in human

17. The English savant Thomas Young (also famous for his work on the wave theory of light) did not begin his attempt to decipher the hieroglyphs until 1814, the French scholar Jean-François Champollion not until 1821. Only in 1824 did the latter publish a provisional solution, which became the basis for all further Egyptological research: see Parkinson, *Cracking codes* (1999).

18. The Dendera zodiac was first pictured in Denon, *Voyage dans l'Égypte* (1802), pl. 130, and later featured as the largest of all the hundreds of engraved plates in [Jomard], *Description de l'Égypte* (1809–28), *Antiquités* vol. 4 (1817), pl. 21, which is reproduced (though much reduced in size) in Gillispie and Dewachter, *Monuments of Egypt* (1987). De Luc (G.-A.), "Réflexions sur les zodiaques" (1802); Guignes, "Sares des Chaldéens" and "L'origine du zodiaque" (both 1809); Buchwald, "Egyptian stars under Paris skies" (2003).

records, or by some other event of comparable magnitude. De Luc had attributed them to violent explosions that had ejected them from the depths of the earth. Saussure, on the other hand, had suggested that they had strayed horizontally and might have been shifted by some kind of "*débâcle*" or violent rush of water. Dolomieu had modified that idea by suggesting that the blocks might have slid down a gently inclined plane surface, which had subsequently been eroded away, leaving them in their strange perched positions. The puzzle remained awkwardly unresolved.

One detailed study of erratics had indeed suggested how they might have been transported without any catastrophic event, but the suggestion did not seem plausible as a general explanation. At the time, its author, Erhard Georg Friedrich Wrede (1766–1826), had been teaching mathematics and natural sciences at a high school [*Gymnasium*] in Berlin. He had studied the granite erratics strewn along the low-lying Baltic coast of his native East Prussia, and in 1804 he had published a causal explanation of them. The granite was utterly unlike anything else on the north German plain, but was known in situ in Scandinavia. So the problem was to explain the transport of thousands of large blocks of granite across the Baltic sea. Wrede had taken his inspiration in part from the geotheory that Lamarck had just published (§7.4). He had suggested that a shift in the earth's axis of rotation could have lowered the general sea level in the northern hemisphere in the quite recent past. If, at an earlier time when it was higher, the climate had also been somewhat

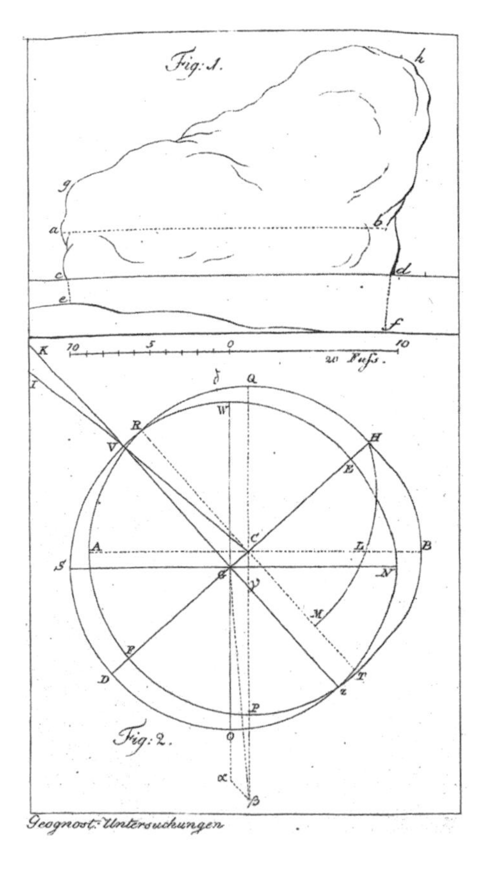

Fig. 10.4. The frontispiece of Wrede's book on the geognosy of the south Baltic coast in East Prussia (1804). The outline sketch (above) of one of the largest erratics illustrates the magnitude of the phenomenon to be explained (the scale shows it as some twenty feet across); this particular block was lying just offshore, submerged to the level of *c-d*, but if the sea had once been higher it could conceivably have been rafted on an ice floe from Scandinavia. The diagram (below) illustrates Wrede's conjecture that a change in the earth's axis of rotation could have effected the necessary change in sea level throughout the northern hemisphere. In the eyes of many other savants, the context of speculative geotheory would have detracted from the plausibility of the specific explanation of erratics. (By permission of the British Library)

colder, the blocks of granite might have drifted across the Baltic from Scandinavia *rafted on ice floes*. It was an attractive idea, for drift ice might have provided sufficient buoyancy to float even the largest blocks; and since they were all stranded on low-lying ground and many were close to the present shoreline, a quite small change of sea level might have been sufficient (Fig. 10.4).[19]

Von Buch on Alpine erratics

Wrede's explanation was quite attractive as an explanation for the erratics on the low-lying north German plain. But it was much less plausible for the otherwise similar blocks perched high in the Alps; furthermore, it was linked to a questionable geotheory. This must have been apparent to Wrede's fellow Berliner the geognost Leopold von Buch (1774–1853), who was the first to make a *detailed* study of the problem of erratics in its more refractory Alpine form, coincidentally around the time that Brongniart and Cuvier were completing their joint work on the Paris Basin.

After being trained under Werner at Freiberg, von Buch had been appointed to a position in the Prussian mines service. As a young man before the turn of the century he had toured through Italy, from the Tyrol to Vesuvius, in the company of Humboldt. Later, he saw for himself the extinct volcanoes of Auvergne, which made him an influential convert to Desmarest's Vulcanist interpretation of basalt (§2.4). Later still he had traveled extensively through Norway and up into Lappland. His travel books and other publications—most of which were translated into French, and some into English too—had established his reputation throughout Europe as an outstanding geognost and physical geographer. In 1806 he had been elected to the Prussian Academy of Sciences in Berlin, and had used his inaugural lecture to promote "*Geologie*". He had presented it as an exciting new science that was disclosing a sequence of "major formative epochs" [*grosse Bildungsepoche*] in the earth's history, leading from inorganic beginnings though successively higher forms of life to its recent culmination in humankind. This was an eloquent expression of the kind of geohistorical synthesis that by that time was being widely adopted among savants. Von Buch was certainly well-informed and up-to-date: as evidence for the history of life, for example, he had cited Lamarck's current work on fossil shells and Cuvier's latest on the bones from the Paris gypsum.[20]

Early in his career in the Prussian mines service, von Buch had been sent to work in Neuchâtel, a small outlying Prussian territory (not yet a part of Switzerland) at the foot of the Jura hills, on the far side of the Swiss plain from the Alps. There he surveyed and reported on mineral resources, but in the course of his fieldwork he

19. Fig. 10.4 is reproduced from Wrede, *Geognostische Untersuchungen* (1804) [London-BL: B.390.(1)], frontispiece (its only illustration), explained on 15–17; his geotheory is related to Lamarck's *Hydrogéologie* (1804) on 55n. The erratic was on the north coast of a small island near Cammin in Pomerania (now Kamien Pomorski in Poland).

20. Buch, "Fortschreiten der Bildungen" (1808); his lecture "Ueber den Gabbro" (1810), a similar piece of *haute vulgarisation*, was mentioned earlier as an illustration of the structural character of geognosy (§2.3). His travels are in *Deutschland und Italien* (1802–9), which printed his letters to Karsten from Auvergne as a long appendix, and in *Norwegen und Lappland* (1810).

could hardly ignore the erratics for which the area was already famous. In 1810 he returned to Neuchâtel to make a closer study of them, and he extended his fieldwork into the Alps to locate their source. The following year he reported his conclusions to the academy in Berlin. Von Buch's paper "On the causes of the spreading of large Alpine blocks" moved persuasively from the specific to the general: from a single remarkable erratic near Neuchâtel to those on the Jura hills as a whole; to their inferred source far away in the Alps; and finally to comparable cases in other parts of the Alps, and the even more widespread distribution of similar erratics on the plains of northern Europe. Partly as a result of the war, this outstandingly important paper was not published until 1815, and it was still longer before a leading French periodical made it more widely known. Its impact on international scientific debate was therefore delayed; but it needs to be described at this point, because it offered a detailed analysis of the effects of an apparently catastrophic event in the geologically recent past, *before* Cuvier's "preliminary discourse" made the notion of some such event widely familiar to the educated public throughout Europe (see §10.3).[21]

On a hillside some 250m above the town and lake of Neuchâtel was (and is) the huge Pierre à Bot: a block of granite the size of a small house, lying on ground composed of Secondary limestone. It was on the lower slopes of the Jura facing the Alps: from it one could look down over the town and its broad lake, then across the low hills of the Vaud, and finally to the high ramparts of the snow-covered Alps rising spectacularly (on a clear day) on the far horizon. Yet the nearest granite known in situ was even further away, in the Mont Blanc massif, some 100km from the Pierre à Bot. The problem of erratic blocks could hardly have been posed in a more dramatic form: how on earth (literally) could the Pierre à Bot have been shifted from somewhere near Mont Blanc to a point high up on the slopes of the Jura? Certainly, no amount of deep time would in itself resolve the puzzle, nor any immediately obvious "actual cause". The Pierre à Bot epitomized the powerful argument for some kind of exceptional event in the quite recent past.[22]

Having posed the problem, von Buch had little difficulty disposing of de Luc's explosion hypothesis, which did not begin to account for the highly specific distribution pattern of the erratics. Nor did any suggestion (extended from Wrede's idea) that they might have drifted from the Alps to the Jura, carried on ice floes or icebergs while the sea level was—hypothetically—far higher in the distant past. It was the observant Saussure who had noticed the important point that erratics were not distributed uniformly. They were most abundant on that part of the Jura that was directly in line with the deep valley of the Rhône, where it cuts through the Alps to emerge on to lower ground at Lac Léman (the Lake of Geneva). Von Buch's careful survey showed, furthermore, that the zone of erratics reached its highest altitude (about 600m above the plain) in that very same part of the Jura, from which it declined on both directions (to about 250m, for example, at the Pierre à Bot above Neuchâtel to the northeast). Far away in the high Alps of the Valais, von Buch located the exact source of the granite erratics at the north end of the Mont Blanc massif, at a point almost directly in line with the part of the Rhône valley just mentioned, and with the part of the Jura where the erratics were most abundant and at

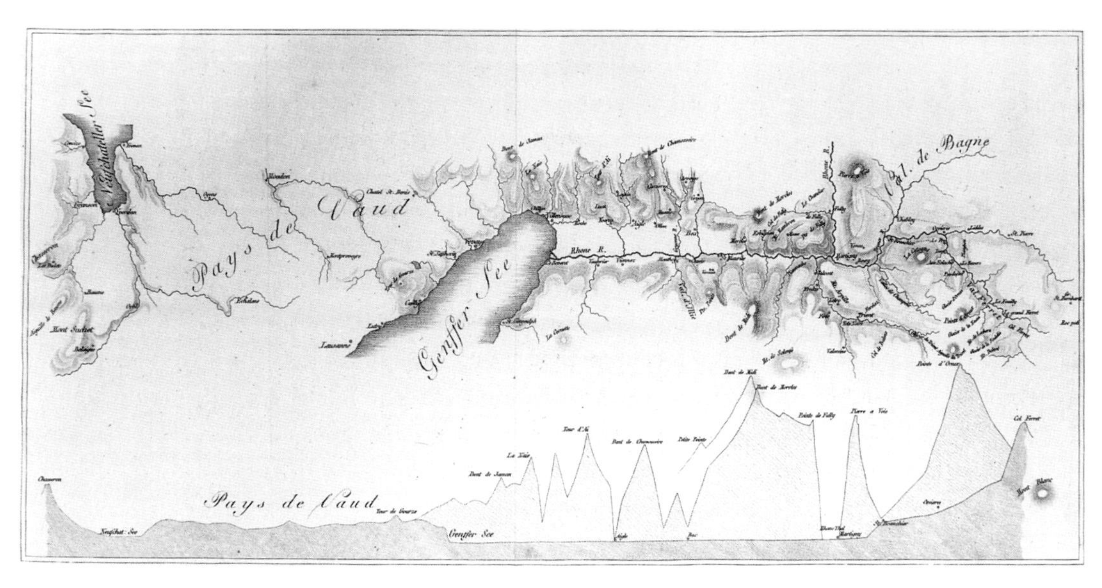

Fig. 10.5. Von Buch's topographical map and longitudinal profile of part of the Rhône valley and Swiss plain, stretching from the high Alps to the Jura hills, illustrating the paper he read in Berlin in 1811 (published in 1815), on the origin of the Alpine erratic blocks. He claimed that the granite erratics had somehow been moved about 100km from their source in the Mont Blanc massif (right)—specifically from the Pointe d'Ornex (left from Val Ferret)—down a long straight stretch of the deep Rhône valley, past the upper end of Lac Léman (the Lake of Geneva, here *Genffer See*), across the low hills of the Pays de Vaud, before being dumped high on the slopes of the limestone hills of the Jura (far left). (Neuchâtel is off the map, near the far end of its lake, *Neufchateller See*, top left). The Val de Bagnes (top right) was the site of a humanly catastrophic mudslide a few years later, which provided a small-scale model for the kind of event that von Buch then envisaged for the movement of the Alpine erratics (see §10.5). The diagrammatic profile (below) matches the map and shows, with great vertical exaggeration, the measured altitudes along the inferred line of movement, and those of nearby Alpine peaks; the total relief is almost 3,000m. (By permission of the Syndics of Cambridge University Library)

greatest height; he was puzzled, however, that they were rarely found on the intervening lower ground of the Vaud country. In other words, the erratics seemed to have been moved some 100km *in a straight line* without, as it were, stopping on the way: an accurate map made the point strikingly clear (Fig. 10.5).[23]

21. Buch, "Verbreitung grosse Alpengeschiebe" (1815), read at the Akademie der Wissenschaften in Berlin on 31 October 1811; abridged (by Brochant) as "Blocs de roches des Alpes" (1818) in the Parisian *Annales de chimie et de physique*.

22. Buch, "Verbreitung grosser Alpengeschiebe" (1815), 161–62; the erratic is now in thick forest, but the view can be seen from nearby. Historians who indulge in cheap sneers at the expense of the "catastrophist" geologists of von Buch's time should be sentenced to walk (not drive) from Neuchâtel up to the Pierre à Bot and back, five times without a break, to impress on them the strength of the evidence, and to deter them from defaming those who can no longer defend themselves. To put a serious point more seriously, those who write about early nineteenth-century catastrophism should at least make the effort to understand what was taken to be the relevant empirical evidence; and by far the most effective way to gain such an understanding is to see some of the evidence *at first hand in the field*, not from books in a library. The value of such *historical* fieldwork, which underlies most of the research described in the present volume, is also expounded briefly in Oldroyd, "Non-written sources" (1999), 409–12.

23. Fig. 10.5 is reproduced from Buch, "Verbreitung grosse Alpengeschiebe" (1815) [Cambridge-UL: P340:3.b.24.1], unnumbered pl. The map is oriented from north-northwest to south-southeast; by modern measurements the Pointe d'Orny is 3,270m above sea level, Lac Léman 372m. The Great Saint Bernard pass, on the main Alpine watershed, is on the far right; Mont Blanc (bottom right) is in its correct position for

This reconstruction was not based on any merely fanciful conjecture about the source of the granite. Following the leads given in Saussure's *Alpine Travels*, von Buch knew that each of the granite massifs in the Alps had a specific variety of granite that could be distinguished from others. So he was confident that the erratics on the Jura could have come only from the Mont Blanc massif, and their trail led unmistakably back to its northern end at the Pointe d'Ornex, towering with its glaciers above the deep Rhône valley. The inference was amply confirmed by other distinctive rocks from the same area—among them Saussure's famous *poudingue* or conglomerate of Vallorcine (§2.4)—which were also found as erratics on the Jura, although less commonly. Much of von Buch's text was devoted to local details establishing the exact sources of all these rocks; cumulatively they amounted to an impressively persuasive case.

Nor were the erratics of the Rhône valley and the Jura the only example of the phenomenon, though it was the one that von Buch studied in greatest detail. Saussure himself had traced the granite erratics perched on the hill of the Salève above Geneva, back up the Arve valley to their source at the other (west) end of the Mont Blanc massif near Chamonix. Reports by other naturalists enabled von Buch to cite three more cases further to the east, making in total five distinct lines along which erratics had been moved northwards from the high Alps, in each case over distances of several tens of kilometers. Finally, he argued that the well-known erratics of northern Europe were a part of the same general phenomenon on an even larger scale: they were strewn in a vast arc stretching from eastern England through the Netherlands, Prussia, and Poland into Russia; yet their only possible source was among the granites and gneisses of Norway and Sweden, on the far side of the North and Baltic Seas, implying that they had been moved several hundred kilometers. At the end of his paper, and just in passing, von Buch tentatively linked this vast phenomenon with the apparently extinct elephants and other fossils that Cuvier had described from Superficial deposits distributed even more widely.[24]

The structure of von Buch's argument shows that he was distinguishing clearly between geohistory and earth physics, between establishing the historicity of an extraordinary event and discovering its physical cause. The geohistorical reality could hardly be disputed, and his own detailed study of the Rhône valley erratics made an overwhelming case for an event of astonishing magnitude in that specific area. On the other hand, he could make scarcely any headway with the problem of explaining the physical cause by which the erratics had been shifted. He likened their movement to the flight of cannon balls: it was as if they had been fired from their Alpine source in trajectories that took them right over the low country of the Vaud, to crash onto the slopes of the Jura beyond. But a rough calculation gave him an "unbelievable" figure for the velocity that would have been required, if this imagery had been taken at all literally (he explained later that the cannon ball had only been intended as a loose analogy). Saussure's conjecture of a *débâcle* of violently moving water seemed almost equally inadequate in the light of another rough calculation. So he left the matter open, noting only that the movement of the erratics must have had a "wholly different cause", and evidently a far more general one.

However, that unresolved conclusion was itself significant. Unlike earlier savants with their comprehensive geotheories, von Buch as a new-style geologist was content to wait for further research to solve the puzzle: like Cuvier's, his was an investigation that was open to being improved.[25]

Hall's mega-tsunami

A few months after von Buch read his paper in Berlin—and without having yet heard about it, above the tumult of the war raging in central Europe—Hutton's old friend Sir James Hall read a rather similar and equally important paper to the Royal Society in Edinburgh. It too started from the problem of the Alpine erratics; but Hall, like Wrede, wanted to integrate his explanation into a context of geotheory. Specifically he wanted to use it to support what he called Hutton's "Plutonic system". That name signaled his recognition that Hutton's geotheory (§3.4) went far beyond the increasingly consensual Vulcanist interpretation of basalt as a volcanic rock (to which von Buch had earlier added his substantial weight) in that it treated a putatively vast store of heat in the earth's *deep* interior (the realm of Pluto) as the causal agency behind major movements of elevation or upheaval in the earth's crust. Ever since Hutton's own deistic metaphysics had been airbrushed out of Playfair's *Illustrations* (§8.4), his Edinburgh supporters had reformulated his system in terms of just three essential features: a dynamic deep interior driven by the immense expansive power of heat; the total adequacy of slowly operating actual causes (which Hall called "diurnal" or everyday) in shaping the earth's external features; and an overall steady-state cyclicity, operating through indefinitely long spans of time. Hall argued that his latest ideas strengthened the case for the Plutonic system, although they entailed abandoning Hutton's (and Playfair's) insistence on invariably gradual processes and the total adequacy of actual causes narrowly defined.[26]

In Hall's view, the essential core of Hutton's system was "the great circle of events" or cyclic process that he had envisaged, which required that the ceaseless action of subaerial erosion be balanced by the repeated elevation of new land

---

the map, although marked confusingly close to the profile. An unfortunate printer's error in von Buch's text put the highest erratics on the Jura at 5,900 feet above the plain, a figure he later corrected to 1,900 feet: see "Alpengeschiebe" (1815), 166, and "Additions au mémoire" (1819), 242n.

24. Buch, "Verbreitung grosser Alpengeschiebe" (1815), 184–86. The other Alpine lines of movement, each marked by erratics of distinctive rock types but not yet mapped in detail, were (1) from the peaks of the Bernese Oberland down the Aare valley past Bern on to the Jura above Biel; (2) from the peaks near the Saint Gotthard pass down the Reuss valley and past Zurich out on to lower ground; and (3) from the peaks of the Glarus down the Limmat valley and out past Winterthur; he also predicted, as a possible further case not yet explored at all, (4) a line down the upper Rhine valley to the Bodensee (Lake of Constance).

25. Buch, "Verbreitung grosser Alpengeschiebe" (1815), 183. There were good reasons why the later solution to the puzzle of erratic blocks—in terms of an "Ice Age" or geologically recent (Pleistocene) period of very extensive glaciation—was far from being as blindingly obvious in von Buch's time as it may seem to modern geologists enjoying the benefit of hindsight (I hope to explore this point historically in a sequel to the present volume).

26. Hall, "Revolutions of the earth's surface" (1814), read at Royal Society of Edinburgh on 16 March, 8 June 1812.

from the floor of the ocean to replace the eroded continents. But in Hall's opinion there was nothing in Hutton's steady-state system that required the elevation to be insensibly gradual: it could be sudden, or at least jerky or intermittent, and Hall claimed that this was just what the empirical evidence suggested. He recalled how, while Hutton was still alive, he himself had been greatly attracted by the older savant's grand theorizing. Yet he had never given up some of the ideas that he imbibed from his reading of Saussure and Pallas: "I have, therefore, been always disposed to combine the doctrines of Hutton with those professed by the gentlemen just named, relative to marine inundations". Unlike Hutton, he had the incalculable advantage of having seen with his own eyes some of Saussure's evidence for such *débâcles* while traveling through the Alps as a young man on his way back from studying the active volcanoes in Italy. "I remember well", he recalled, "to have witnessed . . . where, in high situations, on the face of Jura, I rode through great assemblages of granitic blocks, three or four feet in diameter".[27]

A major problem with Saussure's explanation of erratics was of course their sheer size: as von Buch recognized, it was difficult to imagine how even a mass of violently swirling water could have moved four-foot blocks, let alone the huge Pierre à Bot, any distance at all. In an attempt to alleviate that problem and thereby improve the explanation, Hall adopted Wrede's suggestion that erratics could have been carried on icebergs or ice floes, though he rejected the Prussian's conjecture about a formerly higher sea level. As Hall recognized, this made a distinctive and original synthesis: "I am induced, in attempting to unite the ideas of Saussure with those of Hutton, to retain part of the system proposed by M. Wrede, in so far as to consider the granitic blocks as having been made to float, by means of a mass of ice attached to each".[28]

Hall carefully considered but dismissed the attempts that had been made to explain the transport of erratics by ordinary diurnal causes, as well as de Luc's explosion hypothesis. Instead, he suggested that Saussure's notion of a sudden and violent *débâcle* of water offered a plausible mechanism, provided that icebergs or ice floes had given the erratics the necessary buoyancy. He also suggested that such a mega-tsunami could in turn have been caused—as Pallas had suggested briefly (§3.5)—by some distant volcanic upheaval on the ocean floor. His reasoning was explicitly actualistic. Humanly witnessed tsunamis were his small-scale precedents for imagining far larger incidents of the same kind in the distant past: "the events of Lisbon and of Callao [in Peru], though on a scale comparatively diminutive, help to lead the imagination to the conception of this colossal disaster". Such an event might have been so colossal that no human witnesses would have survived to tell the tale: no textual record or oral tradition of its historical reality was to be expected, and the only possible evidence was geological.

Hall's explanation of erratics, then, was based on a *causal* hypothesis to explain the much smaller—but humanly catastrophic—effects of an ordinary tsunami, which could then be scaled up to account for far larger effects. His hypothesis was that any sudden upheaval of the earth's crust on the ocean floor, at any time, would produce a momentary swelling of the ocean surface, which in turn would generate a huge wave radiating from the epicenter of the quake; ever the hopeful experimentalist, he tried to simulate this effect by exploding some gunpowder underwater and observing the disturbance at the surface. On reaching a shoreline, the wave would first show itself in an ominous withdrawal of the sea—as reported in historic cases of tsunamis—before returning with a vengeance to crash over any low-lying coastal land (Fig. 10.6).[29]

Hall argued that a mega-tsunami could be caused at any time by a proportionately large elevation of part of an ocean floor, which in turn he envisaged as just one phase in the intermittent process by which a whole new continent might be

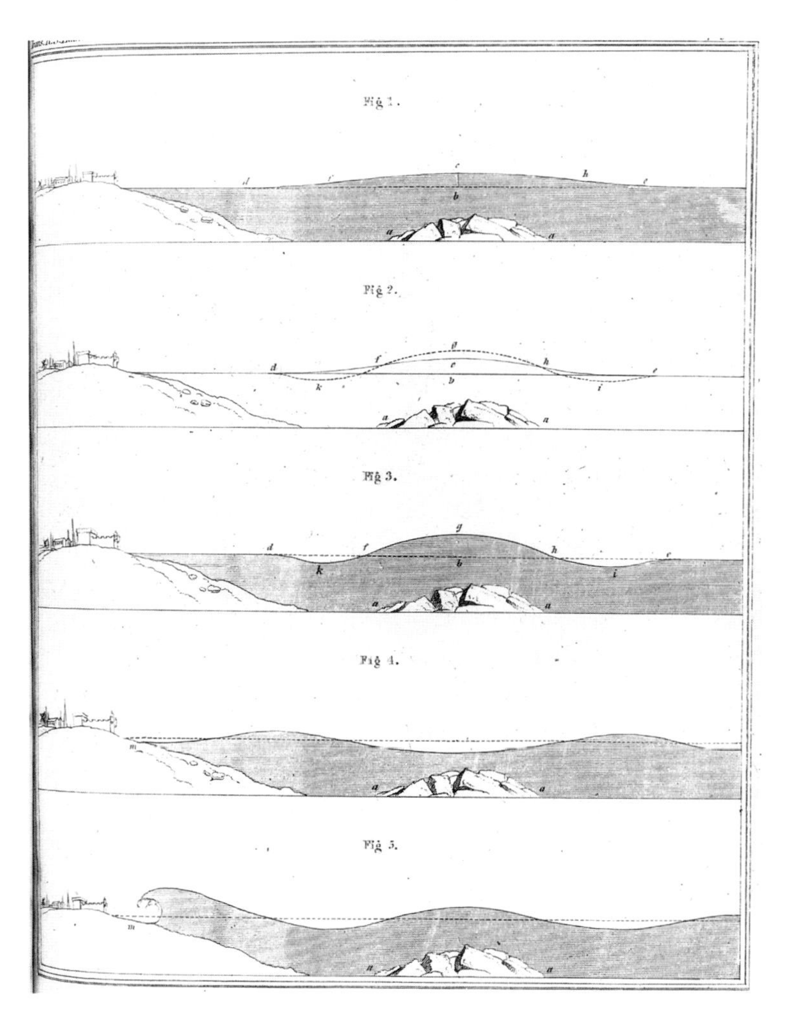

Fig. 10.6. Sir James Hall's diagrammatic sections (published in 1814) illustrating his hypothesis for the generation of a tsunami or "tidal wave" by a sudden upheaval on the ocean floor (figs. 1–3); at any coastline a temporary withdrawal of the sea (fig. 4) would be swiftly followed by a catastrophic inundation (fig. 5). Hall claimed to have sketched the initial form of the disturbance (fig. 1) after observing the result of an experimental underwater explosion. The town (left), with its castle, church, and windmill, indicates that the diagram represents a tsunami more or less on the scale of those recorded at Callao (1746) and Lisbon (1755). But Hall treated it as a model for the possible cause of the far larger (and far earlier) mega-tsunami that he claimed had overwhelmed the region around Edinburgh. Hall's explanation was in the tradition of earth physics (§2.4), invoking an *ahistorical* process that might operate at any time in past, present, or future. (By permission of the British Library)

27. Hall, "Revolutions of the earth's surface" (1814), 143–46.

28. Hall, "Revolutions of the earth's surface" (1814), 145–46, 156–58; he read about Wrede's work in de Luc, *Geological travels* (1810–11), 1: 33–37, describing the latter's earlier fieldwork in northern Europe. Wrede had left Berlin in 1806 to become professor of mathematics at Königsberg in East Prussia (now the Russian enclave of Kaliningrad); there he just missed being a colleague of the great Immanuel Kant, who had died in 1804.

29. Fig. 10.6 is reproduced from Hall, "Revolutions of the earth's surface" (1814) [London-BL: T.C.15.b.7], pl. 6, explained on 154–56. He had outlined his theory briefly, long before, in a footnote to his "Experiments on whinstone" (1799), 67n, in which he had first tried—despite earlier discouragement from Hutton—to simulate experimentally the reactions in rocks at the high temperatures and pressures that Hutton's system required: see also Hall, "Series of experiments" (1812), and Dean, *James Hutton* (1992), 140–43.

elevated. Provided one conceded that the process of elevation might be jerky, or occasionally violent, rather than invariably smooth and insensibly gradual, then Saussurian *débâcles* or mega-tsunamis—which Hall called "*diluvian waves*"—would be natural events that might be expected to occur, however infrequently, at any time in past, present, or future. "We might thus", Hall concluded, "by the help of this diluvian agent, complete the great circle of events, so elegantly pointed out by Dr Hutton, but which the diurnal agents seem quite insufficient to fulfil". This was strong stuff to propose in Edinburgh, in the very building where Hutton's system had first been put forward a quarter-century earlier (§3.4). Hall was currently the president of the Royal Society, and Playfair its secretary, so it is not surprising that he did his best to placate Hutton's hard-line supporters by presenting his own ideas as a minor "deviation" from their conception of "the Huttonian hypothesis". Yet Hall's putative mega-tsunamis, of a magnitude far beyond anything ever witnessed or recorded in human history, were bound to sound heretical in such circles. Indeed, his decision to call them "*diluvian* waves" sounded suspiciously like a concession to those who insisted on the historicity of *the* "diluvial" event, namely the Deluge recorded in Genesis. But Hall was unabashed: as he had told one of his correspondents a few weeks earlier, "I am a great friend to controversy and keep it up here [in Edinburgh] as much as I can".[30]

Hall insisted that his diluvian waves were not only impeccably natural events, but also events with respectable physical credentials: his causal hypothesis for explaining tsunamis—whether small-scale or large—belonged clearly in the tradition of earth physics (§2.4). Furthermore, they were events that were a part of the terrestrial "system" and must therefore have occurred repeatedly in the past, just as they were bound to recur in the future. He emphasized that point by describing Scottish field evidence for the highly folded strata that, in his opinion, recorded far earlier episodes of elevation (and he devoted two of his four plates to depicting them). If, therefore, it happened that there had been a mega-tsunami in the geologically recent past, there was certainly nothing unique about it. However, Hall claimed that just such an event had indeed taken place, not only far away in the Alps and Scandinavia but also close to home: "As the inferences derived from these distant facts are called into question by some gentlemen of the highest authority in this Society, I am happy to have it in my power to produce a set of observations made in this immediate neighbourhood, which seem in a manner no less satisfactory, to lead to similar conclusions."[31]

Three months later, in the second part of his paper—"an account of the diluvian facts in the neighbourhood of Edinburgh"—Hall presented evidence that indeed challenged his critics by its sheer proximity. The most prominent of all the features that he attributed to the action of a geologically recent diluvian wave was none other than Edinburgh's Castle Rock, which towered above the planned and spacious New Town where savants and others like them had their comfortable modern homes. The Rock rose precipitously (crowned by the castle itself) at its west end and highest point, but tapered gently down towards the east as the Old Town, where the lower orders were crowded in old and smelly tenements; and it was flanked north and south by deep valleys not obviously related to any of the present rivers. Hall argued that this strange topography could not be explained as the product of any diurnal agency such as the present river system. However, it found a true but far smaller diurnal analogue in the way that any river in flood could be seen to scour its way past a boulder or other resistant obstacle, and Hall attempted "to apply these principles to the great scale of geology".[32]

The clearest evidence that Hall offered, however, was a short ride on horseback—an easy excursion for any of the Edinburgh savants—to the west of the city, around Corstorphine Hill. This was (and is) a ridge about 2km long and nearly 200m in height, oriented roughly north-south and therefore at right angles to the putative flow indicated by the form of Castle Rock and the Old Town. Hall described in detail how, at many points on the west face of the hill and on its crest, broad pavements of solid bedrock were exposed, deeply scratched with dozens of parallel grooves, each up to several feet in length and all with the same orientation. To the east of the hill, and putatively in its lee, the lower ground was marked by long parallel ridges with the same bearing, some of them with solid rock outcropping at the west end rather like small versions of Castle Rock. Hall's detailed map of the area also showed the locations of fifteen "specimen" sites where the grooved rocks were most clearly exposed: his readers were not to take his word for it and remain merely virtual witnesses, but to go there and see these specimens of the "diluvian facts" for themselves, as it were in a natural open-air museum (Fig. 10.7).[33]

Hall interpreted these grooved rocks as powerful evidence that a huge diluvian wave had indeed swept over the whole region, roughly from west to east, bearing with it a mass of mud, stones, and boulders; the latter had done the scratching. He denied that any "diurnal agent" or actual cause could have been responsible for them: for example, the scratched pavements were up on top of the hill, not on low ground and certainly not near any river. In his Huttonian *Illustrations*, Playfair had publicly doubted that anything like this, which might need to be ascribed to "some other cause than the ordinary *detritus* and wasting of the land", had yet been found; he had advised that "it seems best to wait until the phenomenon is observed, before we seek for the explanation of it". Hall had now explicitly met this challenge. Furthermore, he denied that invoking vast spans of time—the Huttonians' other standard argument—would make any difference, for the observable agency of ordinary erosion was busy effacing the features in question, not generating them:

30. Hall, "Revolutions of the earth's surface" (1814), 159, 166–67; Hall to Alexander Marcet, 7 February 1812 (Edinburgh-NLS, MS 3818, f.49).

31. Hall, "Revolutions of the earth's surface" (1814), 162–63, 167. His pls. 7 and 8 were views of folded strata on the Berwickshire coast, and his "Convolution of certain strata" (1814), read only a month earlier, had focused specifically on this kind of evidence. Playfair was the most prominent of his critics among the society's gentlemen, but not the only one.

32. Hall, "Revolutions of the earth's surface" (1814), 169–76. The valleys on both sides of Edinburgh's Old Town are those spanned by the North and South Bridges; observant visitors who arrive at the city's main rail station (Waverley) can hardly be unaware that it lies in the northern valley, with steep slopes, and lengthy flights of steps, up to the Old Town on one side and to the New on the other.

33. Fig. 10.7 is reproduced from Hall, "Revolutions of the earth's surface" (1814) [London-BL: T.C.15.b.7], pl. 9; the specimen sites are described on 184–91. Spectacular examples of striated rock pavements are still visible at many points on and near the crest of Corstorphine Hill, which is now a public park; but the area to the east is covered with suburban housing and the parallel ridges are hard to discern.

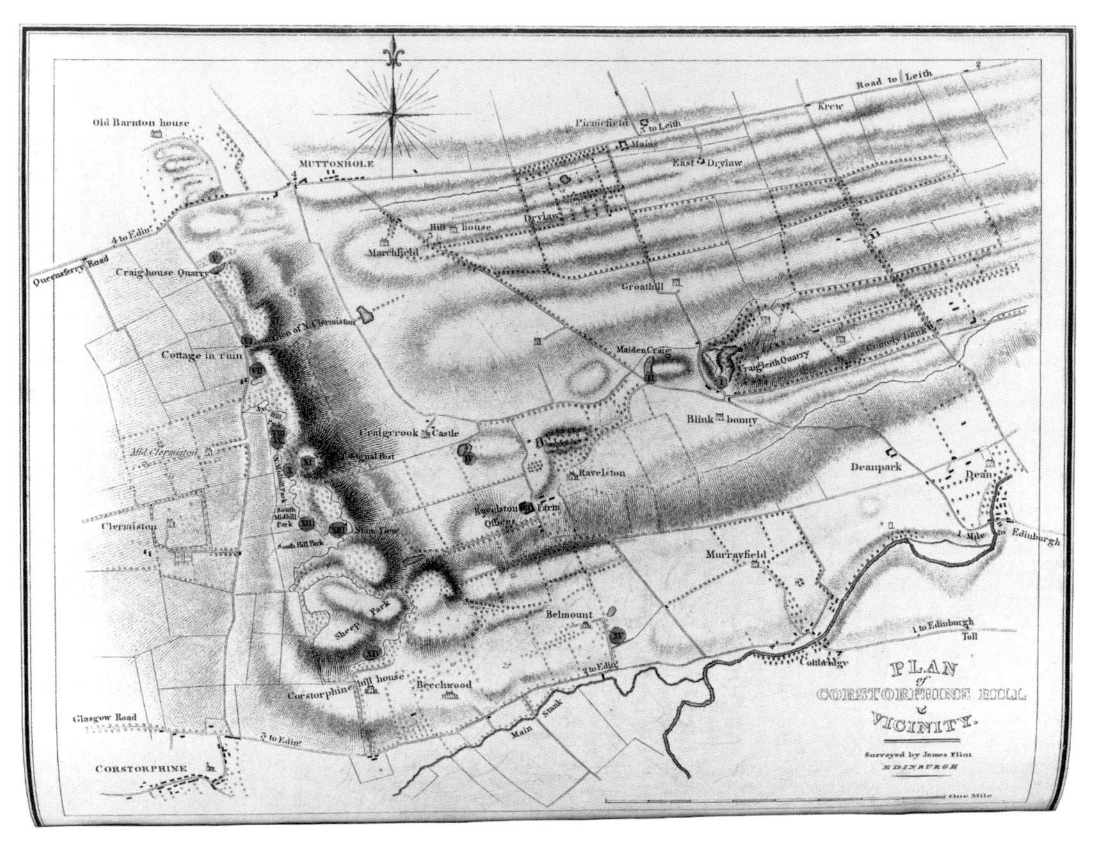

Fig. 10.7. Hall's map of Corstorphine Hill, west of Edinburgh (published in 1814), depicting the topography that he attributed to a "diluvian wave" or mega-tsunami flowing roughly west to east across the whole region. The small dark circles marked fifteen "specimen" sites where deeply grooved bedrock was exposed, which Hall interpreted as the result of boulders being dragged across the surface in a mass of mud at the base of a huge body of water in rapid motion. Craigleith Quarry (the site furthest to the east) was an example of a "crag" of hard rock with a long "tail" to the east, forming a small-scale version of Castle Rock and the Old Town in Edinburgh itself. (By permission of the British Library)

> It is in vain that a vast duration is ascribed to the influence of an agent, unless it can be shewn, that its action has a tendency to produce the alleged result. If it has a tendency to produce a different result, that difference would be augmented in proportion to the duration of the action. Now, the diurnal operations are everywhere found in the act of corroding and altering the forms alluded to; but they are nowhere seen to produce them. This class of [diluvian] facts, on the other hand, all conspire in giving probability to the hypothesis of a diluvian wave, which affords an easy explanation of all the large features of this country [around Edinburgh].[34]

So Hall confronted critics such as Playfair with an awkward example almost on their own doorstep: these were major topographical features that could not plausibly be attributed to any cause observably in action in the area, but they were quite easily explicable if in fact a mega-tsunami had swept over the Edinburgh region in the geologically recent past, bearing with it a mass of mud and boulders that could have acted like a giant abrasive. In a footnote, added to clarify what some in his audience had evidently found obscure, Hall explained that he envisaged a wave lasting no more than a few minutes, like humanly witnessed tsunamis; "but during that short time, I conceive the water to have been urged forward with such force, and to have carried with it so many powerful agents [i.e., the boulders], that it has produced effects equal to the work of ages under other circumstances". Since the grooves in the bedrock looked much the same on the top of Corstorphine Hill as at its base, he inferred that even the top must have been deeply submerged: he guessed that the hill might have been covered by as much water again, making a wave about 1,000 feet high, or some sixteen times the 60-foot tsunami at Lisbon; but in the Alpine case the erratics on the Jura indicated that it would have had to be as much as 2,000 feet high. Hall's estimates were thus edging back towards the magnitude suggested long before, though on different grounds, by Dolomieu (§6.3). But now the necessity of invoking some such exceptional event, however distasteful to hard-line Huttonians, was becoming as overwhelming as the catastrophe itself.[35]

In his conclusion, Hall suggested that accurate local details such as those he had presented, if emulated by naturalists elsewhere, might eventually "furnish the means of ascertaining the direction of diluvial inundations across the great continents . . . by a comparison of directions, these tremendous agents may be traced to their source". Rather as modern seismologists triangulate from records made around the globe to locate the epicenter of an earthquake, Hall looked forward to the possibility of finding the point at which a sudden upheaval might have generated a mega-tsunami affecting widely separated regions. The Alpine and Baltic erratics, and perhaps the Scottish features too, might turn out be the products of a single event of vast magnitude and scope. Yet even if this were so, it would have been just the most recent example of a class of event built into the earth's dynamic system. Hall later made this clear, when he offered to conduct a prominent young member of the Geological Society around his open-air museum:

> I shall be most happy to show the Diluvian facts to Mr. Horner. I must have expressed myself very ill in some parts of the paper to give rise to the opinion which he ascribes to me, that only *one* such cause has acted. On the contrary, it is my firm belief [that] thousands of such actions have taken place in succession; nay, I conceive them to have arisen as the necessary consequence of all the Huttonian elevations.[36]

Had Hall already known about von Buch's work, he would have recognized that his own research agenda was one on which the Prussian had already made an important start. The difference, however, was that for von Buch the putative catastrophe was important primarily as a solution to a long-standing puzzle in physical

34. Hall, "Revolutions of the earth's surface" (1814), 177–78; Playfair, *Illustrations* (1802), 411–12.

35. Hall, "Revolutions of the earth's surface" (1814), 194–95n.

36. Hall, "Revolutions of the earth's surface" (1814), 207–10; Hall to Alexander Marcet, 15 September 1813 (Edinburgh-NLS, MS 3813, f.51), referring to Leonard Horner (1785–1864). Horner had studied at Edinburgh (under Playfair, among others) before starting a career in London in commerce and insurance, but was about to move back to Edinburgh; he later gave Brocchi's Subapennine work (§9.4) an enthusiastic review; much later still, one of his daughters married Charles Lyell.

geography, in terms of a specific and decisive event near the end of a long and complex geohistory. For Hall, in contrast, it was simply a geologically recent—and therefore exceptionally well-preserved—example of a kind of event that was part of the unchanging causal dynamics of the earth. Hall, unlike von Buch, was engaged in earth physics and only incidentally in geohistory.

Hall did in fact have strong interests in at least one problem in *human* history, which he pursued in parallel with his geological research; yet it is significant that even here his approach was primarily causal. At the same time that he was reading his paper on the earth's "revolutions" he was also completing a handsome illustrated book, *The Origin, History, and Principles of Gothic Architecture* (1813). The problem was that of explaining the distinctive style of late-medieval buildings, the beauty of which he had first learned to appreciate as a young man, while returning through France from seeing the very different beauties of the Classical and Renaissance art (and the volcanoes) of Italy. Characteristically, Hall tackled this problem with a causal hypothesis: that the soaring forms of Gothic columns, tracery, and vaulting were all modeled in stone on earlier prototypes in wood; they were enduring embodiments of structural principles first developed (though more crudely and on a smaller scale) in a quite different and more perishable medium. "The whole of this theory has been submitted to an experimental test," he explained, "by the construction of a wicker fabric which is now standing in my garden": an elaborate and convincingly Gothic structure (which he depicted in a fine frontispiece), composed entirely of slender poles of wood bound together in a kind of large-scale basketwork. It was in effect a causal hypothesis to solve a historical problem preserved from the human past; as such it was closely analogous to what he was doing at the same time in his geological investigations, offering a causal explanation for enigmatic natural features preserved from the deeper past of geohistory.[37]

Anyway, by describing a specific new set of "diluvian facts" in persuasive detail, Hall strengthened the argument for the sheer historicity of a geologically recent catastrophe. And by giving the putative class of mega-tsunamis a plausible physical explanation, he certainly undermined the traditional hard-nosed criticism that they were inexplicable, and the hypothesis itself "unphilosophical" or unscientific; or—worse still—that they were imaginary, and the hypothesis perhaps covertly supernaturalist. Nor was his case for a major catastrophe known only to other Edinburgh savants or only to other anglophones. As usual, the effects of the wars were circumvented by the *Bibliothèque Universelle* in Geneva, which promptly published an appreciative summary of Hall's paper for the benefit of Continental savants.[38]

Conclusion

Notwithstanding their important differences, von Buch and Hall jointly reinforced Cuvier's insistence on the sheer historicity of a geologically recent "catastrophe", whatever its exact character might have been and whatever its ultimate physical cause. In contrast, Wrede's earlier suggestion about the possible quiet flotation of erratics on ice floes seemed inadequate to explain the transport of any but those in low-lying areas, though ice might have helped elsewhere to give the boulders some buoyancy in a turbulent flood. Von Buch's detailed study of the erratic blocks high in the Alps and on the Jura established the fact of their long-distance transport almost beyond question, while paradoxically deepening the problem of accounting for it in causal terms. Hall's equally detailed study of the Edinburgh region likewise established the historicity of some kind of extraordinary event; but he also suggested a plausible causal explanation both for the tsunamis witnessed in human history and for the far larger mega-tsunami to which he ascribed the "diluvian facts" he had observed. Both savants inferred an extraordinary event that was at least compatible with the catastrophe of mass extinction that Cuvier claimed on the quite independent evidence of fossil bones.

However, neither von Buch nor Hall made any conjecture, at least in public, about the dating of the extraordinary event in relation to human history, whereas Cuvier claimed that there was textual evidence—however faint and garbled—that it had happened within the memory of early literate cultures. The next section returns to Cuvier, to describe how his "preliminary discourse" tackled the problem of the "last revolution" and how other savants commented on it or adapted it for their own purposes.

10.3 "A GREAT AND SUDDEN REVOLUTION" (1812–16)

The inadequacy of actual causes

Both von Buch and Hall presented their studies of "diluvial" features—in Berlin and Edinburgh respectively—before Cuvier published his *Researches on Fossil Bones* in Paris. But neither was published until later, and their powerful evidence in favor of some kind of catastrophic event in the geologically recent past was unknown to Cuvier when he composed the "preliminary discourse" that introduced his collected papers. Had he been aware of it, he would have seen at once that it greatly strengthened one of the arguments he mounted in that attractively readable essay. In addition to advocating research on the fossils of the younger Secondary formations (Brocchi's "Tertiary") as the highest priority for geology (§9.3), much of Cuvier's discourse explored the relation between the present world and an even more recent part of geohistory, as represented by the Superficial deposits and the fossil bones they contained. These two strands of his argument were in fact closely entwined; here they have been separated, somewhat artificially, in order to show how they contributed to the two rather distinct lines of research described in the previous chapter and in this one.

37. Hall, *Gothic architecture* (1813), 9, 18–19, 27–28, 100–102. Like his hypothesis of diluvian waves, he had first suggested his wickerwork theory many years earlier, in this case in a paper read to the Literary section of the Edinburgh society: Hall, "Gothic architecture" (1798).

38. Hall, "Des révolutions" (1814), in fact paraphrased by the editor, who added the dramatic evidence of the Alpine erratics that Saussure had first described on the Salève above Geneva. The periodical had changed its name from *Britannique* after the end of the wars made its exclusively anglophone orientation no longer necessary.

After presenting himself as nature's "antiquarian" and introducing his main themes (§9.3), Cuvier launched into a review of the evidence for vast "revolutions" in the distant past. As usual, and like other naturalists, he used the word to denote any major change, whether sudden or gradual. He emphasized that there had been not just one catastrophe—which some of his readers might have identified at once as Noah's Flood—but repeated upheavals, which had left traces of their action in the form of tilted and folded strata, massive beds of conglomerate, and so on. His brief survey of the record of the rocks, right back to the Primaries, showed that such revolutions even preceded the first appearance of life on earth and were clearly part of the ordinary course of nature from start to finish. None of this was novel, or even original to Cuvier: he was merely setting his own research in a consensual context, for the sake of ordinary readers who might not be familiar with recent work in geology.[39]

Cuvier next considered "what takes place on earth today", as the obvious potential key to the physical agencies or "causes" responsible for past revolutions: as usual, actualistic comparison was assumed to be the best starting point. Cuvier expressed that principle by means of yet another analogy with human history; significantly, however, he described the claim that actual causes were adequate for explanation—he might have had Desmarest in mind (§10.1)—as the majority view that he himself was about to challenge:

> It has long been thought possible to explain earlier revolutions by these present causes, just as past events in political history are easily [*sic*!] explained when one knows well the passions and intrigues of our [own] times. But we shall see that unhappily this is not so in physical history. The thread of operations is broken; nature has changed course, and none of the agents she employs today would have been sufficient to produce her former effects [*ouvrages*].[40]

Cuvier's assertion that "the thread of operations is broken" has given rise to more argument among historians than perhaps any other phrase in his entire published work. But in context, and particularly in the light of his earlier writings, its meaning is not obscure. None of the processes "that still operate", he argued, is adequate to account for the observed effects; therefore those effects must be attributed to causes of another kind. As usual, he declined to suggest what they might have been, because he was doing geohistory rather than earth-physics. But there is no reason to suppose that he thought they had been in any way contrary to the ordinary ahistorical laws of nature, let alone that they were due to unmediated divine intervention. Anyway, it was a moot point whether, for example, a "deluge" large and violent enough to have ravaged the continents (as Dolomieu had supposed) could properly be said to be of the same "kind" as the humanly catastrophic but far smaller tsunami that had devastated Lisbon; or whether a sudden collapse of continents and consequent emergence of ocean floors as dry land (as envisaged by de Luc) was of the same "kind" as some sudden landslip or the consequent draining of a lake-bed. Yet such huge putative catastrophes had been suggested by those earlier savants as entirely natural events powered by ordinary physical processes, although on an extraordinary scale unparalleled in human history. Such were the respectable scientific precedents that Cuvier was following.[41]

Cuvier clearly felt himself bound to follow those precedents, and to claim that quite extraordinary "causes" must have been responsible for the earth's major revolutions, because the "actual causes" known to him seemed so puny and inadequate. His own firsthand field experience was limited, so it is hardly surprising that he borrowed from contemporary published work on physical geography, reviewing a range of present processes that would later seem utterly inadequate: landslides and cliff erosion as destructive agents; sand dunes, alluvial and coastal deposits, precipitates and coral reefs as constructive ones; volcanoes as a special case but a strictly localized one. Extraterrestrial causes suggested by astronomers—such as the precession of the equinoxes—were no less inadequate, because they were cyclic in character and limited in extent: "these excessively slow movements cannot explain catastrophes, which must necessarily have been sudden". Putatively worldwide changes, such as the progressive diminution of the oceans built into the standard model of geotheory (§3.5) or the gradual global cooling built into Buffon's (§3.2), were inapplicable for the same reason, that "no slow cause can have produced sudden effects". For Cuvier, what had to be explained was altogether more dramatic: "all that is nothing to what has overturned our strata, enveloped in ice large quadrupeds complete with their flesh and skin, brought on to dry land shellfish as well preserved as if they had been fished out alive, and, finally, destroyed whole species and genera."[42]

The role of fossil quadrupeds

Cuvier claimed that the relatively rare and fragmentary fossil remains of terrestrial quadrupeds could—paradoxically—provide more decisive evidence for geohistory than the far more abundant fossil mollusks studied by naturalists such as Lamarck. This was because the living relatives of the former were known far more completely, and could therefore provide a much more reliable baseline for comparison, and more solid evidence for the revolutions that had affected them. Here Cuvier was not only establishing his own intellectual territory but also, more importantly,

39. Cuvier, *Ossemens fossiles* (1812), 1: "Discours", 5–16, trans. Rudwick, *Georges Cuvier* (1997), 187–93.

40. Cuvier, *Ossemens fossiles* (1812), 1: "Discours", 16–17, trans. Rudwick, *Georges Cuvier* (1997), 193.

41. Cuvier's "catastrophist" argument was exactly analogous to the one employed since the late twentieth century by the many scientists who attribute the mass extinctions—including the last of the dinosaurs—at the end of the Cretaceous period (at the "K/T boundary") to the impact of a comet or asteroid: identical in kind to the event that produced Meteor Crater in Arizona and other such features, but vastly greater in magnitude. It is ironic that the modern theory of bolide impacts was fiercely criticized in the 1980s for being "unscientific", just as Cuvier's catastrophism was censured by geologists later in the nineteenth century and by historians through most of the twentieth.

42. Cuvier, *Ossemens fossiles* (1812), 1: "Discours", 24–25, trans. Rudwick, *Georges Cuvier* (1997), 198. His review of "actual causes" was conspicuously lacking in anything that might have suggested the earth's internal powers of crustal elevation, but he was hardly unusual at this time in being unaware of any good evidence that these were observably effective.

sidestepping the traditional argument about putative "living fossils" (§5.1): naturalists could not have the same degree of certainty about changes in marine faunas, simply because knowledge of exotic and deepwater forms was so incomplete.

To make his own argument convincing, however, Cuvier had to confront the possibility that terrestrial quadrupeds as yet unknown alive might still be roaming the poorly explored interiors of the continents, in effect as living fossils. He minimized this possibility by arguing that human occupation of even the fringes of the continents would have yielded reliable knowledge—in the course of time—about at least the larger animals living in the interior; and recent explorers had not in fact sighted herds of mammoths, mastodons or megatheria. But the most effective way for Cuvier to establish this point was to show that all the animals reported or described in the course of human history could be identified as species known to naturalists in his own time. So he plunged into a detailed review of the animals known to the writers of Classical Antiquity. He used his formidable knowledge of ancient literature—doubtless amplified by his work for the French edition of *Asiatick Researches* (§7.4)—to show that every one of the animals recorded in Antiquity could be assigned to some known living species. This proved that the animals of the Old World—even those of Africa—had all become known to Europeans quite early in human history; by analogy, most of those living in the New World ought likewise to have been discovered in the past few centuries, even if the deep interiors of the continents remained poorly explored.

The educated readers whom Cuvier hoped to attract to his Discourse would have found his Classical authors reassuringly familiar and his analysis quite easy to follow despite its technical detail. Cuvier was acting here as a textual critic, for his review required him to discriminate between the soberly factual, the garbled and legendary, and the utterly fabulous. Reports of unicorns and griffins were treated with the same respect and rigor as those of giraffes and zebras: the unicorn, for example, was demythologized into an oryx that had been depicted conventionally in strictly lateral profile. Anyway, Cuvier concluded that all the animals recorded in Antiquity were species known to be still alive; they belonged on the modern side of the great divide. They were not, he claimed, those whose fossil bones he would describe in the body of his work, which must therefore have been destroyed by "general causes" or natural events, not by any human action.

However, although large terrestrial quadrupeds were the most favorable material for resolving the puzzle about the earth's revolutions—the most critical experiment, as it were—they brought problems of their own. Unlike the well-preserved fossil shells that Lamarck had described and classified, Cuvier's skeletons usually had to be pieced together from scattered bones. But he turned this to his advantage, because the necessary reconstructions depended on the validity of his zoological principles, which underlay his main research field of comparative anatomy. Those principles—expressing the functional integration of the animal body—were the basis for his conception of animal species as being well-adapted to their respective modes of life, and for the consequent impossibility of any gradual transmutation from one into another. The role of these zoological principles in the primarily *geological* argument of Cuvier's discourse was strictly instrumental, in that their application simply ensured that the scattered bones were pieced together correctly, to reconstruct a real animal that had once been alive, and not an imaginary chimera or monster. Nonetheless, that role was so important in the argument that Cuvier expounded his principles more clearly here than in any of his overtly zoological publications.[43]

This was the context in which Cuvier made his famous claim to be able in favorable cases to identify the affinities of a fossil quadruped even from a single fragment of bone. Indeed it was "a more certain mark than all those of Zadig"; for Cuvier claimed he could identify totally unknown animals, whereas in Voltaire's story Zadig had merely traced animals already well-known—the king's horse and the queen's bitch—by recognizing and following their tracks after they escaped. To support such confident claims, Cuvier was careful to stress his almost unique qualifications: not only some fifteen years of "assiduous research", but also the advantage of having access, at the Muséum in Paris, to the finest collection of the necessary comparative materials, the skeletons of living species of quadrupeds.[44]

Transformism rejected

At this point Cuvier tackled one of the most serious objections to his interpretation of fossil bones (§10.1): that the contrasts between fossil and living species might simply be the result of transmutation, produced by changed circumstances and the lapse of time, rather than showing the interposition of any catastrophic mass extinction. Cuvier's first line of defense was to emphasize the absence of the intermediate forms that should have been preserved if Lamarck was right: "Why have the entrails of the earth preserved no monuments of such a curious genealogy, unless it is because the species of former times were as constant as ours?" Cuvier reviewed what was known about variation in living species, and claimed that it was strictly limited, certainly in the wild and even when animals were domesticated or transplanted far from their natural habitats. In the extreme case of the varied breeds of dog, the skeletons prepared for him by his brother Frédéric—who was in charge of the menagerie at the Muséum—showed little significant variation except in size. So he concluded that there was no evidence for the unlimited potential for variability that transmutation would require.[45]

Cuvier then claimed that there was nothing to suggest that, in modifying animal form, "time has any more effect than climate". He summarized the detailed work he had done on the great variety of mummified animals that Geoffroy had brought

43. See Russell, *Form and function* (1916), 31–44; Coleman, *Georges Cuvier* (1964), 119–26; Theunissen, "Cuvier's *lois zoologiques*" (1986).

44. Cuvier, *Ossemens fossiles* (1812), 1: "Discours", 58–66, trans. Rudwick, *Georges Cuvier* (1997), 217–22. Cuvier's prose style tended towards the hyperbolic, so it is not surprising that he was later misunderstood to be claiming an almost magical ability to *reconstruct* an entire animal from a single bone, rather than just to *identify* it. In fact he was also careful to restrict his claim to cases where the bone was well-preserved and of the right kind to allow any such identification: for example, a jawbone was far more revealing than a rib (as it would be for a modern vertebrate paleontologist).

45. Cuvier, *Ossemens fossiles* (1812), 1: "Discours", 73–79, trans. Rudwick, *Georges Cuvier* (1997), 226–28; Cuvier (F.), "Races du chien domestique" (1811).

back from Egypt. He had found no detectible differences from their living counterparts: the apparently anomalous ibis had fallen into line after more rigorous study (§7.4). "I am well aware that I am citing monuments of only two or three thousand years ago", he conceded, "but this is as far back as it is possible to reach". Far from betraying an inadequate conception of the magnitude of deep time, the comment reveals Cuvier's acute sense of the kind of reasoning that was needed on this issue. Just as astronomers used the small but detectible short-term movements of the outer planets to calculate orbits measured in centuries, so a long-term transformation in animal form should not be inferred unless its effects could be detected, if not in a brief human lifetime then at least within the total span of known human history. As Cuvier put it, obviously with Lamarck in mind (§7.4, §10.1): "I know that some naturalists rely a lot on the thousands of ages [*siècles*] that they pile up with a stroke of the pen; but in such matters we can hardly judge what a long time would produce, except by multiplying in thought what a lesser time produces". Zero multiplied even by thousands would still be zero.[46]

Cuvier attributed the disappearance of the fossil genera and species not to their transmutation into the forms now known alive but to their extinction in some kind of natural catastrophe. That left him, of course, to explain the *origin* of species (as Darwin rightly defined the problem decades later). Cuvier's provisional solution was, at the time, not unsatisfactory. First, he explicitly repudiated any explanation beyond the strictly natural: "I do not claim that a new creation was needed to produce the existing species; I say only that they did not exist in the same places, and that they must have come from elsewhere". He then posed a simple thought experiment, set not in the deep past but in the imaginable future. Suppose that at some future time Australia were to suffer "a great irruption of the sea", smothering its present surface with detrital deposits. Then its strange endemic fauna of marsupial and egg-laying mammals—newly known in Cuvier's time—would become extinct, and their bones might be fossilized, just as he claimed that in the past his fauna of mammoths and mastodons had been wiped out and their bones preserved. If, at the same time, this future "revolution" happened to create a land bridge through the East Indies, the placental mammals of Asia would soon colonize the new Australian landmass. But if, later, a second revolution were to overwhelm Asia in turn, and destroy its indigenous fauna, future naturalists would puzzle over the origin of what would then be the living Australian elephants and camels, and their relation to the fossil kangaroos that would be found in the detrital deposits there, just as Cuvier and his colleagues puzzled over the origin of the present mammalian fauna and its relation to the fossil mammoths and mastodons. "What New Holland would be, in the conjecture we have just made," he concluded, "Europe, Siberia, and a large part of America are in [present] reality".[47]

In retrospect, of course, it is easy to see that Cuvier's explanation of the origin of present species, in terms of migration from vanished continents, was inadequate, because it led to a potentially infinite regress. It remained fairly satisfactory, however, as long as the question was that of relating a *single* set of fossils to their living counterparts, or as long as the former world remained undifferentiated. And this was just the form in which Cuvier had encountered the problem at the start of his career (§7.1). It was only very recently that it had begun to be deepened and even transformed, as a consequence of the doubly enriched geognosy that he and Brongniart had begun to develop (§9.1). Only in this new and more fully *geohistorical* perspective could it become apparent that—quite apart from the origin of life itself—an explanation would have to be found for the origins of several successive faunas, not just of one. If Cuvier had not yet fully assimilated the implications of his own recent research, he should not be judged too harshly by posterity.[48]

Dating the last revolution

Among living mammals, the human species was of course of supreme interest, and Cuvier turned next to the question of its existence in the former world. He had carefully reviewed the alleged cases of fossil human bones and found that all those he had been able to examine at first hand were spurious. While in Italy, for example, he had visited Pavia to scrutinize those that Spallanzani had collected long ago on the Greek island of Cerigo [Kithira] (§5.4), but he had found not a single genuinely human bone among them. Scheuchzer's famous specimen was no "witness of the Deluge" but a giant salamander; Cuvier had vindicated that interpretation during his visit to Haarlem (§9.3), just in time to be mentioned in the discourse. He also dismissed the alleged finds of human artifacts, such as Lamanon's famous iron key from the Paris gypsum (Fig. 5.16). Some genuine human bones were not true fossils at all, he argued, but remains buried within historical times; and a few other cases were questionable because they had been excavated without due note of their exact situation in relation to the surrounding deposits. Yet there was nothing to suggest that human bones were any less readily preserved than those of other mammals. So Cuvier concluded that "the human species did not exist in the countries where fossil bones are found, at the time [*époque*] of the revolutions that buried those bones". Yet, he added at once, "I do not want to conclude that man did not exist at all before that time", for the earliest human beings might have been confined to limited areas. But apart from the few who survived to repopulate the modern continents, their remains might all lie inaccessibly on the present ocean floors. So in Cuvier's view the former world could almost be equated with the earth's prehuman state, but not precisely.[49]

At this point it is worth mentioning a case that came to light just too late to be included in Cuvier's discourse, but which in effect confirmed this part of his argument. In 1805 a fragmentary but unquestionably human skeleton had been found

46. Cuvier, *Ossemens fossiles* (1812), 1: "Discours", 79–80, trans. Rudwick, *Georges Cuvier* (1997), 228–29. Cuvier did not use the astronomical analogy explicitly at this point, but it is reasonable to infer that he had it in mind, as he certainly did when comparing his own work to Laplace's, in the dedication of *Ossemens fossiles* and in his purple passage about "bursting the limits" of both space and time (§9.3).

47. Cuvier, *Ossemens fossiles* (1812), 1: "Discours", 81–82, trans. Rudwick, *Georges Cuvier* (1997), 229–32.

48. This analysis assesses Cuvier's explanation of the origin of the present fauna as part of his *geological* thinking. It is beyond the scope of this book to review his interpretation of organic change as a fully *biological* problem, which in any case has already been the subject of much thorough historical study: see, for example, Coleman, *Georges Cuvier* (1964), Laurent, *Paléontologie et évolution* (1987), and Corsi, *Age of Lamarck* (1988).

49. Cuvier, *Ossemens fossiles* (1812), 1: "Discours", 82–85, trans. Rudwick, *Georges Cuvier* (1997), 232–34.

embedded in solid rock on the French island of Guadaloupe. In the course of the fighting in the Caribbean, this precious specimen had been captured by the British—like the Rosetta stone in Egypt—and it was brought to London as another prize for the British Museum. In 1814, little more than a year after the publication of Cuvier's discourse and probably as a result of his argument becoming known in Britain, the Guadaloupe skeleton was described to the Royal Society by König, the curator of the museum's natural-history collections—overcoming his alleged aversion to fossils (§9.5)—and his paper was published promptly in the *Philosophical Transactions*. König recognized that its age was ambiguous. It was embedded in a limestone as hard and compact as marble, which by the lithic criterion (§3.5) might have suggested a very high antiquity. But unlike true fossil bones the skeleton was not "petrified", and a chemical analysis by Davy showed that the bones still contained organic material. So König concluded that the skeleton was probably quite recent in origin; he inferred that the human species was created *after* the "general cataclysm" that seemed to have wiped out the truly fossil mammals. Cuvier's inference that there were no genuine human fossils therefore remained unchallenged; he checked the case for himself in 1818, during his first visit to London after peace returned to Europe (Fig. 10.8).[50]

Fig. 10.8. The incomplete human skeleton found on the Caribbean island of Guadaloupe in 1805, brought as a trophy of war to London, and described to the Royal Society in 1814 by Charles König, the curator of the British Museum's natural-history collections. Although the skeleton was embedded in solid rock, König concluded that it was not truly fossilized and that it was probably quite recent in origin: it did not support claims for the high antiquity of the human species. (By permission of the Syndics of Cambridge University Library)

Returning to Cuvier's argument in the discourse: if the human species was no exception to his claims about the drastic effects of the earth's last major revolution, he needed to show that the event had indeed been very recent in relation to the rest of geohistory. He reviewed two distinct kinds of evidence, the physical and the textual, which converged towards the same conclusion. The physical evidence for a geologically recent date was "one of the best demonstrated and least expected results of a sane geology, and a result that is all the more valuable in that it links natural with civil history in an uninterrupted chain". He described how de Luc and Dolomieu had used natural chronometers such as alluvial deposits, peats, screes, dunes, and so forth, all indicating a very recent starting point for the subaerial "actual causes" that produced such features (§6.2, §6.3). Volcanoes were dismissed as indecisive: they could have continued to grow in size whether they were erupting below or above sea level, before or after the revolution in physical geography, so their apparently vast ages (§2.5) had no bearing on the present problem.[51]

Most decisive was the evidence for the growth of deltas within historic times. Cuvier summarized Dolomieu's work on the Nile delta (§6.3); he mentioned de Luc's account of the Rhine delta (§6.2) and confirmed it from his own recent travels in the Netherlands; he added information on the Po delta supplied by a leading French civil engineer, who had been working on flood prevention measures in that recently conquered part of Italy; and he used textual evidence from Herodotus to plot the changes since Antiquity in the delta of the Dnieper, where it entered the Black Sea. "We see sufficiently that nature everywhere maintains the same language", Cuvier concluded; "everywhere she tells us that the present order of things does not reach back very far". That nature expressed itself (or, in French, herself) in a "language" was of course a traditional metaphor, but Cuvier used it in a distinctively historical sense. This was nature's language recording *past* events as well as present ones; it was not—as it had been for Galileo and other natural philosophers—one that just disclosed the unchanging ahistorical mechanisms of nature; and unlike the enigmatic hieroglyphs this was a historical language that could already be deciphered.[52]

Cuvier turned next to the evidence from human history. "What is indeed remarkable", he maintained, was that "mankind everywhere speaks to us like nature": the "history of nations" told the same story of a major physical event, ancient in human terms although recent geologically. So at this point he plunged for the second time into the human records of Antiquity: once again he became a textual critic, not this time of ancient reports of animals but of an exceptional catastrophe near the start of written records. The critic had to distinguish the "real facts" that

50. Fig. 10.8 is reproduced from König, "Fossil human skeleton" (1814) [Cambridge-UL: T340:1.b.85.103], pl. 3 (read on 10 February 1814). The rock contained shells identical to those found on the nearby beaches of coral sand, suggesting that the limey deposit had been naturally cemented into solid rock at a comparatively recent time: the lithic criterion was fallible.

51. The geology to which Cuvier referred was "*saine*" because it was based on field observation, in contrast to the extravagant indoor speculations of geotheory, or "geology" in the older sense (§6.5).

52. Cuvier, *Ossemens fossiles* (1812), 1: "Discours", 85–94, trans. Rudwick, *Georges Cuvier* (1997), 234–39; report on Po delta by Gaspard Riche, baron de Prony, director of the civil engineering school [*École des Ponts-et-Chaussées*] in Paris ("Discours", 117–20).

some ancient texts contained, and "separate out the self-interested fictions that mask their truth".[53]

Cuvier's textual survey was, as before, worldwide and multicultural. That he began with the case of ancient Jewish records might have raised the hopes of some of his readers that he would reveal himself as a follower of de Luc in the interpretation of Genesis and Noah's Flood. In fact, Cuvier's handling of the case could hardly have been more different. He began with the history of the text itself, borrowing from the latest edition of the famous *Introduction to the Old Testament* by the orientalist and biblical scholar Johann Gottfried Eichhorn (1752–1827), one of Blumenbach's most distinguished colleagues at Göttingen; not for the first or last time, Cuvier's ability to read German gave him an edge over his francophone critics. Eichhorn had argued from textual analysis that the Pentateuch (including Genesis as its first book) must date from before the ancient schism between Jews and Samaritans, and Cuvier saw no reason to doubt that its assembly or editing [*rédaction*] had been by Moses himself and therefore dated still further back to the time of the Jewish exile in Egypt. He assessed Moses' reliability in terms of the realpolitik of the time, and concluded that the biblical date for the Flood—some fifteen centuries before Moses' time (and about fifty before Cuvier's)—was probably consistent with the traditions then current in Egypt. In effect, Cuvier treated the Genesis story of "a general catastrophe, an irruption of the waters, an almost total regeneration of mankind" as a second-hand *surrogate* for what the undeciphered Egyptian records might one day reveal. In sum, Cuvier's treatment of the Bible was conventional, not by traditional standards but by those of the best contemporary biblical scholarship, particularly in the German cultural sphere. It was certainly not what would have been expected from a closet literalist.[54]

Cuvier's draft text shows that his reason for putting the biblical case in first place was simply that among Mediterranean cultures the Jewish was the oldest that could easily be read and evaluated. The others followed: "Assyrian" (Babylonian), Egyptian, and Greek records—the first two known only at second hand—were treated in turn in exactly the same way. All added to the testimony of a catastrophic event, despite heavy overlays of myth and legend: "the incoherence of these narratives, which attests to the barbarism and ignorance of all the peoples [i.e., including the Jews!] on the shores of the Mediterranean, indicates equally the recentness of their establishment; and that recentness is itself strong evidence of a major catastrophe". Cuvier then turned to the further parts of Asia. He drew on the scholarship of the *Asiatick Researches* to argue that although the ancient Indian literature was too heavily mythical to yield much of historical value, the last of the revolutions it recorded seemed to date from some five thousand years ago and might be equated with the biblical Flood. Ancient Chinese records were more revealing, although they too required critical assessment of fact, legend, and myth; using the standard French edition of the *Chou-king* [Shu-jing], as well as citing one of Fortia's volumes (§10.1), Cuvier concluded that the Chinese "date their deluge from more or less the same epoch as we do". By contrast, records from the Americas were almost useless, yet even there "some traces of a deluge are believed to be perceptible in crude hieroglyphs" just described from Mexico by Humboldt. Africa was dismissed as useless for historical purposes, since its peoples had "nowhere preserved either annals or traditions".[55]

Cuvier used all this multicultural evidence to argue that the enigmatic event was likely to be authentically historical, precisely because the diverse records of it were clearly *not* all derived from a biblical original. The "Altaic" peoples were so different in every way from the "Caucasians" that, Cuvier suggested, "it is tempting to believe that their ancestors and ours escaped from the great catastrophe on two different shores", and he conjectured that the Africans might have escaped by yet another route. Anyway, the singularity of Noah and his family, in a vessel traditionally carrying as much theological as zoological freight, was tacitly abandoned; once again, Cuvier showed himself to be no biblical literalist. It was indeed the sheer pluralism of the records that in his opinion made their united witness so remarkable:

> Is it possible that it is simply chance that gives such a striking result, and that has the traditional origins of the Assyrian, Indian, and Chinese monarchies dating back more or less forty centuries? Would the ideas of peoples who have had so little connection with each other—whose language, religion, and laws have nothing in common—be in accord on this point, unless they were based on truth?[56]

Winding up his lengthy but well-structured case, Cuvier dismissed a few anomalous reports: he accepted current antiquarian opinion that the famous Egyptian zodiacs did not support the vast antiquity that had been claimed for them (§10.1) and that the quite advanced state of astronomy even in Antiquity was no evidence for an immensely long earlier history for civilized society. He concluded by aligning himself explicitly with what de Luc and Dolomieu had claimed a few years earlier: one of the best established inferences in geology was that "the surface of our globe has been the victim of a great and sudden revolution, the date of which cannot reach back much more than five or six thousand years". Having long vacillated between Dolomieu's transient mega-tsunami and de Luc's idea of a permanent exchange between continents and oceans, Cuvier now chose the latter, imagining how "since that revolution the small number of [human] individuals spared by it have spread out and reproduced on the land newly laid dry"; only since that time had literate cultures arisen and written records been kept.[57]

53. Cuvier, *Ossemens fossiles* (1812), 1: "Discours", 94, trans. Rudwick, *Georges Cuvier* (1997), 239.

54. Cuvier, *Ossemens fossiles* (1812), 1: "Discours", 94–95, trans. Rudwick, *Georges Cuvier* (1997), 239–40; Eichhorn, *Einleitung in das Alte Testament*, 3rd ed. (1803), 1: 140–292.

55. Cuvier, *Ossemens fossiles* (1812), 1: "Discours", 95–106, trans. Rudwick, *Georges Cuvier* (1997), 240–46; Guignes, *Le Chou-king* (1770); Fortia, *Histoire ancienne* (1805–9), 4, 5 (1807); Humboldt, *Vues des Cordillères* (1810).

56. Cuvier, *Ossemens fossiles* (1812), 1: "Discours", 105, trans. Rudwick, *Georges Cuvier* (1997), 245; he derived his anthropological classification from Blumenbach.

57. Cuvier, *Ossemens fossiles* (1812), 1: "Discours", 106–110, trans. Rudwick, *Georges Cuvier* (1997), 246–48. The Dendera zodiac, hitherto known only from a drawing made during Napoleon's expedition, was itself brought to Paris in 1822, and sparked further controversy: Buchwald, "Egyptian stars under Paris skies" (2003).

An anglicized Cuvier

It would be difficult to exaggerate the impact of Cuvier's discourse on the savant world (hence the length at which it has been described in this narrative). Ironically, however, although it was published in Paris at the height of Napoleon's wars, it first became known to a wider public in the land of Napoleon's most tenacious opponents. Jameson's edition of the discourse (§9.3) promptly made Cuvier's argument more generally accessible in Britain than elsewhere, since it became available to anglophone readers as a small book rather than as part of a large and expensive four-volume work. At the same time, however, Jameson distorted Cuvier's intentions in two important ways. The first, already mentioned, was that he gave Cuvier's text the title of "Essay on the Theory of the Earth" (Fig. 9.10). This hinted at the role Jameson intended the discourse to play in his own argument with Hutton's followers in Edinburgh, for it implied that the work belonged to the genre of geotheory rather than being a contribution to, and a program for, research on geohistory. The second way in which Jameson distorted Cuvier's goals related to the latter's argument for "a great and sudden revolution" in the fairly recent past. Appearing only a year after Hall read his paper on "diluvian waves" (§10.2), Jameson's volume confirmed the worry among the Huttonians that on the question of a putative deluge Hall might have sold the pass to their opponents.

In his preface, Jameson attributed to Cuvier's work the religious goals that in fact were characteristic of de Luc's. He made it appear that Cuvier's research was primarily concerned with the relation between the new geology and traditional biblical interpretation. He asserted that since the features described by geologists confirmed "the Mosaic account of the creation of the world", they could be "used as proofs of its author having been inspired"; and he argued that "even the six days of the Mosaic description, are not inconsistent with our theories of the earth", in the opinion of "many of the ablest and most learned scripture critics". Above all, however, Jameson claimed that Cuvier's research was important for establishing the historical reality of the biblical Flood and the recent origin of the human species:

> The deluge, one of the grandest natural events described in the Bible, is equally confirmed, with regard to its extent and the period of its occurrence, by a careful study of the various phenomena observed on and near the earth's surface. The age of the human race, also a most important enquiry, is satisfactorily determined by an appeal to natural appearances; and the pretended great antiquity of some nations, so much insisted on by certain philosophers, is thereby shewn to be entirely unfounded. These enquiries, and particularly what regards the *deluge*, form a principal object of the Essay of Cuvier, now presented to the English reader.[58]

Jameson thus recruited Cuvier's scientific prestige in support of his own (and de Luc's) objectives: to assert that the new geology confirmed the historical accuracy of the Flood story in Genesis and thereby tacitly reinforced the authority of the Bible as a whole. Cuvier did indeed argue that the ancient Jewish record was evidence for a major catastrophic event early in human history. But, as already pointed out, that text had no privileged position in his analysis; the Genesis story featured as just one of many ancient multicultural records of the same kind, all equally garbled and unreliable unless treated with rigorous caution. So while Jameson's edition showed anglophone readers the new scientific evidence for some kind of catastrophe at the dawn of human history, it greatly altered the cultural context of Cuvier's research. Jameson adopted Cuvier's claim that—contrary to what Hutton had asserted—there was strong scientific evidence for a major catastrophe in the geologically recent past, and for an almost equally recent origin for humankind. But unlike Cuvier he used both points primarily to vindicate the reliability of Moses and Genesis, and thereby to confirm the authority of the Bible itself and its place in religious, social, and political life.[59]

Given this cultural context, some of the Huttonians might well have been worried that Hall's hypothesis of diluvian waves would give comfort to their opponents; they might have been expected to resist Jameson's promotion of Cuvier's similar theory in their own city. But the account of it in the influential Whig quarterly *Edinburgh Review*, probably written by Playfair, was quite fair. He criticized Jameson's choice of "Theory" for its title, suggesting that "Considerations on fossil quadrupeds" would have been more accurate; but with that exception his essay was in fact an appreciative but not uncritical review of *Cuvier's* ideas. Playfair even conceded that actual causes were not adequate to explain everything, though he thought they could explain more than Cuvier allowed. He approved Cuvier's case for extinction, though he doubted his conjecture about a sudden deep-freeze to explain the carcasses found in Siberia. He agreed on the absence of genuine human fossils, though he questioned Cuvier's claims for a very recent date for the human species. He conceded that most human *records* stretched only a few millennia into the past, but he thought that some might be genuinely "antediluvian", and he asked only for the "liberty" of putting the origin of human societies somewhat further back than Cuvier seemed to allow. In the end, he dissociated the biblical Flood from any case for a "geological deluge": judging by the account in Genesis, he argued, the former would have been so brief and gentle as to leave no physical trace of its passing; tacitly he allowed the possible reality of the latter at a much earlier date, though he was skeptical about it.[60]

Playfair's important essay on Cuvier's work was duly noted elsewhere in Britain by the intellectual elite who read the *Edinburgh Review*. But it also caught the eye of

58. Cuvier, *Theory of the earth* (1813), Jameson's "preface", v–vii. Most of the *translator's* footnotes merely clarified the text, but one of them reinforced this distortion of Cuvier's work by giving the "short timescale" dates calculated by early chronologers for the creation *of the world*, based on the Hebrew, Samaritan, and Septuagint texts of the Pentateuch (149n). Byron, *Cain* (1822), vi–vii, later and notoriously inverted Jameson's strategy by putting Cuvier's geohistory of prehuman "revolutions" poetically into the mouth of Lucifer!

59. In the absence of international copyright agreements, Cuvier had no formal means of control over Jameson's edition; in the wartime conditions, even informal pressure would have been difficult. There seems to be no documentary evidence to show what he thought of it, or even whether he knew about it in advance of publication.

60. [Playfair?], "Theory of the earth" (1814); the Wellesley Index gives the attribution as probable though not certain. The reviews were formally anonymous, but their authors could usually be identified by knowledgeable readers, or from current gossip.

the editor of the *Bibliothèque Britannique* in Geneva, who promptly translated it for the benefit of his readers throughout Continental Europe (he noted that it allowed him to publish a summary of Cuvier's important work, circumventing his own periodical's normal policy of confining reviews to non-francophone publications). He approved Cuvier's "diluvian chronology" for the recent catastrophe, contrasting it with Playfair's "system of *unassignable antiquity*". Most significantly, however, the editor commented that if Cuvier's critic had been writing at Geneva—or indeed if he had just taken into account the work of his compatriot Hall, who had seen the relevant features (§10.2)—he might have been less skeptical about a recent physical revolution:

> If the Scottish author had been placed as we are, that is to say, having—not just in thought or as *it is said*, but *before one's eyes*—the unquestionable evidence of this débâcle, in the enormous granite blocks lying here and there on the limestone ground near Geneva, at ten or fifteen leagues [50–70km] as the crow flies from the central Alpine chain, where *alone* are found in place the rocks from which these blocks have been detached, then certainly he would not have asked where is the evidence for a débâcle.[61]

Jameson's edition was the first to make Cuvier's discourse available in full in another language, but it was followed by an abridged version in German. Like an earlier abridgement of the report on the Paris Basin (§9.2), this was published in the prestigious *Annalen der Physik*. Gilbert's brief editorial introduction was in stark contrast to Jameson's, in a way that underlines the peculiarity (in both senses) of the British context. Gilbert located the importance of the essay in its relation to geohistory, not to Genesis: Cuvier, he wrote, had used fossil evidence, interpreted in the light of comparative anatomy, to "develop inferences that disclose, with some probability, the way that the present state of the earth's surface has been shaped". He hoped his synopsis would show his readers that recent research outside the German lands was "treating *scientifically* [*wissenschaftlich*] some of the baffling issues in geology that have always and so often given rise to fantastic products". In other words, Cuvier's essay did *not* belong in the genre of geotheory to which Jameson had assigned it, still less in the genre of apologetics, but in the vanguard of truly scientific geology. In addition to his synopsis of Cuvier's work, Gilbert appended summaries of Playfair's Scottish review and of another by two Genevan naturalists (one of them de Luc's nephew and namesake). Such was the international context in which the debate about the significance of Cuvier's work was carried on, towards the end of the Napoleonic wars and at the start of the subsequent peace.[62]

Conclusion

In addition to setting an agenda for research on the Secondary formations and their fossils (§9.3), Cuvier's discourse made a closely argued case for the reality of some kind of physical catastrophe at a much more recent point in geohistory, at or near the dawn of human history. He claimed that none of the "actual causes" known to him was adequate to explain the observable effects of this event, though he implied that the true cause had been equally natural, albeit of unwitnessed magnitude. He argued that its huge impact on the living world was best assessed from the evidence of the fossil quadrupeds that he had made his special study: their living counterparts were far more completely known than those of marine fossils, so that the living fossil argument could be ruled out with much greater confidence. But since fossil quadrupeds were usually preserved in a highly fragmentary form, his anatomical principles were needed to ensure that they were correctly reconstructed.

The consequent anatomical comparisons pointed to a mass extinction of mammals. Cuvier rejected Lamarck's alternative explanation in terms of transmutation: not only did living species not show the kind of variability that that process would require, but also the longest lapse of time available in human records (i.e., since the ancient Egyptians) showed no change whatever in organic forms. Cuvier recognized the need to account not only for the disappearance of species by extinction, but also for their origin. But he circumvented the latter problem, at least in part, by suggesting a model of faunal migration consequent on changes in physical geography in the course of geohistory. He rejected all the alleged cases of human fossils as spurious in one way or another, so that the earth's last revolution was in effect the boundary between the present world of human life and an almost completely prehuman world of extinct animals. (The famous Guadaloupe skeleton was described just too late to be included, but it supported Cuvier's argument.)

Cuvier claimed that the great boundary event of his mass extinction was only a few millennia in the past. This was an approximate date supported not only by the physical evidence of natural chronometers, but also by multicultural textual evidence of some kind of deluge near the dawn of human civilizations. The biblical Flood was just one of many equally faint and garbled records of this event, but their combined testimony was all the stronger for coming from independent sources. However, Jameson's edition of the discourse seriously distorted Cuvier's work by introducing it to anglophone readers as a vindication of the Flood story in Genesis, whereas Continental reviews recognized its primarily geohistorical goals. All in all, Cuvier's discourse was as influential in the argument about the most recent phase of geohistory—the reality or otherwise of a catastrophe at the dawn of human history, and its physical character and cause—as it was in the parallel project of reconstructing the still earlier geohistory preserved in the Tertiary and Secondary formations.

61. [Playfair?], "Essai sur la théorie" (1815): see editorial comments in [pt. 1], 353; [pt. 3], 106n, 113n. In the same year, Playfair's *Illustrations* (1802) was published in full in Paris, as Basset, *Explication de Playfair* (1815), making his case for the adequacy of "actual causes"—and his skepticism about any recent catastrophe—more widely known on the Continent.

62. Cuvier, "Geognostische Betrachtungen" (1816): see Gilbert's introduction (117–19, mispaginated as 1–3). Most of Gilbert's footnotes refer not to Cuvier's ideas but to those expressed in the reviews that followed (159–76).

10.4 BRITAIN BROUGHT BACK INTO EUROPE (1813–16)

Buckland's Oxford lectures

In the last years of the Napoleonic wars, it was becoming a commonplace belief among savants throughout Europe, and also increasingly among the wider public that took an interest in their ideas, that there had been a major physical catastrophe in the geologically recent past. Since it appeared to have been some kind of aqueous event, and the evidence for it was primarily physical rather than textual, it was often called the "geological deluge". Whether it was to be identified with the "biblical deluge"—the Flood supposedly experienced by Noah and recorded by Moses—was a moot point and a controversial one; but the reality and historicity of the physical event did not stand or fall by any such equation. In the wake of what von Buch and Hall, for example, had demonstrated (§10.2), those who wanted to undermine the historical reliability of Genesis, and with it the authority of the Bible as a whole, would now find it difficult to claim that there had been no deluge of any kind. But they could still insist that the geological one had no bearing on any biblical issues, because it had happened long before any humans were around to witness it, let alone to survive and record it.

In the anglophone world, Cuvier's concept of a recent geological catastrophe first became widely known in Jameson's edition, and therefore with that editor's distinctively biblical slant (§10.3). In this form, Cuvier's ideas were adopted enthusiastically by Buckland and Conybeare, who like Jameson had their own local reasons for welcoming the putative conjunction between geology and Genesis. Oxford was the intellectual center of the established Church of England, and its primary social function was to produce well-educated clergymen. The newer natural sciences could not hope to be accepted, even peripherally, into the university's conception of sound knowledge unless they could be shown to be compatible with its traditional studies, primarily in Classical but also in biblical literature. In the case of geology, this would mean demonstrating that the new ideas about the history of the earth, based on rocks and fossils, were not incompatible with the well-established kind of human history based on Classical and biblical texts. The social imperative at this time—in Britain as a whole, but particularly at Oxford—to show that geology was compatible with Genesis has usually been interpreted by later historians in relentlessly negative terms, as a deplorable constraint on the search for scientific truth. But it had a more positive side, as the following narrative will suggest. For it fostered the further transformation of the relevant parts of geology into a *geohistorical* form in which they could be linked to an existing body of knowledge about *human* history.

Most of the early work of the two young Oxford geologists—both Buckland and Conybeare were still in their twenties when Jameson's book was published—was not notably geohistorical. Buckland's lectures, prepared with assistance from Conybeare, simply summarized the geognostic sequence of formations in an international context; their fieldwork was designed to improve the enriched geognosy that Smith called "stratigraphical", both in their own country and in correlation with the Continent (§9.5). However, in preparing his lectures Buckland would also have felt it necessary to deal with the two related issues on which—as Jameson had pointed out in his preface to Cuvier's work—geology was entwined with human history: the historicity of the biblical Flood in relation to the Superficial deposits that were widely regarded as "diluvial" in origin, and the biblical origin of humankind in relation to the sequence of formations.

Far from being reluctant to tackle these thorny problems, Buckland welcomed the opportunity to demonstrate to his colleagues that the science of geology—new at least to them—was compatible with the scriptural foundations of their academic work. Specifically, he would have wanted to reassure his audience that traditional ideas about the recent origin of the human race and the even more recent reality of the Flood had not been undermined, let alone demolished, by the latest discoveries in geology; and that the even more insidious challenge of Hutton's eternalism could be overcome. In the weeks before his first course began, Buckland was asking for Conybeare's help: "pray send me the notes you had begun touching Moses and Huttonianism", he pleaded, adding in another letter, "I suppose you have had no leisure to think of Moses or Creation".[63]

As Buckland's wording suggests, he and his Oxford colleagues were still innocent of the scholarly biblical criticism that had long been flourishing on the Continent. For example, he used the seventeenth-century chronologer Edward Stillingfleet—a contemporary of the early geotheorist Thomas Burnet (§3.1)—as an authoritative source, in a way that Blumenbach's colleague Eichhorn, for example, would have regarded as quite outdated and no longer acceptable in a modern scholar. But since Buckland was starting from such an antiquated position, he would have found the biblical slant that Cuvier's ideas had received in Jameson's edition—published in 1813 with perfect timing for his lectures—a highly congenial way in which to interpret the work of that distinguished savant.[64]

However, Buckland cut no scholarly corners while preparing his lectures. Jameson's edition may have given him a first broad view of Cuvier's ideas, but he then made good use of the full *Fossil Bones*, a copy of which had evidently reached Oxford (or at least London) in spite of the war, and which of course he could read in the original French. He noted, for example, that the fossil bones of elephants were always found in the Superficial deposits, not in the older Secondary formations; paraphrasing a passage from the monograph on the Paris region (§9.2), he noted that "it is therefore probable that these bones have been enveloped by the last general Catastrophe to wh[ich] our Earth has been subjected". This was an inference that he took over from Cuvier without major change, and it became the cornerstone of his own interpretation.[65]

Buckland followed de Luc, not Cuvier, in identifying the "geological deluge" exclusively with the biblical Flood rather than with a broad multicultural range of

63. Buckland to Conybeare, April 1814 (quoted in Gordon, *Life of Buckland* (1894), 14–15).

64. Stillingfleet's *Origines sacrae* (1662) is cited in Buckland, "No Pecora till deluge gravel" (Oxford-MNH, Buckland MSS, note watermarked 1813).

65. Buckland, "Elephant bed" (Oxford-MNH, Buckland MSS, note watermarked 1812), citing Cuvier, *Ossemens fossiles* (1812), 1: "Géographie minéralogique", 236 (describing bones found in digging the Ourcq canal outside Paris); he may have used a copy at the Geological or the Royal Society rather than one at Oxford.

ancient records. In reviewing the evidence for that equation, he had to face the awkward fact—which Cuvier had emphasized—of the total absence of any genuine human fossils, an absence underlined at just this time by König's account of the Guadaloupe skeleton (Fig. 10.8). He solved the problem by inferring that at the time of the Flood the human species had not yet spread widely from the point at which it had been created. He cited Stillingfleet's opinion that that event had been in Asia, so his adoption of the chronologers' short timescale (for humankind, *not* for the earth itself) made the explanation even more plausible: if no more than about sixteen centuries had elapsed between the creation of man and the Flood, early humans would hardly have had time to spread as far as Europe. On the other hand, Buckland claimed that there was *indirect* evidence that humans had indeed been present on earth by the time that the "*Diluvium*", as he called it, was deposited. He noted, on the basis of Cuvier's work, that "*Pecora*", or animals of the cattle family, were known as fossils only from these Superficial deposits. Applying a traditional kind of argument from natural theology (§1.4), he inferred that cattle were divinely designed for human use and intended to be domesticated and would therefore have been placed on earth at the same time as humans. So in effect they could act in the fossil record as surrogates for the missing human remains:

> Non Existence of [truly] fossil Pecora renders the existence of Man improbable—& per contra their presence in Diluvium indicates that Man existed somewhere / Pecora essential to [even] the simplest state of Society, were created with Man—spread more rapidly in the antediluvian World than Man / Checks to rapid extension of human Society in the Savage or Civilised State.[66]

When Buckland gave his lectures for the second time, in 1815, he seems to have dealt more fully with the "diluvial" fossils. In his notes he attributed the "Elephant bed" of the Superficial gravels to "a transient marine Inundation", thus adopting Cuvier's earlier view (derived from Dolomieu), rather than his later idea (derived from de Luc) of a permanent interchange of continents and oceans (§10.3). The former option made the "catastrophe" more readily compatible with the story in Genesis, provided the "inundation" had been a relatively gentle event rather than Hall's (or Dolomieu's) mega-tsunami. On the other hand, it could not have been such a "quiet deluge" as to leave no physical trace at all: that argument had just been used by Playfair, in his anonymous review of Jameson's edition of Cuvier, but it avoided conflict between geology and Genesis at the price of severing any link whatever between human history and geohistory. Buckland had good reason to want to display geology to his colleagues in a more positive light.[67]

A geological triumvirate

The year after Waterloo, Conybeare suggested to Buckland that they should persuade Greenough to join them in "a geological triumvirate" for a major tour on the Continent; an unstated advantage of the plan was perhaps that Greenough's great wealth might relieve the costs falling on the relatively impecunious clergymen.

Buckland wrote to Greenough, urging the potential value of "this triple alliance". Greenough agreed, and the triumvirate crossed the Channel—with a carriage specially modified for carrying a heavy load of specimens—as soon as Buckland had finished his lectures for 1816. Conybeare had suggested an itinerary that would have taken them first to Auvergne, to see the celebrated extinct volcanoes and the controversial evidence of diluvial valley erosion, and then over the Alps into northern Italy to see the Tertiary formations. In the event, however, they headed east into Germany. The change may reflect the weight of Greenough's personality (and his wealth), or perhaps advice given by von Buch, who had been visiting Buckland while the tour was being planned. But anyway, its effect was to give priority to Greenough's preference for mineralogy and the geognosy of the older rocks of central Europe, at the expense of the other geologists' greater interest in the evidence of relatively recent geohistory.[68]

Wherever they went, the triumvirate met savants and saw their collections, much as Brongniart had done on his German tour three years earlier (§9.3). Although in conversation the Englishmen could always use French, the international language, it was an advantage to them that Greenough could also speak German as a result of his time long before as a student at Göttingen. At Gotha they met Schlotheim (§8.1) and inspected his vast and well-arranged collection of fossils, many of them unlike any they knew in England; at Weimar they met Goethe (§1.2) and saw his collections too, though they found him disinclined to talk about geology. When they reached Leipzig, Conybeare, whose first child was only a few months old, decided to return to England. Greenough the confirmed bachelor noted scornfully that his companion was homesick; but Conybeare may also have felt that under Greenough's leadership their tour was proving less interesting to him than he had hoped.

Anyway, Greenough and Buckland traveled onwards to Freiberg, where they met the geognosts attached to the famous mining school and spent much time with Werner in particular. Greenough was not impressed with him, noting that he "knows little of the structure even of Germany" and that he was "no admirer of Cuvier [or] Brongniart". But the great geognost was persuaded to write out a list of formations summarizing his current views, and Buckland later added this to the printed "stratigraphical table" that accompanied his Oxford lectures (§9.5). From Freiberg, the two Englishmen continued eastwards as far as Cracow (at that time part of the Russian empire). Turning south, they visited the mines and mining school at Schemnitz, and in Vienna they saw some of the rich collections brought

66. Buckland, "No Pecora till deluge gravel" (Oxford-MNH, Buckland MSS, note watermarked 1813). The context makes it clear that by "fossil" he meant petrified and preserved in the Secondary rocks. Ironically, his argument was similar to Hutton's use of animal and plant fossils in general as surrogates for the human presence throughout (§3.4), as part of the eternalistic "Huttonianism" that Buckland was concerned to refute.

67. Buckland, "Elephant bed" (Oxford-MNH, Buckland MSS, note watermarked 1815). [Playfair?], "Theory of the earth" (1814); the same argument had been used long before by Buffon (§3.2).

68. Buckland to Greenough, 21 February 1816 (London-UCL, Greenough MSS); Torrens, "Geology in peace time" (1998), gives a detailed and entertaining account of the German portion of the tour, based mainly on Greenough's notebooks.

to the capital from all parts of the Austrian empire. Crossing the Carinthian Alps into Italy then gave Buckland a first taste of high mountain topography.[69]

In northern Italy—much of it now under postwar Austrian rule—Greenough and Buckland did in fact see some of the geologically younger features that might have been in Conybeare's mind before they set out. After visiting Venice, for example, the depleted triumvirate studied the extinct volcanoes of the Euganean Hills [Colli Euganei] near Padua—Italy's answer to France's Auvergne—and then skirted the southern edge of the Alps much as Brongniart was to do four years later (§9.6). Buckland was evidently correlating Tertiary formations with those he knew in England: he noted "the fossil Fish quarries of Monte Bolca [see Fig. 5.7] which are in a formation above & lying on Chalk, & allied to the English Sheppey [i.e., London] Clay & French Calcaire Grossier [Coarse Limestone]". Their route later took them to Piacenza to see Cortesi's vast fossil collection from the Subapennine formation. In the field, Buckland was able to collect some of its superb shells, "many of which resemble those of Hampshire and Sheppey Island"; while doing so he was arrested as a suspected spy—a routine hazard for geologists in the field—and imprisoned briefly at Parma. In Milan he saw Cortesi's original collection and met its curator, Brocchi, who duly presented him with a copy of his great work on the fossil shells (§9.4). It may have been this copy that Leonard Horner used later the same year, when he reviewed Brocchi's book (with copious translated extracts) for the *Edinburgh Review*, thus making the Tertiary formations and fossils of Italy almost as well-known to British geologists as those of the Paris basin.[70]

From Milan, Greenough and Buckland crossed back over the Alps, this time by the Simplon pass into Switzerland (which had just regained its independence). After further fieldwork they traveled home by way of Geneva (which after regaining independence had voted to join Switzerland) and Paris. Greenough was back in London in time for the start of the Geological Society's new season of meetings; Buckland was late for the start of Oxford's new academic year. Their five-month tour had certainly added to Greenough's extensive network of useful contacts and taken him to places that even he had not visited previously. For Buckland, however, the impact of the tour must have been far greater. It was his first time outside the British Isles, and he met savants of many nationalities and several languages. But he also saw a wide range of geological features quite beyond his previous experience, from high mountains and their glaciers to extinct volcanoes. Although the Geological Society had already introduced him to a cosmopolitan circle of Londoners, his first Continental tour seems to have given him a newly international outlook, which distanced him profoundly from most of his Oxford colleagues (though not from Conybeare). In the next few years it was to put his important geological research into a much wider perspective.

Caves and fossil bones

As this brief summary suggests, most of the geological tour that Buckland and his companions made on the Continent in the first full year of peace was quite conventional in character. They wanted to meet foreign geologists and other savants, to look at their collections, and above all to see with their own eyes some of the more famous localities that had been described and discussed during the long years of war. If there was any particular slant to their itinerary, it was perhaps the attention they gave to the Tertiary formations once they reached Italy. Buckland clearly wanted to keep abreast of international research on the relatively recent or Tertiary part of geohistory, to which in a small way he had already contributed in some early fieldwork around London. At one point, however, he and his companions had visited a celebrated region that bore on the distinctly different problem of understanding the most recent phase of all, the enigmatic "last revolution" in the earth's long history.

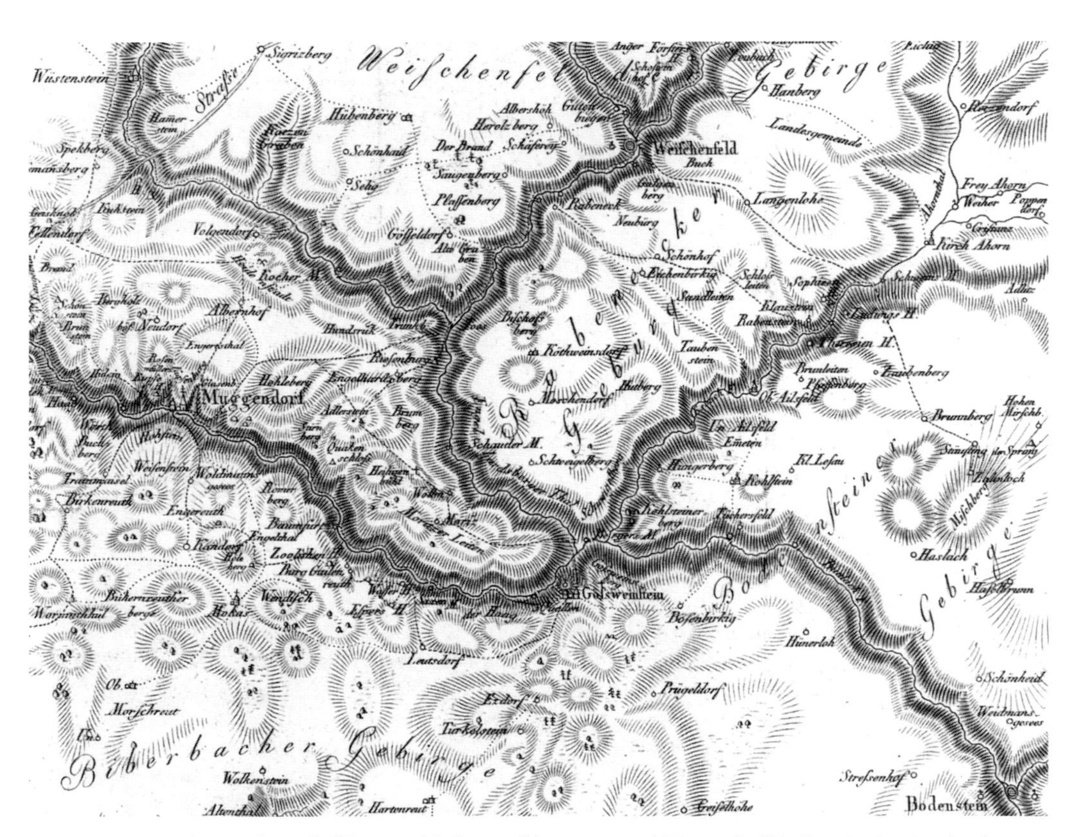

Fig. 10.9. Part of a "Topographical map of the area around Muggendorf" in Bavaria, showing deeply entrenched valleys cutting through a plateau; it illustrated a "pocketbook for amateurs of natural and antiquarian science" (1810) written by Georg August Goldfuss (1782–1848), who at the time was teaching natural history at Erlangen, the nearest university. This local guidebook, which was probably used by Greenough, Conybeare, and Buckland when they visited Muggendorf in 1816, indicates how the bone caves were already a major attraction for scientifically minded tourists: the dashed lines show "paths leading to natural curiosities [*Naturmerkwürdigkeiten*]". The cave near the village of Burg Gailenreuth, the most famous and the richest in fossil bones (see Fig. 10.11), is marked as the "*Zoolithen H[öhle]*" southeast of Muggendorf. The area shown here is about 16km by 12km. (By permission of the British Library)

69. Torrens, "Geology in peace time" (1998). Werner expressed a great desire to visit England, but it was not to be: he died the following year at the age of sixty-eight. Greenough's notes provide no evidence that Jameson was right to claim (§9.3) that Werner was treating fossils in geognosy in a Smithian manner.

70. Buckland to Lady Mary Cole, 3 April 1817 (Cardiff-NMW, De la Beche MSS, 146); Buckland to Brocchi, 1 June 1818 (Bassano-BC, Brocchi MSS, II.33); [Horner], "Brocchi, *Conchiologia*" (1816).

Fig. 10.10. Buckland's sketch of one of the valleys near Muggendorf, later published as a lithograph (1823). It shows two of the caves with their entrances high above the present bed of the little Esbach River *D* and below the ruined castle of Rabenstein *A*. The nearer cave *E* (Kühloch) has its interior sketched as if visible through the solid rock: a sloping passage *F* leads to a large chamber *G* with a bone deposit *H* on its floor. Buckland inferred that the bears that had used the cave as a den had been annihilated by a sudden "deluge", which had also excavated or at least deepened the valley.

While they were in Germany, and still a triumvirate, the English geologists studied the famous bone caves in Bavaria, which Esper had explored over forty years earlier (§5.3) and Rosenmüller more recently (§7.1). That long tradition had made the area around Muggendorf highly attractive to travelers with scientific interests, and the English triumvirate was no exception (Figs. 10.9, 10.10).[71]

Several of the caves had been mined for their fossil bones ever since Esper's time. The most famous and most prolific, near the village of Gailenreuth, was explored by Buckland, Conybeare, and Greenough from its entrance (see Esper's drawing, Fig. 5.13) to its furthest recesses (Fig. 10.11).[72]

The English geologists knew of course that Cuvier had identified the bones as those of a large extinct species of bear (Fig. 10.3); among the specimens they purchased from a local dealer was "a fine Bears head", probably much like the skulls that had been illustrated by Esper and Rosenmüller (Figs. 5.14, 7.2). They were probably aware that Esper had interpreted the bones as the remains of animals swept into the caves from elsewhere at the time of the Flood; but that Rosenmüller, followed by Cuvier, had inferred instead that the bears lived on the spot, using the caves as their dens. There is little direct evidence to show what Buckland, in particular, thought about the matter while he was there, or indeed soon after his return to England; but he may already have begun to formulate privately the interpretation that he later expounded in public (§10.6). This, in effect, combined Esper's ideas with Cuvier's: the caves had been the dens of bears, but the animals had been annihilated on the spot, and their bones preserved in the caves, as a direct result of the "geological deluge". So he may well have made the crucial leap of imagination that transformed his sketch of Gailenreuth cave into a vivid scene in which he unexpectedly encountered the extinct bears as if they were still very much alive and well (Fig. 10.12).[73]

Fig. 10.11. Tourists being shown the interior of Gailenreuth cave, near Muggendorf in Bavaria, by local guides with flaming torches. This engraving, after a drawing by Webster, was based on a sketch made by Buckland during his visit in 1816; it was published in 1822 in his study of bone caves in England and abroad (§10.6). It used the same visual convention as geological sections, showing the cave as it would be seen if the earth were sliced open along a vertical plane. Skulls and other bones are lying on the ground in each chamber, while workmen in the innermost one are excavating a thick bone-bearing deposit below that surface. Buckland, like Rosenmüller and Cuvier before him, interpreted the Bavarian caves as the former dens of an extinct species of bear, representing the last fauna of the former world. (By permission of the Syndics of Cambridge University Library)

71. Fig. 10.9 is reproduced from Goldfuss, *Umgebungen von Muggendorf* (1810) [London-BL: 10231.aa.11], part of last (unnumbered) pl. The book was subtitled a "Taschenbuch für Freunde der Natur- und Alterthumskunde", the latter the *science* of Antiquity pioneered by German historians, notably at Göttingen. Fig. 10.10 is reproduced from Buckland, *Reliquiae diluvianae* (1823), part of pl. 18, explained on 272–73. He may have seen these particular caves in 1816, but the sketch was probably made in 1822 on a later visit (§10.6). As was customary in field sketches, the vertical dimension was exaggerated: the sides of the valley are steep but not in fact precipitate (and are now thickly forested). Rabenstein is 8km east-northeast of Muggendorf.

72. Fig. 10.11 is reproduced from Buckland, "Fossil teeth and bones" (1822) [Cambridge-UL: T340:1.b.85.111], pl. 16. Only one chamber is now accessible, and its entrance is far wider than in Esper's or Buckland's time; all the bones have long since been removed.

73. Fig. 10.12 is reproduced from part of an undated manuscript sketch preserved among Buckland's lecture materials (Oxford-MNH, MS D338). The details of the cave profile, bones, etc., show that it was copied from the sketch he made in 1816 (or from the subsequent drawing or engraving: Fig. 10.11), *not* from his later and significantly modified version (Fig. 10.21); so it may well have been used in Buckland's teaching *before* the discovery of Kirkdale cave turned his attention more fully on to bone caves (§10.6). The inclusion of the ladder suggests that the bears have been imagined transposed into the modern world, not the gentleman into the antediluvian; I am indebted to Ralph O'Connor for the suggestion that the bear was performing for the intruder.

Fig. 10.12. Gailenreuth cave in Bavaria, with Buckland (or one of his companions) in a surprise encounter with some of the extinct cave bears that, he inferred, had once used it as their den: a drawing probably used by Buckland, after his return to Oxford, to enliven his lectures. The imagined juxtaposition expressed his vividly *geohistorical* conception of geology: the fossil bones have been "resurrected" into animals that have time-traveled from the antediluvian world into the world of the Regency gentleman, who seems to have transformed them into *performing* bears and himself into a menagerie showman. (By permission of the Museum of Natural History, Oxford)

Conclusion

Buckland recognized that his science of geology would become acceptable and respectable at Oxford if he could show that its findings were compatible with the university's traditionally text-based culture. This entailed forging a link between the human history based on Classical and biblical texts and the far longer history being disclosed by geological research; more specifically, he had to show that geology did not undermine traditional notions of the origin of humankind and the later Flood recorded in Genesis. In his early lectures, however, these themes were quite subordinate to others that were the main objects of attention at the Geological Society and elsewhere, namely the geognostic or stratigraphical classification of rock formations and, increasingly, their transformation into geohistory.

When, in the first full year of peace, Buckland and Conybeare were joined by Greenough in a lengthy Continental tour, their fieldwork and their contacts with other savants were focused almost entirely on the latter issues, which were at the heart of current international debate among geologists. Only at one point, when they visited the famous bone caves in Bavaria, did their tour impinge on the quite distinct problem of understanding the earth's most recent "revolution", namely the "geological deluge" that was often—though far from universally—identified as Noah's Flood. This firsthand experience was a significant event in Buckland's scientific trajectory, for in the years that followed he made the investigation of the "deluge"—both geological and biblical—a major focus of his research. This is the topic of the final sections of this final chapter.

10.5 TRACING THE GEOLOGICAL DELUGE (1816–22)

Making claims for geology

In the first few years of peace, research on the "stratigraphy" of the Secondary and Tertiary formations, enriching traditional geognosy with the use of fossils, was developing with spectacular rapidity and success throughout Europe. Two concrete bodies of work, on Secondaries and Tertiaries respectively, were treated as exemplary for this new practice: first, Smith's strictly empirical use of "characteristic" fossils to distinguish discrete "strata" or formations (§8.2, §9.5); and second, Prévost's and Brongniart's more subtle and geohistorical use of whole fossil faunas, within which the slowly changing assemblages of species might mark the passage of time (§9.6). In contrast, the thin and variable Superficial deposits overlying such formations were not merely neglected but often scorned as hardly worthy of serious attention. Yet what was almost negligible from a geognostic or stratigraphical perspective might be of crucial importance for geohistory. The Superficial deposits clearly represented some kind of temporal link between the "present world"—as de Luc, followed by Cuvier and many others, had called it—and the "former world" of the "regular" formations described by geognosts. Patchy clays and gravels, the mere detritus of apparently more interesting solid rocks beneath, might be the key to the relation between past and present. But only those for whom *history* was a paramount concern could perceive that these unimpressive deposits might play that pivotal role. It is therefore not surprising that their significance had been championed by naturalists such as de Luc and Cuvier, who for different reasons had wanted to link the prehuman history of the earth forwards onto human history.

In the postwar years Buckland became a major new recruit to this research tradition, which he followed in parallel with his work on the quite separate problems of stratigraphy. His institutional position at Oxford gave him powerful reasons for approaching geology with a historical perspective, to make the new science acceptable and even attractive to his academic colleagues. In his earliest lectures he had treated some of the Superficial deposits as the traces of the biblical Flood or Deluge (§10.4). This diluvial interpretation was of course far from original. It owed its inspiration to de Luc's indefatigable promotion of a binary geotheory (§3.3, §6.2, §6.4), which only ceased at his death in 1817 at the ripe age of ninety; to Cuvier's programmatic "preliminary discourse" as mediated by Jameson (§10.3); and to Hall's theory of repeated "diluvian waves" and von Buch's related ideas about a recent watery *débâcle* (§10.2). However, a new and enhanced position at Oxford gave Buckland an opportunity to formulate his diluvial theory more coherently and to make it more widely known.

Buckland had been appointed Kidd's successor as reader in mineralogy (§9.5), but he wanted to extend his brief to cover the new and exciting outdoor science of geology, which in practice was now quite distinct from the older indoor science of mineralogy. In 1818, after giving his annual lectures the significant title "The Elements of Mineralogy *and Geology*" (emphasis added), he petitioned the Prince Regent, through the chancellor of the university and the prime minister, claiming that it was not feasible to cover both sciences adequately in one course and asking for an additional title and extra salary to give a new and separate one on geology. The petition was successful; although he was not given much extra salary, it certainly heightened his academic profile.[74]

As Oxford's first reader in geology, Buckland now had an opportunity to explain his new science to his colleagues, and to defend its right to a place in the university's field of studies, by giving a formal inaugural lecture. As with his earlier annual lectures, he consulted Conybeare on its topic and contents; in the event, Buckland's inaugural was largely shaped by Conybeare, who later used much of the same material in his introduction to "Conybeare and Phillips" (§9.5). It was Conybeare, for example, who advised Buckland to set geology in a context of natural theology and include plenty of allusions to Classical literature in order to make the science more palatable to their colleagues. In the spring of 1819 Buckland duly gave his inaugural lecture to a large and generally appreciative audience; he published the text the following year with the appropriately Classical title of *Vindiciae Geologicae*. His "*vindiciae*", or claims on behalf of his science, were for the respect that was its due, based on what his subtitle called "the connexion of geology with religion". The connection was with both natural and revealed theology (§1.4), and of course Buckland claimed that it was one of harmony rather than conflict.[75]

Buckland argued that geology, no less than older branches of natural history, gave evidence of divine providence, normally exercised through the "secondary" laws of nature or ordinary physical processes. More important, however, were his claims relating geology to *historical* issues raised by its "connexion" to the biblical records, specifically those relating to the primal Creation and the far more recent Flood. With regard to the first of these, he argued that the evidence for a period before life itself existed—that is, in the Primary rocks, with no trace of fossils—was enough to indicate an origin within time, and hence to refute "the hypothesis of an eternal succession of causes". With a thinly veiled allusion to Hutton, he maintained that eternalism was as incompatible with modern geology as it was with orthodox theology: and he dismissed the Scotsman's geotheory as an outdated hangover from the speculative philosophies of Antiquity:

> If some writers on Geology in later times have professed to see in the earth nothing but the marks of an infinite series of revolutions, without the traces of a beginning; it will be quite

sufficient to answer, that such views are confined to those writers who have presumed to compose theories of the earth, in the infancy of the science, before a sufficient number of facts had been collected.[76]

Buckland sketched the complex and varied geohistory that recent research had disclosed and emphasized that it pointed to an unimaginably vast timescale. Although this had long been taken for granted by naturalists familiar with the field evidence (§2.5), it may have been news to many in Buckland's audience, and to some a source of anxiety about possible conflict with traditional religion. But Buckland allayed that anxiety by citing recent authors of impeccable piety and orthodoxy in support of his position—likewise long taken for granted among savants, at least beyond Britain—that a nonliteral interpretation of the Creation story in Genesis was not merely permissible but positively incumbent on any scholarly exegete. What mattered more was that there was no good reason to doubt the traditional chronology in respect to *human* history: Buckland emphasized that there was no direct evidence for human existence even in the most recent deposits, and certainly not in the ordinary "regular" formations. So his most important claim—echoing Jameson's introduction to Cuvier, and explicitly recruiting the Frenchman in his own support—was that "the two great points then of the low antiquity of the human race, and the universality of a recent deluge, are most satisfactorily confirmed by every thing that has yet been brought to light by Geological investigations". Of these two points, the second received particular emphasis:

> The grand fact of *an universal deluge* at no very remote period is proved on grounds so decisive and incontrovertible, that, had we never heard of such an event from Scripture, or any other authority, Geology of itself must have called in the assistance of some such catastrophe, to explain the phenomena of diluvian action which are universally presented to us, and which are unintelligible without recourse to a deluge exerting its ravages at a period not more ancient than that announced in the Book of Genesis.[77]

For Buckland to claim the global universality of the diluvian effects was bold, perhaps even rash, but in every other way this was a fair summary of what was now

74. Edmonds, "Oxford readership in geology" (1979), prints the petition in full (38–41); on the subsequent success of Buckland's courses, see Rupke, "Oxford's scientific awakening" (1997). He soon began to give himself the title of "professor", perhaps in part because "reader" was not used or widely understood outside Oxford and Cambridge, and certainly not outside Britain. Cambridge had had a succession of professors to cover the sciences of the earth since early in the previous century, endowed by a legacy from Woodward, though most of them had been less than distinguished and many had treated the position as a sinecure.

75. Conybeare to Buckland, December 1818 (Exeter-DRO, Buckland MSS, F548); Buckland, *Vindiciae geologicae* (1820); Edmonds, "*Vindiciae geologicae*" (1991).

76. Buckland, *Vindiciae geologicae* (1820), 21–22.

77. Buckland, *Vindiciae geologicae* (1820), 23–25. One of his most weighty citations was to *Records of the creation* (1816) by John Bird Sumner, the bishop of Chester (and later, archbishop of Canterbury), who argued that the "succession of former worlds" invoked by "geologers" was not improbable, by analogy with the plurality of worlds suggested by astronomers, and that "Mosaic history [is] not inconsistent with geological discoveries" (267–85). It is beyond the scope of the present book to trace the ramifications of the debates in Britain, in the "public realm" of the literate classes, about the interpretation of Genesis in relation to Buckland's presentation of the new geology.

a widespread opinion among geologists, in the light of detailed studies such as those by von Buch and Hall (§10.2), added to Cuvier's case for a mass extinction of an earlier fauna of terrestrial mammals (§10.3). Something very drastic and unusual seemed to have happened in the geologically recent past, and possibly even within the span of recorded human history. The case for such an event was indeed independent of the Flood story in Genesis, as shown by the fact that some respected geologists did *not* equate the "geological deluge" with the biblical one. Of course Buckland had his own reasons for claiming their identity; but the only other point on which many naturalists parted company with him was in his emphasis on the global extent of the diluvial event, and on its having covered even the highest mountains. That was indeed a bold extrapolation from what was known from "facts", even those observed by others, let alone the smaller selection that Buckland had so far seen for himself.

In an appendix, added to his text after he gave the lecture, Buckland summarized the geological evidence for the event that he identified as "the Mosaic Deluge". Most important was his claim that it had been the main cause—or even the sole cause—of what Farey had called "denudations", or the erosion of the solid rocks to form the present topography (see Fig. 9.13). On this point Farey himself was one of Buckland's most forceful critics, although, as so often when he wrote for the *Philosophical Magazine*, he remained notionally anonymous. Farey did not question the reality of many violent episodes in the earth's past history, or that the most recent had been responsible for major erosion. But he rejected Buckland's claim that that event had been recent enough to overlap with human history, and hence its identification with the biblical deluge or any other recorded event. For Farey, as for many others, the geological deluge had been real enough, but it was not the biblical one at all. However, Buckland remained unmoved by such criticism, and continued to treat the equation of the two events as almost self-evident.[78]

The deluge on Buckland's doorstep

By the time Buckland added these comments to his text he had already embarked on fieldwork that was designed to substantiate his diluvial claims. Shortly after giving the lecture, he made a field trip into the English Midlands to the north of Oxford, in the company of Count Breunner, a Viennese amateur who was visiting Oxford at the time (see Fig. 9.18). Buckland must already have noticed that some of the Superficial gravel deposits around Oxford contained, among many other kinds of pebble, smoothly rounded ones composed of a distinctive hard quartzite; and he may have suspected that these might act as mini-erratics, as it were, that would enable him to trace the course of a diluvian wave like the ones that von Buch had detected in the Alps, and Hall around Edinburgh (§10.2). Buckland duly traced the pebbles north into Warwickshire, where he found that they were the main constituent of a solid pudding-stone or conglomerate. This belonged to the "Red Marl" or New Red Sandstone, one of the Secondary formations without fossils that Smith and Greenough had both mapped with an extensive outcrop across the Midlands. Evidently the pebbles had been worn smooth in some ancient environment and then incorporated in an ancient gravel that had later been cemented into a solid rock. At a far more recent time, they must have been eroded out of their original matrix, swept southwards in the putative deluge, and then recycled, as it were, into the diluvial gravel. This was the geohistorical sequence with which Buckland later explained them; but he must have reasoned in this way while still in the field, for he extended his trip to Lickey Hill near Birmingham, where he located the *original* source of the distinctive quartzite itself, in a small "inlier" of Primary rocks, underlying and surrounded by Secondaries.[79]

The paper that Buckland based on this fieldwork was read at the Geological Society the following winter. It formed an instructive contrast to his inaugural lecture in Oxford (though the latter had not yet been published). Both highlighted the geological evidence for a major "deluge" of some kind, but the "connexion of geology with religion" that was the raison d'être of the Oxford lecture was barely alluded to in the more secular atmosphere at the London society: the two presentations were to quite different audiences, with only a handful of Oxford geologists in common. In fact the main title of Buckland's paper might have given other geologists at the society the impression that it was just a standard local study of some Primary and Secondary rocks: a "description" of the quartzite in situ and of its ancient erosion to form the pebbles in the overlying New Red Sandstone. Only in the later part of the paper (and in its subtitle) did Buckland mention his main and more controversial objective, which was to use the recycled quartzite pebbles to trace "the evidence of a recent deluge" all the way from Warwickshire past Oxford and down the Thames valley to London. Significantly, when writing to Greenough beforehand, he referred to the paper as being on "the Pebble question"; sending a preprint to Cuvier, he called it his "Diluvial Gravel paper" (Fig. 10.13).[80]

Most important, perhaps, was Buckland's proposal that the gravels he had traced, and other similar deposits, should be called *Diluvium*, to distinguish them from the later and "postdiluvian" *Alluvium*, the deposits still being formed by the present rivers; the former was of course a theory-laden term, but no more so than, for example, the Freshwater Formation defined by Brongniart and Cuvier (§9.1).

78. Buckland, *Vindiciae geologicae* (1820), 35–38; [Farey?], "Reflections on the Noachian deluge" (1820); Page, *Rise of the diluvial theory* (1963), 88n, mentions the strong internal and stylistic evidence that Farey was the author.

79. Buckland, "Lickey Hill" (1821), read at Geological Society, 3 December 1819 (London-GS, OM/1); Lickey Hill, a fine viewpoint, is about 15km southwest of the center of Birmingham. Buckland was almost certainly guided in the field by a copy of the relevant sheet of either Smith's map or a prepublication version of Greenough's, and possibly also by local informants such as "Honorary" or country members of the Geological Society; as usual, where field notebooks or similar records are no longer extant, such provincials are now "invisible" (§1.2).

80. Fig. 10.13 is reproduced from Buckland, *Reliquiae diluvianae* (1823) [Cambridge-UL: MF.40.10], pl. 27, first published in "Lickey Hill" (1821), pl. 37. One of his local assistants was not "invisible": he thanked the young amateur naturalist Mary Morland (1797–1857)—later his wife—for help in mapping the gravel from above Banbury downstream past Oxford to her father's home in Abingdon (525). The map dates from the heyday of canals and all-weather toll roads, before the earliest railways: the Oxford Canal, for example, took advantage of the low col north of Banbury to link the navigable lower reaches of the Thames to those of the Trent basin to the north. The quoted phrases about the paper are in Buckland to Greenough, 9 November 1819 (London-UCL, Greenough MSS), and Buckland to Pentland, 21 May [1820] (Paris-MHN, MS 627, no. 19; the year is inferred from internal evidence). Joseph Pentland (1797–1873) was a young Irish naturalist who was acting in effect as an honorary research assistant to Cuvier: see Sarjeant and Delair, "Letters of Pentland" (1980).

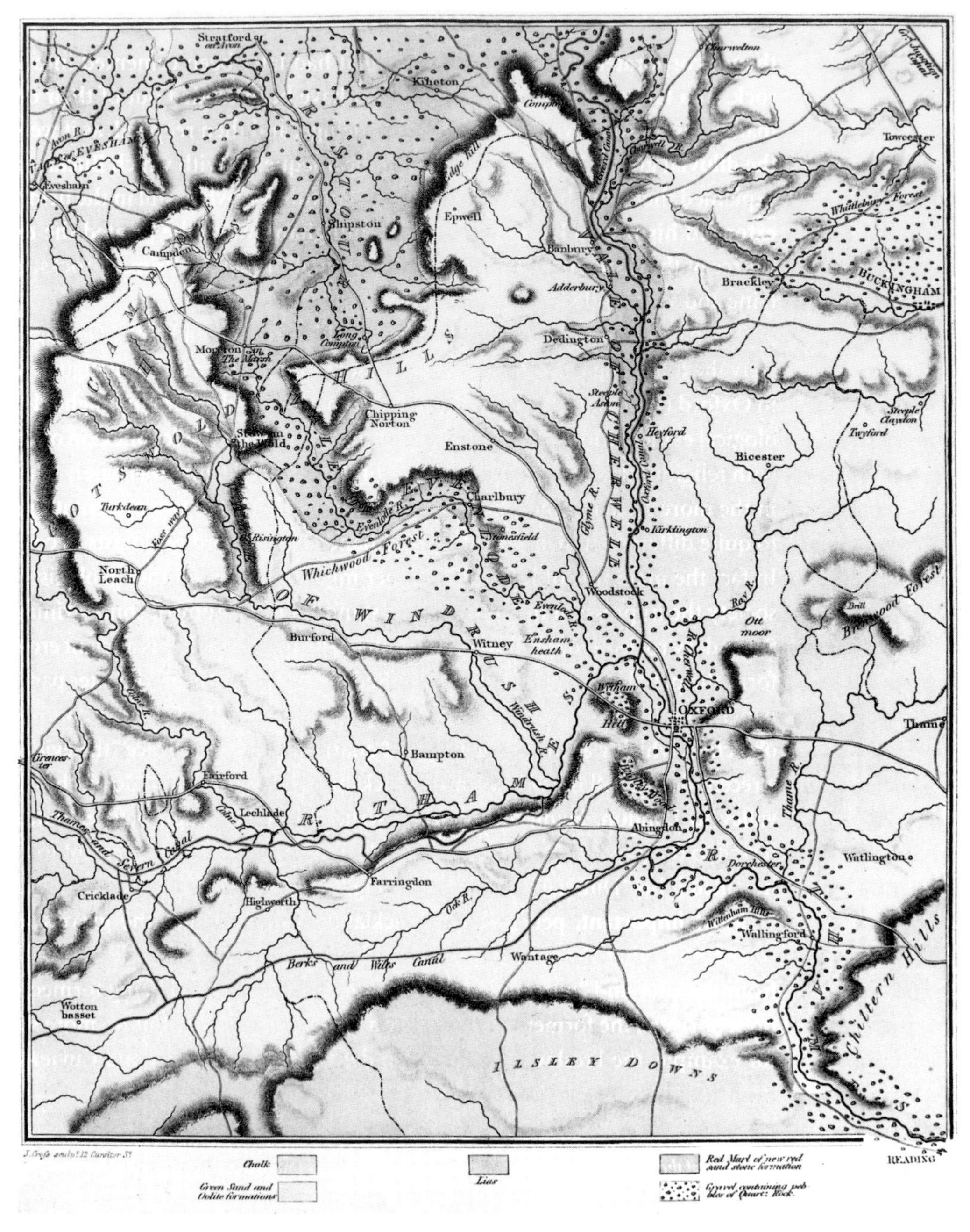

Fig. 10.13. Buckland's map of the Oxford region, published in 1821, plotting the distribution of the "gravel containing pebbles of quartz rock" (stippled), overlying the Secondary formations that both Smith and Greenough had mapped; the formations ranged from the "Red Marl" or New Red Sandstone (in the northwest) up to the Chalk (in the southeast), with the successive scarps of the Cotswold Hills, the low hills near Oxford itself, and the Chiltern Hills. The map showed how the gravels covering the Midland plain (in the northwest) passed right *over* the watershed between two distinct drainage basins in a way that could not be explained in terms of the present rivers: it was found in the valleys of the Cherwell and Evenlode (north of Oxford), both with low cols at their heads—the former exploited by the Oxford Canal—but not in the Windrush and upper Thames valleys (to the west), which had no such cols. Buckland argued that this peculiar distribution could be explained as the result of a transient but violent "diluvian current" that had swept southwards from Warwickshire over the two cols through the Cotswolds, past Oxford, and all the way down the Thames valley, through the narrow gap in the Chilterns (in the southeast corner), past Reading to London, excavating or at least deepening the valleys and depositing in them the diluvial gravel with its telltale quartzite pebbles. (By permission of the Syndics of Cambridge University Library)

He had already used "Diluvium" in his lectures, and one of his auditors, William Thomas Horner Fox-Strangways (1795–1865), had later adopted it in his accounts of the geology around St Petersburg. This young aristocratic diplomat had described to the Geological Society how erratics of clearly Finnish origin were widely scattered across the plains of northern Russia.[81]

Buckland's paper was closely argued. Even such an apparently minor detail as the shape of the hard quartzite pebbles was significant in his analysis. Had he not discovered from his fieldwork that they had been recycled from a solidified gravel of far more ancient origin, their smoothly rounded form might have been taken to imply a very long process of gradual attrition in the beds of ordinary rivers, rather than the transport and deposition of *already rounded* pebbles in the brief and extraordinary event that he inferred. He noted that the pebbles of other kinds of rock, in contrast, were often angular, as would be expected if they had been moved rapidly and briefly: for example, those of the distinctive Red Chalk known in situ away to the northeast, in Lincolnshire, were not rounded at all, although that rock was quite soft.

In the "factual" atmosphere of the Geological Society's meeting room, Buckland was wise to restrict himself to such sober inferences. But in his published paper he included an appendix that went beyond "facts" in a way that was unprecedented in the society's *Transactions.* Here he summarized the evidence for the worldwide event that he inferred from his specific case, and praised Cuvier for "the most enlarged and philosophical view of the state of the question that has ever been taken". He noted that the English gravels contained fossil bones, such as those of "the Siberian rhinoceros" (Fig. 5.10), identical to those that Cuvier had described as belonging to extinct species of mammals. But this fossil evidence was only one part of his argument. Buckland linked Cuvier's name with those of Hall and Saussure; he cited Fox-Strangways on Russia and other recent reports; and he concluded that their combined research supported "the important fact" of the *universality* of the putative deluge. Yet even here he left the equation of this event with the biblical deluge implicit and unargued: "Accumulations of superficial gravel more modern than the most recent of the regular strata are found in all parts of the world, under circumstances of such exact resemblance, that it is impossible not to refer them to one and the same cause, viz. a recent deluge acting universally and at the same period over the surface of the whole globe."[82]

Buckland conceded that his "diluvian current" was only a "hypothesis"—and therefore suspect in the eyes of some of the society's leaders, notably Greenough—but he claimed that it accounted for the "facts" observed in the field, better than any other explanation. Significantly, he referred to the event as the "latest diluvian catastrophe", thus tacitly disclaiming any suggestion that it had been unique and agreeing with Hall that such events had occurred repeatedly in geohistory. In fact,

81. Fox-Strangways, *Strata des environs de St Petersbourg* (1819); "Environs of Petersburg" (1821), read on 16 April 1819; "Geology of Russia" (1822), see map (pl. 1). He eventually inherited his father's title as earl of Ilchester.

82. Buckland, "Lickey Hill" (1821), appendix (538–44), 544.

Fig. 10.14. Kensington Gravel Pits, painted by John Linnell around 1812. These pits were in the gravel terrace, well above the river Thames in London, that represented the furthest extent of a "diluvian current", which Buckland in 1819 traced all the way from the English Midlands to London by means of its telltale quartzite pebbles. Linnell was known for his honest depiction of unromantic scenes, so this painting probably gives a reliable impression of everyday work in a gravel pit, around the time that Buckland was studying the Superficial gravels that he called the "Diluvium". The unmechanized digging gave excellent opportunities for finding fossil bones, which, if sold to collectors, could greatly augment the meager wages of such manual laborers. (By permission of Tate Britain, London)

however, notwithstanding his claims for the magnitude and universality of his deluge, his own fieldwork suggested, on the contrary, a local current of relatively modest size. It had flowed over the tract indicated by the gravels that it had deposited on its way, "bursting in over the lowest point of depression" in the Cotswold Hills rather than totally submerging them (and even at their highest they only rise to about 300m). Still, Buckland had made a strong case, if not for a universal deluge, then at least for a more localized "diluvian current", which had swept all the way from Warwickshire (and perhaps also from Lincolnshire), through gaps in the Cotswolds, past the site of Oxford and down the Thames valley as far as the outskirts of London, where its debris was found as a gravel terrace well above the present level of the river (Fig. 10.14).[83]

A new theory of erratics

Buckland's confidence in his diluvial interpretation of the English gravels would have been reinforced by von Buch's improvement of his own analogous theorizing in the Alps (§10.2). Although his paper had been read in Berlin as early as 1811, it was not published there until after the coming of peace in 1815; and it was not widely noticed outside the German lands until 1818, when André Jean Marie Brochant de Villiers (1772–1840), the director of the School of Mines in Paris under the Restoration regime, translated part of it for the French *Annales de Chimie et de Physique* (one of a proliferating range of scientific periodicals that were springing up all over postwar Europe). Brochant then pressed von Buch to say more about the possible *cause* of the movement of the erratics, and he published the Prussian's comments in the same periodical. As usual among geologists, von Buch took the present to be the key to the past. Just as Hall had used the historic tsunamis at Lisbon and Callao to imagine a mega-tsunami capable of shaping the topography around Edinburgh (§10.2), so von Buch used reports of a catastrophic but localized mudflow in an Alpine valley as a small-scale analogue of an event that could perhaps have shifted large erratics for dozens of miles.[84]

In 1816—only a year after von Buch's original paper on the erratics was published—a huge landslide of boulders, mud, and ice had blocked the upper reaches of the Val de Bagnes in the Swiss Alps (coincidentally, not far from the peak that von Buch had identified as the source of the granite erratics on the Jura: see Fig. 10.5). The water draining the upper valley was ponded back by this natural dam and soon formed a large lake. Then, without warning, the weight of the water had burst the dam. According to the accounts of survivors, collated and checked by the Swiss naturalist Hans Conrad Escher (1767–1823), a huge mass of liquid mud and boulders had swept down the valley, carrying uprooted trees and the debris of the villages in its path: blocks of rock up to a cubic meter in volume had been carried some thirteen miles in thirty-five minutes, or at an average velocity of about thirty-three feet a second. Escher reported on the disaster to the Swiss Society for Natural Sciences, meeting in 1818 in Lausanne, using a three-dimensional model of the area with great effect. At the equivalent gathering the following year, he explicitly used the Val de Bagnes as a small-scale analogue for what might have caused the far wider distribution of the Alpine erratics, while modestly calling his work a mere "pendant" to von Buch's.[85]

Although its consequences for the inhabitants were disastrous, the mudflow in the Val de Bagnes was of immense scientific importance because it showed on a

83. Fig. 10.14 is reproduced from Linnell's oil painting (London-TB, N05776), which was exhibited in London and Liverpool in 1813; see Parris, *Landscape in Britain* (1973), 102–5, no. 245. Linnell was a respected artist in his own right but also, a few years later, a friend and important patron of the then rarely appreciated William Blake. These pits were near Notting Hill Gate, an area now completely urbanized as part of West London. On fossils as a source of income for manual workers at this period, see Knell, *Culture of English geology* (2000). The quoted phrases are from Buckland, "Lickey Hill" (1821), 518–19, 532, 537.

84. Buch, "Transport des blocs de roches" (1818) and "Additions au mémoire" (1819), the latter written in fact by Brochant but summarizing von Buch's informal comments to him.

85. Escher, "Val de Bagnes" (1818), read on 29 July 1818 (see 291n on the model), and "Blocs de roche" (1822), read on 28 June 1819 (see 277–82 on Val de Bagnes). A fine modern dam now ponds back a large lake (for hydroelectric power) in almost the same position in the upper Val de Bagnes, making it easy to imagine the sheer scale of the disaster, which swept boulders as far as Martigny in the Rhône valley (see Fig. 10.5). The Société Helvétique des Sciences Naturelles, or Schweitzerische Gesellschaft für die Naturwissenschaften, was founded in 1817 to bring together savants from around the newly independent (and bilingual) Swiss Confederation; it became the model for the Deutsche Gesellschaft für Naturforscher und Ärzte (founded 1822), which moved similarly around the fragmented German states, and which in turn was the inspiration for the peripatetic British Association for the Advancement of Science (founded 1831) and its still later American namesake.

relatively small scale how a diluvian current might have operated. It proved that water could be an effective means of transport, even for large boulders, provided that they were suspended or borne along rapidly in a dense fluid composed mainly of mud. It therefore seemed to support Hall's conjecture that a mega-tsunami carrying a mass of mud and boulders at its base might have been capable of scouring out new valleys and scratching the bedrock in its path. And, as von Buch and Escher both recognized, the Val de Bagnes mudflow also offered a plausible explanation of how erratic blocks, even those the size of the Pierre à Bot above Neuchâtel, might have been transported from the Alps to the Jura, provided the event had been large enough in scale.

These papers were published in the Genevan *Bibliothèque Universelle* and the Parisian *Annales de Chimie et de Physique*, both of which would certainly have been seen by savants in London. Even if Buckland himself did not read about the Val de Bagnes, it is inconceivable that von Buch did not discuss it with him while he was visiting Oxford at just this time, since it was so obviously relevant to what Buckland was doing. However, Buckland did not mention it in his own published work, either because he considered it inapplicable to his own "diluvian current", which had apparently swept widely across lowland England rather than descending from Alpine heights, or simply because he scrupulously abstained from *any* causal theorizing about the event.[86]

The problem of valley erosion

In the summer of 1820, having given his annual lectures, Buckland made a second Continental tour with Greenough, this time with Breunner making it a triumvirate. His main objective was to construct a sequence of "regular" formations that might be valid throughout Europe (§9.5). But after consulting Cuvier and other savants in Paris, he and his companions gave the extinct volcanoes of Auvergne first priority, as Conybeare had proposed for their trip four years earlier. Buckland had already been primed for this classic and contentious ground by his younger Oxford colleague Charles Giles Brindle Daubeny (1795–1867), who had toured the area the previous summer. As a chemist, Daubeny had been interested mainly in the volcanic rocks themselves and in what they might contribute to the still simmering arguments about basalt; but he had also been struck by the contrast between "ancient" and "modern" volcanic products. He knew of Montlosier's classic work on Auvergne (§6.1) and had, for example, gone to see for himself the famous case in which the Sioule had been diverted by a "modern" lava flow (Fig. 6.2). But he was not convinced by Montlosier's (and Desmarest's) claims that the main valleys had been eroded gradually by the streams that still flow in them. Instead he had adopted something like Dolomieu's alternative (§6.3), inferring that a sudden episode of violent valley erosion had been interposed between the ancient flows and the modern ones. Not surprisingly, in the wake of Buckland's inaugural lecture, Daubeny identified that erosive event as "the Mosaic deluge".[87]

When Buckland himself reached Auvergne and saw the volcanoes and valleys for himself, he added them at once to his tally of diluvial evidence, judging them "the finest thing by far in Europe". He incorporated them subsequently into his lectures, distinguishing the older and newer lavas as "antediluvian" and "postdiluvian"; since he believed that the latter had not been eroded at all since their eruption, they counted as evidence that "modern Causes [i.e., the present streams] will not make Vallies". For much of the rest of this Continental tour, his diluvial research took second place to stratigraphy and other issues, but it was not forgotten. For example, he and Breunner saw some fine fossil bones in a collection at Kremsmünster abbey, not far from Linz in Austria; he told Cuvier they had been found in "the universal Diluvian gravel that contains Elephant, Rhinoceros etc.".[88]

Buckland's fieldwork the following summer, although again mainly stratigraphical in its objectives, also gave him an opportunity to collect evidence for the diluvial erosion of valleys on the south coast of England. The fine coastal cliffs of east Devon and Dorset showed unambiguously that the valleys running down to the sea had been excavated through almost horizontal formations: at least in these cases valleys were evidently not the result of any crustal disturbance. However, that still left open the question whether they had been eroded swiftly by a violent diluvial current or very slowly by the small streams that still flowed in them. Significantly, however, Buckland was so convinced on other grounds that the former

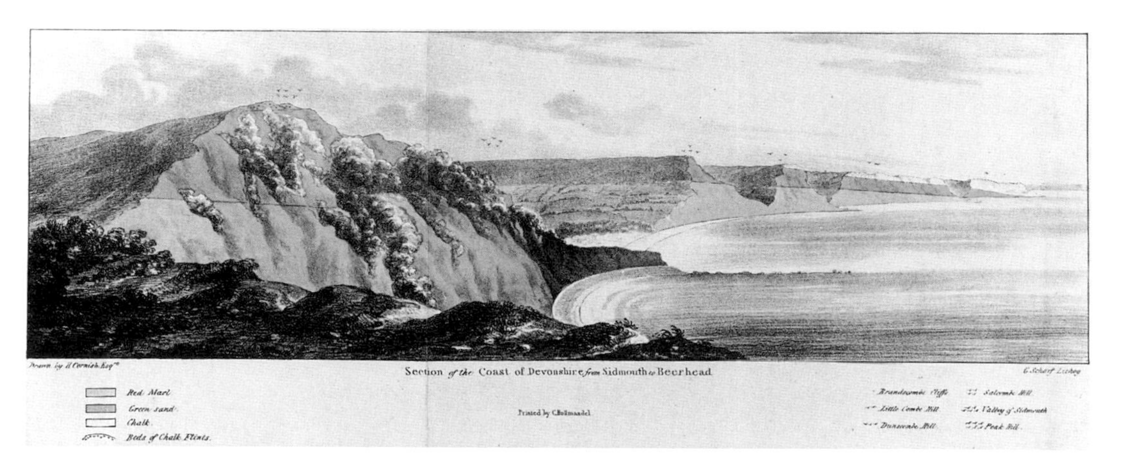

Fig. 10.15. The coastal cliffs of east Devon, on the south coast of England, drawn for Buckland and lithographed for his paper "On the Excavation of Valleys by Diluvial Action" (1822). The valleys have been cut through undisturbed Secondary formations dipping very gently to the east (right); Buckland argued that they must have been scoured out by a sudden and violent "diluvian current", not eroded slowly by the small streams that now flow in them: two of the valleys emerge at the coastline well above sea level. The landmarks are identified by different numbers of seagulls, a common pictorial convention that avoided sullying the beauty of the scene with intrusive numbers or letters. (By permission of the Wellcome Library, London)

86. Buch had already presented the Geological Society with a copy of his original paper (London-GS, OM/1, 4 December 1818).

87. Gordon, *William Buckland* (1894), 37–40; Daubeny to Buckland, 5 March 1820 (Oxford-MNH, Buckland MSS); Daubeny, "Volcanoes of the Auvergne" (1820–21), published after Buckland's return in the recently founded *Edinburgh philosophical journal*, coedited by Jameson.

88. Buckland, lecture note "Deluge Auvergne Volcanic District" [1821?] (Oxford-MNH, Buckland MSS); his denial that there had been any further erosion in Auvergne since the later lava flows were erupted suggests that he had not read Desmarest or Montlosier very carefully. Buckland to Cuvier, 17 October 1820 (Paris-MHN, MS 627, no. 17).

was the correct explanation that in his paper he did not mention the latter even to refute it (Fig. 10.15).[89]

Cuvier's revised discourse

By the time that Buckland presented his paper on valley erosion at the Geological Society, Cuvier had given further powerful support to such diluvial theorizing by publishing a revised version (1821) of his "preliminary discourse" in the new edition of his *Fossil Bones*. After the first edition was published in 1812 (§9.3), Cuvier had resumed work on his great survey, the *Animal Kingdom* (1817). At the same time, however, he had also accumulated material for the revision and enlargement of his many articles on specific kinds of fossil bones. But he had withdrawn from active involvement in geology, leaving the geological implications of his research to others, and mainly to Brongniart (§9.6). So his "preliminary discourse" was largely unaltered, and would have struck well-informed geologists as already rather outdated. Nonetheless, in the peacetime conditions his new edition certainly gained him many new readers, both savants and amateurs, naturalists and the educated public.[90]

In its original language Cuvier's introductory essay continued to be available only as part of an expensive publication, in a volume that it now shared with his revised articles on fossil elephants and other pachyderms. In other languages, however, the discourse was more accessible. Jameson used it promptly for the fourth edition of what he continued to call "Essay on the Theory of the Earth", now greatly bloated with his own ever more verbose "illustrations" or commentary. Jakob Nöggerath (1788–1877), the professor of mineralogy and mining at Bonn, published a German edition equally promptly. In addition, the earlier version had already been given wider currency by a New York edition of Jameson's work (with extra notes on American geology) and a Swedish translation from the original French. So Cuvier's eloquent case for a geologically recent catastrophe, which he claimed had been responsible for the mass extinction of the large mammalian species found in the Superficial deposits, became ever more widely known across the scientific world.[91]

Cuvier's revised discourse hardly referred to the mass of new research on fossil bones that had been done in the intervening years—much of it his own work, or at least inspired by it—although of course this provided the material for his revisions in the rest of the seven volumes that appeared over the next four years. The only important exception was in his review of the evidence for human fossils. Having seen the Guadaloupe skeleton (Fig. 10.8) while he was in London, Cuvier agreed with König that it was probably quite recent in origin. He discounted the human bones that Schlotheim had recently reported from Köstritz in Saxony; although they were apparently mixed with the bones of some of the usual extinct mammals, Schlotheim himself conceded that the case was at best ambiguous and that they could well have been interred at a much later date. Other human bones, found in a cave at Durfort in southern France around the same time, had looked "petrified" and therefore "fossil". But the naturalist who studied them had concluded that they probably dated from historic times—he suggested that they belonged to Gaulish,

Roman, or Saracen victims of war—and were simply encrusted with the stalagmite that was usual in limestone caves: it was a common tourist attraction at "petrifying springs" to leave familiar objects to be "turned into stone" in the same way. All in all, Cuvier saw no reason to modify his conclusion that the human species did not extend far back into geohistory.[92]

In fact Cuvier's introductory essay was more substantially enlarged at just one rather surprising point. He greatly extended his review of the *textual* evidence for the limited span of *human* history; as in the first edition he then used this as evidence for the geologically recent date of the physical event to which he attributed his mass extinction. In other words, the one point at which he tried to reinforce his interpretation of geohistory was on the date of its most recent catastrophe. Cuvier's textual sources of evidence for his "revolution" were—in contrast to Buckland's "Mosaic deluge"—more multicultural than ever: he insisted once more that it had happened at the dawn of human history, and that it was recorded, however obscurely, in several of the earliest human records. Conversely, he was concerned to refute claims that some of those records stretched back much further into the past, with no trace of any disruptive event at all.[93]

This surprising enlargement of Cuvier's earlier incursion into human historiography was probably triggered by the "zodiacomania" that gripped savant Paris just as he was completing his first volume for the press. The antiquity claimed for the celebrated zodiacs from Egypt (§10.3) had been criticized even before his first edition was published: far from being evidence that ancient Egyptian civilization had been flourishing without a break for tens of thousands of years, they were said to date merely from Hellenistic times. But the dispute had revived. Once again it was no mere dry-as-dust dispute among antiquarians, for the zodiacs were again being used to undermine the credibility of the biblical account of human origins by those who opposed the revived political dominance of Catholicism in France under the Restoration. Although Cuvier was far less attached to scriptural traditions than his fellow Protestant Buckland, he was equally sensitive to the misuse of scientific evidence for antireligious purposes. Above all, Cuvier was as much concerned as ever

89. Fig. 10.15 is reproduced from Buckland, *Reliquiae diluvianae* (1823), part of pl. 25 [London-WL: V25115], first published in "Excavation of valleys" (1822), pl. 14; it was drawn by Hubert Cornish and lithographed by George Scharf (the complementary map is pl. 13). Buckland's paper was not read at the Geological Society until 19 April 1822, having been postponed on account of the still greater importance of his analysis of bone caves (§10.6). George Scharf (1788–1860) was a Bavarian who made a career in London as a painter and as a skillful practitioner of the new craft of lithography, in which he was frequently employed by members of the Geological Society.

90. Cuvier, *Ossemens fossiles*, 2nd ed. (1821–24); *Règne animal* (1817).

91. Cuvier, *Theory of the earth*, 4th ed. (1822); *Ansichten von der Urwelt*, 1 (1822), of which the second volume (1826) later made available a selection of Cuvier's work from the rest of *Ossemens fossiles*. On the American (1818) and Swedish (1821) editions, see Smith, *Georges Cuvier* (1993), items 659, 664.

92. Cuvier, *Ossemens fossiles*, 2nd ed., 1 (1821), "discours préliminaire", lxv–lxvi. Schlotheim, *Petrefaktenkunde* (1820), xliii–lxi, translated (by the younger Charpentier) as "Sur les anthropolites" (1820), for the *Bibliothèque universelle*. See Grayson, *Human antiquity* (1983), 97; Köstritz is in the valley of the Elster, near Gera in Saxony. Hombres-Firmas, "Ossemens humains fossiles" (1821), was just too late to be included by Cuvier, but he must have heard of it soon afterwards; Durfort is about 20km southwest of Alès (Gard).

93. Cuvier, *Ossemens fossiles*, 2nd ed., 1 (1821), "discours préliminaire", lxxix–cxv; compare with 1st ed. (1812), 1, "discours", 94–106.

to maintain the historicity of his most recent "revolution", as the natural *cause* of the mass extinction that his research on fossil bones had disclosed. If any of the world's ancient civilizations extended as far back into the past as was being claimed, the veracity of his geohistory would be jeopardized. So he judged it necessary to refute such claims with renewed vigor, and thereby to reassert the strong evidence for a decisive physical break at the geologically recent dawn of human history. This was of course a view that Buckland could appropriate as powerful support for his own narrower concept of a geological deluge that could be identified literally with the biblical one.[94]

Conclusion

Buckland had swiftly become as prominent in the long-running debates about the reality of a geologically recent catastrophe as he was in the more prosaic project of constructing a stratigraphical sequence of general validity. In both areas of research he was not only the most prominent individual at Oxford—which was hardly difficult—but also among the most prominent at the Geological Society in London. Still more importantly, he had become a geologist whom Continental savants, from Cuvier and von Buch downwards, now recognized as a heavyweight in this scientific arena. Buckland had strengthened Cuvier's case for a recent catastrophe of some kind—marking in effect the boundary between the former world and the present—by providing further concrete evidence for violent "diluvian currents", this time in England. Added to Cuvier's evidence from mammalian fossils, von Buch's from erratics and Hall's from scratched bedrock, some such extraordinary agency now seemed far more credible than Playfair's reliance on the ordinary slow and gentle processes observably at work in the present world. And anyway the Val de Bagnes disaster, as described by Escher and interpreted by him and von Buch, showed that in fact the modern world included observable processes that were *not* slow or gentle, but powerful enough to suggest a causal explanation for erratics, at least in mountain regions. All in all, a "geological deluge" was coming to be accepted, almost universally among geologists, as a real feature of recent geohistory.

On the other hand, Buckland's equation of this event with the Flood recorded in Genesis, while echoing Cuvier's more multicultural argument for some such dating, was far more questionable. Many geologists suspected that the geological deluge might be quite distinct from the biblical one: much earlier, more violent, and perhaps truly universal. The final section of this narrative describes research by Buckland that seemed to throw further light on the event, or at least to open an unprecedented window on to a vanished "antediluvian" world.

10.6 A SPY HOLE INTO THE PAST (1821–23)

Kirkdale cave

Before Buckland had time to write up his paper "On the excavation of valleys by diluvial action", his research on the putative deluge gained a huge impetus from an

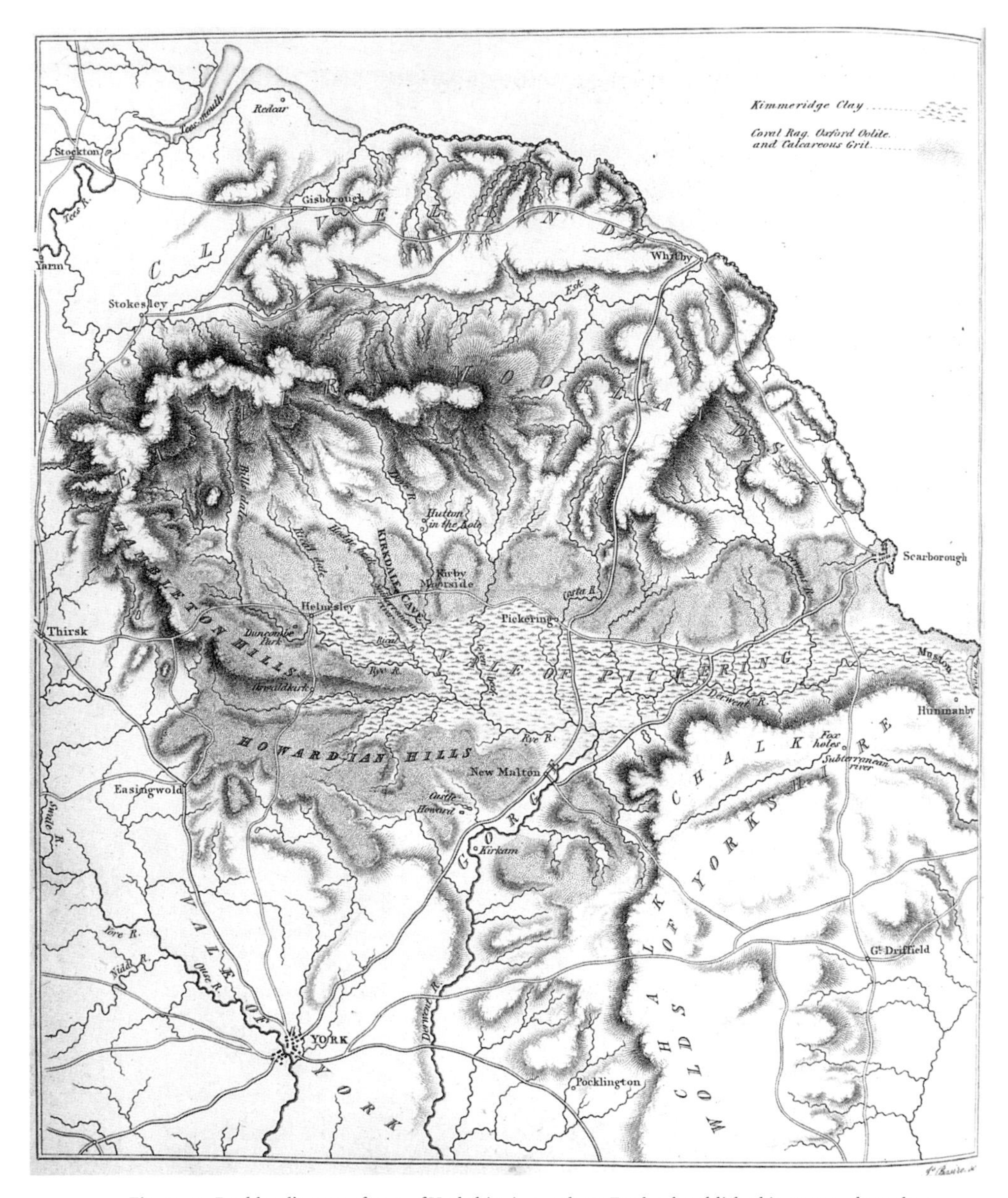

Fig. 10.16. Buckland's map of part of Yorkshire in northern England, published in 1822 to show the geographical and geological situation of Kirkdale cave. The cave is marked on the northern edge of the flat alluvial Vale of Pickering (shown as marshy ground), between the curved scarps of the "Eastern Moorlands" (now the North York Moors) and the Yorkshire Wolds. The vale is drained by the Derwent, which rises very close to the North Sea near Scarborough, but flows *inland*, leaving the vale by a narrow "gorge" northeast of York. Buckland inferred that in antediluvian times the cave had been near the edge of a large lake on the site of the present vale and that the lake had drained away when the gorge was excavated by the waters of the geological deluge. He claimed that the same event had exterminated the animal life of the region, including the hyenas that used the cave as a den. (By permission of the Syndics of Cambridge University Library)

94. Cuvier's discussion of zodiacs (cxv–cxxxiv), like the preceding section, was hugely expanded from the first edition (106–9); see also Negrin, *Georges Cuvier* (1977), 298, and Buchwald, "Egyptian stars under Paris skies" (2003). The argument for their very high antiquity was finally exploded a few years later, as a result of the successful decipherment of the hieroglyphs. Using the demotic inscription on the Rosetta stone as the crucial intermediate between the hieroglyphs and the text in Greek, the French scholar Jean-François Champollion published a preliminary analysis in 1822, and more fully in 1824: see Parkinson, *Cracking codes* (1999). A safely late date could then be assigned to the zodiacs that had threatened Cuvier's peace of mind.

unexpected discovery in northern England. Its beginnings were unspectacular. In the summer of 1821, the entrance to a small cave was uncovered by men working a limestone quarry at Kirkdale, on the edge of the Vale of Pickering in Yorkshire (Fig. 10.16).[95]

Abundant animal bones were found buried in the floor of Kirkdale cave. The quarrymen assumed they were the remains of cattle that had been thrown into the cave after dying in some past epidemic, so the bones were used to fill potholes in the nearby road. There they were noticed by an amateur naturalist from London, who was visiting his native town of Kirby Moorside; he recognized that the bones were not those of cattle. As the news spread an intense bone rush ensued, and local clergymen, country gentlemen, and other collectors scrambled to acquire the best specimens that the quarrymen could find in the still undisturbed parts of the cave floor. Some of the specimens were sent to the Royal College of Surgeons in London, where the curator William Clift (1775–1849)—who as a young man had been John Hunter's assistant—was able to cast his expert eye on them. Their importance was such that without delay Clift mentioned them to Cuvier: "I think *Hyaena*", he wrote, referring to the most common ones, adding that they were larger than the

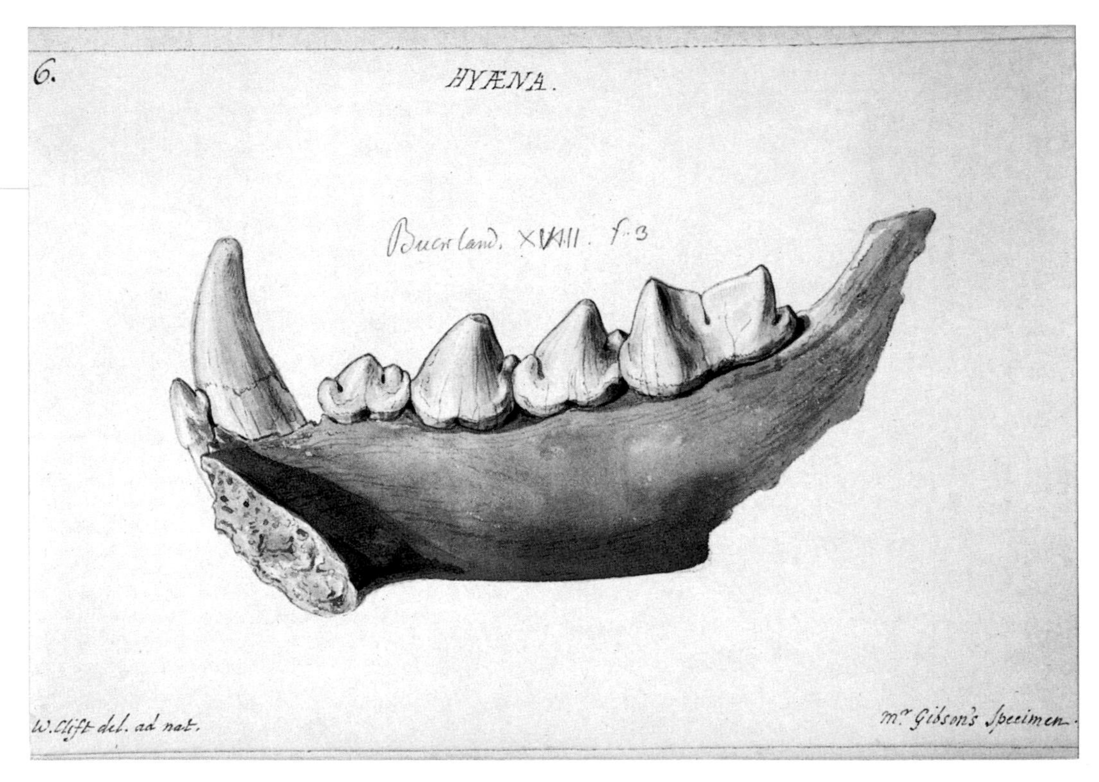

Fig. 10.17. The fragmentary lower jaw of a hyena found in Kirkdale cave in the north of England. This specimen was owned by the collector who had first realized the significance of the bones. It was drawn "from nature" [*ad nat.*] by the anatomist William Clift, who identified the jaw as that of a large species of hyena and sent this picture as a proxy specimen to Cuvier in Paris. As indicated in another hand, the same specimen was later published in Buckland's paper to the Royal Society, in which he interpreted Kirkdale cave as a former den of antediluvian hyenas. (By permission of the Bibliothéque Centrale du Muséum National d'Histoire Naturelle, Paris)

modern species. He then had drawings made of some of the specimens, and sent them to Paris (Fig. 10.17).[96]

Around the same time, Buckland was told about the cave by his colleague the warden of All Souls College (and bishop of Oxford), who showed him some of its bones. Like Clift, Buckland wrote promptly to tell Cuvier about them. As Cuvier was well aware, the bone caves on the Continent seemed to be of two distinct kinds: some contained mainly the bones of carnivores, others the bones of herbivores. The former could be interpreted as true caves, which animals had used as dens; the latter, as fissures into which they had been swept by a deluge or fallen by accident. But Kirkdale did not fit easily into either category, and Buckland conjectured that it might be "a compound one of a fissure falling into or crossing a Cavern"; for although hyena bones were most abundant, there were also those of an extraordinary mixture of other animals. Buckland interpreted the cave as diluvial, though without yet using the word: "It is not easy to conceive that any thing short of the common Calamity of a simultaneous destruction could have brought together into so small a Compass so heterogeneous an Assemblage of Animals as we find here intombed in a common Charnel-House—animals which no habit or Instinct we are acquainted with could ever have associated with a den of Hyaenas."[97]

A few days later, however, perhaps after receiving a report on the cave itself, Buckland modified his opinion. He himself was no expert anatomist, and the specimens at his disposal must already have been identified for him, perhaps by Clift. As he told an amateur naturalist, the daughter of an old friend in Wales, the cave was said to be some two hundred yards (180m) in length, its floor "paved" with hyena bones, but also those of elephants, rhinoceros, horses, oxen, deer, foxes, and rats:

> How the latter [i.e., all but the hyenas] got there is not easy to be conceived unless they be either the wreck of the Hyaenas' larder, or were drifted into a fissure by the Diluvian waters; both are possible causes, but the latter assumes that there was a fissure open at the top, and the account I at present have of it states that the aperture is a Cavern covered all over at the Top with continuous Beds of Lime St[one] and if so, we can only suppose the Bones to be the wreck of Animals that were dragged in for food by the Hyaenas.[98]

Buckland's next step was of course to visit Kirkdale and see the cave for himself. He traveled north as soon as he could, though midwinter was not the best time for

95. Fig. 10.16 is reproduced from Buckland, *Reliquiae diluvianae* (1823) [Cambridge-UL: MF.40.10], pl. 1, first published in "Fossil teeth and bones" (1822), pl. 15. The two scarps correspond to the Cotswolds and Chilterns further south, as shown on Buckland's similar map of the Oxford region (Fig. 10.13); that is, they are composed of the Oolitic (Jurassic) and the Chalk (Cretaceous) formations respectively. There is now little at Kirkdale except its historically important pre-Conquest church and the overgrown quarry and cave that made it briefly famous.

96. Fig. 10.17 is reproduced from Paris-MHN, MS 634(1), "Hyène fossile" dossier; published in Buckland, "Fossil teeth and bones" (1822), pl. 18, fig. 3, and *Reliquiae diluvianae* (1823), pl. 4, fig. 3. Clift to Pentland [i.e., to Cuvier], 11, 29 November 1821 (Paris-MHN, MS 627, nos. 25–27). See also Boylan, *William Buckland* (1984), 112, 365–66.

97. Buckland to Pentland, 18 November 1821 (Paris-MHN, MS 627, no. 22). Buckland, like Clift and other English naturalists, used Pentland as a conveniently anglophone means of communicating with Cuvier.

98. Buckland to Miss [Mary Theresa] Talbot, 26 November 1821 (Cardiff-NMW, MS 162).

Fig. 10.18. Kirkdale cave in Yorkshire, as illustrated in Buckland's paper (1822) to the Royal Society in London. The original entrance had been quarried away, but what remained showed the small size of the cave—the quarryman provided scale—and also its position in the middle of solid strata of limestone. A typical cross section (fig. 2) again showed that the cave was far too small—only four feet wide and four feet high—for any large carcasses to have been swept in; the silt (A) in which the bones were embedded had stalagmitic deposits both below (B) and above it (C, D), and there were corresponding stalactites (E) on the solid roof of the cave. These features were crucial in Buckland's interpretation of the cave as an antediluvian hyena den. (By permission of the Syndics of Cambridge University Library)

fieldwork. What he found was a cave far smaller and less impressive than the famous ones that he had visited in Bavaria five years earlier (§10.4), yet it was clearly of outstanding interest. He confirmed that there was no fissure in its roof through which any of the animals could have fallen. Yet its entrance was so small, and it was so low and narrow throughout, that it was impossible for the carcasses of elephants and rhinoceros to have been swept into it. So the alternative seemed inescapable: it must have been a den of hyenas, which, as scavengers, had dragged into it the already dismembered parts of the larger animals. Furthermore, although most of the floor of the cave had by now been excavated in the hunt for bones, enough remained to show that all of them had been embedded in a silty deposit that was both underlain and overlain by layers of stalagmite (Fig. 10.18).[99]

Buckland wrote to Cuvier immediately on his return to Oxford, telling him that Kirkdale cave was so important that he intended to present his interpretation of it not to the Geological Society but to a wider savant audience at the Royal Society. For it showed "the state of habitation in this country [i.e., England] in the period immediately preceding the Deluge"; it gave further evidence for the reality of that event; and it provided "a natural Chronometer" for both the "Antediluvian Period" and the "Postdiluvian", showing that "the latter must have begun at or about the point of time assigned to it by our common Chronologies". In effect, it allowed the construction of a *geohistorical* narrative that would link the antediluvian world forwards into human history, confirming his own and Cuvier's opinion that the date of the geological deluge was compatible with its being the Flood recorded in Genesis and other ancient human records. It also helped to confirm what Cuvier had long argued, namely that the mammals found so widely in the "diluvial Gravel"—

from Siberia through Europe to North America—"may have been Natives of the Countries in which their remains are found & were not drifted to their present Place of Sepulture from tropical Regions". It followed that "the Antediluvian Climate" might have been much the same as the present, since Cuvier himself had shown that all the fossil species were distinct from those now living in the tropics, and the Siberian elephant and rhinoceros were known to have had thick woolly coats (§10.1).

Buckland explained to Cuvier how he inferred that Kirkdale cave had been "an antediluvian Den of Hyaenas". The bones had not been washed in by the deluge itself, but had been dragged into the den to be chewed and digested at leisure, most of them being gnawed into fragments. All this had happened *before* the deluge itself, which had merely served to bury the bones in a fine silt on the floor of the cave: "the diluvian waters entered it, & being there quiescent deposited their Mud". At the same time, detritus borne by the turbulent floodwaters outside had probably blocked the original entrance of the cave, which had not been occupied subsequently. Instead, the groundwater percolating through the limestone had sealed the bone deposit with a crust of stalagmite "like Cream on a Pan of Milk", its small thickness implying a lapse of time consistent with the kind of date calculated by traditional chronology.[100]

The small thickness of stalagmite *below* the bone layer suggested likewise that a short duration must also be assigned to the antediluvian period. By this Buckland meant that the cave could not have been above sea level for any very long span of time before being occupied by the hyenas, or the stalagmite would have filled it up completely. It had probably been underwater "during the formation of all the more recent strata", that is, those more recent than the Secondary limestone in which the cave had later been formed. So the natural chronometer of stalagmite accumulation showed a geologically brief lapse of time, not only since the putative deluge but also before that event, back to an assumed emergence of the present land as an abode of primal humankind: "the perfect Harmony of all the Circumstances of this Cave and their Confirmations of each other & of the 2 important facts of the Mosaic Deluge & truth of the Antediluvian Mosaic Chronology render these by far the most interesting geological Phenomena I have ever met with."[101]

Buckland wrote to Greenough the following day in much the same way, summarizing "the Kirkdale Story" of a hyena den: "all this was in the Antediluvian Period during the Reign of Messrs. Elephant & Co.", he explained, but "the Cave shews also traces of Diluvian action & of Postdiluvian & is the finest Chronometer I have ever seen". This letter was probably intended to be leaked to Geological Society members, as advance publicity for what Buckland promised would be a "long paper" to

99. Fig. 10.18 is reproduced from Buckland, *Reliquiae diluvianae* (1823) [Cambridge-UL: MF.40.10], pl. 2, figs. 1, 2; first published in "Fossil teeth and bones" (1822), pl. 16. The cave entrance is still recognizably similar, though the quarry, long disused, is now densely wooded.

100. The homely analogy was also a vivid one, since Buckland was clearly thinking of *clotted* cream—still today a delicious specialty of his native region—not the bland modern supermarket variety.

101. Buckland to Cuvier, 12 January 1822 (Paris-MHN, MS 634(2), Kirkdale dossier, nos. 87–90. Buckland seems not to have known that both de Luc and Dolomieu had argued long before that the small thickness of stalagmite on the floors of Alpine caves could be used as a natural chronometer (§6.2, §6.5).

Fig. 10.19. The spotted hyena from the Cape of Good Hope in southern Africa: a portrait from life published by Goldfuss of Erlangen in an illustrated book on living mammals (1809). Another animal of this species in a London menagerie provided dung that matched objects found among the fossil bones at Kirkdale. Buckland also asked Cuvier to report on the bone-gnawing habits of yet another individual in the menagerie attached to the Muséum in Paris, and he himself later made a similar study of one in a traveling show that was passing through Oxford. These observations strongly reinforced his interpretation of Kirkdale cave as a den of antediluvian hyenas of a similar but larger species. (By permission of the British Library)

the Royal Society. Soon his interpretation of the cave had further and unexpected support. In the silty deposit were found not only gnawed bones but also objects that had been neglected by the bone hunters, but which Buckland—a man noted for his robustly pre-Victorian sensibilities—identified as the dung of the hyenas. The distinguished London chemist William Wollaston analyzed them for him and reported that their composition was indeed what would be expected from a diet high in bone material, though he added later, rather archly, that "it may be as well for you and me not to have the reputation of too frequently and too minutely examining faecal products". The objects were also taken to Exeter Change, a commercial menagerie near the Royal Society's premises in London, where the keeper confirmed that they were just like what his spotted Cape hyena produced on a similar diet (Fig. 10.19).[102]

Buckland at the Royal Society

As Buckland promptly reported to Cuvier, his paper had an "enormous" audience when it was read at the Royal Society. His "Account of an assemblage of fossil teeth and bones of elephant, rhinoceros, hippopotamus, bear, tiger, and hyaena, and sixteen other animals, discovered in a cave at Kirkdale" occupied three successive weekly meetings. Banks's long and autocratic reign had ended with his death in 1820, and it was his successor Davy who presided. Davy told Buckland afterwards that "I do not recollect a paper read at the Royal Society which has created so much interest as yours". But Cuvier's opinion was far more important. As soon as the paper had been read, Buckland sent a copy to Paris, asking Cuvier to keep it confidential until it was published in the *Philosophical Transactions*; but told him he was welcome to use it—and its superb lithographed illustrations—for his new edition of *Fossil Bones*. Unfortunately the London animal had died, so Buckland also asked Cuvier to try a dietary experiment on the one in the Muséum's menagerie, to "see whether He gets through either the Astragalus or Os Calcis [of cattle]—my Yorkshire friends never did—though they tried every Corner". In other words, if the Kirkdale bones were indeed the remains of the hyenas' meals, they should show the same differential preservation in proportion to the solidity of the various bones. He also asked Cuvier to send him specimens of the dung of the Muséum's animal, for it was the fossil dung that really clinched his case:

> The discovery of this Dung . . . has supplied the only link that was wanting [i.e., missing] to complete the Evidence of the Cave having been a Den, & being established on such authority as that of Dr Wollaston all the other facts of teeth & gnawed fragments of bone follow of course—so that my otherwise almost incredible Story obtains universal Credit & no small Wonder at the Proof Positive that Hyaenas Elephants Rhinoceros & Hippopotamus were the Antediluvian inhabitants of Yorkshire & generally of England.[103]

In his paper, Buckland again defined the widespread Superficial deposits as "Diluvium". In arguing that they were the products of a "recent and transient inundation" he claimed Cuvier's support, but in fact this was a point on which the Frenchman had been ambivalent (§10.3). Buckland was providing new evidence for one of the two alternatives that Cuvier had considered. The Kirkdale case greatly strengthened the plausibility of a *transient* event, perhaps a mega-tsunami of the kind that Dolomieu had suggested long before, and von Buch and Hall more recently. Conversely, Buckland was tacitly rejecting the kind of sudden but *permanent* interchange between continents and oceans that de Luc had championed for so long. If Buckland's interpretation of Kirkdale was correct, the physical character of the earth's "last great convulsion" would be greatly clarified, even though its physical cause remained uncertain.

102. Fig. 10.19 is reproduced from Goldfuss, *Naturbeschreibung der Säugethiere* (1809) [London-BL: 446.e.4], pl. 32, "painted from life [*ad viv.*]" by Johann Eberhard Ihle. Buckland to Greenough, 13 January 1822 (London-UCL, Greenough MSS); Wollaston to Buckland, 24 June 1822 (London-RS, Buckland MSS, Bu.11). Exeter Change was demolished soon afterwards and part of its site was used for the new National Gallery (1829) in Trafalgar Square.

103. Buckland to Pentland [i.e., to Cuvier], 2 February 1822 (Paris-MHN, MS 634(2), Kirkdale dossier, nos. 95–98); also 11, 23 February 1822; Buckland to Cuvier, 4 March 1822; Buckland's manuscript text and proof copies of the plates, as sent to Cuvier (same dossier, nos. 77–78, 93–94, 99, 103–72). Buckland, "Fossil teeth and bones" (1822) was read on 7, 14, 21 February 1822; Davy to Buckland, 18 March 1822 (London-RS, Buckland MSS, Bu.10).

The core of Buckland's paper was his careful description of the cave and analysis of its varied bones. He set out his evidence for its having been a hyena den for a long period, before the sudden deluge wiped out the fossil fauna and buried the bones on the cave floor. Whether antediluvian conditions at Kirkdale had been colder or warmer than the present climate of Yorkshire was a point on which Buckland sat on the fence. He could afford to do so, because it was peripheral to his main goal. Far more important to him was the proof that the bones had not been swept into England from the tropics—an idea that he claimed was "never till now disproved"—and that the whole fauna had lived on the spot. On this point, however, Buckland was flogging a dead horse. Cuvier and other Continental naturalists had long been convinced that the fossil evidence could not sustain the older idea—going back to Pallas (§5.3) and others—that carcasses and other debris had been swept vast distances around the globe. As for the physical cause of the deluge, Buckland conjectured that it might have been due to a passing comet or some other astronomical event; but he wisely shelved that problem, because he knew he was doing geohistory, not earth physics. This became clear in his verbal reconstruction of the whole ecosystem (to use the modern term) revealed by the Kirkdale fossils, involving herbivores and carnivores as well as the scavenging hyenas. He developed this more informally and vividly, among his friends at the Geological Society's anniversary dinner, only a few days before the formal reading of the paper was due to begin at the Royal Society:

> The hyaenas, gentlemen, preferred the flesh of elephants, rhinoceros, deer, cows, horses etc., but sometimes, unable to procure these, & half starved, they used to come out of the narrow entrance of their cave *in the evening* down to the water's edge of a lake which *must* once have been there, & so helped themselves to some of the innumerable water-rats in wh[ich] the lake abounded.[104]

Buckland's former student the prosaic young lawyer Charles Lyell (1797–1875), in this report of his mentor's table talk, was not only amused but also somewhat bemused, for he was uncertain how seriously to take it all. But Buckland, less inhibited and more imaginative than Lyell, certainly intended it as a plausible reconstruction. Even his details were derived from what he had observed in Yorkshire and what he had learned or already knew of the habits of living hyenas. The lake, for example, was no fanciful conjecture, but vividly imagined as the former state of the broad flat Vale of Pickering that could be seen stretching into the distance from near the cave entrance (see Fig. 10.16). As always, Buckland's method was thoroughly actualistic: the present was always the best key to the past.[105]

Buckland's verbal reconstruction of the time of the hyenas was one scene in a more extended chronological account. Just as he had described his reconstruction informally as a "story", so his formal paper culminated in a "detailed history": both expressions clearly identified his research as *geohistorical.* His narrative reconstruction began in the antediluvian period, before the arrival of the hyenas, when stalagmite accumulated on the cave floor, undisturbed by animal activity. The period of the hyenas' occupation followed, and was only brought to an end when the deluge "extirpated" the whole fauna and buried the bones in mud. The subsequent post-diluvian period had seen the resumption of slow stalagmitic accumulation, the small amount of which was congruent with Cuvier's estimate of only a few millennia, itself modeled on the natural chronometers pioneered by de Luc and Dolomieu. Here at the Royal Society, as earlier at the Geological, Buckland made no overt reference to the "Mosaic" or biblical Flood. It was enough to claim the historicity of the geological deluge as a real natural event; but of course its approximate dating implied that it could in fact be equated with the one recorded in Genesis. To Cuvier, whom he judged to be sympathetic on this point, Buckland had revealed more openly his own delight at finding traditional chronology confirmed by the geological evidence; but among the assembled savants and amateurs at the Royal Society that point could be left unstated.[106]

Buckland's paper on Kirkdale cave was published in the Royal Society's *Philosophical Transactions* only a few months later. Even before it was available, those with scientific interests could read a short summary in the *Annals of Philosophy.* This was promptly translated for the Genevan *Bibliothèque Universelle* and for a German review periodical edited in Weimar. Further extracts followed later in other periodicals, and a fuller translation in the Parisian *Journal de Physique.* In short, the importance of Buckland's work was recognized immediately throughout Europe. Towards the end of 1822 the Council of the Royal Society decided to give him their highest award, specifically for his Kirkdale paper, and he became the first geologist ever to receive the coveted Copley medal. Making the presentation, Davy summarized the scientific importance of what Buckland had done: "by these inquiries, a distinct epoch has, as it were, been established in the history of the revolutions of our globe: a point fixed, from which our researches may be pursued through the immensity of ages, and the records of animate nature, as it were, carried back to the time of the creation."[107]

Davy's citation was significant, because he clearly recognized the innovative character of what Buckland had done. It was a kind of science quite unlike his own mainly chemical researches, in that it was concerned not with discovering perennial "laws of nature" but with reconstructing nature's contingent and unrepeated *history.* Davy, like any savant, took for granted the "immensity" of geohistory as a

104. Buckland's table talk on 1 February 1822, as reported in Lyell to Mantell, 8 February 1822 (Wellington-ATL, Mantell MSS), printed in part in Thackray, *Fellows fight* (2003), 2. The core of Buckland's more formal case for interpreting Kirkdale as a hyena den is reprinted (from his paper) in Rudwick, *Scenes from deep time* (1992), 38–40. He later doubted that the hyenas could have eaten the rats and other small mammals, partly because modern ones refused to do so: see Buckland to Cole, 22 December 1823 (Cardiff-NMW, 172), printed in part in North, "Paviland cave" (1942), 117. The importance of the rest of Buckland's analysis is described in Boylan, "Taphonomy and palaeoecology" (1997).

105. Buckland, "Fossil teeth and bones" (1822), 203n, reconstructs the lake. He, like Cuvier, knew at first hand of an important analogous site in Italy, namely the broad Val d'Arno—with deposits famous for fossil bones—drained by the Arno through its narrow valley upstream from Florence; this showed that the Yorkshire case was not unique.

106. Buckland, "Fossil teeth and bones" (1822), 204–8.

107. Davy, speech on 30 November 1822, printed in *Six discourses* (1827), 51; Royal Society Council, 14 November 1822 (London-RS, Council minutes 10: 12–13). Davy was far from ignorant about geology, having lectured on the science as early as 1805: see Ospovat, "Geological lectures" (1978), and Siegfried and Dott, *Humphry Davy on geology* (1980).

whole, but within that vast panorama Buckland had vividly reconstructed one specific moment or "epoch". In contrast to Buffon's famous *époques*, however, this had not been deduced as the necessary consequence of an overarching physical model (§3.2); he had inferred it from a detailed study of specific historical "records" left by nature itself. Davy saw that this "fixed point" in the immediately antediluvian world could act both as an exemplar of how geohistory should be done, and as a point of reference from which it could be extended back into the deeper past towards the primal "creation". Other naturalists had already explored this kind of geohistorical approach: Davy's reference to "the revolutions of our globe", for example, echoes Cuvier's phrase. But there had been no comparable reconstruction of a whole vanished world: landscape, lake, cave, mammoths, hyenas, water rats, and all.

Bursting the limits of time

Soon after the reading of Buckland's paper, further evidence to support his diluvial interpretation began to accumulate. Until Kirkdale was discovered, he had not given any special attention to the fossil mammals found in the Superficial deposits: he did not have the expert anatomical knowledge to do more than add footnotes, as

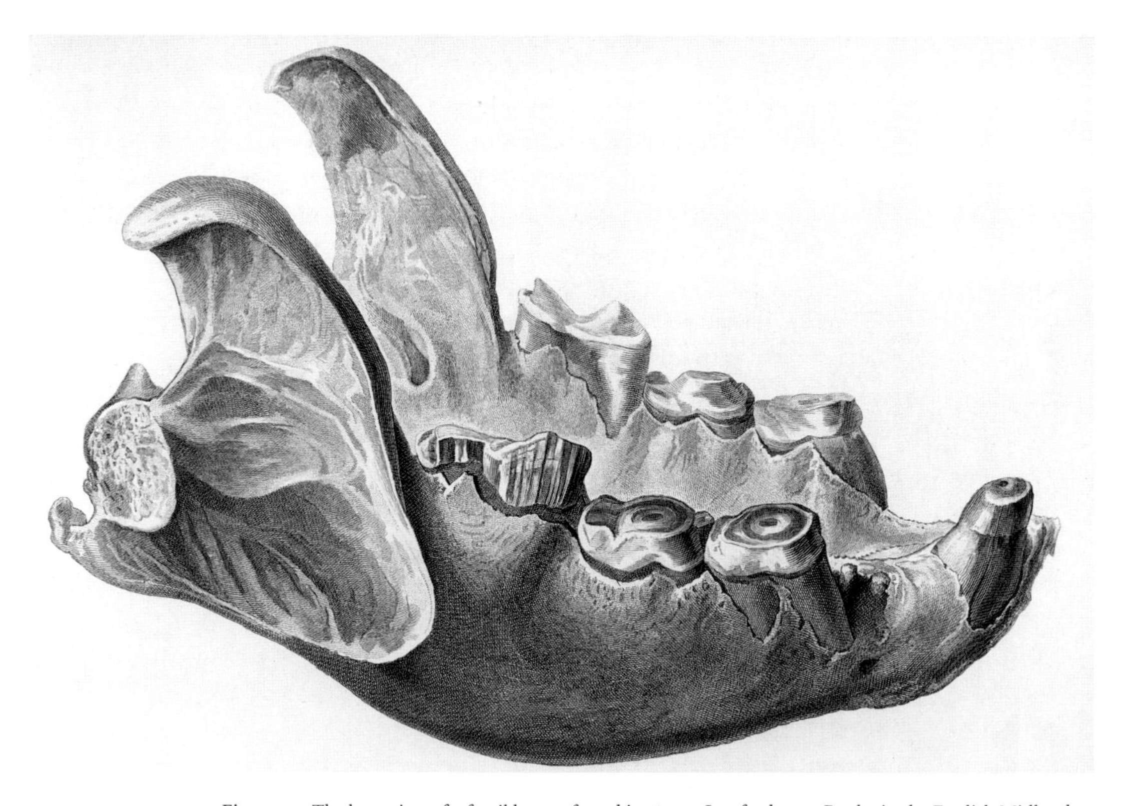

Fig. 10.20. The lower jaw of a fossil hyena, found in 1822 at Lawford, near Rugby in the English Midlands: an engraving based on a drawing by Webster, published in the enlarged version (1823) of Buckland's paper on Kirkdale cave. The jaw belonged to a large and elderly hyena, with heavily worn teeth. Buckland attributed its almost unbroken state—in contrast to the fragmentary and chewed bones found in the putative den at Kirkdale—to its demise by drowning in the diluvian waters that had made its species extinct. (By permission of the Syndics of Cambridge University Library)

it were, to Cuvier's great work, which was now being published in its revised edition. However, when he interpreted Kirkdale cave as a hyena den, Buckland became aware that there had been no previous records of fossil hyenas in England. Yet if his "hyaena story" was correct, they would surely have been part of the antediluvian ecosystem quite generally, not just at a single spot in Yorkshire. This anomaly was resolved on cue when a fine hyena jaw was found in diluvial gravel in another part of England; and its almost perfect preservation and unchewed state was just what would be expected if this individual, unlike some of those previously living at Kirkdale, had perished in the deluge. The new specimen was just in time to be mentioned in the published paper (Fig. 10.20).[108]

The second part of Buckland's paper had been a review of similar bone caves elsewhere in Britain and on the Continent. It was designed to show that Kirkdale was just one instance of features that were so widespread that the diluvial event must be considered universal. But Buckland's account of foreign localities was illustrated only with a drawing of Gailenreuth cave in Bavaria (Fig. 10.11), based on the sketch he had made during his brief visit in 1816. He must have recognized the need for a closer study of that and other caves, in the light of the diluvial theory he had developed since that time, for as soon as his lecturing duties were completed he set off on another tour on the Continent. In addition to studying collections at Bonn, Frankfurt, and Dresden, and discussing their fossil contents with various naturalists, he visited some of the famous bone caves in the Harz and revisited those in Bavaria. In the caves around Muggendorf (see Fig. 10.9) he confirmed what he must have been hoping to find but had overlooked on his earlier visit, namely that a crust of stalagmite had overlain and sealed the bone-bearing deposits, before they were excavated for their fossils. As at Kirkdale, the stalagmite provided a natural chronometer that proved that the caves had been dens—of bears, in this case, not hyenas—long in the past, so that the animals were clearly antediluvian (Fig. 10.21).[109]

After his return to Oxford Buckland had another stroke of luck, when a traveling show passed through Oxford, including in its menagerie another live hyena. With its keeper's collaboration, this gave him the opportunity to do at first hand what he had earlier asked of Cuvier: to observe its eating and defecating habits, studies that his staid Oxford colleagues may have thought barely compatible with the dignity of his academic position. He carefully noted which bones the animal preferred, how it broke them open to extract the nutritious marrow, and how the action of its powerful jaws and teeth literally left their mark on the fragments. He compared what the living hyena discarded with the various fragmentary bones from Kirkdale and found a perfect match. In the enlarged version of his paper, published the following

108. Fig. 10.20 is reproduced from Buckland, *Reliquiae diluvianae*, 2nd ed. (1824) [Cambridge-UL: MF.16.3], pl. 12, engraved by Joseph Basire and first published in the 1st ed. (1823); it was mentioned in "Fossil teeth and bones" (1822), 228n, a postscript dated 24 May 1822.

109. Fig. 10.21 is reproduced from Buckland, *Reliquiae diluvianae* (1823) [Cambridge-UL: MF.40.10], pl. 17, lithographed by Scharf and explained on 271–72. Buckland admitted that his drawing *restored* the stalagmite (C) in the inner chamber to what he believed it had been when the diggings were first started. On this Continental tour, see Boylan, *William Buckland* (1984), 121, 382.

Drawn by T. Webster from a Sketch by W. Buckland. *G. Scharf Lithog, Printed by C. Hullmandel*

Fig. 10.21. Buckland's section of Gailenreuth cave, near Muggendorf in Bavaria: a drawing by Webster based on a sketch made by Buckland during his visit in 1822 (published 1823), showing a layer of stalagmite (C) *overlying* the bone deposits (D, G, I) in each chamber. This silently amended the version he had published in his Royal Society paper the previous year, based on his visit in 1816, which had shown no stalagmite and the bones lying on the surface (Fig. 10.11). As at Kirkdale (Fig. 10.18), the stalagmite implied that the cave had been a bears' den *before* the putative deluge, and thereafter uninhabited and undisturbed until the bone mining began. (By permission of the Syndics of Cambridge University Library)

year, his claim that the cave had been a hyena den was epitomized, with consummate visual rhetoric, in a lithographed plate that depicted past and present side by side (Fig. 10.22).[110]

Buckland's research was characterized memorably by his Oxford colleague Philip Nicholas Shuttleworth, the warden of New College, when he parodied Alexander Pope's famous couplet about Newton: in the new version, "All was darkness once about the Flood/Till Buckland rose, and all was clear—as mud." But it was summarized most perceptively by Conybeare, who, of all the leading English geologists at this time, was arguably the most imaginative as well as the most intelligent. In the wake of Buckland's paper at the Royal Society, Conybeare celebrated the occasion in the jokey manner that was customary in his (and Shuttleworth's) cultural milieu, by writing a poem on Kirkdale cave. In doggerel verses he parodied Buckland's analysis of the cave and its bones, its stalagmite and diluvial mud, and his interpretation of them all as the relics of an antediluvian hyena den. Above all,

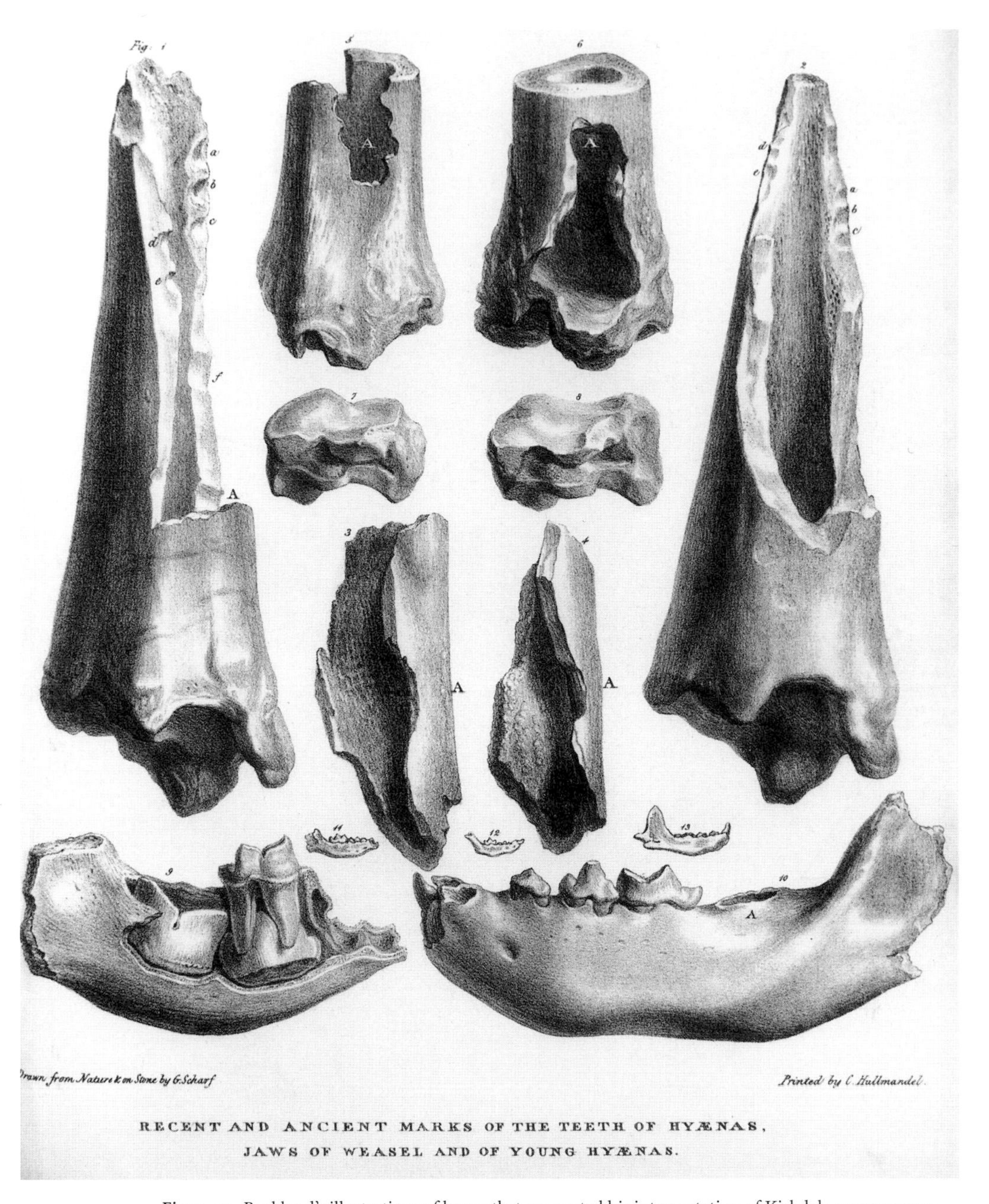

RECENT AND ANCIENT MARKS OF THE TEETH OF HYÆNAS,
JAWS OF WEASEL AND OF YOUNG HYÆNAS.
Printed by C. Hullmandel.

Fig. 10.22. Buckland's illustrations of bones that supported his interpretation of Kirkdale cave as an ancient hyena den: modern specimens (left) were matched symmetrically with fossil ones (right). Toothmarks (*a–e*) on a broken ox tibia chewed by a living hyena (fig. 1) matched those on one from Kirkdale (2); a similar match could be made with splinters left after the extraction of marrow (3, 4), and with holes (A) made by the bite of the hyena (5, 6); but a solid and unnutritious scaphoid (7) "lay all night untouched" by the living animal, and the same bone was found unchewed at Kirkdale (8). A parallel was also found in the jaws of *young* hyenas (9, 10), which was taken as further evidence that the antediluvian ones were living on the spot. The fossil weasel jaws (11–13) were included simply to make full use of the space available on this plate, but they are representative of the smaller animals in the hyenas' putative diet at Kirkdale. (By permission of the Syndics of Cambridge University Library)

110. Fig. 10.22 is reproduced from Buckland, *Reliquiae diluvianae*, 2nd ed. (1824) [Cambridge-UL: MF.16.3], pl. 23, first published in 1st ed. (1823) and explained on 276–78; the plate was drawn by Scharf (except fig. 9, by Clift). On the menagerie at Oxford in December 1822, see Boylan, *William Buckland* (1984), 372.

Mystic Cavern, the gloom of thy rock,
Shedding light on each point that was dark,
Tells the hour by Shrewsbury clock
When old Noah went into the Ark.
By the crust on thy Stalactite floor,
The Post-Adamite ages I've reckoned,
Summed their years, days & hours & more,
And I find it comes right to a second.
Mystic Cavern, thy chasms sublime,
All the chasms of History supply;
What was done ere the birth-day of time,
Thro one other such hole I could spy.

Fig. 10.23. The final stanzas of Conybeare's poem "The Hyaena's Den at Kirkdale near Kirby Moorside in Yorkshire, discovered A.D. 1821", which he distributed in 1822 as a lithographed broadsheet. The "mystic cavern" had yielded a natural chronometer, dating the geological deluge with an accuracy that matched the proverbial "Shrewsbury clock"; it threw light on the obscurity of geohistory, providing a "spy-hole" that could act as an exemplar for other such reconstructions of the still deeper past.

Fig. 10.24. The lithographed caricature that accompanied Conybeare's poem about Kirkdale cave. Buckland, crawling through the narrow entrance, is startled to see by the light of his candle that the antediluvian hyenas are very much alive and well, feasting on the bones of mammals large and small. By bringing the light of science to illuminate the prehistoric scene, Buckland was fulfilling Cuvier's aspiration "to burst the limits of time", making the deep past knowable in the present, even in the absence of human records: in effect, traveling in deep time.

however, Conybeare portrayed his friend's research as *geohistory*. He had his poem written out and lithographed with a matching visual image, making a broadsheet that was distributed widely among British geologists and also sent to favored foreigners such as Cuvier. Words and image alike depicted Buckland as having penetrated from the present world into the antediluvian past, or at least as having found a "spy-hole" through which to see it as it really was. Conybeare's doggerel and caricature were probably designed in conscious emulation of the well-known purple passage in Cuvier's "preliminary discourse" (§9.3), which had been newly reissued the previous year. Conybeare saw Buckland as having fulfilled Cuvier's aspiration "to burst the limits of time", penetrating from human history back into the hitherto dimly known recesses of antediluvian geohistory (Figs. 10.23, 10.24).[111]

Conclusion

Kirkdale cave received the widest publicity in the international savant world, and Buckland's "hyaena story" about it won him the highest scientific honor and recognition that was possible for an Englishman. The point deserves emphasis, because some later geologists and historians have condemned Buckland as a reactionary obscurantist or dismissed him as a buffoon. He was neither. His fanciful jokes and eccentricities, although sometimes puzzling to his more staid colleagues, usually had a serious scientific purpose. At the time of his Kirkdale research he did indeed retain some of the precritical biblical literalism that was still common in his cultural milieu at Oxford. But he was no fundamentalist *avant la lettre*. In his first letter to Cuvier after hearing about the discovery at Kirkdale, he did indeed express the hope that it would demonstrate the congruence between geological evidence and a quite traditional understanding of the early chapters of Genesis. Not only did he believe that the "postdiluvian" period recorded in the cave matched the date for Noah's Flood calculated by traditional chronology; he also evidently thought that his "antediluvian" period would likewise match the traditional dating that carried history back to the story of Creation. But the latter was interpreted as referring to the start of *human* history, allowing prehuman geohistory to extend back through the vastly longer periods recorded by the huge pile of geological formations. Buckland's geology was certainly not constrained in any way by a short timescale for geohistory as a whole; and even for the brief final portion that ran in parallel with human history his opinions coincided with those of Cuvier and many

111. Fig. 10.23 is reproduced from Conybeare's anonymous and undated broadsheet, entitled as noted in the caption; the full fifteen stanzas are printed in Rudwick, *Scenes from deep time* (1992), 40–43. It was first printed, and reliably attributed to Conybeare, in Daubeny, *Fugitive poems* (1869), 92–94, an anthology that illustrates the genre. Fig. 10.24 is reproduced from the same broadsheet, and was probably drawn by a professional artist rather than by Conybeare himself. It may have been inspired by a similarly jocular drawing of the cave bears at Gailenreuth (Fig. 10.12) and been based on a more serious reconstruction of the hyenas in their den—a drawing now apparently lost—that Buckland used as a visual aid in his lectures. A copy of the broadsheet was certainly sent to Cuvier (Paris-MHN, MS 634(1), "Caverne de Kirkdale" dossier) and probably to other foreign savants, as well as to those in Britain. The design was perhaps also inspired by the time-traveling experience of entering the full-size replica of the tomb of the pharaoh Sethos I ("Seti"), famously exhibited at the Egyptian Hall in London in 1821–22 by the traveler and showman Giovanni Belzoni: see Siliotti, *Belzoni's travels* (2001), 67–69 (I am indebted to Ralph O'Connor for this important suggestion).

other savants. His identification of the "geological deluge" as the Flood recorded in Genesis was quite widely criticized; but his work greatly strengthened the case for the physical event itself, the reality of which would now be much harder to refute. Noah's Flood might be sidelined in savant discourse, but the "geological deluge" could not so easily be dismissed.

Above all, Buckland's detailed analysis of Kirkdale cave, and the panoply of other evidence with which he surrounded it, was recognized immediately as an outstanding exemplar of how *geohistory* could and should be done. Conybeare's jokey drawing put into pictorial form the perception that Buckland had fulfilled Cuvier's aspiration "to burst the limits of time", even more fully than the great French naturalist himself. Buckland had reconstructed not just the bodies of extinct animals but a complete scene of "antediluvian" life in its physical environment, not as a fanciful conjecture but by rigorously disciplined inference from detailed evidence, keyed into a knowledge of comparable situations in the present world. In the years that followed, Buckland's work on bone caves received its share of criticism—like any comparable piece of modern scientific research—and his conclusions underwent substantial modification. But its value as an exemplar, and as an inspiration to others, was profound. Just as Brongniart's work on Tertiary stratigraphy demonstrated how that portion of geohistory could link the present to the still deeper and stranger world of the Secondaries (§9.6), so Buckland's cave bones showed how the very last phase of that inconceivably lengthy "former world" could be reconstructed on the threshold of human history. It is therefore appropriate to bring this long narrative to a close at this point, near the start of an astonishingly fruitful decade, because by that time the science of geology had firmly embedded within itself the practice of *geohistory*.

CODA

Retrospect and prospect

RECALLING SAUSSURE

Buckland interpreted Kirkdale and other bone caves—showing with characteristic panache how Cuvier's ambition "to burst the limits of time" could be realized—less than four decades after Saussure became the first savant to stand on the summit of Mont Blanc. The span of time covered in this book fell easily within a single scientific career. Cuvier, for example, would have heard of the Alpine feat while he was a student in Stuttgart; and he was at the height of his powers (in all senses) in Paris at the time that Buckland's antediluvian interpretation of bone caves was taking shape. Saussure's moment of glory has been taken in this book as a historian's golden spike, a symbolic baseline from which the gradual spread of a newly geohistorical outlook could be traced in the practice of the sciences of the earth. This coda summarizes the main features of the synchronic survey of these sciences around the time of Saussure's Alpine feat, as described in Part One, and those of the diachronic narrative in Part Two, which has brought the story as far as Buckland's work. Finally, it just touches on how the debates among those who by then called themselves "geologists" developed and matured during the years that followed Buckland's moment of glory (I hope to trace these later debates in a sequel to the present volume).

The synchronic survey began with a review of the social and cognitive circumstances in which the sciences of the earth were practiced (Chapter 1). At the time that Saussure climbed to the highest point of the highest mountains in Europe (§1.1), he and his like-minded contemporaries called themselves not scientists but "savants" (§1.2). They were respected for their learning and expert knowledge in any of a wide range of sciences (in the broad sense still retained in most languages, e.g., *Wissenschaften*), not only in the natural sciences (the singular "Science" of the modern anglophone world). Savants regarded themselves as members of an informal "Republic of Letters", normally using French as their international language. They were an intellectual elite, but their work was supported in practice by many other people, some but not all of them of lower social status. Most savants were in the modern sense "professionals", in that they earned their living from their savant work, but those who were "amateurs" were esteemed no less and their work was certainly not amateurish. In the natural sciences, new knowledge was generated not only in laboratories but also—and more importantly for the sciences of the earth—in museums and in the field (§1.3). But it was consolidated in social interaction: savants met together both formally and informally, visited each other when they could and wrote letters when they could not, and published their work in books or as articles in a proliferating range of periodicals. They located their work on a mental map of knowledge that broadly distinguished "literary" or textual studies from those that were "philosophical" or concerned with the natural world, including human nature (§1.4). The latter were divided in turn into "natural history", which described and classified natural features and phenomena of all kinds, and "physics" (or "natural philosophy"), which investigated their lawlike regularities and their causes.

The sciences of the earth, like other kinds of natural knowledge, were spread across this mental map in ways that are unfamiliar to modern eyes (Chapter 2). There were four fairly distinct sciences concerned with the earth, three of which fell within natural history because they were concerned primarily with description and classification. "Mineralogy" (§2.1), in a broad (and now obsolete) sense that included the study of fossils, was above all a science of specimens. These were identified and classified indoors, in museums or private "cabinets"; by Saussure's time, major categories such as minerals (in the modern sense), rocks, and fossils (again in the modern sense, objects of organic origin) had been defined and clarified. Specimens of all kinds were bought, sold, and exchanged among collectors; but they were also disseminated in the form of drawings or paintings, often published, which acted as paper "proxies" that made rare or unique specimens mobile across the Republic of Letters. Specimens of minerals, rocks, and fossils, like those of animals and plants, usually carried with them little information from the field, except the localities at which they had been collected, and certainly no clear sense that they might be of quite different relative ages. "Physical geography" (§2.2), in which Saussure was most prominent, was likewise a descriptive science, but it was focused on outdoor work in the field, where large-scale features such as mountains and volcanoes, and their spatial distribution, could be analyzed. These features were transposed into indoor arenas of debate in the form of proxy landscape pictures and as maps. "Geognosy" (§2.3) was also a field science, but one rooted in the underground world of mining. It therefore added the dimension of depth: geognosts tried to discern the three-dimensional structure of solid rock masses [*Gebirge*]. These too were classified, for example by Werner; but even the major categories of "Primary" and "Secondary" rock masses were defined in practice by their structural relations (lying below or above) rather than by the relative ages (older or younger) that might be inferred from those positions. Like mineralogy and physical geography, geognosy was a science of description and classification, not of causes, and certainly not a science of the earth's history.

The fourth and last of the sciences of the earth was earth physics [*physique de la terre*] (§2.4), which belonged in "physics" because it was concerned with the causal origins of what the three natural-history sciences described and classified. It tried to tackle such disparate puzzles as the stoniness of rocks, the emplacement of marine fossils far from the sea, the formation of the rock masses in which they were found, and the origins of valleys and of folded strata. The causal analysis of such features gave earth physics a dimension of time that was absent from or only implicit in the other three sciences of the earth. But although the inferred causes were necessarily temporal, they were as ahistorical as those of (say) chemical reactions, for they were taken to be based on laws of nature that were the same in past, present, and future. By Saussure's time, most savants recognized the overwhelming evidence for a timescale of humanly inconceivable magnitude (§2.5); they were reluctant to put quantitative figures on it, but only because any such estimate was highly speculative. The practice of earth physics was not constrained by the short timescale provided by the traditional science of "chronology", which had calculated the total history of the earth (and indeed of the cosmos) as only a few millennia in duration. Contrary to later historical mythmaking, the sciences of the earth were not engaged in any significant conflict with religious beliefs, and many of the relevant savants were themselves religious people.

What was far more important about the sciences of the earth was that none of them was primarily geohistorical: they were not concerned to reconstruct the particularities of the stages by which the earth had come to be the kind of place that now forms the stage for human lives. However, in Saussure's time there was a distinctive genre of writing about the earth that did, on the face of it, make just such claims (Chapter 3). The genre, for which de Luc coined the neologism "geology", was that of the "theory of the earth" (§3.1). Specific instances of the genre, which were often referred to as rival "systems", have here been termed "geotheories". Using the causal analyses of earth physics as their foundation, they were conjectural attempts to explain how the earth works, accounting causally for its major features and phenomena and relating them to the life of humankind and often also to the purposes of God. But geotheories were rarely geohistorical, because they gave accounts of the earth's temporal development deduced mainly from first principles rather than being induced from the particulars of surviving relics of the past: they described how the earth must (or at least ought to) have developed in the past and how it would necessarily continue to change in the future, given certain initial conditions and unchanging physical principles.

For example, early in his career Buffon (§3.2) treated the earth as a steady-state system that never had been and never would be significantly different from its present condition; its origin was awkwardly tacked on as a cosmic "accident" at the start. But later he developed a strongly directional geotheory: the earth was a body that was cooling gradually from an intensely hot origin (the same cosmic accident) through its present state towards an inevitable end in intense cold. This process was marked by a sequence of temporal milestones or "epochs"; life, for example, had originated at one point in the past and would end at another in the future, bracketed by the constraints of heat and cold. This highlighted the deterministic character of the theory; the temporal development of all other planets and their satellites would be marked by the same sequence of events. De Luc's geotheory (§3.3) was also directional, but was focused on a recent "revolution" of major geographical change, separating an almost undifferentiated prehuman "ancient" or "former world" [*ancien monde*] from the "present world" of human life. He took the former to have been unimaginably lengthy in duration, but he argued that the latter could be dated by "natural chronometers" to no more than a few millennia, which matched the timescale of traditional chronology for humankind (but not for the earth). Hutton's geotheory (§3.4) was radically different, not in its indefinitely vast timescale—which was a commonplace among geotheorists—but in being strictly steady-state and tacitly eternalistic. Like other geotheories, Hutton's was openly hypothetico-deductive; he propounded it *before* undertaking fieldwork to find evidence to confirm it. Finally, the kind of geotheory that is here termed the "standard model" (§3.5) was too general to be identified with any specific savant, though it has often been attributed, misleadingly, to Werner. It was as strongly directional as Buffon's, but its directionality lay in the gradual subsidence of an initially global proto-ocean of complex chemical composition. From this a determinate sequence of precipitates (and their secondary erosion) had formed the series of rock masses described by geognosy, while the continents had gradually emerged as dry land and the oceans were eventually confined within their present limits. Once again, given certain initial conditions, the developmental pathway was determined and inevitable; the standard model, like other geotheories, was not truly geohistorical.

However, in Saussure's time a few savants were beginning to treat the earth as the product of nature's own history (Chapter 4). They were trying to construct narratives of events or states that could not be predicted (or rather, retrodicted) from any assumptions about initial conditions, but only pieced together from detailed analysis of specific relics from the deep past. There were two related sources for this new sense of "geohistory" (§4.1). Ironically, the first was the radically historical perspective of the Judeo-Christian tradition. Rather than being the enemy of progress in the sciences of the earth, as later mythmaking has portrayed it, this orientation fostered the extension of historicity to the previously uncharted vastnesses of prehuman time. The second conceptual source was the secular analogue of biblical religion, namely the work of "erudite" historians and "antiquarians" in the practice of human history, which was expanding at just this time from its traditional focus on written texts to embrace a much wider range of evidence. Most significant was the excavation of Herculaneum and Pompeii, which provided new and far richer materials for the reconstruction of the life of the Roman world in all its particularity.

In Saussure's time a few savants began to transpose methods, metaphors, and analogies from this kind of human history and apply them consciously and deliberately to each of the three descriptive sciences of the earth. The traditional metaphor of fossils as "nature's coins", for example, took on new meaning in conjunction with the growing awareness of the vastness of apparently prehuman geohistory (§4.2). Burtin, who exemplified this transposition of human history into the science of fossil specimens, used it to infer specific revolutions in the deep past, placing even the most recent of these natural events long before recorded human history. Desmarest was the outstanding exponent of the same historical approach in its application to the science of physical geography (§4.3). He reconstructed the immensely long and complex geohistory of Auvergne using the traces of its occasional volcanic eruptions and other features as "nature's documents" or "nature's monuments"; and Lamanon's reconstruction of a great lake on the site of Paris showed that such geohistorical inferences need not be confined to volcanic regions. Soulavie was an outstanding advocate of a similar historicizing of the solid rock masses described by geognosy (§4.4). He reconstructed the long geohistory of Vivarais, using not only its extinct volcanoes but also the fossils in its Secondary formations: these were "nature's archives", and he presented himself as nature's archivist.

Finally, although the causal science of earth physics was intrinsically ahistorical, in practice some of the geotheories that were built on its foundations did incorporate some degree of historicity (§4.5). A steady-state geotheory such as Hutton's allowed no role at all for geohistory, and Buffon's directional system gave it only a marginal place. But the standard model of geotheory often incorporated details that were derived from empirical observation rather than being deduced from first principles. And at the end of the spectrum furthest from Hutton's geotheory, de Luc's was the most geohistorical of all. He made only a perfunctory attempt to account causally for the great revolution at the heart of his system, because he was more concerned to establish its historicity and to determine its approximate date. It is no coincidence that de Luc's system was the most strongly geohistorical, because of all these savants he was the one most explicitly committed to the historical perspective of biblical religion, a perspective that he aspired to extend to the whole of geohistory. In effect, his belief in God's sovereignty translated into a sense of the sheer contingency of geohistory—the sense that at any point events might have taken a different path—which doomed any attempt to deduce geohistory from first principles. So, ironically, it was de Luc who radically undermined the genre of geotheory—which he himself had called "geology"—by beginning to transform it into geohistory.

In the event, a geohistorical orientation developed most effectively through the application of one particular class of evidence, namely fossils (Chapter 5). Fossils were potentially the key, because they were both like and unlike living animals and plants: like them certainly in general terms, but mostly unlike them in detail (§5.1). So they might enable savants to determine whether the deep past had been much like the present, or truly a foreign country where nature did things differently. For there were three alternative explanations for the disparity between living and fossil organisms. The fossil species might still be alive and well, flourishing as "living fossils" in remote and unexplored parts of the world or in the ocean depths, having merely migrated from the regions where their fossil counterparts were found. Or they might have become truly extinct, by some natural process (not just by human agency). Or they might have changed by being "transmuted" (in modern terms, evolved) into the species known alive, without any extinction at all. All three explanations had their supporters, and none was self-evidently superior to the others (and none, it should be noted, was by itself "correct" by the standards of modern knowledge).

By far the commonest fossils, and the most diverse, were those of mollusks and other lowly animals that were inferred to have been marine in origin. Naturalists agreed that many of them, perhaps all of them, differed from their living counterparts at least at the level of species; but there was no agreement on how this should be explained (§5.2). Since the faunas of distant seas and ocean depths were so poorly known, and since "living fossils" were discovered quite frequently, explanations in terms of migration often seemed the most plausible; certainly the case for extinction was far from overwhelming, and that for transformism even more dubious. Fossil plants were clear evidence of the existence of land areas in the distant past, but were otherwise equally inconclusive; large fossil bones were far rarer, and were some of the most highly prized of all fossils (§5.3). Yet they too were inconclusive, because they did not clearly belong to species unknown alive; or if they did, those species might still be lurking as living fossils in the poorly explored interiors of the continents. Finally, evidence about humans, the species of greatest intrinsic interest, was also inconclusive (§5.4). It was quite uncertain whether the human species was a late arrival, as Buffon claimed by making that event his most recent major epoch, and de Luc by dating his major revolution to the dawn of human history; or whether humans had been an essential part of the terrestrial economy from all eternity, as the logic of Hutton's system demanded and as he implied by treating fossil plants and animals as surrogates for the missing human fossils. As far as it went, however, the evidence seemed to favor the first option: fossil human bones and human artifacts were reported occasionally, even from Secondary formations, but in all such cases there were good reasons to doubt their authenticity. Most of geohistory therefore seemed likely to have been truly prehuman.

To summarize: the first part of this book depicts the scientific world of the late eighteenth century, around the time of Saussure's ascent of Mont Blanc, as one of intense activity. The sciences of the earth were certainly not in a state of "preparadigmatic" confusion: research was structured within well-established sciences, each with its own methods, genres, exemplars, and goals. The diversity of terrestrial nature was being described and classified quite effectively, and its processes investigated in temporal and causal terms. A proliferation of conjectural geotheories even offered ambitious accounts of how the earth must have changed in the past and would necessarily change in the future, based on reasoning from the known laws of nature. But attempts to reconstruct the detailed particularities of the earth's history from the concrete relics or traces of the deep past were few and far between: geohistory was little practiced and not highly regarded.

THE CENTRALITY OF GEOHISTORY

The second part of this book has traced in narrative terms how this marginal genre of geohistory came to be an integral part of the mainstream, thereby helping to create a new science of "geology", the first science to claim not only to elucidate nature's perennial laws or regularities but also to reconstruct nature's own unique and unrepeated history. The narrative began when, only two years after the golden spike of Saussure's Alpine feat, political revolution erupted in France (Chapter 6). This did not initially affect the practice of the sciences, but it did highlight one of the meanings of the word "revolution", as applied to nature as much as to society (§6.1): in both spheres, revolutions were major changes, but they might be either smoothly regular or dramatically sudden. In nature, Blumenbach posited a sudden "total revolution" in recent geohistory to account for the apparently complete contrast between fossil and living species. On the other hand Montlosier, following in Desmarest's footsteps, saw in the valleys of Auvergne the traces of continuous erosion, a revolution as gradual and regular as the orbiting of the planets around the sun. Meanwhile the genre of geotheory continued to flourish (§6.2). De Luc revised his system, reluctantly excising its metaphysical and theological components for the sake of publishing it in a leading scientific periodical, and modifying his own word "geology" to denote, in effect, the causal science of earth physics. He now differentiated his ancient or former world into a sequence of distinctive periods, largely as a result of giving more attention to the fossils in the Secondary formations: this made his system even more geohistorical than before. And he trenchantly criticized Hutton for using time itself as a causal explanation and failing to take adequate account of what he termed "actual causes", the processes observably at work in the present world.

As the Revolution in France lurched into its most violent and radical phase of regicide and Terror, the practice of the sciences in Paris—the very center of the scientific world—came almost to a halt, and its main institutional base was abolished (§6.3). Dolomieu was among the many savants who went into actual or internal exile, though he was still able to develop a physical theory of extremely large-scale tidal waves (here termed "mega-tsunamis") and argue geohistorically that the last such revolutionary event had indeed been as recent as de Luc supposed, judging by the natural chronometer of the subsequent growth of the Nile delta. Savants outside France were also affected by the Revolution, and not only by the direct effects of the wars that had broken out across Europe. In an increasingly isolated Britain, for example, Hutton's eternalistic geotheory came under suspicion of impiety and therefore of sedition (§6.4). In the same counterrevolutionary atmosphere, however, de Luc's geotheory gained favor because its explicit claim about the historicity of Noah's Flood was taken to reinforce biblical authority in general. But "Science" was not always or everywhere the underdog to "Religion": Blumenbach found it prudent, amidst religion's "cultured despisers" in Germany, not to disclose his agreement with de Luc's concordance between the two.

After the end of the Terror in the coup d'état of Thermidor, the sciences in France revived spectacularly, with lively new institutions to match (§6.5). Desmarest published a vast survey of geotheories; but their sheer multiplicity led him to conclude that the genre was bankrupt. Saussure, who had hoped to add to their number, instead produced at the end of his life an "agenda" for the sciences of the earth. In the event, this was far more influential than any geotheory would have been, because it suggested a series of more limited and more solvable problems to tackle, some of them clearly geohistorical in orientation. So by the end of the century the meaning of "geology" was shifting from geotheory to earth physics, and thence towards being a useful label for a cluster of potentially soluble puzzles about the earth, combining the causal with the geohistorical.

Two of the most significant puzzles in geology (as it may now be called without anachronism) were the physical character of the earth's most recent major revolution—if in fact there had been one—and the interpretation of fossil bones as potentially the clearest evidence about the history of organisms. Where these puzzles intersected lay the most decisive focal problem of all for the elucidation of geohistory, namely the effect of the last revolution on terrestrial vertebrates, as it might be assessed through a study of the most recent set of fossil bones (Chapter 7). At least two young men had the anatomical talent to replace the older savants who had tackled this problem in earlier years; but only one of them, Cuvier, also had the opportunity to work with the world's finest database of comparative material, in the Muséum in Paris (§7.1). He also benefited from the scientific spoils of the Revolutionary wars, in the form of decisive new specimens (or at least proxies for them). They enabled him to claim, with authority beyond his years, that the Siberian mammoth was distinct from either of the living species of elephants, and that a fossil skeleton of similar size found in Spanish America was an unknown giant edentate, which he named the megatherium. Cuvier's tally of distinctive fossil mammals, many of them of spectacular size, was rapidly enlarged (§7.2), not only by his own efforts, but also by assimilating earlier cases such as the well-known "Ohio animal" (which he later named the mastodon) and by reinterpreting the work of other less expert naturalists such as Jefferson. He inferred that all these fossils represented the distinctively different fauna of the former world, and he attributed its total extinction to the kind of drastic physical revolution proposed by both de Luc and Dolomieu, and indeed by Blumenbach and others. This was just the start of Cuvier's ambitious research project to identify and describe all known fossil "quadrupeds". Above all, he defined himself as "nature's antiquarian", reconstructing the animals of a vanished former world analogous to the vanished Roman life of Pompeii: fossils were to be the key to geohistory.

Cuvier's meteoric rise to scientific prominence, which continued smoothly after Napoleon seized power, enabled him to issue an international appeal for further specimens of fossil bones, or at least paper proxies for them (§7.3). This gave him an invaluable network of scientific informants all over Europe and even beyond it—despite the wars—and brought him a vast "paper museum" that greatly increased the authority of his conclusion that the mammalian fauna of the former world was truly extinct. However, that conclusion was threatened by his older colleague Lamarck, whose steady-state geotheory and concept of universal transformism suggested an alternative explanation (§7.4). Cuvier acknowledged Lamarck's expertise in the description of the rich fauna of fossil shells found around Paris, but not his interpretation of the contrast between fossil and living species in general. The two savants agreed, for example, that the mummified fossil animals brought back by Napoleon's Egyptian expedition were identical to living species; but Cuvier took this as decisive evidence that species were truly stable natural entities, whereas Lamarck claimed there had been insufficient time since ancient Egypt for their gradual transmutation to be detectible. This was a dispute of great importance not only for what Lamarck was newly calling "biology" but also for the new science of geology. In the event, Cuvier's claim that species were stable, and therefore that the fossil mammals were truly extinct, gained ground by the sheer force of cumulative evidence, in the form of an ever expanding range of fossil mammals that he demonstrated were distinct from any living species (§7.5). He also claimed that some of them, from the Secondary formations around Paris itself, were even more distinct from any living species than the obviously more recent cases of the mammoth, mastodon, and megatherium, and therefore represented a still earlier phase in the history of life.

This slender hint of a more differentiated conception of the ancient or former world, already foreshadowed by de Luc, was firmly consolidated when the structural pile of rock formations described by the field science of geognosy began to be combined systematically with the products of the museum science of fossils (Chapter 8). For example, Cuvier's hunch that an age of reptiles had preceded an age of mammals remained merely a hunch, until specific fossils could be related more clearly to specific formations, so that they could be assigned with confidence to distinctly different periods of geohistory (§8.1). This was sketched programmatically by Blumenbach when he suggested—applying yet another analogy from human history—that the earth's "archaeology", a true geohistory, could be traced from the sequence of fossils found in its successive formations. Werner, and the geognosts trained or inspired by him, were beginning to pay more attention to the Secondary formations, in which fossils were often abundant, as well as to the economically important Primary rock masses, which contained none. In the event, however, an "enriched geognosy" (as it has been termed here), in which fossils were added consistently to the other criteria by which formations could be distinguished, was developed most effectively by Smith in England, independently of Continental geognosy (§8.2). Smith's use of what he called "characteristic fossils" to trace the outcrops of successive Secondary formations across England and Wales was unprecedented in its geographical scope and level of detail. But it was still unmistakably geognosy, albeit enriched by the use of fossils: like any other geognost, Smith was concerned above all with the structural "order" of the formations, not with causal questions about either them or their fossils, still less with questions about the geohistory they might represent.

Meanwhile Cuvier developed his distinctively geohistorical approach by sketching for the Parisian educated public an outline of geohistory that incorporated his own latest research on fossil bones (§8.3). On the politically sensitive question of the timescale of geohistory, he steered a sober middle course between two extremes. He emphatically rejected the traditional short timescale of a few millennia, which was enjoying a popular revival in the wake of Napoleon's new pact with the Catholic church. But he also rejected the eternalism that Lamarck was continuing to propound, which was being exploited by the anti-religious party for equally ideological purposes. Cuvier's middle way was to claim, more scientifically, a vast but finite timescale for geohistory as a whole, but also to follow de Luc and Dolomieu in dating its most recent catastrophe to the dawn of human history, implicitly confirming the Flood story in Genesis and thereby linking geohistory to its human epilogue. Cuvier then used another Parisian occasion to censure the genre of geotheory as sterile, on the grounds that its speculations were barely constrained by empirical data (§8.4). In its place he advocated a research agenda—inspired by Saussure's— focused on the soluble problems of the Secondary formations and their fossils. Finally, his report to Napoleon on the natural-history sciences gave him an occasion to bring "geology" in from the cold: having scorned it in its earlier meaning as a synonym for geotheory, he now adopted it in its new meaning as an empirically based science with geohistory as one of its attainable goals. Not by coincidence, this new meaning was also endorsed when the world's first body to be devoted to this kind of science was founded in London, and its founders chose—unexpectedly—to call it the Geological Society.

Cuvier's agenda was immensely fruitful, but it led to a bifurcation of subsequent geological research, which is reflected in the contents of the last two chapters of this book. The first of these equally seminal lines of research aimed at relating the present world to the geohistory recorded in the Secondary formations. Cuvier urged that attention should be focused first on the *youngest* of the Secondaries, because they were nearest the present world both in time and in the kinds of life that their fossils disclosed (Chapter 9). He and the mineralogist Brongniart provided the decisive exemplar when they jointly surveyed the younger Secondary formations around Paris (§9.1). This geognostic research, enriched by close attention to fossils, was a practice that Brongniart may or may not have learned indirectly from Smith, while visiting London during the brief Peace of Amiens; but even if he did, the two Parisians took it far beyond anything that Smith was doing, by doubly enriching it into geohistory. They claimed that formations with fossil shells of marine mollusks alternated with others containing only the shells of freshwater genera. They then translated this geognosy into a geohistory for the Paris region. They reconstructed a complex story in which seas had alternated in the deep past with freshwater lakes or lagoons: it was a geohistory as unpredictable and contingent as the turbulent politics they had both lived through in the past two decades. And it was not to be confined to the Paris region: Brongniart lost no time in extending it to other parts of France and defending their geohistory against its critics (§9.2).

The doubly enriched geognosy that Cuvier developed with Brongniart gave him an indispensable framework on which to plot the history of the quadrupeds, because it fixed the relative ages of his various fossils (§9.3). It confirmed his earlier hunch that an age of reptiles had preceded the first mammalian fauna, which in turn had become *by degrees* more like that of the present world. Having described and illustrated all the fossil bones known to him, Cuvier reprinted his many specialized papers, making the resultant volumes more attractive by adding an essay based on his earlier lectures to the general public. Here he presented himself once more as nature's antiquarian, writing nature's history and making the deep past of the prehuman world knowable in the present, just as Laplace had made the deep space of the solar system knowable to earthbound humans (hence the title of the present book and its epigraph). Cuvier's eloquent essay had an immediate impact throughout Europe and even beyond it. But in Britain his aims and intentions were distorted for local reasons by Jameson, whose editions represented Cuvier as offering geotheory rather than geohistory and as concerned above all to reconcile geology with Genesis. Soon afterwards, however, Napoleon's defeat brought peace at last throughout Europe, enabling savants to travel more freely and ending the intellectual isolation of Britain.

Even before the peace, English geologists took the hint from Cuvier's agenda and began to study their own younger Secondary formations (§9.4). Webster found on the Isle of Wight an alternation of marine and freshwater formations that matched the Parisian sequence. He claimed that even these geologically recent strata had been affected by a major movement of the earth's crust, implying a geohistory far more dynamic than had been generally assumed. Meanwhile, in the foothills of the Apennines in northern Italy, Brocchi described a rich fauna of fossil mollusks in the younger Secondary formations, which he usefully defined as "Tertiary". As Cuvier had predicted, Brocchi's "Subapennine" Tertiary fauna seemed to bridge the gap between Lamarck's Tertiary shells around Paris and the living faunas of present seas, thereby linking the deep geohistorical past still more firmly to the present: Brocchi speculated that such faunas might have changed not in any mass extinctions but gradually by the piecemeal deaths of species with finite life spans.

After the peace made extensive travel and fieldwork possible once more, Brongniart found fossil evidence in the Alps for a spectacular amount of crustal elevation during Tertiary time (§9.5). This made even Webster's evidence look puny and reinforced the new sense of the earth's dynamic history. Meanwhile, in England, Smith managed finally to publish the map, sections, and proxy fossils that substantiated what he called his "stratigraphical" geology, at last bringing his methods fully into the public realm. Despite Smith's own hostility to the Geological Society, two of its younger members, Conybeare and Buckland, soon made the stratigraphy of the Secondary formations of England a standard of reference for the rest of Europe and even, more tentatively, for the rest of the world. This provided a potential framework—though as yet no more—for extending the detailed reconstruction of geohistory back before the time of the Tertiary formations.

Finally, Brongniart's ambitious synthesis of the Tertiary formations throughout Europe—most of it based on his own fieldwork—implied that the corresponding Tertiary era must have been extremely long and a far more significant part of geohistory than had previously been recognized (§9.6). Prompted by his former student Prévost, Brongniart began to estimate the relative ages of Tertiary formations not by a handful of "characteristic fossils" but by a quantitative assessment of whole fossil faunas. Like Brocchi he claimed that such faunas had changed steadily in the course of time, approximating gradually to those of the present world: earlier species had died out and newer ones had replaced them in piecemeal fashion. This research marked the establishment of a fruitful line of research that would be carried much further in subsequent years. It vindicated Cuvier's prediction that the Tertiaries could act as the gateway, as it were, to the even deeper past of the Secondaries (in the newly restricted sense of that term), linking those older formations, with their deeply unfamiliar remains of much stranger creatures, to the familiar world of the present.

The second line of research suggested by Cuvier's agenda tackled the same problem of relating past geohistory to the present world, but it did so at a higher level of temporal resolution. The focus here was on the decisive event—if indeed there had been one—marking the boundary between even the most recent portion of geohistory and the periods of recorded human history (Chapter 10). As in the work on Tertiary geohistory, here too Cuvier's own research led the way (§10.1). His claim that the "last revolution" had been a sudden and drastic event was reinforced by the discovery of a frozen mammoth carcass in Siberia; and his further work on an array of more ambiguous fossil mammals confirmed his view that the whole earlier fauna had been wiped out in a single event. But this interpretation was challenged from three different directions: first, from Desmarest's renewed insistence that the actual cause of ordinary valley erosion had continued smoothly from deep past into present, uninterrupted by any catastrophic event; second, by Lamarck's renewed claims for universal transformism in place of any natural extinction; and third, by the claims of some antiquarians that certain literate civilizations had immensely ancient records stretching back into deep geohistory. On the other hand, Cuvier's concept of a recent mass extinction was strongly supported by von Buch's fieldwork on the puzzling erratic blocks in the Alps, and by Hall's similar inference that a massive mega-tsunami had swept quite recently across Scotland (§10.2). So the essay that Cuvier prefixed to his collected papers on fossil bones made a persuasive case for "a great and sudden revolution" in the geologically recent past (§10.3): he dismissed actual causes as inadequate, rejected transformism, refuted reports of human fossils, and debunked inflated claims for human antiquity. The Flood story in Genesis was reduced to just one of many garbled traces of a real event at the dawn of human history, recorded independently in many diverse cultures; only in Britain did Jameson's editions turn Cuvier into something like a scriptural literalist.

As with research on Tertiary geohistory, here too the return of peace allowed British savants to travel freely and ended their intellectual isolation (§10.4). Inspired by Cuvier but with an initially more literal approach, Buckland promoted geology in the intellectually conservative environment of Oxford by identifying the putative "geological deluge" with the biblical Flood. But he supported the historicity of the deluge itself by reconstructing the course of a "diluvial current" across central and southern England, based on detailed fieldwork (§10.5). Meanwhile von Buch was improving his explanation of the transport of the Alpine erratics, by scaling up a well-documented case of a catastrophic mudslide: so an actual cause seemed likely after all to be adequate to account for some of the most refractory diluvial phenomena. Buckland's extensive travels on the Continent greatly widened his horizons, made him aware of such issues, and in particular gave him firsthand knowledge of the problem of accounting for caves with fossil bones. His geohistorical reconstruction of Cuvier's revolution took a new turn—and fully integrated the field study of bone caves with Cuvier's museum study of the bones themselves—when a small but decisive bone cave was discovered in northern England (§10.6). Buckland interpreted it as the den of antediluvian hyenas and reconstructed the ecosystem of which they had been a part. He enlarged this with further research on other caves on the Continent, expanding his work into a powerful exemplar that fulfilled Cuvier's prescription by showing how the prehuman world could indeed be known in detail, by careful inference from the preserved "relics" of the deep past.

TOWARDS MODERNITY

Brongniart's synthesis of Tertiary stratigraphy and geohistory, and Buckland's reconstruction of immediately antediluvian geohistory, jointly constitute the point at which this narrative has been brought to a close in the early 1820s. This book has not been about a specific scientific problem that reached a clear resolution at a particular time, like, say, the "great Devonian controversy" of the following decade. Instead, it has traced the formulation and consolidation of a novel perspective and a novel method in the sciences of the earth: the injection of *history* into sciences that had previously been either descriptive or causal in their orientation, creating for the first time nature's own history. It has traced how this novel geohistorical approach was derived from transpositions from the human world into the natural: both from the profoundly historical perspective of Judeo-Christian religion and from its secular counterpart in erudite human history and antiquarian research. The former, far from being an obstacle to the perception of the immense timescale of geohistory, facilitated the extension of historicity back into the vastness of deep time. And the latter provided the new practice of geohistory with its crucial conceptual metaphors of nature's documents and archives, coins and monuments, annals and chronologies. By the time Brongniart and Buckland completed their respective exemplars of geohistorical research, the analogy with human history was becoming so familiar to the relevant savants—who now called themselves geologists—that they no longer needed to make it explicit in their expert discourse, though it remained invaluable when they explained their science to the wider public.

So this volume ends not with the resolution or "closure" of any specific problem or controversy, but with the establishment of a method or perspective that by its nature was open-ended. Brongniart's review of Tertiary stratigraphy and Buckland's analysis of bone caves, and the geohistorical interpretations offered by both, signaled not the completion of fruitful lines of research but their start. Brongniart's work called for still more thorough fieldwork, more diligent collecting of fossils and their more accurate identification, and more rigorous reconstructions of the conditions in which they had lived; above all it called for the extension of such methods not only to Tertiary formations in other regions but also to those dating from the still earlier Secondary periods of geohistory. Buckland's work called likewise for scrupulously careful excavation of further bone caves; but it also called for study of other aspects of the most recent phase of geohistory, and above all for the elucidation of the physical character and cause of the putative boundary event between past and present, and a more rigorous assessment of its date in relation to early human history. The published achievements of both geologists were widely praised, but neither was uncontroversial; Buckland's work in particular was criticized by some of his colleagues as forcefully as it was approved by others. But no geologists could ignore what they had done, and their work provided foundations on which others could build.

In the years that followed the publication of these two exemplars of the new geohistorical practice, the rich potential that they offered for further work was rapidly and fruitfully exploited. For a time the two lines of research that they represented continued on their separate trajectories; but eventually they were combined in new geological syntheses of outstanding importance (I hope to trace the course of this work in a sequel to the present volume). Like the research that led to the establishment of the geohistorical approach itself, its further development into a new kind of geotheory, far more geohistorical than the earlier genre and far more soundly based, was not the work of any single savant but of a cluster of many, often in intense interaction with one another. If Cuvier has emerged as the pivotal figure in the history that has been traced in this volume, Lyell was to emerge as a comparable figure in the history of the subsequent decades; but neither had his own way in every respect, and neither deserves uncritical adulation by posterity. Yet jointly, and in conjunction with many other individuals of scarcely less importance, they did produce a body of scientific practice that has continued to be fundamental ever since. The best evidence for this, and the finest tacit tribute that all these earlier figures could have, is that the geologists of the twenty-first century can now take the result of their work completely for granted, and that most of them know little about how it was first achieved.

SOURCES

1. PLACES AND SPECIMENS

This is an unusual, perhaps unprecedented, category with which to begin a list of the sources used in writing a book on the history of the sciences. But it deserves its place. For several years, some historians of the experimental sciences have been demonstrating the value of reconstructing the apparatus and replicating or "re-staging" the experiments of historical figures in order to understand more fully how their hands-on laboratory experience of natural phenomena translated into theoretical conclusions. For a science such as geology, focused more on outdoor work in the field and indoor work in museums (particularly in the period covered in this book: see §1.3), a similar kind of experiential replication—which might be called "re-treading"—is not yet accepted as equally valuable, at least by historians. There is a good but insufficient reason for their suspicion of it. Geologists with historical interests, and even historians of geology, have often used replication in the field or in museums as a way to discover the modern explanation of the objects or features that historical actors described or discussed and then to use what "we now know" as a standard by which to judge their conclusions and even their competence or intelligence. This kind of historically sterile "presentism" deserves to be rejected, as much when it is used to judge past performances in the field or in museums as it is when applied to manuscript or printed texts from the past. However, all historians of the sciences would concede that textual sources need not be read in a presentist way; indeed, most would insist that they should not be. The same is the case with the sources represented by features of historical importance that can be studied in the field or in a museum (for example, a specific rock outcrop or fossil specimen). Just as it is essential for a historian to follow historical actors in what they read and wrote, so it is equally instructive to follow them in what they saw.

This is what I have tried to do, for many years, in all the historical research that lies behind the writing of this book. The "material" sources I have used (as distinct from the more conventional textual ones) are in two categories: features seen in the field and specimens seen in museums. Listed below are the maps I have found helpful in studying the features or regions that were significant in the fieldwork of some of the historical figures mentioned in this book, and in trying to understand and appreciate their interpretations of what they saw, in the light of seeing the same features for myself (unlike the fauna and flora, many geological features have changed relatively little in the past two centuries, once one mentally subtracts the modern overlay of superhighways, power lines, urbanization, etc.). I list topographical rather than geological maps, because the latter are inherently interpretative and their use in the field inexorably intrudes modern concepts and conclusions into what should be an exercise in *historical* understanding. Most of the maps listed are on a fairly small scale, which is best suited to a preliminary appreciation of the issues; larger-scale maps are also invaluable while one is in the field. Some localities are listed more than once, because some sheets overlap substantially. The maps are listed under the countries in which the localities are now situated, not under their political positions two centuries ago.

The second category of material sources, namely the historically decisive specimens that are still on public display in certain museums, should also logically be listed here. But since they have been cited in the footnotes with the same style of abbreviation as the libraries and archives that contain manuscript sources and pictures, the two lists have been combined, and all the abbreviations of this kind are explained in the "manuscripts and pictures" section below.

To repeat the point, both lists—like the textual sources listed in the bibliographies that follow—are confined to features and specimens that I myself have been able to study at first hand. Many others, equally relevant and instructive, might have been included, had I had the time and opportunity to see them.

France

Institut Géographique National (1:100000):

42 *Clermont-Ferrand / Montluçon*: Parc Naturel Régional des Volcans d'Auvergne (Puy de Dôme, Olby, Lac Aidat, Lac Chambon, etc.).

45 *Annecy/Lausanne*: Geneva, Mont Salève, Arve valley, Cascade de l'Arpenaz, Le Buet, Mont Blanc massif, Chamonix, Vallorcine, Dorénaz, Rhône valley and delta into Lac Léman.

49 *Clermont-Ferrand/Aurillac*: Parc Naturel Régional des Volcans d'Auvergne [overlaps largely with sheet 42].

52 *Grenoble/Valence*: Plateau du Coiron, St Vincent-de-Barrès, St Jean-le-Centenier [Vivarais basalts, etc.].

53 *Grenoble/Mont Blanc*: Mont Blanc massif; Crammont.

59 *Privas/Alès*: Ardèche valley, Aubenas, Vals-les-Bains, Antraigues, etc. [Vivarais volcanoes].

66 *Avignon/Montpellier*: Rhône delta into Mediterranean.

90 *Environs de Paris*: Seine valley, Montmartre, Sèvres, Grignon, Fontainebleau, etc. [Paris Basin].

Germany

Kompass Wander- und Radtourenkarte (1:50000):

1026 *Mittleres Erzgebirge*: Freiberg and Erzgebirge.

Fritsch Wanderkarte (1:35000):

123 *Gössweinstein / Pottenstein*: Muggendorf; Gailenreuth and other caves.

Italy

Istituto Geografico Centrale (1:50000):

4 *Massiccio di Monte Bianco*: Mont Blanc massif, Chamonix, Crammont.

Touring Club Italiano, Carta turistica (1:50000):

Parco del Etna: Etna and environs.

Touring Club Italiano, Grande carta stradale (1:200000):

Campania e Basilicata: Vesuvius, Campi Flegrei, Pompeii, Herculaneum [Ercolano].

Toscana: Volterra, Pisa; north and south flanks of Apennines.

Veneto, Friuli, Venezia, Giulia: Val d'Agno, Bolca, Colle Euganei, Po delta.

Netherlands

Michelin Régional (1:200000):

523 *Zuid-Nederland / Pays-Bas Sud*: Rhine delta, Maastricht.

Switzerland

Carte Nationale de la Suisse/Landeskarte der Schweiz (1:100000):

31 *Saane/Sarine*: Neuchâtel.
40 *Le Léman*: Geneva, Mont Salève, Lac Léman.
41 *Col du Pillon*: Rhône valley and delta into Lac Léman, Bex, Dorénaz, Diablerets massif.
45 *Haute-Savoie*: Mont Salève, Arve valley, Cascade de l'Arpenaz, Rochers de Fis.
46 *Val de Bagnes*: Mont Blanc massif, Chamonix, Le Buet, Vallorcine, Val de Bagnes.

United Kingdom

Ordnance Survey, Landranger series (1:50000):

48 *Iona & West Mull*: Staffa (Fingal's Cave).
66 *Edinburgh, Penicuik & North Berwick*: Edinburgh Castle Rock, Corstorphine Hill, Salisbury Crags.
67 *Duns, Dunbar & Eyemouth*: Siccar Point ("Hutton's unconformity" [Berwickshire]).
69 *Isle of Arran*: Goat Fell granite, "Hutton's unconformity" [Arran].
100 *Malton & Pickering, Helmsley & Easingwold*: Vale of Pickering, Kirkdale.
151 *Stratford-upon-Avon, Warwick & Banbury* and 164 *Oxford, Chipping Norton & Bicester*: Thames and Cherwell valleys.
192 *Exeter & Sidmouth* and 193 *Taunton & Lyme Regis, Chard & Bridport*: East Devon and Dorset coastline.
196 *The Solent & Isle of Wight, Southampton & Portsmouth*: Isle of Wight (Needles, Alum Bay, Headon Hill) and Hampshire coastline.

Ordnance Survey of Northern Ireland, Discoverer series (1:50000):

5 *Ballycastle*: Giant's Causeway.

2. MANUSCRIPTS AND PICTURES

Below is a key to the abbreviations used in the footnotes to denote the libraries or archives in which cited manuscripts and reproduced prints and paintings are held. Also listed here, for convenience, are museums that display historically important specimens mentioned in the text (see "Places and Specimens", above).

Amsterdam-UB	Universiteitsbibliotheek, Universiteit van Amsterdam
Bassano-BC	Biblioteca Civica di Bassano del Grappa (Veneto)
Berlin-SB	Staatsbibliothek (Preussischer Kulturbesitz, Handschriftenabteilung), Berlin
Bolca-MF	Museo dei Fossili, Bolca (Verona)
Cambridge-FM	Library of Fitzwilliam Museum, University of Cambridge
Cambridge-SM	Sedgwick Museum, University of Cambridge
Cambridge-UL	University Library, Cambridge
Cardiff-NMW	National Museum of Wales (Department of Geology), Cardiff
Chatsworth-CH	Chatsworth House, Chatsworth (Derbyshire)
Edinburgh-NGS	National Gallery of Scotland, Edinburgh
Edinburgh-NLS	National Library of Scotland, Edinburgh
Edinburgh-RS	Royal Society of Edinburgh
Exeter-DRO	Devon Record Office, Exeter [Buckland papers, formerly on loan but now returned to private ownership]
Genève-BPU	Bibliothèque Publique et Universitaire, Geneva
Genève-MHN	Musée d'Histoire Naturelle, Geneva
Genève-MHS	Musée d'Histoire des Sciences, Geneva

Glasgow-HM	Hunterian Museum, University of Glasgow
Göttingen-NSU	Niedersächsische Staats- und Universitätsbibliothek, Göttingen
Haarlem-HM	Hollandsche Maatschappij der Wetenschappen, Haarlem
Haarlem-TM	Teylers Museum, Haarlem
London-BL	British Library, London
London-BM	British Museum (Department of Prints and Drawings), London
London-GS	Geological Society, London
London-NPG	National Portrait Gallery, London
London-RAS	Royal Astronomical Society, London
London-RCS	Royal College of Surgeons, London
London-RI	Royal Institution, London
London-RS	Royal Society, London
London-TB	Tate Britain [formerly Tate Gallery], London
London-UCL	D. M. S. Watson Library, University College London
London-WL	Wellcome Library, London
Madrid-MCN	Museo Nacional de Ciencias Naturales, Madrid
Milano-AS	Archivio di Stato, Milan
Milano-MSN	Museo di Storia Naturale, Milan
Oxford-MNH	Museum of Natural History [formerly University Museum], University of Oxford
Paris-EM	École des Mines de Paris
Paris-IF	Bibliothèque de l'Institut de France, Paris
Paris-MHN	Bibliothèque Centrale [for MSS] and Galérie de Paléontologie [for specimens], Muséum National d'Histoire Naturelle, Paris
Reggio-MC	Musei Civici, Reggio dell'Emilia (Emilia-Romagna)
Verona-BC	Biblioteca Civica, Verona
Verona-MSN	Museo Civico di Storia Naturale di Verona
Vicenza-BB	Biblioteca Bertoliana, Vicenza (Veneto)
Weimar-GW	Goethes Wohnhaus, Weimar
Wellington-ATL	Alexander Turnbull Library, Wellington (New Zealand)

3. PRINTED SOURCES: PRIMARY

In this bibliography, alphabetization of proper names with "de", "von", "van", etc. generally follows what was customary at the time or is usual in recent historical work: e.g., de La Métherie is listed under L, not D or M; von Buch under B, not V; van Marum under M, not V. De Luc, who was and is commonly cited under both D and L, is here placed, rather arbitrarily, under L. Contemporary abstracts and translations (into any language other than the original) are not listed here, unless they are commented on in a substantial way in the main text. Most of those published in periodicals, except the earliest in date, can be traced easily through the *Royal Society Catalogue of Scientific Papers 1800–1863* (London, 1867–72), which in fact extends back into the 1790s; those published as books are listed in the catalogues of the British Library, Bibliothèque National, Library of Congress, and other major libraries, and in standard reference works such as Louis Agassiz's *Bibliographia Zoologiae et Geologiae* (London, 1848).

The following list is a key to the abbreviations used in this bibliography for items published in periodicals and multi-volume "dictionaries". (References include not only page numbers but also the numbers of *plates*; the latter are often physically separated from the texts and are all too easily missed.)

AAEM	Acta Academiae Electoralis Moguntinae Scientiarum Utilium quae Erfordiae est [Erfurt]
AASP	Acta Academiae Imperialis Scientiarum Petropolitana [St Petersburg]
ACP	Annales de chimie et de physique [Paris]
AKAW	Abhandlungen der Königlichen Akademie der Wissenschaften [Berlin]
AMHN	Annales du Muséum d'Histoire Naturelle [Paris]
AP(H)	Annalen der Physik [Halle]
AP(L)	Annals of philosophy [London]
ASN	Annales des sciences naturelles [Paris]
BB	Bibliothèque Britannique [Geneva]
BC	British Critic [London]
BJ	Bergmännische Journal [Freiberg]
BSP	Bulletin [scientifique], Société Philomathique [Paris]
BU	Bibliothèque universelle [Geneva] [continuation of *BB*]
DGANS	Denkschriften der Vaterländischen Gesellschaft der Aerzte und Naturforscher Schwabens [Tübingen]
DSN	Dictionnaire des sciences naturelles [Paris]
EBH	Ephemeriden über den Berg- und Huttenkunde
EPJ	Edinburgh philosophical journal
ER	Edinburgh review
GA	Göttingische Anzeigen
GMWL	Göttingische Magazin der Wissenschaften und Litteratur
HARIB	Histoire de l'Académie Royale des Inscriptions et des Belle-Lettres [Paris]
HARS	Histoire de l'Académie Royale des Sciences [Paris]
HCATP	Historia et commentationes Academiae Electoralis Scientiarum et Elegantiorum Litterarium Theodora-Palatinae [Mannheim]
JM	Journal des mines [Paris]
JN	Journal du nord [St Petersburg]
JP	Journal de physique, de l'histoire naturelle et des arts [Paris] [full title varies; continuation of *OP*]
MASBB	Mémoires de l'Académie Impériale et Royale des Sciences et des Belles-Lettres de Bruxelles
ME	Magasin encyclopédique [Paris]
MASSP	Mémoires de l'Académie Impériale des Sciences de St Pétersbourg [continuation of *NAASP*]
MIIF	Mémoires de l'Institut Impérial de France, Classe des Sciences Naturelles et Mathématiques [continuation of *MIN*]
MIN	Mémoires de l'Institut National des Sciences Naturelles et des Arts; Sciences mathématiques et physiques [Paris]
MM	Monthly magazine [London]
MNEGN	Magazin für den neuesten Entdeckungen in der gesammte Naturkunde [Berlin] [continuation of *SBGNF*]
MNPN	Magazin für den neueste aus der Physik und Naturgeschichte [Gotha]
MR	Monthly review [London]
MSIS	Memorie di matematica e fisica della Società Italiana delle Scienze [Modena, Verona]
NAASP	Nova acta Academiae Imperialis Scientiarum Petropolitanae [St Petersburg] [continuation of *AASP*]
NCASP	Novi commentarii Academiae Imperialis Scientiarum Petropolitanae [St Petersburg]
NBSP	Nouvelle bulletin, Société Philomathique [Paris] [continuation of *BSP*]
OP	Observations sur la physique, l'histoire naturelle et les arts [Paris] [full title varies]

OSSA	Opusculi scelti sulle scienze e sulle arti [Milan]
PM	Philosophical magazine [London]
PTRS	Philosophical transactions of the Royal Society [London]
SBGNF	Schriften der Berlinischen Gesellschaft Naturforschender Freunde
TAPS	Transactions of the American Philosophical Society [Philadelphia]
TGS	Transactions of the Geological Society [London]
TRIA	Transactions of the Royal Irish Academy [Dublin]
TRSE	Transactions of the Royal Society of Edinburgh
VHMW	Verhandelingen uitgegeeven door de Hollandsche Maatschappij der Weetenschappen te Haarlem
VTTG	Verhandelingen van Teylers Tweede Genootschap [Haarlem]

Abildgaard, P. C. 1796. *Kort beretning om det Kongelige Naturalcabinet i Madrid, med en beskrivelse over et gigantisk skelet af et nyt ubekiendt dyr, som er opgravet i Peru og bevares i dette museum.* Copenhagen.

[Académie des Sciences de Saint Petersbourg]. 1786. *Mémoires présentés à l'Académie Impériale des Sciences pour répondre à la question minéralogique proposée pour le Prix de MDCCLXXXV.* St Petersburg.

Adams, M. 1807. Relation d'un voyage à la Mer Glaciale et découverte des restes d'un mammouth. *JN* 32 (supplément): 633–40, 621–28 [latter pp. misnumbered].

André, Noël [formerly Chrysologue de Gy]. 1806. *Theorie de la surface actuelle de la terre, ou plutôt recherches impartiales sur le temps et l'agent de l'arrangement actuel de la surface de la terre . . .* Paris. *See also* Chrysologue de Gy.

Anonymous. 1753. Chronologie. *Encyclopédie* 3: 390–400.

Anonymous. 1788. [Review of] Theory of the earth... By James Hutton, M.D., F.R.S. Edin. *MR* 79: 36–38.

Anonymous. 1790. [Review of] *Réponse à la question physique proposée par la Société de Teyler* &c . . . by François Xavier Burtin . . . Haarlem, 1790. *MR* 1790 3: 539–44.

Anspach, Margrave of. 1794. Account of some remarkable caves in the Principality of Bayreuth, and of the fossil bones found therein, extracted from a paper sent, with specimens of the bones, as a present to the Royal Society, by his most Serene Highness the Margrave of Anspach. *PTRS* 84: 402–6.

Barruel, Augustin. 1812. *Les Helviennes, ou lettres provinciales philosophiques.* 5th ed. 4 vols. Paris.

Basset, C. A. (ed.). 1815. *Explication de Playfair sur la théorie de la terre par Hutton, et examen comparatif des systèmes géologiques fondés sur le feu et sur l'eau, par M. Murray; en réponse à l'Explication de Playfair.* Paris.

Bayardi, Ottavio Antonio (ed.). 1757–92. *L'Antichità di Ercolano esposte.* 8 vols. Naples.

Bell, William. 1793. Description of the double-horned rhinoceros of Sumatra. *PTRS* 83: 3–6, pls. 2–4.

Berger, J. F. 1811. A sketch of the geology of some parts of Hampshire and Dorsetshire. *TGS* 1: 249–68.

Blainville, Henri Ducrotay de. 1827. *Mémoire sur les bélemnites, considérées zoologiquement et géologiquement.* Paris.

Blumenbach, Johann Friedrich. 1779. *Handbuch der Naturgeschichte.* Göttingen.

———. 1781. *Über den Bildungstrieb und das Zeugungsgeschäft.* Göttingen.

———. 1788. *Handbuch der Naturgeschichte.* 3rd ed. Göttingen.

———. 1790. Beyträge zur Naturgeschichte der Vorwelt. *MNPN* 6 (4): 1–17, pl. 1.

———. 1790. *Beyträge zur Naturgeschichte: erster Theil.* Göttingen.

———. 1791. Ein Wort über die im vorjährigen Oktoberstücke dieses Journal beschriebenen Abdrücke im Bitumindsen-Mergelschiefer. *BJ* (4e Jg.) 1: 151–56.

———. 1795. *De generis humani varietate nativa.* 3rd ed. Göttingen.

———. 1796–1810. *Abbildungen naturhistorischer Gegenstände.* Göttingen.

———. 1797. *Handbuch der Naturgeschichte.* 5th ed. Göttingen.

———. 1799. *Handbuch der Naturgeschichte.* 6th ed. Göttingen.

———. 1801. Specimen archaeologiae telluris, terrarumque in primis Hannoveranarum. *GA* 1801: 1977–84.

———. 1803. *Specimen archaeologiae telluris, terrarumque in primis Hannoveranarum.* Göttingen.

———. 1806–11. *Beyträge zur Naturgeschichte.* 2nd ed. 2 vols. Göttingen.

Bournon, (Comte) de. 1808. *Traité complet de la chaux carbonatée et de l'arragonite, auquel on a joint une introduction à la minéralogie en général, une théorie de la cristallisation, et de son application, ainsi que celle du calcul, à la détermination des formes cristallines et de ces deux substances.* 3 vols. London.

Brander, Gustav. 1766. *Fossilia Hantoniensia collecta, et in Musaeo Brittanico deposita.* London.

Brard, Cyprien Prosper. 1809. Mémoire sur les coquilles fossiles du genre Lymnée qui se trouvent aux environs de Paris, sur les autres coquilles qui les acccompagnent, et sur la nature des pierres qui renferment ces fossiles. *AMHN* 14: 426–40, pl. 27.

———. 1810. Seconde mémoire sur les Lymnées fossiles des environs de Paris, et sur les autres coquilles qui les accompagnent. *AMHN* 15: 406–21, pl. 21.

———. 1811. Troisième mémoire sur les coquilles fossiles des environs de Paris qui appartiennent à des genres fluviatiles ou terrestres. *JP* 72: 448–59.

———. 1812. Quatrième mémoire sur les coquilles fossiles qui appartiennent à des genres fluviatiles et terrestres. *JP* 74: 247–61.

———. 1815. *Histoire des coquilles terrestres et fluviatiles qui vivent aux environs de Paris.* Paris and Geneva.

Breynius, Johann Philip. 1741. Observations, and a description of some mammoth's bones dug up in Siberia, proving them to have belonged to elephants. *PTRS* 40: 124–39.

Brocchi, Giovanni Battista. 1807–8. *Trattato mineralogico e chimico sulle miniere di ferro del Dipartimento del Mella con l'esposizione della costituzione fisica delle montagne metallifere della Val Trompia.* 2 vols. Brescia.

———. 1811. *Memoria mineralogica sulla Valle di Fassa in Tirolo.* Milan.

———. 1814. *Conchiologia fossile subapennina con osservazioni geologiche sugli Apennini e sul suolo adiacente.* 2 vols. Milan.

Brongniart, Alexandre. 1798. Discours du C[itoy]en Brongniart, ingénieur des mines de la République, professeur de minéralogie. *JM* [9]: 177–89.

———. 1800. Essai d'une classification naturelle des reptiles. *BSP* 2: 81–82, 89–91, pl. 6.

———. [1802]. *Essai sur les couleurs obtenues des oxydes métalliques, et fixées par la fusion sur les différens corps vitreux.* [Paris].

———. 1807. *Traité élémentaire de minéralogie, avec des applications aux arts; ouvrage destiné à l'enseignement dans les lycées nationaux.* 2 vols. Paris.

———. 1810. Sur les terrains qui paroissent avoir été formés sous l'eau douce. *AMHN* 15: 357–405, pls. 22–23.

———. 1821. Sur les caractères zoologiques des formations, avec l'application de ces caractères à la détermination de quelques terrains de craie. *AM* 6: 537–72, pls. 7–8.

———. 1823. *Mémoire sur les terrains de sédiment supérieurs calcareo-trappéens du Vicentin, et sur quelques terrains d'Italie, de France, d'Allemagne, etc., qui peuvent se rapporter à la même époque.* Paris.

———. 1844. *Traité des arts céramiques ou des poteries considérées dans leur histoire, leur pratique et leur théorie.* 2 vols. and atlas. Paris.

Brongniart, Alexandre, and D. Riocreux. 1845. *Description méthodique du musée céramique de la Manufacture Royale de Porcelaine de Sèvres.* 2 vols. Paris.

Bruguière, Jean-Guillaume. 1792. *Histoire naturelle des vers,* 1 [text]. *Encyclopédie Méthodique:*

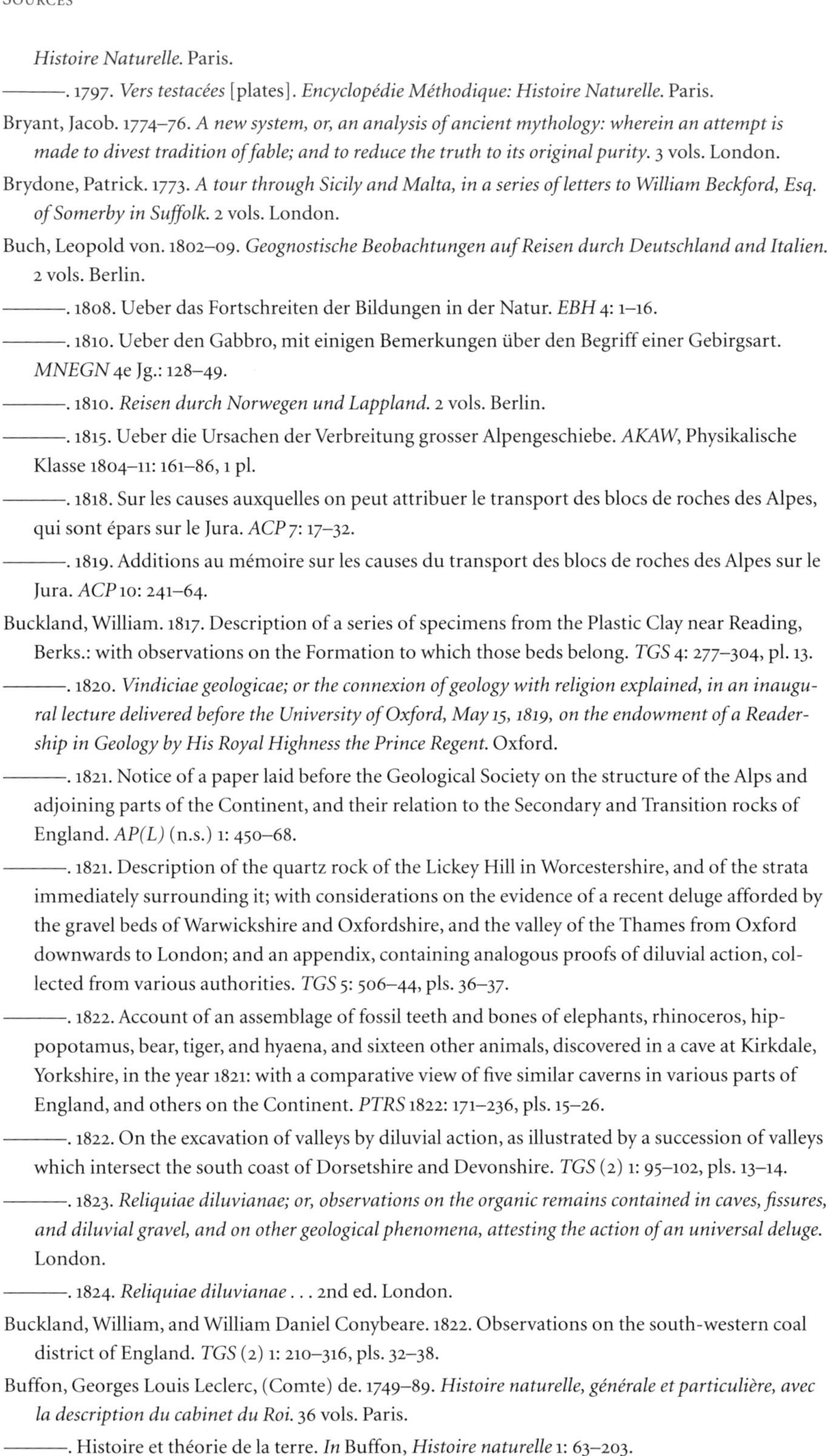

Histoire Naturelle. Paris.

———. 1797. *Vers testacées* [plates]. *Encyclopédie Méthodique: Histoire Naturelle*. Paris.

Bryant, Jacob. 1774–76. *A new system, or, an analysis of ancient mythology: wherein an attempt is made to divest tradition of fable; and to reduce the truth to its original purity*. 3 vols. London.

Brydone, Patrick. 1773. *A tour through Sicily and Malta, in a series of letters to William Beckford, Esq. of Somerby in Suffolk*. 2 vols. London.

Buch, Leopold von. 1802–09. *Geognostische Beobachtungen auf Reisen durch Deutschland and Italien*. 2 vols. Berlin.

———. 1808. Ueber das Fortschreiten der Bildungen in der Natur. *EBH* 4: 1–16.

———. 1810. Ueber den Gabbro, mit einigen Bemerkungen über den Begriff einer Gebirgsart. *MNEGN* 4e Jg.: 128–49.

———. 1810. *Reisen durch Norwegen und Lappland*. 2 vols. Berlin.

———. 1815. Ueber die Ursachen der Verbreitung grosser Alpengeschiebe. *AKAW*, Physikalische Klasse 1804–11: 161–86, 1 pl.

———. 1818. Sur les causes auxquelles on peut attribuer le transport des blocs de roches des Alpes, qui sont épars sur le Jura. *ACP* 7: 17–32.

———. 1819. Additions au mémoire sur les causes du transport des blocs de roches des Alpes sur le Jura. *ACP* 10: 241–64.

Buckland, William. 1817. Description of a series of specimens from the Plastic Clay near Reading, Berks.: with observations on the Formation to which those beds belong. *TGS* 4: 277–304, pl. 13.

———. 1820. *Vindiciae geologicae; or the connexion of geology with religion explained, in an inaugural lecture delivered before the University of Oxford, May 15, 1819, on the endowment of a Readership in Geology by His Royal Highness the Prince Regent*. Oxford.

———. 1821. Notice of a paper laid before the Geological Society on the structure of the Alps and adjoining parts of the Continent, and their relation to the Secondary and Transition rocks of England. *AP(L)* (n.s.) 1: 450–68.

———. 1821. Description of the quartz rock of the Lickey Hill in Worcestershire, and of the strata immediately surrounding it; with considerations on the evidence of a recent deluge afforded by the gravel beds of Warwickshire and Oxfordshire, and the valley of the Thames from Oxford downwards to London; and an appendix, containing analogous proofs of diluvial action, collected from various authorities. *TGS* 5: 506–44, pls. 36–37.

———. 1822. Account of an assemblage of fossil teeth and bones of elephants, rhinoceros, hippopotamus, bear, tiger, and hyaena, and sixteen other animals, discovered in a cave at Kirkdale, Yorkshire, in the year 1821: with a comparative view of five similar caverns in various parts of England, and others on the Continent. *PTRS* 1822: 171–236, pls. 15–26.

———. 1822. On the excavation of valleys by diluvial action, as illustrated by a succession of valleys which intersect the south coast of Dorsetshire and Devonshire. *TGS* (2) 1: 95–102, pls. 13–14.

———. 1823. *Reliquiae diluvianae; or, observations on the organic remains contained in caves, fissures, and diluvial gravel, and on other geological phenomena, attesting the action of an universal deluge*. London.

———. 1824. *Reliquiae diluvianae* . . . 2nd ed. London.

Buckland, William, and William Daniel Conybeare. 1822. Observations on the south-western coal district of England. *TGS* (2) 1: 210–316, pls. 32–38.

Buffon, Georges Louis Leclerc, (Comte) de. 1749–89. *Histoire naturelle, générale et particulière, avec la description du cabinet du Roi*. 36 vols. Paris.

———. Histoire et théorie de la terre. *In* Buffon, *Histoire naturelle* 1: 63–203.

———. 1761. Animaux communs aux deux continens. *In* Buffon, *Histoire naturelle* 9: 97–128.

———. 1775. Recherches sur le refroidissement de la terre & des planètes; fondemens des recherches précédentes sur la température des planètes. *In* Buffon, *Histoire naturelle*, supplément 2: 361–564.

———. 1778. Des époques de la nature [and] Notes justificatives des faits rapportés dans les époques de la nature. *In* Buffon, *Histoire naturelle,* supplément 5: 1–254, 495–599.

———. 1786. Pétrifications et fossiles. *In* Buffon, *Histoire naturelle,* minéraux 4: 156–73.

Burnet, Thomas. 1680–89. *Telluris theoria sacra: orbis nostri originem & mutationes generales, quas aut jam subiit, aut olim subiturus est, complectens.* 2 vols. London.

———. 1692. *Archaeologiae philosophicae: sive doctrina antiqua de rerum originibus.* London.

Burtin, François-Xavier. 1784. *Oryctographie de Bruxelles, ou description des fossiles tant naturels qu'accidentels découverts jusqu'à ce jour dans les environs de ce ville.* Brussels.

———. 1789. *Over de algemeene omkeeringen aan de oppervlakte der aarde, en over de oudheid van onze aardkloot . . . in het oorsponglyk fransch met de nederduitsche vertaaling . . . uitgegeven door Teylers Tweede Genootschap.* Haarlem [repr. *VTTG* 8 (1790)].

———. 1789. *Réponse à la question physique, proposé par la Société de Teyler, sur les révolutions générales, qu'a subies la surface de la terre, et sur l'ancienneté de notre globe.* In *Algemeene omkeeringen,* 1–242.

Byron, George Gordon, Lord. 1822. *Cain; a mystery.* London.

Camper, Adriaan Gilles. 1800. Sur les ossemens fossiles de la Montagne de St Pierre à Maëstricht. *JP* 51: 278–91. pls. 1–2.

Camper, Petrus. 1780. De cranio rhinocerotis Africani, cornu gemino. *AASP* 1777 (2): 193–209, pls. 5–9.

———. 1786. Conjectures relative to the petrifactions found in St Peter's Mountain, near Maestricht. *PTRS* 76: 443–56, pls. 15–16.

———. 1788. Complementa varia Acad. Imper. Scient. Petropolitanae communicanda, ad clar. ac celeb. Pallas. *NAASP* 2: 250–64, pls. 8–9.

Chambers, Ephraim (ed.). 1786. *Cyclopedia, or, an universal dictionary of arts and sciences; with a supplement by Abraham Rees.* 4 vols. London.

Charpentier, Johann Friedrich Wilhelm. 1778. *Mineralogische Geographie der Chursächsischen Lände.* Leipzig.

Chateaubriand, François-Auguste. 1802. *Génie du Christianisme; ou, beautés de la religion chrétienne.* 5 vols. Paris.

Chemnitz, Johann Hieronymus. 1769–95. *Neues systematisches Conchylien-Cabinet.* 16 vols. Nuremberg.

Chrysologue de Gy. 1787. D'une carte physique, minéralogique, civile & ecclesiastique de la Franche-Comté & de ses frontières . . . *OP* 30: 271–84. *See also* André.

Collini, Cosimo Alessandro. 1784. Sur quelques zoolithes du Cabinet d'Histoire Naturelle de S.A.S.E. Palatine & de Bavière, à Mannheim. *HCATP* 5: Phys. 58–103, pls. 1–4.

[Conybeare, William Daniel]. [1822]. *The hyaena's den at Kirkdale near Kirby Moorside in Yorkshire, discovered A.D. 1821.* [folio sheet] n.p.

———. 1823. Memoir illustrative of a general geological map of the principal mountain chains of Europe. *AP(L)* 5: 1–16, 135–49, 210–18, 278–89, 356–59, pl. 19; 6: 214–19.

Conybeare, William Daniel, and William Phillips. 1822. *Outlines of the geology of England and Wales, with an introductory compendium of the general principles of that science, and comparative views of the structure of foreign countries. Part I.* London. [repr. Farnborough (Greg), 1969]

Coupé, Jacques Michel. 1805. Sur l'étude du sol des environs de Paris. *JP* 61: 363–95.

Cuvier, Fréderic. 1807. Du rut. *AMHN* 9: 118–30.

———. 1811. Recherches sur les caractères ostéologiques qui distinguent les principales races du chien domestique. *AMHN* 18: 333–53, pls. 18–20.

Cuvier, Georges. 1795. Discours prononcé par le citoyen Cuvier, à l'ouverture du cours d'anatomie comparée qu'il a fait au Muséum National d'Histoire Naturelle, pour le citoyen Mertrud. *ME* [1e année] 5: 145–55.

———. 1796. Notice sur le squelette d'une très-grande espèce de quadrupède inconnue jusqu'à

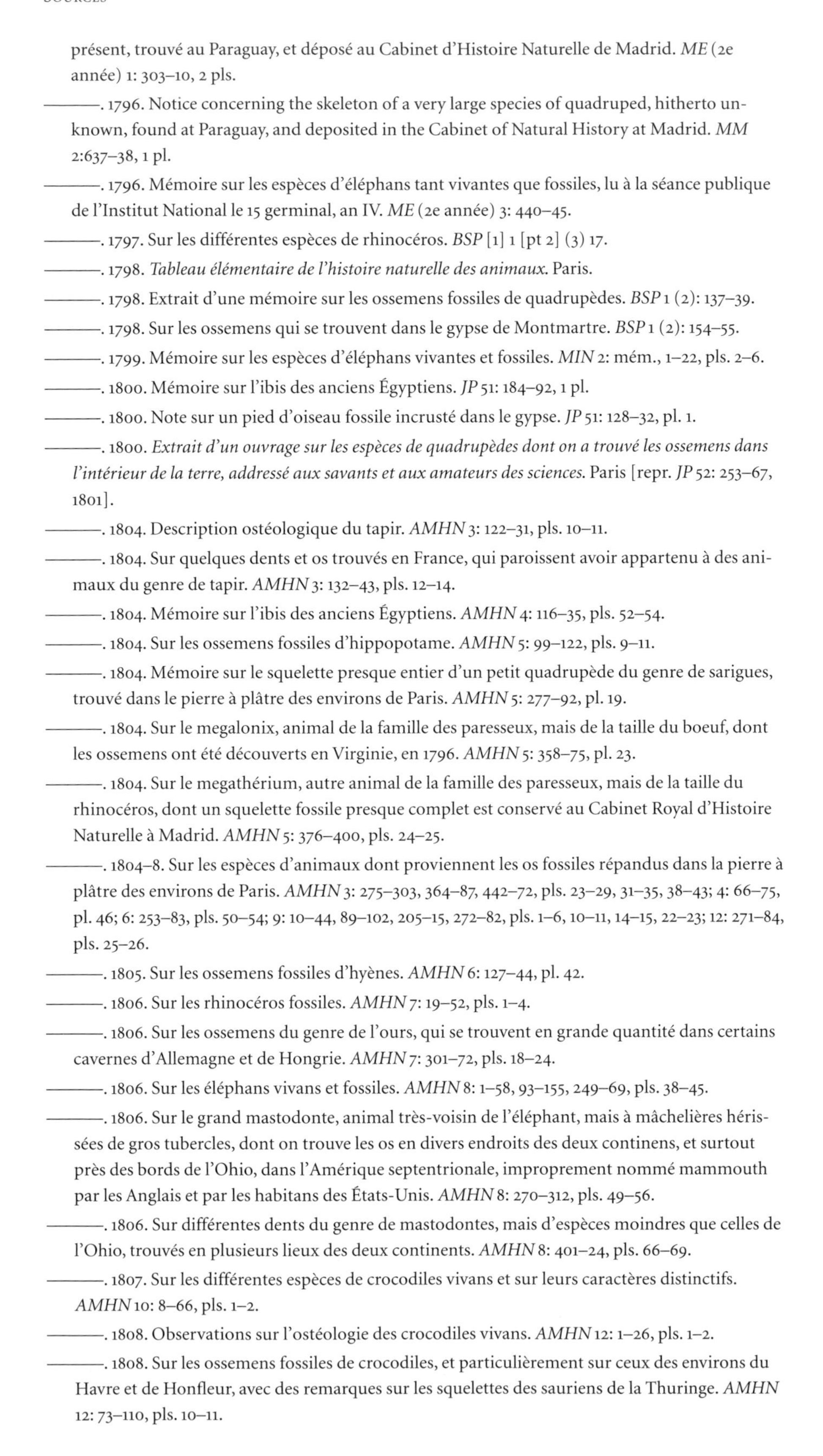

présent, trouvé au Paraguay, et déposé au Cabinet d'Histoire Naturelle de Madrid. *ME* (2e année) 1: 303–10, 2 pls.

———. 1796. Notice concerning the skeleton of a very large species of quadruped, hitherto unknown, found at Paraguay, and deposited in the Cabinet of Natural History at Madrid. *MM* 2:637–38, 1 pl.

———. 1796. Mémoire sur les espèces d'éléphans tant vivantes que fossiles, lu à la séance publique de l'Institut National le 15 germinal, an IV. *ME* (2e année) 3: 440–45.

———. 1797. Sur les différentes espèces de rhinocéros. *BSP* [1] 1 [pt 2] (3) 17.

———. 1798. *Tableau élémentaire de l'histoire naturelle des animaux.* Paris.

———. 1798. Extrait d'une mémoire sur les ossemens fossiles de quadrupèdes. *BSP* 1 (2): 137–39.

———. 1798. Sur les ossemens qui se trouvent dans le gypse de Montmartre. *BSP* 1 (2): 154–55.

———. 1799. Mémoire sur les espèces d'éléphans vivantes et fossiles. *MIN* 2: mém., 1–22, pls. 2–6.

———. 1800. Mémoire sur l'ibis des anciens Égyptiens. *JP* 51: 184–92, 1 pl.

———. 1800. Note sur un pied d'oiseau fossile incrusté dans le gypse. *JP* 51: 128–32, pl. 1.

———. 1800. *Extrait d'un ouvrage sur les espèces de quadrupèdes dont on a trouvé les ossemens dans l'intérieur de la terre, addressé aux savants et aux amateurs des sciences.* Paris [repr. *JP* 52: 253–67, 1801].

———. 1804. Description ostéologique du tapir. *AMHN* 3: 122–31, pls. 10–11.

———. 1804. Sur quelques dents et os trouvés en France, qui paroissent avoir appartenu à des animaux du genre de tapir. *AMHN* 3: 132–43, pls. 12–14.

———. 1804. Mémoire sur l'ibis des anciens Égyptiens. *AMHN* 4: 116–35, pls. 52–54.

———. 1804. Sur les ossemens fossiles d'hippopotame. *AMHN* 5: 99–122, pls. 9–11.

———. 1804. Mémoire sur le squelette presque entier d'un petit quadrupède du genre de sarigues, trouvé dans le pierre à plâtre des environs de Paris. *AMHN* 5: 277–92, pl. 19.

———. 1804. Sur le megalonix, animal de la famille des paresseux, mais de la taille du boeuf, dont les ossemens ont été découverts en Virginie, en 1796. *AMHN* 5: 358–75, pl. 23.

———. 1804. Sur le megathérium, autre animal de la famille des paresseux, mais de la taille du rhinocéros, dont un squelette fossile presque complet est conservé au Cabinet Royal d'Histoire Naturelle à Madrid. *AMHN* 5: 376–400, pls. 24–25.

———. 1804–8. Sur les espèces d'animaux dont proviennent les os fossiles répandus dans la pierre à plâtre des environs de Paris. *AMHN* 3: 275–303, 364–87, 442–72, pls. 23–29, 31–35, 38–43; 4: 66–75, pl. 46; 6: 253–83, pls. 50–54; 9: 10–44, 89–102, 205–15, 272–82, pls. 1–6, 10–11, 14–15, 22–23; 12: 271–84, pls. 25–26.

———. 1805. Sur les ossemens fossiles d'hyènes. *AMHN* 6: 127–44, pl. 42.

———. 1806. Sur les rhinocéros fossiles. *AMHN* 7: 19–52, pls. 1–4.

———. 1806. Sur les ossemens du genre de l'ours, qui se trouvent en grande quantité dans certains cavernes d'Allemagne et de Hongrie. *AMHN* 7: 301–72, pls. 18–24.

———. 1806. Sur les éléphans vivans et fossiles. *AMHN* 8: 1–58, 93–155, 249–69, pls. 38–45.

———. 1806. Sur le grand mastodonte, animal très-voisin de l'éléphant, mais à mâchelières hérissées de gros tubercles, dont on trouve les os en divers endroits des deux continens, et surtout près des bords de l'Ohio, dans l'Amérique septentrionale, improprement nommé mammouth par les Anglais et par les habitans des États-Unis. *AMHN* 8: 270–312, pls. 49–56.

———. 1806. Sur différentes dents du genre de mastodontes, mais d'espèces moindres que celles de l'Ohio, trouvés en plusieurs lieux des deux continents. *AMHN* 8: 401–24, pls. 66–69.

———. 1807. Sur les différentes espèces de crocodiles vivans et sur leurs caractères distinctifs. *AMHN* 10: 8–66, pls. 1–2.

———. 1808. Observations sur l'ostéologie des crocodiles vivans. *AMHN* 12: 1–26, pls. 1–2.

———. 1808. Sur les ossemens fossiles de crocodiles, et particulièrement sur ceux des environs du Havre et de Honfleur, avec des remarques sur les squelettes des sauriens de la Thuringe. *AMHN* 12: 73–110, pls. 10–11.

———. 1808. Sur le grand animal fossile des carrières de Maestricht. *AMHN* 12: 145–76, pls. 19–20.

———. 1808. Sur les os fossiles de ruminans trouvés dans les terrains meubles. *AMHN* 12: 333–98, pls. 32–34.

———. 1809. Sur les brèches osseuses qui remplissent les fentes de rochers à Gibraltar et dans plusieurs autres lieux des côtes de la Méditerranée, et sur les animaux qui en ont fourni les os. *AMHN* 13: 169–206, pls. 15–16.

———. 1809. Sur l'ostéologie du lamantin, sur la place que le lamantin et le dugong doivent occuper dans la méthode naturelle, et sur les os fossiles de lamantin et de phoques. *AMHN* 13: 273–312, pl. 19.

———. 1809. Sur quelques quadrupèdes ovipares fossiles conservés dans les schistes calcaires. *AMHN* 13: 410–37, pls. 30–31.

———. 1809. Sur les ossemens fossiles de tortues. *AMHN* 14: 229–44, pls. 17–18.

———. 1810. *Rapport historique sur le progrès des sciences naturelles depuis 1789, et sur leur état actuel, présenté a Sa Majesté l'Empereur et Roi, en son Conseil d'État, le 6 fevrier 1808, par la Class des sciences physiques et mathématiques de l'Institut, conformément à l'Arrête du Gouvernement du 13 Ventôse an X.* Paris.

———. 1812. *Recherches sur les ossemens fossiles de quadrupèdes, où l'on rétablit les caractères de plusieurs espèces d'animaux que les révolutions du globe paroissent avoir détruites.* 4 vols. Paris.

———. 1813. *Essay on the theory of the earth, with mineralogical notes, and an account of Cuvier's geological discoveries, by Professor Jameson.* Edinburgh.

———. 1815. *Essay on the theory of the earth . . .* 2nd ed. Edinburgh.

———. 1816. Geognostische Betrachtungen, veranlasst durch Untersuchungen der fossilen Knochen vierfüssiger Thiere. *AP(H)* 52: 117–76.

———. 1817. *Le règne animal distribué d'après son organisation, pour servir de base à l'histoire naturelle des animaux et d'introduction à l'anatomie comparée.* 4 vols. Paris.

———. 1817. *Essay on the theory of the earth . . .* 3rd ed. Edinburgh.

———. 1821–24. *Recherches sur les ossemens fossiles de quadrupèdes, où l'on rétablit les caractères de plusieurs animaux que les révolutions du globe ont détruites les espèces.* 2nd ed. 5 vols. in 6. Paris.

———. 1822. *Essay on the theory of the earth . . .* 4th ed. Edinburgh.

———. 1822–26. *Cuviers Ansichten von der Urwelt nach der zweiten Original-Ausgabe verdeutscht und mit Anmerkungen begleitet von Dr Jakob Nöggerath.* 2 vols. Bonn.

———. 1845. *Briefen an C. H. Pfaff aus den Jahren 1788 bis 1792, naturhistorischen, politischen und literarischen Inhalts.* Kiel.

Cuvier, Georges, and Alexandre Brongniart. 1808. Essai sur la géographie minéralogique des environs de Paris. *AMHN* 11: 293–326 [also *JM* 23: 421–58].

———. 1811. Essai sur la géographie minéralogique des environs de Paris. *MIIF* 1810 (1): 1–278, 2 pls, 1 map [also published in book form, Paris, 1811].

———. 1813. Versuch einer mineralogischen Geographie der Gegend um Paris. *AP(H)* 45: 229–76.

———. 1822. *Description géologique des environs de Paris: nouvelle édition, dans laquelle est inséré la description d'un grand nombre de lieux de l'Allemagne, de la Suisse, de l'Italie, etc., qui présentent des terrains analogues à ceux du bassin de Paris.* Paris.

Cuvier, Georges, René-Just Haüy, and Claude Hugues Le Lièvre. 1807. Rapport de l'Institut National (Classe des sciences physiques et mathématiques), sur l'ouvrage de M. André, ayant pour titre: Théorie de la surface actuelle de la terre. *JM* 21: 413–30.

Cuvier, Georges, and Bernard, comte de Lacépède. 1807. Rapport à la Classe des sciences physiques et mathématiques de l'Institut [sur le cadavre d'un animal découvert dans la Mer Glaciale, et intitulé *mammouth*]. *AMHN* 10: 381–86.

Darwin, Erasmus. 1791. *The botanic garden; a poem, in two parts. Part I. containing the economy of vegetation. Part II. The loves of the plants. with philosophical notes.* London.

Daubenton, Louis Jean Marie. 1764. Mémoire sur des os et des dents remarquables par leur grandeur. *HARS* 1762, Mém.: 206–29, pls. 13–14.

Daubeny, Charles Giles Brindle. 1820–21. On the volcanoes of the Auvergne. *EPJ* 3: 359–67; 4: 89–97, 300–15.

———. 1869. *Fugitive poems connected with natural history and physical science.* Oxford.

Davy, Humphry. 1827. *Six discourses delivered before the Royal Society at their anniversary meeting on the award of the Royal and Copley medals; preceded by an address to the Society, on the progress and prospects of science.* London.

Delius, Christoph Traugott. 1773. *Anleitung in der Bergbaukunst nach ihrer Theorie und Ausübung, nebst ein Abhandlung von den Grundsätzen der Berg- und Kammeralwissenschaft für die Kaiserl. Konigl. Schemnitzer Bergakademie.* Vienna.

Deluc, de Luc: *see* Luc.

Denon, Vivant. 1802. *Voyage dans la basse et la haute Égypte, pendant les campagnes du général Bonaparte.* Paris.

Descartes, René. 1644. *Principia philosophiae.* Amsterdam.

Desmarest, Anselme-Gaëtan. 1812. Sur la Gyrogonite. *JM* 32: 341–60, pl. 8.

Desmarest, Nicolas. 1757. Géographie physique. *Encyclopédie* 7: 613–26.

———. 1774. Mémoire sur l'origine et la nature du basalte à grandes colonnes polygones, déterminées par l'histoire naturelle de cette pierre, observée en Auvergne. *HARS* 1771: 705–75, pl. 15.

———. 1777. Mémoire sur le basalte, troisième partie, où l'on traite du basalte des anciens, et où l'on expose l'histoire naturelle des différentes espèces de pierres auxquelles on a donné, en differens temps, le nom de basalte. *HARS* 1773: 599–670.

———. 1779. Extrait d'un mémoire sur la détermination de quelques époques de la nature par les produits des volcans, et sur l'usage de ces époques dans l'étude des volcans. *OP* 13: 115–26.

———. 1794–1811. *Géographie physique.* 4 vols. Paris [*Encyclopédie méthodique*].

———. 1806. Mémoire sur la détermination de trois époques de la nature par les produits des volcans, et sur l'usage qu'on peut faire de ces époques dans l'étude des volcans. *MIN* 6: 219–89, pls. 6–9.

———. 1823. *Carte topographique et minéralogique d'une partie du département du Puy-de-Dôme dans le ci-devant province d'Auvergne où sont determinées la marche et les limites des matières fondues & rejettées par les volcans ainsi que les courants anciens & modernes pour servir aux recherches sur l'histoire naturelle des volcans* [sheet map]. Paris.

Dezallier d'Argenville, Antoine Joseph. 1780. *La conchyliologie, ou histoire naturelle des coquilles de mer, d'eau douce, terrestres et fossiles . . .* 3rd ed. 3 vols. Paris.

Diderot, Denis (ed.). 1749. *Encyclopédie* 1.

Dolomieu, Déodat de. 1783. *Voyage aux Iles de Lipari fait en 1781, ou notices sur les Iles Æoliennes, pour servir à l'histoire des volcans . . .* Paris.

———. 1784. *Mémoire sur les tremblemens de terre de la Calabre pendant l'année 1783.* Rome.

———. 1784. Mémoire sur les volcans éteints du Val di Noto en Sicile. *OP* 25: 191–205.

———. 1788. *Mémoire sur les Iles Ponces, et catalogue raisonné des produits de l'Etna; pour servir à l'histoire des volcans: suivis de la description de l'éruption de l'Etna, du mois de juillet, 1787.* Paris.

———. 1790. Sur la question de l'origine du basalte. *OP* 37: 193–202.

———. 1791. Sur un genre de pierres calcaires très-peu effervescentes avec les acides. *OP* 39: 3–10.

———. 1791. Notes communiqués à Messieurs les naturalistes, qui font le voyage de la Mer de la Sud des contrées voisines au Pôle Austral. *OP* 39: 310–17.

———. 1791–92. Mémoire sur les pierres composées et sur les roches. *OP* 39: 374–407; 40: 41–62, 203–18, 372–403, 481.

———. 1793. Mémoire sur la constitution physique de l'Égypte. *OP* 42: 41–61, 108–26, 194–215.

———. 1794. Distribution méthodique de toutes les matières dont l'accumulation forme les montagnes volcaniques, ou tableau systématique dans lequel peuvent se placer toutes les substances qui ont les relations avec les feux souterrains. *JP* [44]: 102–25; [45]: 81–105.

———. 1794. Mémoire sur les roches composées en général, & particulièrement sur les pétro-siliex, les trapps & les roches de corne, pour servir à la distribution méthodique des produits volcaniques. *JP* [44] 175–200, 241–63.

———. 1795. Passage d'une lettre addressée à l'Agence des Mines. *JM* [2] (9): 59–60.

———. [1797]. Discours sur l'étude de la géologie, prononcé par Déodat Dolomieu, Membre de l'Institut National, à l'ouverture de son cours sur le gissement des minéraux, commencé en ventôse de l'an 5. *JP* [45]: 256–72.

———. 1797. Extrait du resumé succint des observations faites par le citoyen Dolomieu, pendant ses voyages minéralogiques et géologiques des six derniers mois de l'an V. *ME*, 3e année 5: 148–56.

———. 1798. Extrait d'une lettre du citoyen Dolomieu au citoyen Lamétherie. *ME* 4e année, 4: 249–51.

———. 1798. Rapport fait à l'Institut National, par le citoyen Dolomieu, Ingénieur des mines, sur ses voyages de l'an V et de l'an VI. *JM* [7]: 385–402, 405–32.

Drury, Susannah. 1744. *The Giant's Causeway in the county of Antrim in the Kingdom of Ireland* [2 sheet engravings]. [Dublin].

Dupain-Triel, Jean-Louis. 1791. *Recherches géographiques sur les hauteurs des plaines du royaume, sur les mers et leurs côtes presque pour tout le globe, et sur les diverses espèces de montagnes . . .* [Paris].

Ebel, Johann Gottfried. 1808. *Ueber den Bau der Erde in dem Alpengebirge, zwischen 12 Längen- und 2–4 Breitgraden nebst einigen Betrachtungen über die Gebirge und den Bau der Erde überhaupt, mit geognostischen Karten.* 2 vols. and atlas. Zurich.

Eichhorn, Johann Gottfried. 1803. *Einleitung in das Alte Testament.* 3rd ed. 2 vols. Leipzig.

Ellis, John. 1762. An account of an *Encrinus*, or starfish, with a jointed stem, taken on the coast of Barbadoes, which explains to what kind of animal those fossils belong, called Starstones, Asteriae, or Astropoda, which have been found in many parts of this Kingdom. *PTRS* 52: 357–65, pls. 13–14.

Englefield, Henry Charles. 1816. *A description of the principal picturesque beauties, antiquities, and geological phenomena, of the Isle of Wight; with additional observations on the Strata of the island . . . by T. Webster.* London.

Escher, Hans Conrad. 1818. Notice sur le Val de Bagnes en Bas Vallais, et sur la catastrophe qui en a devasté le fond, en juin 1818. *BU* 8: 291–308.

———. 1822. Matériaux pour l'histoire naturelle des blocs de roche, disseminés à la proximité des Alpes. *BU* 21: 259–82.

Esper, Johann Friedrich. 1774. *Ausführliche Nachricht von neuentdeckten Zoolithen unbekannter vierfüssiger Thiere und denen sie enthaltenden, so wie verschiedenen anderen denkwürdigen Grüften der Obergebürgischen Lande des Margrafthums Bayreuth.* Nuremberg [repr. Wiesbaden (Pressler), 1978].

———. 1774. *Description des zoolithes nouvellement découvertes d'animaux quadrupèdes inconnues.* Nuremberg.

———. 1784. Reise zu dem Gailenreuther Osteolithen-Höhlen; aus einem an die Gesellschaft eingeschikten und an des Bayerische Ministerium abgestatteten Bericht gezegen. *SBGNF* 5: 56–106.

Farey, John. 1806. On the stratification of England. *PM* 25: 44–49.

———. 1810. Geological remarks and queries on Messrs Cuvier and Brongniart's memoir on the mineral geography of the environs of Paris. *PM* 35: 113–39.

———. 1811–17. *General view of the agriculture and minerals of Derbyshire, with observations on the means of their improvement.* 3 vols. London. [vol. 1 (1811) repr. Peak District Mines Historical Society, 1989.]

[———?]. 1820. Reflections on the Noachian Deluge, and on the attempts lately made at Oxford, for connecting the same with present geological phenomena. *PM* 56: 10–14.

Faujas de Saint-Fond, Barthélemy. 1778. *Recherches sur les volcans éteints du Vivarais et du Velay; avec un discours sur les volcans brûlans, des mémoires analytiques sur les schorls, la zéolite, le basalte, la pouzzolane, les laves & les différentes substances qui s'y trouvent engagées, &c.* Grenoble and Paris.

———. [1797]. Lettre à Lamétherie sur les dents d'éléphans, d'hippopotame, et autres quadrupèdes, trouvés à dix-huit pieds de profondeur, dans une carrière, à une lieue à l'ouest de la ville de Orleans. *JP* 45 [1797; incorrectly dated 1794]: 445–48.

———. 1799. *Histoire naturelle de la Montagne de Saint-Pierre de Maestricht.* Paris.

———. 1803. Mémoire sur deux espèces de boeufs dont on trouve les crânes fossiles en Allemagne, en France, en Angleterre, dans le nord de l'Amérique et dans d'autres contrées. *AMHN* 2: 188–200, pls. 43–44.

———. 1803–9. *Essai de géologie, ou mémoires pour servir à l'histoire naturelle du globe.* Paris.

———. 1806. De coquilles fossiles des environs de Mayence. *AMHN* 8: 372–82, pl. 58.

Ferber, Johann Jakob. 1773. *Briefen aus Wälschland über natürliche Merkwürdigkeiten dieses Landes; an den Herausgeber derselben Ignatz Edlen von Born.* Prague.

———. 1776. *Travels through Italy, in the years 1771 and 1772, described in a series of letters to Baron Born, on the natural history, particularly the mountains and volcanos, of that country.* London.

———. 1786–88. Réflexions sur l'ancienneté relative des roches et des couches terreuses qui composent la croute du globe terrestre. *AASP* 6(2): 185–213; *NAASP* [1(2)]: 297–322; 2(2): 163–80.

Ferguson, Adam. 1766. *An essay on the history of civil society.* Edinburgh.

Fischer [von Waldheim], Gotthelf. 1802–03. *Das Nationalmuseum der Naturgeschichte zu Paris von seinen ersten Ursprung bis zu seinen jetzigen Glanze.* 2 vols. Frankfurt-am-Main.

Flourens, Pierre. 1856. *Receuil des éloges historiques.* 3 vols. Paris.

Fortia d'Urban, Agricole. 1805–09. *Mémoires pour servir à l'histoire ancienne du globe terrestre.* 10 vols. [volume titles vary]. Paris.

Fortis, Alberto. 1774. *Viaggio in Dalmazia.* Venice.

———. 1786. Extrait d'une lettre . . . sur différentes petrifications. *OP* 28: 161–68.

Fox-Strangways, William T. H. 1819. *Strata des environs de St Petersbourg en ordre de position géologique* [folio sheet]. St Petersburg.

———. 1821. Geological sketch of the environs of Petersburg. *TGS* 5: 392–458, pls. 28–31.

———. 1822. An outline of the geology of Russia. *TGS* (2) 1: 1–39, pls. 1–2.

Frere, John. 1800. Account of flint weapons discovered at Hoxne in Suffolk. *Archaeologia* 13: 204–5, pls. 14–15.

Fréret, Nicolas. 1751. De l'accroissement ou élévation du sol de l'Égypte par le débordement du Nil. *HARIB* 16: mém., 333–77.

Füchsel, Georg Christian. 1761. Historia terrae et maris, ex historia Thuringiae, per montium descriptionen eruta. *AAEM* 2: 44–254, pl. 5.

Garriga, José. 1796. *Descripción de un quadrúpedo muy corpulento y raro, que se conserva en el Real Gabinete de Historia Natural de Madrid.* Madrid [repr. López Piñero, *Bru de Ramón* (1996), 331–48, 5 pls.].

Geoffroy Saint-Hilaire, Étienne, and Georges Cuvier. 1795. Sur le *Rhinocéros bicorne. ME* [1e année] 1: 326–28.

———. 1795. Sur les espèces d'éléphans. *BSP* 1 (1): 90.

———. 1795. Mémoire sur une nouvelle division des mammifères, et sur les principes qui doivent servir de base dans cette sorte de travail . . . *ME* [1e année] 2: 164–90.

———. 1795. Histoire naturelle des orang-outangs. *ME* [1e année] 3: 451–63, 1 pl.

Geological Society. 1808. *Geological inquiries.* [London].

Gmelin, Johann Friedrich. 1777–79. *Des Ritters Carl von Linné . . . vollstandiges Natursystem des Mineralreichs; nach der zwölften lateinischen Ausgabe in einer freyen und vermehrten Uebersetzung von Johann Friedrich Gmelin.* 4 vols. Nuremberg.

Goldfuss, Georg August. 1809. *Vergleichende Naturbeschreibung der Säugethiere.* Erlangen.

———. 1810. *Die Umgebungen von Muggendorf: Taschenbuch für Freunde der Natur- und Alterthumskunde.* Erlangen.

Greenough, George Bellas. 1819. *A critical examination of the first principles of geology; in a series of essays.* London.

———. 1820. *A geological map of England and Wales* [sheet map]. London.

Guettard, Jean Étienne. 1761. Mémoire sur les encrinites et les pierres étoilées, dans lequel on traitera aussi des entroques, &c. *HARS* 1755: mém. 224–63, pls. 8–10.

———. 1768–86. *Mémoires sur différentes parties de la physique, de l'histoire naturelle, des sciences et des arts &c.* 5 vols. in 6. Paris.

Guignes, Joseph de. 1770. *Le Chou-king: un des livres sacrés du Chinois.* Paris.

———. 1808. *Voyages à Peking, Manille et l'Ile de France, faits dans l'intervalle des années 1784 à 1801.* 3 vols. and atlas. Paris.

———. 1809. Observations sur les sares des Chaldéens, et sur le nombre incroyable d'années qu'on assigne aux règnes de leurs premiers rois. *HARIB* 47: mém. 345–77.

———. 1809. Mémoire concernant l'origine du zodiaque du calendrier des orientaux, et celle de différentes constellations de leur ciel astronomique. *HARIB* 47: mém. 378–434.

Haidinger, Karl. 1787. *Systematische Eintheilung der Gebirgsarten: eine Abhandlung welcher am 30ten Dezember 1785 von der kais. Akademie zu St Petersburg der Preis zuerkannt wurde.* Vienna.

Hall, James. 1794. Observations on the formation of granite. *TRSE* 3(1): 8–12.

———. 1798. On the origins and principles of Gothic architecture. *TRSE* 4(2): 3–27, pls. 1–6.

———. 1799. Experiments on whinstone and lava. *TRSE* 5 (1): 43–75.

———. 1812. Account of a series of experiments, shewing the effects of compression in modifying the action of heat. *TRSE* 6 (1): 71–185, pls. 1–5.

———. 1813. *Essay on the origin, history and principles of Gothic architecture.* London and Edinburgh.

———. 1814. On the vertical position and convolution of certain strata, and their relation with granite. *TRSE* 7(1): 79–108, pls. 1–5.

———. 1814. On the revolutions of the earth's surface. *TRSE* 7(1): 139–211, pls. 6–9.

———. 1814. Des révolutions de la surface du globe. *BU* 56: 125–49, 250–61.

Hamilton, William. 1776. *Campi Phlegraei: Observations on the volcanos of the Two Sicilies, as they have been communicated to the Royal Society of London . . .* Naples [pls. repr. Carlo Knight, *Les fureurs de Vésuve*, Paris (Gallimard), 1992].

———. 1777. Account of the discoveries at Pompeii. *Archaeologia* 4: 160–75, pls. 6–18.

———. 1779. *Supplement to the Campi Phlegraei, being an account of the great eruption of Mount Vesuvius in the month of August 1779.* Naples.

———. 1783. An account of the earthquakes which happened in Italy, from February to May 1783. *PTRS* 63: 169–208.

———. 1791–95. *Collection of engravings from ancient vases mostly of pure Greek workmanship discoverd in sepulchres in the Kingdom of the Two Sicilies but chiefly in the neighbourhood of Naples during the course of the years 1789 and 1790.* 5 vols. Naples.

Haüy, René Just. 1801. *Traité de minéralogie.* Paris.

Henry, Robert. 1771–93. *The history of Great Britain, from the first invasion of it by the Romans under Julius Caesar.* 6 vols. London.

Héron de Villefosse, A. M. 1804. Considérations sur les fossiles, et particulièrement sur ceux que présente le pays de Hanovre; ou extrait raisonné d'un ouvrage de M. Blumenbach, ayant pour titre Specimen archaeologiae telluris . . . *JM* 16: 5–36.

Hombres-Firmas, L. A. d'. 1821. Notice sur les ossemens humains fossiles. *JP* 92: 227–33.

Home, Everard. 1799. Some observations on the structure of the teeth of graminivorous quadrupeds; particularly those of the elephant and *Sus Æthiopicus. PTRS* [89]: 237–58, pls. 13–21.

———. 1801. Observations on the structure, and mode of growth, of the grinding teeth of the wild boar, and animal incognitum. *PTRS* [91]: 319–32, pls. 20–23.

Hooke, Robert. 1705. *The posthumous works of Robert Hooke . . . containing his Cutlerian lectures, and other discourses read at the meetings of the illustrious Royal Society . . .* London.

[Horner, Leonard]. 1816. [Review of] G. B. Brocchi, *Conchiologia fossile subapennina. ER* 26: 156–80.

Howard, Philip. 1797. *The scriptural history of the earth and of mankind, compared with the cosmogonies, chronologies, and original traditions of ancient nations; an abstract and review of several modern systems; with an attempt to explain philosophically, the Mosaical account of the Creation and Deluge, and to deduce from this last event the causes of the actual structure of the earth.* London.

Humboldt, Alexander von. 1801. Esquisse d'un tableau géologique de l'Amérique méridionale. *JP* 53: 30–60.

———. 1810. *Vues des Cordillères, et monumens des peuples indigènes de l'Amérique.* Paris.

Humboldt, Alexander von, and Aimé Bonpland. 1805–34. *Voyages aux régions equinoxiales, faits en 1799, 1800, 1801, 1803, 1804, et 1805.* 30 vols. Paris.

Hume, David. 1757. *Four dissertations.* London.

———. 1778. *The history of England, from the invasion of Julius Caesar to the Revolution in 1688: a new edition . . .* London.

Hunter, John. 1794. Observations on the fossil bones presented to the Royal Society by his most Serene Highness the Margrave of Anspach, &c. *PTRS* [84]: 407–17, pls. 19–20.

———. 1859. *Observations and reflections on geology, intended to serve as an introduction to the catalogue of extraneous fossils.* London.

———. 1861. *Essays and observations on natural history, anatomy, physiology, and geology . . . arranged and revised, with notes . . . by Richard Owen.* 2 vols. London.

Hunter, William. 1769. Observations on the bones, commonly supposed to be elephant bones, which have been found near the river Ohio in America. *PTRS* 58: 34–45, pl. 4.

[Hutton, James]. 1785. *Abstract of a dissertation . . . concerning the system of the earth, its duration, and stability.* Edinburgh [repr. White and Eyles, *James Hutton*, 1970].

———. 1788. The theory of rain. *TRSE* 1: 41–86.

———. 1788. Theory of the earth; or an investigation of the laws observable in the composition, dissolution and restoration of the land upon the globe. *TRSE* 1: 209–304, pls. 1–2 [repr. White and Eyles, *James Hutton*, 1970].

———. 1790. Dissertation on written language as a sign of speech. *TRSE* 2 (1): 5–15.

———. 1790. Huttons Theorie der Erde; oder Untersuchung der Gesetze, die bey Entstehung, Auflösung und Wiederherstellung des Landes auf unserm Planeten bemerklich sind. Ein Auszug . . . *MNPN* 6 (4): 17–27.

———. 1794. *An investigation of the principles of knowledge . . .* 3 vols. Edinburgh [repr. Bristol (Thoemmes), ed. P. and J. Jones, 1999].

———. 1794. Observations on granite. *TRSE* 3(2): 77–85 [repr. White and Eyles, *James Hutton*, 1970].

———. 1795. *Theory of the earth, with proofs and illustrations.* 2 vols. Edinburgh [repr. Codicote (Wheldon and Wesley), 1959].

Imrie, (Major). 1798. A short mineralogical description of the Mountain of Gibraltar. *TRSE* 4(2): 191–202.

Jameson, Robert. 1800. *Mineralogy of the Scottish Isles; with mineralogical observations made in a tour through different parts of the mainland of Scotland, and dissertations upon peat and kelp.* 2 vols. Edinburgh.

———. 1802. *Mineralogische Reisen durch Schottland und die Schottischen Inseln; aus den Englischen übersetzt und von einem Auszuge aus Herrn Bergrath Werners Geognosie, die Lehre von den Gebirgsarten betreffend, als Einleitung begleitet von Heinrich Wilhelm Meuder.* Leipzig.

———. 1804–8. *System of mineralogy: comprehending oryctognosie, geognosie, mineralogical chemistry, mineralogical geography, and oeconomical mineralogy.* 3 vols. Edinburgh [vol. 3, *Elements of geognosy*, 1808, repr. New York (Hafner), 1976].

Jefferson, Thomas. 1787. *Notes on the state of Virginia, illustrated with a map, including the states of Virginia, Maryland, Delaware and Pennsylvania.* London.

———. 1799. A memoir on the discovery of certain bones of a quadruped of the clawed kind in the western parts of Virginia. *TAPS* 4: 246–60.

[Jomard, E. F.] (ed.). 1809–28. *Description de l'Égypte, ou receuil des observations et des recherches qui ont été faites en Égypte pendant l'expédition de l'armée française.* 20 vols. Paris.

Kant, Immanuel. 1755. *Allgemeine Naturgeschichte und Theorie des Himmels oder Versuch von der Verfassung und dem mechanische Ursprunge des ganzes Weltgebäudes nach Newtonischen Grundsätzen abgehandelt.* Königsberg [repr. Leipzig (Engelmann: Oswalds Klassiker 12), 1890; trans. *Universal natural history* . . ., Ann Arbor (University of Michigan Press), 1969].

Karg, Joseph Maximilian. 1805. Ueber den Steinbruch zu Oeningen bey Stein am Rheine, und dessen Petrefacte. *DGANS* 1: 1–74, pls. 1–2.

Karsten, Dietrich Ludwig Gustav. 1791. *Tabellarische Übersicht der mineralogisch-einfachen Fossilien.* Berlin.

Kay, John. 1838. *A series of original portraits and caricature etchings, with biographical sketches and illustrative anecdotes.* 2 vols. Edinburgh.

Kirwan, Richard. 1794. Examination of the supposed igneous origin of stony substances. *TRIA* 5: 51–87.

———. 1797. On the primitive state of the globe and its subsequent catastrophe. *TRIA* 6: 233–308.

———. 1799. *Geological essays.* London.

Knorr, Georg Wolfgang. 1766–67. *Deliciae naturae selectae, oder auserlesenes Naturalien-Cabinet . . ./Délices physiques choisies* . . . Nuremberg.

Knorr, Georg Wolfgang, and J. C. Emmanuel Walch. 1755–75. *Sammlung der Merkwürdigkeiten der Natur und Naturgeschichte der Versteinerungen.* 4 vols. Nuremberg.

König, Charles. 1814. On a fossil human skeleton from Guadaloupe. *PTRS* 1814 (1): 107–20, pl. 3.

Labaume, A. (ed.). 1805. *Recherches asiatiques, ou mémoires de la société établie au Bengale pour faire des recherches sur l'histoire et les antiquités, les arts, les sciences, et la littérature de l'Asie* . . . 2 vols. Paris.

La Borde, Jean-Benjamin, and Béat Fidèle Antoine Zurlauben. 1780–88. *Tableaux de la Suisse, ou voyage pittoresque fait dans les XIII cantons et états alliés au Corps Helvétique.* 4 vols. Paris.

Lacépède, Bernard, comte de, and Georges Cuvier. 1801–05. *La ménagerie du Muséum National d'Histoire Naturelle, ou les animaux vivants, peints d'après nature, sur vélin, par le citoyen Maréchal* . . . Paris.

Lacépède, Bernard, comte de, Georges Cuvier, and Jean-Baptiste de Lamarck. 1802. Rapport des professeurs du Muséum, sur les collections d'histoire naturelle rapportées d'Égypte par E. Geoffroy. *AMHN* 1: 234–41.

Lamanon, Robert de Paul de. 1781. Mémoire sur un os d'une grosseur énorme qu'on a trouvé dans une couche de glaise au milieu de Paris; et en général sur les ossemens fossiles qui ont appartenus à des grands animaux. *OP* 17: 393–405, pl. 2.

———. 1782. Description de divers fossiles trouvés dans les carrières de Montmartre près Paris, & vues générales sur la formations des pierres gypseuses. *OP* 19: 173–94, pls. 1–3.

Lamarck, Jean-Baptiste. 1794. *Recherches sur les causes des principaux faits physiques* . . . Paris.

———. 1797. *Mémoires de physique et d'histoire naturelle, établis des bases de raisonnement indépendantes de toute théorie* . . . Paris.

———. 1801. *Systême des animaux sans vertèbres, ou tableau général des classes, des ordres et des genres de ces animaux* . . . Paris.

———. 1802. *Hydrogéologie ou recherches sur l'influence qu'ont les eaux sur la surface du globe terrestre* . . . Paris [trans. A. V. Carozzi, *Hydrogeology*, Urbana (University of Illinois Press), 1964].

———. 1802. *Recherches sur les corps vivans.* Paris [repr. Paris (Fayard), 1986].

———. 1802–9. Mémoires sur les fossiles des environs de Paris, comprenant la détermination des espèces qui appartiennent aux animaux marins sans vertèbres, et dont la plupart sont figurés dans la collection des vélins du Muséum. *AMHN* 1: 299–312 [and many further brief installments, *AMHN* 1–14, with many pls. from 6: pls. 43–46 onwards].

———. 1804. Sur un nouvelle espèce de Trigonie, et sur une nouvelle d'huître découvertes dans le voyage du capitaine Baudin. *AMHN* 4: 351–59, pl. 67.

———. 1805. Considérations sur quelques faits applicable au théorie du globe, observés par M. Peron dans son voyage aux terres australes, et sur quelques questions géologiques qui naissent de la connoissance de ces faits. *AMHN* 6: 26–52.

———. 1809. *Philosophie zoologique, ou exposition des considérations relatives à l'histoire naturelle des animaux; à la diversité de leur organisation et des facultés qu'ils en obtiennent; aux causes physiques qui maintiennent en eux la vie et donnent aux mouvemens qu'il exécutent; enfin, à celles qui produisent, les unes le sentiment, et les autres l'intelligence de ceux qui en sont doués.* 2 vols. Paris.

La Métherie, Jean-Claude de. 1786. Discours préliminaire, contenant un précis des nouvelle découvertes. *OP* 28: 1–53.

———. 1787. Discours préliminaire. *OP* 30: 1–45.

———. 1790. Discours préliminaire. *OP* 36: 1–46.

———. 1791. Discours préliminaire. *OP* 38: 3–51.

———. 1797. *Théorie de la terre: second édition, corrigée, et augmentée d'une minéralogie.* 5 vols. Paris.

———. 1798. Discours préliminaire. *JP* 46: 3–134.

———. 1799. Discours préliminaire. *JP* 48: 3–99.

———. 1802. Idées de Werner sur quelques points de la géognosie: extraits de ses conversations. *JP* 55: 443–50.

———. 1804. *Considérations sur les êtres organisés.* 2 vols. Paris.

———. 1811. Observations sur les terrains qui paroissent avoir été formés sous l'eau douce. *JP* 72: 460–70.

Laplace, Pierre Simon. 1796. *Exposition du systême du monde.* Paris.

———. 1799–1805. *Traité de mécanique céleste.* 5 vols. Paris.

Launay, Louis. 1780. Mémoire sur l'origine des fossiles accidentels des Provinces Belgiques; précédé d'un discours sur la théorie de la terre. *MASBB* 2: 509–82.

———. 1786. Essai sur l'histoire naturelle des roches, précédé d'une exposé systematique des terres et des pierres . . . *In* [Académie de Saint Petersbourg], *Mémoires présentés à l'Académie,* 101 pp.

Lavoisier, Antoine-Laurent. 1793. Observations générales sur les couches modernes horizontales, qui ont été déposées par la mer, et sur les conséquences qu'on peut tirer de leur dispositions, relativement à l'ancienneté du globe terrestre. *HARS* 1789: mém., 351–71, pls. 1–7.

Le Grand d'Aussy, Pierre Jean-Baptiste. 1788. *Voyage d'Auvergne.* Paris.

Lehmann, Johann Gottlob. 1756. *Versuch einer Geschichte des Flötz-Gebürgen, betreffend deren Entstehung, Lage, darinne befindliche Metallen, Mineralien und Fossilien, grösstentheils aus eigenen Wahrnehmungen, chymischen und physicalischen Versuchen, und aus denen Grundsätzen der Natur-Lehre hergeleitet.* Berlin.

Leibniz, Gottfried Wilhelm. 1749. *Protogaea, sive de prima facie telluris et antiquissimae historiae vestigiis in ipsis naturae monumentis dissertatio* . . . Göttingen.

Léman, Sébastien. 1812. Note sur la gyrogonite. *NBSP* 3: 108–10.

Luc, Guillaume-Antoine de. 1802. Réflexions sur les zodiaques trouvés dans la Haute Égypte. *BB* 20: 94–104.

Luc, Jean-André de. 1772. *Recherches sur l'atmosphere* . . . Geneva.

———. 1778. *Lettres physiques et morales sur les montagnes et sur l'histoire de la terre et de l'homme: addressées à la Reine de la Grande-Bretagne.* The Hague.

———. 1779. *Lettres physiques et morales sur l'histoire de la terre et de l'homme: addressées à la Reine de la Grande-Bretagne.* 5 vols. in 6. The Hague and Paris.

———. 1788. Mémoire sur la question, "Que doit-on penser de la *gradation* que plusieurs philosophes, tant anciens que modernes, ont admise entre les *êtres naturels*; et jusqu'à quel point pouvons-nous parvenir à nous assurer de la réalité de cette *gradation*, et de l'ordre que la nature y observe?" *VHMW* 25: 457–98.

———. 1790–91. Letters to Dr James Hutton, F.R.S. Edinburgh, on his theory of the earth. *MR* 2: 206–27, 582–601; 3: 573–86; 5: 564–85.

———. 1790–93. Lettres à M. de La Métherie. *OP* 36–43 [31 installments].

———. 1793–95. Geological letters, addressed to Professor Blumenbach. *BC* 2: 231–38, 351–58; 3: 110–18, 226–37, 467–78, 589–98; 4: 212–18, 328–36, 447–57, 569–78; 5: 197–207, 316–26.

———. 1793–96. Herrn de Luc's geologische Briefen an Hrn. Prof. Blumenbach. *MNPN* 8 (4): 1–41; 9 (1): 1–123; 9 (4): 1–49; 10 (3): 1–20; 10 (4): 1–104; 11 (1): 1–71.

———. 1798. *Lettres sur l'histoire physique de la terre, addressées à M. le professeur Blumenbach; renfermant de nouvelles preuves géologiques et historiques de la mission divine de Moyse.* Paris.

———. 1803. *Abrégé des principes et des faits concernant la cosmologie et la géologie.* Brunswick.

———. 1809. *Traité élémentaire de géologie.* Paris.

———. 1810–11. *Geological travels.* 3 vols. London.

———. 1813. *Geological travels in some parts of France, Switzerland and Germany.* 2 vols. London.

Luc, Jacques-François de. 1762. *Observations sur les savans incrédules, et sur quelqu'uns de leurs écrits.* Geneva.

Maclure, William. 1809. Observations on the geology of the United States, explanatory of a geological map. *TAPS* 6: 411–28, 1 pl. [trans. in *JP* 69: 201–15; 72: 137–65, pl. 1 (1809–11)].

Marsden, William. 1783. *The history of Sumatra, containing an account of the government, laws, customs, and manners of the native inhabitants, with a description of the natural productions and a relation of the ancient political state of that island.* London.

Martini, Friedrich Heinrich Wilhelm, and Johann Hieronymus Chemnitz. 1768–95. *Neues systematisches Conchylien-Cabinet geordnet und beschreiben* . . . 11 vols. Nuremberg.

Marum, Martinus van. 1790. Beschrijving der beenderen van een kop van eenen fisch, gevonden in den St. Pietersberg bij Maastricht, en geplaatst in Teylers Museum. *VTTG* 8: 383–89, pls. 1–2.

Mecheln, Christian von. 1790. *Voyage de M. de Saussure à la cime du Mont Blanc au mois d'août MDCCLXXXVII* [two sheet engravings]. Basel.

———. 1791. *Explication des renvois de l'estampe enluminée qui représente la vallée de Chamouni, et les montagnes adjacentes; executée d'après le relief de M. Exchaquet.* Basel.

Merck, Johann Heinrich. 1782. *Lettre à Monsieur de Cruse: sur les os fossiles d'éléphans et de rhinocéros qui se trouvent dans le pays de Hesse-Darmstadt.* Darmstadt.

———. 1784. *Seconde lettre à Monsieur de Cruse . . . sur les os fossiles d'éléphans et de rhinocéros qui se trouvent en Allemagne et particulièrement dans le pays de Hesse-Darmstadt.* Darmstadt.

———. 1786. *Troisième lettre sur les os fossiles d'éléphans et de rhinocéros qui se trouvent en Allemagne et particulièrement dans le pays de Hesse-Darmstadt: addressée à Monsieur Forster.* Darmstadt.

Michaelis, Christian Friedrich. 1785. Ueber ein Thiergeschlect der Urwelt. *GMWL* 4e Jg. 2: 25–48.

Michell, John. 1760. Conjectures concerning the cause, and observations upon the phenomena of earthquakes; particularly of that great earthquake of the First of November, 1755, which proved so fatal to the city of Lisbon, and whose effects were felt as far as Africa, and more or less throughout Europe. *PTRS* 51: 566–634, pl. 13.

Monnet, Antoine Grimoald. 1780. *Atlas et description minéralogique de la France entrepris par ordre du Roi.* Paris.

Montesquieu, Charles de Secondat. 1748. *De l'esprit des lois, ou, du rapport que les lois doivent avoir avec la constitution de chaque gouvernement* . . . 2 vols. Geneva.

Montlosier, François-Dominique de Reynaud. 1789 [repr. 1802]. *Essai sur la théorie des volcans d'Auvergne.* [Paris].

[Murray, John]. 1802. *A comparative view of the Huttonian and Neptunian systems of geology: in answer to the Illustrations of the Huttonian Theory of the Earth, by Professor Playfair*. Edinburgh.

Olivi, Giuseppe. 1792. *Zoologia Adriatica, ossia catalogo ragionato degli animali del Golfo e delle Lagune de Venezia; preceduto da une dissertazione sulla storia fisica e naturale del Golfo; e accompagnato da memorie, ed osservazioni fisica storia naturale economia*. Bassano.

Paccard, Michel Gabriel. [1786]. [Prospectus for] *Premier voyage à la cime de la plus haute montagne de l'ancien continent, le Mont Blanc, par le docteur Michel Paccard, médecin dans les Alpes & Chamonix, le 8 aout 1786*. [Geneva?].

Paine, Thomas. 1794–95. *The age of reason: being an investigation of true and fabulous theology*. Paris, London.

Pallas, Peter Simon. 1769. De ossibus Siberiae fossilibus craniis praesertium rhinocerotum atque buffalorum, observationes. *NCASP* 13: 436–77, pls. 9–12.

———. 1771–76. *Reisen durch verschiedene Provinzen des Russischen Reichs*. 4 vols. St Petersburg.

———. 1778. Observations sur la formation des montagnes & les changemens arrivés au globe, particulièrement à l'égard de l'Empire de Russe. *AASP* 1777: 21–64.

Parkinson, James. 1804–11. *Organic remains of a former world: an examination of the mineralized remains of the vegetables and animals of the antediluvian world; generally termed extraneous fossils*. 3 vols. London.

———. 1811. Observations on some of the strata in the neighbourhood of London, and on the fossil remains contained in them. *TGS* 1: 324–54.

Peale, Charles Willson. 1802. Extrait d'une lettre de M. Peales [*sic*], directeur du Muséum d'histoire naturelle de Philadelphie, au citoyen Geoffroy. *AMHN* 1: 251–53.

Peale, Rembrandt. 1802. *Account of the skeleton of the mammoth, a non-descript carnivorous animal of immense size, found in America*. London.

Pennant, Thomas. 1771. *Synopsis of quadrupeds*. Chester.

———. 1774–76. *A tour of Scotland, MDCCLXXII*. Chester, London.

Péron, M.-François. 1804. Sur quelques faits zoologiques applicables à la théorie du globe. *JP* 59: 463–79.

Phillips, John. 1844. *Memoirs of William Smith, Ll.D.* London [repr. Bath (Bath Royal Literary and Scientific Institution), ed. Hugh Torrens, 2003].

Phillips, William. 1815. *An outline of mineralogy and geology, intended for the use of those who may desire to become acquainted with the elements of those sciences; especially of young persons*. London.

———. 1816. *Outlines of mineralogy and geology*. 2nd ed. London.

———. 1818. *A selection of facts from the best authorities, arranged so as to form an outline of the geology of England and Wales*. London.

Pinel, Philippe. 1793. Nouvelles observations sur la structure & la conformation des os de la tête de l'éléphant. *JP* 43: 47–60, 1 pl.

Pini, Ermenegildo. 1790. Saggio di una nuova teoria della terra. *OSSA* 13: 361–89.

———. 1790–92. Sulle rivoluzioni del globo terrestre provenienti dall'azione dell'acque: memoria geologica. *MSIS* 5: 163–258; 6: 389–500.

Playfair, John. 1802. *Illustrations of the Huttonian theory of the earth*. Edinburgh [repr. New York (Dover), 1956].

———. 1805. Biographical account of the late Dr James Hutton, F.R.S. Edin. *TRSE* 5(3): 39–99.

[———?]. 1814. [Review of] Essay on the Theory of the Earth: translated from the French of M. Cuvier . . . by Robert Kerr . . . with mineralogical notes . . . by Professor Jameson, Edinburgh, 1813. *ER* 22: 454–75.

[———?]. 1815. [Trans. of review of] Essai sur la théorie de la terre . . . Edimbourg, 1813. *BU* 59: 353–68; 60: 3–23, 105–26.

Poiret, Jean Louis Marie. 1800–1803. Mémoire sur la tourbe pyriteuse du département de l'Aisne; sur sa formation; les différentes substances qu'elle contient, et ses rapports avec la théorie de la terre. *JP* 51: 292–304; 53: 5–19; 55: 189–96; 57: 249–59.

———. 1804. Mémoire sur la formation des tourbes. *JP* 59: 81–91.

———. 1805. Dissertation sur l'étude et les principes de la géologie. *JP* 61: 5–17.

Prévost, Constant. 1820. Sur la constitution physique et géognostique du bassin à l'ouverture duquel est située la ville de Vienne. *JP* 91: 347–67, 460–73, 1 pl.

Prévost, Constant, and Anselme-Gaëtan Desmarest. 1809. Sur les empreintes de corps marins trouvées à Montmartre, dans plusieurs couches de la masse inférieure de la formation gypseuse. *JM* 25: 215–26.

Razumovsky, Grigory Kirillovich. 1789. *Histoire naturelle de Jorat et de ses environs; et celle des trois lacs de Neuchâtel, Morat et Bienne* . . . Lausanne.

[Reale Accademia di Napoli]. 1784. *Istoria de'fenomeni del tremuto avvenuto nelle Calabrie, e nel Valdemone nell'anno 1783.* Naples.

Recupero, Giuseppe. 1815. *Storia naturale e generale dell'Etna* [ed. Agatino Recupero]. 2 vols. Catania.

Reuss, Franz Ambros. 1801–06. *Lehrbuch der Mineralogie.* 4 vols. in 8. Leipzig.

Richard de Saint-Non, Jean Claude. 1781–86. *Voyage pittoresque ou description des Royaumes de Naples et de Sicilie.* 4 vols. in 5. Paris.

Robertson, William. 1769. *The history of the reign of Charles V, with a view of the progress of society in Europe, from the subversion of the Roman Empire, to the beginning of the sixteenth century.* London.

Rosenmüller, Johann Christian. 1794. *Quaedam de ossibus fossilibus animalis cuiusdam, historiam eius et cognitionem accuratiorem illustrantia.* Leipzig.

———. 1795. *Beiträge zur Geschichte und nähern Kenntnis fossiler Knochen.* Leipzig.

———. 1796. *Abbildungen und Bescheibungen merkwürdiger Hölen um Muggendorf im Bayreuthischen Oberlands für Freunde der Natur und Kunst.* Erlangen.

———. 1804. *Abbildungen und Beschreibungen der fossilen Knochen des Höhlenbären.* Weimar.

———. 1804. *Die Merkwürdigkeiten der Gegend um Muggendorf.* Berlin.

Sale, George, et al. (eds.). 1747–48. *An universal history, from the earliest account of time.* 2nd ed. 20 vols. London.

———. 1759–66. *The modern part of an universal history* . . . 44 vols. London.

Saussure, Horace-Bénédict de. 1779–96. *Voyages dans les Alpes, précédés d'un essai sur l'histoire naturelle des environs de Genève.* 4 vols. Neuchâtel.

———. 1787. *Relation abrégée d'un voyage à la cime du Mont Blanc en août 1787.* Geneva.

———. 1796. Agenda, ou tableau général des observations et des recherches dont les résultats doivent servir de base à la théorie de la terre. *JM* 4: 1–70.

Saussure, Nicolas-Théodore de. 1792. Analyse de la dolomie. *OP* 40: 161–72.

Scheuchzer, Johann Jakob. 1709. *Herbarium diluvianum.* Zurich.

———. 1726. *Homo diluvii testis et theoscopos.* Zurich.

———. 1731. *Physica sacra* . . . 4 vols. Augsburg and Ulm.

Schleiermacher, Friedrich. 1799. *Über die Religion: Reden an die Gebildeten unter ihren Verächtern.* Berlin [trans. J. Oman, *On religion*, repr. New York (Harper and Row), 1958].

Schlotheim, Ernst Friedrich von. 1804. *Beschreibung merkwürdiger Kräuter-Abdrücke und Pflanzen-Versteinerungen: ein Beitrag zur Flora der Vorwelt.* Gotha.

———. 1820. *Die Petrefaktenkunde auf ihrem jetzigen Standpunkte durch die Beschreibung seiner Sammlung versteinerter und fossiler Überreste des Thier- und Pflanzenreichs der Vorwelt erläutert.* Gotha.

Schröter, Johann Samuel. 1779–88. *Lithologisches Real- und Verballexicon, in welchem nicht nur die Synonymien der deutschen, lateinischen, französischen und holländischen Sprachen angeführt und erläutert, sondern auch alle Steine und Versteinerungen ausführlich beschrieben werden.* 8 vols. Frankfurt-am-Main.

Smith, William. 1801. *Prospectus of a work, entitled, Accurate delineations of the natural order of the various strata that are found in different parts of England and Wales; with practical observations thereon.* London.

———. 1815. *A delineation of the strata of England and Wales with part of Scotland . . .* [sheet map]. London. [repr. (at reduced scale), Keyworth (British Geological Survey), 2003]

———. 1815. *A memoir to the map and delineation of the strata of England and Wales, with part of Scotland.* London.

———. 1816–19. *Strata identified by organized fossils, containing prints on coloured paper of the most characteristic specimens in each stratum.* London.

———. 1817. *Stratigraphical system of organized fossils, with reference to the specimens of the original geological collection in the British Museum; explaining their state of preservation and their use in identifying the British strata.* London.

———. 1819. *Section of the strata, through Hampshire and Wiltshire to Bath, on the road from Bath to Salisbury* [folio sheet]. London.

Soulavie, Jean Louis Giraud-. 1780. La géographie de la nature, ou distribution naturelle des trois règnes sur la terre: description d'une carte du Vivarais, dressée en relief, où cette distribution est enluminée selon la nature du sol & les variétés des êtres organisés: méthode pour rendre par des reliefs la forme du sol d'une Province dont on écrit l'Histoire Naturelle, pour l'enluminer selon la nature du terrein. *OP* 16: 63–73.

———. 1780–84. *Histoire naturelle de la France méridionale, ou recherches sur la minéralogie du Vivarais . . . de l'Auvergne . . . de la Provence . . .* 7 vols. Nîmes and Paris.

———. 1783. Description des couches superposées de laves du volcan de Boutaresse en Auvergne, et observations sur une planche travaillée par le main de l'homme, trouvée sous des coulées de laves. *OP* 22: 289–94.

———. 1786. Les classes naturelles des minéraux et les époques de la nature correspondantes à chaque classe . . . *In* [Académie de Saint Petersbourg], *Mémoires présentés à l'Académie,* 161 pp., 1 pl.

———. 1790–93. *Mémoires du Maréchal Duc de Richelieu, pour servir à l'histoire du cours de Louis XIV, de la minorité et du règne de Louis XV, etc., etc.* 9 vols. Paris.

Sowerby, James [and James de Carle Sowerby]. 1812–46. *The mineral conchology of Great Britain.* 7 vols. London.

Spallanzani, Lazzaro. 1786. Osservazioni fisiche instituite nell'Isola di Citera oggidì detta Cerigo. *MSIS* 3: 439–64, 1 pl.

———. 1792–97. *Viaggi alle Due Sicilie e in alcuna parti dell'Apennino.* 6 vols. Pavia.

———. 1795–97. *Voyages dans les Deux Siciles et dans quelques parties des Apennins.* 5 vols. Bern.

Steno, Nicolaus. 1669. *De solido intra solidum naturaliter contento dissertationis prodromus.* Florence.

Stillingfleet, Edward. 1662. *Origines sacrae, or a rational account of the grounds of Christian faith, as to the truth and divine authority of the Scriptures, and the matters therein contained.* London.

Sumner, John Bird. 1816. *A treatise on the records of the Creation, and on the moral attributes of the Creator; with particular reference to Jewish history, and to the consistency of the principle of population with the wisdom and goodness of the Deity.* 2 vols. London.

Tarner, George Edward. 1916. *Axioms and postulates of athetic philosophy.* Cambridge.

———. 1924. *A letter to the Vice-Chancellors of both universities.* Cambridge.

Thiéry, Luc Vincent. 1787. *Guide des amateurs et des étrangers voyageurs à Paris, ou description raisonnée de cette ville, & de tout ce qu'elles contiennent de remarquable . . .* 2 vols. Paris.

Tilesius [Wilhelm Gottlieb von Tilenau]. 1815. De skeleto mammonteo Sibirico ad Maris Glacialis littora anno 1807 effosso, cui praemissae elephantini generis specierum distinctiones. *MASSP* 5: 406–513, pls. 10–11.

Toulmin, George Hoggart. 1780. *The antiquity and duration of the world.* London.

———. 1789. *The eternity of the universe.* London.

Trebra, Friedrich Wilhelm Heinrich von. 1785. *Erfahrungen vom Innern der Gebirge nach Beobachtungen gesammelt.* Dessau and Leipzig.

Viallon, (M.). 1792. Lettre à J. C. Delamétherie [sur la géologie]. *OP* 40: 224–28.

[Volta, Giovanni Serafino]. 1796[–1809]. *Ittiolitologia Veronese del Museo Bozziano ora annesso a quello di conte Giovambattista Gazola e di altri gabinetti di fossili Veronesi con la versione latina.* Verona.

Warner, Richard. 1801. *The history of Bath.* Bath.

Webster, Thomas. 1814. On the freshwater formations in the Isle of Wight, with some observations on the strata over the Chalk in the south-east part of England. *TGS* 2: 161–254, pls. 9–11.

Werner, Abraham Gottlob. 1774. *Von den äusserlichen Kennzeichen der Fossilien.* Leipzig.

———. 1787. *Kurze Klassifikation und Bescheibung der verschiedenen Gebirgsarten.* Dresden.

———. 1788. Von den verschiedenen Graden der Festigkeit des Gesteins, als dem Hauptgrunde der Hauptverschiedenheiten des Häuerarbeiten. *BJ* 1: 4–21.

———. 1788. Werners Bekanntmachung einer von ihm am Scheibenberger Hügel über die Entstehung des Basalts gemachten Entdeckung . . . *BJ* 2: 845–55.

———. 1791. *Neue Theorie von der Entstehung der Gänge, mit Anwendung auf dem Bergbau besonders den Freibergischen.* Freiberg.

———. 1802. *Nouvelle theorie de la formation des filons; application de cette théorie à l'exploitation des mines particulièrement de celles de Freiberg.* Paris.

Whitehurst, John. 1778. *An inquiry into the original state and formation of the earth; deduced from the facts and the laws of nature.* London.

———. 1786. *An inquiry into the original state* . . . 2nd ed. London.

Williams, John. 1789. *The natural history of the mineral kingdom.* 2 vols. Edinburgh.

Wistar, Caspar. 1799. A description of the bones deposited, by the President [Jefferson], in the Museum of the Society, and represented in the annexed plates. *TAPS* 4: 526–31, 2 pls.

Woodward, John. 1695. *An essay toward a natural history of the earth and terrestrial bodies, especially minerals, as also of the seas, rivers and springs; with an account of the universal deluge, and of the effects that it had upon the earth.* London.

Wrede, Erhard Georg Friedrich. 1804. *Geognostische Untersuchungen über die Südbaltischen Länder, besonders über das untere Odergebiet; nebst einer Betrachtung über die allmählige Veränderung des Wasserstandes auf der nordlichen Halbkügel der Erde und deren physische Ursachen.* Berlin.

4. PRINTED SOURCES: SECONDARY

Biographical sources are only listed here if they add substantially to information in Charles C. Gillispie (ed.), *Dictionary of Scientific Biography* [*DSB*], 16 vols., New York (Scribner's), 1970–80; in J. C. Poggendorff, *Biographisch-litterarisches Handwörterbuch zur Geschichte der Exacten Wissenschaften*, 2 vols., Leipzig, 1863; or in the relevant national biographical dictionaries and other standard sources. It is worth noting that secondary sources in the *history* of the sciences often enjoy a much longer useful half-life than most papers in the natural sciences themselves. Not listed here, nor cited in the text, are the many recent popular books on the history of the earth sciences, most of which are wholly parasitic on more scholarly secondary sources, and many of which perpetuate historical myths that have long been rejected by historians.

The following list is a key to the abbreviations used in this bibliography for items published in periodicals (those with single-word titles are not abbreviated, and are not included in this list).

AGS	Annales Guébhard-Séverine
AH	Art history

AHR	Agricultural history review
AHSS	Annales: histoire, science sociales
AIV	Atti dell'Istituto Veneto di Scienze, Lettere e Arti
AM	Annales des mines
AMHN	Archives du Muséum d'Histoire Naturelle
AmS	American scientist
ANH	Archives of natural history [continuation of *JSBNH*]
ARCS	Annals of the Royal College of Surgeons
AS	Annals of science
ASG	Archives des sciences, Genève
BAAPG	Bulletin of the American Association of Petroleum Geologists
BBMNH	Bulletin of the British Museum (Natural History)
BGSA	Bulletin of the Geological Society of America
BJHS	British journal of the history of science
BJRL	Bulletin of the John Rylands Library
BMHNP	Bulletin du Muséum d'Histoire Naturelle de Paris
BSGF	Bulletin de la Société Géologique de France
BW	Berichte zur Wissenschaftsgeschichte
CRAS	Comptes-Rendus de l'Académie des Sciences, Paris
CRP	Comptes Rendus Palevol [Académie des Sciences, Paris]
DHVS	Documents pour l'histoire du vocabulaire scientifique
DSB	Dictionary of scientific biography
EDS	Études sur le XVIIIe siècle
EF	Erlanger Forschungen
ES	Engineering and science
ESH	Earth sciences history
GSAM	Geological Society of America, Memoirs
GSASP	Geological Society of America, special papers
HB	Histoire et biologie
HN	Histoire et nature
HS	History of science
HSPS	Historical studies in the physical sciences
HT	History and theory
IAJ	Irish astronomical journal
JGB	Jahrbuch der Geologischen Bundesanstalt [Vienna]
JGE	Journal of geological education
JHB	Journal of the history of biology
JHC	Journal of the history of collections
JHI	Journal of the history of ideas
JSBNH	Journal of the Society for the Bibliography of Natural History
MG	Mercian geologist
MMNHN	Mémoires du Muséum National d'Histoire Naturelle, Paris
MP	Miscellanea Paleontologica [Verona]
MSPHN	Mémoires de la Société de Physique et d'Histoire Naturelle de Genève
MSVSN	Mémoires de la Société Vaudoise des Sciences Naturelles
NRRSL	Notes and records of the Royal Society of London
OE	Open earth

PAPS	Proceedings of the American Philosophical Society
PBNS	Proceedings of the Bristol Naturalists' Society
PGA	Proceedings of the Geologists' Association
PGSL	Proceedings of the Geological Society of London
PNGL	Publicaties van het Natuurhistorische Genootschap van Limburg
PRSE	Proceedings of the Royal Society of Edinburgh
PYGS	Proceedings of the Yorkshire Geological Society
RDAAS	Reports and Transactions of the Devon Association for the Advancement of Science
RHMC	Revue d'histoire moderne et contemporaine
RHS	Revue d'histoire des sciences
RS	Ricerche storiche
RSH	Revue des sciences humaines
SAC	Sussex archaeological collections
SC	Science in context
SHPBBS	Studies in the history and philosophy of the biological and biomedical sciences
SHPS	Studies in the history and philosophy of science
SJT	Scottish journal of theology
SSS	Social studies of science
SVEC	Studies on Voltaire and the eighteenth century
TAPS	Transactions of the American Philosophical Society
TCFHG	Travaux du Comité Français pour l'Histoire de la Géologie (COFRHIGEO)
TGGNWT	Tijdschrift voor de geschiedenis der geneeskunde, natuurwetenschappen, wiskunde en techniek
TRGSC	Transactions of the Royal Geological Society of Cornwall
VDMR	Virginia Division of Mineral Resources
ZDGG	Zeitschrift der Deutsche Geologische Gesellschaft

Aguillon, L. 1889. L'École des Mines de Paris: notice historique. *AM* (8) 15: 433–686.

Albrecht, Helmuth, and Roland Ladwig (eds.). 2002. *Abraham Gottlob Werner und die Begründung der Geowissenschaften.* Freiberg (Technische Universität Bergakademie Freiberg).

Allchin, D. 1994. James Hutton and phlogiston. *AS* 51: 615–35.

Allen, Don Cameron. 1949. *The legend of Noah: Renaissance rationalism in art, science, and letters.* Urbana (University of Illinois).

Anglesea, Martyn, and John Preston. 1980. 'A philosophical landscape': Susanna Drury and the Giant's Causeway. *AH* 3: 252–73, figs. 8–18.

Anonymous (ed.). 1893. *Centenaire de la fondation du Muséum d'Histoire Naturelle, 10 juin 1793–10 juin 1893: volume commemoratif publié par les professeurs du Muséum.* Paris (Imprimerie Nationale).

Anonymous (ed.). 1967. *Abraham Gottlob Werner: Gedenkschrift aus Anlass der Wiederkehr seines Todestages nach 150 Jahren am 30. Juni 1967.* Leipzig (VEB).

Anonymous (ed.). 1974. *Approches des Lumières: mélanges offerts à Jean Fabre.* Paris (Klincksieck).

Anonymous (ed.). 1978. *'Teyler' 1778–1978: studies en bijdragen over Teylers Stichting naar aanleiding van het tweede eeuwfeest.* Haarlem (Teylers Stichting).

Anonymous (ed.). 1994. *Montbéliard sans frontieres: colloque international de Montbéliard, 8 et 9 octobre 1993.* Montbéliard (Société d'Émulation de Montbéliard).

Anonymous (ed.). 2003. L'Essor de la géologie française au XIXe siècle. *Géochronique* no. 88: 16–42.

Appel, Toby A. 1987. *The Cuvier-Geoffroy debate: French biology in the decades before Darwin.* New York and London (Oxford University Press).

Archinard, Margarida. 1975. De Luc et la recherche barométrique. *Gesnerus* 32: 235–47.

———. 1979. *Collection de Saussure*. Geneva (Musée d'Art et d'Histoire).

Ashworth, William B., Jr. 1984. *Theories of the earth, 1644–1830: the history of a genre*. Kansas City, Missouri (Linda Hall Library).

Aufrère, Léon. 1952. *Soulavie et son secret: un conflit entre l'actualisme et la créationisme: le temps géomorphologique*. Paris (Hermann).

Bailey, E. B. 1967. *James Hutton: the founder of modern geology*. Amsterdam, London, and New York (Elsevier).

Baker, Keith Michael. 1988. Revolution. *In* Lucas, *French revolution*, 2: 41–62.

Banks, R. E. R., B. Elliott, J. G. Hawkes, D. King-Hele and G. Ll. Lucas (eds.). 1994. *Sir Joseph Banks: a global perspective*. London (Royal Botanic Gardens).

Barr, James. 1985. Why the world was created in 4004 B.C.: Archbishop Ussher and biblical chronology. *BJRL* 67: 575–608.

Barton, Ruth. 2003. 'Men of science': language, identity and professionalization in the mid-Victorian scientific community. *HS* 41: 73–119.

Bedini, S. A. 1985. Thomas Jefferson and American vertebrate paleontology. *VDMR* Publ. 61.

Belaval, Yvon, and Dominique Bourel (eds.). 1986. *Le siècle des Lumières et la bible*. Paris (Beauchesne).

Bell, W. J. 1949. A box of old bones: a note on the identification of the mastodon, 1766–1806. *PAPS* 93: 169–77.

Berti, Giampietro. 1987. *Un naturalista dell'ancien régime alla restaurazione: Giambattista Brocchi*. Bassano (G. B. Verci).

———. 1987. Aspetti della formazione scientifica e intellettuale di Giambattista Brocchi. *In* Marini, *Brocchi*, 13–40.

Blanckaert, Claude, Claudine Cohen, Petro Corsi, and Jean-Louis Fischer (eds.). 1997. *Le Muséum au premier siècle de son histoire*. Paris (Muséum National d'Histoire Naturelle).

Bloch, Olivier, Bernard Balan, and Paulette Carrive (eds.). 1988. *Entre forme et histoire: la formation de la notion de développement à l'âge classique*. Paris (Meridien Klincksieck).

Bödecker, Hans Erich, Georg G. Iggers, Jonathan B. Knudsen, and Peter H. Reill (eds.). 1986. *Aufklärung und Geschichte: Studien zur Deutschen Geschichtswissenschaft im 18. Jahrhunderts*. Göttingen (Vandenhoeck and Ruprecht)

Bologna, Ferdinando. 1990. The rediscovery of Herculaneum and Pompeii in the artistic culture of Europe in the eighteenth century. *In* Conticello, *Rediscovering Pompeii*, 79–91.

Bork, Kennard B. 1990. Constant Prévost (1787–1856): the life and contributions of a French uniformitarian. *JGE* 38: 21–27.

Bots, Hans, and Rob Visser (eds.). 2003. *La correspondance, 1785–1787, de Petrus Camper (1722–1789) et son fils Adriaan Camper (1759–1820)*. Amsterdam and Utrecht (APA-Holland University Press).

Bots, Hans, and Françoise Waquet (eds.). 1994. *Commercium litterarium: la communication dans la Republique des Lettres 1600–1750*. Amsterdam (APA-Holland).

Bourguet, Marie-Noëlle. 1997. La collecte du monde: voyage et histoire naturelle (fin XVIIème siècle–début XIXème siècle). *In* Blanckaert et al., *Le Muséum*, 163–96.

Bourguet, Marie-Noëlle, and Christian Licoppe. 1997. Voyages, mesures et instruments: une nouvelle expérience du monde au siècle des Lumières. *AHSS* 52: 1115–51.

Bourrouilh-LeJan, Françoise G. 2000. Déodat de Gratet de Dolomieu (1750–1801), vie et oeuvre d'un géologue européen, naturaliste et lithologiste. *CRAS* 330 (IIA): 83–95.

Boyer, Ferdinand. 1971. Les conquêtes scientifiques de la Convention en Belgique et dans les pays rhénans (1794–1795). *RHMC* 18: 354–74.

———. 1973. Le Muséum d'Histoire Naturelle à Paris et l'Europe des sciences sous la Convention. *RHS* 26: 251–57.

Boylan, Patrick J. 1970. An unpublished portrait of Dean William Buckland, 1784–1856. *JSBNH* 5: 350–54.

———. 1984. *William Buckland, 1784–1856: scientific institutions, vertebrate palaeontology, and Quaternary geology.* Leicester (University of Leicester [Ph.D. dissertation]).

———. 1997. William Buckland (1784–1856) and the foundations of taphonomy and palaeoecology. *ANH* 24: 361–72.

Bret, Patrice (ed.). 1999. *L'Expédition d'Égypte: une enterprise des Lumières, 1798–1801.* Paris (Tec et Doc).

Brianta, Donata. 2000. Education and training in the mining industry, 1750–1860: European models and the Italian case. *AS* 57: 267–300.

Broadie, Alexander (ed.). 2003. *The Cambridge companion to the Scottish Enlightenment.* Cambridge (Cambridge University Press).

Broc, Numa. 1991. *Les montagnes au siècle des Lumières.* Paris (Comité des Travaux Historiques et Scientifiques) [repr. of *Les montagnes vues par les géographes et les naturalistes de langue française au XVIIIe siècle*, Paris, 1969].

Brock, Michael G., and Mark C. Curthoys (eds.). 1997. *Nineteenth-century Oxford* [*History of the University of Oxford* 6]. Oxford (Clarendon).

Broman, Thomas. 2000. Periodical literature. *In* Frasca-Spada and Jardine, *Books and the sciences*, 225–38.

Brooke, John H. 1991. *Science and religion: some historical perspectives.* Cambridge (Cambridge University Press).

———. 2003. Science and religion. *In* Porter, *Eighteenth-century science*, 741–61.

Brown, T. G., and Gavin De Beer. 1957. *The first ascent of Mont Blanc, published on the occasion of the centenary of the Alpine Club.* London (Alpine Club).

Browne, Janet. 1983. *The secular ark: studies in the history of biogeography.* New Haven (Yale University Press).

———. 2003. Noah's Flood, the Ark, and the shaping of early modern natural history. *In* Lindberg and Numbers, *Science and Christianity*, 111–38.

Bruijn, J. G. de. 1977. *Inventaris van de prijsvragen uitgescheven door de Hollandsche Maatschappij der Wetenschappen, 1753–1917.* Haarlem (Hollandsche Maatschappij der Wetenschappen).

Brush, Stephen G. 1988. *The history of modern science: a guide to the second scientific revolution, 1800–1950.* Ames: Iowa University Press.

Buchwald, Jed Z. 2003. Egyptian stars under Paris skies. *ES* 66: 20–31.

Buffetaut, E., J. M. Mazin, and E. Salmon (eds.). 1982. *Actes du symposium paléontologique G. Cuvier.* Montbéliard (Ville de Montbéliard).

Bultingaire, L. 1932. Iconographie de Georges Cuvier. *AMHN* (6) 9: 1–12, pls. 1–11.

Burke, Peter (ed.). 1991. *New perspectives on historical writing.* Cambridge (Polity).

Burkhardt, Richard W., Jr. 1977. *The spirit of system: Lamarck and evolutionary biology.* Cambridge (Harvard University Press).

———. 1997. La Ménagerie et la vie au Muséum. *In* Blanckaert et al., *Le Muséum*, 481–508.

———. 1997. Unpacking Baudin: models of scientific practice in the age of Lamarck. *In* Laurent, *Jean-Baptiste Lamarck*, 497–514.

Burn, Lucilla. 1997. Hamilton and the Greekness of Greek vases. *JHC* 9: 187–89.

Buyssens, Danielle. 2001. Saussure mémorable dans les arts et les sciences. *In* Sigrist, *Saussure*, 1–22.

Bynum, W. F., and Roy Porter (eds.). 1985. *William Hunter and the eighteenth-century medical world.* Cambridge (Cambridge University Press).

Cannon, Susan Faye. 1978. *Science in culture: the early Victorian period.* New York (Dawson and Science History).

Carozzi, Albert V. (ed.). 1968. *Telliamed, or conversations between an Indian philosopher and a French missionary, on the diminution of the sea.* Urbana (University of Illinois Press).

———. 1988. La géologie [des savants genevois]: de l'histoire de la terre selon le récit de Moïse aux premiers essais sur la structure des Alpes et à la géologie experimentale, 1778–1878. *In* Trembley, *Savants genevois*, 203–65.

———. 1989. Forty years of thinking in front of the Alps: Saussure's (1796) unpublished theory of the earth. *ESH* 8: 123–40.

———. 1992. De Maillet's Telliamed: the diminution of the sea or the fall portion of a complete cosmic eustatic cycle. *In* Dott, *Eustasy*, 17–24.

———. 1998. Découverte d'une grande découverte: Horace-Bénédict de Saussure et les refoulements horizontaux en sens contraire dans la formation des Alpes. *In* Carozzi et al., *Les plis du temps*, 223–367.

———. 2000. *Manuscripts and publications of Horace-Bénédict de Saussure on the origin of basalt.* Geneva (Zoé).

———. 2001. Du dogme neptuniste au concept de refoulements horizontaux: les étapes d'une reflexion géologique. *In* Sigrist, *Saussure*, 83–108.

Carozzi, Albert V., and Gerda Bouvier. 1994. The scientific library of Horace-Bénédict de Saussure (1797): annotated catalog of an 18th-century bibliographic and historic treasure. *MSPHN* 46, 201 pp.

Carozzi, Albert V., and Marguerite Carozzi. 1991. Reevaluation of Pallas' theory of the earth (1778). *ASG* 44 (1), 105 pp.

Carozzi, Albert V., Bernard Crettaz, and David Ripoll (eds.). 1998. *Les plis du temps: mythe, science et H.-B. de Saussure.* Geneva (Musée d'Ethnographie).

Carozzi, Albert V., and John K. Newman (eds.). 2003. *Lectures on physical geography given in 1775 by Horace-Bénédict de Saussure at the Academy of Geneva.* Geneva (Zoé).

Carozzi, Albert V., and Donald H. Zenger. 1981. Sur un genre de pierres calcaires très-peu effervescentes avec les acides, & phophorescentes par la collision. Déodat de Dolomieu 1791. Translation with notes on Dolomieu's paper reporting his discovery of dolomite. *JGE* 29: 4–10.

Carozzi, Marguerite. 1986. From the concept of salient and re-entrant angles by Louis Bourguet to Nicolas Desmarest's description of meandering rivers. *ASG* 39: 25–51.

———. 2000. H.-B. de Saussure: James Hutton's obsession. *ASG* 53: 77–158.

Cartwright, David E. 1997. Robert Paul de Lamanon: an unlucky naturalist. *AS* 54: 585–96.

Chaldecott, J. A. 1968. Scientific activities in Paris in 1791: evidence from the diaries of Sir James Hall for 1791, and other contemporary records. *AS* 24: 21–52.

Ciancio, Luca. 1995. *Autopsie della terra: illuminismo e geologia in Alberto Fortis (1741–1803).* Florence (Olschki).

Clark, William. 2000. On the bureaucratic plots of the research library. *In* Frasca-Spada and Jardine, *Books and the sciences*, 190–206.

———, Jan Golinski, and Simon Schaffer (eds.). 1999. *The sciences in Enlightened Europe.* Chicago (University of Chicago Press).

Cohen, Claudine. 1994. *Le destin du mammouth.* Paris (Seuil) [trans. *The fate of the mammoth*, Chicago (University of Chicago Press), 2002].

———. 1996. Leibniz's *Protogaea*: patronage, mining, and evidence for a history of the earth. *In* Marchand and Lunbeck, *Proof and persuasion*, 125–43.

———. 1997. Stratégies et rhétorique de la preuve dans les *Recherches sur les ossements fossiles de quadrupèdes* de Cuvier. *In* Blanckaert et al., *Le Muséum*, 523–39.

Cohn, Norman. 1996. *Noah's Flood: the Genesis story in Western thought.* New Haven (Yale University Press).

Coleman, William. 1964. *Georges Cuvier zoologist: a study in the history of evolution theory.* Cambridge (Harvard University Press).

Constantine, David. 2001. *Fields of fire: a life of Sir William Hamilton.* London (Weidenfeld and Nicolson).

Conticello, B. (ed.). 1990. *Riscoprire Pompei / Rediscovering Pompeii.* Rome ("L'Erma" di Bretschneider).

Conybeare, F. C. (ed.). 1905. *Letters and exercises of the Elizabethan schoolmaster John Conybeare.* London (Henry Frowde).

Cooper, Alix. 1998. From the Alps to Egypt (and back again): Dolomieu, scientific voyaging, and the construction of the field in eighteenth-century natural history. *In* Smith, *Making space for science*, 39–63.

Cooter, Roger, and Stephen Pumphrey. 1994. Separate spheres and public places: reflections on the history of science popularization and science in popular culture. *HS* 32: 237–67.

Cormack, Lesley B. 1997. *Charting an empire: geography at the English universities, 1580–1620.* Chicago (University of Chicago Press).

Corsi, Pietro. 1988. *The age of Lamarck: evolutionary theories in France, 1790–1830.* Berkeley (University of California Press) [revised ed. of *Oltre il mito: Lamarck e la scienze naturali del suo tempo*, Bologna (Il Mulino), 1983].

———. 2001. *Lamarck: genèse et enjeux du transformisme, 1770–1830.* Paris (CNRS) [trans. of *Oltre il mito*, with new "annexes" (329–83) on his auditors].

Cox, L. R. 1942. New light on William Smith and his work. *PYGS* 25: 1–99.

Craig, Gordon Y. (ed.). 1978. *James Hutton's theory of the earth: the lost drawings* [with portfolio of facsimiles]. Edinburgh (Scottish Academic Press).

Cremo, Michael A., and Richard L. Thompson. 1993. *Forbidden archeology: the hidden history of the human race.* 2 vols. San Diego (Bhaktivedanta Institute).

Cunningham, Andrew R. 1988. Getting the game right: some plain words on the identity and invention of "science". *SHPS* 19: 365–89.

Cunningham, Andrew R., and Nick Jardine (eds.). 1990. *Romanticism and the sciences.* Cambridge (Cambridge University Press).

Cunningham, Andrew R., and Perry Williams. 1993. De-centring the big picture: *The origins of modern science* and the modern origins of science. *BJHS* 26: 407–32.

Curi, Ettore (ed.). 1999. *Scienza, tecnica e "pubblico bene" nell'opera di Giovanni Arduino (1714–1795).* Verona (Accademia di Agricoltura, Scienze e Lettere).

Czerkas, Sylvia J., and Everett C. Olson (eds.). 1987. *Dinosaurs past and present.* 2 vols. [Seattle] (University of Washington Press).

Dance, S. Peter. 1986. *A history of shell collecting.* Leiden (E. J. Brill).

Darby, H. C. and Harold Fullard. 1970. *The new Cambridge modern history atlas.* Cambridge (Cambridge University Press).

Darnton, Robert. 1979. *The business of Enlightenment: a publishing history of the* Encyclopédie, *1775–1800.* Cambridge (Harvard University Press).

———. 1984. *The great cat massacre and other episodes in French cultural history.* New York (Basic Books).

———. 1991. History of reading. *In* Burke, *New perspectives*, 140–67.

———. 2001. Epistemological angst: from encyclopedism to advertising. *In* Frängsmyr, *Structure of knowledge*, 53–75.

Darnton, Robert, and Daniel Roche (eds.). 1989. *Revolution in print: the press in France, 1775–1800.* Berkeley (University of California Press).

Daston, Lorraine J. 1988. The factual sensibility. *Isis* 79: 452–70.

———. 1991. The ideal and reality of the republic of letters in the Enlightenment. *SC* 4: 367–86.

Daudin, Henri. 1926. *Cuvier et Lamarck: les classes zoologiques et l'idée de série animale (1790–1830).* Paris (Félix Alcan) [repr. 1983].

Davies, Gordon L. 1967. George Hoggart Toulmin and the Huttonian theory of the earth. *BGSA* 78: 121–24.

———. 1969. *The earth in decay: a history of British geomorphology 1578–1878.* London (Macdonald).

Dean, Bashford. 1904. A reference to the origin of species in an early letter (1796) signed by both Lamarck and Geoffroy. *Science* (n.s.) 19: 798–800.

Dean, Dennis R. 1979. The word geology. *AS* 36: 35–43.

———. 1981. The age of the earth controversy: beginnings to Hutton. *AS* 38: 435–56.

———. 1992. *James Hutton and the history of geology*. Ithaca (Cornell University Press).

———. 1997. *James Hutton in the field and in the study, being an augmented reprinting of vol. III of Hutton's* Theory of the Earth. Delmar [New York] (Scholars' Facsimiles and Reprints).

De Beer, Gavin. 1960. *The sciences were never at war*. London (Nelson).

Decrouez, Danielle, and Edouard Lanterno. 1998. La collection géologique d'Horace-Bénédict de Saussure. *In* Carozzi et al., *Les plis du temps*, 211–21.

Dehérain, Henri. 1908. *Catalogue des manuscrits du fonds Cuvier (travaux et correspondence scientifique) conservés à la Bibliothèque de l'Institut de France*. Paris (Honoré Champion).

Delécraz, Christian. 1998. Les reliefs de montagne au XVIIIe siècle, entre science et mythe. *In* Carozzi et al., *Les plis du temps*, 125–41.

Desjardins-Menegali, Marie-Hélène. 1982. Georges Cuvier à Fécamp (1788–1795): les prémices d'une carrière. *In* Buffetaut et al., *Symposium Cuvier*, 143–56.

Donovan, Arthur. 1978. James Hutton, Joseph Black, and the chemical theory of heat. *Ambix* 25: 176–190.

Donovan, Arthur, and Joseph Prentiss. 1980. James Hutton's medical dissertation. *TAPS* 70 (57 pp.).

Dott, Robert H., Jr. (ed.). 1992. Eustasy: the historical ups and downs of a major geological concept. *GSAM* 180.

Dougherty, F. W. P. 1986. Der Begriff der Naturgeschichte nach J. F. Blumenbach anhand seiner Korrespondenz mit Jean-André Deluc: ein Beitrag zur Wissenschaftsgeschichte bei der Entdeckung der Geschichtlichkeit ihres Gegenstandes. *BW* 9: 95–107.

Drake, Ellen Tan. 1996. *Restless genius: Robert Hooke and his earthly thoughts*. New York and Oxford (Oxford University Press).

Durant, Graham P., and W. D. Ian Rolfe. 1984. William Hunter (1718–1783) as natural historian: his "geological" interests. *ESH* 3: 9–24.

Eddy, M. D. 2001. Geology, mineralogy, and time in John Walker's University of Edinburgh natural history lectures. *HS* 39: 95–119.

Edmonds, J. M. 1978. Patronage and privilege in education: a Devon boy goes to school, 1798. *RDAAS* 110: 95–111.

———. 1979. The founding of the Oxford Readership in geology. *NRRSL* 34: 33–51.

———. 1991. *Vindiciae geologicae*, published 1820: the inaugural lecture of William Buckland. *ANH* 18: 255–68.

Edmonds, J. M., and J. A. Douglas. 1976. William Buckland, F.R.S. (1784–1856) and an Oxford geological lecture, 1823. *NRRSL* 30: 141–67.

Edwards, Nicholas. 1971. Thomas Webster (circa 1772–1944). *JSBNH* 5: 468–73.

Ellenberger, François. 1973. La thèse de doctorat de James Hutton et la rénovation perpetuelle du monde. *AGS* 49: 497–533.

———. 1979. Aux sources de la géologie française: guide de voyage à l'usage de l'historien des sciences de la terre sur l'itinéraire Paris-Auvergne-Marseille. *HN* 15, [29 pp.].

———. 1983. La dispute des lignites du Soissonais. *TCFHG* (2) 1: 1–22.

———. 1983. Documents pour une histoire du vocabulaire de la géologie: 1) le terme de *géographie souterraine*. *DHVS* 4: 35–42.

———. 1985. Recherches et réflexions sur la naissance de la cartographie géologique, en Europe et plus particulièrement en France. *HN* 22/23: 3–54.

———. 1987. Les causes actuelles en géologie: origine de cette expression, la légende et la réalité. *BSGF* (8) 3: 199–206.

———. 1989. Étude du terme révolution. *DHVS* 9: 69–90.

———. 1992. Les sciences de la terre avant Buffon: un bref coup d'oeil historique. *In* Gayon, *Buffon 88*, 327–42.

———. 1994. *Histoire de la géologie* 2: *La grande éclosion et ses prémices, 1660–1810*. Paris (Tec et Doc) [trans. *History of geology* 2, Rotterdam (A. A. Balkema), 1996].

Ellenberger, François, and Gabriel Gohau. 1981. A l'aurore de la stratigraphie paléontologique: Jean-André de Luc, son influence sur Cuvier. *RHS* 34: 217–57.

Engelhardt, W. von. 1980. Carl von Linné und das Reich der Steine. *In* Goerke et al., *Carl von Linné*, 81–96.

Erwin, Douglas H., and Scott L. Wing (eds.). 2000. *Deep time: paleobiology's perspective. Paleobiology* 26 (4), supplement.

Eyles, Joan M. 1967. William Smith: the sale of his collection to the British Museum. *AS* 23: 177–202.

———. 1969. William Smith (1769–1839): a chronology of significant dates in his life. *PGSL* 1969–70: 173–76.

———. 1969. William Smith: some aspects of his life and work. *In* Schneer, *History of geology*, 142–58.

———. 1969. William Smith (1769–1839): a bibliography of his published writings, maps, and geological sections, printed and lithographed. *JSBNH* 5: 87–109.

———. 1979. William Smith: great discoverer or mere fossil collector? *OE* 2: 11–13.

———. 1985. Sir Joseph Banks, William Smith, and the French geologists. *In* Wheeler and Price, *Linnaeus to Darwin*, 37–50.

Eyles, V. A. 1950. Note on the original publication of Hutton's *Theory of the earth*, and on the subsequent forms in which it was issued. *PRSE* 63B: 377–86.

———. 1955. A bibliographical note on the earliest printed version of James Hutton's *Theory of the earth*, its form and date of publication. *JSBNH* 3: 105–08.

Eyles, V. A., and Joan M. Eyles. 1938. On the different issues of the first geological map of England and Wales. *AS* 3: 190–212.

Feldman, Theodore S. 1990. Late Enlightenment meteorology. *In* Frängsmyr et al., *Quantifying spirit*, 143–77.

Findlen, Paula. 1994. *Possessing nature: museums, collecting, and scientific culture in early modern Italy*. Berkeley (University of California Press).

Flügel, Helmut W. 2003. Carl Maria Haidingers und Abraham Gottlob Werners 'Klassifikation' des 'Gebirgsarten' von 1787. *JGB* 143: 535–41.

Ford, Trevor D. 1960. White Watson (1760–1835) and his geological sections. *PGA* 71: 349–63.

———. 1967. The first detailed sections across England, by John Farey, 1806–8. *MG* 2: 41–49.

Ford, Trevor D., and Hugh S. Torrens. 1989. John Farey (1766–1826), an unrecognised polymath. *In* Farey, *Derbyshire* [orig. pub. 1811], 44 pp. [repr. Torrens, *British geology*, 2002, art. VI].

Foucault, Michel. 1966. *Les mots et les choses: une archéologie des sciences humaines*. Paris (Gallimard) [trans. *The order of things*, London (Tavistock), 1970)].

Frängsmyr, Tore. 1969. *Geologi och skapelsetro: föreställingar om jordens historia från Hiärne till Bergman*. Stockholm (Almqvist & Wiksell).

———. 1983. *Linnaeus: the man and his work*. Berkeley (University of California Press).

——— (ed.). 2001. *The structure of knowledge: classifications of science and learning since the Renaissance*. Berkeley (University of California Press).

Frängsmyr, Tore, J. L. Heilbron, and Robin E. Rider (eds.). 1990. *The quantifying spirit in the eighteenth century*. Berkeley (University of California Press).

Frasca-Spada, Marina, and Nick Jardine (eds.). 2000. *Books and the sciences in history*. Cambridge (Cambridge University Press).

Freshfield, Douglas W. 1920. *The life of Horace Benedict de Saussure*. London (Edward Arnold).

Frigo, Margherita, and Lorenzo Sorbini. 1997. *600 fossili per Napoleone*. Verona (Museo Civico di Storia Naturale di Verona).

Fritscher, Bernhard. 1991. *Vulkanismusstreit und geochemie: die Bedeutung der Chemie und des Experiments in der Vulkanismus-Neptunismus-Kontroverse.* Stuttgart (Franz Steiner).

Fritscher, Bernhard, and Fergus Henderson (eds.). 1998. *Toward a history of mineralogy, petrology, and geochemistry.* Munich (Institut für Geschichte der Naturwissenschaften).

Forbes, R. J., E. Lefebvre, and J. G. de Bruijn (eds.). 1969–76. *Martinus van Marum: life and work.* 6 vols. Haarlem (Hollandsche Maatschappij der Wetenschappen).

Fuller, John G. C. M. 1969. The industrial basis of stratigraphy: John Strachey, 1671–1743, and William Smith, 1769–1839. *BAAPG* 53: 2256–73.

———. 1992. The invention and first use of stratigraphic cross-sections by John Strachey, F.R.S. (1671–1743). *ANH* 19: 69–90.

———. 1995. *'Strata Smith' and his stratigraphic cross sections, 1819.* Denver (American Association of Petroleum Geologists); London (Geological Society of London).

———. 2001. Before the hills in order stood: the beginning of the geology of time in England. *In* Lewis and Knell, *Age of the earth,* 15–23.

Gascoigne, John. 1998. *Science in the service of empire: Joseph Banks, the British state, and the uses of science in the age of revolution.* Cambridge (Cambridge University Press).

———. 1999. The Royal Society and the emergence of science as an instrument of state policy. *BJHS* 32: 171–84.

Gaudant, Jean. 1993. L'exploration géologique du Bassin Parisien: quelques pionniers, le plus souvent méconnus. *HN* 30: 17–39, pls. 1–3.

———. 1999. La querelle des trois abbés (1793–1795): le débat entre Domenico Testa, Alberto Fortis et Giovanni Serafino Volta sur la signification des poissons petrifiés du Monte Bolca (Italie). *MP* 8: 159–206.

Gayon, Jean (ed.). 1992. *Buffon 88: Actes du colloque international, Paris, Montbard, Dijon, Paris.* Paris (Éditions du Muséum).

Geikie, Archibald. 1905. *The founders of geology.* 2nd ed. London (Macmillan) [repr. New York (Dover), 1962].

Gerstner, Patsy A. 1968. Hutton's theory of the earth and his theory of matter. *Isis* 59: 26–31.

———. 1971. The reaction to James Hutton's use of heat as a geological agent. *BJHS* 5: 353–62.

Geus, Armin. 1997. Specimens and visual representations of animals and plants extinct in historical time. *In* Mazzolini, *Non-verbal communication,* 391–409.

Ghiselin, Michael T., and Alan E. Leviton (eds.). 2000. *Cultures and institutions of natural history: essays in the history and philosophy of science.* San Francisco (California Academy of Sciences).

Giglia, Gaetano, Carlo Maccagni, and Nicoletta Morello (eds.). 1995. *Rocks, fossils, and history.* Florence (Festina Lente).

Gillispie, Charles Coulston. 1951. *Genesis and geology: a study in the relations of scientific thought, natural theology, and social opinion in Great Britain, 1790–1850.* Cambridge (Harvard University Press).

———. 1980. *Science and polity in France at the end of the Old Regime.* Princeton (Princeton University Press).

Gillispie, Charles Coulston, and Michael Dewachter (eds.). 1987. *Monuments of Egypt, the Napoleonic edition: the complete archeological plates from La description de l'Égypte.* Princeton (Princeton Architectural Press).

Goerke, H., H. Querner, C. Andree, W. von Engelhardt, and V. H. Heywood. 1980. *Carl von Linné: Beiträge über Zeitgeist, Werk und Wirkungsgeschichte.* Göttingen (Vandenhoeck und Ruprecht).

Gohau, Gabriel. 1988. Naissance de la géologie historique. *In* Bloch et al., *Entre forme et histoire,* 127–43.

———. 1990. *Les sciences de la terre aux XVIIe et XVIIIe siècles: naissance de la géologie.* Paris (Albin Michel).

———. 1992. La "théorie de la terre", de 1749. *In* Gayon, *Buffon 88,* 343–52.

———. 1997. L'Hydrogéologie et l'histoire de la géologie. *In* Laurent, *Jean-Baptiste Lamarck*, 137–47.

——— (ed.). 1997. *De la géologie à son histoire: ouvrage édité en hommage à François Ellenberger.* Paris (C.T.H.S.).

——— (ed.). 1998. *Buffon: des époques de la nature.* Paris (Diderot).

———. 2003. *Naissance de la géologie historique: la terre, des 'théories' à l'histoire.* Paris (Vuibert and ADAPT).

Goldgar, Anne. 1995. *Impolite learning: conduct and community in the Republic of Letters, 1680–1750.* New Haven (Yale University Press).

Golinski, Jan. 1998. *Making natural knowledge: constructivism and the history of science.* Cambridge (Cambridge University Press).

———. 1999. Barometers of change: meteorological instruments as machines of Enlightenment. *In* Clark et al., *Sciences in Enlightened Europe*, 69–93.

———. 2003. Chemistry. *In* Porter, *Eighteenth-century science*, 375–96.

Goodman, David, and Colin Russell (eds.). 1991. *The rise of scientific Europe, 1500–1800.* London (Hodder & Stoughton).

Goodman, Dena. 1994. *The republic of letters: a cultural history of the French Enlightenment.* Ithaca (Cornell University Press).

Gordon, [Elizabeth Oke]. 1894. *The life and correspondence of William Buckland, D.D., F.R.S.* London (Murray).

Gorst, Martin. 2001. *Aeons: the search for the beginning of time.* London (Fourth Estate).

Gould, Stephen Jay. 1987. *Time's arrow, time's cycle: myth and metaphor in the discovery of geological time.* Cambridge (Harvard University Press).

———. 2000. *The lying stones of Marrakesh: penultimate reflections on natural history.* London (Jonathan Cape).

———. 2000. The proof of Lavoisier's plates. *In* Gould, *Lying stones of Marrakesh*, 91–114.

Grafton, Anthony T. 1975. Joseph Scaliger and historical chronology: the rise and fall of a discipline. *HT* 14: 156–85.

———. 1983. *Joseph Scaliger: a study in the history of Classical scholarship.* Oxford (Clarendon).

———. 1991. *Defenders of the text: the traditions of scholarship in an age of science 1450–1800.* Cambridge (Harvard University Press).

Grandchamp, Philippe. 1994. Deux exposés des doctrines de Cuvier antérieurs au "Discours préliminaire": les cours de géologie professés au Collège de France en 1805 et 1808. *TCFHG* (3) 8: 13–26.

———. 1999. Un essai inédit de classification des terrains: le cours de géognosie professé en 1813 par Alexandre Brongniart à la Faculté des Sciences de Paris. *TCFHG* (3) 13: 99–113.

Grant, R. 1979. Hutton's theory of the earth. *In* Jordanova and Porter, *Images of the earth*, 23–38.

Grayson, Donald K. 1983. *The establishment of human antiquity.* New York (Academic Press).

Greene, Mott T. 1992. *Natural knowledge in pre-Classical antiquity.* Baltimore (Johns Hopkins University Press).

Grell, Chantal, and Catherine Volpilhac-Auger (eds.). 1994. *Nicolas Fréret, légende et verité.* Oxford (Voltaire Foundation).

Grossklaus, Götz, and Ernst Oldenmeyer (eds.). 1983. *Natur als Gegenwelt: Beiträge zur Kulturgeschichte der Natur.* Karlsrühe (Von Loeper).

Grote, Andreas (ed.). 1994. *Macrocosmos in microcosmo: die Welt in der Stube: zur Geschichte des Sammelns 1450 bis 1800.* Opladen (Leske & Budrich).

Guntau, Martin. 1984. *Abraham Gottlob Werner.* Leipzig (B. G. Teubner).

———. 1995. The beginnings of lithostratigraphic and biostratigraphic thinking in Germany. *In* Giglia et al., *Rocks, fossils, and history*, 149–55.

———. 1996. The natural history of the earth. *In* Jardine et al., *Cultures of natural history*, 211–29.

———. 2002. Zu den Prinzipien der Klassifikation natürlicher Objekte in den Vorstellungen von Abraham Gottlob Werner. *In* Albrecht and Ladwig, *Abraham Gottlob Werner*, 79–87.

Guntau, Martin, and Wolfgang Mühlfriedel. 1968. Die Bedeutung von Abraham Gottlob Werner für die Mineralogie und die Geologie. *Geologie* 17: 1096–1115.

Haber, Francis C. 1959. *The age of the world, Moses to Darwin.* Baltimore (Johns Hopkins University Press).

Hackmann, W. D. 1973. *John and Jonathan Cuthbertson: the invention of the eighteenth-century plate electrical machine.* Leiden (National Museum of the History of Science).

Hahn, Roger. 1971. *Anatomy of a scientific institution: the Paris Academy of Sciences, 1666–1803.* Berkeley (University of California Press).

———. 1997. Du Jardin du Roi au Muséum: les carrières de Fourcroy et de Lacépède. *In* Blanckaert et al., *Le Muséum*, 31–41.

Hamm, Ernst P. 1990. *Goethe on granite.* Toronto (University of Toronto [PhD dissertation]).

———. 1997. Knowledge from underground: Leibniz mines the Enlightenment. *ESH* 16: 77–99.

———. 2001. Unpacking Goethe's collections: the public and private in natural-historical collecting. *BJHS* 34: 275–300.

Hampton, John. 1955. *Nicolas-Antoine Boulanger et la science de son temps.* Geneva (E. Droz).

Hamy, Ernest-Théodore. 1893. Les derniers jours du Jardin du Roi et la fondation du Muséum d'Histoire Naturelle. *In* Anonymous, *Centenaire du Muséum*, 1–162.

Hardin, Clyde L. 1966. The scientific work of the Reverend John Michell. *AS* 22: 27–47.

Haskell, Francis. 1993. *History and its images: art and the interpretation of the past.* New Haven (Yale University Press).

———. 1993. *The paper museum of Cassiano dal Pozzo: a catalogue raisonné . . .* London (Harvey Miller).

Hay, Denys. 1977. *Annalists and historians: Western historiography from the eighth to the eighteenth centuries.* London (Methuen).

Heesen, Anke te. 2004. From natural historical investment to state service: collectors and collections of the Berlin Society of Friends of Nature Research, c. 1800. *HS* 42: 113–31.

Heilbron, J. L. 1979. *Electricity in the 17th and 18th centuries: a study of early modern physics.* Berkeley (University of California Press).

———. 1982. *Elements of early modern physics.* Berkeley (University of California Press).

———. 2002. Physics and its history at Göttingen around 1800. *In* Rupke, *Göttingen*, 50–71.

Heimann, P. M., and J. E. McGuire. 1971. Newtonian forces and Lockean powers: concepts of matter in eighteenth-century thought. *HSPS* 3:233–306.

Heller, Florian. 1972. Die Forschungen in der Zoolithenhöhle bei Burggaillenreuth von Esper bis zur Gegenwart. *EF* (B) 5: 7–52.

Herries Davies, Gordon L.: *see* Davies, Gordon L.

Himbergen, E. J. van. 1978. De prijsvragen van de Tweede Genootschap 1778–1978. *In* Anonymous, *Teyler 1778–1978*, 37–55.

Hineline, Mark L. 1993. *The visual culture of the earth sciences 1863–1970.* San Diego (University of California San Diego [Ph.D. dissertation]) [University Microfilms no. 94–20724].

Hobsbawm, E. J. 1962. *The age of revolution 1789–1848.* London (Weidenfeld and Nicholson).

Hobson, Marian. 1992. "Nexus effectivus" and "nexus finalis": causality in the *Inégalité* and in the *Essai sur les origines des langues. In* Hobson and Leigh, *Rousseau and the eighteenth century*, 225–50.

Hobson, Marian, and J. T. A. Leigh (eds.). 1992. *Rousseau and the eighteenth century: essays in memory of R. A. Leigh.* Oxford (Voltaire Foundation).

Hodge, Jonathan. 1992. Two cosmogonies (theory of the earth and theory of generation), and the unity of Buffon's thought. *In* Gayon, *Buffon 88*, 241–54.

Hofstetter, Robert. 1959. Les rôles respectifs de Bru, Cuvier et Garriga dans les premières études concernant *Megatherium. BMHNP* 31: 536–45.

Hölder, Helmut. 1985. Goethe als Geologe. *ZDGG* 136: 1–21.

Holmes, Frederic L. 2000. The "revolution in chemistry and physics": overthrow of a reigning paradigm or competition between contemporary research programs? *Isis* 91: 735–53.

Hooykaas, R. 1959. *Natural law and divine miracle: a historical-critical study of the principle of uniformity in geology, biology, and theology.* Leiden (E. J. Brill).

———. 1966. James Hutton und die Ewigkeit der Welt. *Gesnerus* 23: 55–66.

———. 1970. *Continuité et discontinuité en géologie et biologie.* Paris (Seuil).

Hope, V. (ed.). 1984. *Philosophers of the Scottish Enlightenment.* Edinburgh (Edinburgh University Press).

Huta, Carole. 1998. Jean Senebier (1742–1809): un dialogue entre l'ombre et la lumière; l'art d'observer à la fin du XVIIIe siècle. *RHS* 51: 93–105.

Iggers, Georg G. 1986. The European context of eighteenth-century German Enlightenment historiography. *In* Bödecker et al., *Aufklärung und Geschichte*, 225–45.

Iliffe, Rob. 2003. Science and voyages of discovery. *In* Porter, *Eighteenth-century science*, 618–45.

Impey, Oliver, and Arthur MacGregor 1985. *The origins of museums: the Cabinet of Curiosities in seventeenth-century Europe.* Oxford (Clarendon).

Inkster, Ian, and Jack Morrell (eds.). 1983. *Metropolis and province: studies in British culture 1780–1850.* London (Hutchinson).

Jackson, Donald D. (ed.). 1978. *Letters of the Lewis and Clark expedition, with related documents, 1783–1854.* Urbana (University of Illinois Press).

Jacob, Margaret C., and Wijnand W. Mijnhardt (eds.). 1992. *The Dutch republic in the eighteenth century: decline, enlightenment, revolution.* Ithaca (Cornell University Press).

Jahn, Melvin E. 1969. Some notes on Dr Scheuchzer and on *Homo diluvii testis. In* Schneer, *History of geology*, 193–213.

Jahn, Melvin E, and Daniel J. Woolf. 1963. *The lying stones of Dr. Johann Bartholomew Beringer, being his Lithographiae Wirceburgensis.* Berkeley (University of California Press).

Jankovic, Vladimir. 2000. The place of nature and the nature of place: the chorographic challenge to the history of British provincial science. *HS* 38: 79–113.

Jardine, Nick. 2000. Uses and abuses of anachronism in the history of the sciences. *HS* 38: 251–70.

Jardine, Nick, J. A. Secord, and E. C. Spary (eds.). 1996. *Cultures of natural history.* Cambridge (Cambridge University Press).

Jenkins, Ian. 1996. "Contemporary minds": William Hamilton's affair with antiquity. *In* Jenkins and Sloan, *Vases and volcanoes*, 40–64.

Jenkins, Ian, and Kim Sloan (eds.). 1996. *Vases and volcanoes: Sir William Hamilton and his collections.* London (British Museum).

Johns, Adrian. 1994. The ideal of scientific collaboration: the "man of science" and the diffusion of knowledge. *In* Bots and Waquet, *Commercium litterarium*, 3–22.

———. 1998. *The nature of the book: print and knowledge in the making.* Chicago (University of Chicago Press).

———. 2003. Print and public science. *In* Porter, *Eighteenth-century science*, 536–60.

Jones, Jean. 1985. Hutton's agricultural research and his life as a farmer. *AS* 42: 573–601.

Jones, Jean, Hugh S. Torrens, and Eric Robinson. 1994. The correspondence between James Hutton (1726–1797) and James Watt (1736–1819) with two letters from Hutton to George Clerk-Maxwell (1715–1784); part I. *AS* 51: 637–53.

Jones, Peter. 1984. An outline of the philosophy of James Hutton (1726–97). *In* Hope, *Philosophers of Scottish Enlightenment*, 182–210.

Joost, Ulrich, and Albrecht Schöne (eds.). 1983–92. *Georg Christoph Lichtenberg Briefwechsel: im Auftrag der Akademie der Wissenschaften zu Göttingen.* 4 vols. Munich (Beck).

Jordanova, L. J. 1984. *Lamarck.* Oxford (Oxford University Press).

Jordanova, L. J., and Roy S. Porter (eds.). 1979. *Images of the earth: essays in the history of the environmental sciences.* Chalfont St Giles (British Society for the History of Science) [repr. 1997].

Keller, Susanne B. 1998. Sections and views: visual representation in eighteenth-century earthquake studies. *BJHS* 31: 129–59.

Kemp, Martin. 2000. *Visualizations: the* Nature *book of art and science.* Oxford (Oxford University Press).

Klonk, Charlotte. 1996. *Science and the perception of nature: British landscape art in the late eighteenth and early nineteenth centuries.* New Haven (Yale University Press).

———. 2003. Science, art, and the representation of the natural world. *In* Porter, *Eighteenth-century science*, 584–617.

Knell, Simon J. 2000. *The culture of English geology, 1815–1851.* Aldershot (Ashgate).

Koerner, Lisbet. 1996. Purposes of Linnaean travel: a preliminary research report. *In* Miller and Reill, *Visions of empire*, 117–52.

———. 1996. Carl Linnaeus in his time and place. *In* Jardine et al., *Cultures of natural history*, 145–62.

Kölbl-Ebert, Martina. 2002. British geology in the early nineteenth century: a conglomerate with a female matrix. *ESH* 21: 3–25.

———. 2003. George Bellas Greenough (1778–1855): a lawyer in geologist's clothes. *PGA* 114: 247–54.

Konvitz, Josef. 1987. *Cartography in France, 1660–1848: science, engineering, and statecraft.* Chicago (University of Chicago Press).

Kuhn, Thomas S. 1970. *The structure of scientific revolutions.* 2nd ed. Chicago (University of Chicago Press).

———. 1979. *The essential tension: selected studies in scientific tradition and change.* Chicago (University of Chicago Press).

Lacroix, A. 1921. *Déodat de Dolomieu: sa vie aventureuse, sa captivité, ses oeuvres, sa correspondance.* Paris (Perrin).

Laissus, Yves. 1964. Les cabinets d'histoire naturelle. *In* Taton, *Enseignement et diffusion*, 659–712.

———. 1995. *Le Muséum d'Histoire Naturelle.* Paris (Gallimard).

——— (ed.). 1998. *Il y a 200 ans: les savants en Égypte.* Paris (Nathan).

Landon, H. C. Robbins. 1977. *Haydn: the years of "The Creation".* London (Thames and Hudson).

Latour, Bruno. 1990. Drawing things together. *In* Lynch and Woolgar, *Representation*, 19–68.

Laudan, Rachel. 1976. William Smith: stratigraphy without palaeontology. *Centaurus* 20: 210–26.

———. 1977–78. The problem of consolidation in the Huttonian tradition. *Lychnos* 1977–78: 195–206.

———. 1982. Tensions in the concept of geology: natural history or natural philosophy? *ESH* 1: 7–13.

———. 1987. *From mineralogy to geology: the foundations of a science, 1650–1830.* Chicago (University of Chicago Press).

Launay, Louis de. 1940. *Une grande famille de savants: les Brongniart.* Paris (G. Rapilly).

Laurent, Goulven. 1987. *Paléontologie et évolution en France de 1800 à 1860: une histoire des idées de Cuvier et Lamarck à Darwin.* Paris (Comité des Travaux Historiques).

———. 1993. Ami Boué (1794–1881): sa vie et son oeuvre. *TCFHG* (3) 7: 19–30.

——— (ed.). 1997. *Jean-Baptiste Lamarck 1744–1829.* Paris (Comité des Travaux Historiques et Scientifiques).

———. 2002. Jean-Guillaume Bruguière (1750–1798) et les débats de la paléontologie des invertèbres. *TCFHG* (3) 16: 37–46.

Leclaire, Lucien. 1992. L' 'Histoire naturelle des mineraux' ou Buffon géologue universaliste. *In* Gayon, *Buffon 88*, 353–69.

Lejoix, Anne. 1997. Alexandre Brongniart: scholar and member of the Institut de France. *In* Préaud, *Sèvres porcelain*, 24–41.

Lennon, Thomas M., John M. Nicholas, and John W. Davis (eds.). 1982. *Problems of Cartesianism.* Kingston and Montreal (McGill-Queens University Press).

Lenoir, Timothy. 1982. *The strategy of life: teleology and mechanics in nineteenth-century German biology.* Chicago (University of Chicago Press).

Lepenies, Wolf. 1976. *Das Ende der Naturgeschichte: Wandel kultureller Selbstverständigkeiten in den Wissenschaften des 18. und 19. Jahrhunderts.* Munich and Vienna (Hanser).

Levere, T. H. and G. L'E. Turner. 2002. *Discussing chemistry and steam: the minutes of a coffee house philosophical society 1780–1787.* Oxford: Oxford University Press.

Levine, Joseph M. 1987. *Humanism and history: origins of modern English historiography.* Ithaca (Cornell University Press).

———. 1994. Strife in the Republic of Letters. *In* Bots and Waquet, *Commercium litterarium,* 301–19.

Lewis, C. L. E., and S. J. Knell (eds.). 2001. *The age of the earth: from 4004 B.C. to A.D. 2002.* London (Geological Society).

Lindberg, David C., and Ronald L. Numbers (eds.). 1986. *God and nature: historical essays on the encounter between Christianity and science.* Berkeley and Los Angeles (University of California Press).

——— (eds). 2003. *When science and Christianity meet.* Chicago (University of Chicago Press).

Lindenfeld, David F. 1997. *The practical imagination: the German sciences of state in the nineteenth century.* Chicago (University of Chicago Press).

Lippincott, Kristen (ed.). 1999. *The story of time.* London (National Maritime Museum).

Livingstone, D. N. 1986. Preadamites: the history of an idea from heresy to orthodoxy. *SJT* 40: 41–66.

———. 1992. The preadamite theory and the marriage of science and religion. *TAPS* 82 (3).

———. 2002. *Science, space, and hermeneutics.* Heidelberg (Department of Geography, University of Heidelberg).

López Piñero, José M. 1988. Juan Bautista Bru (1740–1799) and the description of the genus *Megatherium. JHB* 21: 147–63.

———. 1996. *Juan Bautista Bru de Ramón: el Atlas Zoológica, el megaterio y las téchnicas de pesca Valencianas, 1742–1799.* Valencia (Ayuntamiento de Valencia).

Löwenbrück, Anna-Ruth. 1986. Johann David Michaelis et les débuts de la critique biblique. *In* Belaval and Bourel, *Lumières et la bible,* 113–28.

Lucas, Colin (ed.). 1988. *The French Revolution and the creation of modern political culture.* Oxford (Pergamon).

Lynch, Michael, and Steve Woolgar (eds.). 1990. *Representation in scientific practice.* Cambridge (MIT Press).

MacArthur, C. W. P. 1990. Dr Jean-François Berger of Geneva: from the Travelling Fund to the Wollaston Donation. *ANH* 17: 97–119.

McCarthy, James. 1997. The biblical chronology of James Ussher. *IAJ* 24: 73–82.

McClellan, James E., III. 1979. The scientific press in transition: Rozier's Journal and the scientific societies in the 1770s. *AS* 36: 425–49.

———. 1985. *Science reorganized: scientific societies in the eighteenth century.* New York (Columbia University Press).

———. 2003. Scientific institutions and the organization of science. *In* Porter, *Eighteenth-century science,* 87–106.

McClelland, Charles E. 1980. *State, society, and university in Germany, 1700–1914.* Cambridge (Cambridge University Press).

McIntyre, Donald B., and Alan McKirdy. 1997. *James Hutton: the founder of modern geology.* Edinburgh (H.M. Stationery Office).

McMahon, Darrin M. 2001. *Enemies of the Enlightenment: the French counter-Enlightenment and the making of modernity.* New York (Oxford University Press).

McPhee, John. 1981. *Basin and range.* New York (Farrar, Strauss, Giroux).

Marchand, Suzanne, and Elizabeth Lunbeck (eds.). 1996. *Proof and persuasion: essays on authority, objectivity, and evidence.* Amsterdam (Brepols).

Marini, Paola (ed.). 1987. *L'Opera scientifica di Giambattista Brocchi (1772–1826).* Bassano del Grappa (Città di Bassano).

Mauriès, Patrick. 2002. *Cabinets of curiosities.* London (Thames & Hudson).

Mazzolini, Renato G. (ed.). 1997. *Non-verbal communication in science prior to 1900.* Florence (Olschki).

Mazon, A. 1893. *Histoire de Soulavie (naturaliste, diplomate, historien).* Paris (Fischbauer).

Mendyk, Stan A. E. 1989. *"Speculum Britanniae": regional study, antiquarianism, and science in Britain to 1700.* Toronto (University of Toronto Press).

Mijnhardt, Wijnand Wilhelm. 1987. *Tot heil van't menschdom: culturele genootschappen in Nederland, 1750–1815.* Amsterdam (Rodopi).

Millasseau, Silvie. 1997. Brongniart as taxonomist and museologist: the significance of the Musée Céramique at Sèvres. *In* Préaud, *Sèvres porcelain,* 122–47.

Miller, David P. 1996. Joseph Banks, empire, and "centres of calculation" in late Hanoverian London. *In* Miller and Reill, *Visions of empire,* 21–37.

———. 1999. The usefulness of natural philosophy: the Royal Society and the culture of practical utility in the later eighteenth century. *BJHS* 32: 185–201.

Miller, David P., and H. D. Reill (eds.). 1996. *Visions of empire: voyages, botany, and representations of nature.* Cambridge (Cambridge University Press).

Momigliano, Arnaldo. 1966. *Studies in historiography.* London (Weidenfeld & Nicolson).

———. 1990. *The Classical foundations of modern historiography.* Berkeley (University of California Press).

Montalenti, Giuseppe, and Paolo Rossi (eds.). 1982. *Lazzaro Spallanzani e la biologia del Settecento: theorie, esperimenti, istituzioni scientifiche.* Florence (Leo S. Olschki).

Morello, Nicoletta. 1982. Lazzaro Spallanzani geopaleontologo: dall'origine delle sorgenti alla vulcanologia. *In* Montalenti and Rossi, *Lazzaro Spallanzani,* 271–81.

——— (ed.). 1998. *Volcanoes and history.* Genoa (Glauco Brigati).

Morrell, Jack, and Arnold Thackray. 1981. *Gentlemen of science: early years of the British Association for the Advancement of Science.* Oxford (Clarendon).

Morris, A. D. 1989. *James Parkinson: his life and times.* Boston, Basel, Berlin (Birkhäuser).

Murat, Laure, and Nicolas Weill. 1998. *L'Expédition d'Égypte: le rêve orientale de Bonaparte.* Paris (Gallimard).

Nadault de Buffon, Henri. 1860. *Correspondence inédit de Buffon.* 2 vols. Paris (L. Hachette).

Negrin, Howard E. 1977. *Georges Cuvier: administrator and educator.* New York (New York University [Ph.D. dissertation]) [University Microfilms no. 78–3124].

Newcomb, Sally. 1990. Contributions of British experimentalists to the discipline of geology: 1780–1820. *PAPS* 134: 161–225.

———. 1998. Laboratory variables in late eighteenth-century geology. *In* Fritscher and Henderson, *History of mineralogy,* 81–99.

Newth, D. R. 1952. Lamarck in 1800: a lecture on the invertebrate animals, and a note on fossils taken from the *Système des animaux sans vertèbres* by J. B. Lamarck. *AS* 8: 229–54.

North, F. J. 1942. Paviland cave, the "Red Lady", the Deluge, and William Buckland. *AS* 5: 91–128.

———. 1956. W. D. Conybeare: his geological contemporaries and Bristol associations. *PBNS* 29: 133–46.

Oldroyd, D. R. 1979. Historicism and the rise of historical geology. *HS* 17: 191–213, 227–57 [repr. *Sciences of the earth* (1998), art. XII].

———. 1998. *Sciences of the earth: studies in the history of mineralogy and geology.* Aldershot (Ashgate).

———. 1999. Non-written sources in the study of the history of geology: pros and cons, in the light of the views of Collingwood and Foucault. *AS* 56: 395–415.

———. 2003. A Manichean view of the history of geology [essay-review of Şengör, "Is the present the key to the past?" (2001)]. *AS* 60: 423–36.

Olmi, Giuseppe. 1992. *L'Inventario del mondo: catalogazione della natura e luoghi del sapere nella prima età moderna.* Bologna (Il Mulino).

———. 1997. From the marvellous to the commonplace. *In* Mazzolini, *Non-verbal communication*, 235–78.

Ophir, Adi, and Steven Shapin. 1991. The place of knowledge: a methodological survey. *SC* 4: 3–21.

O'Rourke, J. E. 1978. A comparison of James Hutton's *Principles of knowledge* and *Theory of the earth. Isis* 69: 5–20.

Ospovat, Alexander M. 1969. Reflections on A. G. Werner's "Kurze Klassifikation". *In* Schneer, *History of geology*, 242–56.

———. 1971. *Abraham Gottlob Werner: Short classification and description of the various rocks.* New York (Hafner).

———. 1978. Four hitherto unpublished geological lectures given by Sir Humphry Davy in 1805. *TRGSC* 21 (1), 96 pp.

———. 1980. The importance of regional geology in the geological theories of Abraham Gottlob Werner: a contrary opinion. *AS* 37: 433–40.

Outram, Dorinda. 1978. The language of natural power: the "Éloges" of Georges Cuvier and the public language of nineteenth-century science. *HS* 16: 153–78.

——— (ed). 1980. *The letters of Georges Cuvier: a summary calendar of manuscript and printed materials*... Chalfont St Giles (British Society for the History of Science).

———. 1982. Storia naturale e politica nella correspondenza tra Georges Cuvier e Giovanni Fabbroni. *RS* 13: 185–235.

———. 1983. Cosmopolitan correspondence: a calendar of the letters of Georges Cuvier (1769–1832). *Archives* 16: 47–53.

———. 1983. The ordeal of vocation: the Paris Academy of Sciences and the Terror, 1793–95. *HS* 21: 251–73.

———. 1984. *Georges Cuvier: vocation, science, and authority in post-Revolutionary France.* Manchester (Manchester University Press).

———. 1986. Uncertain legislator: Georges Cuvier's laws of nature in their intellectual context. *JHB* 19: 323–68.

Padian, Kevin. 1987. The case of the bat-winged pterosaur: typological taxonomy and the influence of pictorial representation on scientific perception. *In* Czerkas and Olson, *Dinosaurs past and present*, 2: 64–81.

Page, Leroy E. 1963. *The rise of the diluvial theory in British geological thought.* Norman [Okla.] (University of Oklahoma [Ph.D. dissertation]) [University Microfilms no. 64–215]

———. 1969. Diluvialism and its critics in Great Britain in the early nineteenth century. *In* Schneer, *History of geology*, 257–71.

Palm, L. C. 1987. Van Marums contacten: reizen en correspondentie. *In* Wiechmann and Palm, *Martinus van Marum*, 223–44.

Palmer, Katherine V. W. 1977. *The unpublished vélins of Lamarck (1802–1809): illustrations of fossils of the Paris Basin Eocene.* Ithaca (Paleontological Research Institution).

Pancaldi, Giuliano. 1983. *Darwin in Italia: impresa scientifica e frontiere culturali.* Bologna (Il Mulino) [trans. *Darwin in Italy*, Bloomington (Indiana University Press), 1991].

———. 1987. La teoria delle specie di Giambattista Brocchi. *In* Marini, *Brocchi*, 41–53.

Parkinson, Richard. 1999. *Cracking codes: the Rosetta stone and decipherment.* London (British Museum).

Parris, Leslie. 1973. *Landscape in Britain, c. 1750–1850*. London (Tate Gallery).

Phillips, Mark Salber. 1996. Reconsiderations on history and antiquarianism: Arnaldo Momigliano and the historiography of eighteenth-century Britain. *JHI* 58: 297–316.

———. 2000. *Society and sentiment: genres of historical writing in Britain, 1740–1820*. Princeton (Princeton University Press).

Pichard, Georges. 1992. Robert de Paul de Lamanon (1752–87): entre théorie de la terre et géologie. *TCFHG* (3) 6: 29–67.

Pickstone, John V. 1994. Museological science? The place of the analytical/comparative in nineteenth-century science, technology, and medicine. *HS* 32: 111–38.

———. 2001. *Ways of knowing: a new history of science, technology, and medicine*. Chicago (University of Chicago Press).

Pinault Sorensen, Madeleine. 1990. *Le peintre et l'histoire naturelle*. Paris (Flammarion) [trans. *The painter as naturalist*, Paris (Flammarion), 1991].

Plan, Danielle. 1909. *Un génévois d'autrefois: Henri-Albert Gosse, 1753–1816*. Paris (Fischbacher); Geneva (Kundig).

Pomian, Krzysztof. 1987. *Collectionneurs, amateurs et curieux: Paris, Venise, XVIe–XVIIIe siècle*. Paris (Gallimard) [trans. *Collectors and curiosities*, Cambridge (Polity Press), 1990].

Popkin, Richard H. 1987. *Isaac La Peyrère (1596–1676): his life, work, and influence*. Leiden and New York (Brill).

———. 2003. *The history of scepticism from Savonarola to Bayle*. Revised ed. [1st ed. 1979]. New York (Cambridge University Press).

Porter, Roy S. 1977. *The making of geology: earth science in Britain, 1660–1815*. Cambridge (Cambridge University Press).

———. 1978. George Hoggart Toulmin and James Hutton: a fresh look. *BGSA* 89: 1256–58.

———. 1978. George Hoggart Toulmin's theory of man and the earth in the light of the development of British geology. *AS* 35: 339–52.

———. (ed.). 2003. *Eighteenth-century science*. Cambridge (Cambridge University Press).

Porter, Theodore M. 1981. The promotion of mining and the advancement of science: the chemical revolution of mineralogy. *AS* 38: 543–70.

Préaud, Tamara (ed.). 1997. *The Sèvres porcelain manufactory: Alexandre Brongniart and the triumph of art and industry, 1800–1847*. New Haven (Yale University Press).

———. 1997. Brongniart as administrator. *In* Préaud, *Sèvres porcelain*, 42–51.

———. 1997. Brongniart as technician. *In* Préaud, *Sèvres porcelain*, 64–73.

Price, David. 1989. John Woodward and a surviving British geological collection from the early eighteenth century. *JHC* 1: 79–95.

Pyenson, Lewis, and Susan Sheets-Pyenson. 1999. *Servants of nature: a history of scientific institutions, enterprises, and sensibilities*. London (HarperCollins).

Rappaport, Rhoda. 1968. Lavoisier's geologic activities 1763–1792. *Isis* 58: 375–84.

———. 1969. The geological atlas of Guettard, Lavoisier, and Monnet: conflicting views of the nature of geology. *In* Schneer, *History of geology*, 272–87.

———. 1969. The early disputes between Lavoisier and Monnet, 1777–1781. *BJHS* 4: 233–44.

———. 1973. Lavoisier's theory of the earth. *BJHS* 6: 247–60.

———. 1978. Geology and orthodoxy: the case of Noah's Flood in eighteenth-century thought. *BJHS* 11: 1–18.

———. 1982. Borrowed words: problems of vocabulary in eighteenth-century geology. *BJHS* 15: 27–44.

———. 1994. Baron d'Holbach's campaign for German (and Swedish) science. *SVEC* 323: 225–46.

———. 1997. *When geologists were historians, 1665–1750*. Ithaca (Cornell University Press).

———. 2003. The earth sciences. *In* Porter, *Eighteenth-century science*, 417–35.

Regteren Altena, C. O. van. 1956. Achtienden eeuwse verzamelaars van fossilien te Maastricht en het lot hunner collecties. *PNGL* 9: 83–112.

Rice, Howard C. 1951. Jefferson's gift of fossils to the Museum of Natural History in Paris. *PAPS* 95: 597–627.

Richards, Robert J. 2000. Kant and Blumenbach on the *Bildungstrieb*: a historical misunderstanding. *SHPBBS* 31C: 11–32.

Richet, Pascal. 1999. *L'Age du monde: à la découverte de l'immensité du temps.* Paris (Seuil) [repr. *The age of the world*, Chicago (University of Chicago Press), forthcoming].

Ripoll, David. 2001. L'iconographie des *Voyages dans les Alpes. In* Sigrist, *Saussure*, 315–35.

Roberts, Gerrylynn K. 1991. Establishing science in eighteenth-century central Europe. *In* Goodman and Russell, *Rise of scientific Europe*, 361–86.

Roberts, Lissa. 1999. Going Dutch: situating science in the Dutch Enlightenment. *In* Clark et al., *Sciences in Enlightened Europe*, 350–88.

Roberts, Michael B. 1998. Geology and Genesis unearthed. *Churchman* 112: 225–55.

Roche, Daniel. 1988. *Les republicains des lettres: gens de culture et Lumières au XVIIIe siècle.* Paris (Fayard).

Roger, Jacques. 1953. Un manuscrit inédit perdu et retrouvé: *Les anecdotes de la nature*, de Nicolas-Antoine Boulanger. *RSH* 71: 231–54 [repr. *Histoire des sciences* (1995), 313–44].

———. 1962. Buffon, *Les époques de la nature*: édition critique. *MMNHN* (C) 10 (clii + 343 pp.) [repr. 1988].

———. 1963. *Les sciences de la vie dans la pensée française du XVIIIe siècle: la génération des animaux de Descartes à l'Encyclopédie.* Paris (Colin) [trans. *Life sciences in eighteenth-century French thought*, Stanford (Stanford University Press), 1997].

———. 1973. La théorie de la terre au XVIIe siècle. *RHS* 26: 23–48 [repr. *Histoire des sciences* (1995), 129–54].

———. 1974. Le feu et l'histoire: James Hutton et la naissance de la géologie. *In* Anonymous, *Approches des Lumières*, 415–29 [repr. *Histoire des sciences* (1995), 155–69].

———. 1982. The Cartesian model and its role in eighteenth-century "theory of the earth". *In* Lennon et al., *Problems of Cartesianism*, 95–125.

———. 1988. L'Europe savante 1700–1850. *In* Trembley, *Savants genevois*, 23–54.

———. 1988. La place de Buffon dans l'histoire des sciences de la terre. *TCFHG* (2) 2: 81–88.

———. 1989. *Buffon: un philosophe au Jardin du Roi.* Paris (Fayard) [trans. *Buffon: a life in natural history*, Ithaca (Cornell University Press), 1997].

———. 1992. Buffon et l'introduction de l'histoire dans l' 'Histoire naturelle'. *In* Gayon, *Buffon 88*, 193–205.

———. 1995. *Pour une histoire des sciences à part entière* [ed. Claude Blanckaert]. Paris (Albin Michel).

Rogerson, John. 1984. *Old Testament criticism in the nineteenth century in England and Germany.* London (SPCK).

Rolfe, W. D. Ian. 1983. William Hunter (1718–1783) on Irish "elk" and Stubbs's *Moose. ANH* 11: 263–90.

———. 1985. William and John Hunter: breaking the Great Chain of Being. *In* Bynum and Porter, *William Hunter*, 297–319.

Roller, Duane H. D. (ed.). 1971. *Perspectives in the history of science and technology.* Norman (Oklahoma University Press).

Rossi, Paolo. 1979. *I segni del tempo: storia della terra e storia delle nazioni da Hooke a Vico.* Milan (Feltrinelli) [trans. *The dark abyss of time*, Chicago (University of Chicago Press), 1984].

Rowlinson, J. S. 1998. "Our common room in Geneva" and the early exploration of the Alps of Savoy. *NRRSL* 52: 221–35.

Rudwick, Martin J. S. 1962. Hutton and Werner compared: George Greenough's geological tour of Scotland in 1805. *BJHS* 1: 117–35.

———. 1963. The foundation of the Geological Society of London: its scheme for cooperative research and its struggle for independence. *BJHS* 1: 325–55.

———. 1970. The glacial theory. *HS* 8: 136–57 [repr. *New science of geology* (2004), art. XIV].

———. 1971. Uniformity and progression: reflections on the structure of geological theory in the age of Lyell. *In* Roller, *Perspectives in the history of science*, 209–27 [repr. *Lyell and Darwin* (2005), art. I].

———. 1972. *The meaning of fossils: episodes in the history of palaeontology.* London (Macdonald); New York (American Elsevier) [2nd ed. 1976, repr. Chicago (University of Chicago Press), 1985].

———. 1976. The emergence of a visual language for geological science 1760–1840. *HS* 14: 149–95 [repr. *New science of geology* (2004), art. V].

———. 1982. Charles Darwin in London: the integration of public and private science. *Isis* 73: 186–206 [repr. *Lyell and Darwin* (2005), art. IX].

———. 1985. *The great Devonian controversy: the shaping of scientific knowledge among gentlemanly specialists.* Chicago (University of Chicago Press).

———. 1986. The shape and meaning of earth history. *In* Lindberg and Numbers, *God and nature*, 296–301 [repr. *New science of geology* (2004), art. II].

———. 1990. The emergence of a new science [essay-review of Laudan, *Mineralogy to geology*, 1987]. *Minerva* 28: 386–97 [repr. *New science of geology* (2004), art. IV].

———. 1992. *Scenes from deep time: early pictorial representations of the prehistoric world.* Chicago (University of Chicago Press).

———. 1996. Cuvier and Brongniart, William Smith, and the reconstruction of geohistory. *ESH* 15: 25–36 [repr. *New science of geology* (2004), art. VII].

———. 1996. Minerals, strata, and fossils. *In* Jardine et al., *Cultures of natural history*, 266–86 [repr. *New science of geology* (2004), art. III].

———. 1997. *Georges Cuvier, fossil bones, and geological catastrophes.* Chicago (University of Chicago Press).

———. 1997. *Recherches sur les ossements fossiles*: Georges Cuvier et le collecte des alliés internationaux. *In* Blanckaert et al., *Le Muséum*, 591–606 [trans. *New science of geology*, art. VIII].

———. 1999. Geologists' time: a brief history. *In* Lippincott, *Story of time*, 250–53 [repr. *New science of geology* (2004), art. I].

———. 2000. Georges Cuvier's paper museum of fossil bones. *ANH* 27: 51–68 [repr. *New science of geology* (2004), art. IX].

———. 2001. Jean-André de Luc and nature's chronology. *In* Lewis and Knell, *Age of the earth*, 51–60 [repr. *New science of geology* (2004), art. VI].

———. 2004. *The new science of geology: studies in the earth sciences in the age of revolution.* Aldershot (Ashgate).

———. 2005. *Lyell and Darwin, geologists: studies in the earth sciences in the age of reform.* Aldershot (Ashgate).

Rupke, Nicolaas. 1983. *The great chain of history: William Buckland and the English school of geology.* Oxford (Clarendon).

———. 1990. Caves, fossils, and the history of the earth. *In* Cunningham and Jardine, *Romanticism and the sciences*, 241–59.

———. 1994. C. C. Gillispie's *Genesis and geology. Isis* 85: 261–70.

———. 1997. Oxford's scientific awakening and the role of geology. *In* Brock and Curthoys, *Nineteenth-century Oxford*, 543–62.

——— (ed.). 2002. *Göttingen and the development of the natural sciences.* Göttingen (Wallstein).

Russell, E. S. 1916. *Form and function: a contribution to the history of morphology.* London (Murray) [repr. Chicago (University of Chicago Press), 1982].

Sadrin, Paul. 1986. Nicolas-Antoine Boulanger (1722–1759), ou avant nous le déluge. *SVEC* 240, 266 pp.

Sarjeant, William A. S., and Justin B. Delair. 1980. An Irish naturalist in Cuvier's laboratory: the letters of Joseph Pentland 1820–1832. *BBMNH* (Hist.) 6: 245–319.

Schaer, Roland. 1993. *L'Invention des musées.* Paris (Gallimard).

Schaffer, Simon. 1980. Herschel in Bedlam: natural history and stellar astronomy. *BJHS* 13: 211–39.

Schmidt, Benjamin, and Pamela Smith (eds.). Forthcoming. *Knowledge and its making in early modern Europe.* Chicago (University of Chicago Press).

Schnapp, Alain. 1993. *La conquête du passé: aux origines de l'archéologie.* Paris (Carré) [trans. *The discovery of the past,* London (British Museum), 1996].

Schneer, Cecil J. (ed.). 1969. *Toward a history of geology.* Cambridge (MIT Press).

———. 1979. *Two hundred years of geology in America.* Hanover (University Press of New Hampshire).

Semonin, Paul. 2000. *American monster: how the nation's first prehistoric creature became a symbol of national identity.* New York (New York University Press).

Şengör, A. M. Cêlal. 2001. Is the present the key to the past or the past the key to the present? James Hutton and Adam Smith versus Abraham Gottlob Werner and Karl Marx in interpreting history. *GSASP* 355, 51 pp.

Seta, Cesare de. 1992. *L'Italia del Grand Tour: da Montaigne a Goethe.* Naples (Electa).

Shapin, Steven. 1984. Pump and circumstance: Robert Boyle's literary technology. *SSS* 14: 481–520.

———. 1989. The invisible technician. *AmS* 77: 554–63.

———. 1994. *A social history of truth: civility and science in seventeenth-century England.* Chicago (University of Chicago Press).

Siegfried, Robert, and Robert H. Dott, Jr. 1980. *Humphry Davy on geology: the 1805 lectures for the general audience.* Madison (University of Wisconsin Press).

Sigrist, René (ed.). 2001. *H.-B. de Saussure (1740–1799): un regard sur la terre.* Geneva and Paris (Georg).

———. 2001. La géographie de Saussure à l'horizon des savants du XVIIIe siècle. *In* Sigrist, *Saussure,* 215–48.

Siliotti, Alberto. 2001. *Belzoni's travels: narrative of the operations and recent discoveries in Egypt and Nubia.* London (British Museum Press).

Simpson, A. D. C. 1982. *Joseph Black 1728–1799: a commemorative symposium.* Edinburgh (Royal Scottish Museum).

Simpson, G. G. 1942. The beginnings of vertebrate paleontology in North America. *PAPS* 86: 130–88.

Sloan, Philip R. 1992. L'hypothétisme de Buffon: sa place dans la philosophie des sciences du dix-huitième siècle. *In* Gayon, *Buffon 88,* 207–22.

———. 1992. Organic molecules revisited. *In* Gayon, *Buffon 88,* 415–38.

———. 1997. Le Muséum de Paris vient à Londres. *In* Blanckaert et al., *Le Muséum,* 607–34.

Smith, Crosbie (ed.). 1998. *Making space for science: territorial themes in the shaping of knowledge.* Basingstoke (Macmillan).

Smith, Jean Chandler. 1993. *Georges Cuvier: annotated bibliography of his published works.* Blue Ridge Summit [Penn.] (Smithsonian Institution Press).

Snelders, H. A. M. 1992. Professors, amateurs, and learned societies: the organization of the natural sciences. *In* Jacob and Mijnhardt, *Dutch republic,* 308–22.

Snobelen, Stephen. 1999. Isaac Newton, heretic: the strategies of a Nicodemite. *BJHS* 32: 381–419.

Spadafora, David. 1990. *The idea of progress in eighteenth-century Britain.* New Haven (Yale University Press).

Spallanzani, Maria Franca. 1985. *La collezione naturalistica di Lazzaro Spallanzani: i modi e i tempi della sua formazione.* Reggio Emilia (Comune di Reggio nell'Emilia).

———. 1994. Vom "Studiolo" zum Laboratorium: die "piccolo raccolta di naturali produzioni" des Lazzaro Spallanzani (1729–99). *In* Grote, *Macrocosmos*, 679–94.

Spary, E. C. 1999. The "Nature" of Enlightenment. *In* Clark et al., *Sciences in Enlightened Europe*, 272–304.

———. 2000. *Utopia's garden: French natural history from Old Regime to Revolution.* Chicago (University of Chicago Press).

Stengers, Jean. 1974. Buffon et la Sorbonne. *EDS* 1: 97–127.

Stewart, Larry. 1999. Other centres of calculation, or, where the Royal Society didn't count: commerce, coffee-houses, and natural philosophy in early modern Britain. *BJHS* 32: 133–53.

Taquet, Philippe. 1994. Georges Cuvier, ses liens scientifiques européens. *In* Anonymous, *Montbéliard sans frontières*, 287–309.

———. 1997. Lamarck en 1809, un auteur soucieux de la diffusion de son oeuvre auprès du public. *In* Laurent, *Jean-Baptiste Lamarck*, 95–118.

———. 1998. Les premiers pas d'un naturaliste sur les sentiers du Wurtemberg: recit inédit d'un jeune étudiant nommé Georges Cuvier. *Geodiversitas* 20: 285–318.

Taquet, Philippe, and Kevin Padian. 2004. The earliest known restoration of a pterosaur and the philosophical origins of Cuvier's *Ossemens fossiles. CRP* 3: 157–75.

Taton, René (ed.). 1964. *Enseignement et diffusion des sciences en france au XVIIIe siècle.* Paris (Hermann).

Taylor, Kenneth L. 1968. *Nicolas Desmarest (1725–1815): scientist and industrial technologist.* Cambridge (Harvard University [Ph.D. dissertation]).

———. 1979. Geology in 1776: some notes on the character of an incipient science. *In* Schneer, *Two hundred years of geology*, 75–90.

———. 1982. The beginnings of a French geological identity. *HN* 19/20: 65–82.

———. 1988. Les lois naturelles dans la géologie du XVIIIème siècle: recherches préliminaires. *TCFHG* (3) 2: 1–28.

———. 1992. The historical rehabilitation of theories of the earth. *Compass* 69: 334–45.

———. 1992. The *Époques de la nature* and geology during Buffon's later years. *In* Gayon, *Buffon 88*, 371–85.

———. 1994. New light on geological mapping in Auvergne during the eighteenth century: the Pasumot-Desmarest collaboration. *RHS* 47: 129–36.

———. 1995. Nicolas Desmarest and Italian geology. *In* Giglia et al., *Rocks, fossils, and history*, 95–109.

———. 1998. Volcanoes as accidents: how "natural" were volcanoes to 18th-century naturalists? *In* Morello, *Volcanoes and history*, 595–618.

———. 2001. Buffon, Desmarest, and the ordering of geological events in *époques. In* Lewis and Knell, *Age of the earth*, 39–49.

———. 2001. The beginnings of a geological naturalist: Desmarest, the printed word, and nature. *ESH* 20: 44–61.

———. 2002. Two ways of imagining the earth at the close of the 18th century: descriptive and theoretical traditions in early geology. *In* Albrecht and Ladwig, *Werner*, 369–78.

———. 2002. Un commentaire anonyme inédit sur les observation et les idées de William Hamilton. *TCFHG* (3) 15: 1–35.

Thackray, John C. 1976. James Parkinson's *Organic remains of a former world* (1804–1811). *JSBNH* 7: 451–66.

——— (ed.). 2003. *To see the Fellows fight: eye witness accounts of meetings of the Geological Society of London and its Club, 1822–1868.* British Society for the History of Science (monograph no. 12).

Théodoridès, Jean. 1969. Une lettre inédite de Georges Cuvier à la Gesellschaft Naturforschender Freunde zu Berlin (1800). *HB* 2: 58–68.

Theunissen, Bert. 1980. De briefwisseling tussen A. G. Camper en G. Cuvier. *TGGNWT* 3: 155–77.

———. 1984. A. G. Camper, Cuvier en het Mosasaurusvraagstuk: een case-study van Cuviers paleontologie. *TGGNWT* 7: 65–78.

———. 1986. The relevance of Cuvier's *lois zoologiques* for his palaeontological work. *AS* 43: 543–56.

———. 1987. Martinus van Marum, 1750–1837: "Ten nutte en ten genoegen der ingezetenen". *In* Wiechmann and Palm, *Martinus van Marum*, 11–32.

Thomson, Duncan. 1997. *Raeburn: the art of Sir Henry Raeburn 1756–1823*. Edinburgh (Scottish National Portrait Gallery).

Tomkeieff, S. I. 1962. Unconformity, an historical study. *PGA* 73: 383–417.

Torrens, Hugh S. 1979. Geological communication in the Bath area in the last half of the eighteenth century. *In* Jordanova and Porter, *Images of the earth*, 215–247 [repr. *British geology* (2002), art. III].

———. 1989. In commemoration of the 150th anniversary of the death of William Smith (1769–1839). *TCFHG* (3) 3: 57–63.

———. 1994. Patronage and problems: Banks and the earth sciences. *In* Banks et al., *Sir Joseph Banks*, 49–75 [repr. *British geology* (2002), art. V].

———. 1995. Mary Anning (1799–1847) of Lyme: "the greatest fossilist the world ever knew". *BJHS* 28: 257–84.

———. 1997. Le "nouvel art de prospection minière" de William Smith et le "projet de houillère de Brewham": un essai malencontreux de recherche de charbon dans le sud-ouest de l'Angleterre, entre 1803 et 1810. *In* Gohau, *Géologie à son histoire*, 101–18 [repr. *British geology* (2002), art. IV].

———. 1998. Geology in peace time: an English visit to study German mineralogy and geology (and visit Goethe, Werner, and Raumer) in 1816. *In* Fritscher and Henderson, *History of mineralogy*, 147–75.

———. 1999. Coal hunting at Bexhill 1805–1811: how the new science of stratigraphy was ignored. *SAC* 136: 177–91 [repr. *British geology* (2002), art. VII].

———. 2001. Timeless order: William Smith (1769–1839) and the search for raw materials 1800–1820. *In* Lewis and Knell, *Age of the earth*, 61–83 [repr. in Phillips (J.), *Memoirs of William Smith* (1844), repr. 2003, 155–92].

———. 2002. *The practice of British geology, 1750–1850*. Aldershot (Ashgate).

———. 2003. Introduction to the life and times of William Smith. *In* Phillips (J.), *Memoirs of William Smith* (1844), repr. 2003, xi–xxxviii.

Toulmin, Stephen, and June Goodfield. 1965. *The discovery of time*. London (Hutchinson) [repr. Chicago (University of Chicago Press), 1977].

Trabant, Jürgen. 2001. New perspectives on an old academic question. *In* Trabant and Ward, *Origin of language*, 1–17.

Trabant, Jürgen, and Sean Ward (eds.). 2001. *New essays on the origin of language*. Berlin (Mouton de Gruyter).

Trembley, Jacques (ed.). 1988. *Les savants genevois dans l'Europe intellectuelle du XVIIe au milieu du XIXe siècle*. Geneva (Journal de Genève).

Tunbridge, Paul A. 1971. Jean André de Luc, F.R.S. (1727–1817). *NRRSL* 26: 15–33.

Turner, Anthony. 1987. *Early scientific instruments: Europe 1400–1800*. London (Sotheby's).

Uglow, Jennifer. 2002. *The Lunar men: the friends who made the future, 1730–1810*. London (Faber).

Vaccari, Ezio. 1992. Geologia e attività mineraria in Italia settentrionale tra settecento ed ottocento: l'influenza della "scuola di Freiberg" su alcuni scienziati Italiani. *Nuncius* 7: 93–107.

———. 1993. I manoscritti di uno scienziato Veneto del settecento: notizie storiche e catalogo del fondo "Giovanni Arduino" della Biblioteca Civica di Verona. *AIV* 151: 271–373.

———. 1993. *Giovanni Arduino (1714–1795): il contributo di uno scienziato Veneto al dibattito settecentesco sulle scienze della terra*. Florence (Olschki).

———. 1998. Mineralogy and mining in Italy between the eighteenth and nineteenth centuries: the extent of Wernerian influences from Turin to Naples. *In* Fritscher and Henderson, *History of mineralogy*, 107–30.

———. 1998. Lazzaro Spallanzani and his geological travel to the "Due Sicilie": the volcanology of the Aeolian Islands. *In* Morello, *Volcanoes and history*, 621–51.

———. 1998. Quelques réflexions sur les instructions scientifiques destinées aux géologues voyageurs aux dix-huitième et dix-neuvième siècles. *TCFHG* (3) 12: 39–57.

———. 1999. La 'classificazione' delle montagne nel Settecento e la teoria litostratigrafica di Giovanni Arduino. *In* Curi, *Scienza di Arduino*, 47–80.

———. 2000. Mining and knowledge of the earth in eighteenth-century Italy. *AS* 57: 163–80.

———. 2000. The museum and the academy: geology and mineralogy in the Accademia dei Fisiocritici of Siena during the 18th century. *In* Ghiselin and Leviton, *Institutions of natural history*, 5–25.

———. 2001. European views on terrestrial chronology from Descartes to the mid-eighteenth century. *In* Lewis and Knell, *Age of the earth*, 25–37.

Vaj, Daniela. 2001. Saussure à la découverte de l'Italie. *In* Sigrist, *Saussure*, 269–301.

Van Riper, A. Bowdoin. 1993. *Men among the mammoths: Victorian science and the discovery of human prehistory*. Chicago (University of Chicago Press).

Vergani, Raffaelo. 1987. Brocchi funzionario alle miniere: da Regno Italico alla dominazione Austriaca. *In* Marini, *Brocchi*, 67–77.

Vidler, Alec. 1961. *The church in an age of revolution: 1789 to the present day*. Harmondsworth (Penguin).

Viénot, J. 1932. *Le Napoléon de l'intelligence: Georges Cuvier*. Paris (Fischbacher).

Visconti, Agnese. 1987. Il ruolo delle raccolte del Museo Reale di Storia Naturale di Milano per la realizzazione della *Conchiologia fossile Subapennina* di Giambattista Brocchi. *In* Marini, *Brocchi*, 143–53.

Visser, Robert Paul Willem. 1985. *The zoological work of Petrus Camper (1722–1789)*. Amsterdam (Rodopi).

Wagenbreth, Otfried. 1967. Abraham Gottlob Werners System der Geologie, Petrographie und Lagerstättenlehre. *In* Anonymous, *Werner Gedenkschrift*, 83–148.

———. 1968. Die Paläontologie in Abraham Gottlob Werners geologischen System. *Bergakademie* 20: 32–36.

Wagner, Monika. 1983. Das Gletschererlebnis: visuelle Natureignung im frühen Tourismus. *In* Grossklaus and Oldenmeyer, *Natur als Gegenwelt*, 235–63.

Wakefield, André. 2002. The cameralist tradition in Freiberg. *In* Albrecht and Ladwig, *Werner*, 379–88.

———. Forthcoming. The fiscal logic of Enlightened German science. *In* Schmidt and Smith, *Knowledge and its making*, [forthcoming].

Wegmann, Eugène. 1967. Évolution des idées sur le déplacement des lignes de rivage: origines en Fennoscandie. *MSVSN* 14: 129–90.

———. 1969. Changing ideas about moving shorelines. *In* Schneer, *History of geology*, 386–414.

Weindling, Paul. 1979. Geological controversy and its historiography: the prehistory of the Geological Society of London. *In* Jordanova and Porter, *Images of the earth*, 248–71.

———. 1983. The British Mineralogical Society: a case study in science and social improvement. *In* Inkster and Morrell, *Metropolis and province*, 120–150.

Wellnhofer, Peter. 1982. Cuvier and his influence on the interpretation of the first known pterosaur. *In* Buffetaut et al., *Symposium Cuvier*, 535–48.

Wheeler, Alwyne, and James H. Price (eds.). 1985. *From Linnaeus to Darwin: commentaries on the history of biology and geology*. London (Society for the History of Natural History).

Wiechmann, A., and L. C. Palm (eds). 1987. *Een elektriserend geleerde: Martinus van Marum, 1750–1837*. Haarlem (Enschedé).

Wiechmann, A., and Lydia Touret. 1987. Frappez, frappez toujours! Van Marum als verzamelaar en bezieler van het geleerd bedrijf in Haarlem. *In* Wiechmann and Palm, *Martinus van Marum*, 103–53.

Wilcox, Donald J. 1987. *The measure of times past: pre-Newtonian chronologies and the rhetoric of relative time.* Chicago (University of Chicago Press).

Withers, Charles W. J. 1997. On Georgics and geology: James Hutton's "Elements of agriculture" and agricultural science in eighteenth-century Scotland. *AHR* 42: 38–48.

Wood, Paul. 2003. Science in the Scottish Enlightenment. *In* Broadie, *Scottish Enlightenment*, 95–116.

Wood Jones, F. 1953. John Hunter as a geologist. *ARCS* 12: 219–44.

Woodward, Horace B. 1907. *The history of the Geological Society of London.* London (Geological Society).

Wyatt, John F. 1995. George Bellas Greenough: a Romantic geologist. *ANH* 22: 61–71.

Yeo, Richard. 2001. *Encyclopaedic visions: scientific dictionaries and Enlightenment culture.* Cambridge (Cambridge University Press).

———. 2003. Classifying the sciences. *In* Porter, *Eighteenth-century science*, 241–66.

INDEX

A page number in *italics* refers to a figure caption. Page numbers 639 and above refer to the summary in the "Coda", and may be helpful as a first point of reference.

CONTENTS

ILLUSTRATIONS

Saussure's century than the first ascent of Everest was in the twentieth; as remarkable in its time as the first successful expeditions to the north and south poles.[2]

Saussure and his party were not quite the first to stand on the summit of Mont Blanc. He was guided there by Jacques Balmat, a villager from Chamonix, who had reached the peak the previous summer with the local physician, Michel-Gabriel Paccard. Balmat had already been dubbed "le Mont Blanc" for his achievement; and when Saussure wrote an account of his own ascent, he praised the local pair for their courage in climbing to the summit "without even being certain that men could live in the places they aspired to reach". There had in fact been a series of attempts on the mountain during the previous decade, by Saussure himself among others, and some of the villagers had become quite experienced at scaling the slopes of ice and snow. After his first success, Balmat had been commissioned by Saussure to try again in 1787, and he had duly made a second ascent in July, accompanied this time by two other villagers. Saussure had been on his way to Chamonix when they did so, but had not got there in time to go with them. Then the weather had closed in for almost a month, and he had had to wait in patience for another opportunity. When at last the clouds cleared, he had been ready to make it a truly scientific expedition.

Saussure and his party found the ascent arduous. After a relatively simple first day's climb over grassy slopes and easy rocks, they camped 779 *toises* or fathoms (about 1,500m) above Chamonix. But the second day took them up across a glacier, where they had to negotiate dangerous crevasses—one man fell in, but was saved by

A
B
C
M. T. Bourrit fecit
Vue du Mont-Blanc et de la Route par la quelle on a atteint sa Cime,
A. Cime du Mt. Blanc. B. Dome du Gouté. C. Aiguille du Gouté. D.E. Arve et Vallée de Chamouny. * * Places ou l'on a campé en montant.

Fig. 1.2. The north face of Mont Blanc, as seen from the far side of the valley of the Arve, in which Chamonix lies: an engraving based on a drawing by Marc-Théodore Bourrit, published in a later volume of Saussure's *Alpine Travels* (1796). His route to the peak (A) is shown, with asterisks to denote their camp sites on the way up. (By permission of the Syndics of Cambridge University Library)

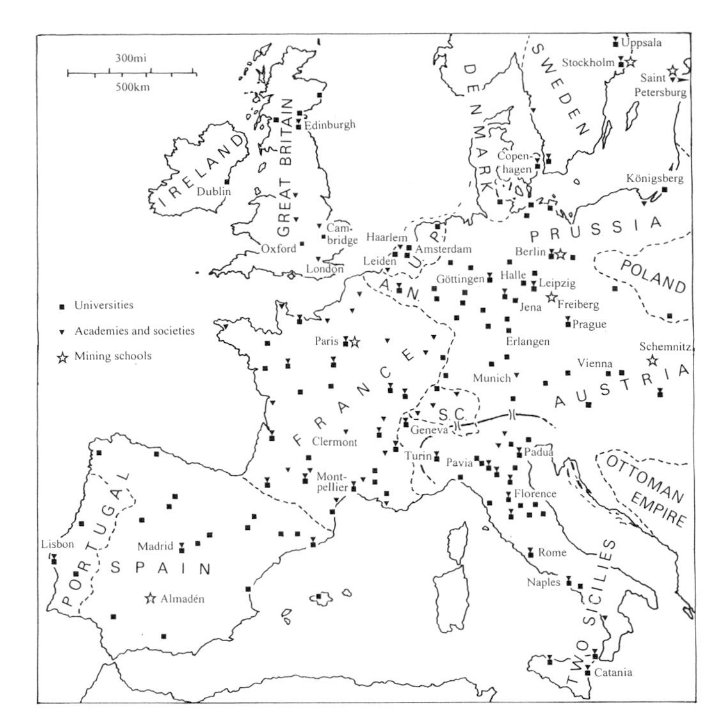

The relation between theology and the natural sciences, like other aspects of the map of knowledge, was not just an abstract issue, but was embodied in concrete social and institutional forms. At specific times and in specific places, there was indeed ecclesiastical suspicion or even hostility towards certain claims to natural knowledge. But this was more on the grounds of their heterodox theological implications than because they contradicted a literalistic reading of biblical texts. For example, recurrent speculations about the natural origin of life, and more particularly of the human species, were often suspect. The suspicion arose, however, not just because such ideas seemed to contradict the literal sense of the creation narratives in Genesis; anyway, biblical scholars—and through them, the educated laity too—were well aware, for example, of the multiple meanings of the Hebrew words translated as "create" and "day". What often made this kind of speculation unacceptable was that it was seen as undermining the theistic concept of creation itself, with its radical distinction between God and *all* kinds of "creature"—from atoms to humans—and all the personal, social, and political implications that flowed from that emphasis on divine transcendence. In other words, certain claims in natural philosophy were objectionable to the religious because they were also tacit claims about the proper conduct and ultimate meaning of human lives: it was a case of one religion pitted against another, even if one of them was often called antireligious or even atheistic (see also §2.5).

Most importantly, however, the relations between scientific savants and ecclesiastical authorities varied greatly from one cultural environment to another. The highly centralized and absolutist French state, with Catholicism as its established form of religion, tended to equate any philosophical speculation with potential sedition. Yet even here the official censorship was generally either lax or relaxed: those responsible for checking the credentials of new books were often themselves savants, and might be sympathetic to their authors' scholarly goals; and anyway, in case of difficulty, books could be published in Amsterdam or Geneva and then shipped back to France, or else be printed in France but issued under the false imprints of such safely foreign cities. In England, by contrast, the political and ecclesiastical establishments were intertwined in such a way that it was usually taken for granted that sound scientific work was more likely to support religious practice than to subvert it; the doctrinally liberal "latitudinarian" party in the established Anglican church was generally able to call the tune. In the German and Italian lands the sheer multiplicity of sovereign states generated a political and cultural pluralism that ensured that most kinds of scholarly work could find a congenial home somewhere: even within the Catholic world, for example, what was unacceptable in the Papal States might be published not far away in the Republic of Venice.[58]

56. See for example Löwenbrück, "Michaelis et la critique biblique" (1986), on the famous Göttingen savant Johann David Michaelis (1717–91), whose less famous son Christian Friedrich worked on fossil bones (§5.3); more generally, Rogerson, *Old Testament criticism* (1984), chap. 1.

57. Brooke, "Science and religion" (2003), reviews these issues, and particularly *natural* theology, for the eighteenth century.

58. For France, see for example Darnton, *Business of enlightenment* (1979), and *Great cat massacre* (1984), chap. 4; and Darnton and Roche, *Revolution in print* (1989).

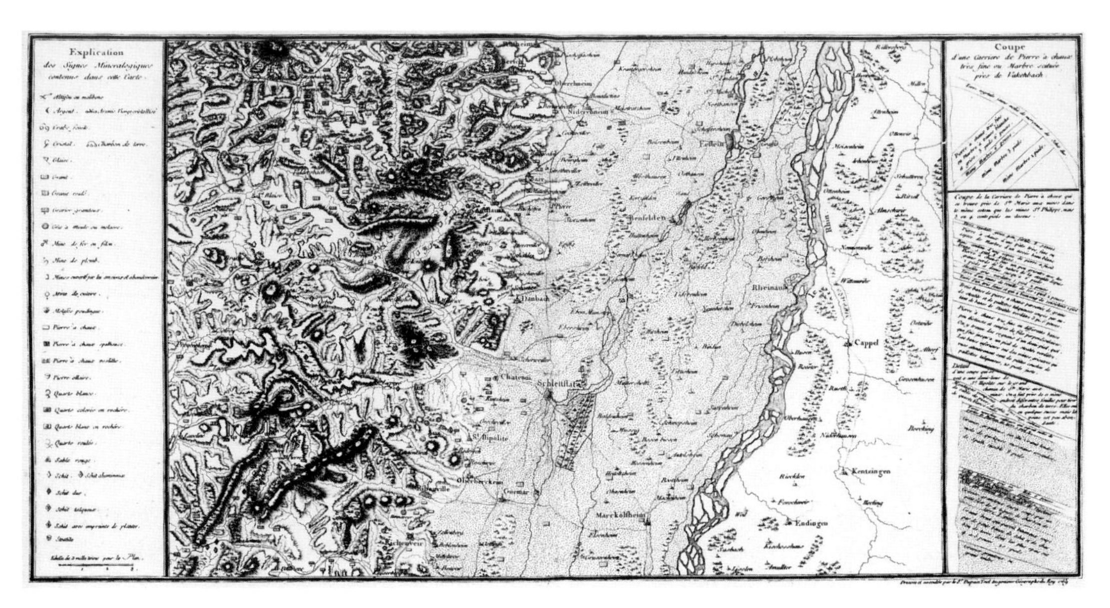

Fig. 2.11. A sheet of the *Atlas Minéralogique*, dated 1769, showing the hills of the Vosges in Alsace, and the braided river Rhine forming the eastern frontier of France; the town of Schlettstat (now Sélestat) south of Strasbourg is near the center of the map. The topography is plotted with much detail and accuracy, although the hills are depicted only by relatively crude hachuring; the alluvial plain of the Rhine is stippled. The map was based on the earlier topographical survey made for strategic and economic reasons of state; what made the atlas "mineralogical" was the inclusion of many spot symbols (see the key in the left margin) denoting the localities of useful or interesting minerals and rocks. The third dimension (see §2.3) was represented only by small sections of strata at specific localities (in the right margin). This sheet was based on some of the first joint fieldwork by Guettard and Lavoisier. This reproduction shows the general format of the maps in the atlas; Fig. 2.12 shows part of another sheet in greater detail.

physical geography, such as those of the *Atlas*, depicted topography in the cartographical style of the period. In particular, topographical relief was usually shown by hachuring, not contours. Even in the hands of a skilled engraver working from good sketches by the field surveyor, the resultant impression of the form of hills and valleys usually left much to be desired from the point of view of a physical geographer; in other respects, however, the sheets of the *Atlas* were impressive (Fig. 2.11).[30]

The way that the "mineral" component of the survey was recorded on the sheets of the *Atlas* reflects accurately the goals of the science of physical geography. The convention of spot symbols was long established in cartography and used on ordinary maps to show churches, castles, towns of varied status, and so on; here it was simply extended to show noteworthy and distinctive minerals, rocks, and even

28. This was the second and more ambitious Cassini survey (1747–88), with 180 sheets at a scale of 1:86, 400: see Konvitz, *Cartography in France* (1987), chap. 1; his fig. 6 charts its progress; Broc, *Montagnes* (1991), figs. 12, 17, reproduces samples. Maps of a comparable standard did not begin to be produced in Britain, for example, until the threat of Revolutionary war on the Continent caused the military Board of Ordnance to set up an Ordnance Survey for that purpose.

29. Rappaport, "Geological atlas" and "Lavoisier and Monnet" (both 1969); Konvitz, *Cartography in France* (1987), chap. 4. By 1780, 31 sheets out of a projected 214 had appeared, and the final count was only 45. The engraver was Dupain-Triel (see Fig. 4.12). Monnet's MS records of some of his fieldwork for the *Atlas* are in Paris-EM, MS 9, 10.

30. Fig. 2.11 is reproduced from Monnet, *Atlas minéralogique* (1780), unnumbered sheet.

birge]. This expressed their structural relation, namely that the Primaries everywhere underlay the Secondaries and seemed to be the foundation of the earth's crust; and also a general contrast between them, namely that the Secondaries were usually bedded rocks lying in a structural sequence that was distinct, at least in any given region, whereas the Primaries rarely had any clear sequence. Furthermore, the word *Urgebirge* implied a belief that the Primaries were "primitive" or fundamental in a temporal sense; Werner cautiously called them "apparently primitive" [*uranfänglich*]. In French this was expressed less ambiguously: following Rouelle's lead, the rock masses were usually termed the "*ancienne terre*" and the "*nouvelle terre*", terrains old and new.[42]

The French terms, and more ambiguously the others, indicate that relative age was indeed one of the criteria by which these two great categories were to be distinguished. Yet relative age was inferred from structural position, and in practice that directly observable criterion was treated as more important: geognosts usually referred to rock masses as being "above" or "below" others, and only rarely as "younger" or "older". When, for example, the Swedish geognost Johann Jakob Ferber (1743–90) reported on Arduino's classification—and thereby made it more widely known to savants throughout Europe—he explained how the major groups of *Gebirge* were distinguished "by the arrangement of their beds below or above one another, and the different age and origin inferred therefrom". It was the structural position of the Primaries, as the apparent physical foundation of the earth's crust, that entitled them to be treated as primary; and it was the structural position of the Secondaries, clearly overlying the Primaries and sometimes manifestly composed of materials derived from their erosion, that made them secondary, rather than the lesser age that those observed features implied. Anyway, "Primary" and "Secondary" remained terms of classification, not geohistory: the inferred difference of age was just one taxonomic criterion among many, and for most purposes the others were more important.[43]

The Primaries were generally hard rocks, usually found in upland or mountain regions and lacking any trace of fossils; the Secondaries were often softer, they were

41. Laudan, *Mineralogy to geology* (1987), chap. 5, draws a sharp distinction between Werner's classification and others in natural history, on the grounds that Werner's posited gradations between species and used geological time as a major taxonomic criterion. But the nature and reality of the divisions between species was also controversial among botanists and zoologists (hence their lively debates about hybrids and "sports"); and the supposed introduction of time did not alter the fundamental goal of Werner's work, which was to *classify* the varied rock masses (see below).

42. The *Ur*- prefix was ambiguous, because it denoted some kind of fundamental status, but not necessarily a temporal one: in the case of botany, Goethe suggested a hypothetical "original plant" [*Urpflanze*] to represent the archetype or common structural ground plan on which the diversity of plants was constructed, rather than their temporal origin or ancestral form. Werner later proposed a "Transition" category [*Übergangsgebirge*] that bridged the gap between the Primaries and Secondaries (§8.1), but this did not alter the basic character of his classification, any more than the gradations that he allowed between many of the "species" within them. "Primary", "Transition", "Secondary", etc., are given initial capitals throughout this account, to indicate that they were technical terms, like their indirect modern equivalents such as "Carboniferous" and "Jurassic".

43. Ferber, *Briefen aus Wälschland* (1773), 38. French and English editions followed in 1776; however, the latter (*Tour through Italy*) missed the crucial structural point by translating Ferber's phrase only as "according to the difference of their beds and their presumptive antiquity and origin" (36). On Arduino himself, see Vaccari, "La 'classificazione' delle montagne" (1999).

A further distinction between Primaries and Secondaries is well illustrated by this example. As already mentioned, the Secondary rock masses were usually composed of more or less distinct and parallel-surfaced beds or strata (see also Fig. 2.16), apparently deposited in succession (just how they might have been deposited, and from what medium, were separate problems). The Primary rock masses, on the other hand, had more diverse and often obscure structural relations with one another. Some, such as gneiss and schist, seemed to be layered or bedded—Saussure thought the granite near the summit of Mont Blanc was another case in point (§1.1)—but others were "massive" and without any trace of bedding. Some seemed to lie side by side, rather than being either above or below one another; and mineral veins usually cut right through rock masses of some contrasting kind. These confusing structures made the study of the Primaries more difficult than that of the Secondaries. However, it was also literally more rewarding, for most of the economically important minerals—particularly the ores of the precious and other nonferrous metals—were found in veins in the "hard-rock" terrains of the Primaries, which for that reason were sometimes called "veined rock masses" [*Ganggebirge*]. Coal seams and some iron ores were almost the only valuable materials found in the bedded rock masses of the Secondaries (which were also important, however, as sources of workaday materials such as building stone, brick-earth, etc.). Consequently, geognosts paid far more attention to the Primaries than to the Secondaries; the latter were often tacitly dismissed as being of little interest, either to the practical manager of mines or to the *wissenschaftlich* savant.

By comparison with the Primary and Secondary *Gebirge*, the other two high-level categories in geognostic classification were of relatively minor importance. Most of the *alluvial* or *Superficial* deposits were not rock masses at all, at least in the colloquial meaning of "rocks": they were the loose gravels, sands, silts, and muds usually found in river valleys and estuaries and on low-lying plains (see Fig. 2.18), though sometimes they formed low hills. Their origin was inferred to be very recent, relative even to the Secondary rock masses. More important, however, was the observable criterion that they were clearly derivative: Werner called them appropriately the "washed-out" [*Ausgeschwemmte*] deposits. River gravels, for example, might include easily identifiable pebbles of any of the Primary and Secondary rocks in the region drained by the river and its tributaries. They were therefore sometimes called *Tertiary* deposits.[45]

The fourth and last high-level category in geognostic classification was also treated as relatively minor in significance. The main diagnostic feature of the

44. Fig. 2.18 is reproduced from Dupain-Triel, *Recherches géographiques* (1791), pl. 3, fig. [1]. The modern distinction between "hard-rock" and "soft-rock" terrains (and hard-rock and soft-rock geologists) has no simple correlation with age, because it depends on the contingent geohistory of specific regions. Pudding stone or *poudingue* was so called because its embedded pebbles resembled, for example, the fruit and nuts in traditional Christmas puddings.

45. The word Tertiary was not being used here in its modern stratigraphical sense. However, Arduino had used "*monti terziari*" in a way that did approximate to that modern meaning, namely for what others regarded as upper and flat-lying beds of Secondary formations (as shown, though without that label, in Fig. 2.32), containing distinctive fossils such as nummulites. He then distinguished the alluvial deposits on the valley floors and plains—not "monti" at all—as a *fourth* category or "*quatro ordine*" (the indirect forerunner of modern "Quaternary"): see Vaccari, *Giovanni Arduino* (1993), chap. 3.

De Veyrenc Del.
CRATERE DE LA MONTAGNE DE LA COUPE, AU COLET D'AISA,
Avec un Courant de Lave qui donne naissance a un pavé de basalte prismatique.
Cl. Fessard Sculp.

Fig. 2.23. The cratered volcanic cone of the Coupe d'Aizac, near Vals in Vivarais, "with a lava current that gives birth to a pavement of prismatic basalt" at the edge of the stream: an engraving in Faujas's *Extinct Volcanoes* (1778) that strikingly supported his claim for the volcanic origin of this particular prismatic basalt, and also by implication for others elsewhere. The scene is given scale and human significance by gentlemanly figures, with the carriage that has brought them as real witnesses to this remote spot. (By permission of the Syndics of Cambridge University Library)

Werner's Neptunist case was supported by contrasts between basalts on the one hand and what were agreed to be lavas on the other. For example, the spectacular phenomenon of regular prismatic jointing in many basalts, as at the Giant's Causeway and the Isle of Staffa (Figs. 1.15, 2.8), was said to be unknown in any true lava produced by an active volcano; and its similarity—however crude—to the roughly hexagonal cracks that were commonly seen when a muddy pool dried out reinforced the argument that basalt had originally been some kind of precipitate that had likewise dried out after it was formed.

However, this was a case in which fieldwork in physical geography yielded what was claimed to be decisive evidence to refute this Neptunist interpretation. For example, Barthélemy Faujas de Saint Fond (1741–1819), one of the naturalists attached to the royal museum in Paris, traveled extensively in central France, in the remote uplands of the provinces of Vivarais and Velay, and examined in detail some of its recently discovered extinct volcanoes.

62. Fig 2.22 is reproduced from Faujas, *Volcans éteints* (1778) [Cambridge-UL: MA.40.1], pl. 5, engraved by de Veyrenc; explained on 277–80 as the "volcan de Maillas". The hill, above what is now Saint-Jean-le-Centenier (Ardèche), between Aubenas and Montélimar, is not in fact the isolated mass that it appears to be in this view, but a spur projecting from the broad basaltic Plateau de Coiron in the background. The Neptunist case was put forward in, for example, Werner, "Entstehung des Basalts" (1788); particularly striking, in his opinion, was the Scheibenberg, a prominent basalt-capped hill in the Erzgebirge.

Tom. I.
Pl. IV.
Vüe de la Montagne du Nant d'Arpenaz entre Maglan et Salanche en Faucigny.

Fig. 2.25. A spectacular case of strongly curved strata, exposed at Nant d'Arpenaz in the valley of the Arve downstream from Chamonix: an engraving, based on a painting by Bourrit, published by Saussure in his *Alpine Travels* (1779). He claimed that the rocks in the cliff were just the lower part of an even larger S-shaped fold, the top of which was continued (here in foreshortened view) in the distant cliff on the right. To give his readers a sense of the vast scale of the phenomenon, he noted that the waterfall alone was about 800 feet (250m) high. This striking instance of hard rocks apparently folded like putty became a constant empirical reference point in Saussure's long search for an adequate causal explanation for the origin of mountains. (By permission of the Syndics of Cambridge University Library)

result of huge lateral compressive movements of the earth's crust, like the crumpling of a tablecloth, operating over vast spans of time (Fig. 2.25).[66]

It therefore became necessary to infer—as Saussure did—that the earth's crust had somehow been buckled, not just in the earth's infancy, but since the deposition of at least some of the Secondary formations. That implied that not all mountains were "primitive" or original features of the earth's crust; at least some must have been formed during the earth's later "revolutions". However, that inference did little to solve the causal problem. What deep-seated agency could have been capable of causing whole mountains made of Secondary formations to be heaved up on end? Or what could have caused the deep foundations of those mountains to be so undermined that they had collapsed? Or what huge lateral force could have buckled solid formations like a gigantic crumpled tablecloth? These were serious problems; worse, they were problems for which any proposals seemed unavoidably speculative and untestable, since they concerned putative processes or agencies hidden in the unobservable depths of the earth.

figure of about three million years on the whole sequence, and he toyed with figures up to ten million. Even his much lower published estimate was of course more than enough to breach the traditional short timescale, so he is unlikely to have kept his larger estimates to himself in order to avoid criticism from religious authorities. Rather, it was a matter of scientific caution: he could at least offer experimental evidence in favor of the lower figures, but nothing more concrete than a hunch to justify the higher ones.

Nor was Buffon unusual in this respect, even with his higher figures. As an order of magnitude, his three million years were in line with the guesstimates of others such as Werner and Lavoisier, and about as large as most savants would have regarded as justified by the evidence available. It is significant that even thirty years earlier the editor of *Telliamed* had silently reduced the few "billions" of years (for each vast cosmic cycle) in de Maillet's manuscript down to a few "millions" in the published text, evidently to bring that extremely fanciful theory somewhat more within imaginative reach of its readers.[90]

Encounters with theologians

After the publication of *Nature's Epochs*, Buffon did have a brush with the Sorbonne, the theological faculty in Paris, but it was no more than a little local difficulty (on an earlier encounter, see §3.2). The great public acclaim for his work obliged the theologians to react officially to his handling of the Genesis narrative, which expanded the biblical "days" into vast spans of time and treated the text as one intended for the understanding of common people rather than scholars. Although Buffon was not at all original in this, the sheer popularity of his work made it seem a threat to the Sorbonne's authority: in effect, Buffon the philosopher was trespassing on the intellectual turf that properly belonged, in their opinion, to theologians like themselves. He gave them a perfunctory statement of orthodoxy, but he was scornful of their action and openly derided it. He was far too powerful—and protected by the king—for the Sorbonne to dare censure him formally. The whole episode had no impact on Buffon's work or that of other savants, many of whom were themselves clergymen.[91]

Soulavie's corresponding little local difficulty had the same specific character. He became the target of malicious attacks by a fellow priest and native of Vivarais, the ex-Jesuit Augustin Barruel, but apparently more out of jealousy and resentment at Soulavie's worldly success than from any deep zeal for orthodoxy. In his pamphlet "Genesis according to Mr Soulavie" (promptly reprinted in his popular "Provincial philosophical letters" from Vivarais), Barruel parodied Soulavie as claiming that 356,913,750 years had elapsed (by 1780) since the formation of the

90. Buffon, "Époques de la nature" (1778), 67–70. See Roger, *Buffon philosophe* (1989), 537–43, and "Buffon: Époques" (1962), lx–lxvii. On de Maillet, see the modern edition by Carozzi, *De Maillet: Telliamed* (1968), 182, and 381n52, and Rappaport, *Geologists were historians* (1997), 229.

91. Stengers, "Buffon et la Sorbonne" (1974), describes the episode in detail, using inter alia the archival records of the Sorbonne's deliberations; see also Roger, *Buffon philosophe* (1989), 554–56.

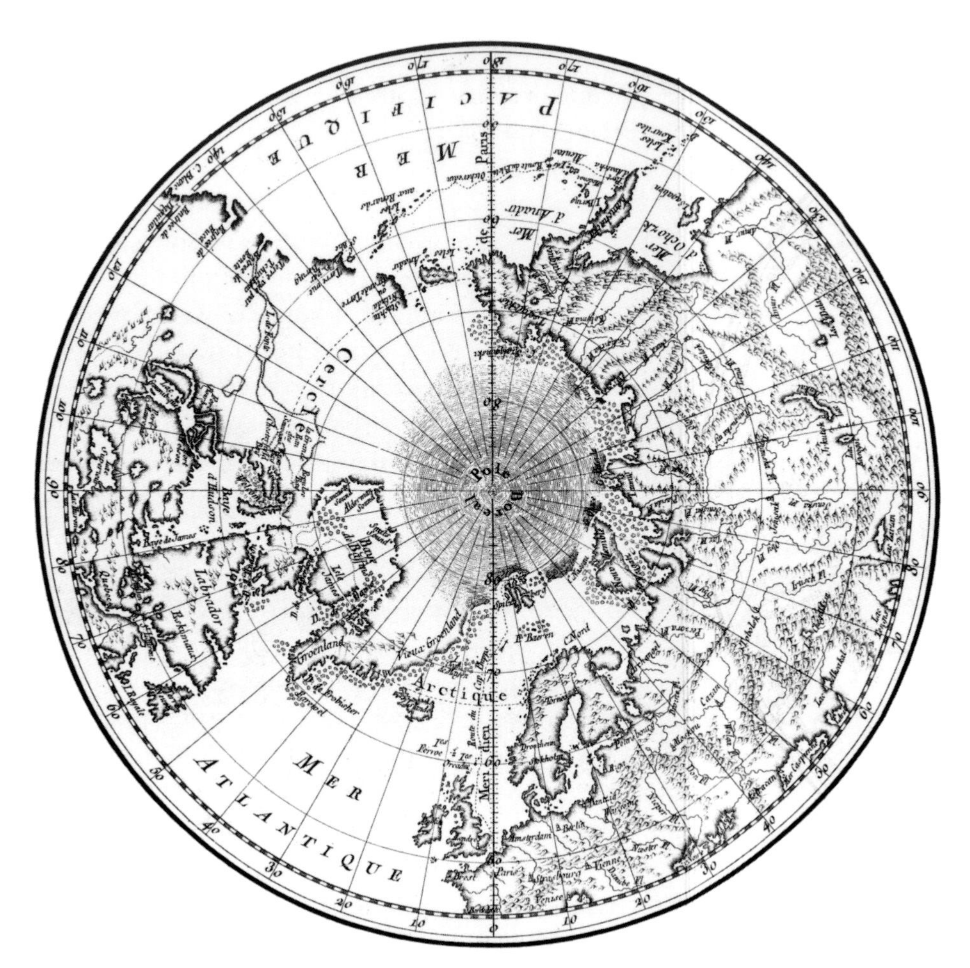

Fig. 3.3. Buffon's map of the earth's northern latitudes, in polar projection, from *Nature's Epochs* (1778), illustrating his theory of the progressive cooling of the globe. The vast areas of the polar ice-sheets indicated the encroaching arctic regime: "those that have already enveloped the vicinity of [Novaya] Zemlya", for example, were just one tongue of the huge polar ice cap, "throwing a shadow on this part of the earth for ever lost to us". The map also served to locate the finds of fossil remains of apparently tropical mammals at high latitudes, for example in eastern Siberia, near the mouth of the River Lena on the Arctic Ocean: they proved that those regions had formerly enjoyed far warmer climates than at present. As the map shows, the geography of Siberia was much better known at this time than that of North America, as a result of extensive explorations funded by the Russian imperial government. The map extends down to the latitude of 45°, which allows it to include Paris. (By permission of the Syndics of Cambridge University Library)

17. The "generation" of life from nonliving materials, by *some* kind of natural process, was a commonplace of Enlightenment theorizing. Buffon, with his well-developed theory of "*molécules organiques*" and "*moules intérieurs*", was a prominent contributor to these debates; but his ideas in this field have only tenuous links with geotheory, except analogically, and lie beyond the scope of this book. See the classic work of Roger, *Sciences de la vie* (1963), and more briefly in *Buffon philosophe* (1989), esp. chaps. 9, 10; also Sloan, "Organic molecules revisited" (1992), Hodge, "Two cosmogonies" (1992), and essays by other Buffon scholars in Gayon, *Buffon 88* (1992), part 5.

18. For Buffon, the heat of volcanoes had no connection with the earth's internal heat: see Taylor, "Geology during Buffon's later years" (1992).

19. Fig. 3.3 is reproduced from Buffon, "Époques de la nature" (1778) [Cambridge-UL: MG.7.20], pl. opp. 615; quotation on 603–4. The matching map (not reproduced here) showed the Antarctic, likewise apparently enveloped in a vast ice cap, but in fact known at this time from little more than its encircling zone of dangerous icebergs. Cook's famous second voyage (1772–75) had just proved conclusively that no *temperate* southern continent existed.

We admire the power by which the human mind has measured the movements of the globes, which nature seemed to have concealed forever from our view; genius and science have burst the limits of space, and observations interpreted by reason have unveiled the mechanism of the world. Would there not also be some glory for man to know how to burst the limits of time, and, by observations, to recover the history of this world, and the succession of events that preceded the birth of the human species?

Georges Cuvier, *Researches on Fossil Bones*, 1812

In memory of Susan Abrams
(1945–2003)

Martin J. S. Rudwick is research associate in the Department of History and Philosophy of Science at the University of Cambridge and professor emeritus of history at the University of California, San Diego. He is the author of *The Meaning of Fossils, The Great Devonian Controversy, Scenes from Deep Time,* and *Georges Cuvier,* all published by the University of Chicago Press.

The University of Chicago Press, Chicago 60637
The University of Chicago Press, Ltd., London
 Published 2005
Printed in the China
Published with the generous support of the Getty Grant Foundation.
14 13 12 11 10 09 08 07 06 05   1 2 3 4 5

ISBN: 0-226-73111-1 (cloth)

Library of Congress Cataloging-in-Publication Data

Rudwick, M. J. S.
Bursting the limits of time : the reconstruction of geohistory in the age of revolution / Martin J. S. Rudwick.
p. cm.
"Based on the Tarner Lectures delivered at Trinity College, Cambridge, in 1996".
Includes bibliographical references and index.
ISBN 0-226-73111-1 (cloth : alk. paper)
1. Science—Europe—History—18th century. 2. Geology—Europe—History—18th century. I. Title.
Q127.E8R83 2005
551'.094'09034—dc22

2004022007

∞ The paper used in this publication meets the minimum requirements of the American National Standard for Information Sciences—Permanence of Paper for Printed Library Materials, ANSI Z39.48-1992.

Martin J. S. Rudwick

Bursting THE LIMITS OF TIME

The Reconstruction of Geohistory in the Age of Revolution

Based on the Tarner Lectures delivered at Trinity College, Cambridge, in 1996

The University of Chicago Press · Chicago & London

Bursting the Limits of Time

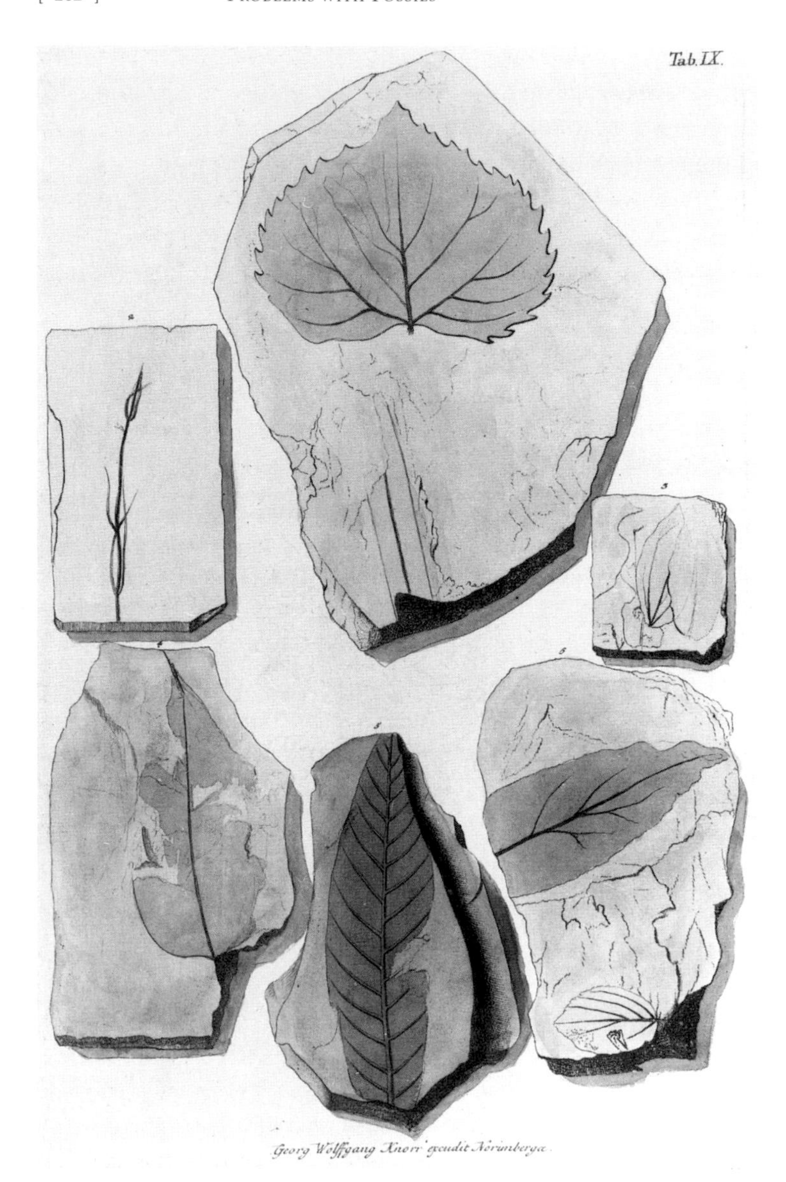

also some of the well-known but puzzling forms that were quite unknown alive, such as ammonites and belemnites (§5.2). Sharks' teeth and the rare remains of turtles completed this unquestionably *marine* fauna. Faujas considered himself qualified to assert that at least some of the fossils had true and exact "analogues" alive in the present world, and he listed forty-one examples from Maastricht and other localities. He noted that his colleague Lamarck, who since Bruguière's untimely death had begun to study fossil and living mollusks in detail, agreed with him on this point. The clear implication was that in time, with further exploration, *all* the fossil forms might turn out to be alive in the present world. Like La Métherie, Faujas concluded that the main task was to explain why so many of the analogues of the fossil species were now living in quite different climates; a major shift in climatic distribution was the big problem, not extinction.[36]

Faujas's supreme example was the famous "Maastricht animal", which Camper had identified as a toothed whale (§5.2; see Fig. 2.6). But Faujas claimed that before his death Camper had changed his mind and decided that it was a crocodile. This was Faujas's own view, and one that he supported with an anatomical comparison with museum specimens of crocodiles of living species. Above all, however, Faujas made his case with the finest known example of the enigmatic fossil, which had been the most highly prized of all the specimens that he had acquired

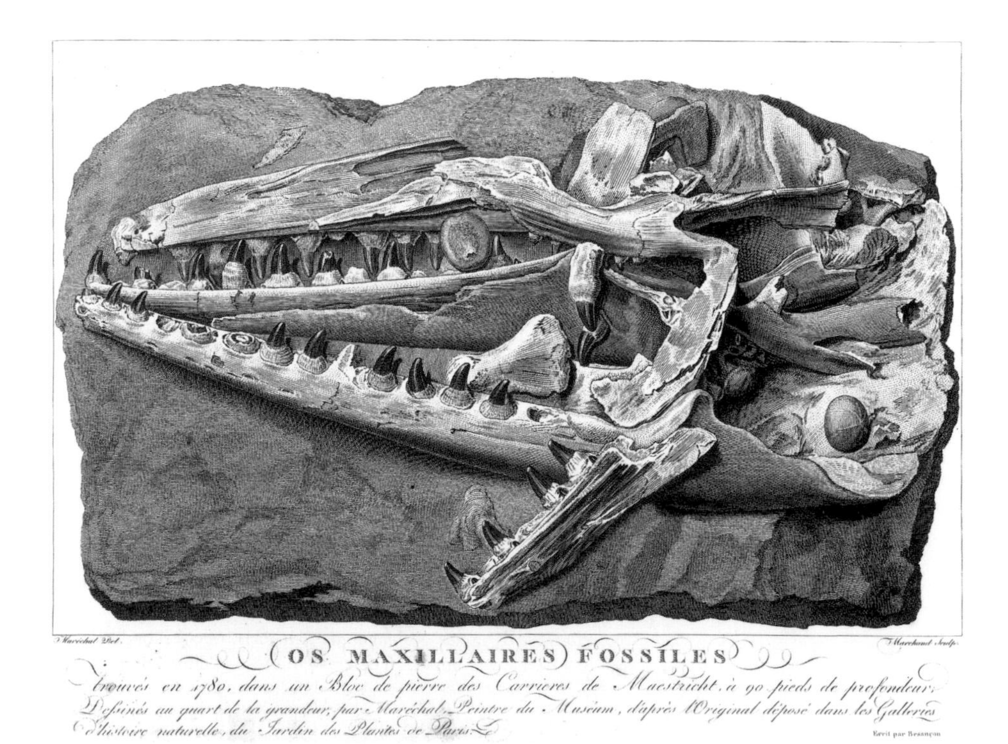
Maréchal Del.
Marchand Sculp.
(OS MAXILLAIRES) FOSSILES
trouvés en 1780, dans un Bloc de pierre des Carrieres de Maestricht, à 90 pieds de profondeur;
Dessinés au quart de la grandeur, par Maréchal, Peintre du Muséum, d'après l'Original déposé dans les Galleries
d'histoire naturelle, du Jardin des Plantes de Paris.
Ecrit par Besançon

Fig. 7.7. The finest specimen (over a meter in length) of the jaws of the famous fossil animal from Maastricht; it was acquired by Faujas in 1794 in the newly captured Dutch town, taken to Paris and placed in the Muséum, and treated as the centerpiece of his lavish descriptive account of all the Maastricht fossils (1799). Camper had identified the huge animal as a toothed whale, but Faujas claimed it was a gigantic crocodile. The two oval objects are sea urchins, preserved by chance among the bones and witnessing to the marine origin of the deposit. (By permission of the British Library)

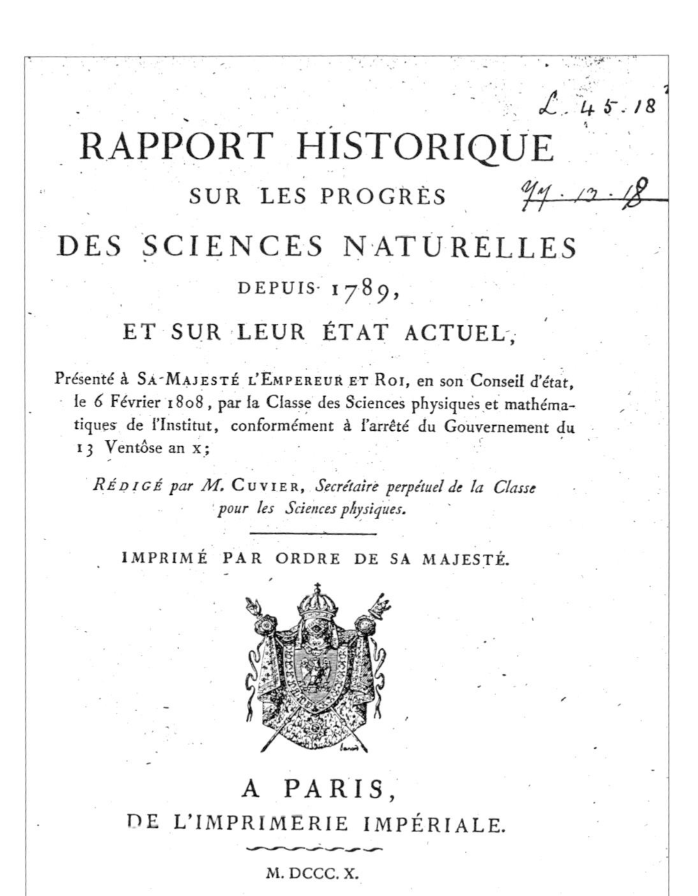